D1259760

Crystal Structure
Analysis for
Chemists and Biologists

Methods in Stereochemical Analysis

Series Editor
Alan P. Marchand, Denton, Texas, USA

Crystal Structure Analysis for Chemists and Biologists

Jenny P. Glusker

with

Mitchell Lewis
Miriam Rossi

VCH

Jenny P. Glusker
Fox Chase Cancer Center
Institute for Cancer Research
7701 Burholme Avenue
Philadelphia, PA 19111

Mitchell Lewis
Department of Biochemistry
 and Biophysics
University of Pennsylvania
School of Medicine
36th and Hamilton Walk
Philadelphia, PA 19104

Miriam Rossi
Chemistry Department
Vassar College
Poughkeepsie, NY 12601

This book is printed on acid-free paper. ∞

Library of Congress Cataloging-in-Publication Data

Glusker, Jenny Pickworth.
   Crystal structure analysis for chemists and biologists / Jenny P.
Glusker, Mitchell Lewis, Miriam Rossi.
      p.   cm.
   Includes bibliographical references and index.
   ISBN 0-89573-273-4
   1. X-ray crystallography.   I. Lewis, Mitchell.   II. Rossi, Miriam.
III. Title.
QD945.G583   1994
548'.83 — dc20   92-7886
                                                        CIP

©1994 VCH Publishers, Inc.

Printed in the United States of America

ISBN 0-89573-273-4 VCH Publishers

Printing History:
10 9 8 7 6 5 4 3

Published jointly by

VCH Publishers, Inc.
220 East 23rd Street
New York, NY 10010-4606

VCH Verlagsgesellschaft mbH
P.O. Box 10 11 61
69451 Weinheim, Germany

VCH Publishers (UK) Ltd.
8 Wellington Court
Cambridge CB1 1HZ
United Kingdom

This volume is dedicated to
Kenneth N. Trueblood,
who excels in crystallography,
in teaching, and as a friend.

Series Foreword

Methods in Stereochemical Analysis provides a forum for critical and timely reviews that deal with the applications of physical methods for determining conformation, configuration, and stereochemistry. The term "stereochemical analysis" is interpreted in its broadest sense, encompassing organic, inorganic, and organometallic compounds, as well as molecules of biochemical and biological significance. The methods include, but are not restricted to, *spectroscopic techniques* (e.g., NMR, infrared, UV-visible, Raman, mass, and optical spectroscopy), *physical techniques* (e.g., calorimetry, photochemical, kinetic, and "direct" methods such as X-ray crystallography, neutron and electron diffraction), and *applied theoretical approaches* to stereochemical analysis.

In establishing the series, the editor and members of the advisory board seek to attract contributions of the highest scientific caliber from outstanding investigators who are actively pursuing research on stereochemical applications of these various techniques and/or applied theoretical approaches. The editor and board members envision contributions in the form either of a monograph or of a multiauthor treatise with individual chapters contributed by a number of outstanding research scientists. Regardless of format, the editor and board members prefer that the contribution consists of critical and timely reviews that place the author's own work in perspective with regard to other important literature in the field, while at the same time retaining the highly personal character of his or her individual contribution. Indeed, rather than necessarily comprehensive, reviews should be *critical* and *timely.*

Whatever merit the resulting volumes possess necessarily must derive from the excellence of the individual contributions. Accordingly, the editor welcomes suggestions from members of the scientific community of potential topics for inclusion in the series, and of names of potential contributors. The editor also welcomes suggestions of a critical nature, which might assist him in better fulfilling the stated objectives. It seems fitting, therefore, that the series be dedicated to its readership among members of the scientific community, for ultimately *they* will gauge the degree to which the series fulfills its objectives.

Alan P. Marchand, Editor
Denton, Texas

Preface

This volume has been written for those chemists and biochemists who may never themselves do X-ray diffraction analyses of crystals, but who need to be able to understand the results of such studies on structures of immediate interest to them. The fields of structural biology and chemistry have blossomed in the years since X-ray diffraction was discovered in 1912. For example, the three-dimensional structures of benzene, graphite, the alkali halides, the boron hydrides, the rare gas halides, penicillin, vitamin B_{12}, hemoglobin, lysozyme, transfer RNA, and the common-cold rhinovirus have been determined and, in each case, the results have greatly increased our understanding of fundamental chemistry and biochemistry.

Dorothy Hodgkin,[1] who obtained a Nobel Prize in 1965 for her X-ray diffraction work, wrote, "a great advantage of X-ray analysis as a method of chemical structure analysis is its power to show some totally unexpected and surprising structure with, at the same time, complete certainty." Her X-ray diffraction studies are the reason that we now know that penicillin has a β-lactam structure and that vitamin B_{12} contains what is now known as a "corrin" ring system. The results of all X-ray diffraction studies are used by chemists and biochemists, and these scientists need to be able to appreciate the significance and extent to which such results are useful and precise.

We have written this book with two main purposes in mind. One aim is to acquaint chemists and biochemists with the general principles of crystal structure analysis so that they can critically appraise articles in the crystallographic literature, and extract, with a reasonable comprehension of the precision of the results of the experiment, any structural information they are interested in. The second aim is to make the reader aware of the vast amount of structural information that has resulted from this method of analysis, to inform him or her of how to access the results in the most useful way, and to indicate the manner by which these types of data have enhanced our understanding of chemistry and biochemistry. It is our aim to place the method in context with other methods of structure analysis, such as solution studies (nuclear magnetic resonance and infrared analyses) and molecular modelling.

As is necessary for a book of reasonable size, only an overview can be given here; more complete information has to be found by consulting the references. For a similar reason it has been necessary to select a few experimental results of interest and, regretfully, not to mention other equally excellent studies. Further details on methods may be found in any of the excellent textbooks on the subject,[2-12] while experimental results[13, 14] are found in the current scientific literature.

Our overall aims are:

- To show what a crystal is and how molecules or ions pack in it; and how it forms faces, cleaves, and has useful physical properties (see Chapters 1, 2, and 5).
- To describe methods for obtaining good crystals suitable for an X-ray structural analysis (see Chapter 2).
- To show what diffraction is and that it can be observed on an everyday basis. The relationship between optical diffraction and X-ray diffraction is described (see Chapter 3).
- To show how diffraction of X-rays by a crystal may be used to determine the atomic arrangement within that crystal (see Chapter 6).
- To show how the X-ray diffraction pattern of a crystal is measured, and how the precision of these measurements may be assessed (see Chapter 7).
- To show how the phase problem is solved in order to be able to calculate an electron density map (see Chapter 8).
- To describe electron density maps and their interpretations and misinterpretations (see Chapter 9).
- To describe the meaning of the parameters obtained by the crystal-structure refinement, and to compare X-ray diffraction results with those from neutron diffraction and other physical techniques (see Chapters 11 and 12).
- To assess the precision of the distances, angles, and torsion angles determined by an X-ray structure analysis, and to explain why some are more precise than others even if they both are derived from structures with the same R-value (see Chapters 10 and 11).
- To describe torsion angles and conformation, and their significance (see Chapter 12).
- To explain thermal motion and temperature effects, and the diffraction consequences of disorder and imperfections in crystals (see Chapter 13).
- To describe anomalous scattering and its use in determining absolute configuration and in phasing (see Chapter 14).
- To show how molecules and ions pack in crystals and in any symmetry that can be found in this packing (see Chapters 4 and 15).
- To describe methods to compare molecules and assess conformational variability (see Chapter 16).
- To describe what has been learned about how molecules, and more specifically, functional groups within them, "recognize" other molecules or groups. This leads to information on what biological "receptors" might look like (see Chapters 15 and 17).
- To show what can be learned from X-ray diffraction analyses about chemical reactions and the intermediates in their reaction pathways (see Chapter 18).

We thank those who have helped us, particularly Dr. John J. Stezowski for his help with Chapters 2 and 13, and for reading other chapters, and Drs. Margaret J. Adams, Harold Greenwald, Kenneth N. Trueblood, and Betty K. Patterson, who read the entire manuscript. We also thank Esther Steinbrecher for extensive help with references and with the first names of many scientists. Many others have contributed by helping with typing, reading various portions, figure preparations, and reference listings. We also thank Dr. Alan Marchand, who patiently encouraged us throughout the writing of this book. Our appreciation to Carol Afshar, Karen Albert, Pat Bateman, Bud Carrell, Francesco Caruso, Jonah Erlebacher, Maria Flocco, Philip George, Ann Glusker, Carrie Hafer, Henry Katz, Amy Kaufman Katz, Kathy Lewis, Dan Lipton, Frank Manion, Ann Nista, Eileen Pytko, Liat Shimoni, Sheryl Silverman, Joel Sussman, Vesselin Tomov, Dave Zacharias, and also NIH grant CA-10925.

Jenny P. Glusker
Mitchell Lewis
Miriam Rossi

April 1994

References

1. Hodgkin, D. C. (1965) The X-ray analysis of complicated molecules. *Science,* **150**, 979–988 (Lecture, 11 December 1964 for the Nobel Prize in Chemistry).

2. Blundell, T. L. and Johnson, L. N. *Protein Crystallography.* Academic Press: New York, London, San Francisco (1976).

3. Bunn, C. W. *Chemical Crystallography. An Introduction to Optical and X-ray Methods.* Clarendon Press: Oxford (1946).

4. Dunitz, J. D. *X-ray Analysis and the Structure of Organic Molecules.* Cornell University Press: London, Ithaca, NY (1979).

5. Glusker, J. P. and Trueblood, K. N. *Crystal Structure Analysis. A Primer.* Second edition. Oxford University Press: New York, Oxford (1985).

6. Ladd, M. F. C. and Palmer, R. A. *Structure Determination by X-ray Crystallography.* Second edition. Plenum Press: New York, London (1985).

7. Luger, P. *Modern X-ray Analysis on Single Crystals.* Walter de Gruyter: Berlin, New York (1980).

8. Stout, G. H. and Jensen, L. H. *X-ray Structure Determination. A Practical Guide.* Macmillan: New York (1968); Second edition, John Wiley: New York, Chichester, Brisbane, Toronto, Singapore.

9. Vainshtein, B. K., Fridkin, V. M. and Indenbom, V. L. *Modern Crystallography II. Structure of Crystals.* Springer-Verlag: Berlin, Heidelberg, New York (1982).

10. Giacovazzo, C. (ed.) *Fundamentals of Crystallography.* International Union of Crystallography/Oxford University Press: New York (1992).

11. Hahn, T. (ed.) *International Tables for Crystallography. Brief Teaching Edition of Volume A. Space-group Symmetry.* Reidel: Dordrecht (1985).

12. Taylor, C. A. (ed.) 19 pamphlets on various aspects of crystallography for teachers. IUCr Commission on Crystallographic Teaching. University of Cardiff: Cardiff (1981–1984). Available from Polycrystal Book Service.

13. Mak, T. C. W., and Zhou, G.-D. *Crystallography in Modern Chemistry. A Resource Book of Crystal Structures.* John Wiley: New York (1992).

14. Bränden, C-I., and Tooze, J. *Introduction to Protein Structures.* Garland: New York, London (1991).

Contents

8. Estimation of Relative Phase Angles 281

9. Electron-Density Maps 345

10. Least-Squares Refinement of the Structure 389

11. Interpreting x, y, and z (Atomic Coordinates) 413

CHAPTER

1

Introduction To Crystal Structure Analysis

The purpose of this book is to introduce the reader to the methods of molecular structure determination by X-ray diffraction of crystals. This will enable the reader to appreciate those results published in the scientific literature not only in terms of the structural parameters derived, but also in terms of their precision. We will also demonstrate how useful these results can be to the chemist and biochemist.

1.1 How can we "see" atoms?

It is not possible to "see" atoms and molecules because these components of matter are too small for us to view by the usual methods. Ideally, a "supermicroscope" should be used to allow us to see molecules. While a microscope can be built (using the fact that light can be focused by a lens), the **resolution** is not high enough for us to "see" atoms. We cannot see the details of an object with a microscope unless these details are separated by at least half the wavelength of the radiation used to view them.[1,2] Visible light has wavelengths of 4 to 8 $\times$ 10^{-5} cm, and is, therefore, of no use when we are trying to view atoms that are separated in molecules by distances of the order of 10^{-8} cm (0.1 nm). The appropriate radiation to use to view atoms would be **X rays** with wavelengths in the nanometer range. Thus, the supermicroscope that we might build to view atoms would have to employ X rays rather than visible light, and it would give us a wonderful chance to step briefly into the nanometer world of molecules, seeing their shapes and how they interact with each other! This is the aim of the studies described in this book.

In an ordinary optical or electron microscope, radiation is scattered by the object that one wishes to "see" at higher magnification. This radiation is recombined by the lens system, resulting in an image of the scattering matter, appropriately magnified. In such a microscope the

1

flow of radiation (light), after being scattered by the object, is continuous through the lens system and beyond. When the scattered waves are recombined by the lens, the relationships between the **phases** (the relative locations of peaks and troughs[3] of the various scattered waves) have been maintained. Such phase relationships are illustrated by the surfers in Figure 1.1, where one sees a series of waves in phase, out of phase and partially out of phase. This figure illustrates that sets of waves can produce very different effects, depending on their relative phases.

On a much smaller dimensional scale than that just described for visible light, X rays, which have very short wavelengths, are scattered by the electrons in atoms. The scattered X rays cannot, however, normally be focused by any currently known experimental technique because no convenient material that can focus X rays (in the way that glass lenses focus visible light) has yet been devised. Unfortunately, this prohibits the building of an X-ray "supermicroscope," and therefore another technique must be used to view atoms. Studies with longer-wavelength X rays[4] or with a scanning tunneling microscope[5] enable us to infer the shapes of molecules, but only those on the surface of a structure.

There is, however, an alternative (but still indirect) way to view these molecules. It involves studies of crystalline solids and the use of the phenomenon of **diffraction**. The radiation used is either X rays, with a wavelength on the order of 10^{-8} cm, or neutrons of similar wavelengths. The result of analyses by these diffraction techniques, described in this volume, is a complete three-dimensional elucidation of the arrangement of atoms in the crystal under study. The information is obtained as atomic positional coordinates and atomic displacement parameters. The coordinates indicate the position of each atom in a repeat unit within the crystal, while the displacement parameters indicate the extent of atomic motion or disorder in the molecule. From atomic coordinates, it is possible to calculate, with high precision, interatomic distances and angles of the atomic components of the crystal and to learn about the shape (**conformation**) of molecules in the crystalline state.

Since X rays are simply a shorter form of electromagnetic radiation than visible light, their behavior can be described by use of the theory of Christiaan Huygens.[6] Thus, the effects that occur when a series of waves impinge on an object of similar dimensions can be calculated from a consideration of how the waves are scattered by that object. Each wave has an undulating displacement (the **amplitude**) and a distance (the **wavelength**) between crests. The displacement is periodic in time and/or space. In order to assess the disturbance of a beam of light by an object, it is necessary to know the relative phases of all the waves scattered by it (Figure 1.1) as these relative phases have a profound effect on the intensities of the scattered beams. In an X-ray experiment the scattered beams are captured on photographic film, a scintillation counter, or some other detector. These X-ray beams (waves) impinge on the detector with

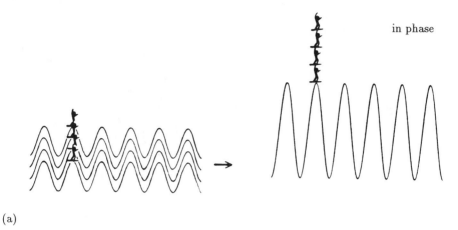

in phase

(a)

out of phase

(b)

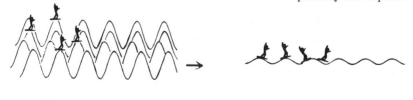

partially out of phase

(c)

FIGURE 1.1. The importance of relative phases on the result of the summation of waves. Sets of waves of equal amplitude and wavelength are shown; only the relative phases of the waves differ. (a) Waves exactly in phase (the wave crests coincide so that the waves reinforce each other), (b) exactly out of phase (troughs and crests coincide giving no resultant wave), and (c) partially out of phase. On the left is shown the set of waves and on the right the result of interference [constructive in (a) or destructive in (b)], with surfers to accentuate the result.

intensities and relative phases determined by the atomic arrangement in the crystal. X-ray detection devices, however, are sensitive only to the intensity of the wave, which is proportional to the square of the amplitude of the wave; they are not, unfortunately, sensitive to the the phase of the wave relative to other scattered waves. In contrast, a microscope lens, which is composed of glass and therefore refracts visible light waves, collects the waves scattered by the object and combines them all again with due appreciation for their relative phases. Therefore, by the analogy between a microscope and X-ray diffraction (Figure 1.2), if one uses X-ray diffraction to "view" atomic structures, it is necessary to obtain a measure of the relative phases of diffracted X-ray beams. Then

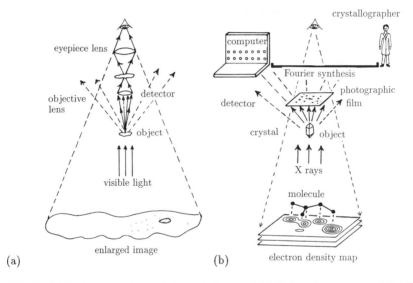

FIGURE 1.2. A comparison of the techniques of (a) light microscopy and (b) X-ray diffraction for "viewing" a magnified image of a small object. In the light microscope, scattered light is focused, first by the objective lens and then by the eyepiece lens, so that a magnified image is obtained. X rays, which are scattered by the electrons in atoms, are needed if smaller objects, such as atoms, are to be viewed. Unfortunately no lens is available to focus scattered X rays. Therefore the second stage in the action of a light microscope cannot be carried out with X rays. Although the intensity pattern of scattered X rays may be detected directly, the relative phases of these scattered X rays are lost; these phases must be derived in a different way. The strategy used is to simulate the action of the lens by mathematical techniques. Since X rays are scattered by electrons in the crystal, the result of this mathematical simulation is an electron-density map with peaks at atomic positions. [From: Glusker, J. P., and Trueblood, K. N. *Crystal Structure Analysis. A Primer.* 2nd edn. Oxford University Press: New York, Oxford (1985)]

they must be recombined with the correct phase relationships in order to obtain a view of the atomic structure of the crystal. This lack of relative phase information has been termed the **phase problem**.

In order to understand the techniques of crystal structure analysis, the reader must appreciate the significance of the terms crystal and diffraction. One can then learn how to measure the X-ray diffraction pattern of a crystal and find experimentally, with high precision, the arrangement of atoms in the crystal under study. The first part of this book describes experimental techniques used to determine the arrangement of atoms in a crystal. The second part deals with the chemical and biochemical information that can be learned from the results of such analyses.

1.1.1 Early studies of crystals

One of the prerequisites of this method of molecular structure determination is a suitable crystal.[7] The word **crystal** is derived from the Greek $\kappa\rho\upsilon\sigma\tau\alpha\lambda\lambda o\varsigma$, meaning clear ice (*crystallum* in Latin). This name was applied to a clear transparent form of quartz (rock crystal) because it was thought that it had been formed from water by intense cold. Sodium chloride (table salt) forms crystals in the shape of little cubes. This is well known to anyone who has taken a magnifying glass to the dinner table. Photographs of some naturally occurring quartz crystals, and of a large crystal of ammonium dihydrogen phosphate are shown in Figure 1.3.

Our understanding of the internal structures of crystals began in the early seventeenth century. In 1611 Johannes Kepler[8] wrote an essay on the hexagonal snowflake. He tried to explain the shape of a snowflake

(a)

(b)

FIGURE 1.3. Crystals of (a) quartz, and (b) ammonium dihydrogen phosphate. Note the symmetry of face development, especially in (b).

and considered how it might be built up from the packing of spherical particles. He noted that the closest packing arrangement of equal-sized balls in two dimensions is hexagonal. He felt this model must be related in some way to the shape of the snow flake, but was not able to account for the dendritic (multibranched) appearance of most snowflakes. The concept of internal periodicity in crystals was investigated further by many others, notably René Descartes, Domenico Guglielmini, Robert Hooke, and Christiaan Huygens later in the seventeenth century. Hooke[9] continued the research of Kepler on crystals and suggested that alum, salt, and saltpeter crystals could all be considered to be composed of "spheroids." He used such spheroids to account for some of the crystal shapes he observed, and described these spheroid assemblies in his book, *Micrographia*, published in 1665.

The crystals of quartz in Figure 1.3(a) have flat sides (**faces of a crystal**). Although quartz crystals vary in size and shape from specimen to specimen, it was pointed out by Nicolaus Steno[10] (Niels Stenson) in 1669 that the angles between certain pairs of faces are characteristic of the substance (quartz), even though the size of such faces depend on the conditions that have occurred during crystal growth. Further investigations were made by Jean Baptiste Louis Romé Delisle (de l'Isle),[11] René Just Haüy,[12,13] and Torbern Bergman[14] who were interested in the relationship between the external shape of a crystal and its internal structure. They inferred, from observations of the manner by which new faces appear when crystals are broken, that crystals were composed of small polyhedral particles, which they called "molécules intégrantes," related to Hooke's spheroids and to what were later called **unit cells**. These were considered to stack in regular arrays, like bricks, to form crystals. This concept of the smallest component of a crystal came at the time that John Dalton[15] was formulating his atomic theory of molecular structure. Although Dalton was aware of the studies of the crystallographers of the time, he did not use their ideas in formulating his theory. His work was based, instead, on the combining weights of elements rather than on their possible volumes and the shapes of the molecules formed from the atoms.

The macromolecules of interest to biochemists and biologists also form beautiful crystals provided that the appropriate conditions for their crystallization can be found. Crystals of hemoglobin[16,17] were described as early as 1830. The first crystals of an enzyme were obtained by James B. Sumner[18] when he crystallized urease in 1926. Then, in 1930, John H. Northrop[19] crystallized pepsin, trypsin, and chymotrypsin. While several early crystallographers recorded diffraction patterns of such proteins, their experimental conditions led only to very weak intensities, and it was believed that the structures of the proteins studied, such as hemoglobin, were too complicated to be profitably investigated by X-ray diffraction. It was not until 1934 that John Desmond Bernal and Dorothy Crowfoot (Hodgkin)[20] found that crystalline proteins would give good diffraction patterns provided that, during the diffraction experiment, the crystals were suspended in their mother liquor rather than being removed and left to dry. Although these scientists showed that crystals of pepsin, when mounted in this way, gave clear X-ray diffraction patterns, it was many years before the extensive diffraction patterns of proteins could be analyzed correctly so that their molecular structures could be found. Even more amazing than this work on proteins was the crystallization of living material, tobacco mosaic virus (TMV), by Wendell Stanley.[21] Later, Sir Frederick C. Bawden and Norman W. Pirie,[22] J. D. Bernal and Isidor Fankuchen[23] showed by X-ray diffraction how TMV particles pack in a regularly-repeating way in three dimensions. This was also demonstrated by electron microscopy (see Chapter 2, Figure 2.1, for an electron micrograph of tobacco necrosis virus crystals).

1.1.2 The regularly repeating internal structures of crystals

The word crystal can have various meanings to different people. The X-ray crystallographer defines a crystal as a solid that has, in all three dimensions, a regularly repeating internal arrangement of atoms. It is not sufficient to describe a solid as crystalline based on macroscopic observable properties such as flat faces; it must have a regular internal structure at the molecular level to satisfy the definition just given. A glass goblet is not crystalline in the scientific sense we use here, even though it may be described commercially as "crystal." The glass from which the goblet is made does not have a regularly repeating internal structure that we expect for a crystalline solid. No matter how much glass is ground or polished to produce flat faces, it will not become a crystal.

The internal regularity of a crystal may, as shown by the early studies just described, be represented by a unit cell, which is the basic building unit. This unit cell is considered to be repeated continuously in three dimensions to form the crystal. In reality, the crystal is composed of the atoms that make up its structure. One should remember that the edges of the unit cell are imaginary and are merely chosen, for the convenience of the scientist, to define the shape and size of the repeating portion of the crystal structure. Thus the unit cell can be considered as the building block from which the crystal is constructed, analogous to (nonimaginary) bricks that are repeated in a similar way to construct a (two-dimensional) wall. The faces of the crystals may be considered to be the ends of the block formed from the orderly aggregation of many such unit cells. Often these faces are formed by a stepwise arrangement of the building units. Therefore, a careful measurement of the angles between crystal faces can lead to an assessment of the unit cell shape, but not its absolute size. Such derivations of unit cell shape formed the basis of the nineteenth-century studies of crystalline minerals. The concept of the internal periodicity of crystals was put on a firm mathematical basis at that time by Auguste Bravais,[24,25] who derived the number of possible unit cell shapes.

These ideas were viewed from a different perspective by Ludwig August Seeber,[26] who realized that, in order to explain physical properties such as the thermal expansion and elasticity of crystals, it is necessary to have a better model than a microscopic bricklike structure — one closer to reality. He proposed a model of atoms with mutual distances determined by the balance of attractive and repulsive forces between them, but containing the regularity that Haüy and others had so elegantly deduced earlier. These notions were verified when William J. Pope and William Barlow[27] together proposed atomic arrangements in salts such as sodium chloride that involved an equilibrium of positively and negatively charged ions. It was Pope who urged William Lawrence **Bragg** to determine the structure of sodium and potassium chloride crystals by X-ray diffraction techniques,[28] since he believed that the atomic arrangement in crystals of these alkali halides was a cubic close-packed structure in which no individual "molecules" of sodium chloride would be found. The crystal would consist of closely packed Na^+ and Cl^- ions. Bragg showed that this hypothesis was correct.[29]

1.2 Diffraction of light

Since the possibility of determining the structure of sodium chloride by X-ray diffraction techniques has just been introduced to the reader, it is now necessary to explain the meaning of the term diffraction. Diffraction occurs when radiation passes through an aperture that has dimensions similar to the wavelength of the radiation; as a result, some of the radiation appears to be slightly deflected and light appears in the area expected, for geometric reasons, to be in shadow. If there are several slits, fringes of light are observed.

Early studies of diffraction were made with **diffraction gratings** prepared by scratching fine, parallel lines on glass. Such gratings were used by Francesco Maria Grimaldi[30] in the seventeenth century and Joseph von Fraunhofer[31] in the early nineteenth century. They showed that diffraction involves the constructive and destructive **interference** of scattered radiation. The scattered radiation is recombined by the summation of waves with appropriate phases, as proposed by Huygens[6] and described in considerable detail by Friedrich Magnus Schwerd[32] in 1835. These scientists realized that the relative phases of the scattered rays (Figure 1.1) will determine the overall extent of their interference with each other, and, hence, the intensity of the diffracted beam. This important discovery will be discussed in more detail in Chapter 3.

Diffraction need not, however, just be investigated in a physics laboratory; it is an everyday phenomenon. For example, when one looks through a silk umbrella or a fine fabric curtain at a distant street light at night, the small light will have additional spots of light regularly disposed around it. Diffraction may even be seen through your eyelashes if you view a small distant light in the dark. The phenomenon can be demonstrated very effectively by use of fine mesh sieves, such as those used for separating powders. As shown in Figure 1.4, when a small distant light is viewed through one of these sieves, there appears not just one spot of light but several. Alternatively a fine laser beam projected onto a screen, with a sieve interposed, will give the same effect. These are experiments that the reader can easily do. The spots that we see around the central spot (the direct beam) are the diffracted rays and these are regularly arranged. This arrangement of spots of light, each with a measurable position and intensity, is referred to as the **diffraction pattern** of the sieve. The spacing between these spots is inversely related to the spacing between the wires of the sieve grid and is also a function of the distance between the sieve and detector (eye-to-screen distance). A coarser sieve (wider distance between the wires of the grid) gives a finer diffraction pattern (shorter distance between spots at the detecting system), while a finer sieve gives a wider spacing between spots in the diffraction pattern. All objects may individually diffract radiation of the appropriate wavelengths, but the diffraction effect is reinforced when there is a regular periodicity of structure (as in a sieve or a crystal). This periodicity greatly increases the intensity of the diffraction and makes it more readily observable.

Now that we have introduced crystals and diffraction, we can discuss how these two are related. In the experiment described above, a sieve was used to diffract visible light. If the scale of this experiment is reduced

FIGURE 1.4. Diffraction produced by the passage of visible light through a fine sieve. A small light source, such as a microscope light with a pinhole aperture in front of it (or sunlight reflected off a car bumper) is viewed through the sieve (e.g., 0.0015 inches between wires). The result is called the diffraction pattern. The finer the mesh of the sieve, the further apart are the diffracted beam positions hitting the detection system (photographic film in this case). The reader is encouraged to try this experiment.

several orders of magnitude, we can perform an analogous experiment. The sieve is replaced by a crystal with its much smaller regularly repeating internal order, and the visible light is replaced by X rays which have a correspondingly smaller wavelength. The resulting diffraction patterns are similar, although they require different detection methods.

1.3 Diffraction of X rays by crystals

The diffraction of X rays by crystals was first demonstrated by Walther Friedrich, Paul Knipping, and Max von Laue,[33] seventeen years after Wilhelm Konrad Röntgen[34] discovered X rays in 1895. Although X rays were known to be highly penetrating, their nature was still a mystery. The question posed at that time was "are X rays particles, or are they electromagnetic radiation, like visible light?" The definitive experiment used to demonstrate a wavelike nature for radiation was whether or not the radiation could be diffracted. Since diffraction is an interference phenomenon, a property of waves, it follows that if X rays could be diffracted, they must be waves. Charles Glover Barkla[35] had already shown in 1911 that X rays are scattered by matter. Bernhard Walter and Robert Wichard Pohl[36,37] had estimated that, if X rays were waves, then their wavelengths must be of the order of 10^{-9} cm. The definitive experiment was done by von Laue and co-workers, who wrote (translated)[33] "If X rays really are electromagnetic waves, it would seem likely that the structure of the crystal lattice would give rise to interference phenomena, resulting from the atoms being excited to either free or constrained vibration. In fact, the interference phenomena should be similar in nature to those known for optical grating spectra."

Relevant to this story of the diffraction of X rays by crystals are the experiments then underway to determine Avogadro's number N, the number of molecules in a gram mole. This is an important fundamental constant to the chemist. It had been measured in several ways[38] and found to be approximately 6×10^{23}. For example, for sodium chloride crystals, a knowledge of the density (2.16 g cm^{-3}) and the formula weight (58.4 g) leads to an estimate of the interatomic distance between Na$^+$ and Cl$^-$ ions. A cubic crystal, 1 mm on an edge, will weigh 2.16×10^{-3} g. Since 6×10^{23} NaCl units weigh 58.4 g mole $^{-1}$, this 1 mm cube contains approximately 4×10^{-5} moles or 22×10^{18} pairs of Na$^+$ and Cl$^-$ ions. Since the crystal edge is 1 mm, there are $\sqrt[3]{22 \times 10^{18}} \approx 3 \times 10^6$ ion pairs along this 1 mm distance. The separation between ion pairs is therefore about $[10^{-1}/(3\times10^6)]$ cm $= 3 \times 10^{-8}$ cm. Thus, von Laue realized that if X rays were wavelike, with a wavelength much shorter than that of visible light, they should be diffracted by crystals containing an internal periodicity of appropriate dimensions. Later, it was found that "soft" (i.e., less penetrating) X rays with longer wavelengths (of the order of 10^{-7} cm), could be diffracted (at grazing incidence) by diffraction gratings and, as a result, their wavelengths could be measured directly. Then, by determining interatomic distances in crystals of known structures (so that the unit cell dimensions obtained could be interpreted in terms of structure) and from a knowledge of the crystal density, it is possible to obtain an exact and direct determination of Avogadro's number.[39,40]

Friedrich and Knipping[33] performed the first X-ray diffraction experiment with a copper sulfate crystal and demonstrated the wavelike nature of X rays. A beam of X rays was directed onto the crystal, and a piece of photographic film was placed behind the crystal to detect any diffracted X rays. From this arrangement of X-ray source, crystal and detector (Figure 1.5), they obtained a diffraction pattern and concluded that X rays must be electromagnetic radiation with wavelengths of the order 10^{-8} cm (a **unit of length** defined as 1 **Ångström unit, 1 Å**).

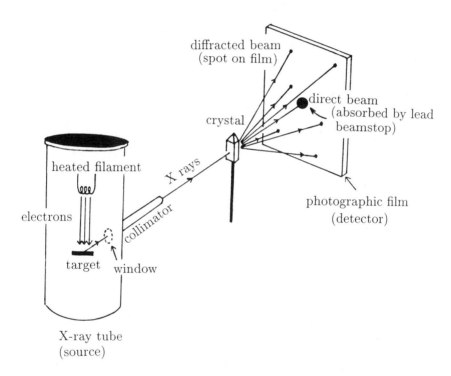

FIGURE 1.5. The experimental setup used by Friedrich and Knipping[33] to measure X-ray diffraction intensities. The important components consisted of an X-ray source to provide a finely collimated X-ray beam, a crystal to scatter X rays, and a detection system, such as photographic film, to measure the directions and intensities of the diffracted beams. The intensities so measured are related to the squares of the amplitudes of the scattered beams, but information on the relative phases of these scattered beams is lost. This same general experimental setup is currently used, although the source of X rays and the detection system are now much more sophisticated.

This experimental setup (Figure 1.5) is essentially the same as that used today. The equipment now is much more elaborate, but the essential components are still an X-ray source, a crystal that is bathed in the incident beam, and a detector to measure the intensities of the diffracted X-ray beams. Details of present-day apparatus will be described later in Chapter 7.

During the twentieth century, as different types of diffraction have been found, we have come to understand that no firm distinction can be made between waves and particles. For example, neutrons[41,42] and electrons,[43] which would be considered particles according to classical mechanics, can be diffracted. With great perspicacity, William Henry Bragg wrote[44] in 1912, "The problem then becomes, it seems to me, not to decide between two theories of X rays, but to find, as I have said elsewhere, one theory which possesses the capacities of both."

Several types of diffraction by crystals are now studied. Neutron diffraction[45] can be used with great effectiveness to give information on molecular structure. These results complement those from X-ray diffraction studies, because there are different mechanisms for the scattering of X rays and of neutrons by the various atoms. X rays are scattered by electrons, while neutrons are scattered by atomic nuclei. Neutron diffraction is important for the determination of the locations of hydrogen atoms which, because of their low electron count, are poor X-ray scatterers. Electron diffraction,[46] while requiring much smaller crystals and therefore being potentially useful for the study of macromolecules, produces diffraction patterns that are more complicated. Their interpretation is hampered by the fact that the diffracted electron beams are rediffracted within the crystal much more than are X-ray beams. This has limited the practical use of electron diffraction in the determination of atomic arrangements in crystals to studies of surface structure.

1.4 Use of X-ray diffraction to find atomic arrangements

Once it had been shown that crystals diffract X rays, the relationship between the observed effect and the experimental conditions was put on a sound mathematical basis by Max von Laue, Paul P. Ewald and many others.[47] X-ray diffraction by crystals represents the interference between X rays scattered by the electrons in the various atoms at various locations within the unit cell. It must, however, be stressed again that any molecule or ion can diffract X rays or neutrons. It is only when this diffraction is reinforced by the repetition of the unit cell in the crystal that diffraction by atoms is a conveniently observable effect, for example as spots of differing intensity on photographic film. Of particular interest to chemists and biochemists is the work by W. L. Bragg,[29] who demonstrated that measurement of the diffraction patterns gives information on the distribution of electron density in the unit cell, (i.e., the arrangement of atoms within this unit cell).

As an example of diffraction by atoms (ions) in crystals, the structure and resulting diffraction pattern of sodium chloride are shown in Figure 1.6 and compared with potassium chloride.[29] The diffraction pattern of potassium chloride has a different spacing between spots (diffracted X rays hitting the photographic film) from the diffraction pattern of sodium chloride. This is because the unit cells are different sizes, because a potassium ion is larger than a sodium ion. The reciprocal relationship between crystal periodicity and the diffraction pattern of that crystal (as mentioned earlier for sieves) causes the spots in the diffraction pattern of potassium chloride to be closer together than those for sodium chloride (Figure 1.6).

The question at this stage is "How does one derive the atomic arrangement in a crystal, such as that of sodium chloride or potassium chloride, from the intensities in their respective diffraction patterns?" The answer is that the diffracted X-ray beams, which have amplitudes represented by the square roots of their measured intensities, must be recombined in a manner similar to that achieved by a lens in an optical microscope. This recombination is done by a mathematical calculation called a **Fourier synthesis.** The recombination cannot be done directly because the phase relations among the different diffracted X-ray beams usually cannot be measured. If the phases, however, can be estimated by one of the methods described in Chapter 8, an approximate image of the arrangement of atoms in the crystal can be obtained.

Throughout this book we will illustrate results of a single crystal structure analysis; but there is always a problem in displaying a three-dimensional image in two dimensions. Therefore we use stereodiagrams, as in Figure 1.6(a). These are pairs of illustrations of the same image of the three-dimensional structure, but with one rotated 6° about a vertical axis in order to mimic the directions of two eyes that produce stereovision. By visual training, or by use of stereoviewers, these two images can be made to blend into one "three-dimensional" image.

1.4.1 19th century chemical studies of molecular structure

A historical account of some of the early structures studied by X-ray diffraction techniques will illustrate how the structural principles, laboriously derived by organic and inorganic chemists in the nineteenth century, were verified in the twentieth century.

The problem of determining the shape of a molecule has been the subject of research for a long time.[48] This was addressed at the beginning of the nineteenth century by John Dalton.[15] He proposed that the combining proportions of atoms in the molecules that make up compounds could be explained by assuming that each element had a different weight (its atomic weight). By chemical analysis of simple compounds he could determine their formulæ and propose the connectivities of the component atoms. Attempts to describe the shapes of molecules continued to interest

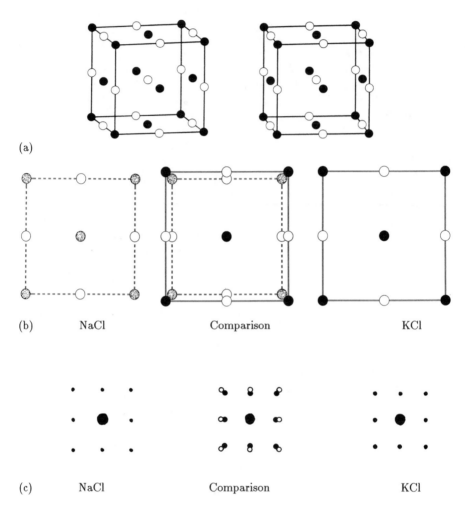

(a)

(b)   NaCl   Comparison   KCl

(c)   NaCl   Comparison   KCl

FIGURE 1.6. Diffraction of X rays by the crystalline alkali halides.[29] Sodium chloride has a cubic unit cell, $a = 5.628$ Å, and potassium chloride has the same arrangement of ions, but $a = 6.291$ Å, larger because a potassium ion is larger than a sodium ion. (a) Stereoview of the contents of one unit cell with alkali metal cations as filled circles and chloride ions as open circles. (b) Comparison of the unit cells with potassium ions (filled circles), sodium ions (stippled circles), and chloride ions (open circles). (c) Comparison of the X-ray diffraction patterns of sodium chloride (left) and potassium chloride (right). In the middle is a superposition of the two diffraction patterns (NaCl, open circles; KCl, filled circles). Since the unit cell of potassium chloride is larger than that of sodium chloride by 12%, the spots on the diffraction pattern are correspondingly 12% closer together. This comparison emphasizes the reciprocal relationship between the dimensions of a repeating object and its diffraction pattern.

chemists throughout the nineteenth century. The proposal of a tetrahe-drally bound carbon atom by Jacobus Hendricus van't Hoff[49] and Joseph Achille Le Bel[50] in 1874 had a profound effect on our understanding of the three-dimensional structure of organic molecules. From their investigations, van't Hoff and Le Bel suggested that the atoms attached to a saturated carbon atom are arranged at the corners of a regular tetrahedron with the saturated carbon atom in the middle. Le Bel derived his ideas from the work of Louis Pasteur, who had surmised that crystalline form may be related to the chemical and physical properties of a compound. The ideas of van't Hoff were derived from the more traditional organic chemists, particularly Friedrich August Kekulé. Both van't Hoff and Le Bel[51] were trying to find out why some compounds had more isomers[52] than were expected from the chemical formulæ. Organic chemists were then able to build three-dimensional models of organic molecules. Emil Fischer[53] described Adolph von Baeyer in the winter of 1890–1891 sitting at a table during a meal constructing "carbon atom models from bread crumbs and toothpicks." These three-dimensional molecular models were the forerunners of the models we build today using molecular modeling kits, or the models that can be viewed on a screen using computer graphics.

The concept of the tetrahedral arrangement of bonds about a carbon atom could not, however, be used to explain the structure of benzene, which has the formula C_6H_6. It was difficult to determine the arrangement of atoms in benzene simply from the proportions of atom. Kekulé[54] suggested that the structure of benzene could be considered as a hexagon of connected carbon atoms, each with one hydrogen atom attached to it. This implied alternating single and double bonds to satisfy the four valences of carbon. The manner by which Kekulé arrived at this structural formula has been the subject of much discussion. The validity of the story that he had a dream in which a snake grabbed its own tail has been questioned,[55–58] but it illustrates the problems faced by scientists at that time in deducing the correct connectivity of atoms.

On the other hand, cyclohexane, unlike benzene, is saturated, and its structure can readily be explained on the basis of a tetrahedral carbon atom. Hermann Sachse[59] in 1890 deduced that cyclohexane is a cyclic hexamer. It is, however, nonplanar with several forms for the puckered rings existing in solution in an equilibrium. These ideas, surprisingly, lay dormant for many years until Ernst Mohr[60] reintroduced them in 1915.

From then on, many more structural principles were derived by chemists and their knowledge of structure was systematized. The absolute configurations of optically active organic compounds were all related to each other by Emil Fischer[61] and Martin A. Rosanoff,[62] who used naturally occurring sugars as standards. In 1893 the inorganic chemist Alfred Werner,[63] again considering the number of isomers of a given compound, deduced that tripositive cobalt has an octahedral arrangement of groups around it and showed that bidentate ligands can give isomers that show optical activity. Two other significant concepts in structural studies (discussed in Chapter 12) are Baeyer's theory of strain[64] and Victor Meyer's principle of steric hindrance.[65] The packing of ions was studied by William Barlow in England and Gustav Tammann in Germany. William Barlow and William J. Pope,[66] as mentioned earlier, considered ions in crystals to be hard spheres touching each other. They correctly deduced the crystal structures of sodium and potassium chlorides from a model of closely packed spheres, long before these crystal structures were determined by X-ray diffraction.

1.4.2 The impact of X-ray diffraction on structure analysis

During this period of deduction of molecular structure by chemists came the discovery by von Laue[33] in 1912 of X-ray diffraction by crystals. This was accompanied by the idea of W. L. Bragg[29] that diffraction could be used to determine the atomic arrangement in crystals. As a result, all the structural studies of the nineteenth-century organic chemists, described above, could be verified not only with respect to the way the atoms were connected and their disposition in space, but also with respect to the precise angles and distances between them. John Monteath Robertson wrote on this subject,[67] "Perhaps the most striking feature of the results is the amazing verification which they afford of the stereochemical conceptions of organic chemistry. Of course it may be argued that these fundamental formulas did not stand in any need of verification — they were firmly established by chemical methods long before the diffraction of x rays was discovered, indeed before the discovery of x rays at all." He goes on to say "The greatest difference lies in the exact metrical representation of the structures which has now been achieved. The interatomic distances appear as constants which are definitely characteristic of certain types of binding between the atoms."

The crystal structure of diamond[68] was one of the first to be investigated by X-ray diffraction. Each carbon atom is surrounded equiangularly (tetrahedrally) by four other carbon atoms, just as van't Hoff and Le Bel[49,50] had predicted in 1874. But the X-ray experiment gave more information. By then, the wavelength of the X rays had been determined. The symmetry of the packing of the carbon atoms in the unit cell of these cubic crystals and the unit cell dimensions together defined the C—C distance that was calculated from these data to be 1.544 Å. At last, the dimensions of covalent molecules could be measured on an absolute scale! Furthermore, when a more extensive picture was made for several unit cells, it could be seen that diamond has a rigid three-dimensional structure (Figure 1.7), which may account for its high stability and hardness. Six-membered rings, equivalent to the carbon atom arrangement in cyclohexane, may be picked out of the network of carbon atoms. Even larger rings can also be picked out and

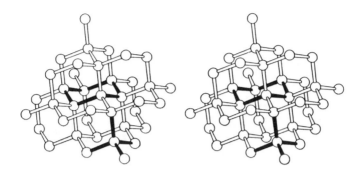

FIGURE 1.7. The crystal structure of diamond.[68] Stereoview showing several unit cells. Each carbon atom is tetrahedrally surrounded by four others. The arrangement of carbon atoms found in cyclohexane and in tetramethylmethane (shown with black bonds) can also be picked out from this rigid structure.

these shapes are stable for the saturated carbon-rings found in certain larger organic molecules.

The structure of graphite,[69,70] an allotrope of diamond, posed more problems for the crystallographer. It was difficult to interpret the diffraction pattern, and it was several years before this structure was determined. Graphite crystallizes in a hexagonal unit cell. The carbon atoms are arranged in flat sheets with C—C bond lengths of 1.42 Å and nonbonded interplanar spacings of 3.36 Å (Figure 1.8). This three-dimensional structure accounts for the physical properties of graphite. Unlike diamond, graphite can be used as a lubricant because its cleaves readily between the planes of the hexagons. Such structure also accounts for the difficulty in obtaining good crystals of graphite. The slippage of the planes of atoms relative to one another causes disorder from unit cell to unit cell.

The determination of the structure of hexamethylbenzene[71] (Figure 1.9) was an outstanding achievement by Kathleen Lonsdale in 1928. It established the hexagonal symmetry of hexamethylbenzene and, by inference, of benzene. This symmetry implies that benzene does not consist of a Kekulé-like structure, possibly in equilibrium with other Kekulé structures, but has a single chemical structure. The molecule consists of a flat, regular hexagonal arrangement of carbon atoms with the six C—C bonds of equal length. This was interpreted as demonstrating electron delocalization. Benzene is a liquid at room temperature, and so its crys-

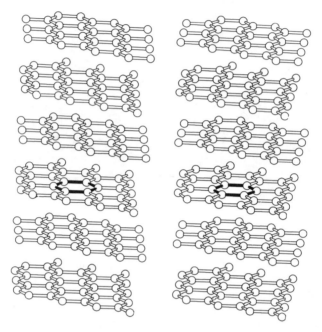

FIGURE 1.8. The crystal structure of graphite.[69,70] This consists of hexagonal sheets of carbon atoms with weak forces between the sheets. Note that the arrangement of the carbon atoms in benzene (black bonds) is a portion of this structure.

talline hexamethyl derivative was chosen for X-ray diffraction study at that time. The structure of benzene itself was determined[72] by X-ray diffraction techniques in 1954. The diffraction experiment was conducted at $-3°C$. Since benzene crystals melt at $5.5°C$, the crystal was nearly melting when studied, and the atoms (and molecules) were relatively mobile in the crystal. As a result, the C—C bond lengths were poorly determined. A few years later Durward W. J. Cruickshank and co-workers[73] showed how the interatomic distances could be corrected for thermal motion, giving values that agreed with results from spectroscopic observations.[74] These concepts will be discussed in more detail in the chapter on thermal motion and disorder, Chapter 13. Neutron diffraction[75] and other studies later confirmed that all six C—C bonds were of equal length. Deductions about the crystal structures of polycyclic aromatic compounds were made by W. H. Bragg.[76,77] He showed that the sizes of the three-dimensional packing units (unit cells) of naphthalene and anthracene differ mainly in one dimension and that this difference, 2.9 Å, is approximately the width of a hexagon in graphite. Bragg thus deduced that these planar polycyclic molecules lie with their long axes along that one direction.

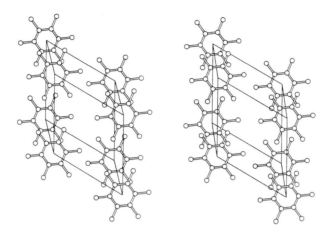

FIGURE 1.9. Stereoview of the crystal structure of hexamethylbenzene.[71] Hydrogen atoms are omitted. Unit-cell edges are indicated.

The structures of cyclohexane derivatives were also determined by X-ray diffraction techniques. Measurements on the β-isomers of hexachloro-cyclohexane and hexabromocyclohexane[78] verified the ideas of Sachse[59] regarding cyclohexane. The determination of the locations of the halogen atoms in the unit cell showed that these molecules exist in the chair conformation with substituents in so-called *equatorial* positions as shown in Figure 1.10.

A significant contribution to our understanding of the structure of steroids was made long before it was possible to determine the atomic coordinates of these more complicated molecules. Bernal and his co-workers[79] measured not only the unit cell dimensions of many steroids but also their **refractive indices**. The unit cell dimensions and the directions of the largest and smallest refractive indices with respect to the edges of the unit cell indicated the orientation of the steroid molecules in their unit cells. They showed that steroid molecules have dimensions 17–20 Å long, 7–8 Å wide, and 5 Å thick.[79] It must be stressed that Bernal did not use unit cell dimensions alone to determine the dimensions of a steroid, but used them in combination with refractive index data that showed the directions of the long and short axes of the molecules in the crystals. When a model was built using the current formula for steroids proposed by Heinrich Wieland and Adolf Windaus with C—C bonds of 1.5 Å and tetrahedral C—C—C angles, the model in Figure 1.11 (for-

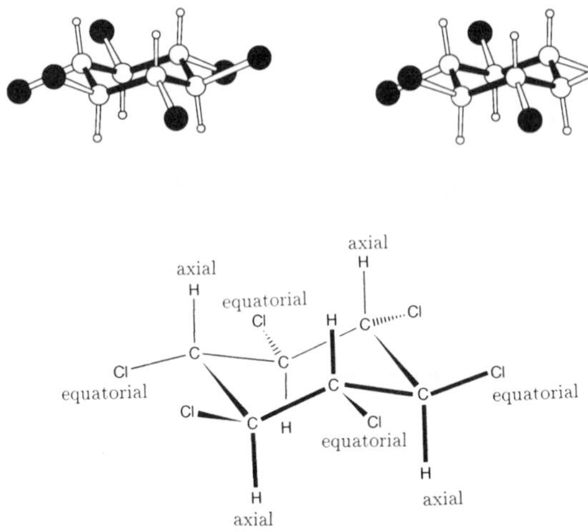

(a)

(b)

FIGURE 1.10. Stereoview of hexachlorocyclohexane,[78] $C_6H_6Cl_6$, in the chair form with all chlorine substituents (filled circles) in *equatorial* positions (most nearly in the plane of the overall ring system, the "equator"). This implies that the hydrogen atoms are all in *axial* positions (perpendicular to the plane of the overall ring system). (a) Stereoview and (b) designations of *axial* and *equatorial*.

mula **I**) had dimensions $18.0 \times 7.0 \times 8.5$ Å³. This would not fit into the measurements derived by Bernal from his X-ray diffraction studies. The same difficulty arose with respect to formula **II**. Further studies of steroid-molecule dimensions were made using the Langmuir trough. These gave the surface area of a monomolecular layer. If the amount of material in the monolayer is known, molecular cross-section areas can be determined.[80] Measurements on monolayers of several steroids reinforced Bernal's conclusions. Cross sections of 37–44 Å² were obtained that agreed with the values for the two smaller steroid dimensions obtained from the X-ray and optical experiments (35–40 Å²). In response to these various experimental data, an alternate chemical formula for steroids (formula **III**), derived from chrysene, was suggested by Otto Rosenheim and Harold King.[81] This formula was slightly modified (formula **IV**) by Heinrich Wieland and Elisabeth Dane.[82] In this way, the structure of the ring system in steroids was established. Bernal and co-workers[83,84] continued these studies and reported measurements of unit cell dimensions and refractive indices for 105 steroids. They clearly demonstrated how the steroid molecules packed together in the crystal with the formation of hydrogen bonds. Further corroboration was provided later by the determination of the crystal structure of cholesteryl iodide,[85] shown

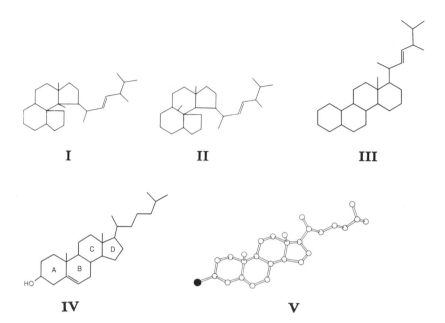

I   II   III

IV   V

FIGURE 1.11. The derivation of the chemical formulæ of steroids resulting from Bernal's X-ray crystallographic work.[79] **I** and **II** had been put forward as possible formulæ, but **III** seemed to provide a better fit to the experimental data on molecular size from refractive index data combined with unit cell dimensions.[81] Bernal listed sizes $8.5 \times 7.0 \times 18$ Å^3 for **I**, $11.0 \times 7.5 \times 15$ Å^3 for **II**, $7.5 \times 4.5 \times 20$ Å^3 for **III** (from models built to scale), and $7.2 \times 5.0 \times 17$–20 Å^3 from observed data from X-ray studies and refractive index measurements. The correct formula, which is very similar to formula **III**, is shown in formula **IV** (cholesterol).[82] The crystal structure of cholesteryl iodide[85] is shown as **V**. (The value of the parameter x reported for C15 apparently should be 0.585, not that reported in ref. 85.) Iodine atom, filled circle.

as **V** in Figure 1.11. Today a wealth of information is available in a huge compendium of data listing the results of X-ray analyses of steroid structures.[86] These volumes were referred to by one organic chemist as a "mail-order catalog of steroid structures!"

As techniques for structure determination by X-ray diffraction evolved, molecules of even greater complexity could be studied. Two molecules of special biological interest were penicillin,[87] the important antibiotic discovered by Alexander Fleming,[88] and vitamin B_{12},[89] used in the treatment of pernicious anemia. Several possible structures had been proposed for penicillin, but the β-lactam structure was considered unlikely by Robert Robinson and John Cornforth from the point of view of organic chemists.[90] To their surprise, X-ray diffraction studies of the potassium and rubidium salts of benzylpenicillin[87] proved that penicillin had this β-lactam structure (Figure 1.12). Vitamin B_{12}[89] (Figure 1.13) has an even more complex structure. In fact, it was the largest molecule of unknown chemical formula to be studied at that time. Sir Lawrence Bragg[91] referred to its structure determination as breaking "the sound barrier." By the 1960s, it became possible to determine structures of biological macromolecules, although at lower resolution. The first protein structure to be determined was that of myoglobin by John Kendrew and co-workers.[92] This work was followed by a similar determination of the structure of hemoglobin.[93,94] Since that time, many protein and nucleic acid structures have been studied to atomic resolution.

Summary

1. A **crystal** is a solid with a regularly repeating arrangement of atoms in its internal structure.
2. X rays are used because their wavelengths are of the order of the distances between atoms (10^{-8} cm $= 1$ Å or 1 **Ångström unit**).

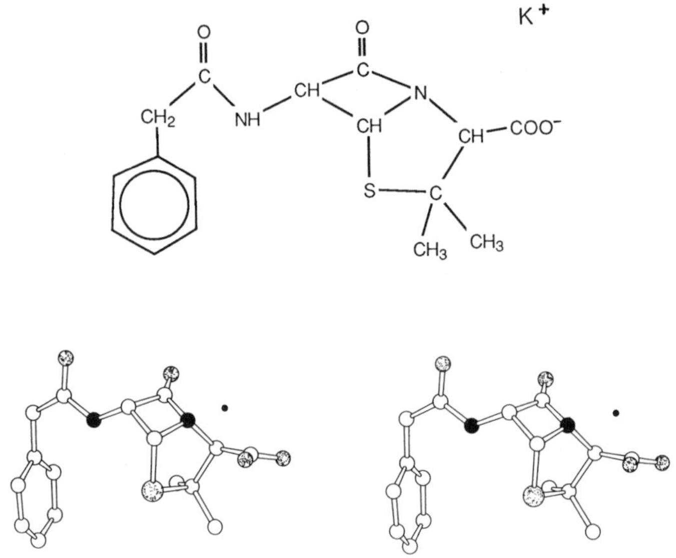

FIGURE 1.12. (a) Chemical formula and (b) stereoview of the potassium salt of the benzyl derivative of the antibiotic penicillin.[87] This crystallographic analysis established this chemical formula of penicillin showing that it contained, unexpectedly, a β-lactam (four-membered) ring. Circles representing atoms: S, large stippled; O, small stippled; N, small black; C, small white; K^+, very small black; hydrogen atoms omitted.

3. Each molecule diffracts X rays, but the regularly repeating internal structure of a crystal causes reinforcement of the **diffraction** effect making it become readily measurable.

4. Unlike visible light, X rays cannot be focused, so we use diffraction of X rays by crystals to "see" molecules, but, to do this, the microscope lens must be replaced by a mathematical computation (a **Fourier synthesis**).

5. X-ray diffraction methods yield data that can result in precise determinations of the positions of atoms in a crystal. The molecular geometry can then be calculated with high precision.

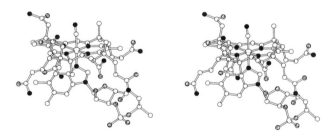

(a)

(b)

FIGURE 1.13. (a) Chemical formula and (b) stereoview of the anti-pernicious anæmia factor, vitamin B_{12} cyanocobalamin (Me = $-CH_3$, Ac = $-CH_2COOH$, Pr = $-CH_2CH_2COOH$). This crystallographic study[89] established the chemical formula of the vitamin (nitrogen atoms black, oxygen atoms stippled).

Glossary

Amplitude: The height of a wave. It is the maximum numerical value of a periodic function, such as a cosine wave function, measured from its mean value. Thus it is half the peak-to-valley displacement. The square of the amplitude is the intensity of the wave.

Ångström unit: The unit of length commonly used in reporting the results of crystal structure analyses. $1 \text{ Å} = 10^{-8}$ cm. It is named after Anders Jonas Ångström, a spectroscopist from Upsala, Sweden.

Bragg: There are two crystallographers called Bragg mentioned here. W. H. Bragg, the father, pioneered instrumental methods of measuring X-ray diffraction. W. L. Bragg, the son, developed methods for analyzing the experimental data in terms of the atomic structure of the crystal.

Conformation: One of the various shapes that a molecule can adopt as a result of rotation about single bonds.

Crystal: A solid with a regularly repeating internal arrangement of atoms.

Diffraction: When radiation passes through a narrow aperture, some waves appear to be deflected to a small extent so that some light appears in the area normally considered to be geometric shadow. This bending of light is best explained as the interference of secondary waves generated in the area of the slit. Each such generated (secondary) wave traveling in a given direction will be slightly out of phase with neighboring waves by an amount that depends on the wavelength and the angle of deviation of the secondary wave from the direct beam. The shorter the wavelength or the larger the angle, the more a wave is out of phase with its neighbor. The intensity of the beam in a given direction may be determined, according to the theory of Huygens, by a superposition of all the waves scattered in that direction. The more that waves are out of phase, the weaker the resultant intensity of the combined waves.

Diffraction grating: A system of close, equidistant, parallel lines, ruled on a polished surface, that causes diffraction to occur when light impinges on it.

Diffraction pattern: The intensity pattern obtained when light is diffracted by an object with a spacing of the order of the wavelength of that light. For a crystal, this is the experimentally measured values of intensities, diffracting angle (direction), and order of diffraction for each diffracted beam obtained when the crystal is placed in a narrow beam of radiation such as X rays or neutrons.

Faces of a crystal: The smooth, flat surfaces of a crystal that intersect giving sharp edges and that show definite symmetrical relationships leading to information on the symmetry of the internal structure of the crystal.

Fourier synthesis: A method of summing waves (such as scattered X rays) to obtain a periodic function (such as a representation of the electron density in a crystal). (See Chapter 6 glossary for a more detailed definition.)

Interference: The mutual effect of two waves traveling in the same direction on each other. If one wave enhances the intensity of the other the interference is said to be constructive; if one wave decreases the intensity of the other the interference is said to be destructive.

Phase: The point, expressed as a fraction of the wavelength, to which the crest of a given wave has advanced in relation to a standard position, e.g., the starting point or origin of the unit cell. The relative phase of two waves of the same wavelength traveling in the same direction is the difference in the positions of their crests, measured as a fraction of the wavelength.

Phase problem: The problem of determining the phase angle (relative to a chosen origin) that is to be associated with each diffracted wave that is combined to give an electron-density map. The measured intensities of diffracted beams give only the squares of the amplitudes, but the relative phases cannot normally be determined experimentally (see Chapter 8). The determination of the relative phases of the Bragg reflections is crucial to the calculation of the correct electron density map.

Refractive index: The ratio of the velocity of light in a vacuum to its lower velocity in a material under study. The refractive index gives an indication of the ability of a substance to bend (refract) light. When a colorless substance is immersed in a colorless liquid with the same refractive index, the crystal becomes invisible.

Resolution of a crystal structure: The process of distinguishing two close objects as separate entities rather than as a single, blurred object.

Unit cell: The basic building block of a crystal. (See Chapter 2 glossary for a more detailed definition).

Units of length:

Meter:	1 m	=	1	m =	10^2	cm =	10^{10}	Å.
Millimeter:	1 mm	=	10^{-3}	m =	10^{-1}	cm =	10^7	Å.
Micrometer:	1 μm	=	10^{-6}	m =	10^{-4}	cm =	10^4	Å.
Nanometer:	1 nm	=	10^{-9}	m =	10^{-7}	cm =	10	Å.
Ångström:	1 Å	=	10^{-10}	m =	10^{-8}	cm =	1	Å.
Picometer:	1 pm	=	10^{-12}	m =	10^{-10}	cm =	10^{-2}	Å.

X rays: Electromagnetic radiation with a wavelength of the order of 10^{-8} cm emitted when a metal target is bombarded by fast electrons. X rays with wavelengths characteristic of the elemental composition of the metal target are emitted when an electron that has been displaced from an inner shell of an atom is replaced by another electron that falls in from an outer shell with loss of energy. This energy is detected as X rays of a specific energy (wavelength).

Wavelength: The distance between two similar points, for example the crests of a cosine wave.

References

1. Abbé, E. Beiträge zur Theorie des Mikroskops und der mikroskopischen Wahrnehmung. I. Die Konstruction von Mikroskopen auf Grund der Theorie; II. Die dioptrischen Bedingungen der Leistung des Mikroskops; III. Die physikalischen Bedingungen für die Abbildung feiner Structuren; IV. Das optische Vermögen des Mikroskops. [Contributions to the theory of the microscope and microscopic observations. I. Construction of a microscope on the basis of this theory. II. The dioptric conditions for the working of a microscope. III. Physical conditions for the imaging of small structures. IV. The optical properties of the microscope.] *Archiv für Mikroskopische Anatomie* 9, 413–468 (1873).
2. Porter, A. B. On the diffraction theory of microscopic vision. *Phil. Mag.* 11, 154–166 (1906).

3. Waser, J. Pictorial representation of the Fourier method of X-ray crystallography. *J. Chem. Educ.* **45**, 446–451 (1968).

4. Sayre, D., Kirz, J., Feder, R., Kim, B., and Spiller, E. Potential operating region for ultrasoft X-ray microscopy of biological materials. *Science* **196**, 1339–1340 (1977).

5. Binnig, G., Rohrer, H., Gerber, C., and Weibel, E. Tunneling through a controllable vacuum gap. *J. Appl. Phys.* **40**, 178–180 (1982).

6. Huygens, C. *Traité de la lumière où sont expliquées les causes de ce qui luy arrive dans la réflexion et dans la réfraction. Et particuliérement dans l'étrange réfraction du cristal d'Islande.* [Treatise on light in which the causes of the events that result in reflection and refraction are explained. Of particular interest is the unusual refraction of Iceland spar.] Pierre van der Aa: Leiden (1690). **English translation**: Thompson, S. P. Macmillan: London (1912).

7. Gramaccioli, C. M. *Il Meraviglioso Mondo dei Cristalli.* [The Marvelous World of Crystals.] Calderini: Bologna (1986).

8. Kepler, J. *Strena seu de Nive Sexangula.* Godefridum Tampach: Francofurti ad Mœnum (1611). **English translation** of excerpts: Silverman, J. S. *A New Year's Present; on Hexagonal Snow.* In: *Crystal Form and Structure* (**Ed.**, Schneer, C. J.) pp. 16–17. Dowden, Hutchinson & Ross: Stroudsburg, PA (1977). Also: Hardie, C. *The Six-cornered Snowflake.* Clarendon Press: Oxford (1966).

9. Hooke, R. *Micrographia: or some Physiological Descriptions of Minute Bodies made by Magnifying Glasses. With Observations and Inquiries thereupon.* J. Martyn and J. Allestry (for the Royal Society): London (1665). Reprinted as: Gunther, R. T. *Early Science in Oxford. Vol. XIII. The Life and Work of Robert Hooke (Part V). Micrographia, 1665.* Oxford University Press: Oxford (1938).

10. Steno, N. *De Solido intra Solidum Naturaliter Contento. (Dissertationis prodromus).* [The prodromus to a dissertation concerning solids naturally contained within solids.] Stellæ: Florentiæ (1669). **English translation**: 1671. Oldenburgh, H. [Published by J. Winter: London (1671)] Also in: Ostwald's *Klassiker der exacten Wissenschaften,* Number **209**. Akademie Verlagsgesellschaft: Frankfurt/M (1967). Also: *The prodromus of Nicolaus Steno's dissertation concerning a solid body enclosed by the process of nature within a solid.* **English translation**: Winter, J. G. Macmillan: New York (1916).

11. Romé de Lisle, J. B. L. *Essai de Cristallographie, ou Description des Figures géométriques, propres à différents Corps du Règne Minéral, connus vulgairement sous le Nom de Cristaux.* [Essay on crystallography, or a description of geometric figures, characteristic of different materials in the mineral kingdom, known commonly under the name of crystals.] 1st edn. Didot, Knapen and Delaguette: Paris (1772). 2nd edn. *Cristallographie, ou Description des Formes propres à tous les Corps du Règne minéral, dans l'Etat de Combination saline, pierreuse ou métallique.* 4 vols. Impr. de Monsieur: Paris (1783).

12. Haüy, R. J. *Traité Élémentaire de Physique.* [Elementary treatise on physics.] 2 vols. Delanee and Lesueur: Paris (1804). **English translation,** abridged: In: *Crystal Form and Structure.* (**Ed.**, Schneer, C. J.). pp. 18–20. Dowden, Hutchinson & Ross: Stroudsburg, PA (1977).

13. Haüy, R. J. (Abbé Haüy) *Essai d'une Théorie sur la Structure des Crystaux appliquée à plusieurs genres de substances crystallisées.* [Essay on a theory concerning the structure of crystals applied to several types of crystallized materials.] Gogué & Née de la Rochelle: Paris (1783). Additional treatises were published. (Citoyen Haüy) *Traité de Minéralogie.* Vol. 1. Louis: Paris (1801)) and Bachelier & Huzard: Paris (1822).

14. Bergman, T. Variæ Crystallorum Formæ e Spatha Ortis explicatæ. [Various crystal shapes explained by the position of heavenly bodies.] *Nova Acta Reg. Soc. Sci. Upsal.* 1, 150–155 (1773). **English translation**: Cullen, E. *Physical and Chemical Essays.* 3 volumes. J. Murray: London (1784).

15. Dalton, J. *A New System of Chemical Philosophy.* 2 vols. S. Russel: Manchester (1808).

16. Baumgärtner, K. H. *Beobachtungen über die Nerven und das Blut.* [Observations on nerves and blood.] Groos: Freiburg (1830).

17. Reichert, E. T., and Brown, A. P. *The Differentiation and Specificity of Corresponding Proteins and Other Vital Substances in Relation to Biological Classification and Organic Evolution: The Crystallography of Hemoglobin.* Publication No. 116. Carnegie Institute of Washington: Washington, DC (1909).

18. Sumner, J. B. The isolation and crystallization of the enzyme urease. Preliminary paper. *J. Biol. Chem.* **69**, 435–441 (1926).

19. Northrop, J. H. Crystalline pepsin: I. Isolation and tests of purity; II. General properties and experimental methods. *J. Gen. Physiol.* **13**, 739–780 (1930).

20. Bernal, J. D., and Crowfoot, D. X-ray photographs of crystalline pepsin. *Nature (London)* **133**, 794–795 (1934).

21. Stanley, W. M. Isolation of a crystalline protein possessing the properties of tobacco-mosaic virus. *Science* **81**, 644–645 (1935).

22. Bawden, F. C., and Pirie, N. W. The isolation and some properties of liquid crystalline substances from solanaceous plants infected with three strains of tobacco mosaic virus. *Proc. Roy. Soc. (London)* **B123**, 274–320 (1937).

23. Bernal, J. D., and Fankuchen, I. X-ray and crystallographic studies of plant virus preparations. I. Introduction and preparation of specimens. II. Modes of aggregation of the virus particles. III. *J. Gen. Physiol.* **25**, 111–165 (1941).

24. Bravais, A. Études cristallographiques, Part I: Du cristal considéré comme un simple assemblage de points. Part II. Du cristal considéré comme un assemblage de molécules polyatomiques. *Journal de l'École Polytechnique (Paris)* (cahier 34) pp. 101–194, 194–247. Bachelier: Paris (1851). **English translation**: [Part I. The crystal considered as a simple assemblage of points. Part II. The crystal considered as an assemblage of polyatomic molecules.] Bonnabaud, F. pp. 165–170, pp. 194–202 of above reference.

25. Bravais, A. Mémoire sur les systèmes formés par des points distribuées régulièrement sur un plan ou dans l'espace. *Journal de l'École Polytechnique (Paris)* **19** (cahier 33), 1–128. Printed vol: Bachelier: Paris (1850). **English translation**: Shaler, A. J. On the systems formed by points regularly distributed on a plane or in space. Crystallographic Society of America Memoir # 1 (Monograph # 4) New York (1949).

26. Seeber, L. A. Versuch einer Erklärung des inneren Baues der festen Körper. [Attempts to explain the internal structure of a solid body.] *Gilbert's Annalen der Physik* **76**, 229–248, 349–372 (1824).

27. Barlow, W. Probable nature of the internal symmetry of crystals. *Nature (London)* **29**, 186–188, 205–207 (1883).

28. Ewald, P. P. (**Ed.**) *Fifty Years of X-ray Diffraction.* p. 62. Oosthoek: Utrecht (1962).

29. Bragg, W. L. The structure of some crystals as indicated by their diffraction of X rays. *Proc. Roy. Soc. (London)* **A89**, 248–277 (1913).

30. Grimaldi, F. M. *Physico-Mathesis de Lumine, Coloribus, et Iride.* [Physical insights into light, colors, and the spectrum (rainbow)] Hæredis Victorii Benatii: Bononiae [Bologna] (1665).

31. Fraunhofer, J. von. Kurze Berichte von den Resultaten neuer Versuche über die Gesetze des Licht, und die Theorie Derselben. [Short communications on the results of new experiments on the principles of light and its theory.] *Denkschriften der königlichen Akademie der Wissenschaften zu München* **8**, p. 1. (1821–1822); *Gilberts Annalen der Physik* **74**, 337–378 (1823). **English translation**: Ames, J. S. *Prismatic and Diffraction Spectra. Memoirs by Joseph von Fraunhofer.* pp. 11–38. Harper: New York, London (1898).

32. Schwerd, F. M. *Die Beugungserscheinungen, aus den Fundamentalgesetzen der Undulationstheorie analytisch entwickelt.* [Diffraction derived from the fundamental laws of wave theory.] Schwan and Goetz: Mannheim (1835).

33. Friedrich, W., Knipping, P., and Laue, M. Interferenz-Erscheinungen bei Röntgenstrahlen. [Interference phenomena with X rays.] *Sitzungsberichte der mathematisch-physikalischen Klasse der Königlichen Bayerischen Akademie der Wissenschaften zu München*, pp. 303–322 (1912). **English translation**: Stezowski, J. J. In: *Structural Crystallography in Chemistry and Biology.* (**Ed.**, Glusker, J. P.) pp. 23–39. Hutchinson & Ross: Stroudsburg, PA (1981). [Laue became von Laue after this article was published].

34. Röntgen, W. C. Über eine neue Art von Strahlen. *Sitzungsberichte der Würzburger Physikalischen-Medizinischen Gesellschaft*, pp. 132–141 (1895). **English translation**: Stanton, A. On a new kind of ray. *Science* **3**, 227–231 (1896); *Nature (London)* **53**, 274–276 (1896).

35. Barkla, C. G. The spectra of fluorescent Röntgen radiations. *Phil. Mag., Ser. 6* **22**, 396–412 (1911).

36. Walter, B., and Pohl, R. Zur Frage der Beugung der Röntgenstrahlen. [Concerning the question of X-ray diffraction.] *Annalen der Physik* **25**, 715–724 (1908).

37. Walter, B., and Pohl, R. Weitere Versuche über die Beugung der Röntgenstrahlen. [Further experiments on X-ray diffraction.] *Annalen der Physik* **29**, 331–354 (1909).

38. Condon, E. U., and Shortley, G. H., Jr. *The Theory of Atomic Spectra.* Cambridge University Press: New York, London (1935).

39. Birge, R. T. Probable values of the general physical constants. *Phys. Rev. Supplement* **1**, 1–73 (1929).

40. Bijvoet, J. M., Kolkmeyer, N. H., and McGillavry, C. H. *X-ray Analysis of Crystals.* Interscience: New York; Butterworths: London (1951). Originally: *Röntgenanalyse van Kristallen.* 2nd edn., revised. D. B. Centen: Amsterdam (1948).

41. Elsasser, W. M. Sur la diffraction des neutrons lents par les substances cristallines. [On the diffraction of slow neutrons by crystalline materials.] *Comptes Rendus, Acad. Sci. (Paris)* **202**, 1029–1030 (1936).

42. Mitchell, D. P., and Powers, P. N. Bragg reflection of slow neutrons. *Phys. Rev.* **50**, 486–487 (1936).

43. Davisson, C., and Germer, L. H. The scattering of electrons by a single crystal of nickel. *Nature (London)* **119**, 558–560 (1927).

44. Bragg, W. H. X rays and crystals. (Letter to the editor). *Nature (London)* **90**, 360 (1912).

45. Bacon, G. E. *Neutron Diffraction.* 2nd edn. Oxford University Press: Oxford (1962).

46. Vainshtein, B. K. *Structure Analysis by Electron Diffraction.* **English translation:** Feigl, E., and Spink, J. A. Macmillan: New York (1964).

47. Ewald, P. P. Zur Begründung der Kristalloptik. Teil III. Die Kristalloptik der Röntgenstrahlen. [Explanation of crystal optics. Part III. The crystal optics of X rays.] *Annalen der Physik* **54**, 519–556 (1917).

48. Burke, J. G. *Origins of the Science of Crystals.* University of California Press: Berkeley, Los Angeles (1966).

49. van't Hoff, J. H. *Voorstel tot uitbreiding der tegen woordig in de scheikunde gebruike structuur-formules in de ruimte, benevens een daarme samehangende vermogen en chemische constitutie van organische verbindingen.* [A suggestion concerning the extension into space of the structural formulas at present used in chemistry and a note upon the relation between the optical activity and the chemical constitution of organic compounds.] J. Greven: Utrecht (1874). **French translation:** Sur les formules de structure dans l'espace. *Archives Néerlandaises des Sciences Exactes et Naturelles,* **9**, 445–454 (1874). **English translation:** *Arrangement of Atoms in Space.* Longmans, Green: London (1898); also in: *A Source Book in Chemistry 1400–1900.* (Eds., Leicester, H. M., and Klickstein, H. S.) p. 445–458. McGraw-Hill: New York, Toronto, London (1952).

50. Le Bel, J. A. Sur les relations qui existent entre les formules atomiques des corps organiques, et le pouvoir rotatoire de leurs dissolutions. [On the relations which exist between the atomic formulas of organic compounds and the rotatory power of their solutions.] *Société chimique de France Bulletin,* **22**, 337–347 (1874). **English translation:** In: *Classics in the Theory of Chemical Combinations.* Benfy, O. T. pp. 151–171. Dover: New York (1963); also in: *A Source Book in Chemistry 1400–1900.* (Eds., Leicester, H. M., and Klickstein, H. S.) p. 459–462. McGraw-Hill: New York, Toronto, London (1952).

51. Ramsay, O. B. (**Ed.**). *Van't Hoff–Le Bel Centennial.* ACS Symposium Series 12. American Chemical Society: Washington, DC (1975).

52. Meister, O. Sitzung der Chemischen Harmonika vom 2. Novbr. [Meeting of the "Chemical Harmonika Group" on November 2nd.] [A group of theoretically minded German chemists who met regularly to discuss fundamental problems of chemistry and published once a year. Thanks to Theo Hahn for this information.] *Ber. Deutsch. Chem. Gesell.* **2**, 619–621 (1869).

53. Fischer, E. *Aus meinem Leben.* [From my life.] p. 134. Springer: Berlin (1922).

54. Kekulé, A. Ueber die Konstitution und die Metamorphosen der chemischen Verbindungen und über die chemische Natur des Kohlenstoffs. [The constitution and metamorphoses of chemical compounds and the chemical nature of carbon.] *Liebig's Annalen der Chemie.* **106**, 129–159 (1858); Untersuchungen über aromatische Verbindungen. [Studies of aromatic compounds.] *Liebig's Annalen der Chemie und Pharmacie.* **137**, 129–196 (1865); Sur la constitution des substances aromatiques. [On the constitution of aromatic materials.] *Bulletin de la Société chimique Française* **3**, 98–110 (1865).

55. Verkade, P. E. August Kekulé. In: *Theoretical Organic Chemistry.* Papers presented to the Kekulé Symposium organized by the Chemical Society (London, September, 1958). pp. ix–xvii. Butterworth: London (1959).

56. Lipeles, E. S. Profiles in Chemistry: Friedrich August Kekule *J. Chem. Educ.* **58**, 624–625 (1981).

57. Wotiz, J. H., and Rudofsky, S. Kekulé's dreams: fact or fiction? *Chem. Britain* **20**, 720–723 (1984).

58. Ramsay, O. B., and Rocke, A. J. Kekulé's dreams. Separating the fiction from the fact. *Chem. Britain* **20**, 1093–1094 (1984).

59. Sachse, H. Über die geometrischen Isomerien der Hexamethylenderivate. [Concerning geometrical isomers of hexamethylene derivatives.] *Ber. Deutsch. Chem. Gesell.* **23**, 1363–1370 (1890).

60. Mohr, E. Die Baeyersche Spannungstheorie und die Struktur des Diamants. [Baeyer's tension theory and the structure of diamond.] *Chemisches Zentralblatt* **2**, 1065 (1915).

61. Fischer, E. Synthesen in der Zuckergruppe I, II. [Syntheses in sugar groups I, II.] *Ber. Deutsch. Chem. Gesell.* **23**, 2114–2141 (1890); **27**, 3189–3232 (1894).

62. Rosanoff, M. A. On Fischer's classification of stereo-isomers. *J. Amer. Chem. Soc.* **28**, 114–121 (1906).

63. Werner, A. Beitrag zur Konstitution anorganischer Verbindungen. [Contribution concerning the constitution of inorganic compounds.] *Z. Anorg. Allgemeine Chem.* **3**, 267–342 (1893). **English translation**: In: Kauffman, G. B. *Classics in Coordination Chemistry, Part I: The Selected Papers of Alfred Werner.* pp. 5–88. Dover: New York (1968).

64. Baeyer, A. Über Polyacetylenverbindungen. [Concerning polyacetylene compounds.] *Ber. Deutsch. Chem. Gesell.* **18**, 2269–2281 (1885). **English translation**: In: Leicester, H. M. and Klickstein, H. S. *A Source Book in Chemistry 1400–1900.* p. 465–467. McGraw-Hill: New York, Toronto, London (1952).

65. Meyer, V. Ergebnisse und Ziele der stereochemischen Forschung. [Results and goals of stereochemical research.] *Ber. Deutsch. Chem. Gesell.* **23**, 567–630 (1890).

66. Barlow, W., and Pope, W. J. The relation between the crystalline form and chemical constitution of simple inorganic substances. *Trans. Chem. Soc. (London)* **91**, 1150–1214 (1907).

67. Robertson, J. M. Metrical representation of some organic structures by quantitative X-ray analysis. *Chem. Rev.* **16**, 417–437 (1935).

68. Bragg, W. H., and Bragg, W. L. The structure of the diamond. *Nature (London)* **91**, 557 (1913).

69. Bernal, J. D. The structure of graphite. *Proc. Roy. Soc. (London)* **A106**, 749 (1924).

70. Hassel, O., and Mark, H. Über die Kristallstruktur des Graphits. [The crystal structure of graphite.] *Z. Physik* **25**, 317–337 (1924).

71. Lonsdale, K. The structure of the benzene ring. *Nature (London)* **122**, 810 (1928).

72. Cox, E. G., and Smith, J. A. S. Crystal structure of benzene at –3°C. *Nature (London)* **173**, 75 (1954).

73. Cox, E. G., Cruickshank, D. W. J., and Smith, J. A. S. The crystal structure of benzene at –3°. *Proc. Roy. Soc. (London)* **A247**, 1–21 (1958).

74. Stoicheff, B. P. High resolution Raman spectroscopy of gases. II. Rotational spectra of C_6H_6 and C_6D_6 and internuclear distances in the benzene molecule. *Can. J. Phys.* **32**, 339–346 (1954).

75. Bacon, G. E., Curry, N. A., and Wilson, S. A. A crystallographic study of solid benzene by neutron diffraction. *Proc. Roy. Soc. (London)* **A279**, 98–110 (1964).

76. Bragg, W. H. The structure of organic compounds. *Proc. Phys. Soc. (London)* **34**, 33–50 (1921).

77. Bragg, W. H. Crystalline structure of anthracene. *Proc. Phys. Soc. (London)* **35**, 167 (1922).

78. Dickinson, R. G., and Bilicke, C. The crystal structures of beta benzene hexabromide and hexachloride. *J. Amer. Chem. Soc.* **50**, 764–770 (1928).

79. Bernal, J. D. Carbon skeleton of the sterols. *Chemistry and Industry* **51**, 466 (1932).

80. Adam, N. K., and Rosenheim, O. The structure of surface films. Part XIII. Sterols and their derivatives. *Proc. Roy. Soc. (London)* **A126**, 25–34 (1929).

81. Rosenheim, O., and King, H. The ring system of sterols and bile acids. *Chemistry and Industry* **51**, 464–466 (1932).

82. Wieland, H., and Dane, E. Untersuchungen über die Konstitution der Gallensäuren. Über den Aufbau des Gesamtgerüstes und über die Natur von Ring D. [The constitution of bile acids. I. The structure of the entire framework and the nature of ring D.] *Hoppe-Seylers Z. Physiol. Chem.* **216**, 91–104 (1933).

83. Bernal, J. D., and Crowfoot, D. X-ray crystallographic data on the sex hormones, œstrone, androsterone, testosterone and progesterone and related substances. *Z. Krist.* **A93**, 464–480 (1936).

84. Bernal, J. D., Crowfoot, D., and Fankuchen, I. X-ray crystallography and the chemistry of the steroids, Part 1. *Phil. Trans. Roy. Soc. (London)* **A239**, 135–182 (1940).

85. Carlisle, C. H., and Crowfoot, D. The crystal structure of cholesteryl iodide. *Proc. Roy. Soc.* (*London*) **A184**, 64–83 (1945).

86. Duax, W. L., and Norton, D. A. (**Eds.**) *Atlas of Steroid Structure.* Vol. 1. Plenum Press: New York, Washington, London (1975). *Atlas of Steroid Structure.* Vol. 2. (**Eds.**, Griffin, J. F., Duax, W. L., and Weeks, C. M.) Plenum: New York, Washington, London (1984).

87. Crowfoot, D., Bunn, C. W., Rogers-Low, B. W., and Turner-Jones, A. The X-ray crystallographic investigation of the structure of penicillin. In: *The Chemistry of Penicillin.* (**Eds.**, Clarke, H. T., Johnson, J. R., and Robinson, R.) Ch. XI, pp. 310–367. Princeton University Press: Princeton (1949).

88. Fleming, A. On the antibacterial action of cultures of a penicillium with special reference to their use in isolation in *B. influenzæ*. *Brit. J. Exp. Path.* **10**, 226–236 (1929).

89. Hodgkin, D. C., Pickworth, J., Robertson, J. H., Trueblood, K. N., Prosen, R. J., and White, J. G. The crystal structure of the hexacarboxylic acid derived from B_{12} and the molecular structure of the vitamin. *Nature* (*London*) **176**, 325–328 (1955).

90. Perutz, M. Forty years of friendship with Dorothy. In: *Structures of Molecules of Biological Interest.* (**Eds.**, Dodson, G. G., Glusker, J. P., and Sayre, D.) Oxford University Press: Oxford (1981).

91. Bragg, W. L. The growing power of X-ray analysis. In: *Fifty Years of X-ray Diffraction.* (**Ed.**, Ewald, P. P.). Oosthoek: Utrecht. Chapter IV, part 8, pp. 120–135 (1962).

92. Kendrew, J. C., Dickerson, R. E., Strandberg, B. E., Hart, R. G., and Davies, D. R. Structure of myoglobin. A three-dimensional Fourier synthesis at 2 Å resolution. *Nature* (*London*) **185**, 422–427 (1960).

93. Perutz, M. F., Muirhead, H., Cox, J. M., Goaman, L. C. G., Mathews, F. S., McGandy, E. L., and Webb, L. E. Three-dimensional Fourier synthesis of horse oxyhæmoglobin at 2.8 Å resolution: (1) X-ray analysis. *Nature* (*London*) **219**, 29–32 (1968).

94. Perutz, M. F. Stereochemistry of cooperative effects in hæmoglobin. *Nature* (*London*) **228**, 726–734 (1970).

2

Crystals

The chemist or biochemist who wants to determine the crystal structure of a molecule must first obtain diffraction-quality crystals. In this chapter we will describe some of the general methods commonly used for obtaining crystals and what can be learned from their unit cells.

We are all familiar with crystals, such as quartz, amethyst, and iron pyrite, which occur naturally in rock formations, often as well-formed crystalline specimens. Many are collected and prized for their æsthetic beauty; the development of faces on these crystalline minerals gives rise to the shapes that are appreciated by those who are artistically inclined. Instead of merely admiring these minerals, early scientists made precise measurements of their external structure, such as the angles between the faces, the constancy of which implies that there is an internal regularity in crystals, as described in Chapter 1. Such observations led to a preliminary understanding of the regularly repeating internal structure of a crystal. The established basic scientific definition of a crystal is that it is *a solid composed of a regularly repeated arrangement of atoms*. This regular arrangement of atoms is the cause of the well-developed faces on good crystalline specimens.

2.1 Visualizing the internal regularity of crystals

Methods available for the direct viewing of molecules in crystals have improved dramatically in the past forty years. The notion that a crystal is composed of molecules or ions laid down in a very regular, periodic manner is now well supported by physical evidence. For example, electron microscopy can be used to view certain molecules if they are sufficiently large. This technique has been applied successfully to crystalline macromolecules. Examples have been provided by Ralph W. G. Wyckoff, who, in the 1950s, took elegant electron micrographs of crystalline viruses[1] (Figure 2.1). The photograph shown in this figure depicts the develop-

FIGURE 2.1. An electron micrograph of the rhombic form of tobacco necrosis virus,[1] showing the regular arrangement of spherical particles (molecules) in the crystal and also the formation of crystal faces. In this micrograph the viral particles each have a diameter of approximately 250 Å. (Micrograph courtesy R. W. G. Wyckoff.)

ment of crystal faces from the packing of what appear to be roughly spherical molecules. This is similar to Kepler's proposal,[2] centuries before, that simple crystals are composed of "spheroids." An electron micrograph of a hepadnavirus (Figure 2.2) again shows the regularity of packing of the particles. Similarly, an electron micrograph of T4 bacteriophage crystals[3] (Figure 2.3) illustrates the regularity (and the occasional irregularity) of the packing. Some of the bacteriophage in this photograph are aligned with their heads facing in the opposite direction to those of the majority; others are merely slightly misaligned. This, we might surmise, is how molecules pack in crystals — generally with one distinct orientation, but also with occasional **disorder**.

It is now also possible to view individual atoms on the surfaces of crystals. One example is by field-ion microscopy, which was developed by Erwin W. Müller.[4] The first images of individual heavy atoms supported on films of light atoms were produced by the use of a dark-field scanning

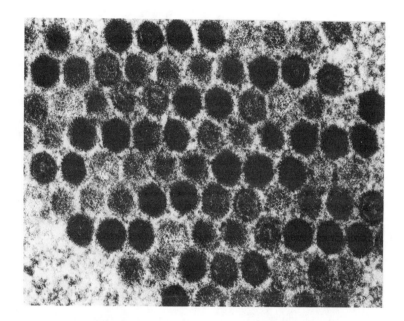

FIGURE 2.2. Electron micrograph of a hepadnavirus in a duck hepatocyte nucleus showing the regularity of packing of the particles in two dimensions (hexagonal close packing). Each particle is about 700 Å in diameter. Such crystalline arrays of viral particles are typical of adenovirus infections. (Photograph courtesy Sally Shepardson and Jesse Summers).

electron microscope.[5] Moreover, since electron microscopy, unlike X-ray diffraction, does not depend on the periodicity of the structure, it is possible to obtain images of crystal defects in addition to those of well-ordered areas.[6,7] The crystals used must be thin (generally 50–100 Å thick), the crystal axes must be aligned accurately with respect to the incident beam direction and the image must be recorded out of focus by a carefully calculated amount. Certain niobium oxides have been studied in this way to high resolution by John M. Cowley and Sumio Iijima.[7,8] Recently pictures of ordered molecules bound on surfaces have been obtained by a scanning tunneling microscope.[9,10] The method involves a needlelike device capable of scanning the surface from a distance of few Å. A small voltage is applied to the needle, and a current passes between the needle and sample; the size of the current is a function of the distance between needle and sample. Such a micrograph of gallium arsenide is shown in Figure 2.4.[11]

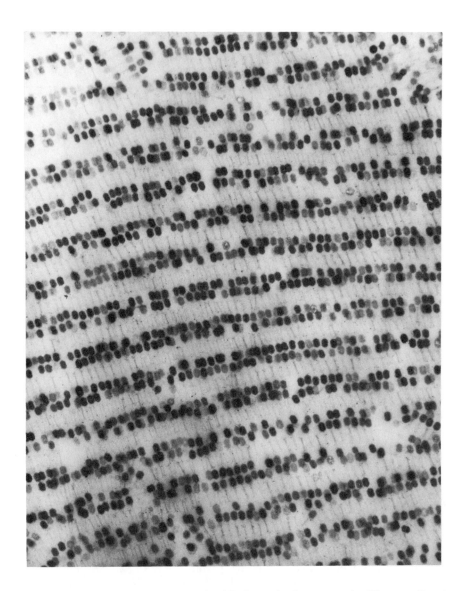

FIGURE 2.3. An electron micrograph of T4 bacteriophage crystals. The overall periodicity is evident; note, however, that some of the bacteriophage are out of alignment with the others. Some are even completely misoriented with respect to their neighbors. This type of disorder is also found for molecules in some crystals. (Micrograph courtesy L. H. Khairallah and the *Journal of Molecular Biology*.[3])

Ga   Ga   Ga   Ga   Ga   Ga

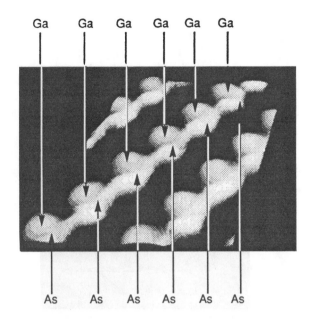

As   As   As   As   As   As

(a)

(b)

FIGURE 2.4. Combined scanning tunneling microscope images of the gallium arsenide (110) surface. (a) A computer scan of a photograph showing the identification of atoms. The image of the gallium ions was acquired by tunneling into empty states that are localized around surface Ga atoms. The image of the arsenide ions was acquired by tunneling out of filled states that are localized around surface As atoms. The original photograph provided [reproduced in (b)] showed Ga in blue and As in red. (Photograph courtesy R. M. Feenstra, IBM, Yorktown Heights, NY.)[11]

2.2 The unit cell

The micrographs in Figures 2.1 to 2.3 demonstrate well the internal periodicity of solid structures. For example, a repeat distance of approximately 250 Å can be found in Figure 2.1, and one of approximately 700 Å in Figure 2.2. It is usual to select out from a regular array of molecules or ions in a crystal, one small unit, called the **unit cell** (Figure 2.5). This unit cell repeats in a regular manner in each direction. The lengths of the sides of this unit cell and the angles between these sides are all that is needed to define it.

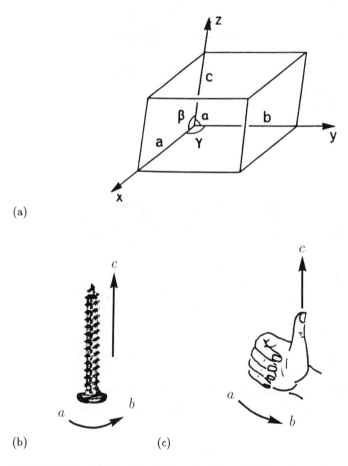

(a)

(b)   (c)

FIGURE 2.5. (a) A unit cell showing the axial lengths and interaxial angles. The directions of the axes are given in a right-handed axial system with **a** along the x direction and the angle γ between **a** and **b**. As **a** is moved to **b**, the screw (b) or thumb (c) proceeds in the c-direction, in a right-handed manner.

By convention a unit cell is characterized by three vectors, **a**, **b**, and **c**, and the interaxial angles between them, α, β, and γ. Also by convention, α is the angle between **b** and **c**, β is between **a** and **c**, and γ is between **a** and **b**. These unit-cell vectors form the edges of a **parallelepiped** where directions are given in a **right-handed axial system**. There are often many ways of selecting a unit cell. It is conventional to choose a unit cell that is the simplest, with the axial (repeat) lengths as short as possible and interaxial angles as near to 90° as possible. The three-dimensional shape of a unit cell can often be derived from the way that faces form on a crystal. In Chapter 3 we will show how X-ray diffraction gives complete information on the unit cell dimensions.

The unit cell is an imaginary building block that helps us to appreciate the internal periodicity of a crystal. We must remember that a crystal is actually composed of an **aggregation** of molecules or ions rather than of unit cells and that the unit cell is merely a convenient mathematical device for describing this aggregation. It greatly simplifies our description of the atomic structure, however, because it

1. divides the crystal into identical small components composed of specific arrangements of atoms. These components are closely packed; and
2. requires only a finite number of parameters to describe these arrangements of atoms.

Once the concept of a unit cell is understood, the reader will appreciate that the problem of determining the structure of a crystal has been reduced to that of determining the structure of the contents of one unit cell. Every other unit cell will be like the first, so that the entire crystal structure will be built up.

Millions of unit cells make up a crystal. The larger a molecule the smaller the number of unit cells found in a fixed crystal volume. The unit cell of sodium chloride is a cube, $a = 5.6$ Å in each direction; there are approximately 10^{19} unit cells in a crystal 1 mm on an edge (volume 10^{21} Å^3). The enzyme D-xylose isomerase[12,13] crystallizes with a unit cell ($a = 93.9$ Å, $b = 99.6$ Å, $c = 102.9$ Å) with a volume of 10^6 Å^3. It contains 10^{15} unit cells per mm^3. Thus the same volume crystal can contain 10^{19} unit cells of sodium chloride but only 10^{15} unit cells of xylose isomerase, four orders of magnitude fewer. The greater number of unit cells in crystals of small molecules accounts, at least in part, for the generally higher intensities obtained in the X-ray diffraction data of crystals of small molecules versus those of proteins or other large molecules. In practice, only the contents of one unit cell and its immediate molecular surroundings are considered when the molecular structure of a crystal is reported. The crystallographer selects out this simplified part of the crystal structure for description.

2.3 The process of crystal growth

The formation of a crystal depends both on the solubility of the compound (a thermodynamic property) and on nucleation and growth roles (kinetic properties). Crystal growth involves a phase change from liquid or gas to solid, such as the **precipitation** of a solute from solution or the formation of a crystalline solid by condensation of a gas. Conditions favoring these processes include supersaturation of a solution (resulting eventually in precipitation) and the use of temperature gradients that cause condensation of molecules from a gas or freezing of a liquid (solidification). Crystal growth may be roughly divided into three stages. The first stage is nucleation, in which a few molecules or ions form a stable aggregate. The second stage is growth, i.e., the orderly addition of further molecules or ions in a regular manner. In the final stage, termination, growth ceases.

To understand crystal growth the reader should spend some time looking through a microscope, watching the evaporation of solvent from a solution in a shallow, uncovered watchglass.[14-18] Small crystals may grow slowly and then enlarge to give well-formed specimens. Alternatively, many very tiny crystals may appear, glistening in the microscope light. These are crystals that have come out of solution quickly. Consequently, all of the material in the solution may be used up before any crystal can become very large, or crystals may coalesce, forming misaligned multicrystalline aggregates. Branching formations of **dendrites**, somewhat like very fine snowflakes, or tiny **spherulites**, looking like prickly sea urchins, may also form. The presence of such aggregates indicates that it may be possible to grow larger crystals by adjusting the experimental conditions. Another observation may be a gum, gel, glass, amorphous powder, or precipitate with no crystallinity. Such results can be frustrating because there is no clue how the experimental conditions have to be adjusted in order for crystallization to occur.

2.3.1 Nucleation

The initial stage in crystallization is the formation of aggregates to which more material will add. This process is called **nucleation**. A submicroscopic nucleus is formed by the chance association of several molecules in solution. These molecules have approached each other in appropriate orientations and have formed an aggregate. Further material can then be adsorbed and aligned on the surface of this nucleus, giving rise to ordered growth, and the formation of a crystal. The probability that this crystal will grow depends on the concentration of the solute, the temperature, the nature of the chemical species that causes precipitation of material from solution (the **precipitant**), the pH, and other factors. The aim is to obtain a highly ordered, repetitive array of components (atoms, ions, molecules), in other words, a crystal.

The number (concentration) of nuclei that are formed has an important effect on the course of the subsequent crystallization. To obtain a crystal suitable for X-ray diffraction studies, conditions should be such that only a few nuclei are formed. Failure to obtain crystals at all may indicate a difficulty in obtaining stable nuclei. On the other hand, if too many nuclei are formed, masses of microcrystals, unsuitable for single-crystal X-ray diffraction studies, result.[18,19]

2.3.2 Deposition on an initial nucleus

Charles W. Bunn wrote about crystal formation:[15] "One of the interesting questions about a newborn nucleus is, 'How quickly does it put on weight?' " Growth is a dynamic process; a molecule or ion approaches the surface and may either remain there or leave. When additional components reach the surface, they must interact with this surface in the appropriate orientation if they are to remain and support the formation of a crystal. The greater the number of specific interactions that each component forms as it settles on the growing crystal, the more tightly it will be bound and the lower the energy and the greater the stability of the conglomerate or crystal. A molecule or ion is particularly likely to remain if it is bound at steps or ledges such as the concave regions at the protrusions and pits on the surface of a crystal. In this case there is the possibility of a larger number of interactions than would occur on a flat surface [see Figure 2.6(a)]. It is easier for a molecule to add to and remain on the steps of partially formed layers than it is for it to attach itself to an edge or initiate the formation of a new layer.

Frederick Charles Frank[20] showed that crystal **defects**, that is, distortions from the regular order in crystals, such as are found in **dislocations**, provide the steps and ledges required for strong binding of a newly arrived molecule. An example is a **screw dislocation** [see Figure 2.6(b)].[20-24] If a portion of a crystal is moved so that a ledge is formed, as shown in this figure, new molecules may add and bind well to it. The dislocation has an approximately spiral structure that is maintained as more molecules are added, and provides a matrix for continuous growth of the crystal, making it unnecessary ever to form new layers. Screw dislocations have been observed in electron micrographs of paraffin crystals.[25] Defects like these may heal as the crystal grows, and this mechanism for facilitating crystal growth may be lost. Growth facilitated by screw dislocations is not the only mechanism for crystal growth. It is possible that large, irregularly shaped molecules also provide good sites for the deposition of fresh molecules. For example, new layers of macromolecules may form more readily than do new layers of simpler molecules.[15]

We noted the importance of obtaining an appropriate number of nuclei for suitable crystallization. The fate of dislocations upon which growth occurs is important. High-quality crystals are obtained when the rate of deposition of material from solution and the healing of defects

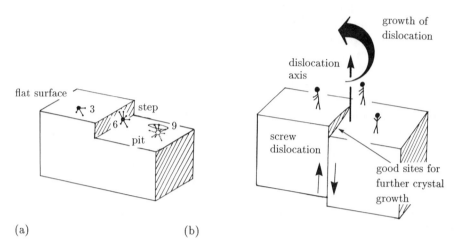

(a)   (b)

FIGURE 2.6. Examples of the number of interactions formed by a hypothetical molecule as it approaches the surface of a growing crystal. (a) When the molecule lands on the surface of a crystal, it is more likely to remain if several interactions between it and the crystal are formed. This is illustrated for a flat surface (3 interactions), a ledge (6 interactions), and a hole (9 or more interactions). Therefore ledges and holes will be eliminated by the deposition of molecules in them during crystal growth. (b) If there is a dislocation in the crystal that causes formation of a ledge (as shown here for a screw dislocation), molecules adding to the ledge tend to remain. In this case, however, the ledge is not eliminated; the screw dislocation continues to grow up the dislocation axis (marked with the thin vertical arrow) and, as demonstrated by figures going around the spiral, is self-perpetuating through the crystal.

are appropriately balanced. A high deposition rate may lead to defects in crystal, rapid growth in too many directions, and the formation of dendritic (that is, multibranched) crystals. When the healing rate is too fast, the desired growth sites may disappear, and the growth rate may be slowed or cease altogether. Since a reasonable rate of crystal growth appears to depend on the presence of defects, it is difficult to grow a perfect crystal with no internal disorder. The rate of diffusion of molecules into the region of crystal growth will influence the growth rate.[26,27] When a new molecule is bound to a crystal face, energy is released and transferred to neighboring molecules. The rate at which this energy is dissipated is yet another factor in crystal growth. Crystal growth also depends on the availability of material to add to the growing surface. When material is deposited on a crystal face there is a temporary depletion of material in that region of the solution. Thus, while crystal growth is often faster at lower temperatures, if the solution becomes too viscous at low temperatures, the growth rate will be decreased because molecular travel to the growing crystal face is impeded.

Crystals may also be grown from molten material.[28] One technique involves placing a seed crystal on the surface of the melt maintained just above its melting point. The crystal is pulled slowly from the melt as it grows. Another method involves lowering the molten material through a controlled temperature gradient; crystallization occurs at the appropriate location on this gradient.[29-31]

2.3.3 Development of crystal faces

One of the first things one may notice about a crystal is the development and arrangement of its faces (i.e., the smooth planes that form the crystal surfaces). The shape or form of a crystal, including its faces and the angles between them, is called **morphology**.[17,32,33] The word **habit** is used for the appearance of a crystal that has developed under certain crystallization conditions.[34-36] Variation of these conditions (e.g., the solvent used) may lead to crystals with a different habit.[37-43] Naphthalene, which normally grows as bulky crystals, will give very thin plates if grown rapidly. The habit of a crystal occasionally gives information on the inherent molecular arrangement within it. Some flat molecules, if they stack one on the other, form crystals with an **acicular** or needlelike habit. This is analogous to stacks of books.

The formation of the faces on a crystal depends on the internal atomic structure and the interaction of the surface of the crystal with other molecules during its growth. There is a well-defined equilibrium shape for a crystal grown under given conditions from a selected solvent. The shape is influenced by the need to attain a minimum total surface free energy for the volume of crystal. For this reason, a small drop of liquid is spherical. Molecules in the drop of liquid have lower energy in the bulk of a liquid than they do if they lie on its surface. In the solid state, a crystal might tend towards a spherical habit if all faces were to develop at exactly equal rates. The rates of development of various crystal faces are, however, generally different, and these rates are modified by experimental conditions. Significantly, however, they determine the overall shape of the crystal, which illustrates the relation between the thermodynamic and kinetic factors in crystal growth. A crystal is bounded by the faces that have grown most slowly.[37-43] If the crystal grows more rapidly in a direction perpendicular to small corner faces than in a direction perpendicular to the other faces, the small corner faces will eventually disappear, as shown in Figure 2.7.

2.3.4 Polymorphism

There may be more than one arrangement of components in a crystal that have similar overall energies. Changes in experimental conditions may

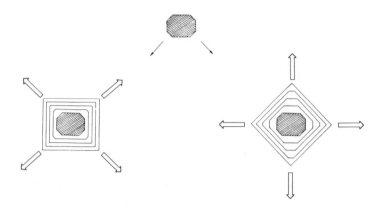

FIGURE 2.7. A crystal is bounded by its slowest-growing faces. Changes in shape, as a function of time, are shown for two sets of crystal faces as the original crystal (shaded) grows larger. The directions of fastest growth are indicated by white arrows. On the left, the crystal progresses to a rectangular shape and, on the right, to a diamond shape. This diagram illustrates that crystal faces eventually disappear if they lie perpendicular to the direction of fastest growth (white arrows).

produce crystals with different internal arrangements of atoms and different unit cell dimensions. **Polymorphism** is the existence of two or more crystalline forms of a given compound (dimorphism if only two forms). The polymorphs have crystal structures with different packings of the same kinds of molecules. Polymorphism is distinct from the differences in crystal habit of one crystal form; the word habit implies that crystals that have different appearances are actually composed of identical unit cells and atomic arrangements. In polymorphs, the shape and size of the unit cells and the arrangements of their contents differ. It is conceivable that most compounds can, under appropriate experimental conditions, crystallize as polymorphs. This phenomenon is discussed in more detail in the chapter on crystal packing (Chapter 15).

2.3.5 Isomorphism

Sometimes different compounds give apparently identical crystals. **Isomorphism** is the similarity of crystal shape, unit cell dimensions, and structure between substances of nearly, but not completely, identical chemical composition. It is derived from the Greek words – *isos* meaning equal and *morphe* for form or shape. The arrangements of atoms in the isomorphous crystals are identical, but the identity of one or more atoms in this arrangement has been changed. For example, sulfur in a sulfate may often be replaced by selenium, to give an isomorphous selenate. Ideally, isomorphous compounds are so closely similar in composition that they can form a continuous series of solid solutions. Eilhard Mitscherlich described this phenomenon in 1819 and wrote[44,45] (translated)[46] "An equal number of atoms, if they are bound in the same way, produce similar crystal forms, and the crystal form depends not on the nature of the atoms but on the number and method of combination."

He had discovered this phenomenon while studying the phosphates and arsenates of potassium (KH_2PO_4 and KH_2AsO_4).[45] He noted that crystals of these two compounds had such a strong resemblance to each other that "it was quite impossible to find any real difference." The isomorphism of potassium perchlorate and potassium permanganate led to the correct determination of the atomic weight for manganese because the formula of the permanganate was deduced from that of the isomorphous perchlorate.[47] Similarly atomic weights were assigned to selenium and chromium on the basis of the isomorphism of potassium sulfate, potassium selenate, and potassium chromate.[47] Some pairs of compounds containing elements of the same group in the Periodic Table may not, however, be isomorphous because of differing ionic sizes.

The best known examples of isomorphism are provided by the alums. These have the general formula $(M_1)_2(SO_4)\cdot(M_3)_2(SO_4)_3\cdot24H_2O$, where M_1 is a monovalent cation such as potassium or ammonium and M_3 is a trivalent cation such as aluminum or chromium. Various alums have different colors, and therefore the effects of isomorphism may be readily visible. For example, chrome alum crystals ($M_1 = K$, $M_3 = Cr$) can be grown from aqueous solution as purple octahedra. On the other hand, potash alum ($M_1 = K$, $M_3 = Al$) is colorless but will grow over a deep purple chrome alum octahedral crystal. This gives a large octahedral crystal with a purple interior. Readers are urged to try this experiment.[14] In such mixed crystals of alums, the only difference is the replacement of aluminum by chromium in the crystal structure; the ionic replacement is possible because the sizes of the ions are similar. This overgrowth of one isomorph on another is not a necessary condition for isomorphism, but is commonly observed.

Isomorphous replacement is now employed in the determination of the structures of biological macromolecules. These molecules crystallize with 50% or more of the crystal volume filled with solvent molecules. Murray Vernon King, working with David Harker, conceived the idea of soaking protein crystals in solutions of compounds containing a heavy atom.[48] These heavy-atom compounds are diffused into the crystals through the solvent channels and settle on preferred sites on protein molecules. The diffraction patterns of the unperturbed crystal (described as "native") and the heavy-atom derivative are then compared in such a way that an electron-density map for the protein results. The method of isomorphous replacement, and the manner by which it is used to derive relative phases, are described in detail in Chapter 8.

2.4 Growing crystals from solution: some practical advice

The question the scientist seeking a crystal structure determination will ask is, "What are the experimental conditions necessary for obtaining diffraction-quality crystals?" There are many factors involved in the growth of large, well-defined crystals, as mentioned previously. The type of growth observed depends on the balance among these in each case. As discussed, for a solution to yield good crystals, only a limited number of nuclei should be formed. The degree of supersaturation of the solution should support continued, ordered growth.[38] In practice, the experiment is usually most effective if crystallization occurs over a period of one to several days. Solutions of the material being crystallized, preferably saturated or nearly so, should be filtered to remove most nuclei and then left undisturbed. The aim is to obtain crystals that are at least 0.1 to 0.3 mm in each dimension.

The growth of diffraction-quality crystals is most frequently accomplished by preparing a solution that is nearly saturated with the compound to be crystallized. Conditions are then changed very slowly to achieve modest supersaturation while controlling the number of nucleation sites.[49-54] In this manner a few crystals of suitable size for diffraction experiments can be obtained. For example, when one grows crystals of table salt by evaporation, the solution must be kept left undisturbed. In this way several large crystals may be obtained. If, however, the solution is disturbed (by picking it up to examine it for crystals, for example) or if seed particles settle on the surface, a large number of small crystals usually form because many nucleation sites have been introduced.

If crystals do not grow as expected, crystallization may be promoted by adding nuclei, thst is, small seed crystals of the same material. Many a worried chemist, stroking his beard while bent over a crystallizing dish, has introduced seeds to the solution in the dish, and nucleated crystal formation. Alternatively, the organic chemist has often scratched the interior of a glass test tube with a glass rod to try to induce crystallization by distributing nuclei throughout the solution. Boiling chips may serve the same purpose via cavitation, that is, by virtue of the pores in the chip. Crystals of unrelated compounds having a similar unit cell shape have also been used as nuclei for crystallization.[38] Unfortunately, the organic chemist often employs techniques (such as allowing heated solutions to cool too quickly) that are not likely to give diffraction-quality crystals. Beautiful crystals may be obtained at the end of a synthesis and then, to the horror of the crystallographer, they are crushed in order to dry the sample for a chemical analysis or for use in a subsequent reaction. One chemist remarked that he had to forget all he learned in his courses and had to start learning new techniques when trying to grow diffraction-quality crystals. If a chemist obtains crystals of a compound of interest, it is best to investigate their suitability for crystallographic analysis before proceeding further with other experiments.

Proteins generally are available in small quantities and are often difficult to crystallize. Therefore the strategy is to use a combinatorial method[55] in which selected combinations of different organic solvents, different *p*H ranges, different salt concentrations, and different precipitating agents are tested for their ability to aid in production of diffraction-quality crystals. Very small amounts of protein solution (10–30 mg/ml), for example, 2 μl (also sometimes called 2 λ) are used. In this way the best conditions for crystallization can be refined. Recipes for growing crystals are usually only specific for a particular protein.

When crystallizing proteins, it is important to keep them hydrated, to control *p*H and temperature,[49,53] and to change conditions slowly. Initiation of crystallization may be achieved by the use of precipitating agents or by *p*H changes so that only a limited degree of supersaturation results. In aqueous solution, a protein, with many hydrophilic groups on its surface, is covered with water molecules. If ions are added, they will pull off some of these water molecules to re-establish osmotic equilibrium. This may temporarily leave sites on the protein that are free to bind other protein molecules, thereby facilitating protein aggregation. Organic solvents may have the same effect and have also been used to concentrate proteins.

The most common precipitating agents used in protein crystallization are those highly soluble inorganic salts (ammonium sulfate or sodium chloride, for example) and organic polyethers (polyethyleneglycols of a selected molecular weight range) that do not denature the protein. If diffraction-quality crystals are to be grown, the protein must be sufficiently pure, concentrated (*ca.* 5–100 mg/ml), and free from solid contaminants. Analogous techniques are used for other types of macromolecules, such as nucleic acids.

Some ways of growing crystals from solution are illustrated in Figure 2.8. It is best to obtain a large amount of material, at least 100 mg, so that a variety of crystallization experiments may be tried and the best conditions for crystallization determined. Often, however, such a large quantity of material is not available.

2.4.1 Crystallization by controlled temperature change

A commonly used crystallization method involves controlled temperature change. Slow cooling of a saturated solution can be effective in producing crystals if the compound is more soluble at higher temperatures; alternatively, slow warming can be applied if the compound is less soluble at higher temperatures (as is the case for most proteins). The rate of the temperature change must be carefully controlled. If microcrystals form, the rate was too rapid and the system needs to be insulated by, for example, placing it in a water bath (to increase the thermal inertia) or using equipment in which the temperature can be programmed to change more slowly.

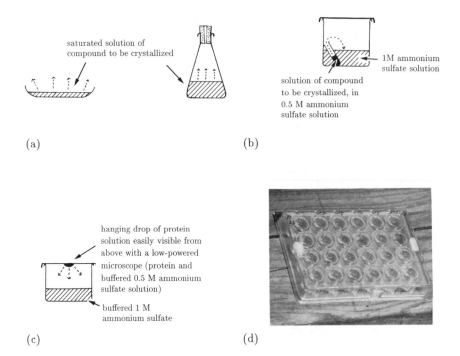

saturated solution of
compound to be crystallized

(a)

1M ammonium
sulfate solution

solution of compound
to be crystallized, in
0.5 M ammonium
sulfate solution

(b)

hanging drop of protein
solution easily visible from
above with a low-powered
microscope (protein and
buffered 0.5 M ammonium
sulfate solution)

buffered 1 M
ammonium sulfate

(c)   (d)

FIGURE 2.8. Some methods for growing crystals from solution. (a) Slow solvent evaporation, primarily used for small molecules. (b) Vapor diffusion. (c) The hanging drop method, primarily used for macromolecules. (d) A crystallization plate used for hanging drops; there are slightly different crystallization conditions in each well, varied systematically with respect to, for example, pH and ionic strength.

2.4.2 Crystallization by slow solvent evaporation

Slow solvent evaporation is a valuable method for producing crystals of high quality; it is used primarily for small molecules. The rate of growth is manipulated by adjusting conditions (initial concentration, temperature, evaporation rate) so that crystals are obtained in a few days, not months or minutes. The rate of evaporation can be adjusted by covering the solution with a watch glass or a piece of film (aluminum or Parafilm®) with a number of small holes punched in it. The solution is then left to stand undisturbed.

Single-component solvent evaporation methods depend on the removal of solvent in order to obtain the **saturated solution** required for crystallization to occur. If single-solvent solutions do not yield diffraction-quality crystals, mixtures of solvents may be tried. Multi-component solvent evaporation methods depend on the difference in the solubility of the solute in various solvents. A solvent system is selected in which the solute is more soluble in the component with the higher vapor pressure. As the solution is left to evaporate, the volume of the solution is reduced and, because these solvents evaporate at different rates, the composition of the mixture of solvents changes. At the same time, the solubility of the solute is reduced as the proportion of solvent in which it is more soluble decreases. Concentration gradients that depend on evaporation rates can be set up in order to control crystal growth.

2.4.3 Crystallization by vapor diffusion

In the vapor diffusion method, a solution of the solute in solvent is placed in a small, open container that, in turn, is placed in a larger container with a small amount of a miscible, volatile nonsolvent. The larger container is then tightly closed. As solvent equilibrium is approached, the nonsolvent diffuses through the vapor phase into the solution, and saturation or supersaturation is achieved, perhaps producing crystals. Subtle changes in solubility can be tested by using multicomponent solvent systems with different percentage compositions for the solution and for the outer reservoir. An alternative method is to allow the two liquid phases (the solution and a nonsolvent) to diffuse directly into each other. The densities and miscibility of the two liquid phases are chosen so that they create a concentration gradient in the vessel. Crystals appear as the solution and nonsolvent diffuse into each other.

2.4.4 Crystallization by hanging or sitting drop methods

Adaptations of vapor diffusion and solvent gradient effects are the most common methods used for growing protein crystals. The hanging-drop or sitting-drop methods have proven to be particularly productive of crystals and are generally tried first when any soluble protein is slated for crystallization experiments. Because the available amounts of purified proteins (or other macromolecules) are often very small (*ca.* 5–20 mg), microcrystallization techniques have to be employed. In the hanging-drop vapor diffusion technique, a 5–15 μl drop of a concentrated protein solution containing precipitants (nonsoluble) is suspended from a silicone-treated microscope cover glass placed over a spot plate reservoir containing about 500 μl of a more concentrated solution of the precipitants. The cover glass forms a tight seal over the reservoir. As equilibrium between the liquids in the reservoir and the hanging drop is approached, protein crystals may form.[53,56] Some examples of methodologies are given in Figure 2.8. Dialysis techniques, using membranes to separate the protein solution from a solution containing more precipitant, can produce similar effects.[57] Experiments are now in progress to test the effects of gravity (which causes flow currents through the solution) on the quality of the crystals grown.[58–60]

2.4.5 Epitaxy

While crystals may grow on small nuclei formed by the aggregation of molecules of the compound under study, it is often found that dust particles, of unspecified composition, also cause nucleation. This nucleation may be by epitaxial growth, a phenomenon that has been successfully used in the preparation of certain crystals. **Epitaxy**[38] is the oriented growth of one compound on a crystal of an entirely different substance (Figure 2.9). Essentially the first crystal is acting as a nucleus for the further growth of the second. Repeat distances in the two crystalline compounds should be similar, hence the requirement for interfacial orientation. Epitaxy has important applications in modern electronics. Oriented gallium arsenide crystals[61] are grown on silicon crystals that have been etched by agents, such as hydrofluoric acid, to produce an appropriate surface for epitaxial growth. Because different faces on silicon crystals are etched at different rates, the electronic properties of the layers of material produced by epitaxial growth can be controlled by this process.

Alexander McPherson and Paul J. Shlichta[38,62] have suggested using insoluble minerals as heterogeneous nuclei for the crystallization of macromolecules. They obtained excellent protein crystals, which could be cleaved from the mineral nucleus and used for X-ray diffraction studies. The mineral is introduced into a supersaturated solution of the material to be crystallized. As supersaturation increases, nucleation occurs on a specific face of the mineral nucleus, and a crystal begins to grow. The orientation and periodicity of the molecules on the nucleus surface promote an oriented overgrowth that has a similar periodicity.

There need not be a perfect match of unit cell dimensions for such epitaxy to occur. Several unit cell faces of one compound may be needed

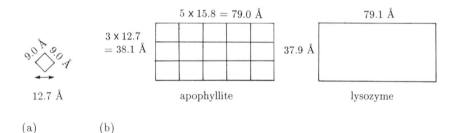

FIGURE 2.9. Epitaxy of lysozyme on the mineral apophyllite. (a) The diagonal across the *ab* face of apophyllite ($a = b = 9.0$ Å, $c = 15.8$ Å) is $9.00 \times (\sqrt{2}) = 12.7$ Å. (b) Fifteen unit cells of apophyllite, 3×12.7 Å by 5×15.8 Å, are almost the same size in cross-section of one unit cell of lysozyme (79.1×37.9 Å).

to match one unit cell face of the other. For example, crystals of lysozyme (large unit cell dimensions) grew on the mineral apophyllite (small unit cell dimensions). The protein unit cell covered several unit cells of the mineral, and the fit was nearly exact (Figure 2.9).[63,64] Thus, an inorganic matrix may influence the deposition of certain crystalline proteins on its surface.[65-67] This has general effects in nature. The crystallization of minerals in bones and shells, for example, appears to be affected by the nature of the organic surfaces on which they grow.

It is, however, not always easy to grow crystals. When crystals of a chosen compound defy the best efforts of the crystallographer to grow them, it is sometimes possible to prepare molecular complexes that crystallize more readily, where one component of the complex is the compound of interest. An example is provided by Margaret Etter and co-workers.[68] They demonstrated that, if a compound contains an acidic hydrogen atom that can readily form hydrogen bonds, then formation of a complex with triphenylphosphine oxide may lead to diffraction-quality crystals of the complex.

2.5 The shapes of crystals

Naturally occurring crystalline minerals often show well-developed, flat faces, as described previously, so that much can be learned about the shape of a unit cell from an examination of the geometrical arrangement of the faces of a crystal. The size of a crystal is a function of the amount of material available and the time over which crystallization has been allowed to proceed. The interfacial angles, however, are a function of the internal atomic arrangement in the crystal, and of the shape and relative dimensions of the unit cell. The prominence of faces is governed by the population density of the atoms in particular planes; therefore, those faces that contain the most closely spaced atoms occur most readily.

2.5.1 Measurement of angles between crystal faces

The first reported crystal measurements were of the angles between faces. This was generally done using a contact **goniometer** or an optical goniometer (from the Greek: *gonia* = angle).[69] A contact goniometer [Figure 2.10(a)] consists of two straight edges hinged together and placed so they are in contact with two crystal faces. The angle between the two straight edges is equal to the interfacial angle. The more sophisticated reflecting (or optical) goniometer [Figure 2.10(b)] was introduced by William Hyde Wollaston.[70] It consists of a rotatable mounting on which the crystal can be aligned with a set of crystal faces parallel to the axis of rotation. First the crystal is rotated until light from a distant light source is reflected from one of the crystal faces onto a fixed detector. It is rotated again until a second face reflects the beam of light. The angle between these two positions is then recorded as the **interfacial angle**.

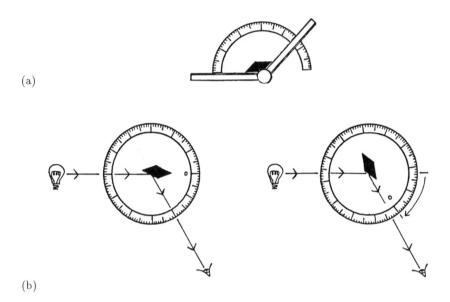

(a)

(b)

FIGURE 2.10. Methods of measuring the angles between crystal faces. (a) A contact goniometer in which the angle between two arms in contact with the faces can be recorded by direct measurement. (b) A reflecting goniometer contains a light source and measures the angles by which a crystal has to be rotated (shown by the arrow) so that reflections of the incident light from different crystal faces can be measured. Measurements for two crystal faces are diagrammed.

Interfacial angles (angles between crystal faces) are characteristic of the crystal form being studied and, in certain cases, may aid in identification of the material. When calcite crystals are broken, they form rhombohedra with interfacial angles of 75°. Haüy discovered this when, to his chagrin, he dropped a valuable Iceland spar (calcite) crystal. He found that the particles into which it broke were always the same type of rhombohedra,[71,72] with the same shape (but not necessarily the same size). This occurred no matter what external form the original crystal had.

These observations led to the concept of a unit cell. Haüy was able to build realistic models of calcite crystals by stacking rhombohedral building blocks of uniform size (each with interfacial angles of 75°).[73] Clearly, the interfacial angles are important dimensions of the exteriors of crystals. The **Law of the Constancy of Interfacial Angles**[73] was first proposed by Steno. It states that: in all crystals of a given kind, that is, with the same composition and the same crystal form, the angles between corresponding analogous faces have a constant value. For example, when calcite crystals are broken up, they form rhombohedra with interfacial angles of 75°. This Law explains why the relative sizes of crystal faces are less characteristic of the material than are the angles between faces.

2.5.2 Indexing crystal faces

It is important to describe each crystal face in a numerical way if data on different crystals or from different laboratories are to be compared. The method used to describe crystal faces is derived from the **Law of Rational Indices**, proposed by Haüy and Arnould Carangeot.[69,71] This Law states that each face of a crystal may be described, by reference to its intercepts on three noncollinear axes, by three small whole numbers (that is, by three rational indices).[71] From this law, William Whewell introduced a specific way of designating crystal faces by such indices, and William Hallowes Miller[74] popularized it. The integers that characterize crystal faces are called **Miller indices** h, k, and l. When this method is used to describe crystal faces, it is rare to find h, k, or l larger than 6, even in crystals with complicated shapes. An example of the buildup of unit cells to give crystals with different faces is shown in Figure 2.11.

The geometrical construction for determining the indices of a crystal face [Figure 2.12(a)] is as follows: From a point inside the crystal three noncollinear axes are chosen. These axes, x, y, and z are assigned, by convention, in a right-handed system (see Figure 2.5). If we have chosen these axes with care (or by trial and error), we can orient their coordinate system so that it coincides with the unit cell axes with a along x, b along y, and c along z. Further verification of the choice of unit cell axes follows

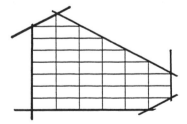 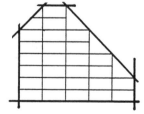

FIGURE 2.11. Diagrams showing the development of what amounts to crystal faces (shown with heavy lines) as unit cells build up. Each of these faces can be given a numerical indexing as shown in Figure 2.12.

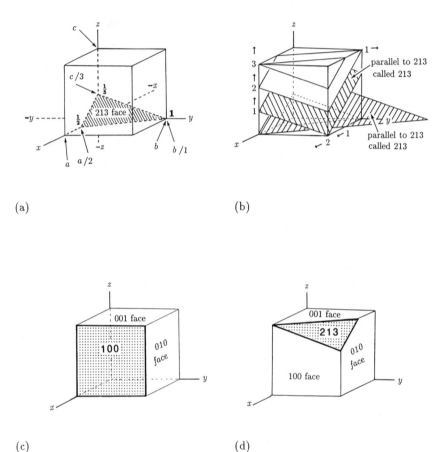

(a)

(b)

(c)

(d)

FIGURE 2.12. The indexing of crystal faces. (a) The derivation of the integers used to describe the crystal faces (Miller indices). The (213) face is marked as an example. It intersects the axes of the unit cell at $a/2$, $b/1$, and $c/3$. (b) Several parallel planes can be represented by these Miller indices. They intersect at, for example, a, $2b$, $2c/3$; $3a/2$, $3b$, c. (c) A crystal with its (100) face (stippled), intersecting the x axis at $x = a$ and the y and z axes at infinity (that is, parallel to them). (d) A crystal with a (213) face stippled.

from any symmetry in the crystal and this will be described in Chapters 4 and 5. If we assume one corner of the unit cell lies at $x = a$, $y = b$, and $z = c$ along these three axes, then a crystal face, designated hkl, will make intercepts on the three axes at $x = a/h$, $y = b/k$, and $z = c/l$. When the indices are negative, they are designated $\bar{h}$, $\bar{k}$, $\bar{l}$ (read h–bar, etc., for

American or bar–h, etc., for British practice). Planes with these indices cut the unit cell axes at $-a/h$, $-b/k$, $-c/l$, respectively. If the values of h, k, and l are small for all observed crystal faces, then a reasonable unit cell has probably been chosen. For reasons of symmetry, planes in hexagonal crystals are conveniently described by four axes, three in a plane at 120° to each other. This leads to four indices, $hkil$, where $i = -(h + k)$, and equivalent crystal faces will have similar indices.

An alternate way of considering crystal planes, indexed as described earlier, is to consider the number of times a set of parallel planes intersects each crystal axis within one unit cell. The set of hkl planes cuts the a axis at h positions, the b axis at k positions, and the c axis at l positions [Figure 2.12(b)]. Ralph Steadman suggests[75] "To determine hkl ⋯ move along the whole length of the unit cell vector **a**. Count the number of spaces crossed, and this is h. By spaces we mean spaces between planes." In Figure 2.12 $hkl = 426 = 213$, because, in usual crystallographic practice,[76] any common divisor is factored out. In X-ray diffraction, described in Chapter 3, (426) is used to denote the second-order diffracted beam from a (213) plane.

In summary, we choose the unit cell so that the faces can be described by small values for h, k, and l. This is done so that the faces with the simplest indices [(100), (010), and (001)] are parallel to the sides of the unit cell, thereby keeping the symmetry of the crystal in mind. More details on unit cells will be given in Chapter 4.

When the faces have been indexed and angles between them measured with fair precision, they can be represented in two (instead of three) dimensions by a stereographic projection (Figure 2.13).[77–79] This will demonstrate, in a diagram that can be drawn on a sheet of paper, the three-dimensional information on any symmetry in the development of crystal faces. To construct a stereographic projection, the crystal is surrounded by an imaginary sphere where the coordinate centers of the crystal and this surrounding sphere are the same. Lines perpendicular to each crystal face are drawn outwards, touching the surface of the sphere. Since a two-dimensional plot is easier to work with than the surface of a sphere, the results are transferred to a two-dimensional plot as follows: (1) An equatorial plane is drawn through the center of the sphere. (2) The points from the crystal planes are projected onto the surface of this sphere and joined to the opposite pole, that is, those points above (north of) the equatorial plane are joined to the lower (south) pole. (3) The points of intersection of these lines with the equatorial plane are marked with filled circles for the northern hemisphere and open circles for the southern hemisphere [see examples in Figure 2.14(b) to (h)]. The stereographic projection is useful for demonstrating the symmetry of the crystal, without the need for a three-dimensional model. Inspection of a stereographic projection will often indicate whether crystals of different habits belong to the same crystal form. Some of the many types of crys-

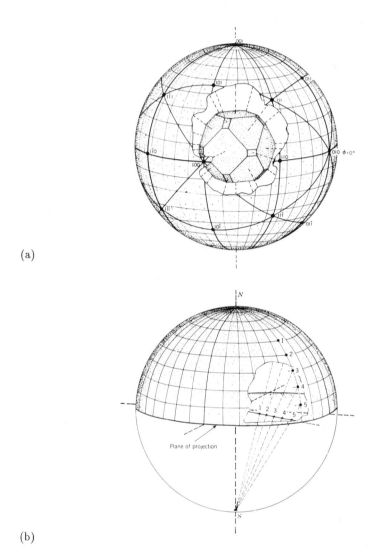

(a)

(b)

FIGURE 2.13. The stereographic projection for representing, in two dimensions, the arrangement and directions of faces in a three-dimensional crystal. (a) The crystal is surrounded by a sphere. The points at which normals (perpendiculars) to the faces hit this sphere are noted by points labeled 100, 110, etc., the same as the crystal faces they represent. (b) Each point (1 to 5, for example) representing a crystal face is joined to the opposite pole ("south" if the point is in the northern hemisphere).

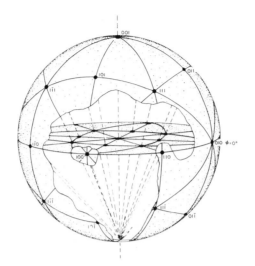

(c)

(d)

FIGURE 2.13 (cont'd). An equatorial plane (plane of projection) is selected, and the points where these lines to the pole touch the equatorial plane are noted (labeled 1 to 5). (c) The points of intersection with the equatorial plane are joined by lines or arcs. (d) The resulting diagram, viewed on the equatorial plane, shows the symmetry of the crystal, cubic in this case. (Courtesy Ernest E. Wahlstrom and John Wiley & Sons, Inc. From: Wahlstrom, E. E. *Optical Crystallography.* 5th edn. John Wiley: New York, Chichester, Brisbane, Toronto (1979). © John Wiley & Sons, Inc.)

tal shapes that can be formed from a cubic unit cell, together with their stereographic projections, are shown in Figure 2.14.

Impurities often have a profound effect on crystal habit. Adsorption of an impurity on faces of a crystal may retard the growth of certain crystal faces, and these will therefore become prominent in the final crystal, as described earlier (see Figure 2.7).[80] For example, sodium chloride[81]

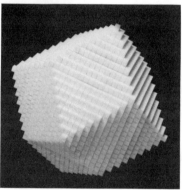

(a)

FIGURE 2.14. Some of the various shapes (habits) of crystals that can result from a crystal structure with a cubic unit cell. (a) Models of a cube, octahedron, and rhombic dodecahedron, composed of cubic unit cells. (b) Tetrahedron, (c) cube, (d) octahedron, (e) rhombic dodecahedron, (f) tetrahexahedron, (g) icositetrahedron, and (h) irregular dodecahedron (pyritohedron). The corresponding stereographic projections are also shown, with filled circles for the northern hemisphere and open circles for the southern hemisphere (see Figure 2.13). (Models courtesy Amy Kaufman).

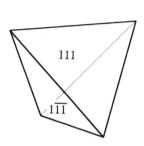

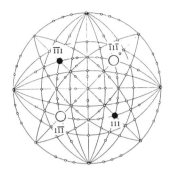

(b) Tetrahedron

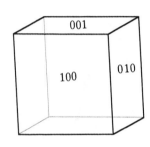

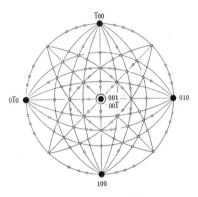

(c) Cube

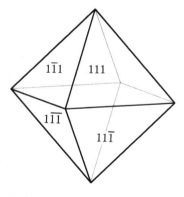

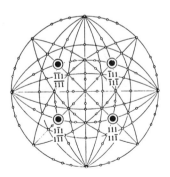

(d) Octahedron

FIGURE 2.14 (cont'd).

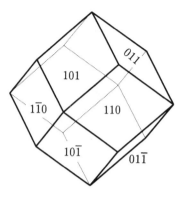
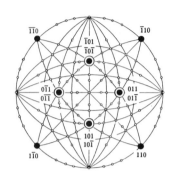

(e) Rhombic dodecahedron

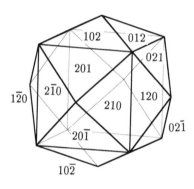
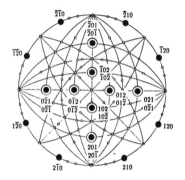

(f) Tetrahexahedron

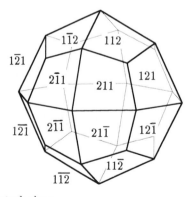
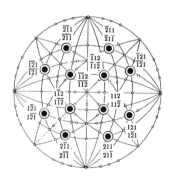

(g) Icositetrahedron

FIGURE 2.14 (cont'd).

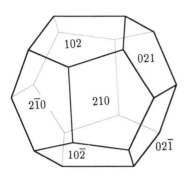
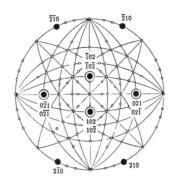

(h) Pyritohedron
FIGURE 2.14 (cont'd).

crystallizes as a cube with (100), (010), and (001) as the most commonly developed faces. There are equal numbers of sodium and chloride ions on the surface. This alternation of positive and negative ions on the faces of a crystal is normally encountered for salts. On the other hand, in the presence of urea or other compounds that interact with either positive or negative ions, octahedra grow with well-developed (111) faces. These faces have ions of only one type (sodium or chloride) on their surfaces (Figure 2.15). Although the unit cells are the same for both the cubic and the octahedral crystals, different faces have developed. The latter are (111), (11$\bar{1}$), (1$\bar{1}$1), ($\bar{1}$11), (1$\bar{1}\bar{1}$), ($\bar{1}$1$\bar{1}$), ($\bar{1}\bar{1}$1), ($\bar{1}\bar{1}\bar{1}$), the eight faces of an octahedron rather than (100), (010), (001), ($\bar{1}$00), (0$\bar{1}$0), (00$\bar{1}$), the six faces of a cube. The numerical listings of crystal faces given above are rather clumsy and may be abbreviated {111} and {100}, respectively, where (*hkl*) with round parentheses means a set of parallel planes and {*hkl*} with braces denotes all the sets of planes *hkl* related by symmetry {100} and {111}, respectively for the faces of a cube and octahedron. Square brackets [*hkl*] are used to indicate a direction in the crystal, designated such that [100], [010] and [001] are the directions of the three unit cell edges. Thus, [001] describes a line (the *c* axis) connecting (001) to the origin.

2.6 The crystal lattice: a mathematical concept

We have discussed the unit cell, which has imaginary boundaries. For simplification, each unit cell in a crystal can be represented by a single point. The result is called a **crystal lattice**, which is an imaginary three-dimensional arrangement of points such that the view in a given direction from each point in this lattice is identical to the view in the same direction from any other lattice point. In other words, it is an array of points that repeat, in the same regular manner in three directions, the way in which the unit cell contents repeat in the crystal. The term crystal lattice is specific and refers only to this arrangement of points (each representing the unit cell contents); the term should not be used to denote the entire atomic arrangement. As Steadman says[75] "the ruthless stripping away of the atoms of a structure to leave behind the dead lattice may make for simplicity, but it removes all trace of the real crystal" He makes the reader aware, however, of what the lattice is. In fact, he stresses: "A crystal *structure* is made of *atoms*. A crystal *lattice* is made of *points*. A crystal *system* is a *set of axes*." In other words, the structure is an ordered array of atoms, ions, or molecules. By contrast, the lattice is an ordered array of imaginary points, each point being merely a representation of the unit cell contents.

The relationship between the crystal lattice, the unit cell contents, and the crystal structure is shown in two dimensions in Figure 2.16. This raises the concept of **convolution**, which is the equivalent of folding one function into another. It is a method for converting a unit of a pattern into many identical copies arranged on a lattice; all that need be defined is the unit of a pattern and the lattice; the convolution gives the rest.

If the contents of one unit cell are laid on each of the crystal lattice points, the result (Figure 2.16) is a representation of the entire crystal

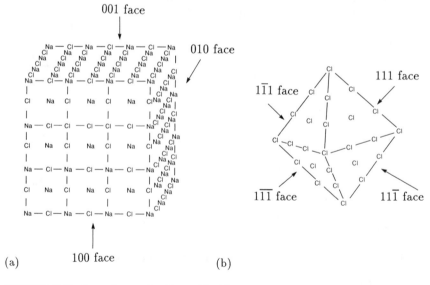

FIGURE 2.15. Some faces of sodium chloride crystals, labeled for a unit cell oriented as in Figure 2.5. Sodium cations are denoted Na, and chloride anions are denoted Cl. (a) When a crystal grows as a cube, its faces have equal numbers of sodium and chloride ions on their surfaces. (b) When the crystals grow as octahedra, the faces contain only one type of ion, either sodium or chloride.

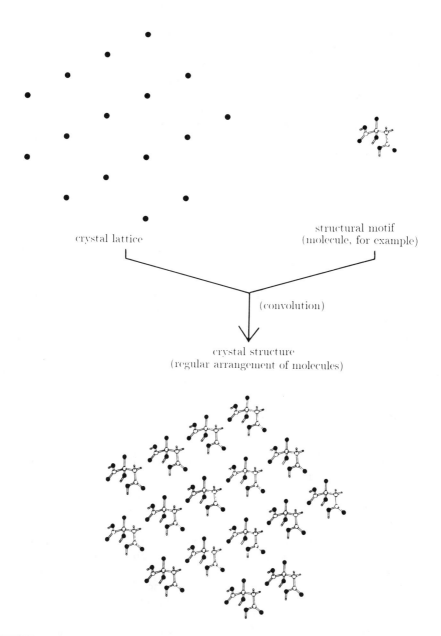

FIGURE 2.16. The crystal structure is composed of a crystal lattice (represented by points) and a structural motif (the atomic contents of one unit cell). The convolution of these two items will give the entire crystal structure.

structure. Kenneth C. Holmes and David M. Blow[82] wrote: "A general statement of the operation of convolution of two functions is: set down the origin of the first function in every possible position of the second, multiply the value of the first function in each position by the value of the second at that point and take the sum of all such possible operations ⋯ The concept of convolution is useful because it considers a periodic structure such as a crystal to be made from two separable parts: the lattice, and a density function describing the component molecules." This is shown numerically, for those interested, in Figure 2.17. The crystal, therefore, can be described mathematically in terms of a crystal lattice of defined dimensions and an arrangement of atoms that fills one unit cell.

Summary

1. **Crystals** are defined as solids with a regularly repeated internal arrangement of atoms.
2. **Crystals** develop **faces** in a manner that depends on the internal structure of the crystal, the concentration of atoms on a face, and the relative rate of growth of that face.
3. For diffraction purposes crystals are generally grown by a very slow precipitation from a saturated solution. The aim is to obtain a crystal about 0.1 – 0.3 mm in all dimensions.
4. Crystals are **isomorphous** if they have approximately the same unit cell dimensions, have the same arrangement of atoms in the unit cell, and differ only by the identities of a small number of atoms in this arrangement.
5. Compounds are polymorphic if they form several different types of crystals with different unit cell dimensions and different atomic arrangements.
6. The indexing of crystal faces with respect to the unit cell axes is a method used for describing crystal habit. This can be represented diagrammatically by a stereographic projection.
7. A crystal may be described as the **convolution** of a **crystal lattice** with the contents of one **unit cell**.

Glossary

Acicular crystals: Long, needle-shaped crystals (from the Greek: *aci*, a point).

Aggregation: The clustering together of a small group of ions or molecules

Constancy of Interfacial Angles, Law of: In all crystals of a given type from a given compound, the angles between corresponding faces have a constant value. This law applies to one particular form if the crystal is polymorphous.

Convolution: The convolution of two mathematical functions $f(y)$ and $g(y)$ is $C(x) = \int_y f(y)g(x-y)dy$. To calculate this function, set the origin of the first function in turn on every position of the second, multiply the value of the first function in each position by the value of the second at that point and take the sum of all such possible operations. Thus, the convolution of two functions $f(y)$ and $g(y)$ at the

$$Convolution = \int f(y)g(x-y)dy \approx \sum f(y)g(x-y)$$

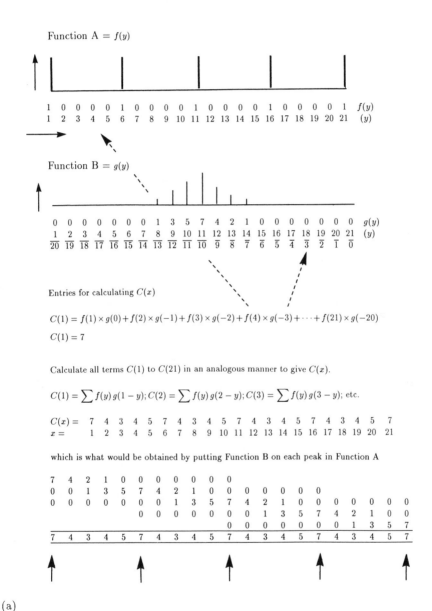

Function A = $f(y)$

| 1 | 0 | 0 | 0 | 0 | 1 | 0 | 0 | 0 | 0 | 1 | 0 | 0 | 0 | 0 | 1 | 0 | 0 | 0 | 0 | 1 | $f(y)$ |
| 1 | 2 | 3 | 4 | 5 | 6 | 7 | 8 | 9 | 10 | 11 | 12 | 13 | 14 | 15 | 16 | 17 | 18 | 19 | 20 | 21 | (y) |

Function B = $g(y)$

0	0	0	0	0	0	0	1	3	5	7	4	2	1	0	0	0	0	0	0	0	$g(y)$
1	2	3	4	5	6	7	8	9	10	11	12	13	14	15	16	17	18	19	20	21	(y)
$\overline{20}$	$\overline{19}$	$\overline{18}$	$\overline{17}$	$\overline{16}$	$\overline{15}$	$\overline{14}$	$\overline{13}$	$\overline{12}$	$\overline{11}$	$\overline{10}$	$\overline{9}$	$\overline{8}$	$\overline{7}$	$\overline{6}$	$\overline{5}$	$\overline{4}$	$\overline{3}$	$\overline{2}$	$\overline{1}$	$\overline{0}$	

Entries for calculating $C(x)$

$$C(1) = f(1) \times g(0) + f(2) \times g(-1) + f(3) \times g(-2) + f(4) \times g(-3) + \cdots + f(21) \times g(-20)$$

$$C(1) = 7$$

Calculate all terms $C(1)$ to $C(21)$ in an analogous manner to give $C(x)$.

$$C(1) = \sum f(y)\,g(1-y);\, C(2) = \sum f(y)\,g(2-y);\, C(3) = \sum f(y)\,g(3-y);\, \text{etc.}$$

| $C(x) =$ | 7 | 4 | 3 | 4 | 5 | 7 | 4 | 3 | 4 | 5 | 7 | 4 | 3 | 4 | 5 | 7 | 4 | 3 | 4 | 5 | 7 |
| $x =$ | | 1 | 2 | 3 | 4 | 5 | 6 | 7 | 8 | 9 | 10 | 11 | 12 | 13 | 14 | 15 | 16 | 17 | 18 | 19 | 20 | 21 |

which is what would be obtained by putting Function B on each peak in Function A

7	4	2	1	0	0	0	0	0	0	0										
0	0	1	3	5	7	4	2	1	0	0	0	0	0	0	0					
0	0	0	0	0	0	0	1	3	5	7	4	2	1	0	0	0	0	0	0	0
					0	0	0	0	0	0	0	1	3	5	7	4	2	1	0	0
										0	0	0	0	0	0	0	1	3	5	7
7	4	3	4	5	7	4	3	4	5	7	4	3	4	5	7	4	3	4	5	7

(a)

FIGURE 2.17. A convolution. (a) The mathematics.

Convolution of Function A

and Function B

gives Function C.

(b)

FIGURE 2.17 (cont'd). (b) Graphical view (solid lines) of the same convolution. Broken lines indicate the components B.

point x_o is obtained by multiplying together the values of $f(y)$ and $g(x_o - y)$ for each set of possible values of x, and then summing all the products. This is repeated for each value of x in order to obtain the entire convolution. This function is also called the folding or *Faltung* of two functions. An example is worked out numerically in Figure 2.17.

Crystal lattice: Crystals are composed of groups of atoms repeated with the same orientation at regular intervals in three dimensions. Each such group of atoms may be replaced by a representative point, and the collection of points so formed is called the crystal lattice. The term crystal lattice is specific and should not be used to denote the entire atomic arrangement.

Defects: Irregularities in the precise periodicity of the arrangement of unit cells in crystals, i.e., imperfections in the crystal lattice.

Dendrites: Multibranched crystals (from the Greek: *dendron*, tree).

Dislocations: Displacements in the packing of unit cells an otherwise normal crystal structure resulting in a distortion in the internal periodicity. Dislocations provide good sites for further crystal growth.

Disorder: A disturbance to the regular organization of an entity (here a crystal).

Epitaxy: The oriented overgrowth of one crystalline material on the surface of another. Generally there is some match of periodicity between the two.

Goniometer: An instrument for measuring angles, such as those between crystal faces of a crystal grown under specific conditions.

Habit: The appearance of a crystal grown under specific conditions. The habit is evident in the relative development of the different faces of a crystal for a given material.

Interfacial angle: The angle between adjoining faces of a crystal.

Isomorphism: Similarity of crystal shape, unit cell dimensions, and structure between substances of similar chemical composition. Generally only the identity of one atom in the chemical formula is changed. Ideally, the substances are so closely similar that they may form a continuous series of solid solutions.

Miller indices: Integers (h,k,l), reduced to the least common denominator, used to describe any plane through a crystal. The plane with Miller indices h,k and l makes intercepts a/h, b/k, c/l with the unit-cell axes a, b, and c.

Morphology: The general shape of a crystal of a particular material.

Nucleation: The action of a tiny crystal, dust particle or other solid material in starting crystallization.

Parallelepiped: A six-sided figure each side of which is a parallelogram; that is, opposite sides are parallel to each other.

Polymorphism: The occurrence of two or more crystalline forms of the same substance that differ in their crystal structures (have different atomic arrangements).

Precipitant: A chemical used to promote protein crystallization, but not denaturation. Examples are highly soluble inorganic salts (ammonium sulfate or sodium chloride), and organic polyethers (polyethyleneglycols of a selected molecular weight range).

Precipitation: The act of separation of a solid mass from solution.

Rational Indices, Law of: A rational number is an integer or the quotient of two integers. The Law of Rational Indices states that all of the faces of a crystal may be described, with respect to their intercepts on three noncolinear axes, by three small whole numbers.

Right-handed axial system: A system of axes, usually orthogonal (at right angles to each other), that are designated x, y, and z, in such an order that x is converted to y and y is converted to z in the same manner that a right-handed screw (moving clockwise into a piece of wood) would proceed (with x to y as the clockwise motion and z the direction into the wood).

Saturated solution: A solution is said to become saturated when the solute and the solution are at equilibrium.

Screw dislocation: A dislocation formed in a crystal upon nucleation, with an approximately spiral structure that is maintained as more molecules are added and the crystal grows.

Spherulites: Solids composed of radiating, branched fibers, such as found in some vitreous volcanic rocks, for example, obsidian. These fibers are packed to give an approximately spherical shape.

Unit cell: The basic building block of a crystal, repeated infinitely in three dimensions. It is characterized by three vectors, **a**, **b**, and **c**, that form the edges of a parallelepiped and the angles between them, α, β, and γ (with γ between **a** and **b**, α between **b** and **c**, and β between **a** and **c**).

References

1. Labaw, L. W., and Wyckoff, R. W. G. The electron microscopy of tobacco necrosis virus crystals. *J. Ultrastruct. Res.* **2**, 8–15 (1958).

2. Kepler, J. *Strena seu De Nive Sexangula.* Godefridum Tampach: Francofurti ad Mœnum (1611). **English translation** (of excerpts): Silverman, J. S. *A New Year's Present; on Hexagonal Snow.* In: *Crystal Form and Structure.* (**Ed.**, Schneer, C. J.) pp. 16–17. Dowden, Hutchinson & Ross: Stroudsburg, PA (1977). Also: Hardie, C. *The Six-cornered Snowflake.* Clarendon Press: Oxford (1966).

3. Speyer, J. F., and Khairallah, L. H. Crystalline T4 bacteriophage. *J. Molec. Biol.* **76**, 415–417 (1973).

4. Müller, E. W. Atoms visualized. *Sci. Amer.* **196(6)**, 113–122 (1957).

5. Crewe, A. V., Wall, J., and Langmore, J. P. Visibility of single atoms. *Science* **168**, 1338–1340 (1970).

6. Allen, F. M., Smith, B. K., and Buseck, P. R. Direct observation of dissociated dislocations in garnet. *Science* **238**, 1695–1697 (1987).

7. Cowley, J. M. *Diffraction Physics.* 2nd edn. North-Holland: Amsterdam, Oxford (1981).

8. Iijima, S. Direct observation of lattice defects in H-Nb_2O_5 by high resolution electron microscopy. *Acta Cryst.* **A29**, 18–24 (1973).

9. Binnig, G., Rohrer, H., Gerber, C., and Weibel, E. Tunneling through a controllable vacuum gap. *J. Appl. Phys.* **40**, 178–180 (1982).

10. Binnig, G., and Rohrer, H. The scanning tunneling microscope. *Sci. Amer.* **253(2)**, 50–56 (1985).

11. Feenstra, R. M., Stroscio, J. A., Tersoff, J., and Fein, A. P. Atom-selective imaging of the GaAs (110) surface. *Phys. Rev. Lett.* **58**, 1192–1195 (1987).

12. Berman, H. M., Rubin, B. H., Carrell, H. L., and Glusker, J. P. Crystallographic studies of D-xylose isomerase. *J. Biol. Chem.* **249**, 3983–3984 (1974).

13. Carrell, H. L., Rubin, B. H., Hurley, T. J., and Glusker, J. P. X-ray crystal structure of D-xylose isomerase at 4-Å resolution *J. Biol. Chem.* **259**, 3230–3236 (1984).

14. Holden, A., and Singer, P. *Crystals and Crystal Growing.* Anchor Books, Doubleday: Garden City, NY (1960).

15. Bunn, C. *Crystals: Their Role in Nature and in Science.* Academic Press: New York, London (1964).

16. Newnham, R. E., and Markgraf, S. A. *Classic Crystals: A Book of Models.* Materials Research Laboratory, The Pennsylvania State University: University Park, PA 16801 (1986).

17. Schneer, C. J. (**Ed.**) *Crystal Form and Structure.* Dowden, Hutchinson & Ross: Stroudsburg, PA (1977).

18. Bunn, C. W. *Chemical Crystallography. An Introduction to Optical and X-ray Methods.* 1st edn. 1946. 2nd edn, 1961. Clarendon Press: Oxford (1961).

19. Strickland-Constable, R. F. *Kinetics and Mechanism of Crystallization.* Academic Press: London, New York (1968).

20. Frank, F. C. On the equations of motion of crystal dislocations. *Proc. Phys. Soc. (London)* **A62**, 131–135 (1949).

21. Frank, F. C. The influence of dislocations on crystal growth. *Disc. Faraday Soc.* **5**, 48–54 (1949). Also: *Crystal Growth.* Gurney and Jackson: London, Edinburgh (1949).

22. Burton, W. K., Cabrera, N., and Frank, F. C. The growth of crystals and equilibrium structure of their surfaces. *Phil. Trans. Roy. Soc. (London)* **A243**, 299–358 (1951).

23. Burton, W. K., and Cabrera, N. Crystal growth and surface structure. Part I. Part II. *Disc. Faraday Soc.* **5**, 33–39, 40–48 (1949). Also: *Crystal Growth.* Gurney and Jackson: London, Edinburgh (1949).

24. van der Merwe, J. H. Misfitting monolayers and oriented overgrowth. *Disc. Faraday Soc.* **5**, 201–214 (1949). Also: *Crystal Growth.* Gurney and Jackson: London, Edinburgh (1949).

25. Dawson, I. M., and Vand, V. The observation of spiral growth-steps in *n*-paraffin single crystals in the electron microscope. *Proc. Roy. Soc. (London)* **A206**, 555–562 (1951).

26. Feher, G., and Kam, Z. Nucleation and growth of protein crystals: general principles and assays. *Methods in Enzymology. Diffraction Methods for Biological Macromolecules. Part A.* (**Eds.**, Wyckoff, H. W., Hirs, C. H. W., and Timasheff, S. N.) **114**, 77–112 (1985).

27. Leubner, I. H. Crystal formation (nucleation) under kinetically controlled and diffusion-controlled growth conditions. *J. Phys. Chem.* **91**, 6069–6073 (1987).

28. Czochralski, J. Ein neues Verfahren zur Messung der Kristallisationsgeschwindigkeit der Metalle. [A new method for the measurement of the velocity of crystallization of the metals.] *Z. Phys. Chem.* **92**, 219–221 (1917); Dislozierte Reflexion in Dienste der Metallkunde. [The use of directed reflection in metallography.] *Z. Anorg. Allgemeine Chem.* **144**, 131–141 (1925).

29. Bridgman, P. Effect of tension on the transverse and longitudinal resistance of metals. *Proc. Amer. Acad. Arts Sci.* **60**, 423–49 (1925).

30. Stockbarger, D. C. Production of large single crystals of lithium fluoride. *Rev. Sci. Instruments* **7**, 133–136 (1936).

31. Verneuil, A. Mémoire sur la reproduction artificielle du rubis par fusion. [Note on the artificial production of rubies by fusion.] *Compt. Rendus, Acad. Sci. (Paris).* **135**, 791–794 (1902); *Annales de Chimie et de Physique.* Huitième série. **III**, 20–48 (1904).

32. Buerger, M. J. The relative importance of the several faces of a crystal. *Amer. Mineralogist* **32**, 593–606 (1947).

33. Donnay, J. D. H., and Harker, D. A new law of crystal morphology extending the law of Bravais. *Amer. Mineralogist* **22**, 446–467 (1937).

34. Kitaigorodskii, A. I. *Organic Chemical Crystallography.* (**English translation**: Consultants Bureau: New York) (Russian edition, 1955. Academy of Sciences Press: Moscow) (1961).

35. Groth, P. *Physikalische Kristallographie.* [Physical Crystallography.] 4th edn. Engelmann: Leipzig (1905).

36. Groth, P. *Chemische Kristallographie.* [Chemical Crystallography.] 5 vols. Engelmann: Leipzig (1906-1919).

37. Wells, A. F. Crystal habit and internal structure. – I. *Phil. Mag.* **37**, 184–199, 217–236 (1946).

38. McPherson, A., and Shlichta, P. Heterogeneous and epitaxial nucleation of protein crystals on mineral surfaces. *Science* **239**, 385–387 (1988).

39. Gibbs, J. W. On the equilibrium of heterogeneous substances. *Transactions of the Connecticut Academy of Sciences.* (1878) In: *The Scientific Papers of J. Willard Gibbs. Vol. 1., Thermodynamics.* p. 320. Longmans & Green: London (1906).

40. Curie, P. Sur la formation des cristaux et sur les constantes capillaires de leurs différentes faces. *Bull. soc. min. de France* **8**, 145–150 (1885). **English translation**: Schneer, C. J. On the formation of crystals and on the capillary constants of their different faces. *J. Chem. Educ.* **47**, 636–637 (1970).

41. Hartman, P., and Perdok, W. G. On the relations between structure and morphology of crystals. I. II. III. *Acta Cryst.* **8**, 49–52, 521–529 (1955).

42. Wulff, G. Zur Frage der Geschwindigkeit des Wachstums und der Auflösung der Kristallflächen. *Z. Krist.* **34**, 449–530 (1901). **English translation**: Watson, D. On the question of the rate of growth and of dissolution of crystal faces. In: *Crystal Form and Structure.* (**Ed.**, Schneer, C. J.) pp. 16–17. Dowden, Hutchinson & Ross: Stroudsburg, PA (1977).

43. Ohara, M., and Reid, R. C. *Modeling Crystal Growth Rates from Solution.* Prentice-Hall: Englewood Cliffs (1973).

44. Mitscherlich, E. Über die Kristallisation der Salze in denen das Metall der Basis mit zwei Proportionen Sauerstoff verbunden ist. [Concerning the crystallization of salts in which the base metal is bound to two proportions of oxygen.] *Abhandlungen der Königlichen Akademie der Wissenschaften in Berlin.* pp. 427–437 (1818–1819).

45. Mitscherlich, E. Sur la relation qui existe entre la forme cristalline et les proportions chimiques. I. Mémoire sur les arseniates et les phosphates. [On the relationship that exists between crystal form and chemical proportions. I. Note on arsenates and phosphates.] *Ann. Chim. Phys.* **19**, 350–419 (1822).

46. Leicester, H. M., and Klickstein, H. S. *A Source Book in Chemistry. 1400-1900.* McGraw-Hill: New York, Toronto, London (1952).

47. Hartshorne, N. H., and Stuart, A. *Crystals and the Polarizing Microscope. A Handbook for Chemists and Others.* 2nd edn. Edward Arnold: London (1950).

48. King, M. V. Referred to by Wood, E. A. In: The development of X-ray diffraction in the U.S.A. From the beginning of World War II to 1961. In: *Fifty Years of X-ray Diffraction.* (**Ed.**, Ewald, P. P.) N. V. A. Oosthoek: Utrecht (1962).

49. McPherson, A. *The Preparation and Analysis of Protein Crystals.* Wiley: New York (1982).

50. McPherson, A. Crystallization of macromolecules: general principles. *Methods in Enzymology. Diffraction Methods for Biological Macromolecules. Part A.* (**Eds.**, Wyckoff, H. W., Hirs, C. H. W. and Timasheff, S. N.) **114**, 112–120 (1985).

51. Ducruix, A., and Giegé, R. *Crystallization of Nucleic Acids and Proteins: A Practical Approach.* IRL Press: Oxford, New York, Tokyo (1991).

52. Stout, G. H., and Jensen, L. H. *X-ray Structure Determination. A Practical Guide.* Macmillan: New York (1968); 2nd edn. John Wiley: New York, Chichester, Brisbane, Toronto, Singapore (1989).

53. Blundell, T. L., and Johnson, L. N. *Protein Crystallography.* Academic Press: New York, London, San Francisco (1976).

54. Gilman, J. J. *The Art and Science of Growing Crystals.* Wiley: New York (1963).

55. Carter, C.W., Jr., and Carter, C.W. Protein crystallization using incomplete factorial experiments. *J. Biol. Chem.* **254**, 12219–12223 (1979).

56. Wyckoff, H. W., Hirs, C. H. W., and Timasheff, S. N. (**Eds.**) *Methods in Enzymology. Diffraction Methods for Biological Macromolecules. Part A.* **114**, *Part B.* **115**. Academic Press: New York (1985).

57. Zeppezauer, M., Eklund, H., and Zeppezauer, E. S. Micro diffusion cells for the growth of single protein crystals by means of equilibrium dialysis. *Arch. Biochem. Biophys.* **126**, 564–573 (1968).

58. Littke, W., and John, C. Protein single crystal growth under microgravity. *Science* **225**, 203–204 (1984).

59. DeLucas, L. J., Suddath, F. L., Snyder, R., Naumann, R., Broom, M. B., Pusey, M., Yost, V., Herren, B., Carter, D., Nelson, B., Meehan, E. J., McPherson, A., and Bugg, C. E. Preliminary investigations of protein crystal growth using the Space Shuttle. *J. Crystal Growth* **76**, 681–693 (1986).

60. DeLucas, L. J., Smith, C. D., Smith, H. W., Vijay-Kumar, S., Senadhi, S. E., Ealick, S. E., Carter, D. C., Snyder, R. S., Weber, P. C., Salemme, F. R., Ohlendorf, D. H., Einspahr, H. M., Clancy, L. L., Navia, M. A., McKeever, B. M., Nagabhoshan, T. L., Nelson, G., McPherson, A., Koszelak, S., Taylor, G., Stammers, D., Powell, K., Darby, G. and Bugg, C. E. Protein crystal growth in microgravity. *Science* **246**, 651–654 (1989).

61. Masselink, W. T., Henderson, T., Klem, J., Fischer, R., Pearah, P., Morkoc, H., Hafich, M., Wang, P. D., and Robinson, G. Y. Optical properties of gallium arsenide on (100)-silicon using molecular beam epitaxy. *Appl. Phys. Lett.* **45**, 1309–1311 (1984).

62. McPherson, A., and Shlichta, P. J. Facilitation of the growth of protein crystals by heterogeneous/epitaxial nucleation. *J. Crystal Growth* **85**, 206–214 (1987).

63. Steinrauf, L. K. Preliminary X-ray data for some new crystalline forms of β-lactoglobulin and hen egg-white lysozyme. *Acta Cryst.* **12**, 77–79 (1959).

64. Taylor, W. H., and Náray-Szabó, S. The structure of apophyllite. *Z. Krist.* **77**, 146–158 (1931).

65. Berman, A., Addadi, L., and Weiner, S. Interaction of sea-urchin skeleton macro-molecules with growing calcite crystals — a study of intracrystalline proteins. *Nature (London)* **331**, 546–548 (1988).

66. Mann, S. Molecular recognition in biomineralization. *Nature (London)* **332**, 119–124 (1988).

67. Weiner, S. Organization of extracellularly mineralized tissues: a comparative study of biological crystal growth. *CRC Crit. Rev. Biochem.* **20**, 365–408 (1986).

68. Etter, M. C., and Baures, P. W. Triphenylphosphine oxide as a crystallization aid. *J. Amer. Chem. Soc.* **110**, 639–640 (1988).

69. Carangeot, A. Goniomètre, ou mésure–angle. [Goniometry or angle measurement.] *Observations sur la Physique, sur l'Histoire Naturelle, et les Arts.* (*J. de Phys.*) **22**, 193–197 (1783).

70. Wollaston, W. H. The description of a reflective goniometer. *Phil. Trans. Roy. Soc. (London)* **99** 253–258 (1809).

71. Haüy, R. J. *Traité de Minéralogie.* [Treatise on mineralogy.] 5 volumes. Louis: Paris (1801).

72. Haüy, R. J. *Traité de Cristallographie.* [Treatise on crystallography.] 3 volumes. Bachelier & Huzard: Paris (1822).

73. Burke, J. G. *Origins of the Science of Crystals.* University of California Press: Berkeley, Los Angeles (1966).

74. Miller, W. H. *A Treatise on Crystallography.* Deighton: Cambridge, England, and Parker: London (1839).

75. Steadman, R. *Crystallography.* Van Nostrand Reinhold: New York, Cincinnati, Toronto, London, Melbourne (1982).

76. Klug, H. P., and Alexander, L. E. *X-ray Diffraction Procedures for Polycrystalline and Amorphous Materials.* 2nd edn. Wiley-Interscience: New York (1974).

77. Penfield, S. L. The stereographic projection and its possibilities from a graphical standpoint. *Amer. J. Sci.* **11**, 1–24, 115–144 (1901).

78. Ladd, M. F. C., and Palmer, R. A. *Structure Determination by X-ray Crystallography.* 2nd edn. Plenum: New York, London (1985).

79. Wahlstrom, E. E. *Optical Crystallography.* 5th edn. (1st edn., 1943). John Wiley: New York, Chichester, Brisbane, Toronto (1979).

80. Bunn, C. W. Absorption, oriented overgrowth and mixed crystal formation. *Proc. Roy. Soc. (London)* **A141**, 567–593 (1933).

81. Barlow, W., and Pope, W. J. The relation between the crystalline form and the chemical constitution of simple inorganic substances. *Trans. Chem. Soc.* **91**, 1150–1214 (1907).

82. Holmes, K. C., and Blow, D. M. *The Use of X-ray Diffraction in the Study of Protein and Nucleic Acid Structure.* Revised reprint from *Methods of Biochemical Analysis.* (**Ed.**, Glick, D.) **13**, 113–239 (1966). Interscience (John Wiley): New York, London, Sydney (1966). Reprinted: Krieger: Melbourne, FL (1979).

CHAPTER

3

Diffraction By Crystals

In this chapter we discuss the phenomenon of X-ray diffraction that results from the scattering of the X rays by the electrons surrounding atomic nuclei in crystals. Because there is no lens available to recombine X rays after they have been scattered in this way, mathematical techniques, analogous to wave recombination by a lens, are employed in order to obtain an image. To assist the reader in understanding how this image of the scattering matter is obtained from X-ray diffraction studies, we will review the description of the microscope and the analogies with diffraction methods, a continuation of the description in Chapter 1.

Methods for the experimental measurement of the intensities of the diffracted beams will be described in Chapter 7, and methods for deriving relative phases of these beams for recombination will be discussed in Chapter 8. The result of these mathematical calculations, which simulate the action of a lens, will be all the information that is needed for the calculation of a three-dimensional electron–density map. This is a map in which peaks are situated at or near atomic positions. In this way, measurements of the diffraction pattern lead to an image of the molecules or ions and their arrangement in the crystal under study. Details of the calculation of the electron density maps that reveal the atomic arrangement will be described in Chapter 9. For more information on each aspect of diffraction by crystals, the reader is referred to the many texts on the subject listed in the Preface to this book.

3.1 Obtaining a magnified image: the microscope

We see images of objects because the lenses in our eyes recombine the scattered visible radiation; the signals generated in this process are detected by the retina and then interpreted, very efficiently, in the brain. Unfortunately, many objects that we would like to examine are too small to be viewed without the aid of a device, such as a hand lens, that will magnify their image. The compound microscope, as discussed in Chapter 1, contains two lenses that recombine light scattered by an object. The first lens, that closest to the object of study, is the objective lens, and it produces an image that is further magnified by a second lens, the eye-piece. In this way we are able to "see" a highly magnified image of very small objects. An optical microscope, however, is not powerful enough to allow us to view atoms; they are too small. An instrument using radiation with a shorter wavelength, X rays, for example, would have to be constructed in order to let us do this.

There are two parts to the process of magnifying an image by means of a lens, as shown in Figure 1.2 (Chapter 1). The first part involves the scattering of light by the object, and the second part involves the recombination (focusing) of these scattered waves to give a magnified image. When we attempt to view atoms by X-ray diffraction methods, we carry out the first part of the action of an X-ray microscope by analyzing the pattern of scattered X rays. This is generally done by causing a fine beam of X rays to impinge on a crystal and measuring the intensity variation of the scattered X rays around the crystal. We cannot carry out the second part of the action of a microscope (in which a lens focuses scattered radiation) because no material with a refractive index greater than 1.0 for X rays has yet been found, and this is needed to form an X-ray lens.[1] To overcome this problem, the crystallographer has had to resort to mathematical methods in order to recombine the waves. Thus the lens that is used for X rays is a mathematical one, produced computationally, but giving the same results that would be expected of a real lens with visible light. The manner in which a lens takes the relative phases of light beams into account when visible light is focused in a microscope is analogous to the manner in which the relative phases of X rays are recombined in an X-ray diffraction experiment. This is the subject of Chapter 8 and constitutes what is called the phase problem.

3.2 Light and X rays as wave motions

Our present-day understanding of the nature of light as a wave motion (see Figure 3.1) stems from the work of Christiaan Huygens,[2] who considered light to be a wave disturbance. This idea was in contrast to that of Isaac Newton, who developed the notion that light is a stream of particles. In the description given by Huygens, as the wave disturbance proceeds its forward surface, at any instant in time, constitutes what is known as the **wave front**. A principle, now called the Huygens–Fresnel principle, states that, at any instant in time the unobstructed part of any wave front becomes a source of secondary wavelets that have the same wavelength as the original wave and these secondary wavelets spread out to give a new wave front, as shown in Figure 3.2. This description forms the basis of the discussion of diffraction that follows. Newton's ideas,

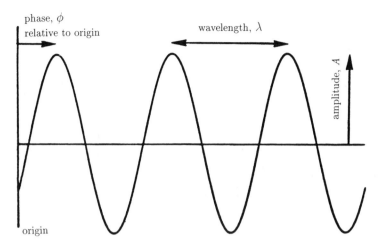

FIGURE 3.1. The amplitude A, wavelength λ, and relative phase ϕ (relative to a chosen origin) of a sinusoidal wave. The vertical axis represents the extent of deviation from a base line. The horizontal axis, and hence the phase, represents the time (or distance) required for the wave crest to reach a defined point.

however, cannot be neglected, in view of the fact that neutrons (which are considered to be particles) are also diffracted by crystals.

3.2.1 Diffraction of visible light by small holes and slits

The pattern of radiation scattered by any object is called its diffraction pattern.[3-5] When objects are large compared with the wavelength of the light that is falling on them, relatively sharply defined shadows are cast, as expected, since light is considered to travel in straight lines. But when the wavelength of the light and the size of the object are of the same order of magnitude, an observable amount of light spreads into the area expected to be in shadow. This effect is particularly noticeable as the light passes the edges of an opaque object or through a narrow slit. For example, in the early seventeenth century, Francesco Maria Grimaldi[5] observed that a fine thread does not cast as sharp a shadow as would be expected. Fringes of parallel dark and bright bands are produced at the edges of the shadow. This effect he named diffraction.

The term diffraction is used whenever the straight-line path of a wave front traveling through a uniform medium is caused to change direction. Note that this is different from refraction, which is the bending of light that occurs when the nature of the medium changes. Diffraction is found to occur whenever a wave front is obstructed in some way.[6] Henry S. Lipson[7] wrote: "An object in the path of a beam of light alters the beam so that information about the object is impressed upon the light waves; this is called the diffraction of the light waves by the object. A perfect

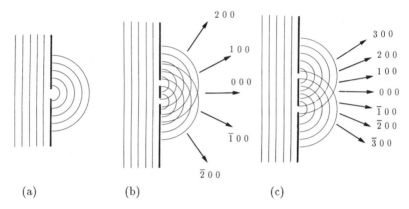

FIGURE 3.2. The spreading of light at an aperture to give a diffraction pattern. The equally spaced circles and lines represent the crests of waves (their maximum amplitude), and these waves reinforce each other when they intersect. Angles where this reinforcement occurs (the diffracted beams) are indicated by arrows. The order of diffraction is the number of wavelengths in the path difference between beams diffracted by the two slits (000 for the direct beam, 100 for the first order, etc., and $\bar{1}$00, etc. for negative directions). (a) Diffraction at a single slit (aperture). (b,c) Diffraction by two slits, (b) close together and (c) further apart. Note that in (b), where the slits are closer to each other, the directions of wave reinforcement (diffracted beams) are at higher angles from the horizontal (for the same order of diffraction, see arrows) than in (c), where the apertures are further apart.

image of the object can be obtained only if the whole of the information carried by the waves is used." This information consists of the amplitude, wavelength (periodicity), and relative phase of each wave (see Figure 3.1). All three must be known to obtain the required image.

Interference between waves was demonstrated by Thomas Young[8,9] at the beginning of the nineteenth century by an experiment that can be tried by the reader. Augustin Jean Fresnel[10] did a similar experiment independently, and the two scientists corresponded in a friendly fashion with each other about their results. In a modern-day version of this experiment, two pinholes are punched 2–3 mm apart in a piece of metal (aluminum foil is good) and a distant street light is viewed at night with the pinholes held close to an eye. Young's experiment had, as a result of Newton's theory of light, been expected to give two lines when projected onto a screen. The actual result is a series of light and dark bands, called fringes.

Interference between waves occurs when they are traveling in the same direction. They do not appreciably affect each other when they are traveling in different directions. When his son asked "Can waves cross?"

Sir Lawrence Bragg quickly replied,[11] "Of course they can, otherwise you could not see your brother across the table, while I was looking at your mother."

But how do different waves traveling in the same direction interfere with each other? If the two waves have the same wavelength, the effects of constructive or destructive interference, shown in Figure 1.1 (Chapter 1) and Figure 3.3, result. This interference can be demonstrated by throwing two stones simultaneously into water (see Figure 3.2). The secondary peaks have higher peaks than the original when two primary waves coincide, and have deeper valleys than the original when two valleys coincide. If a peak and a valley coincide, they cancel each other out.

The important feature that determines the consequences of interference of two waves of the same wavelength is their **path difference** (see Figure 3.3), that is, the fraction of a wavelength one wave is out of phase

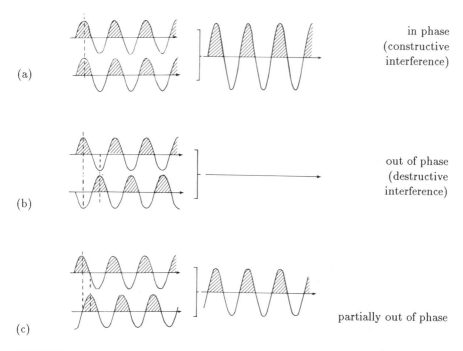

FIGURE 3.3. Two waves traveling in the same direction with the same amplitude and wavelength are (a) in phase (path difference of zero or an integral number of wavelengths), (b) out of phase (path difference half a wavelength), and (c) partially out of phase (path difference $\lambda/4$ in this case). This Figure illustrates another version of the situations shown in Figure 1.1 (Chapter 1).

with another. Young wrote: "When two undulations, from different origins, coincide perfectly or very nearly in direction, their joint effect is a combination of the motions belonging to each."[8] This means that the displacements of the wave motions from the horizontal axis add to each other to give the combined, diffracted beam, as illustrated in Figure 3.3. The terms interference and diffraction are used somewhat indiscriminately when describing these effects. According to Richard P. Feynman,[12] the difference between them is merely a question of usage.

The principle of Huygens and Fresnel[6] explains what happens when light is diffracted by an object such as the fine thread that Grimaldi used or the narrow slit that Young used. In diffraction, secondary waves traveling in the same direction are combined with one another, and the amplitude (hence the **intensity**) of the diffracted beam is determined by the superposition of all such secondary waves with due attention to their relative phases. Each secondary wave will generally be slightly out of phase with its neighbors traveling in the same direction. The extent of this phase difference depends on the angular deviation of the direction of the diffracted beam from that of the direct beam and on the wavelength of the light (Figures 3.2 and 3.3). The larger this angle and the shorter the wavelength of the radiation, the more one wave is out of phase with its neighbor. At some angles the scattered waves are all in phase, and the diffracted beam is strong [Figure 3.3(a)]. At other angles they cancel each other out [Figure 3.3(b)], and there is no apparent diffracted beam. At still other angles there is partial interference and the diffracted beam has intermediate intensity [Figure 3.3(c)]. Since a single slit has width, waves traveling in the same direction from the two edges will interfere with each other in the manner just described. The amount that these waves are out of phase will be greater for the wider slit and therefore the diffracted beam will be narrower. This gives a diffraction pattern for a single slit, and its intensity profile is referred to as the **envelope** (Figures 3.4 and 3.5).

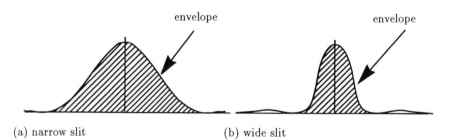

(a) narrow slit   (b) wide slit

FIGURE 3.4. Scattering by a single slit. (a) Diffraction by a narrow slit and (b) the diffraction pattern of a slit that is wider than that in (a). In both cases the intensity variation shown is referred to as the "envelope." The zero point of the horizontal axis represents the direction of the direct beam (cf. Figure 3.5).

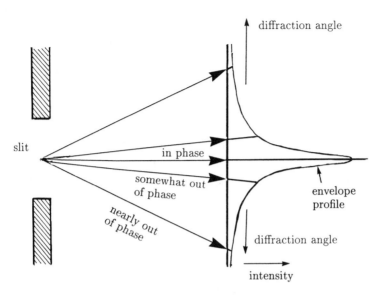

FIGURE 3.5. An illustration of the reason that the diffraction pattern of a single slit has width to its profile (the envelope profile, compare with Figure 3.4). This is because of varying degrees of interference between waves traveling in the directions indicated in this figure. Waves traveling in the direction of the direct beam (if the incident light is perpendicular to the slit) are in phase and give a maximum intensity. At angles other than that of the direct incident beam, the scattered waves are more out of phase. Therefore their intensities decrease because of increased interference. The result is a diffraction pattern with the envelope profile shown here.

The diffraction pattern also gives a quantitative measure of the dimensions of the object causing it. As described in Chapter 1 and implied in Figure 3.2, there is an inverse relationship between the angular spread in the diffraction pattern and the corresponding dimension of the diffracting object. The narrower the slit, the wider the angular spread. Conversely, a wider slit gives a narrower diffraction pattern,[3] as shown in Figure 3.4. The thread and the slit have, interestingly, similar diffraction patterns. In other words, diffraction by a solid object is similar to diffraction by a hole of the same shape.

3.2.2 Diffraction of visible light by regular arrays

So far we have only considered the diffraction pattern of a single slit and have shown that the intensity variation is bell shaped; this is the envelope profile with a width inversely proportional to the width of the slit. Now we will consider what happens to the diffraction pattern when more slits are lined up parallel to the first to give the equivalent of a diffraction grating. This is a two-dimensional analogy to the buildup of a crystal from unit cells. The diffraction pattern of a regular array retains intensity only at isolated regularly spaced positions, as shown in Figure 3.6; these are **sampling regions** of the single slit diffraction pattern.

The diffraction pattern of a series of slits may be considered to be composed of two components. The "envelope profile" results from the interference of waves scattered by an individual slit. The "sampling regions" result from the interference of waves scattered from equivalent points in different slits. Therefore, the locations of the sampling regions give a measure, by a reciprocal relationship, of the spacing between the slits. It is also found that the lines in the diffraction pattern become increasingly sharp the greater the number of slits.

In summary: In diffraction of any form of electromagnetic radiation by a regularly repeating series of identical objects (e.g., a series of slits),

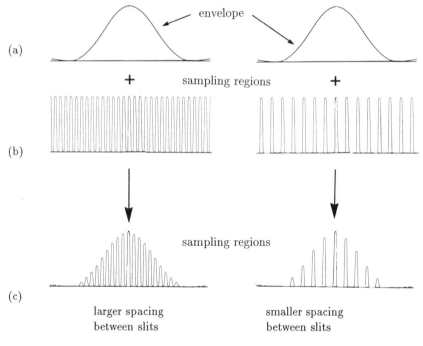

FIGURE 3.6. The diffraction patterns of series of slits. Shown in (a) is the diffraction pattern of a single slit. The sampling regions (from the grating periodicity) are shown in (b). The diffraction pattern is the combination of the envelope profile and the sampling regions, that is, the sampling of the envelope profile at sampling regions only. The results for a series of parallel, regularly spaced slits are shown in (c). The spacing between the slits is greater in the diagrams on the left than in those on the right.

the width and the intensity profile of the envelope are determined by the intensity profile of the diffraction pattern of a single object (e.g., a single slit). The positions in this envelope profile that are sampled to give intensity in the diffraction pattern are determined, in a reciprocal way, by the spacings between the regularities of the arrangement (e.g., the spacings between slits). These features are illustrated in Figure 3.7 by the diffraction patterns of masks consisting of holes in various arrangements, punched into a sheet impervious to light. The effect of different sizes of holes is seen by comparing Figure 3.7(a,b,c) with (d,e,f). The effect of different shapes of holes is illustrated by a comparison of Figure 3.7(a,b,c) with (g,h,i). The effects of different spacings between holes can be seen by a comparison of Figure 3.7(b,e,h) with (c,f,i). The diffraction patterns of more complex patterns are shown in Figure 3.7(k,l).

3.3 Diffraction of X rays by atoms in crystals

Just as visible light can be diffracted by very small objects, X rays are diffracted by electrons, and neutrons are diffracted by the nuclei of atoms. The effect is the same in each case, but the scale of the experiment is reduced to atomic dimensions for X-ray and neutron diffraction. The physical explanation given for diffraction by atoms[13] is this: when X rays hit an atom, the rapidly oscillating electric field of the radiation sets the electrons of the atom into oscillation about their nuclei. This oscillation has the same frequency as that of the incident radiation. The result is that the electron acts as an oscillating dipole that serves as a source of secondary radiation with the frequency of the incident beam. This phenomenon is referred to as **coherent scattering** because there is no wavelength change. Joseph John Thomson[13] showed that when radiation is scattered by an electron, there is a phase shift of 180° in the radiation, in the sense that the electric field in the scattered wave at a given point is opposite to that of the direct (incident) wave at that point. This phase change, diagrammed in Figure 3.8, is the same for scattering by all atoms in the crystal, in the absence of unusual effects (such as anomalous scattering, to be described later, in Chapter 14).

3.3.1 Bragg equation: directions and indices of beams

Now we consider diffraction in terms of the path differences of waves on scattering. This will lead to an understanding of the effect on the diffraction pattern of the periodicity of the crystal from unit cell to unit cell. There are two methods that have been used to consider this, one by Max von Laue and the other by W. L. Bragg.[14,15]

Laue realized[14] in 1912 that the path length differences PD_1, PD_2, PD_3 for waves diffracted by atoms separated by one crystal lattice translation must be an integral number of wavelengths for diffraction (i.e., reinforcement) to occur; further, he showed that this condition must be true

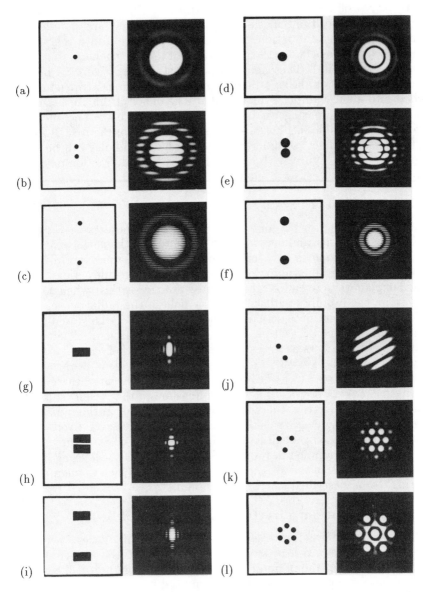

FIGURE 3.7. Examples of diffraction patterns from a variety of masks. The mask with the holes in it is to the left and its diffraction pattern is to the right. The diffraction pattern of (a) a single round hole in a mask, (b) two round holes, and (c) two round holes further apart. Note that in (b) and (c) the diffraction pattern is sampled (in lines) with a spacing that is inversely proportional to the distance between

FIGURE 3.7. (cont'd). Caption starts on opposite page.

holes in the mask. The diffraction effects when the sizes of the holes in the masks are increased are shown in (d), (e), and (f). The holes are larger than in (a), (b), or (c), but the spacings between their centers are the same. Recall that the experimental diffraction pattern is now more compact, illustrating the reciprocal relationship between the size of an object and its diffraction pattern. Since the spacings of holes in the masks are the same in (a), (b), and (c), and in (d), (e), and (f), the distances between sampling regions are also the same. The effect of changing the shape of the holes in the mask is shown in (g), (h), and (i), where the holes are rectangular in shape. Again, the reciprocal relationship between dimensions in real space and in the diffraction pattern is shown. The wider part of the hole gives a narrower diffraction pattern. The spacings between holes are the same as in (a) to (f). Finally, the effect of different arrangements of holes on the diffraction pattern. In (j) there are two holes in the mask, and the resulting diffraction pattern is similar to that in (b), although one mask is rotated with respect to the other. In (k) the effect of three holes, equivalent to three superpositions of the diffraction pattern in (j), each at 120° to each other is seen. In (l) the diffraction pattern of six holes is shown. (Reprinted from G. Harburn, C. A. Taylor and T. R. Welberry: *Atlas of Optical Transforms.* Copyright © 1975 by G. Bell & Sons Ltd. Used by permission of the authors and the publisher, Cornell University Press.)

simultaneously in all three dimensions for the three crystal lattice translations.

$$\text{Laue equations}: PD_1 = h_1\lambda, PD_2 = h_2\lambda, PD_3 = h_3\lambda \qquad (3.1)$$

The condition for reinforcement, when the path difference is an integral number of wavelengths, is shown in Figure 3.9(a). The equations so derived are known as the **Laue equations**.

W. L. Bragg,[15] in the summer of 1913, showed that scattered radiation from a crystal behaves as if the diffracted beam were "reflected" from a plane passing through points of the crystal lattice in a manner that makes these crystal-lattice planes analogous to mirrors. Such **lattice planes** are directions in the crystal lattice that are rich in lattice points. Examples are shown in Figure 3.9(b). Each diffracted beam is considered as a "**Bragg reflection**" from a crystal lattice plane [Figure 3.10], so that the angle of incidence of radiation, $(90° - \theta_{hkl})$, equals the angle of reflection as shown in Figure 3.10(a) and (b). It is convenient to consider the diffracted beam to deviate by the angle $2\theta_{hkl}$ from the main incident beam as shown in Figure 3.10(b). This leads to a value of $90° - \theta_{hkl}$ for the angle between the nth **order of diffraction** of X rays of wavelength λ and the normal to a set of crystal lattice planes. It then follows that, if the perpendicular spacing between the crystal lattice

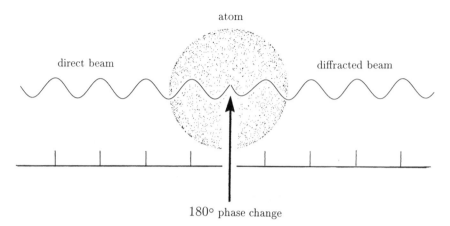

atom

direct beam

diffracted beam

180° phase change

FIGURE 3.8. Diagram illustrating that the phase shift on scattering of X rays is 180° (half a wavelength).

planes is d_{hkl}, the condition that the path difference of waves diffracted by adjacent crystal lattice planes be an integral number of wavelengths leads to:

$$\text{Bragg's Law}: n\lambda = 2d_{hkl}\sin\theta_{hkl}. \qquad (3.2)$$

where n is an integer. Equation 3.2 is known as **Bragg's Law** (see Figure 3.10). Thus, if λ and θ_{hkl} are known, values of d_{hkl}/n may be determined. Sir Lawrence Bragg wrote[16] modestly "$\cdots$ I have always felt the association of my name with it [Bragg's Law] to be an easily earned honour, because it is merely the familiar optical relation giving the colours reflected by thin films, in another guise."

The Laue equations and the Bragg equation are equivalent; they are just different ways of describing the same phenomenon.[17,18] Thus, if an X-ray beam strikes a crystal, coherent diffraction will occur when, and only when, the Bragg Equation (3.2) is satisfied, that is, λ, d_{hkl} and θ_{hkl}, have the relationships to each other defined by this equation. For X rays of a single wavelength and for a specific orientation of the crystal, very few Bragg reflections satisfy this condition (unless the unit cell is very large). Therefore it is usual to oscillate or rotate the crystal through an angular range or to use multiple-wavelength, in order to bring more diffracted beams into the necessary diffraction geometry for measurement.

Since the spacing between lattice planes d_{hkl} is a function of the unit cell dimensions and the indices h,k,l of those crystal planes, Bragg was able to equate the integers h_1, h_2, h_3 from Equation 3.1 of the Laue equations to the Miller indices h,k,l of the lattice planes causing the hkl Bragg reflection (see Chapter 2). Thus the path differences, which must be an integral number of wavelengths for reinforcement, are related to the

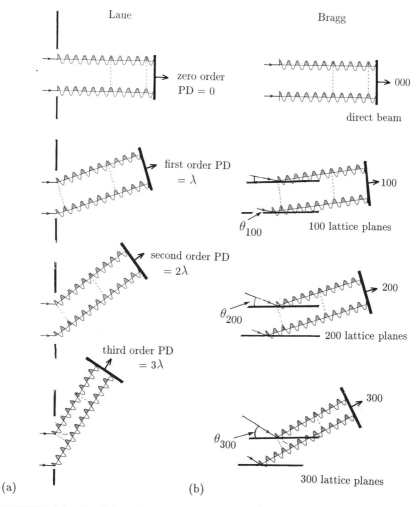

FIGURE 3.9. Conditions for diffraction. (Left) Max von Laue considered the path difference between a scattered and another wave scattered one unit cell away. If the difference is an integral number of wavelengths, reinforcement occurs, and a stronger diffracted beam is observed. Examples are shown for the direct beam, the first, second and third orders. (Right) W. L. Bragg realized that these same conditions could be considered as reflection from lattice planes. In addition, he realized that the orders of reflection [0,1,2,3] are equal to the indices $h00$ of the lattice planes causing that Bragg reflection. Note that the direction of a Bragg reflection is inclined at an angle $2\theta_{hkl}$ to the direct beam. (a) Laue considerations of path differences and (b) Bragg considerations of reflections from lattice planes. In both cases the overall result (diffraction) is the same.

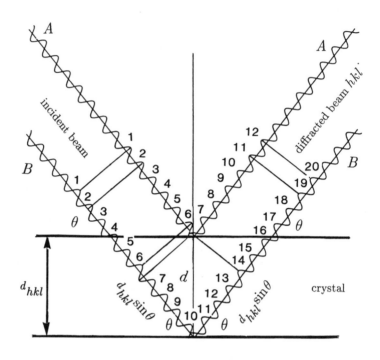

beam A PD = 0
beam B PD = 8λ

(a)

FIGURE 3.10. (a) The geometry of diffraction showing the path differences (8λ) between beams diffracted from two adjacent crystal lattice planes. The total path difference is $2d\sin\theta = n\lambda$.

wavelength of the radiation and the Miller indices of the crystal lattice planes involved. The method of using a set of diffracted beams to obtain d_{hkl} values, and then the unit cell dimensions, is shown in Table 3.1.

The relationship between the crystal structure and its diffraction pattern is diagrammed in Figure 3.11. Since diffraction by a crystal may be considered as reflection from a lattice plane, "reflection" has come to be used to denote a diffracted beam. We have concentrated in this chapter on the analogy to slit diffraction, so we will call it a Bragg reflection or a "diffracted beam." Also, the line drawn perpendicular to the lattice planes causing a Bragg reflection will define the angles of incidence and reflection and is known as the **diffraction vector** [see Figure 3.10(b)].

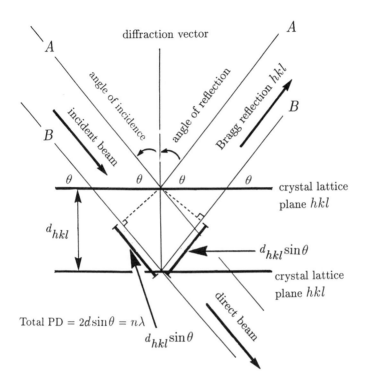

FIGURE 3.10. (b) Geometry of diffraction and its relationship to Bragg's Law. Note the definition of θ.

3.3.2 Scattering of X rays or neutrons by atoms

Since X rays are scattered by electrons, the amplitude of the wave scattered by an atom[19,20] is proportional to how many electrons there are around that atom, that is, its atomic number Z. **Atomic scattering factors** f_j are expressed as the ratio of the scattering of an atom to the scattering by a single electron under the same conditions.[19-24] Values of f_j for X rays fall off in value as a function of the **scattering angle**, $2\theta_{hkl}$ (Figure 3.12), because the atom has a size that is comparable to the X-ray wavelength. Examples are shown in Figure 3.13(a) and Table 3.2. A zero value for the scattering angle means that the incident beam is not deflected at all in that direction, and the various regions of the atom all scatter in phase. At higher scattering angles there is a decrease in scattering power that is a result of the size of the atom; this size causes

---

TABLE 3.1. Obtaining unit cell dimensions from d_{hkl} values.

The general equation for the spacings between lattice planes d_{hkl} for a crystal with unit-cell dimensions a, b, c, α, β, γ is given by:

$$d_{hkl} = X/Y,$$

where X and Y are:

$$X = [1 - \cos^2 \alpha - \cos^2 \beta - \cos^2 \gamma + 2 \cos \alpha \cos \beta \cos \gamma]^{\frac{1}{2}}$$

$$Y = \left[\left(\frac{h}{a} \right)^2 \sin^2 \alpha + \left(\frac{k}{b} \right)^2 \sin^2 \beta + \left(\frac{l}{c} \right)^2 \sin^2 \gamma \right.$$

$$- \frac{2kl}{bc}(\cos \alpha - \cos \beta \cos \gamma) - \frac{2lh}{ca}(\cos \beta - \cos \gamma \cos \alpha)$$

$$\left. - \frac{2hk}{ab}(\cos \gamma - \cos \alpha \cos \beta) \right]^{1/2}. \qquad (3.1.1)$$

In systems of higher symmetry, this equation is greatly simplified. For example, if $\alpha = \beta = \gamma = 90°$:

$$d_{hkl} = 1 \Big/ \sqrt{\frac{h^2}{a^2} + \frac{k^2}{b^2} + \frac{l^2}{c^2}}. \qquad (3.1.2)$$

Unit cell dimensions (22.67, 7.67, 9.77 Å) may be determined, with radiation of a known wavelength, $\lambda = 1.5418$ Å, from the values of $2\theta_{hkl}$ for reflections of known hkl indices, as shown below:

h	k	l	2θ (degrees)	θ (degrees)	sin θ	nλ/2 sin θ (Å)	Spacing
10	0	0	39.74	19.87	0.3400	22.67	d_{100}
22	0	0	96.82	48.41	0.7480	22.67	d_{100}
0	4	0	47.42	23.71	0.4022	7.67	d_{010}
0	0	10	104.14	52.07	0.7887	9.77	d_{001}

some destructive interference between waves scattered by the various regions of the atoms. The wider the electron cloud around an atom, the greater the falloff in diffracted X-ray intensity at high scattering angles. Values of X-ray scattering factors for individual atoms as a function of $2\theta_{hkl}$ may be derived theoretically[19-24] and are listed in *International*

CRYSTAL SPACE   DIFFRACTION SPACE

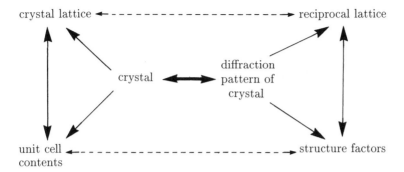

FIGURE 3.11. Relationships between a crystal (in crystal space) and its reciprocal lattice (in diffraction space).

Tables for X-ray Crystallography.[25] Examples are shown in Figure 3.13(a) and Table 3.2.

Neutrons, on the other hand, are scattered by the nuclei of atoms. Wavelengths for neutrons of the order of 1 Å are used for crystal diffraction. The nucleus is minuscule, of the order of 10^{-12} cm in diameter, compared with the larger size of electron clouds that are four orders of magnitude larger (10^{-8} cm). Therefore, scattering by different parts of the nucleus is not a consideration and as a result, there is almost no fall-off in the scattering of neutrons as a function of scattering angle. The amplitude of scattering by neutrons is *not* a function of atomic number, as shown in Table 3.2; it is approximately a sum of two factors. One is the scattering by an impenetrable nucleus with scattering amplitude proportional to $A^{1/3}$, where A is the atomic mass. The second factor depends on the energy of the interaction of the neutron with the atomic nucleus, and it can have the opposite sign to that of the first factor. Therefore, unlike the case for X rays, neutron scattering factors for atoms may even be negative, as for one isotope of hydrogen, implying that the phase change on scattering is 0°, rather than 180°, as normally found. The fact that scattering amplitudes for neutrons bear no resemblance to the values for X-ray scattering [Figure 3.13(b)] makes neutron diffraction useful for distinguishing between isotopes or atoms with similar atomic numbers (see Table 3.2).

3.4 The reciprocal lattice

The crystal lattice, described in Chapter 2 (Figure 2.16), provides information on the three-dimensional periodicity of the internal structure of the crystal, but no information on the detailed arrange-

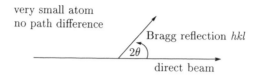

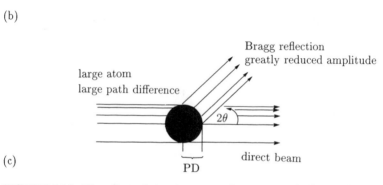

(a)

(b)

(c)

FIGURE 3.12. The effect of the size ment of an atom, relative to the wavelength of the incident radiation, on the intensities of diffracted beams as the scattering angle 2θ increases. As the phase difference between waves scattered by the different outer regions of an atom increases, so does the falloff in intensity as a function of 2θ (or $\sin\theta/\lambda$). (a) A very small atom (e.g., the nucleus of an atom, detected by neutron diffraction). (b) A larger atom, and (c) a very large atom. The falloff in intensity is greatest in (c) because the path difference (PD) for the largest size atom is greatest.

ment of atoms within a unit of that periodicity in the repeating crystal structure. Within this crystal lattice there are lattice planes passing through crystal lattice points [Figure 3.14(a)]. The three-dimensional periodicity of the crystal lattice then gives an infinite number of such lattice planes, each separated by distances that are integral multiples of d_{hkl} from each other.

A three-dimensional lattice, reciprocal to the crystal lattice, is very useful in analyses of X-ray diffraction patterns; it is called the **reciprocal lattice**. Earlier in this chapter the diffraction pattern of a series of regularly spaced slits was considered to be composed of an "envelope profile," the diffraction pattern of a single slit, and "sampling regions,"

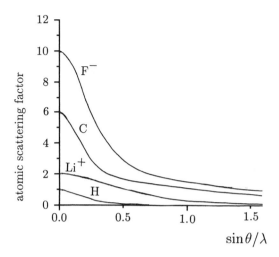

(a) X rays

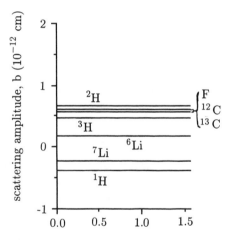

(b) neutrons

FIGURE 3.13. Comparison of scattering curves for X rays and neutrons for selected atoms. (a) Atomic scattering curves for X rays and (b) for neutrons for selected atoms (H, Li, C, F) as a function of $\sin\theta/\lambda$, that is, the angle of scattering and the wavelength. The amount of scattering of X rays by an atom is approximately proportional to the atomic number; the amount of scattering for neutrons does not follow any such proportionality, and the scattering factors for neutrons can be very different for isotopes with the same atomic number but different atomic weights.

TABLE 3.2. Some scattering factors for neutrons and X rays.

Element	Isotope	X rays $\sin\theta/\lambda = 0$	X rays $\sin\theta/\lambda = 0.5/\text{Å}$	Neutrons* $b\,(10^{-12}\ cm)$	Neutrons** (normalized to -1.00 for ^1H)
H	^{1}H ($=$D)	1.0	0.07	-0.38	-1.00
	^{2}H (=D)	1.0	0.07	0.65	1.71
Li	^{6}Li	3.0	1.0	$0.18+0.025i$	$0.71+0.066i$
	^{7}Li	3.0	1.0	-0.25	-0.66
C	^{12}C	6.0	1.7	0.66	1.74
	^{13}C	6.0	1.7	0.60	1.58
O	^{16}O	8.0	2.3	0.58	1.53
Fe	^{54}Fe	26.0	11.5	0.42	1.11
	^{56}Fe	26.0	11.5	1.01	2.66
	^{57}Fe	26.0	11.5	0.23	0.61
Co	^{59}Co	27.0	12.2	0.25	0.66
U	^{238}U	92.0	53.0	0.85	2.24

* *The quantity b is the neutron coherent scattering amplitude. In the entry for Li, $i = \sqrt{-1}$. The anomalous factor for ^{6}Li involves a phase shift of about $8°$ [$(0.025/0.18) = \tan 8°$] relative to most other nuclei.*

** *Neutron scattering amplitudes normalized arbitrarily to the value -1.0 for ^{1}H in order to illustrate the small range of scattering amplitudes observed as compared with that observed in X-ray scattering (2.7 for neutrons versus 92 for X rays).*

inversely related to the distances between the slits. The X-ray diffraction pattern of a crystal can be considered in a similar way. A photograph of an X-ray diffraction pattern consists of isolated spots, not a continuum. This is shown in Figure 3.15. The evident intensity variation across the photograph may be considered as the envelope profile. The spots on the photograph indicate where diffracted beams have hit the photographic film; they are positioned at the sampling regions. In this case the envelope profile is the diffraction pattern of all the atoms in one unit cell.

The sampling regions are arranged on a lattice that is reciprocal to the crystal lattice[26] and hence is called the reciprocal lattice. In other words, in addition to the crystal lattice in real space, there is another lattice in diffraction space, the reciprocal lattice. This concept, initially proposed by Josiah Willard Gibbs in 1884,[27] was introduced to crystallography by Ewald[26,28-30] in 1913. It is the reciprocal lattice points that provide sampling regions in the diffraction pattern of the contents of one unit cell.

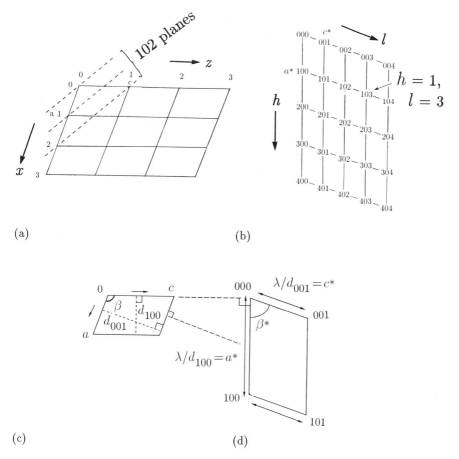

(a)   (b)

(c)   (d)

FIGURE 3.14. The reciprocal lattice and its relationship to the crystal lattice. (a) A crystal lattice is a series of points (see Figure 2.16, Chapter 2). Some sets of planes (hkl) are drawn here by connecting the lattice points. (b) The reciprocal lattice is constructed from the spacing between crystal lattice planes, d_{hkl}. In the reciprocal lattice the point hkl is drawn at a distance $1/d_{hkl}$ from the origin (the direct beam, 000), and in the direction of the perpendicular between lattice planes. (c) Details of the geometry of the crystal lattice and (d) its reciprocal lattice. a^* perpendicular to the (100) plane, length $1/d_{100} = d_{100}^*$, b^* perpendicular to the (010) plane, length $1/d_{010} = d_{010}^*$, c^* perpendicular to the (001) plane, length $1/d_{001} = d_{001}^*$.

$$d_{hkl}^{*2} = h^2 a^{*2} + k^2 b^{*2} + l^2 c^{*2} + 2kl b^* c^* \cos \alpha^* + 2lh c^* a^* \cos \beta^* + 2hk a^* b^* \cos \gamma^*,$$

where $a^* = bc \sin \alpha / V$, etc., and $\cos \alpha^* = (\cos \beta \cos \gamma - \cos \alpha)/(\sin \beta \sin \gamma)$, etc.

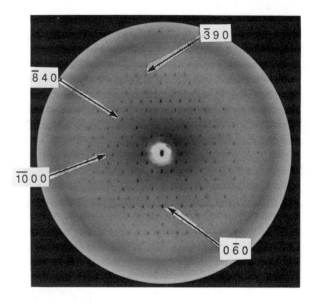

FIGURE 3.15. An X-ray diffraction photograph of the lac repressor protein with some Bragg reflections indexed (hkl). Note the intensity variation among different Bragg reflections. Unit cell dimensions are $a = 164$, $b = 162$, $c = 75$ Å, $\beta = 125°$. In the indexing, h is chosen to point horizontally to the right, and k is vertical and points up.

To construct the reciprocal of a lattice,[31] select any one lattice point as the origin and then look for a series of equally spaced parallel planes from among the lattice points, for example, the 102 series of planes shown in Figure 3.14(a). If the distance between the crystal lattice planes is d_{hkl} (in Å), then the distance of the reciprocal lattice point from its origin in a direction perpendicular to the planes is (depending on the preference of the user) $1/d_{hkl}$ (in Å^{-1}) or, more commonly, λ/d_{hkl} (dimensionless) [Figure 3.14(b)]. This operation is repeated for each possible set of planes. The result is a new lattice, the reciprocal lattice. The crystal lattice and the reciprocal lattice are reciprocal to each other and both are real entities; the geometrical construction just described will convert either into the other. Each set of crystal lattice planes in a structure, specified by the Miller indices h, k and l [Figure 3.14(a)], is represented by a single reciprocal lattice point that is designated hkl [Figure 3.14(b)]. Details of the geometries of the two lattices are given in Figures 3.14(c) and (d).

The X-ray diffraction pattern of a crystal (by analogy to the example of a series of slits, described earlier) is the sampling at the reciprocal lattice points of the X-ray diffraction pattern of the contents of a single unit cell. It is only necessary to find the atomic arrangement in one unit cell, which can be derived from the overall intensity variation in the diffraction pattern, and then this atomic arrangement is repeated in a manner defined by the crystal lattice[32] to give the entire crystal structure. The diffraction pattern of a single molecule is too weak to be observable. When, however, it is reinforced in a crystal because of the repeating unit cells, it can be readily observed at the reciprocal lattice points. That is why X-ray crystallographers obtain their diffraction data as isolated spots on photographic film or as pulses at certain directions on an electronic detection device. The directions in space and the intensities of these diffracted beams are measured. The spatial arrangement of diffracted beams is determined by the geometry of the crystal lattice (Figure 3.14), while the intensities of these beams are determined by the arrangement of atoms within one unit cell.

The concept that the X-ray diffraction pattern of a crystal is a sampling of the diffraction pattern of the contents of a single unit cell is illustrated in Figure 3.16. The X-ray diffraction pattern of vertically-stretched fibers of B-DNA is shown in Figure 3.16(a). Diffraction patterns like this led to the determination of the structure for DNA that has been so important to molecular biology.[33,34] Regularity in a helix leads to a cross in the middle of the diffraction pattern, as shown in Figure 3.16(b), a measure of the pitch (angle) of the helix and the length of one helical repeat (about 27 Å). The more widely spaced blurs at the top and bottom of the diffraction pattern indicate smaller spacings in the DNA structure (because of the reciprocal relationship between atomic and diffraction space). These blurs represent the spacing between the nucleic-acid bases which stack perpendicular to the helix axis at a shorter distance apart, 3.4 Å.

The sampling of the diffraction pattern is illustrated in Figure 3.16(c) for the diffraction pattern of a crystalline oligonucleotide, with a structure similar to that derived for DNA. The crystal gives well-separated spots, unlike the blurs obtained for stretched fibers, shown in in Figure 3.16(a). A comparison of these two diffraction patterns shows the similarities in the overall intensities in both and the principle that the diffraction pattern of a crystal is a sampling of the diffraction pattern of a single object. It immediately leads to the deduction that the helix axis of the crystalline oligonucleotide is vertical with respect to the diffraction photograph in Figure 3.16(c), as for the fibers of DNA in Figure 3.16(a).

Different arrangements of atoms give different diffraction patterns. What does a given diffraction pattern tell us about the atomic arrangement in the crystal that gave that diffraction pattern? Figure 3.16 gives one example where the overall atomic arrangement was clear. In general, however, the answer is complicated, but it is possible to suggest an atomic arrangement. Optical masks, such as those in Figure 3.7, can be obtained by preparing the appropriate array of holes in this suggested

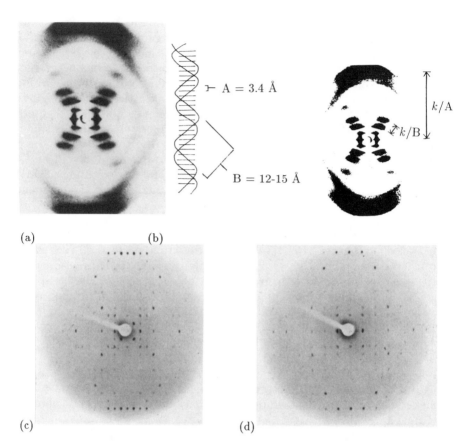

(a)   (b)

(c)   (d)

FIGURE 3.16. Samplings of a diffraction pattern at reciprocal lattice points. (a) The diffraction pattern of stretched fibers of DNA. (Courtesy Robert Langridge). (b) The repetitious features of the structure of DNA showing the vertical distances between bases (3.4 Å) giving the blurs in the upper and lower regions of the diffraction photograph. The distances between the backbone atoms in the helical molecule are larger, hence the cross in the middle of the photograph with smaller spacings. (c) and (d) Two views of the diffraction pattern of a crystalline oligonucleotide (with some information to be explained later). This is a 17° precession photograph of (c) the $h0l$ and (d) the $0kl$ zones of the diffraction pattern of the decamer, CGATCGATCG. Pictures were taken at 4° C. The film-to-crystal distance was 100 mm. Crystals are orthorhombic, space group $P2_12_12_1$ with unit cell dimensions $a = 38.93$, $b = 39.63$ and $c = 33.30$ Å with the c-axis vertical. The c-axis is the axis of a B-type double helix with strong reflections noted around 3.4 Å due to the stacking of the bases. On the $h0l$-photograph, notice the other strong reflections which form a cross pattern indicative of a helix. (Photograph and legend courtesy K. Grzeskowiak, K. Yanagi, G. Prive, and R. E. Dickerson.) (Space group information is given in Chapter 4.)

arrangement in an opaque sheet, using visible light to prepare the diffraction pattern. The photograph of the optical diffraction pattern of a "trial structure" (like that in Figure 3.7) is compared, both in location and intensities of diffraction spots, with the photographs of the experimentally obtained X-ray diffraction.[35-38] The method, developed by Charles W. Bunn,[36] proved useful in establishing the crystal structure of sodium benzylpenicillin.[39] This method is rarely practicable for the determination of most crystal structures, and many better ways of finding an atomic arrangement from the X-ray diffraction pattern have been devised (see Chapter 8).

3.5 The Ewald sphere

How do we know the relationship between the crystal and the diffraction pattern that we will obtain from it? Often it is easier to think of the diffraction experiment with respect to a reciprocal lattice rather than the crystal lattice planes. The reciprocal lattice is a real physical property of a crystal, and rotation of the crystal will cause a rotation of the reciprocal lattice.

A geometrical description of diffraction which encompasses Bragg's law and the Laue equations was originally proposed by P. P. Ewald.[26] The advantage of this description, the Ewald construction (Figure 3.17), is that it allows us to calculate which Bragg reflections will be observed if we know the orientation of the crystal with respect to the incident beam. In Chapter 7 we will discuss how to measure the intensities (and therefore the amplitudes) of the diffracted rays. But, before we can measure the intensity of a Bragg reflection, we need to know where and from what direction to make the measurement. The Ewald construction provides a geometrical relationship between the orientation of the crystal and the direction of the X-ray beams diffracted by it.

For simplicity we consider here a two-dimensional system. The two-dimensional crystal lattice will have a corresponding two-dimensional reciprocal lattice. As is illustrated in Figure 3.10, the Bragg condition for diffraction occurs when a set of lattice planes, with defined d_{hkl} spacings, are inclined with respect to the incident beam by an angle θ_{hkl} and the diffracted beam (a Bragg reflection) occurs at an angle $2\theta_{hkl}$ from the incident beam as illustrated in Figure 3.17. The diffraction vector is perpendicular to the lattice planes and its length is inversely related to the spacing of the planes, $1/d_{hkl} = \sin \theta_{hkl} / \lambda$.

In the Ewald construction (Figure 3.17), a circle with a radius proportional to $1/\lambda$ and centered at C, called the Ewald circle, is drawn. In three dimensions it is referred to as the **Ewald sphere** or the sphere of reflection. The reciprocal lattice, drawn on the same scale as that of the Ewald sphere, is then placed with its origin centered at O. The crystal, centered at C, can be physically oriented so that the required reciprocal lattice point can be made to intersect the surface of the Ewald sphere.

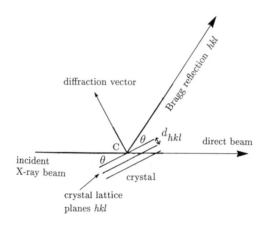

(a)

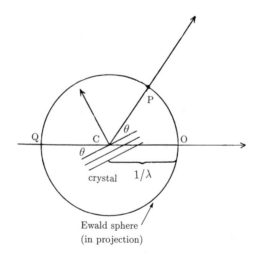

(b)

FIGURE 3.17. The construction of an Ewald sphere of reflection, illustrated in two dimensions (the Ewald circle). (a) Bragg's Law and the formation of a Bragg reflection *hkl*. The crystal lattice planes *hkl* are shown. (b) Construction of an Ewald circle, radius $1/\lambda$, with the crystal at the center C and Q–C–O as the incident beam direction.

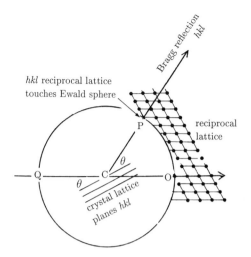

FIGURE 3.17. (c) Orientation of the reciprocal lattice with its origin (*hkl* = 000) at O. If one reciprocal lattice point *hkl* touches the surface of the Ewald circle (sphere), a Bragg reflection *hkl* will be formed.

In Ewald's construction, when the magnitude of the wavelength of the X rays is constant, the vectors describing the incident beam **CO** and the diffracted beam **CP** are equal in length and proportional to the reciprocal of the wavelength. The vectors **CO** and **CP** form an isosceles triangle in which the base **OP** is parallel to the diffraction vector.

As the crystal is rotated, so is its crystal lattice and its reciprocal lattice. If, during the rotation of the crystal a reciprocal lattice point touches the circumference of the Ewald circle (the surface of the Ewald sphere), Bragg's Law and the Laue conditions are satisfied. The result is a Bragg reflection in the direction **CP**, with values of h, k, and l corresponding both to *hkl* values for the reciprocal lattice point and for the crystal lattice planes.

Generally the Bragg equation is useful in describing the directions $2\theta_{hkl}$ of diffracted beams, but provides no information to help us to understand the magnitudes of the intensities obtained. The concepts to be described in detail in Chapter 6 are much more informative about the intensities of the diffracted beams.

Summary

1. Visible light is diffracted by small holes and slits in a manner that can be described by Huygens' wave theory in terms of interference between scattered light. In a similar way X-ray diffraction is the interference between X-ray beams scattered by the electrons surrounding atomic nuclei.

2. When two waves traveling in the same direction interfere with each other, the resultant wave is represented as the sum of the displacements.

3. The diffraction pattern of a repeating object, such as a diffraction grating or a crystal structure, may be considered to be composed of two parts: the diffraction pattern of the repeating object (described as the envelope profile) and the diffraction that occurs as a result of the periodicity of the crystal lattice (described as sampling regions, that is, the regions of the envelope profile that are sampled to give the diffraction pattern).

4. In X-ray diffraction the **envelope profile** is the diffraction pattern of the atomic contents of the unit cell.

5. The **reciprocal lattice**, derived from the crystal lattice, represents the sampling points (see section 3, above). There is a reciprocal relationship between the dimensions of the sampling regions and the repeat in real space (the crystal lattice).

6. Atoms scatter X rays by an amount that depends on the atomic number and the scattering angle. The extent of this scattering is expressed by **atomic scattering factors**. Their values are decreased at high scattering angles by atomic vibrations or disorder.

7. The diffraction pattern of a crystal is the resultant of interference between beams scattered by all the various atoms in the crystal.

8. The angles at which diffraction occurs depend, in an inverse manner, on the periodicity of the crystal lattice. Such diffraction may be considered in terms of path differences (Laue) or reflection from lattice planes (Bragg). The **Bragg equation**, $n\lambda = 2d\sin\theta$, describes the positions of diffracted beams.

9. The **Ewald sphere** is a construction that allows one to determine which Bragg reflections will be observed for a given orientation of the crystal and the incident X-ray beam.

Glossary

Atomic scattering factor: The scattering power of an atom for X rays is defined relative to the scattering of X rays by a single electron under the same conditions. This scattering of X rays depends on the number of extranuclear electrons in the atom and the angle of scattering. At higher scattering angles, and as the vibration of the atoms is increased, the amount of scattering is reduced. Values for scattering factors are computed from theoretical wave functions for free atoms.

Bragg's Law, the Bragg equation: In diffraction of X rays by crystals, each diffracted beam can be considered to be reflected from a set of parallel lattice planes. If the angle between the diffracted X-ray beam (wavelength λ) and the normal (perpendicular) to a set of crystal lattice planes is $90° - \theta_{hkl}$, and if the perpendicular spacing of the lattice planes is d_{hkl}, then:

$$\lambda = 2d_{hkl}\sin\theta_{hkl}. \tag{3.2}$$

This is Bragg's Law, first put forward by W. L. Bragg. When X rays strike a crystal they will be diffracted when, and only when, this equation is satisfied. This equation identifies the orders of X-ray diffraction (hkl) with the Miller indices of the lattice planes (hkl) that cause that Bragg reflection.

Bragg reflection: Diffraction of X rays by a crystal has been shown by W. L. Bragg to be equivalent to **reflection** from a lattice plane in the crystal. While the term **"diffracted beam"** is correct, the term **"reflection"** is also used, although the term **"Bragg reflection"** is preferable.

Coherent scattering: Scattering of radiation in which the fluctuation of the electromagnetic field of an incident X-ray beam causes an electron to oscillate about its nucleus at the same frequency as the beam. This oscillating dipole results in a secondary scattered wave with the same wavelength, but a phase change of 180°. This is the type of scattering considered here.

Diffraction vector: A vector perpendicular to the lattice planes *hkl* causing a Bragg reflection. The diffraction vector bisects the directions of the incident and diffracted beams and lies in their plane.

Envelope: The shape of the plot of intensity versus scattering angle for the diffraction pattern of an isolated object.

Ewald sphere, sphere of reflection: A geometrical construction used for predicting conditions for diffraction by a crystal in terms of its reciprocal lattice rather than its crystal lattice. It is a sphere, of radius $1/\lambda$ (for a reciprocal lattice with dimensions $d^* = \lambda/d$). The diameter of this Ewald sphere lies in the direction of the incident beam. The reciprocal lattice is placed with its origin at the point where the incident beam emerges from the sphere. Whenever a reciprocal lattice point touches the surface of the Ewald sphere, a Bragg reflection with the indices of that reciprocal lattice point will result. Thus, if we know the orientation of the crystal, and hence of its reciprocal lattice, with respect to the incident beam, it is possible to predict which reciprocal lattice points are in the surface of this sphere and hence which planes in the crystal are in a reflecting position.

Intensity: The amount of radiation scattered in a given direction. It is proportional to the square of the amplitude of the wave proceeding in that direction.

Interference: The mutual effect of two waves on each other.

Lattice planes: Planes through at least three noncolinear crystal lattice points.

Laue equations: Equations that, like the Bragg equation, express the conditions for diffraction in terms of the path difference of scattered waves. Laue considered the path length differences of waves that are diffracted by two atoms one lattice translation apart. These path differences must be an integral number of wavelengths for diffraction (that is, reinforcement) to occur. This condition must be true simultaneously in all three dimensions.

$$\text{Laue equations}: PD_1 = h_1\lambda, PD_2 = h_2\lambda, PD_3 = h_3\lambda \tag{3.1}$$

where λ is the wavelength of the radiation, PD represents the path difference and h is the order of diffraction.

Order of diffraction: An integer associated with a given interference fringe of a diffraction pattern. The first order arises as a result of a path difference of one wavelength. Similarly, the *n*th order corresponds to a path difference of *n* wavelengths.

Path difference: This term is used in diffraction to describe the difference in distance that two beams travel when "scattered" from different points in the crystal.

As a result of such path differences, two beams may or may not be in phase and therefore interfere with each other to different extents.

Reciprocal lattice: The lattice with axes $\mathbf{a}^*$, $\mathbf{b}^*$, $\mathbf{c}^*$, related to the crystal lattice or direct lattice axes $\mathbf{a}$, $\mathbf{b}$, $\mathbf{c}$ in such a way that $\mathbf{a}^*$ is perpendicular to $\mathbf{b}$ and $\mathbf{c}$; $\mathbf{b}^*$ is perpendicular to $\mathbf{a}$ and $\mathbf{c}$; and $\mathbf{c}^*$ is perpendicular to $\mathbf{a}$ and $\mathbf{b}$. Quantitatively, $\mathbf{a}^*$, $\mathbf{b}^*$ and $\mathbf{c}^*$ are related to $\mathbf{a}$, $\mathbf{b}$, and $\mathbf{c}$ by, for example,

$$\mathbf{a}* = \frac{\mathbf{b} \times \mathbf{c}}{(\mathbf{a} \cdot \mathbf{b} \times \mathbf{c})}. \tag{3.3}$$

Rows of points (zone axes) in the direct lattice are perpendicular to nets (planes) of the reciprocal lattice, and vice versa. The repeat distance between points in a particular row of the reciprocal lattice is inversely proportional to the interplanar spacing between the nets of the crystal lattice that are normal to this row of points $(d^* = \lambda/d)$.

Sampling regions: Positions in the envelope (q.v.) of the diffraction pattern of an isolated object that have intensity as a result of diffraction by a regular periodic packing of the object.

Scattering angle: The angle at which a scattered wave deviates from the direct beam. Conventionally in X-ray diffraction, as a result of Bragg's Law, it is designated $2\theta_{hkl}$.

Wave front: A surface that consists of all points reached at a given instant in time by a vibrational disturbance as it moves through any material.

References

1. Michette, A. G. No X-ray lens. *Nature (London)* **353**, 510 (1991).
2. Huygens, C. *Traité de la Lumière.* (*Tractatus de Lumine.*) 1678, reprinted Leiden (1690). **English translation:** *Treatise on Light.* Thompson, S. P. Chicago. Macmillan: London (1912). Also, Dover Publications: New York (1962).
3. Jenkins, F. A., and White, H. E. *Fundamentals of Optics.* 4th edn. McGraw-Hill: New York and other international cities (1976).
4. Greenler, R. G., and Hable, J. W. Colors in spider webs. *Amer. Scientist* **77**, 369–373 (1989).
5. Grimaldi, F. M. *Physico-Mathesis de Lumine, Coloribus, et Iride.* [Physical insights into light, colors, and the spectrum (rainbow).] Hæredis Victorii Benatii: Bononiae [Bologna] (1665).
6. Hecht, E. *Optics.* 2nd edn. Addison-Wesley: Reading, MA, and other international cities (1987).
7. Lipson, H. S. *Crystals and X rays.* The Wykeham Science Series # 13. Wykeham Publications: London, Winchester (1970).
8. Young, T. Experiments and calculations relative to physical optics. *Phil. Trans. Roy Soc. (London)* **93**, November 24 (1803).
9. Young, T. Lecture XXXIX. On the nature of light and colours. *A Course of Lectures on Natural Philosophy and the Mechanical Arts.* pp. 457–471 (1807).

10. Fresnel, A. J. Supplément au deuxième mémoire sur la diffraction de la lumière. Prèsenté à l'Académie des Sciences dans la séance du 15 juillet 1816. [Supplement to the second paper on the diffraction of light.] In: *Œvres Complètes d'Augustin Fresnel.* (**Eds.**, de Senarmont, H., Verdet, E. and Fresnel, L.) Vol. 1., pp. 129–170. Imprimerie Impériale: Paris (1866).

11. Bragg, S. L. A personal view by his elder son. In: *Selections and Reflections: The Legacy of Sir Lawrence Bragg.* (**Eds.**, Thomas, J. M., and Phillips, D.) The Royal Institution of Great Britain: Middlesex, UK (1990).

12. Feynman, R. P., Leighton, R. B., and Sands, M. *The Feynman Lectures on Physics. Mainly Mechanics, Radiation and Heat.* Addison-Wesley: Reading, MA, Palo Alto, London (1963).

13. Thomson, J. J. *Conduction of Electricity through Gases.* p. 268 (1903). 2nd edn. p. 321 (1906). 3rd edn., with Thomson, G. P., in two vols. Vol. II, p. 256 (1928 and 1933).

14. Friedrich, W., Knipping, P., and Laue, M. Interferenz-Erscheinungen bei Röntgenstrahlen. [Interference phenomena with X rays.] *Sitzungsberichte der mathematisch-physikalischen Klasse der Königlichen Bayerischen Akademie der Wissenschaften zu München*, pp. 303–322 (1912). **English translation**: Stezowski, J. J. In: *Structural Crystallography in Chemistry and Biology.* (**Ed.**, Glusker, J. P.) pp. 23–39. Hutchinson & Ross: Stroudsburg, PA (1981).

15. Bragg, W. L. The diffraction of short electromagnetic waves by a crystal. *Proc. Camb. Phil. Soc.* **17**, 43–57 (1913).

16. Bragg, L. *The Development of X-ray Analysis.* (**Eds.**, Phillips, D. C., and Lipson, H. F.) p. 24, Hafner Press (Macmillan): New York (1975).

17. Stout, G. H., and Jensen, L. H. *X-ray Structure Determination. A Practical Guide.* Macmillan: New York (1968); 2nd edn. John Wiley: New York, Chichester, Brisbane, Toronto, Singapore (1989).

18. Glusker, J. P., and Trueblood, K. N. *Crystal Structure Analysis: A Primer.* 2nd edn. Oxford University Press: New York, Oxford (1985).

19. James, R. W. *Optical Principles of the Diffraction of X rays.* Bell: London (1954).

20. Hartree, D. R. The atomic structure factor in the intensity of reflexion of X rays by crystals. *Phil. Mag.* **50**, 289–306 (1925).

21. James, R. W., and Brindley, G. W. Some numerical calculations of atomic scattering factors. *Phil. Mag.* **12**, 81–112 (1932).

22. Viervoll, H., and Ögrim, O. An extended table of atomic scattering factors. *Acta Cryst.* **2**, 277–279 (1949).

23. Cromer, D. T., and Waber, J. T. Scattering factors computed from relativistic Dirac–Slater functions. *Acta Cryst.* **18**, 104–109 (1965).

24. Stewart, R. F., Davidson, E. R. and Simpson, W. T. Coherent X-ray scattering for the hydrogen atom in the hydrogen molecule. *J. Chem. Phys.* **42**, 3175–3187 (1965).

25. *International Tables for X-ray Crystallography, Vol. III. Physical and Chemical Tables.* (**Eds.**, MacGillavry, C. H., and Rieck, G. D.) Atomic scattering factors. pp. 201–245. Kynoch Press: Birmingham, New York (1962).

26. Ewald, P. P. Das 'reziproke Gitter' in der Strukturtheorie. Teil I: Das Reziproke eines einfachen Gitters. [The "reciprocal lattice" in the theory of structure. Part I. The reciprocal of a primitive lattice.] *Z. Krist.* **56**, 129–156 (1921).

27. Gibbs, J. W. *Vector Analysis*. Reprinted in his collected works. 1st edn. (1906). 2nd edn. (1928).

28. Ewald, P. P. Zur Theorie der Interferenzen der Röntgenstrahlen in Kristallen. [The theory of the interference of Röntgen rays in crystals.] *Physik. Z.* **14**, 465–472 (1913).

29. Ewald, P. P. Bemerkung zu der Arbeit von M. Laue. Die dreizählig-symmetrischen Röntgenstrahlenaufnahmen an regulären Kristallen. [Note upon M. Laue's paper: "the three-fold symmetry of Röntgen ray photographs from regular crystals."] *Physik. Z.* **14**, 1038–1040 (1913).

30. Cruickshank, D. W. J., Juretschke, H. I., and Kato, N. *P. P. Ewald and his Dynamical Theory of X-ray Diffraction*. Oxford University Press: Oxford (1992).

31. Kettle, S. F. A., and Norrby, L. J. The Brillouin zone — an interface between spectroscopy and crystallography. *J. Chem. Educ.* **67**, 1022–1028 (1990).

32. Holmes, K. C., and Blow, D. M. *The Use of X-ray Diffraction in the Study of Protein and Nucleic Acid Structure*. Revised reprint from *Methods of Biochemical Analysis*. (**Ed.**, Glick, D.) **13**, 113–239 (1966). Interscience (John Wiley): New York, London, Sydney (1966). Reprinted: Krieger: Melbourne, FL (1979).

33. Franklin, R. E., and Gosling, R. G. Molecular configuration in sodium thymonucleate. *Nature (London)* **171**, 740–741 (1953).

34. Watson, J. D. & Crick, F. H. C. A structure for deoxyribose nucleic acid. *Nature (London)* **171**, 737–738 (1953).

35. Bragg, L. [W. L.] Lightning calculations with light. *Nature (London)* **154**, 69–72 (1944).

36. Bunn, C. W. *Chemical Crystallography. An Introduction to Optical and X-ray Methods*. 1st edn. 1946. 2nd edn, 1961. Clarendon Press: Oxford (1961).

37. Taylor, C. A., and Lipson, H. *Optical Transforms: their Preparation and Application to X-ray Diffraction Problems*. G. Bell: London (1964).

38. Harburn, G., Taylor, C. A., and Welberry, T. R. *Atlas of Optical Transforms*. Bell: London and Cornell University Press: Ithaca, NY (1975).

39. Crowfoot, D., Bunn, C. W., Rogers-Low, B. W., and Turner-Jones, A. The X-ray crystallographic investigation of the structure of penicillin. In: *The Chemistry of Penicillin*. (**Eds.** Clarke, H. T., Johnson, J. R., and Robinson, R.). Ch. XI, pp. 310–367. Princeton University Press: Princeton (1949).

CHAPTER

4

Symmetry in Crystals and Their Diffraction Patterns

People instinctively appreciate symmetry.[1-4] For example, any two-dimensional periodicity in a wallpaper pattern is readily noticed. The repeat unit with the highest symmetry possible is usually the one that is selected; The rectangular repeating unit shown in Figure 4.1(a) is a reasonable choice even though it is not the simplest repeating unit. Similarly, any symmetry in the arrangement of crystal faces is usually obvious to a viewer. This external symmetry may give information about the symmetry of the internal three-dimensional structure of the crystal (the crystal lattice, for example). We concentrate in this chapter on a description of symmetry as found in molecules and in crystals.

A study of the symmetry of a crystal is very important when choosing a unit cell for that crystal. The choice of the **origin of a unit cell** is arbitrary. As a result of the study of symmetry it is apparent that it is best to choose this origin in such a way that the symmetry present in the crystal are taken advantage of. There are many ways that a unit cell can be chosen from among an arrangement of molecules, illustrated in Figure 4.1(b). First, one must find the magnitudes and directions of the three unit cell edges. This is often obvious if one looks for repeats in the structure. The unit cell with the highest symmetry is the best choice from among the various possibilities, as shown in Figure 4.1. The symmetry of the atomic arrangement in the crystal is more important in the choice of unit cell than is the molecular content. There does not have to be an atom at the corner of the unit cell. It is also not necessary for one entire molecule to lie in one unit cell; it may span several unit cells.[5] One unit cell, however, should have a volume equal to or larger than the molecular volume (or the volume of a repeating unit of an infinite polymer) because of the repetitions from unit cell to unit cell.

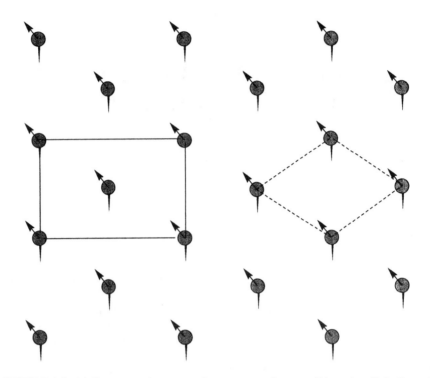

FIGURE 4.1. (a) Symmetry in a repeating pattern. One possible unit cell, indicated by broken lines contains one object per unit cell and is described as "primitive." Another possible unit cell, indicated by solid lines, contains two objects per unit cell and is said to be "centered" (and therefore "nonprimitive"). Thus, while the simplest unit cell is not rectangular, a larger unit cell with higher symmetry, which in this case is rectangular, can readily be picked out from the arrangement of objects.

A **symmetry operation** is the actual event of converting one item into another apparently identical item. This symmetry operation can take place about a point, a line, or a plane of symmetry. Thus, an object is symmetrical when a symmetry operation applied to it gives an object indistinguishable from the original. The point, line or plane about which the symmetry operation is performed is defined[6,7] as a **symmetry element**. For example, an **axis of symmetry** is a symmetry element that describes a line about which all parts of an object are symmetrically disposed. We are all familiar with the vertical reflection plane (or near reflection plane) in the human face. We are also familiar with the six-fold axis of symmetry that lies perpendicular to the plane of a benzene molecule and passes through the center of it (Figure 4.2).

FIGURE 4.1 (cont'd). (b) Diagram illustrating choices of a two-dimensional unit cell from a crystal structure. Each choice encompasses the equivalent of one molecule. (Diagram courtesy S. Silverman.)

These ideas on symmetry are explained by A. Michael Glazer in the following way:[8] "Many scientists tend to mix up the concepts of 'operation' and 'element,' although strictly speaking they are different ideas. In our example [benzene], there is one sixfold symmetry element (the axis of rotation), but six sixfold operations (six successive rotations of 60° to bring the object back to its initial state)."

A collection or set of elements (such as symmetry elements) that are related to each other is called a **group**.[6,9] In this book, the groups we will be discussing are point groups and space groups, both of which contain symmetry elements. When symmetry operations are applied about a single point that remains fixed in space, the result is a **point group**; when symmetry operations are applied to points arranged periodically on a crystal lattice, the result is a **space group**. There are 32 unique crystallographic point groups and 230 unique crystallographic space groups.[10]

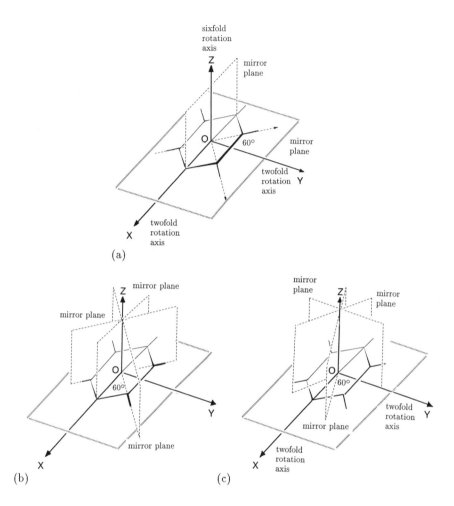

FIGURE 4.2. The symmetry of benzene. This molecule is flat, with a mirror plane through all atoms, and a sixfold rotation axis OZ perpendicular to this molecular plane, passing through the center of the ring, as shown in (a). Twofold rotation axes lie in the plane of the ring and pass through its center; they lie along the HC···CH directions or else bisect the C–C bonds. There are six such twofold rotation axes and six mirror planes perpendicular to them, as shown in (b) and (c).

4.1 Symmetries of isolated objects

A *point group* is defined as *a set of symmetry operations that leave unmoved at least one point* within the object to which these symmetry operations are being applied. No **translation** operations (that is, simple movements along a straight line) can be included in this description;

otherwise any chosen point would have to move, contrary to the definition just given. In a point group, the point or points that remain(s) stationary lie on all the symmetry elements that make up that particular group (because otherwise they would not remain stationary during these symmetry operations). In one dimension, only a reflection symmetry operation is consistent with point symmetry. In two dimensions, rotation about a point and reflection across a line, are both possibilities. The possible symmetry operations in three dimensions are rotation about a line, reflection across a plane and inversion about a point.

There are four symmetry operations that are important in point-group symmetry and these are shown in Figures 4.3 to 4.6. Rotation about an n-fold **rotation axis** is the first considered here. An object possesses an n-fold rotational symmetry if a rotation of $360°/n$ about a particular axis leaves the object indistinguishable from the original. By convention we make rotations in a counterclockwise direction with respect to our line of sight (as in the method of drawing such angles). This symmetry is evident every 180° ($360°/2$; $n = 2$) for a twofold axis (see Figure 4.3), every 120° ($360°/3$; $n = 3$) for a threefold axis, and every 90° ($360°/4$) for a fourfold axis. For example, when two objects or two parts of a single object are related to each other by a twofold axis, the coordinates of one can be superimposed on those of the other by a rotation of 180° about the symmetry axis. If the twofold axis of symmetry passes through the origin and is coincident with the z axis, then the first object at x, y, z is converted by this twofold rotation axis to the second object (or part of the object) at $-x$, $-y$, z.

Consider the case of benzene, for which the symmetry is indicated in Figure 4.2. In the symmetry operations just described, one sixfold rotation rotates the diagram 60°, two sixfold rotations rotate it to 120°, three rotate it to 180° (as for a twofold rotation), four rotate it to 240° $= -120°$, five rotate it to 300° $= -60°$, and six rotate it to 360° $= 0°$ (as if nothing had happened). This last case (0°) is called the **identity operation**. Glazer has a very graphic description of the symmetry of this molecule,[8] and it can be followed by examining Figure 4.2: "Suppose that you were to look away from this book for a moment, and that somehow I could then contrive to rotate this figure [of benzene] through 60° about the axis OZ through its centre. On looking back at the page, you would not be aware that anything had changed, even though *I* know that a change was made. We could then repeat the game over and over again, always with the same result. Note, however, that if the process of rotation by 60° is carried out an integral multiple of six times, the object really *is* back in its original position (but only *I* know that!)"

Another type of symmetry operation is reflection across a **mirror plane** (Figure 4.4). Reflection across mirror planes converts an object into its mirror image. Reflection symmetry occurs when a plane can be constructed such that an object or molecule on one side of the plane is

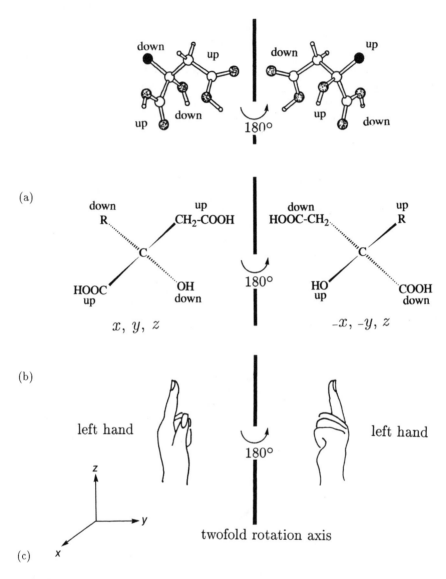

FIGURE 4.3. Two molecules related to each other by a twofold rotation axis. One molecule is rotated half a turn, 180°, about this twofold axis, to give a second molecule. A rotation axis maintains the handedness of the molecule (a left hand is converted to another left hand). The location of groups above the plane of the paper ("up") and below the plane of the paper ("down") are shown in this and subsequent Figures. (a) View of the two molecules. (b) Formulæ of the two molecules. (c) Representations of the two molecules as hands.

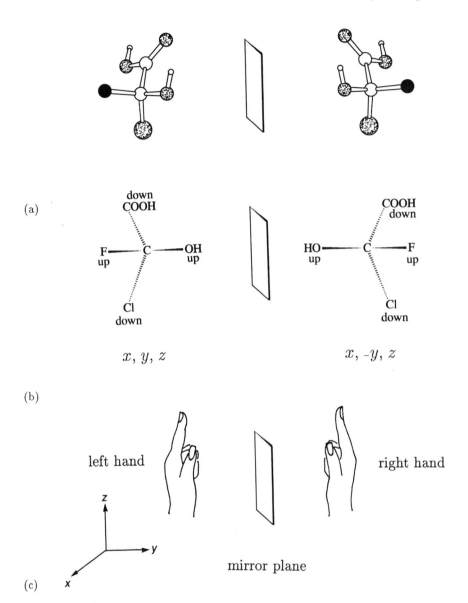

FIGURE 4.4. Two molecules related by a mirror plane. A mirror plane converts one molecule into another with the opposite handedness. For example, a left hand is converted to a right hand. (a) View of the two molecules. (b) Formulæ of the two molecules. (c) Representations of the two molecules as hands with a mirror between them.

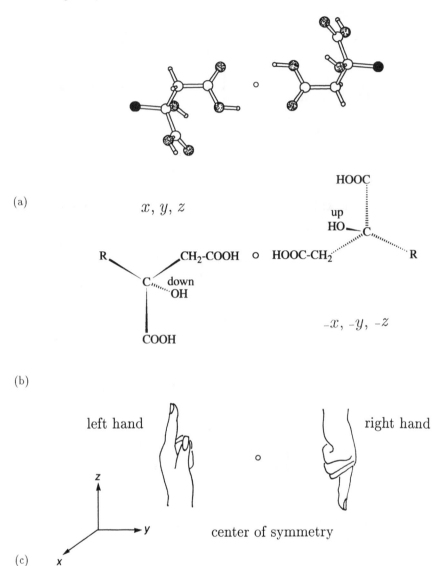

FIGURE 4.5. Two molecules related to each other by a center of symmetry. Each point on the first molecule lies as far in one direction from the center of symmetry as the equivalent point on the second molecule lies in exactly the opposite direction. As a result, a center of symmetry converts a left hand into a right hand and vice versa. (a) View of the two molecules. (b) Formulæ of the two molecules. (c) Representations of the two molecules as hands. The small circle indicates the location of the center of symmetry.

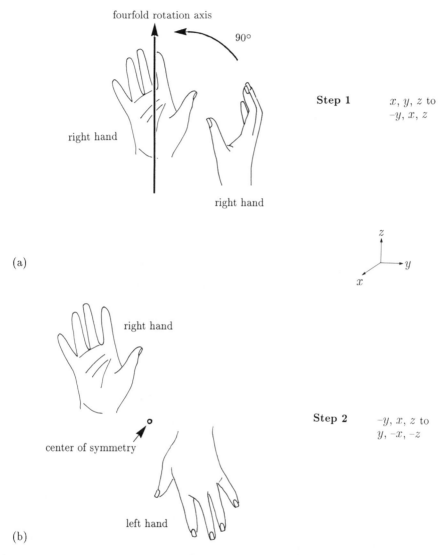

FIGURE 4.6. A rotatory-inversion axis involves a rotation and then an inversion across a center of symmetry. Since, by the definition of a point group, one point remains unmoved, this must be the point through which the rotatory-inversion axis passes and it must lie on the inversion center (center of symmetry). The effect of a fourfold rotation-inversion axis is shown in two steps. By this symmetry operation a right hand is converted to a left hand, and an atom at x,y,z is moved to $y,-x,-z$. (a) The fourfold rotation, and (b) the inversion through a center of symmetry.

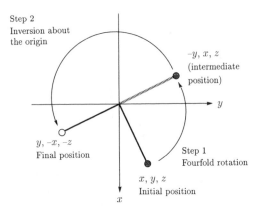

FIGURE 4.6. (c) A rotatory-inversion axis. The view from above, where filled circles lie at $+z$ and open circles at $-z$. Two steps are involved: (1) a fourfold rotation, and (2) inversion about the origin.

related to another as if it were reflected by a mirror between them. A mirror plane parallel to the yz plane and passing through the origin will convert an object at x, y, z into one at $-x$, y, z. A mirror plane of symmetry is designated by the lower case letter m. Naturally occurring biological molecules, such as sugars and nearly all amino acids, have a handedness; generally only D-sugars and L-amino acids are found. None of these asymmetrical biological molecules can have any reflection symmetry, because if they did, one part of them would be related to another as a left hand is to a right hand.

The third type of symmetry operation considered here involves **inversion**, the act of turning an object inside out. If a glove is taken as an example, the reader can verify that inversion will also convert a left hand to a right hand. The point about which inversion of an object occurs is called a **center of symmetry** (or a **center of inversion**, Figure 4.5). If an object lies at a distance r from a center of symmetry, its symmetry-related object lies at an equal distance $(-r)$ in the opposite direction. This means that if a center of symmetry lies at the origin of a unit cell, it converts an object at x,y,z to one at $-x,-y,-z$. This type of symmetry operation will convert a left-handed molecule into a right-handed molecule, and vice versa.

The fourth type of symmetry operation combines rotational symmetry with inversion symmetry to produce what is called a **rotatory-inversion axis**, designated $\bar{n}$ (Figure 4.6). It consists of rotation about a line combined with inversion about a specific point on that line. For example, the operation of fourfold rotation-inversion is done by rotating an object at x,y,z through an angle of 90° about the z axis to produce an object with coordinates $-y$, x, z. Subsequent inversion about the origin will convert these coordinates to y, $-x$, $-z$. Like mirror planes and centers of symmetry, a rotatory-inversion axis converts a left hand into a right hand.

Integral numbers[10] are used to describe the crystal symmetry: 2, 3, 4, 6 for rotation axes refer to rotations of $180° = 360/2°$, $120° = 360/3°$, $90° = 360/4°$, and $60° = 360/6°$, respectively, about an axis. The symbol $\overline{1}$ refers to a center of symmetry; mirror planes are designated m or $\overline{2}$; and rotatory-inversion axes are written as $\overline{3}$, $\overline{4}$, $\overline{6}$.

It is very important in structural chemistry to know whether or not a left-handed molecule is converted to a right-handed molecule. It is customary to divide symmetry operations into those that do and those that do not invert a left hand to a right hand. "**Proper symmetry operations**," such as translation and rotation, maintain the handedness of an object and are symmetry operations that can relate **chiral objects** to each other without changing their chirality. Reflection and inversion are "**improper symmetry operations**;" they invert the handedness of an object, and thus convert a chiral object into its mirror image. Thus, chiral objects (such as D-sugars or L-amino acids) can *never* be assigned point groups having improper symmetry operations, because then they would be superimposable on their own mirror image.

The point group of a crystal can often be determined from the external symmetry of its morphology, that is, the arrangement of the faces on the surface of the crystal. It can also often be determined from various physical properties of the crystal, as will be described later in Chapter 5. When the point group has been determined, it is designated by Schoenflies[11] or Hermann–Mauguin[12] symbols. The chemist is familiar with the former and the crystallographer uses the latter. A table of conversions between the two is given in Table 4.1, since Hermann–Mauguin symbols will be used here.[10,13] The symmetry of the principal axis is denoted first, its symbol being the number, n, of the axis. An example is 4, denoting a fourfold axis, which rotates an object 90°. A mirror plane has as its symbol, m. If the mirror plane is perpendicular to a rotation axis, the symbol $4/m$ or $\frac{4}{m}$ is used. The point group of a crystal can usually be determined by examining its stereographic projection (see Figure 2.13, Chapter 2) if this is appropriately oriented to display the symmetry. Paul Groth,[14–18] at the beginning of this century, catalogued a large number of shapes of crystals and their symmetries by careful observation and illustration of the symmetries of their faces. He heightened a general interest in the study of the symmetry to be found in crystals.

Consider the symmetry of a crystal that has a cubic unit cell; it has three fourfold axes perpendicular to each other, six axes of twofold symmetry, and four axes of threefold symmetry, as shown in Figure 4.7. The threefold axes are important in defining a cube, even though the

TABLE 4.1. Schoenflies and Hermann–Mauguin symbols (see References 8 and 11).

Schoenflies	*Hermann–Mauguin*	*Schoenflies*	*Hermann–Mauguin*

1. Rotation axes ($C = cyclic$).

C_1	1	C_4	4
C_2	2	C_6	6
C_3	3		

2. Rotation axes plus plane of symmetry normal to those axes ($s = Spiegelung$).

$C_1^h = C_s$	m	C_4^h	$4/m$
C_2^h	$2/m$	C_6^h	$6/m$
C_3^h	$3/m = \overline{6}$		

3. Plane of symmetry through an axis.

C_2^v	$mm2$	C_4^v	$4mm$
C_3^v	$3m$	C_6^v	$6mm$

4. Rotatory-inversion axes ($S = Sphenoidisch$).

$S_2 = C_i$	$\overline{1}$	$S_6 = C_3^d$	$\overline{3}$
S_4	$\overline{4}$		

5. Rotation axis, degree n, with n twofold axes normal to it ($D = Di\ddot{e}dergruppe$).

$D_2 = V$	222	D_4	422
D_3	32	D_6	622

6. Rotation axis with twofold axes normal to it plus a horizontal plane.

$D_2^h = V_h$	mmm	D_4^h	$4/mmm$
D_3^h	$\overline{6}m = 3/mm$	D_6^h	$6/mmm$

7. D group with additional diagonal in the vertical plane.

D_2^d	$\overline{4}2m$	D_3^d	$\overline{3}m$

8. Cubic group ($T = $ tetrahedral, $O = $ octahedral)

T	23	T_d	$\overline{4}3m$
T_h	$m3$		
O	432	O_h	$m3m$

fourfold axes are more immediately obvious. A particularly interesting example of an unusually shaped crystal that derives from a cubic unit cell is iron pyrite, a naturally occurring mineral. Crystals of pyrite have five-sided faces, although they are not completely regular. Haüy[19] realized that the unit cell of pyrite is cubic and showed that the five-sided faces correspond to the 210 direction, as described in Chapter 2 where it is illustrated in stereographic projection [Figure 2.14(h)].

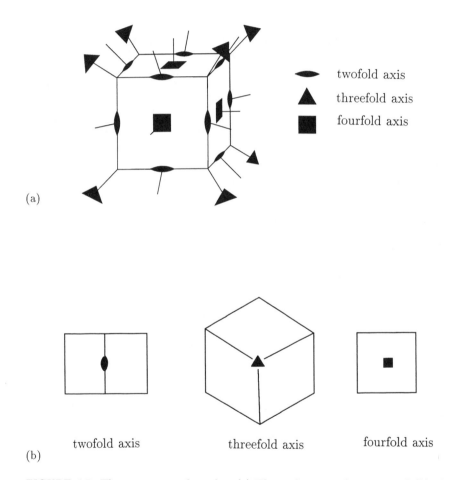

(a)

- twofold axis
▲ threefold axis
■ fourfold axis

twofold axis   threefold axis   fourfold axis

(b)

FIGURE 4.7. The symmetry of a cube. (a) The various rotation axes and (b) view down one of each type of rotation axis. The reader is encouraged to pick up a cube and examine its symmetry.

4.2 Symmetries of regularly repeating objects

Chemists are interested in isolated molecules, and therefore are generally familiar with point groups[7] which apply to isolated objects. These isolated objects can have, for example, n-fold rotation axes, where n may be *any* integer. On the other hand, chemists are usually much less familiar with space group theory which is applied to crystals and involves arrays that repeat in a very regular manner. There is a great difference between the point group of a molecule in a crystal which itself has three-dimensional internal order and that of the isolated molecule. Thus a crystallographic point group is a group of symmetry elements that, upon operation, leave one point in the crystal invariant or unmoved, usually the origin, and that transform a given crystal lattice into one indistinguishable from it. Only a few point group symmetries are consistent with the requirement that the crystal lattice be unchanged after the symmetry operation. Only two-, three-, four- and sixfold rotation axes are allowed, while n-fold axes (where n is any other integer such as five, see Figure 4.8) are not allowed. The requirement for lattice periodicity in a crystal has reduced the number of possible point groups from infinity to 32.

The symmetry classes of crystals were mathematically derived by Johann Friedrich Christian Hessel[20] in 1830. He found that by combining rotation axes (2, 3, 4, 6), mirror planes (m), centers of symmetry ($\bar{1}$), and rotatory inversion axes ($\bar{3}, \bar{4}, \bar{6}$), it was possible to derive 32 point groups. They involve seven different types of lattice symmetry. These, classified by their lattice symmetry rather than their shape, comprise the seven **crystal systems** (Tables 4.2 and 4.3). A primitive unit cell, denoted P, is the simplest possible unit cell and it contains only one lattice point. In some cases the selection of a lattice containing more than one crystal lattice point per unit cell will give a lattice of higher symmetry than that found when there is only one lattice point per unit cell [see Figure 4.1(a)]. The additional lattice point may lie in the center of the unit cell (I for *Innenzentrierte*, German), at the center of one face (C, A or B), or at the centers of all faces (F).

Bravais in 1849 showed that there are only 14 ways that identical points can be arranged in space subject to the condition that each point has the same number of neighbors at the same distances and in the same directions.[21] Moritz Ludwig Frankenheim, in an extension of this study, showed that this number, 14, could also be used to describe the total number of distinct three-dimensional crystal lattices.[22] These are referred to as the 14 **Bravais lattices** (Figure 4.9), and they represent combinations of the seven crystal systems and the four lattice centering types (P, C, F, I). Rhombohedral and hexagonal lattices are primitive, but the letter R is used for the former.

Three additional symmetry operations, in addition to the four types already described, may be applied to an arrangement of atoms while

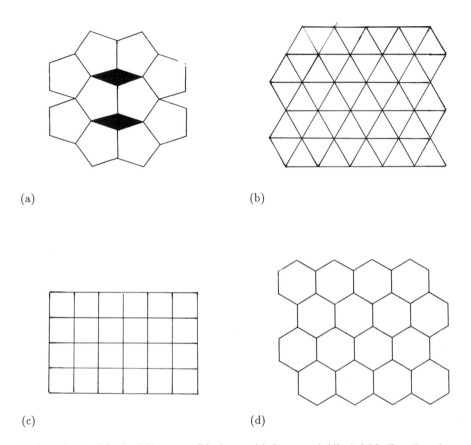

FIGURE 4.8. (a) Fivefold versus (b) three-, (c) four-, and (d) sixfold tiles. Regular three-, four-, and six-sided tiles pack well together. Note that regular five-sided tiles do not fill space, as indicated by the shaded areas of space between them.

still maintaining the internal periodicity of the crystal lattice. These are translations (Figure 4.10), which are simply movements in a straight line; **screw axes** (Figure 4.11), which are rotation axes combined with translations; and **glide planes** (Figure 4.12), which are reflections across mirror planes combined with translations.

The regular arrangement of molecules or ions in a unit cell can be described in terms of translational symmetry. One set of units of translation in a crystal structure are the unit-cell dimensions, that is, the vectors between points in the crystal lattice (**a, b** or **c**). Other translational symmetry operations use fractions of these translations. Screw

TABLE 4.2. The Seven Crystal Systems.

Crystal system (7)	Characteristic symmetry	Lattice (Laue) symmetry	Axial and angular constraints from symmetry*
1. Triclinic	Identity or inversion (onefold rotation or rotatory-inversion axis) in any direction	$\bar{1}$	$a \neq b \neq c$ $\alpha \neq \beta \neq \gamma$
2. Monoclinic	A single 2-fold rotation or rotatory-inversion axis along **b**	$2/m$	$a \neq b \neq c$ $\alpha = \gamma = 90°$, $\beta \neq 90°$
3. Orthorhombic	Three mutually perpendicular 2-fold rotation or rotatory-inversion axes along **a**, **b** and **c**	mmm	$a \neq b \neq c$ $\alpha = \beta = \gamma = 90°$
4. Tetragonal	A single 4-fold rotation or rotatory-inversion axis along **c**	$4/mmm$	$a = b \neq c$ $\alpha = \beta = \gamma = 90°$
5. Cubic	Four 3-fold axes along **a+b+c**, **−a+b+c**, **a−b+c**, **−a−b+c**	$m3m$	$a = b = c$ $\alpha = \beta = \gamma = 90°$
6. Trigonal	A single 3-fold rotation or rotatory-inversion axis along **a+b+c**	$\bar{3}m$	$a = b = c$ $\alpha = \beta = \gamma \neq 90°$ $\gamma < 120°$
7. Hexagonal	A single 6-fold rotation or rotatory-inversion axis (along **c**)	$6/mmm$	$a = b \neq c$ $\alpha = \beta = 90°$ $\gamma = 120°$

* These follow from the definition given in the characteristic symmetry column, rather than being the direct definition of the crystal system. Note that the symbol $\neq$ means that values are not equal for symmetry reasons; they may, however, accidentally be equal.

TABLE 4.3. The 14 Bravais Lattices, 32 Crystallographic Point Groups (Crystal Classes) and Some Space Groups.

Crystal system (7)	Bravais Lattices (14)*	Crystallographic Point Groups (32)**	Common and/or representative space groups (out of 230)
1. Triclinic	P	$1, \bar{1}$ (C_1, C_i)	$P1, P\bar{1}$
2. Monoclinic	P, C	$2, m, 2/m$ (C_2, C_S, C_{2h})	$P2_1$ $C2/c, P2_1/c$
3. Orthorhombic	P, C, I, F	$222, mm2, mmm$ (D_2, C_{2v}, D_{2h})	$P2_12_12_1, Pbca$ $I222, I2_12_12_1$
4. Tetragonal	P, I	$4, \bar{4}, 4/m, 422,$ $4mm, \bar{4}2m, 4/mmm$ $(C_4, S_4, C_{4h}, D_4,$ $C_{4v}, D_{2d}, D_{4h})$	$P4, P4_1, I\bar{4}$ $P4_12_12, I4_1/amd$
5. Cubic	P, I, F	$23, m3, 432,$ $\bar{4}3m, m3m$ (T, T_h, O, T_d, O_h)	$Fm3m, P23$ $Pa3, I2_13$
6. Trigonal	R	$3, \bar{3}, 32, 3m, \bar{3}m$ $(C_3, C_{3i}, D_3, C_{3v}, D_{3d})$	$P3, P3_121, R3c$
7. Hexagonal	P	$6, \bar{6}, 6/m, 622,$ $6mm, \bar{6}m2, 6/mmm$ $(C_6, C_{3h}, C_{6h}, D_6$ $C_{6v}, D_{3h}, D_{6h})$	$P6_1, P6_222$

* P = primitive, C = centered on the (001) faces, I = body-centered, F = all face-centered, R = primitive rhombohedral.

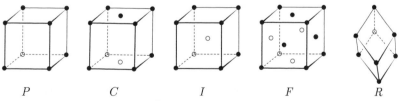

$$P \qquad C \qquad I \qquad F \qquad R$$

** Hermann–Mauguin system followed by the Schoenflies system of notation.

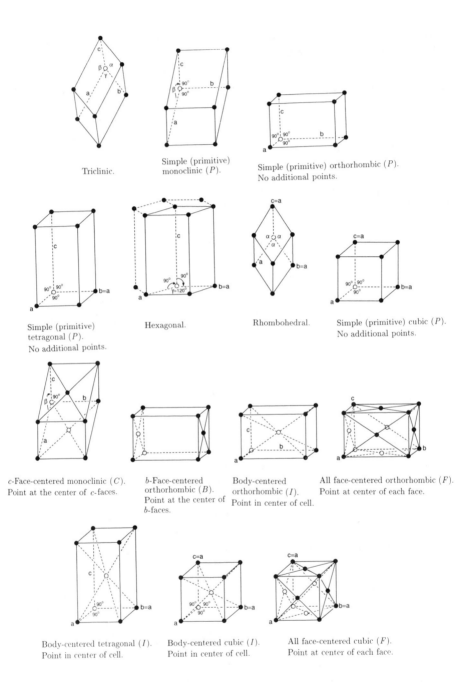

Triclinic.

Simple (primitive) monoclinic (*P*).

Simple (primitive) orthorhombic (*P*). No additional points.

Simple (primitive) tetragonal (*P*). No additional points.

Hexagonal.

Rhombohedral.

Simple (primitive) cubic (*P*). No additional points.

c-Face-centered monoclinic (*C*). Point at the center of *c*-faces.

b-Face-centered orthorhombic (*B*). Point at the center of *b*-faces.

Body-centered orthorhombic (*I*). Point in center of cell.

All face-centered orthorhombic (*F*). Point at center of each face.

Body-centered tetragonal (*I*). Point in center of cell.

Body-centered cubic (*I*). Point in center of cell.

All face-centered cubic (*F*). Point at center of each face.

FIGURE 4.9. The 14 Bravais lattices (7 primitive, 7 nonprimitive).

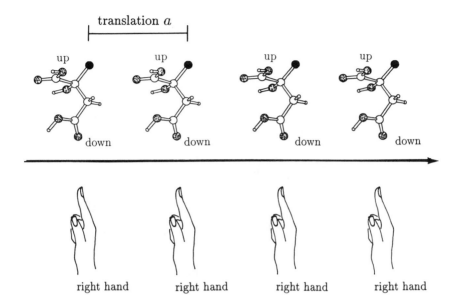

FIGURE 4.10. Translational symmetry. Each molecule or hand is identical, but moved (translated) along the translational vector **a**.

axes are designated as n-fold rotation axes with the fraction of the unit-cell translation indicated by the subscript q, giving n_q. For example, a fourfold screw axis has a rotation of 90° and is denoted 4_1 if the translation is $\frac{1}{4}a$, 4_2 if the translation is $\frac{2}{4}a$, and 4_3 if the translation is $\frac{3}{4}a$. As a consequence, 4_1 and 4_3 are enantiomorphic, related as left and right hands (that is, involving a translation of $+\frac{1}{4}a$ and $-\frac{1}{4}a = +\frac{3}{4}a$, respectively). A glide plane with a translation along an axis of $\frac{1}{2}a$, or $\frac{1}{2}b$, or $\frac{1}{2}c$, is designated a, b, c, respectively, while if the translation is along a diagonal $\frac{1}{2}(b+c)$, $\frac{1}{2}(c+a)$, or $\frac{1}{2}(a+b)$, the glide plane is denoted n. If the translation is $\frac{1}{4}(b\pm c)$, $\frac{1}{4}(c\pm a)$, or $\frac{1}{4}(a\pm b)$, the glide plane is denoted d. The relationship between symmetry operations with translations and those without them is shown in Figure 4.13.

A *space group* is defined as *the set of transformations (symmetry elements) that leave a three-dimensionally periodic, discrete set of labeled points (atoms) unchanged.* Purely integral lattice translations, however, are generally not included in this specific definition, as they are already implied by the lattice concept. A comprehensive scheme of crystal symmetry is shown in Tables 4.2 and 4.3. The 14 Bravais lattices are combined with the 32 crystallographic point groups plus additional symmetry elements with translational components to give the 230 space groups, originally formulated by Leonhard Sohncke,[23] and derived independently

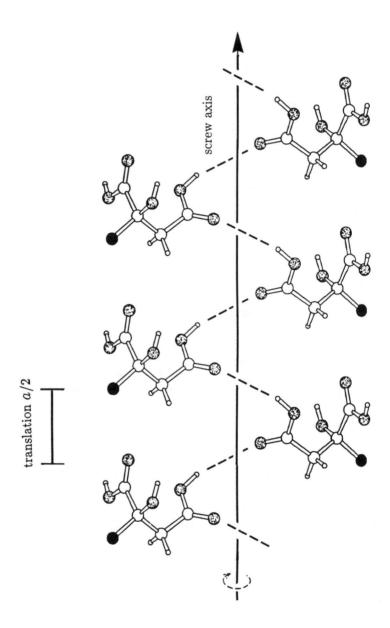

FIGURE 4.11. A twofold screw axis. One molecule is translated half a unit cell (*a* in this example) along the screw axis direction, and then a twofold rotation is applied to it. Two such operations give the original molecule translated to the next unit cell (*a* away). All molecules have the same handedness.

translation $a/2$

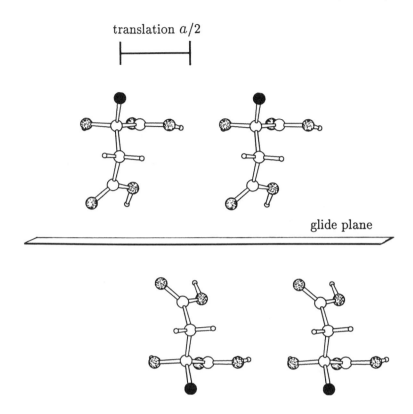

glide plane

FIGURE 4.12. A glide plane causes a reflection across a mirror plane plus a translation of half the unit cell edge (translation $= a/2$) A left-handed molecule is converted to a right-handed molecule by this symmetry operation. Two such operations give the original molecule translated to the next unit cell (translation $= a$).

by Evgraf Stepanovich von Fedorov,[24,25] Artur Moritz Schoenflies,[11] and William Barlow.[26,27] Space group theory is one of the few theories of mathematics considered complete. The rules have also extended to color symmetry[28] in which the objects have different colors.

Plane groups are the groups of symmetry elements that produce regularly repeating patterns in two dimensions, for example, wallpaper. As shown in Figure 4.8, there are only five types of lattices that repeat regularly in two dimensions without leaving any gaps between them; these are square, rectangular, centered rectangular, oblique, and hexagonal. There are only 17 ways of combining identical objects in these two-dimensional lattices while, at the same time maintaining the lattice. The reader is urged to study the 17 plane groups described in the *International Tables*,[10] that is, the 17 symmetry variations of wallpaper. This exercise will make clear the way the plane groups are formed and so,

twofold rotation axis

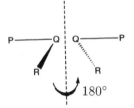

a twofold rotation axis
and a translation $a/2$
gives a screw axis

mirror plane

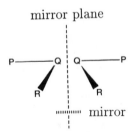

a mirror plane
and a translation $a/2$
gives a glide plane

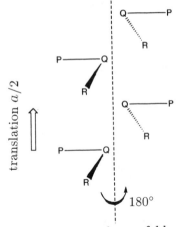

Relationship of a twofold
rotation axis to a twofold
screw axis

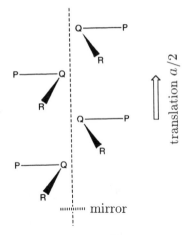

Relationship of a mirror plane
to a glide plane

FIGURE 4.13. The relationships between symmetry operations as a function of their translational properties. Shown above are symmetry operations with no translations involved, and below, the analogous symmetry operations that involve translation. (Left) A twofold rotation axis, combined with a translation of $a/2$, gives a twofold screw axis. (Right) A mirror plane, combined with a translation of $a/2$, gives a glide plane.

by analogy, space groups. When three dimensions are considered, rather than two, it is found that there are 230, and only 230, possible space groups. Crystals each contain an arrangement of atoms corresponding to one of these 230 space groups.

The unit on which the space group operations act to produce the entire crystal structure (an infinitely extending pattern) is called the **asymmetric unit**. This is the simplest unit that can be selected (consistent with the space group symmetry). The complete set of operations of the space group may be described by the **equivalent positions**; these convert a unit at x,y,z into all other units in the unit cell. As a result of space group theory, the description of the contents of a unit cell can be reduced to the description of an asymmetric unit and the space group operations that will generate the entire crystal structure.

The first tabulations of space groups were published by Paul Niggli,[29] Ralph W. G. Wyckoff,[30] and William Thomas Astbury and Kathleen Yardley (Lonsdale).[31] The tabulation generally used now by crystallographers is that published in *International Tables*.[10] All of the preceding ideas on symmetry just discussed are incorporated in these space group tables. The 230 space groups are arranged in order of increasing symmetry, through the 32-point groups; the Bravais lattice is listed for each.

Space groups are designated by Hermann–Mauguin symbols,[12] starting with a letter indicating the type of centering (i.e., the translations implied in the Bravais lattice): P for primitive (simple), A, B, C for (100), (010), (001) face centered, F for all face centered, and I for body centered. Then follows a set of characters describing the various symmetry elements. Some examples of space groups are given in Table 4.4.

The arrangement of atoms in two crystal structures, one in the common space group $P2_12_12_1$ and the other in the common space group $P2_1/a$ are shown in Figure 4.14. The symmetry relationships between molecules are shown. The space group $P2_12_12_1$ describes a primitive orthorhombic unit cell (P) with three mutually perpendicular nonintersecting twofold screw axes (2_1) in the unit cell. This is consistent with the symmetry mmm (see Tables 4.2 and 4.3) of an orthorhombic lattice. The space group $P2_1/a$ describes a primitive monoclinic unit cell with a twofold screw axis along the unique axis (b) and a glide plane perpendicular to this axis with a translation of $a/2$. A knowledge of the space group symmetry is essential for the structure determination because only the contents of the asymmetric unit, combined with the relationships between positions of atoms listed for the space group, are needed to derive a picture of the entire atomic contents of the crystal. The space group $P2_12_12_1$ has only one asymmetric unit that needs to be studied. The symmetry operations: x,y,z; $\frac{1}{2}-x,-y,\frac{1}{2}+z$; $\frac{1}{2}+x,\frac{1}{2}-y, -z$; $-x,\frac{1}{2}+y, \frac{1}{2}-z$, give the rest of the crystal structure. Interestingly, about 65% of all organic structures are found in one of eight space groups,[32] out of a total of 230 (see Table 4.4 for some of these).

4.3 Symmetry in the diffraction pattern

Not all of the Bragg reflections, $I(hkl)$, are unique in intensity. Georges Friedel[33] noted that the intensity distribution in the diffraction pattern is centrosymmetric (diagrammed in Figure 4.15).

$$\text{Friedel's law}: I(hkl) = I(\overline{hkl}). \tag{4.1}$$

As a result, there are Bragg reflections with different (but related) indices that have identical intensities, as shown in an $hk0$ X-ray diffraction photograph (Figure 3.15, Chapter 3), where $I(hk0)$ equals $I(\overline{hk}0)$. The only exception to **Friedel's Law** is found if atoms scatter radiation anomalously, as described in Chapter 14.

Any symmetry in the packing of objects is related (in a reciprocal way) to symmetry in its diffraction pattern, and this symmetry in the diffraction pattern can be used to determine the crystal symmetry (see Tables 4.2 and 4.3). This is of great importance to the X-ray crystallographer because this is the way the space group of a crystal is determined.

Any symmetry in the intensities in the diffraction pattern other than that implied by Friedel's Law is called **Laue symmetry** (because it can be displayed on Laue X-ray diffraction photographs of an appropriately aligned crystal, see Figure 4.16). Friedel's Law implies that there is a center of symmetry in the diffraction pattern. Therefore the Laue symmetry displayed by the diffraction pattern is the point-group symmetry of the crystal with an additional center of symmetry (if this does not already exist). If a crystal is monoclinic then, the intensities $I(hkl)$ and $I(\overline{hk}\overline{l})$ are the same, although $I(hkl)$ does not equal $I(\overline{h}kl)$. Orthorhombic

---

TABLE 4.4. Some common space groups and the equivalent positions of objects in them.

$P\overline{1}$	*triclinic*	$x, y, z; -x, -y, -z$
$P2_1$	*monoclinic*	$x, y, z; -x, \frac{1}{2}+y, -z$
$P2_1/a$	*monoclinic*	$x, y, z; -x, -y, -z; \frac{1}{2}-x, \frac{1}{2}+y, -z; \frac{1}{2}+x, \frac{1}{2}-y, z$
$C2/c$	*monoclinic*	$x, y, z; -x, -y, -z; -x, y, \frac{1}{2}-z; x, -y, \frac{1}{2}+z;$ $\frac{1}{2}+x, \frac{1}{2}+y, z; \frac{1}{2}-x, \frac{1}{2}-y, -z;$ $\frac{1}{2}-x, \frac{1}{2}+y, \frac{1}{2}-z; \frac{1}{2}+x, \frac{1}{2}-y, \frac{1}{2}+z$
$P2_12_12_1$	*orthorhombic*	$x, y, z; \frac{1}{2}-x, -y, \frac{1}{2}+z; \frac{1}{2}+x, \frac{1}{2}-y, -z; -x, \frac{1}{2}+y, \frac{1}{2}-z$

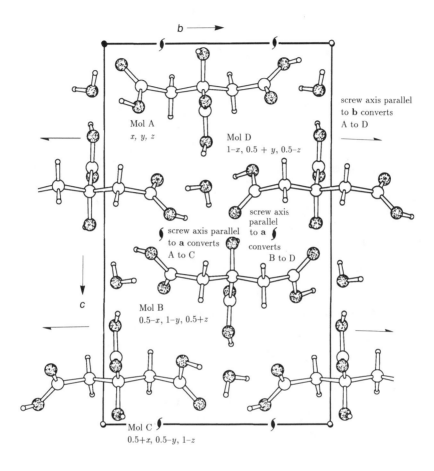

FIGURE 4.14. The symmetry operations involved the packing of molecules in two crystalline forms of citric acid (which has no asymmetric carbon atom). (a) Citric acid monohydrate with a primitive orthorhombic unit cell,[34] space group $P2_12_12_1$, with atoms at $x,y,z;$ $\frac{1}{2}-x,-y,\frac{1}{2}+z;$ $\frac{1}{2}+x,\frac{1}{2}-y,-z;$ $-x,\frac{1}{2}+y,\frac{1}{2}-z.$ All molecules, A, B, C, and D have the same conformational handedness.

crystals, with three mutually perpendicular twofold rotation axes, have intensities $I(hkl) = I(\overline{h}kl) = I(h\overline{k}l) = I(hk\overline{l})$. Therefore only one of these sets needs to be measured; the other Bragg reflections related in this way have the same intensity. If, by chance, the crystal is monoclinic with the β angle equal to 90°, the Laue symmetry in the X-ray diffraction pattern will show that $I(hkl)$ does not equal $I(\overline{h}kl)$. This violates the

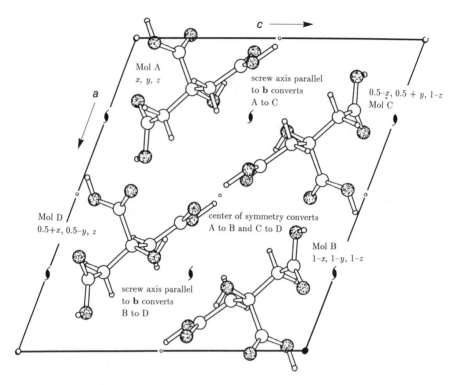

FIGURE 4.14. (b). Anhydrous citric acid with a primitive monoclinic unit cell[35,36] space group $P2_1/a$, with atoms at x,y,z; $-x,-y,-z$; $\frac{1}{2}-x,\frac{1}{2}+y,-z$; $\frac{1}{2}+x,\frac{1}{2}-y,z$. There is a twofold screw axis parallel to **b**. There is also a glide plane parallel to the plane of the paper at $b = \frac{1}{4}$ with translation $a/2$. Molecules A and C are (arbitrarily) conformationally of one handedness, and molecules B and D are conformationally of the opposite handedness.

Laue symmetry of an orthorhombic crystal (see Figure 4.17). Thus, Laue symmetry, not unit cell dimensions, give a measure of the crystal system. As was stressed earlier, it is the symmetry of the diffraction pattern that tells us that a crystal (lattice) is orthorhombic, not the fact that $\alpha = \beta = \gamma = 90°$.

4.4 Space group determination from diffraction patterns

The symmetry of the crystal is indicated in its diffraction pattern. **Systematic absences** in the diffraction pattern show that there are translational symmetry elements relating components in the unit cell. The translational component of the symmetry elements causes selective and predictable destructive interference to occur when the specific translation in the arrangement of atoms are simple fractions of the normal lattice

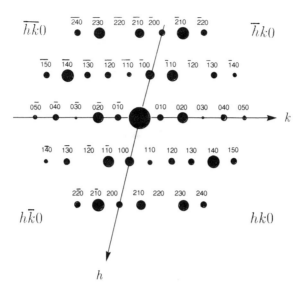

FIGURE 4.15. The symmetry of intensities implied by Friedel's Law. Shown are the *hk0* Bragg reflections. Note the symmetry of the intensities for *hk0* and $\overline{hk}0$ and for $\overline{h}k0$ and $h\overline{k}0$. This gives a twodimensional view of Friedel's Law which states that intensities of Bragg reflections *hkl* and $\overline{hkl}$ should be the same.

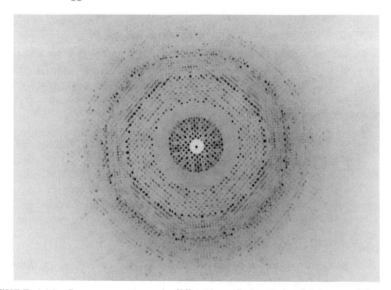

FIGURE 4.16. Laue symmetry. A diffraction photograph of tetragonal lysozyme, viewed down its unique axis (c). Note the fourfold symmetry of the diffraction pattern.

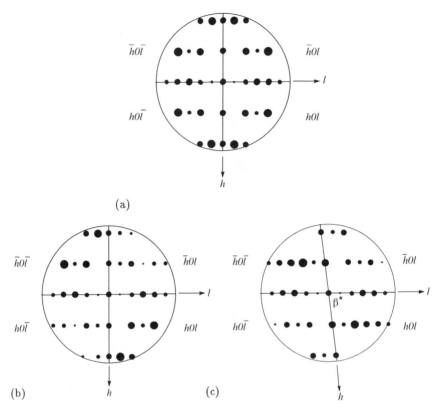

(a)

(b)   (c)

FIGURE 4.17. Laue symmetry in monoclinic and orthorhombic cases in addition to Friedel symmetry $[I(hkl) = I(\overline{hkl})]$. (a) Orthorhombic, $[I(hkl) = I(h\overline{k}l) = I(\overline{h}kl)]$. (b) Monoclinic, $[I(hkl) = I(h\overline{k}l) \neq I(\overline{h}kl)]$ with $\beta = 90°$, and (c) with $\beta \neq 90°$. The nonequivalence of the $h0l$ and $\overline{h}0l$ intensities in (b) indicates that the crystal is monoclinic, not orthorhombic.

translations. These are the systematic absences (also called "extinctions"). The intensity of each member of a particular group of Bragg reflections may be zero, for example, $hk0$ when $h + k$ is odd, as shown in Figure 4.18. These indicate symmetry in the crystal structure, and some specific examples are listed in Table 4.6. Consider a twofold screw axis, parallel to the unit cell translation, **a**. This will convert an atom at x to one at $x + \frac{1}{2}$. This repetition every $\frac{1}{2}\mathbf{a}$ implies apparent halving of the spacing between x values for atoms (it is not necessary to consider y and z in this argument). Therefore, in this case, a Bragg reflection $h00$ will be absent if h is odd, provided all atoms in the structure are repeated at $a/2$. The spacing in reciprocal space has been doubled, $2h$, because the repeat unit in real space has been halved, $a/2$. Space groups are listed in Table 4.5.

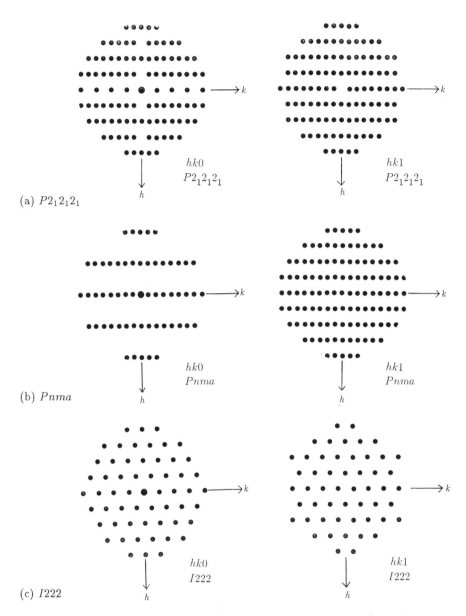

FIGURE 4.18. Space group determinations from systematic absences in Bragg reflections. *hk*0 and *hk*1 diffraction patterns are shown. In each case the unit-cell dimensions are the same, but different Bragg reflections are systematically absent. (a) $P2_12_12_1$, (b) *Pnma*, and (c) *I*222.

TABLE 4.5. Examples of crystal symmetry determinations from systematic absences in Bragg reflections. All Bragg reflections of the type listed below must be absent.

(a) Bravais lattice

absent reflections	deduction on symmetry	symbol	translation
no restrictions	*primitive*	P	none
$k + l$ odd	*centered on A-face* (100)	A	$b/2 + c/2$
$l + h$ odd	*centered on B-face* (010)	B	$c/2 + a/2$
$h + k$ odd	*centered on C-face* (001)	C	$a/2 + b/2$
h, k, l not all odd or all even	*centered on all faces*	F	$(a+b)/2$, $(b+c)/2$ and $(a+c)/2$
$h + k + l$ odd	*body-centered*	I	$(a + b + c/2)$

(b) Symmetry elements

absent reflections	deduction on symmetry	symbol	translation
$h00$, h odd	*2-fold screw axis along a*	2_1	$a/2$
$0k0$, k odd	*2-fold screw axis along b*	2_1	$b/2$
$00l$, l odd	*2-fold screw axis along c*	2_1	$c/2$
$00l$, $l = 3n + 1$ or $3n + 2$	*3-fold screw axis along c*	3_1 or 3_2	$c/3, 2c/3$
$0kl$, k odd	*glide plane perpendicular to a*	b	$b/2$
$0kl$, l odd	*glide plane perpendicular to a*	c	$c/2$

A careful tabulation of systematic absences in the X-ray diffraction pattern will allow the space group of the crystal to be assigned (Table 4.6). This, after the determination of unit cell dimensions, is the next piece of information that the X-ray crystallographer obtains. Space group determination is done by systematically considering each class of Bragg reflection ($h00$, $0k0$, $00l$, $hk0$, $h0l$, $0kl$, hkl) and checking that examples of reflections in these classes with appreciable intensities for odd and even values of each index either are observed, or are consistently absent. Sometimes two or more space groups fit the conditions imposed by crystal symmetry on the Bragg reflections, giving rise to a **space group ambiguity**. Some examples in Chapter 5 demonstrate physical methods for settling a space group ambiguity if, as is often the case, the two possibilities differ in their centricity, one atomic arrangement having a center of symmetry, while the other does not.

Our purpose here has been to give the reader a general idea of symmetry. The manner in which the X-ray crystallographer considers symmetry

TABLE 4.6. Space groups and the symmetry of objects in them.

Non-centrosymmetric, chiral molecules

1	Triclinic	C_1	$P1$
3–5	Monoclinic	C_2	$P2, P2_1, C2$
16–24	Orthorhombic	D_2	$P222, P222_1, P2_12_12, P2_12_12_1, C222_1, C222, F222, I222, I2_12_12_1$
75–80	Tetragonal	C_4	$P4, P4_1, P4_2, P4_3, I4, I4_1$
89–98		D_4	$P422, P42_12, P4_122, P4_12_12, P4_222, P4_22_12, P4_322, P4_32_12, I422, I4_122$
143–146	Trigonal	C_3	$P3, P3_1, P3_2, R3$
149–155		D_3	$P312, P321, P3_112, P3_121, P3_212, P3_221, R32$
168–173	Hexagonal	C_6	$P6, P6_1, P6_5, P6_2, P6_4, P6_3$
177–182		D_6	$P622, P6_122, P6_522, P6_222, P6_422, P6_322$
195–199	Cubic	T	$P23, F23, I23, P2_13, I2_13$
207–214		O	$P432, P4_232, F432, F4_132, I432, P4_332, P4_132, I4_132$

Non-centrosymmetric, both enantiomers

6–9	Monoclinic	C_s	Pm, Pc, Cm, Cc
25–46	Orthorhombic	C_{2v}	$Pmm2, Pmc2_1, Pcc2, Pma2, Pca2_1, Pnc2, Pmn2_1, Pba2, Pna2_1, Pnn2,$ $Cmm2, Cmc2_1, Ccc2, Amm2, Abm2, Ama2, Aba2, Fmm2, Fdd2, Imm2,$ $Iba2, Ima2$
81–82	Tetragonal	S_4	$P\bar{4}, I\bar{4}$
99–110		C_{4v}	$P4mm, P4bm, P4_2cm, P4_2nm, P4cc, P4nc, P4_2mc, P4_2bc, I4mm, I4cm,$ $I4_1md, I4_1cd$
111–122		D_{2d}	$P\bar{4}2m, P\bar{4}2c, P\bar{4}2_1m, P\bar{4}2_1c, P\bar{4}m2, P\bar{4}c2, P\bar{4}b2, P\bar{4}n2, I\bar{4}m2, I\bar{4}c2,$ $I\bar{4}2m, I\bar{4}2d$
147–148	Trigonal	C_{3i}	$P\bar{3}, R\bar{3}$
156–161		C_{3v}	$P3m1, P31m, P3c1, P31c, R3m, R3c$
174	Hexagonal	C_{3h}	$P\bar{6}$
183–186		C_{6v}	$P6mm, P6cc, P6_3cm, P6_3mc$
187–190		D_{3h}	$P\bar{6}m2, P\bar{6}c2, P\bar{6}2m, P\bar{6}2c$
215–220	Cubic	T_d	$P\bar{4}3m, F\bar{4}3m, I\bar{4}3m, P\bar{4}3n, F\bar{4}3c, I\bar{4}3d$

Centrosymmetric, both enantiomers

2	Triclinic	C_i	$P\bar{1}$
10–15	Monoclinic	C_{2h}	$P2/m, P2_1/m, C2/m, P2/c, P2_1/c, C2/c$
47–74	Orthorhombic	D_{2h}	$Pmmm, Pnnn, Pccm, Pban, Pmma, Pnna, Pmna, Pcca, Pbam, Pccn,$ $Pbcm, Pnnm, Pmmn, Pbcn, Pbca, Pnma, Cmcm, Cmca, Cmmm, Cccm,$ $Cmma, Ccca, Fmmm, Fddd, Immm, Ibam, Ibca, Imma$
83–88	Tetragonal	C_{4h}	$P4/m, P4_2/m, P4/n, P4_2/n, I4/m, I4_1/a$
123–142		D_{4h}	$P4/mmm, P4/mcc, P4/nbm, P4/nnc, P4/mbm, P4/mnc, P4/nmm,$ $P4/ncc, P4_2/mmc, P4_2/mcm, P4_2/nbc, P4_2/nnm, P4_2/mbc, P4_2/mnm,$ $P4_2/nmc, P4_2/ncm, I4/mmm, I4/mcm, I4_1/amd, I4_1/acd$
162–167	Trigonal	D_{3d}	$P\bar{3}1m, P\bar{3}1c, P\bar{3}m1, P\bar{3}c1, R\bar{3}m, R\bar{3}c$
175–176	Hexagonal	C_{6h}	$P6/m, P6_3/m$
191–194		D_{6h}	$P6/mmm, P6/mcc, P6_3/mcm, P6_3/mmc$
200–206	Cubic	T_h	$Pm\bar{3}, Pn\bar{3}, Fm\bar{3}, Fd\bar{3}, Im\bar{3}, Pa\bar{3}, Ia\bar{3}$
221–230		O_h	$Pm\bar{3}m, Pn\bar{3}n, Pm\bar{3}n, Pn\bar{3}m, Fm\bar{3}m, Fm\bar{3}c, Fd\bar{3}m, Fd\bar{3}c, Im\bar{3}m, Ia\bar{3}d$

is to be found in *International Tables*.[10] Further details of this fascinating branch of crystallography may be obtained from books by Charles Bunn,[37] Frank Coles Phillips,[13] and Ralph Steadman,[38] as well as the plane and space groups sections of the *International Tables*.[10]

Summary

1. A crystal may be described (see Chapter 2) as the convolution of a lattice with the contents of one unit cell.
2. Crystals are described, in terms of their external symmetry, as belonging to one of 32 symmetry classes (point groups).
3. A crystal may be described, in terms of the symmetry of its internal arrangement of atoms, as belonging to one of the 230 space groups.
4. The entire arrangement of atoms in a crystal may be described in terms of the unit cell dimensions, the coordinates of each atom in the asymmetric unit, and the space group symmetry.

Glossary

Asymmetric unit: The smallest part of a crystal structure from which the complete structure can be derived by use of the space-group symmetry operations (including translations). The asymmetric unit may consist of only one molecule or ion, part of a molecule, or of several molecules that are not related by crystallographic symmetry.

Axis of symmetry: When an object is rotated by less than 360 ° about an axis passing through it and the result is an object indistinguishable from the first, then the original object is said to possess an axis of symmetry. If such a situation occurs for a rotation of $360°/n$, the object is said to have an n-fold axis of symmetry.

Bravais lattice: One of the 14 possible arrays of points repeated periodically in three-dimensional space such that the arrangement of points about any one of the points is identical in every respect to that about any other point in the array.

Center of symmetry (or center of inversion): A point through which an inversion operation (q.v.) is performed, converting an object into its enantiomorph.

Chiral objects: A chiral object or structure cannot be superimposed on its mirror image to give complete equivalence (Greek: *cheir* = hand), (see Chapter 14).

Crystal system: The seven crystal systems, classified in terms of their symmetry and corresponding to the seven fundamental shapes for unit cells consistent with the 14 Bravais lattices.

Equivalent positions: The complete set of positions produced by the operation of the symmetry elements of the space group upon any general position.

Friedel's Law: Georges Friedel noted that $|F(hkl)|^2$ values for centrosymmetrically related Bragg reflections are equal, even for an acentric crystal structure.

$$\text{Friedel's law}: I(hkl) = I(\overline{hkl}) \text{ or } |F(hkl)|^2 = |F(\overline{hkl})|^2 . \qquad (4.1)$$

This law only holds under conditions where anomalous scattering can be ignored.

Glide plane: A glide plane involves reflection across a plane combined with a translation. For an axial glide plane, denoted a, b or c, the translational component is $\frac{a}{2}$, $\frac{b}{2}$, or $\frac{c}{2}$, respectively. For a c-glide plane, if the mirror plane is perpendicular to **b** and the translation is parallel to **c**, an object at x,y,z is converted into its enantiomorph at $x,-y,\frac{1}{2}+z$. Other types of glide planes include a diagonal glide, denoted n, with translations $\frac{a}{2} + \frac{b}{2}$, or $\frac{a}{2} + \frac{c}{2}$, or $\frac{b}{2} + \frac{c}{2}$, and a diamond glide, denoted d, with translations $\frac{a}{4} + \frac{b}{4}$, or $\frac{a}{4} + \frac{c}{4}$, or $\frac{b}{4} + \frac{c}{4}$.

Group (mathematical): A collection or a set of symmetry elements that obey certain mathematical conditions that interrelate these elements. The conditions are that one element is the identity element, that the product of any two elements is also an element, and that the order in which symmetry elements are combined does not affect the result. For every element there exists another in the group that is called the inverse of the first; when these two are multiplied together, the product is the identity element. Some symmetry elements are their own inverses; for example, a twofold axis applied twice gives the same result as the identity operation.

Identity operation: This symmetry operation leaves an object totally unmoved.

Improper symmetry operation: A symmetry operation that converts a right-handed object into a left-handed object. Such operations include mirror planes, centers of symmetry, and rotatory-inversion axes.

Inversion: Inversion involves converting an object into its enantiomorph by projecting through a center of symmetry (also called a center of inversion) and extending it an equal distance beyond the center. An object at x,y,z is converted through a center of inversion (q.v.) at $0,0,0$ into the enantiomorph at $-x,-y,-z$.

Laue symmetry: Symmetry in the intensities of the diffraction pattern beyond that expected from Friedel's Law. The Laue symmetry of the diffraction pattern of a crystal is the point-group symmetry of the crystal plus, as Friedel noted, a center of symmetry. There are 11 Laue symmetry groups.

Mirror plane: A mirror plane converts an object into its mirror image which lies as far behind the mirror plane as the original object is in front. If the mirror plane is perpendicular to **b**, it converts an object at x,y,z into its enantiomorph at $x,-y,z$.

Origin of a unit cell: The point in a unit cell (usually one corner) from which the x, y, and z axes originate. It is designated $0,0,0$ (for its values of x, y, z).

Plane groups: The groups of symmetry elements that produce regularly repeating patterns in two dimensions. There are 17 plane groups (listed in the *International Tables*), such as, the 17 symmetry variations of wallpaper.

Point group: A group of symmetry operations that leave unmoved at least one point within the object to which they apply. Symmetry elements include simple rotation and rotatory–inversion axes; the latter include the center of symmetry and the mirror plane. Since one point remains invariant, all rotation axes must pass through this point and all mirror planes must contain it. A point group is used to describe isolated objects, such as single molecules or real crystals.

Proper symmetry operation: A symmetry operation that maintains the handedness of the object. Such operations include translations, rotation axes, and screw axes.

Rotation axis: If rotation about an axis causes an object to appear identical to the original every $360°/n$, then that axis is an n-fold rotation axis.

Rotatory–inversion axis: Rotation of an object by $360°/n$, about this axis and then inversion through a center of symmetry to give a mirror-image form of the original object.

Screw axis: A screw axis, n_r, involves rotation by $360°/n$ about an axis coupled with a translation parallel to that axis by r/n of the unit cell in that direction. A twofold screw axis through the origin of the unit cell and parallel to **b** converts an object at x,y,z into one at $-x,\frac{1}{2}+y,-z$. The enantiomorphic identity of the object is not changed by this symmetry operation.

Space group: A space group is a set of transformations that leave a triply periodic, discrete set of labeled points (atoms) unchanged, excluding those transformations that are purely integral lattice translations. The space groups can be derived by the addition of translational symmetry operations to the 32 point groups appropriate for structures arranged on lattices (simple translations, screw axes, and glide planes). Hence a space group is the group of operations that converts one molecule or asymmetric unit into an infinitely extending three-dimensional pattern. There are 230 such groups, which can be identified, although sometimes with some ambiguity, from systematic absences in the diffraction pattern.

Space group ambiguity: Sometimes more than one space group fits a given set of systematic absences in intensity. In such a case, other physical methods have to be used to establish the true space group.

Symmetry element: A point, line, plane or other geometrical entity about which a symmetry operation is performed.

Symmetry operation: A symmetry operation or a series of symmetry operations converts an object into an exact replica of itself. In crystal structures, the possible symmetry operations are axes of rotation and rotatory inversion, screw axes, and glide planes, as well as lattice translations. Proper operations, which convert an object into a replica of itself, are translation and rotation. Improper operations, which convert an object into the mirror image of its replica, are reflection and inversion.

Systematically absent Bragg reflections: Bragg reflections that have no intensity, because of translational components of any symmetry in the unit-cell contents and which have h, k, and l values that are systematic in terms of oddness or evenness. These absences depend only upon symmetry in the atomic arrangement in the crystal, and they can be used to derive the space group. For example, all reflections for which $h + k$ is odd may be absent, showing that the unit cell is C-face centered.

Translation: A motion in which all points of a body move in the same direction (i.e., along the same or parallel lines).

References

1. Weyl, H. *Symmetry.* Princeton University Press: Princeton (1952).
2. MacGillavry, C. H. *Symmetry Aspects of M. C. Escher's Periodic Drawings.* Bohn, Scheltema and Holkema: Utrecht (now Kluwer: Dordrecht) (1976).

3. Heilbronner, E., and Dunitz, J. D. *Reflections on Symmetry in Chemistry ··· and Elsewhere.* VHCA: Basel and VCH: Weinheim (1993).

4. Hargittai, I. and Hargittai, M. *Symmetry through the Eyes of a Chemist.* VCH Publishers: New York (1987).

5. Gao, Q. and Craven, B. M. Conformation of the oleate chains in crystals of cholesterol oleate at 123 K. *J. Lipid Res.* **27**, 1214–1221 (1986).

6. Buerger, M. J. *X-ray Crystallography. An Introduction to the Investigation of Crystals by their Diffraction of Monochromatic X-radiation.* J. Wiley: New York (1942). Reprinted. Krieger: Melbourne, FL (1980).

7. Buerger, M. J. *Elementary Crystallography.* Wiley and Sons, Inc.: New York (1956).

8. Glazer, A. M. *The Structure of Crystals.* Adam Hilger: Bristol, UK (1987); Taylor and Francis: Philadelphia, PA (1987).

9. Cotton, F. A. *Chemical Applications of Group Theory.* 2nd edn. Wiley–Interscience: New York, London, Sydney (1971).

10. *International Tables for X-ray Crystallography, Vol. 1. Symmetry Groups.* (**Eds.**, Henry, N. F. M., and Lonsdale, K.) Kynoch Press: Birmingham (1952) and *Volume A. Space-group symmetry.* (**Ed.**, Hahn, T.) 2nd edn., revised. D. Reidel: Dordrecht, Boston, Lancaster, Tokyo (1987). *International Tables for Crystallography. Brief Teaching Edition of Volume A. Space-group Symmetry.* (**Ed.**, Hahn, T.) International Union for Crystallography. D. Reidel: Dordrecht, Boston, Lancaster (1985).

11. Schoenflies, A. *Krystallsysteme und Krystallstruktur.* [Crystal systems and crystal structures.] B. G. Teubner: Leipzig (1891). (2nd edn. 1923.) Reprinted. Springer: Berlin (1984).

12. Bragg, W. H., von Laue, M., and Hermann, C. (**Eds.**) *Internationale Tabellen zur Bestimmung von Kristallstrukturen. Erster Band. Gruppentheoretische Tafeln.* [International Tables for the Determination of Crystal Structures. First Volume. Tables on the Theory of Groups.] [In German, English and French.] Borntraeger: Berlin (1935).

13. Phillips, F. C. *An Introduction to Crystallography.* 3rd edn. Longmans, Green: London (1963). 4th edn. Oliver and Boyd: Edinburgh (1971).

14. Groth, P. *Chemische Krystallographie. Erster Teil. Elemente — Anorganische Verbindungen ohne Salzcharakter — Einfache und Complexe Halogenide, Cyanide und Azide der Metalle, nebst den Zugehörigen Alkylverbindungen.* [Chemical crystallography. First part. Elements — nonionic inorganic compounds — simple and complex metallic halides, cyanides and azides, together with their accompanying alkyl derivatives.] W. Engelmann: Leipzig (1906).

15. Groth, P. *Chemische Krystallographie. Zweiter Teil. Die Anorganischen Oxo- und Sulfosalze.* [Chemical crystallography. Second part. Inorganic oxy- and sulfosalts.] W. Engelmann: Leipzig (1908).

16. Groth, P. *Chemische Krystallographie. Dritter Teil. Aliphatische und Hydroaromatische Kohlenstoffverbindungen.* [Chemical crystallography. Third part. Aliphatic and hydroaromatic carbon compounds.] W. Engelmann: Leipzig (1910).

17. Groth, P. *Chemische Krystallographie. Vierter Teil. Aromatische Kohlenstof-fverbindungen mit einem Benzolringe.* [Chemical crystallography. Fourth part. Aromatic hydrocarbons with only one benzene ring.] W. Engelmann: Leipzig (1917).

18. Groth, P. *Chemische Krystallographie. Fünfter Teil (Schluss). Aromatische Kohlenstoffverbindungen mit mehreren Benzolringen Heterocyclische Verbindungen.* [Chemical crystallography. Fifth part (conclusion). Aromatic hydrocarbons with multiple benzene rings. Heterocyclic compounds.] W. Engelmann: Leipzig (1919).

19. Haüy, R. J. *Traité Élémentaire de Physique.* [Elementary treatise on physics.] 2 vols. Delanee and Lesueur: Paris (1804). **English translation,** abridged: In: *Crystal Form and Structure.* (**Ed.,** Schneer, C. J.). pp. 18–20. Dowden, Hutchinson & Ross: Stroudsburg, PA (1977).

20. Hessel, J. F. C. *Krystallometrie oder Krystallonomie und Krystallographie.* [Crystallometry, or crystallonomy and crystallography.] In: Gehler's *Physikalisches Wörterbuch* **5**, 1023–1360. Schwickert: Leipzig (1830). Reprinted in Ostwald's *Klassiker der exacten Wissenschaften.* **89**, 41–124. W. Engelmann: Leipzig (1897).

21. Bravais, A. Mémoire sur les polyèdres de forme symétrique. [Note on polyhedra with symmetric forms.] *J. de Math. (Liouville)* **14**, 137–180 (1849).

22. Frankenheim, M. L. Die Lehre von der Cohärenz, umfassend die Elasticität der Gase; die Elasticität und Cohäsion der flüssigen und festen Körper und die Krystallkunde nebst vielen neuen Tabellen über alle Theile der Cohäsionslehre, in'd besondere über die Elasticität und die Festigkeit. [Coherence theory, including the elasticity of gases; the elasticity and cohesion of liquid and solid bodies and the crystallography as well as many new tables concerning all aspects of cohesion theory, particularly elasticity and hardness.] August Schulz: Breslau (1835).

23. Sohncke, L. *Entwicklung einer Theorie der Krystallstruktur.* [Development of a theory of crystal structures.] B. G. Teubner: Leipzig (1879).

24. Fedorov, E. S. Simmetriia pravil'nykh sistem figur. *Zap. Min. Obshch.* [Die Symmetrie der Figuren regelmässiger Systeme.] [The symmetry of regular systems of figures.] (in Russian). A. Jakobson: St. Petersburg (1890) *Trans. Mineral. Soc.* **28**, 1–146 (1891). **English translation**: Harker, D. and Harker, K. *Symmetry of Crystals.* American Crystallographic Association Monograph Number 7: New York (1971).

25. Fedorov, E. von. II. Zusammenstellung der krystallographischen Resultate des Herrn Schoenflies und der meinigen. [Survey of the crystallographic results of Schoenflies and myself.] *Z. Krist. Mineral.* **20**, 25–75 (1892).

26. Barlow, W. A mechanical cause of homogeneity of structure and symmetry. *Sci. Proc. Roy. Dublin Soc.* **8**, 527–690 (1897).

27. Barlow, W. Über die geometrischen Eigenschaften homogener starrer Strukturen und ihre Anwendung auf Kristalle. [Concerning the geometrical properties of homogeneous glassy structures and their application to crystals.] *Z. Krist. Mineral.* **23**, 1–63 (1894).

28. Harker, D. The three-colored three-dimensional space groups. *Acta Cryst.* **A37**, 286–292 (1981).

29. Niggli, P. *Geometrische Kristallographie des Diskontinuums.* [The geometrical crystallography of a discontinuum.] Bornträger: Leipzig (1919). Reprint. Sändig: Wiesbaden (1973). [See *Fifty Years of Crystallography* (Ed. Ewald) for a discussion of this title. JPG]

30. Wyckoff, R. W. G. *The Analytical Expression of the Results of the Theory of Space Groups.* Publication 38. Carnegie Institution of Washington: Washington, DC. 180 pp. (1922). 2nd edn. 1930.

31. Astbury, W. T., and Yardley, K. [Lonsdale]. Tabulated data for the examination of the 230 space groups by homogeneous X rays. *Phil. Trans. Roy. Soc. (London)* **A244**, 221–257 (1924).

32. Stout, G. H., and Jensen, L. H. *X-ray Structure Determination. A Practical Guide.* Macmillan: London (1968). 2nd edn. John Wiley: New York, Chichester, Brisbane, Toronto, Singapore (1989).

33. Friedel, G. Sur les symétries cristallines que peut révéler la diffraction des rayons Röntgen. [Concerning symmetries of crystals that can be revealed by X-ray diffraction.] *Comptes Rendus, Acad. Sci. (Paris)* **157**, 1533–1536 (1913).

34. Roelofsen, G., and Kanters, J. A. Citric acid monohydrate, $C_6H_8O_7 \cdot H_2O$. *Cryst. Struct. Commun.* **1**, 23–26 (1972).

35. Nordman, C. E., Weldon, A. S., and Patterson, A. L. X-ray crystal analysis of the substrates of aconitase. II. Anhydrous citric acid. *Acta Cryst.* **13**, 418–426 (1960).

36. Glusker, J. P., Minkin, J. A., and Patterson, A. L. X-ray crystal analysis of the substrates of aconitase. IX. A refinement of the structure of anhydrous citric acid. *Acta Cryst.* **B25**, 1066–1072 (1969).

37. Bunn, C. W. *Chemical Crystallography. An Introduction to Optical and X-ray Methods.* 1st edn. 1946. 2nd edn, 1961. Clarendon Press: Oxford (1961).

38. Steadman, R. *Crystallography.* Van Nostrand Reinhold: New York, Cincinnati, Toronto, London, Melbourne (1982).

5

Physical Properties of Crystals

What can we learn about the internal arrangement of atoms, ions, or molecules from the appearance and physical properties of a crystal? The macroscopic physical properties of crystals are related to the arrangements of atoms within them. A careful examination of these physical properties leads to much useful information on the symmetry of the atomic arrangement, on the shape of the unit cell, and on the overall arrangement of molecules within the unit cell. Selected examples of such studies are described in this Chapter.

5.1 Mechanical properties of crystals

Certain mechanical properties of crystals[1-5] are immediately obvious. For example, crystals may show differing degrees of **hardness**, which means that they are resistant to scratching by other materials. The hardness of a crystal is measured on a scale,[6] devised by Friedrich Mohs, that ranks compounds according to their ability to scratch another. Diamond, with a numerical rating of 10, is the hardest material and will scratch all other materials. Talc, on the other hand, is very soft, and has a numerical rating of 1, since it is scratched by most other materials. The hardness of a crystal is a measure of the cohesiveness of the molecules or ions comprising it. The hardest crystals have high melting points and contain atoms or molecules with very strong forces between them. Diamond contains carbon atoms joined in an infinite network by three-dimensional hydrogen bonds; talc has the formula $Mg_3Si_4O_{10}(OH)_2$ and consists of sheets of magnesium ions surrounded by six oxygen atoms, sandwiching sheets of tetrahedra of silicon bonded to four oxygen atoms. These sheets are only loosely held together by van der Waals forces[7] so that talc is soft.

Another simple mechanical property of some crystals is their ability to be split readily in one or more specific directions to give smooth surfaces. This property is called **cleavage**, and led Haüy, as described in Chapter 2, to realize that crystals of a given material have characteristic interfacial angles. The cohesiveness of readily cleaved crystals is not the same in all directions, and the intermolecular forces are weak and easily broken in the direction of cleavage. If a crystal is cut along a cleavage plane, much less force is needed to separate the crystal into two parts than is required to cut it in another direction. Such cleavage gives information on the packing in the crystal. For example, the mineral, mica, has a crystal structure composed of two-dimensional layers of silicate groups. These layers are only held together weakly and therefore mica can be split into thin sheets. Another example is provided by graphite (Figure 1.8, Chapter 1) which is readily cleaved parallel to {0001} [{hkil}] faces (see Section 2.5.2). This makes it slippery and useful as a lubricant. Cubic ionic compounds with the general formula AX, such as sodium chloride or cesium chloride,[8] have been shown to cleave parallel to {100}. An example of a crystal with ready cleavage is shown in Figure 5.1. When crystals were ground into spheres for diffraction data measurement, hemispheres were generally obtained because the spheres split by cleavage.[9]

Atmospheric conditions, particularly humidity and temperature, may have a profound effect on the appearance of some crystals that contain water (or other solvents) in their crystal structures. Efflorescence (from the Latin: *to blossom*) is the change of a crystal to a powder upon ex-

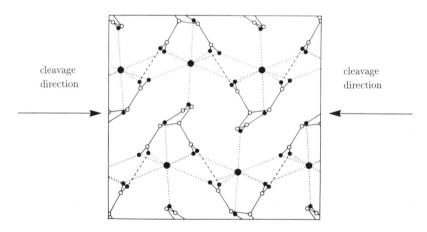

cleavage direction

cleavage direction

FIGURE 5.1. Crystal structure of potassium isocitric lactone, showing the direction of cleavage. Potassium ions large black circles, oxygen atoms small filled circles, carbon atoms open circles, hydrogen atoms omitted. $K^+ \cdots O$ dotted lines, hydrogen bonds dashed lines.

posure to air. It results from the loss of water or some other solvent of crystallization, initially from the surface of the crystal, as a result of the greater vapor pressure of the hydrate compared with the partial pressure of water vapor in the air. As layer after layer of the crystal is destroyed in this way, the inner layers of the crystal become accessible to the effect. As a result, transparent crystals that initially had well defined faces, become powdery and opaque as the process proceeds. Washing soda (sodium carbonate decahydrate $Na_2CO_3 \cdot 10H_2O$) provides an excellent example of an efflorescent material. Under very dry conditions copper sulfate pentahydrate $CuSO_4 \cdot 5H_2O$ will also effloresce. By contrast, other types of crystals may attract, rather than lose, water. Deliquescence (from the Latin: *to become liquid*) is a process by which a crystal absorbs moisture from the air and dissolves in the liquid so absorbed, forming a solution; in other words, the crystal is converted to a puddle. An example is provided by calcium chloride. The extent to which deliquescence occurs depends, again, on the partial pressure of water vapor in the surrounding atmosphere (and, in part, on the solubility of the compound in water).

5.2 Optical properties that indicate symmetry in a crystal

Crystals may be transparent, opaque, colored, or colorless, depending on the way in which they interact with visible light. A beam of light may be reflected from their surfaces, may pass right through them, or may be absorbed by them. Light is a transverse electromagnetic wave motion with a magnetic and an electric vector, perpendicular to each other and to the direction of propagation of the light. It is the electric vector of light that is detected in vision, and this vector can assume any orientation in a plane perpendicular to the direction of propagation of the light (Figure 5.2). When visible light passes through a transparent material, the electric field of the light "polarizes" the atoms by displacing their positive and negative charges and so producing electric dipoles.

The **polarization of an atom** in a crystal is both a function of the electric field the light impinging on it, and of the instantaneous dipole moments of atoms in the neighborhood of that atom.[10,11] The light then produces forced oscillations in the electric dipoles so produced. The oscillations of these induced dipoles are perpendicular to the direction in which the light wave moves and will cause radiation of light with a maximum intensity perpendicular to the direction of vibration. The extent to which light can induce the charge separations in each type of atom will determine how the light is affected by a crystal.[12,13] This polarization of atoms causes light to travel more slowly through the material and is the reason that the refractive index is not unity.

The interaction of crystals with light depends both on the arrangement of atoms in their crystal structure and on the nature of these atoms.[7] The optical properties of crystals indicate the directions of symmetry axes in the crystal and, in certain cases, provide useful preliminary information

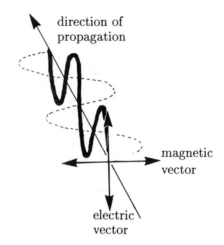

FIGURE 5.2. Light as a wave motion.

on the arrangement of molecules in the unit cell. For detailed information the reader should consult physics texts on the subject.[14]

Any property of a material that is equal in all directions is described as **isotropic** (from the Greek: *isos*, equal; *tropos*, responding to). If a material is isotropic, it appears the same from any direction. Most gases, liquids, and amorphous solids, but only a few types of crystals — those that belong to the cubic system — are isotropic. Such isotropy is, of course, an important property of window glass, otherwise a distorted image might be obtained when looking through it.

The vast majority of crystals are **anisotropic**, that is, their properties are not the same in all directions within the crystal. Light is said to be **plane polarized** (also called linearly polarized) when the electric field oscillates in a straight line. When, on the other hand, the electric field vector travels around a circle, then the light is said to have circular polarization, and when the ends of the vector travel in an ellipse, it is said to be elliptically polarized.

Several interesting effects are observed when light is shone on anisotropic crystals, such as the spectacular property of **pleochroism**, in which the crystals appear to have different colors when viewed from different directions under transmitted white light (dichroism if there are only two colors). The variation of the absorption of light with direction within the crystal will give information on molecular orientation in organic crystals. The absorption of light is greatest when light is vibrating along the bonds of the chromophoric groups rather than perpendicular to them.

5.2.1 Refractive indices of crystals

An important measure of the extent to which light interacts with the atomic arrangement in a crystal is provided by the refractive index. This describes the extent to which the crystal slows down light traveling through it, compared to the maximum velocity of light when it travels in a vacuum. The refractive index is the ratio of the velocity of light in a vacuum to its velocity in the material under study. The value for dry air is essentially unity (1.000274). The velocity of light is, however, considerably reduced in crystals. Values for their refractive indices generally lie in the range 1.3 to 2.4, but depend both on the wavelength of the light and on the atomic arrangement within the crystal.

Since light has a different velocity in different media, the direction of the light beam is altered as it passes from one medium to another, an effect referred to as **refraction** (from the Latin: *refractus*, past participle of *refringere*, to break off). This *break off* is familiar to anyone who has observed a stick emerging from water with an apparent bend in it. Elizabeth A. Wood[15] points out that this can be understood by thinking of the result of the slowing down of the velocity of one wheel of a car compared to that of the other. The car will turn in the direction of lower velocity. The example, illustrated in Figure 5.3, involved a pair of wheels, joined by an axle, rolling down a gentle slope across a concrete pavement and onto grass at an angle. She writes: "the more the first wheel is retarded by the grass (lower velocity), the greater will be the turning in the direction of travel (higher index of refraction)." As a light

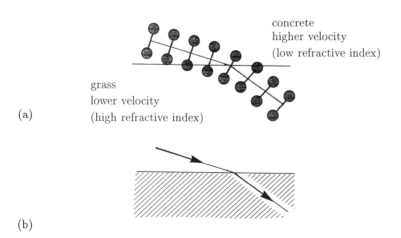

(a)

(b)

FIGURE 5.3. Refraction illustrated by (a) movements of car wheels and (b) passage of light into a different medium.

wave enters a crystal at an angle, the portion of the wavefront inside the crystal lags behind that portion still in air. As a result, just as for the wheels, the direction of the wavefront (and hence of light) changes.

The ratio of the sines of the angles (from the normal) of a beam of light on each side of an interface is the inverse of the ratio of the refractive indices of the materials through which it is traveling [Figure 5.4(a)]. This law was discovered by Willebrord Snell about 1621. Never formally published, it is known as **Snell's Law.**

If the angle of incidence of light is greater than a certain angle (the **critical angle**), the radiation may be reflected back into the incident medium, a phenomenon called **total internal reflection.** Internal re-

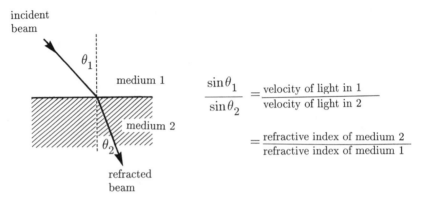

$$\frac{\sin\theta_1}{\sin\theta_2} = \frac{\text{velocity of light in 1}}{\text{velocity of light in 2}}$$

$$= \frac{\text{refractive index of medium 2}}{\text{refractive index of medium 1}}$$

(a) Snell's law:

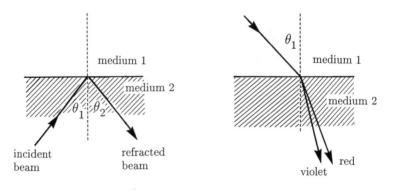

(b) Total internal reflection: (c) Dispersion:

FIGURE 5.4. (a) Snell's Law of Refraction, (b) total internal reflection, and (c) dispersion (the refractive index for violet light is greater than that for red light, therefore the deviation θ_2 is greater for red light).

flection [Figure 5.4(b)] occurs when light strikes the interface from the side of the material with the higher refractive index. Because the components of white light have different wavelengths, and therefore different refractive indices in a particular crystal, the light will be split into its component colors, which travel at different rates and are bent to different degrees at the interfaces [Figure 5.4(c)]. This important phenomenon, the variation of the refractive index of light with its wavelength, is called **dispersion**. These phenomena (refraction, total internal reflection, and dispersion) are combined when a rainbow is formed after a rainstorm. The bow is obtained by two refractions and one internal reflection of sunlight through raindrops (violet at 40.6°, red at 42.4° from the horizon). A secondary, weaker bow may also be seen. This results from one additional internal reflection and, therefore, has its colors reversed (red at 50.4°, violet at 53.6°).

These three effects of light on crystals are evident in gemstones, such as diamonds,[16,17] which are cut with careful attention to their refractive indices so that the extent to which they catch visible light is maximized. The **brilliance** of diamonds is due to a combination of refraction, internal reflection and dispersion of light; the latter, dispersion, gives rise to the so-called **"fire"** of diamonds. These effects have been carefully balanced in the cutting. Light is bent when it enters a diamond, and gem faces are cut so that light will be reflected inside it. Diamond has a high refractive index, 2.4, and a low critical angle, 24.5°, for example, for yellow light, meaning that when yellow light passes into a diamond and hits a second face internally at an angle from the normal greater than 24.5° (an angle of 65.5° or less with the outer surface of the crystal), it cannot pass from the diamond back into air but is reflected back into the gemstone. Also, diamonds disperse light; that is, the refractive indices for red light and violet light are different, 2.409 and 2.465, respectively, so that the gemstone acts like a prism in separating the colors, and its dispersion is 0.056 (the difference). The greater the dispersion, the better the spectrum of colors that is obtained. The faces of gemstones are cut so that when the light returns to the observer, it has already been reflected and dispersed many times within the crystal, as shown in Figure 5.5. This is the reason that diamonds appear brilliant and generate the lovely optical effects for which they are so famous.

If a crystal belongs to the cubic system, the velocity of light through it (and therefore its refractive index) is isotropic (the same in all directions). The refractive index of such an isotropic crystal is measured by observing it when it is immersed in a colorless liquid of matching refractive index (obtained by mixing appropriate liquids of known refractive indices in which the crystal is insoluble). When the refractive index of the surrounding mixture of liquids exactly matches that of the crystal, the latter becomes invisible. The refractive index of the liquid mixture can be measured and is equal to the refractive index of the crystal.

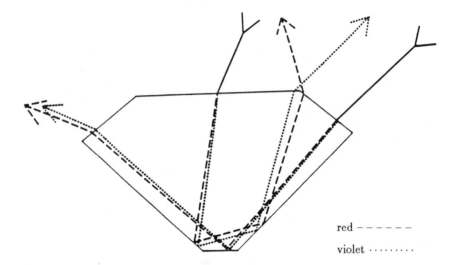

red − − − − −

violet · · · · · · · ·

FIGURE 5.5. The "fire" of diamonds. The refractive indices of diamond are 2.409 for red light and 2.465 for violet light. Light passing through a suitably cut diamond will be refracted and dispersed in a manner such as that shown here.

5.2.2 Birefringence of crystals

Crystals are anisotropic unless they belong to the cubic system. When light enters a crystal with an anisotropic atomic arrangement, an interesting phenomenon, illustrated in Figure 5.6, occurs. The light is split into two components that travel with different velocities along different paths as a result of different extents of refraction. This effect is called **double refraction** or **birefringence** (two refractive indices or refrangibilities), and its presence indicates that the internal atomic arrangement within the crystal is anisotropic.

Birefringence was first noted by Erasmus Bartholinus[18,19] in the naturally occurring mineral Iceland spar (also called calcareous spar or calcite) acquired from a sailor who found them in the Bay of Róerford in Iceland. As shown in Figure 5.6, when you look at a mark on paper through a calcite crystal in a direction perpendicular to one of its rhombohedral faces, you will see two images of the mark. If you position the crystal over the mark and rotate the crystal, one of the images will not move (normal or ordinary refraction) while the second image will move (but keep its orientation) at the same rate as the crystal (extraordinary refraction). There are, thus, two types of refracted rays identifiable by this experiment, and it is possible to tell which refracted ray is which. One, called the **ordinary ray,** obeys the ordinary laws of refraction, and the other, called the **extraordinary ray,** obeys

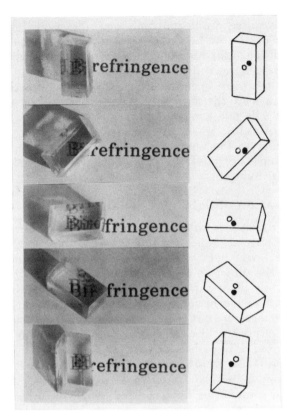

FIGURE 5.6. View through an Iceland spar (calcite) crystal, looking at the word "birefringence" with the first few letters covered by the crystal. Note that the relationship of the two images changes as the calcite crystal is rotated. Follow the line of lettering in the photograph to identify the ordinary ray image (black circles in the diagram on the right) versus the extraordinary ray image (open circles in the diagram on the right). The refractive indices of the ordinary and extraordinary rays are $n_O = 1.6585$, and $n_E = 1.4862$ (for sodium light which is yellow), respectively. The difference, $n_E - n_O = -0.1723$, provides a measure of the birefringence of calcite.

a different law of refraction in which the velocity of light is dependent on its direction of travel. These two types of rays have different refractive indices, n_O for the ordinary ray, and n_E for the extraordinary ray. The difference $n_E - n_O$ between the two types or rays is used as a measure of birefringence, just as the differences in refractive indices of red and violet light gives a measure of dispersion. The birefringence of calcite, shown in Figure 5.6, is very noticeable because there is a large difference in the refractive indices of the two types of rays.

Bartholinus noticed[18] that, on viewing a calcite crystal from all possible directions, there is one (but only one) direction through a calcite crystal that an object can be viewed so that one image of the object (not two) is formed. This direction is called the **optic axis** and is the direction in a birefringent crystal along which the ordinary and extraordinary rays travel at the same speed (meaning that the refractive indices for both types of rays are the same in these directions). The presence of one such axis that will give a single image is the reason that the crystal is described as **uniaxial**. If a plate-like section is cut from a calcite crystal perpendicular to the unique crystallographic axis c (the optic axis in these hexagonal crystals), the uniaxial character of calcite is evident (only one image along c).

5.2.3 Polarization of light by crystals

When light is reflected from a crystal face or split by a birefringent crystal, it acquires special properties, and is said to be polarized. Normally the electric vector of light can assume any orientation in a plane perpendicular to the direction of propagation (Figure 5.2). After light is reflected from any plane surface, such as a crystal face, the electric vector of the light is no longer equally distributed in the plane normal to the direction of propagation, but is concentrated in one direction. This light, for which the electric field oscillates along a straight line, is described as plane–polarized or linearly polarized. Such polarized light is normally difficult to distinguish from ordinary light, but it interacts in different ways with certain materials. For example, so-called polarizing sunglasses reduce glare by eliminating any reflected light, which has the property of being plane-polarized.[20] Note that here we have employed two different uses for the word "polarized:" (1) polarization of light refers to light waves vibrating in one plane only, while (2) polarization of atoms refers to the separation of positive and negative charges around an atom.

This **polarization of light** upon reflection from a surface was discovered and named by Étienne Malus in 1808.[10] He observed this effect when he was idly looking through a crystal of Iceland spar at the setting sun reflected off the windows of the Palais du Luxembourg. The sunlight provided an intense, distant source of light, and the windows provided a reflecting surface. As Malus rotated the crystal while looking through it, he observed two images of the sun; these images alternately disappeared every 90° of rotation of the crystal. He deduced that, after being reflected from the windows, the light had acquired a new property, polarization. Malus was a follower of Newton's ideas on light and thought the light particles were small magnets that became oriented with respect to the earth's poles on reflection. The term polarization that Malus introduced for the effect has persisted even though the wave theory of light was established soon after.

That same evening Malus tested his deduction that light is polarized on reflection by studying the properties of candlelight reflected off the surface of water in a bowl. He found that the candlelight had interesting properties if it had first passed through a calcite crystal so that two rays were produced. When the ordinary ray is reflected from the surface of the water, the extraordinary ray is not reflected at all and vice versa. The explanation is that the plane-polarized ordinary and extraordinary rays, formed when light passes through an anisotropic crystal, vibrate in planes perpendicular to each other. When reflected a second time, these rays only show intensity when their plane of polarization is appropriately oriented with respect to reflection (see Figure 5.7).

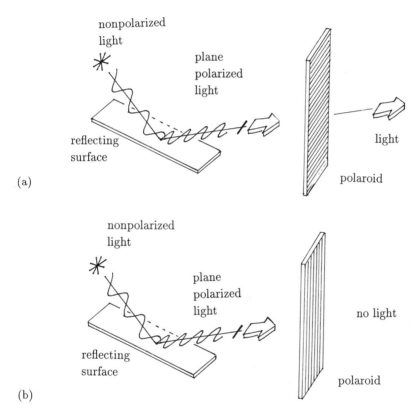

FIGURE 5.7. Light is plane polarized on reflection and is transmitted through a detector of polarization (polaroid shown here) only if the orientation of the detector is correct for the passage of plane-polarized light. This orientation is correct as in (a), but not in (b). The experiment of Malus with candlelight on a bowl of water made use of the same principles, with a calcite crystal replacing the polaroid.

In 1812 Jean Baptiste Biot[21,22] and Sir David Brewster[23,24] independently discovered that some crystals have two optic axes along which the two rays travel at the same speed. Such crystals are designated **biaxial**. When a beam of light passes into a biaxial crystal, it is split into two rays, both of which behave like the extraordinary ray of a uniaxial crystal, that is, both are refracted, travel in different directions with different velocities, and both are plane polarized, vibrating in planes at right angles to each other. Tetragonal, hexagonal, rhombohedral crystals are uniaxial with one optic axis, while triclinic, monoclinic, and orthorhombic crystals are biaxial, with two optic axes. Only cubic crystals (or isotropic noncrystalline materials) are completely isotropic, so that only one image is seen, no matter what the direction of view is.

What does it mean if there are two rays formed by an anisotropic crystal? The refractive index is different for light plane polarized in one direction from one plane polarized in a direction perpendicular to the first. Effectively, because of the atomic arrangement in the crystal, the electrons in the molecules respond better to vibrations in the direction parallel to the axes of the molecule than they do in a direction perpendicular to this direction. For example, if the crystal structure contains long flat molecules, then plane-polarized light that has its electric vector parallel to the molecular planes will respond differently from light plane polarized perpendicular to the molecular plane.

The formation of two rays in a birefringent crystal is diagrammed in Figure 5.8. If the incident light is plane polarized with its electric vector perpendicular to the optic axis, it will travel through in a normal way that can be represented by a progression of a wave front composed of spherical wavelets [Figure 5.8(a)]. This is the ordinary ray, shown in Figure 5.9(a). On the other hand, if the light entering the crystal has its plane of polarization in the plane of the optic axis, the waves passing through the crystal are no longer spherical, but, because of the variation of refractive indices with direction, spread out as ellipsoidal waves with the optic axis as a major axis of the ellipsoid [Figure 5.8(b)]. The wavefront then proceeds in a different direction through the crystal than that of the ordinary ray and hence it is called the extraordinary ray [Figure 5.9(b)]. It emerges parallel to the original (ordinary) beam, but displaced from it, hence two images. The two rays are plane polarized in planes perpendicular to each other. The ordinary ray vibrates in a plane perpendicular to the plane containing the optic axis and the incident ray. The extraordinary ray vibrates in a plane containing the incident ray and the optic axis. Thus, when unpolarized light hits a birefringent crystal, it is separated into two rays, because the components of unpolarized radiation are separated in a birefringent crystal.[12,13,20]

In a uniaxial crystal, for any light traveling along **c** (vibrating in any direction perpendicular to its propagation direction) there is only one refractive index and therefore only one image. The extraordinary ray behaves differently. The value of n_E, the refractive index of the extraordinary ray (see Figure 5.6), varies from its minimum value for vibration parallel to the optic axis (where the extraordinary ray has its maximum velocity) to its maximum value for vibration perpendicular to the optic axis (lowest n_E). The value listed for n_E listed is always the extreme index of refraction for the extraordinary ray, that is, when it vibrates parallel to **c** (travels in a direction perpendicular to **c**). Thus, in any direction that is not along the optic axis, there are two refractive indices, n_O and n_E. If light vibrates in a direction perpendicular to **c**, the refractive index is n_O (a constant with respect to direction). If light vibrates parallel to **c**, the refractive index is n_E (which varies from n_E to n_O as a function of direction).

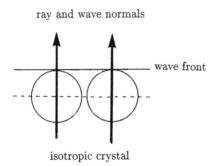

FIGURE 5.8. (a) Waves passing through an isotropic crystal are spherical, whereas those passing through a birefringent crystal (b) behave as ellipsoidal wavelets.

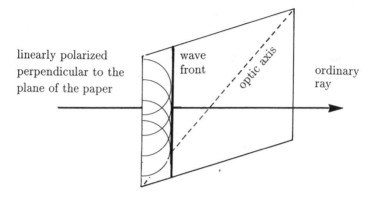

(a)

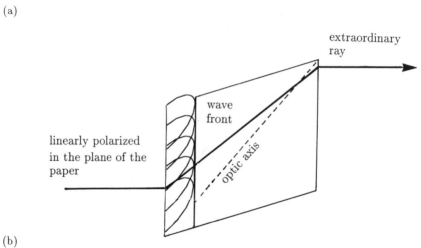

(b)

FIGURE 5.9. Passage of light through a birefringent crystal. (a) The component of the incident light plane polarized perpendicular to the plane of the paper moves as an ordinary ray. On the other hand, (b) the component of the incident light plane polarized in the plane of the paper moves as the extraordinary ray. The result is two types of rays pass through the crystal and two images are obtained (see Figure 5.6).

5.2.4 The optical indicatrix

The **optical indicatrix** is a useful construction for visualizing the variation in the refractive index of a crystal as a function of spatial direction.[25,26] It shows directions (with respect to unit-cell edges) where the refractive index is greatest and where it is the smallest. It is a three-dimensional ellipsoid with a shape defined by the ends of vectors, each with the same fixed origin. The length of each vector is proportional to the refractive index of the light vibrating in the direction of the vector (perpendicular to the direction of travel of the light). As discussed earlier (see Figure 5.2), the directions of vibration are the important quantities in studying refractive indices. When there is some symmetry constraint on the choice of unit cell axes, the optical indicatrix has a principal axis in the direction required by the symmetry.

By the optical methods just described, crystals may be divided into three groups:

1. those that are isotropic,
2. those that are anisotropic with one unique crystallographic axis (uniaxial crystals, and
3. those that have lower symmetry so that no axes are equivalent (biaxial crystals).

The shape of the optical indicatrix gives a measure of the directionality of the refractive indices. If the crystal is cubic, its optical indicatrix is represented by a sphere. If the crystal is uniaxial, the optical indicatrix is an ellipsoid of revolution (obtained by rotating an ellipsoid about a major axis). The unique axis of the ellipsoid of revolution, the direction of the optic axis, represents the refractive index of the extraordinary ray. The circular section perpendicular to the unique axis represents the refractive index of the ordinary ray. Uniaxial crystals are divided into two groups; for those that are positively birefringent, the refractive index for the ordinary ray n_O is smaller than that for the extraordinary ray n_E, while for those that are negatively birefringent, n_O is larger than n_E. Calcite (n_E less than n_O) is negatively birefringent, while quartz (n_E greater than n_O) is positively birefringent (see Table 5.1). The optical indicatrices of these two minerals, shown in Figure 5.10, are oblate (flattened at the poles) for calcite (where n_E is less than n_O), and prolate (stretched at the poles) for quartz (where n_E is greater than n_O).

If the crystal is biaxial, there are three principal refractive indices (α, β, and γ, with α less than β less than γ). The optical indicatrix has

TABLE 5.1. Refractive indices of some uniaxial crystals.

Compound	Formula	n_O (ω)	n_E (ϵ)	$n_E - n_O$ ($\epsilon - \omega$)	Sign
Cinnabar	HgS	2.854	3.201	0.347	+
Calcite	$CaCO_3$	1.658	1.486	0.172	−
Calomel	Hg_2Cl_2	1.973	2.656	0.683	+
Quartz	SiO_2	1.544	1.553	0.009	+

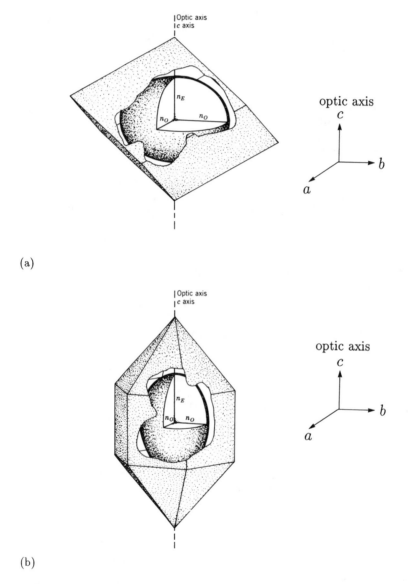

(a)

(b)

FIGURE 5.10. Some crystals, each with their optical indicatrices drawn within them. (a) A uniaxial negative crystal (calcite) showing orientation of the indicatrix. (b) A uniaxial positive crystal (quartz) showing orientation of the indicatrix.

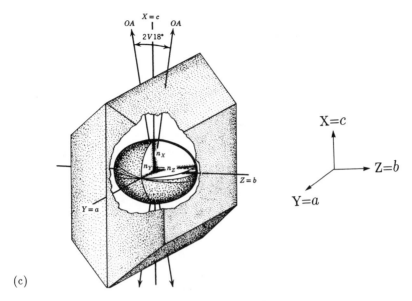

(c)

FIGURE 5.10. (c) A biaxial negative orthorhombic crystal in which [for convenience, in this case], $X = c$, $Y = a$, and $Z = b$. The optic plane is parallel to (100). ((a), (b) and (c) courtesy Ernest E. Wahlstrom and John Wiley & Sons, Inc. From: Wahlstrom, E. E. *Optical Crystallography.* 5th edn. John Wiley: New York, Chichester, Brisbane, Toronto (1979). © John Wiley & Sons, Inc.).[12]

three unequal mutually perpendicular semiaxes with lengths proportional to the refractive indices. The birefringence of such a crystal is defined as $\gamma - \alpha$. Biaxial crystals are considered positive if β is closer to α than γ, and negative if it is closer to γ than α. An optic axis is always perpendicular to a circular section of the optical indicatrix. Thus, for a uniaxial crystal with one optic axis, there is one circular section perpendicular to it, while for a biaxial crystal, the two optic axes are perpendicular to the two circular sections. These two circular sections correspond to directions where there is no double refraction, that is, directions along which the ordinary and extraordinary rays travel at the same speed.

5.2.5 Structural information from refractive indices

The measurement of the refractive index of a birefringent crystal can give information on the orientation of molecules, or portions of them, in a crystal. When crystals are optically anisotropic, plane-polarized light is used to determine the magnitudes of the refractive indices in different directions. Experimentally it is found that a tetragonal (uniaxial) crystal has a different refractive index when the measurement is made with light vibrating parallel to its fourfold axis than when it is vibrating perpendicular to this axis. If the crystal has lower symmetry and is orthorhombic, monoclinic, or triclinic, there are three refractive indices. On the other hand, if the crystal has cubic symmetry, the refractive indices are the same in all directions and only one value will be found.

If the orientation of the principal optical directions can be found with respect to the unit cell vectors of the crystal, the orientations of the molecules in the unit cell, especially those with very anisotropic shapes and considerable unsaturation, may be found. This information was useful in the determination of the shape and size of the steroid nucleus (see Figure 1.11, Chapter 1). The relationship of molecular shape to refractive index is listed in Table 5.2, and the different refractive indices of naphthalene[27] are shown in Figure 5.11.

Each atom in a crystal structure makes a specific contribution (called its **atomic refractivity**) to the refractive index. The refractivities of cations are generally less than those of anions, presumably because the electrons are held more tightly in the former. Each atom becomes an electric dipole as the electric field of the light interacts with it. Double refraction results from the mutual interactions of neighboring dipoles. Induced magnetic dipoles do not influence each other very much. Ionic refractivities were used as a measure of the polarizabilities of the ions. The birefringence of calcite, $CaCO_3$, was analyzed by W. L. Bragg.[28] Calcite crystals are composed of alternate layers of calcium and carbonate ions (Figure 5.12). Bragg assumed that the refractivity of the carbonate group was mainly due to its three oxygen atoms, which are 2.25 Å apart in a plane; the refractivity of a calcium cation is small and was neglected. When light passes through calcite, the three oxygen atoms of the carbonate groups are polarized more strongly when the electric field is in their plane than when it is perpendicular to it. The higher refractive index is perpendicular to the trigonal axis of calcite, because the electric vector is then in the plane of the carbonate groups.

TABLE 5.2. Relationship of refractive indices, n to molecular shape.

Shape of object	*Refractive index*
1. *Equal in all dimensions*	*Isotropic*
2. *Parallel rods*	*Positively birefringent (n_E greater than n_O). Rods parallel to the single vibration direction of greatest refractive index*
3. *Plate shaped*	*Negatively birefringent (n_E less than n_O). Plates perpendicular to the single vibration direction of least refractive index*

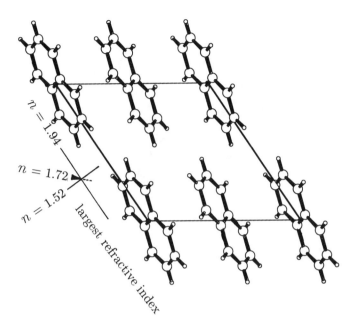

FIGURE 5.11. Refractive indices and molecular stacking in a crystal of naphthalene, showing how the largest refractive index corresponds to the direction of the greatest density of atoms. The edges of the unit cell are shown.

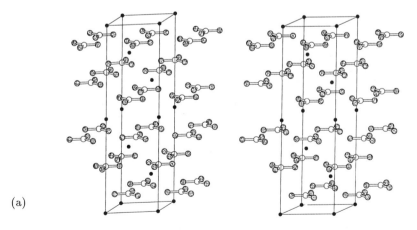

(a)

FIGURE 5.12. Analysis of the birefringence of calcite, $CaCO_3$, in terms of the atomic arrangement in crystals. (a) Stereoview of packing in the crystal. Oxygen atoms stippled, calcium ions small filled circles.

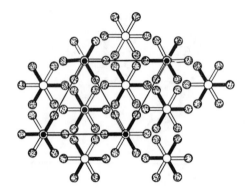

(b)

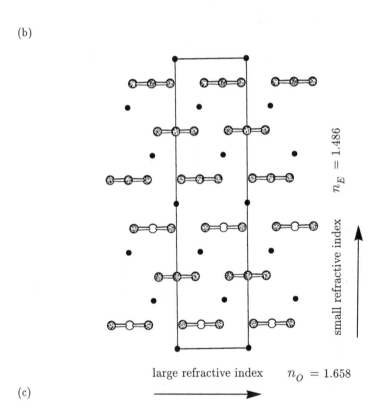

large refractive index $\qquad$ $n_O = 1.658$

(c)

FIGURE 5.12. (b) Calcite crystal structure. Space group $R\bar{3}c$, hexagonal unit cell dimensions $a = b = 4.9898$, $c = 17.060$ Å. View down the unique axis, c, and (c) view perpendicular to this, with the unique axis pointing up.

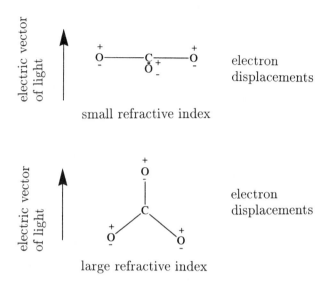

(d)

FIGURE 5.12. (d) Explanation of the different refractive indices in terms of electron displacements.

The biochemist uses the results of birefringence to test whether or not a protein has crystallized. If the material in a test tube is held up to the light and shaken, the presence of birefringent microcrystals of protein may be revealed by a **Schlieren effect** (from the German word for streaks or striae). These streaks are observed when light is viewed across the medium since it has varying refractive index as a result of the presence of the small, randomly oriented protein crystals. If the protein crystals belong to the cubic system, and therefore are isotropic, this Schlieren effect is not seen,[29] and the presence of crystals is established by observing them with a microscope.

5.2.6 Examination of crystals under the optical microscope

Birefringence was used by William Nicol[30] in 1829 to make a polarizing device so that the optical properties of other materials could be studied under a microscope. What he did was devise a way of getting rid of the ordinary ray, so that only the extraordinary ray is available. Nicol cut two calcite crystals and glued them together (thereby making what is called a Nicol prism). The light (the extraordinary ray) that emerges from the Nicol prism (the *polarizer*) is plane polarized (vibrates only in one plane). If a second Nicol prism (called the *analyzer*) is placed in a similar orientation to the first, the extraordinary ray emerging from the first Nicol prism will pass through the second, still vibrating in the same plane. If the second Nicol prism is rotated, only a component (depending on the angle of rotation) will be transmitted. If the Nicol prisms are oriented at 90° to each other, an arrangement referred to as **crossed Nicol prisms,** no light will pass through (see Figure 5.13).

Polaroid®, the material of which most polarizers are currently made, is composed of a clear sheet of polyvinyl alcohol. It is stretched in one direction to align the molecules and dyed with iodine molecules that align themselves along the polymer molecules. This polaroid film will transmit light only in a direction perpendicular to that in which the film was stretched. Polaroid was first made in 1928 by Edwin Herbert Land from quinine sulfate periodide.[20,31] The initial idea for this came from the observation of William Bird Herapath that when iodine is dropped in the urine of a dog that has been fed on quinine, shiny green crystals (now referred to as *herapathite*) with strongly polarizing properties are formed.

When an isotropic crystal is placed between crossed Nicol prisms, there is extinction, which means that nothing is seen, and the field is perfectly dark. On the other hand, if an optically anisotropic, birefringent crystal is similarly viewed, it will appear colored except at certain positions of rotation, usually 90° apart. At these positions, extinction occurs because the vibration directions of the light transmitted through the

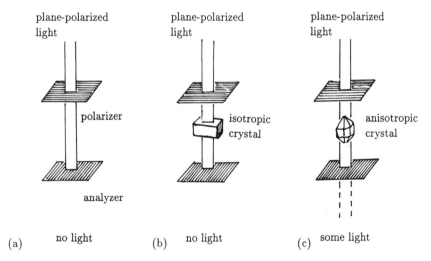

FIGURE 5.13. Crossed Nicol prisms. (a) No crystal, (b) an isotropic crystal between the polarizer and the analyzer and (c) an anisotropic crystal which rotates the plane of plane polarized light to some extent as shown.

Nicol prisms coincide with those of the crystal (ordinary and extraordinary rays) [see Figure 5.13(c)]. In other positions the color is determined by the birefringence of the crystal, the direction of the light and the thickness of the crystal. This method can be used to determine whether the crystal is single or not, since two or more misaligned crystals (twins) do not extinguish the light completely.

Another method for studying the optical properties of crystals is with convergent light, such as is found with a high-power condenser lens on a petrographic microscope. By removing the eyepiece of the microscope and looking straight down the tube, an **interference figure** can be obtained. When a convergent beam of white polarized light is passed through a uniaxial crystal (cut perpendicular to the optic axis and lying between crossed Nicols), an interference figure consisting of concentric circles of interference colors superimposed on a black Maltese cross (an **isogyre**) with its center on the optic axis is obtained. On the other hand, if the crystal is biaxial (with two optic axes), the interference figure is more complicated. If the crystal is thick enough, it is possible to observe colored concentric circles centered on the optic axis directions. The optic axial angle can then be measured by noting the distance between these circles. Thus it is possible from the nature of the interference figures to determine whether the crystal is uniaxial or biaxial (Figure 5.14). Further details on the methods of optical crystallography can be found in the texts on the subject by Elizabeth A. Wood,[15] by Ernest E. Wahlstrom,[12] and by Norman H. Hartshorne and Alan Stuart.[32]

5.2.7 Optical activity of crystals

A substance capable of rotating the plane of polarization of plane-polarized light has **optical rotatory power** and is **optically active**. Optical activity is a property of substances that have a shape such that they cannot be superimposed on their mirror image and appear identical.

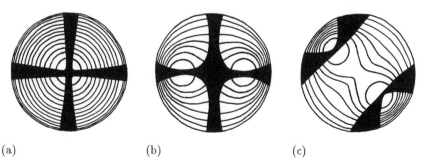

(a)   (b)   (c)

FIGURE 5.14. Idealized interference figures for (a) a uniaxial crystal, such as calcite, looking down the optic axis, (b and c) biaxial crystals with the polarizers in different orientations (parallel to the page edge and at 45° to it, respectively).

The asymmetry may reside either in the molecules forming the crystal or in the molecular arrangement (and hence the development of faces) in the crystal. Optical activity is readily observed by passing a beam of plane-polarized light through materials. If the emergent plane-polarized beam is no longer vibrating parallel to the original direction of vibration, but its plane of vibration has been rotated clockwise or counterclockwise during its transmission through the substance, the substance is said to be optically active. The effects are designated (**dextrorotatory**), or (**levorotatory**), depending on the direction of rotation as viewed by looking towards the light source (clockwise, or counterclockwise, respectively), as shown in Figure 5.15. If the emergent beam is unchanged, the sample as a whole is optically inactive.

In 1811 François Arago[33] discovered optical rotation in hexagonal crystals of quartz. Some crystals rotated the plane of polarization of polarized light to the left and some to the right. The angle of rotation was found to be proportional to the thickness of the crystal. Since the repeating unit of quartz, SiO_2, is symmetrical, he reasoned that the observed optical activity indicated some asymmetry in the molecular packing in the crystal. X-ray diffraction studies have since shown that crystalline quartz consists of helical arrangements of SiO_4 tetrahedra, so that every oxygen atom is common to two tetrahedra, the helices lying along the direction of the optic axis.[34] Levorotatory α-quartz crystallizes in the space group $P3_121$, and its crystal structure contains right-handed helices of SiO_4 tetrahedra, illustrated in Figure 5.16(a).[34] Similarly, sodium chlorate, with pyramidal chlorate ions, shows no optical activity in solution, but forms crystals that can rotate the plane of polarized light. On the other hand, Biot in 1812 found that a similar rotation of the plane of polarized light was obtained for solutions of tartaric acid.[21,22] Since the

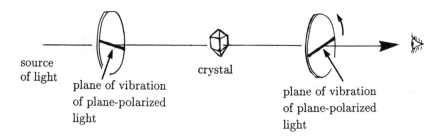

source of light

plane of vibration of plane-polarized light

crystal

plane of vibration of plane-polarized light

FIGURE 5.15. The sense of optical rotation, that is, the direction in which the plane of polarization moves. In the case illustrated here the plane of vibration of plane polarized light is moved in a counterclockwise direction when viewed towards the source, so that the crystal is labelled levorotatory. If it is rotated in the opposite direction the crystal would be said to be dextrorotatory.

tartaric acid molecules were in solution and presumably not aggregated in any regular way, he deduced that he was measuring a property of the molecules, not their aggregates.

Thus there are two classes of optically active crystals. One, like quartz and sodium chlorate, only shows the effect in the crystalline state. Here the ability to rotate the plane of polarized light is related to the fact that the atoms in the crystal are arranged as right-handed or left-handed spirals or other asymmetric shapes. The second class of optically active crystals contains molecules or ions, such as certain tartrates, that are themselves asymmetric; the effect persists even when the crystal melts or is dissolved. More details on this subject are given in Chapter 14.

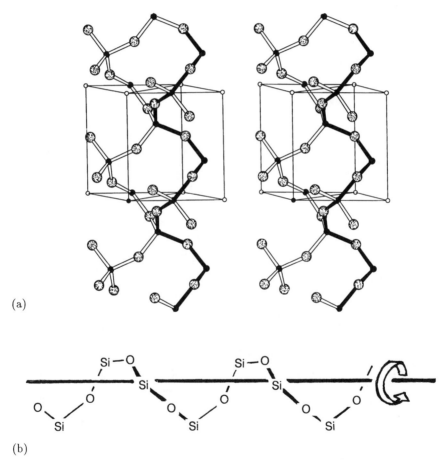

(a)

(b)

FIGURE 5.16. Structure of quartz, SiO_2, showing (a) a stereoview and (b) an indication of atomic identities in the helical packing in levorotatory quartz.

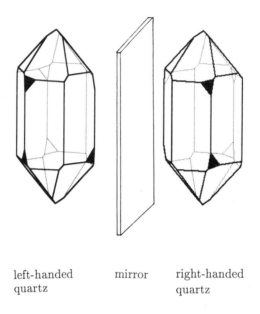

left-handed   mirror   right-handed
quartz   quartz

(c)

FIGURE 5.16 (cont'd). (c) Right-handed and left-handed quartz crystals, with hemi-hedral faces marked in black. On the left is levorotatory α quartz and on the right is dextrorotatory α quartz.

A careful examination of quartz crystals, which have optical activity by virtue of their internal structure, reveals small faces (facets) that give the crystal an apparent lower symmetry. These crystals are called **hemihedral,** a half set of facets, as opposed to **holohedral**, a full set of facets. As a result, there are two types of hexagonal quartz crystals that are hemihedral; these are mirror images (enantiomorphic forms, or **enantiomorphs**) [Figure 5.16(c)]. John Herschel[35] noted that the direction of rotation of the plane of polarization of light by a quartz crystal was dependent on which of the two possible enantiomorphic types of crystal was being studied. This study was continued by Louis Pasteur,[36,37] who showed that the hemihedry of sodium ammonium (+)-tartrate was consistent with its dextrorotatory power. He was even able to use tweezers to separate racemic sodium ammonium tartrate crystals into two groups depending on the handedness of the hemihedry. When dissolved in water, one group gave a dextrorotatory solution, while the other group gave a levorotatory solution.

5.3 Electrical and magnetic effects in crystals

Noncentrosymmetric crystals may generate measurable anisotropic electrical effects in response to external forces such as temperature changes or direct pressure on the crystal.[38] The **piezoelectric effect**[39,40] is the generation of electric charge in certain crystals along their **polar axis** when they are subjected to stress. The polar axis is an axis that has distinguishable ends; that is, positive and negative directions can be identified along this axis. The polarization produced is proportional to strain. Crystals can also display thermochromic and piezochromic effects, that is, color changes associated with temperature changes and pressure changes respectively.

5.3.1 Piezoelectricity

Certain crystals develop electric charges on their faces in response to mechanical stress. The effect, first noted by Jacques and Pierre Curie in 1880,[39] is found only for noncentrosymmetric crystals and may be used as a test for a lack of a center of symmetry in a crystal. It is called the piezoelectric effect, the prefix *piezo* (pronounced *pee-ayzo*) being derived from the Greek *to press*. Good examples of the piezoelectric effect are provided by crystals of quartz or Rochelle salt (sodium potassium tartrate tetrahydrate).[41,42] If a crystal of Rochelle salt is placed between metal plates attached to an electric current sensor such as a small light bulb, and the top of the crystal is cushioned with some cardboard, then if the top of the crystal is struck with a mallet, a voltage will develop on opposite surfaces (the direct effect).[43] This voltage is proportional to the applied force (Figure 5.17). In the reverse process (the converse effect), the dimensions of the crystal will change slightly if a voltage is applied in a specific direction. Thus the piezoelectric effect works both ways[44] and is a simple example of microscopic reversibility. The direct effect is linear with respect to strain, while the converse effect is linear with respect to electrical field. Walter Guyton Cady[45] notes that it is "only for historical reasons that the term direct is applied to one rather than the other of these two effects." Of the 32 crystal classes, 20 (lacking a center of symmetry) are piezoelectric. Only one noncentrosymmetric crystal class (432) is nonpiezoelectric.

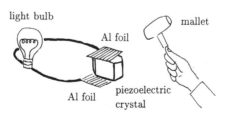

FIGURE 5.17. Measurement of the piezoelectric effect. A crystal is placed between metal plates and these are struck with a mallet. The changes developed as a result cause a current to flow.

A piezoelectric crystal contains moveable ions that, in changing position, cause a change in the shape of the crystal. As pressure is applied to a piezoelectric crystal, the crystal changes shape slightly by the movement of ions. Some of the positive and negative ions move in different directions. When the pressure is released, the ions move back to their original positions and the current moves in the opposite direction. Conversely, if the piezoelectric crystal is placed in an electric field, the positive and negative ions move toward the opposite electrodes, thereby changing the shape of the crystal.

The piezoelectric effect has practical applications. It can be used to produce selected amounts of electric charge by controlling how the piezoelectric crystal is deformed. An alternating pressure can therefore be converted to an alternating voltage, as in a phonograph pickup. Conversely, an alternating voltage can be converted into an alternating pressure as in the generation of supersonic waves. A quartz plate cut with a defined orientation and dimension has a natural frequency of expansion and contraction, that is, a natural frequency of vibration. The frequency is a function of the thickness of the quartz plate and is very high, millions of vibrations per second. If the applied field is made to alternate with this same frequency, the quartz plate will respond with a vigorous mechanical resonant vibration. Then the applied electric oscillation will be augmented by the piezoelectric effect. This provides a way of converting mechanical energy into electrical energy and is particularly effective at high frequencies of stress. For example, quartz crystals have been used for crystal-controlled watches and clocks. Quartz has other interesting properties, such as **triboluminescence**.[46,47] If quartz rocks are scratched together in the dark, it is possible to observe a flash of light. The source of this light is an electric arc resulting from the development of charges on the surface of the quartz.

Cady[45] in World War II realized that such a mechanical resonance of a vibrating crystal could be used in frequency control. This discovery had an important influence on radio communications.[48,49] Alternating electric fields, such as those generated by the radio tubes of the time, were applied to plates of piezoelectric crystals and the expansions and contractions of the plates were caused to react on electrical circuits. If the natural frequency of the mechanical vibration of the quartz plate coincided with the frequency of oscillation of the electric circuit, resonance between the two took place and energy was acquired by the mechanical oscillators. Later, Rochelle salt and barium titanate, which are each both ferroelectric and piezoelectric, were used.[50-52] In ferroelectric crystals, the polarization or dipole moment is reversed or reoriented upon application of an electric field. Ferroelasticity is another property displayed by some crystals in which stress can cause the interconversion between two stable orientational states. These physical properties of crystals are of great use in modern technology.

5.3.2 Pyroelectricity

Temperature changes can cause certain crystals to develop charges at their opposite ends. The **pyroelectric** effect, described and named by Sir David Brewster,[53] is the development of electric charges on the surfaces of certain noncentrosymmetric and optically active crystals that have a polar axis, in response to a temperature change. This effect is shown in 10 noncentrosymmetric point groups — those that possess a unique direction (polar axis). Traders in precious gemstones noted, for example, that the mineral tourmaline, on heating on coals to test its hardness, attracted pieces of paper or cinders.[54-56] On heating, one end of tourmaline crystals (see Figure 5.18) called the "analogous pole" becomes positively charged and the other end, called the "antilogous pole" becomes negatively charged. If the crystal is passed through a flame to remove these charges, it develops charges in the opposite sense on cooling.[57] The locations of positive and negative charge can be found[58-60] by blowing certain colored powders through fine muslin at the crystal. Yellow sulfur powder (itself negatively charged) will be attracted to the positively charged end of the pyroelectric crystal and other, differently colored powders (such as red carmine or red lead, Pb_3O_4, positively charged) will be attracted to the negative end of the crystal (Figure 5.18).

The pyroelectric effect results from the electric charge separation resulting from the stress caused by the temperature change. A small potential difference, sometimes too small to measure, develops across

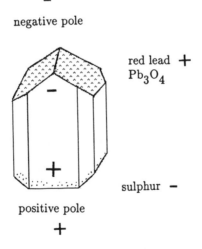

FIGURE 5.18. The pyroelectric effect. Charges develop on opposite faces on heating a pyroelectric crystal. These changes can be located by suitable powders, as shown.

the crystal analogous to the piezoelectric effect described earlier. This temperature-dependent effect results in a reversible change in dipole moment (one end acquires a positive charge and the other end a negative charge). The electric field is generally concentrated at a point and therefore if a dust particle settles on the crystal another dust particle will settle on top of it and the result will be a filamentous growth on the surface of the crystal.

A scheme for the careful optical examination of crystals, incorporating the different ways in which they can interact with light, is summarized in Table 5.3.

TABLE 5.3. Symmetry information from physical studies of crystals.

---

1. **Morphology:** *Measure the angles between crystal faces if they are well developed. Drawing a stereographic projection (see Chapter 2) may help. Note any cleavage directions and any tendency to form twins. Sometimes etching of the crystal faces will reveal their symmetry.*

2. **Color:** *Note any color change with direction of view (pleochroism).*

3. **Rotation of crystal between crossed Nicol prisms:** *If a sample is dark at all angles of rotation, it is optically isotropic and is either a cubic crystal or amorphous. If a crystal appears alternately dark and light as it is rotated, it is optically anisotropic (triclinic, monoclinic, orthorhombic, tetragonal, trigonal, hexagonal).*

4. **Refractive index measurements:** *Uniaxial (two refractive indices) (tetragonal, trigonal, hexagonal). Biaxial (three refractive indices) (triclinic, monoclinic, orthorhombic). Note the directions of the refractive indices with respect to the crystal shape and to the unit cell edge directions (if known).*

5. **Interference figures:** *These will indicate whether the crystal is optically positive or negative and, if biaxial, the angle between the optic axes.*

6. **Optical activity:** *The crystal must belong to a noncentrosymmetric space group.*

7. **Pyroelectricity:** *The observation of a pyroelectric effect implies a noncentrosymmetric space group. It can only exist if there is a unique polar axis in the point group of the crystal. If no effect is observed, it is presumed (but not certain) that the crystal possesses a center of symmetry.*

8. **Piezoelectricity:** *This effect is only observed if the crystal is noncentrosymmetric.*

5.3.3 Nonlinear optical effects

The fact that the optical properties of materials can be altered by a strong electric field was noted by John Kerr[61,62] in 1875. For example, while glass is normally isotropic, when an electric field is applied, the refractive index of glass becomes anisotropic so that it has different values along the field direction and perpendicular to it. A similar behavior was noted by Friedrich Pockels in 1893 for certain crystals[63-65] that did not have a center of symmetry in their structure.

With the advent of laser technology, and the resulting intense and coherent beams, our ability to study the interaction of crystals with light has been greatly expanded. Just as an electric field can modify the optical properties of a material, so can an intense laser beam. The electric field of the light waves is comparable in magnitude to the electric field that binds electrons in atoms and molecules.

We mentioned earlier that light, by virtue of its electric field, induces a slight separation between positive and negative charges on each molecule in the material through which it is traveling. A small electric field forms as a result of this charge separation, opposing the incident light and thereby reducing the speed of light in the material. This reduction in the refractive index is not apparently dependent on the intensity of ordinary light. By contrast, laser light, which has an extremely high intensity, can produce an electric field that has a much more profound effect on the material than does ordinary light.

Peter Alden Franken in 1961 found that when a beam of red light from a ruby laser is passed through a quartz crystal, the emerging light has two components, one red and the second ultraviolet with exactly twice the frequency of the incident light.[66] This frequency doubling is described as the production of a **second harmonic**[67-71] (a term used in acoustics). Even higher order effects, third-order harmonics, for example, are also found to occur with certain materials. Substances showing these properties are referred to as **nonlinear optical materials**. The nonlinear optical effect is important because some of its properties can be used as an optical switch in the design of optical computers.

When a molecule is polarized in response to a local electrical field E, its dielectric polarization can be expressed as:

$$P = \alpha E + \beta E^2 + \gamma E^3 + \cdots \tag{5.1}$$

The linear coefficient α is associated with refraction and the refractive index, described earlier. The other terms represent nonlinear optical terms. If the crystal is centrosymmetric $\beta = 0$. Noncentrosymmetric crystals with a dipole moment show second harmonic generation or frequency doubling, that is, the conversion of coherent light, frequency ω, into light of frequency 2ω. For example, potassium dihydrogen phosphate crystals, when illuminated with light of wavelength 1064 nm, also give off light of wavelength 532 nm. Inorganic crystals, such as potassium dihydrogen phosphate and lithium niobate, are expensive and fragile and become cloudy and so are of limited use as second harmonic generators. Organic molecules show the effect if they contain "polar groups," such as amino and nitro groups, that pack in a noncentrosymmetric manner in the crystal. Large effects are found for molecules with asymmetric conjugated π-electron systems, such as is found in *p*-nitroaniline.[72,73] An organic compound with electron-rich (D) and electron-deficient (A) substituents gives an asymmetric charge distribution and high β (in Equation 5.1) in the D $\rightarrow$ A direction.[74] Unfortunately, those organic materials that have larger nonlinear optical coefficients are, in general, not sufficiently photochemically resistant for use in nonlinear optical devices.

Second harmonic generation has been used as a test for a noncentrosymmetric crystal.[75,76] An example is provided by 9-methyl-10-chloromethylanthracene which crystallizes with disorder in the positions of the methyl and chloromethyl group and therefore it was not clear whether the space group was centrosymmetric or noncentrosymmetric. The measurement of a strong second harmonic confirmed the latter choice.

The **photorefractive effect**[77] was first noted by Arthur Ashkin and co-workers in 1966 when they were analyzing second harmonic effects in lithium niobate ($LiNbO_3$) and lithium tantalate ($LiTaO_3$). A laser beam directed through the crystal caused, after a few minutes of normal transmission, a scattering of the light in all directions. The refractive index of the material had been altered by the intense laser light, as a result of a small change in the positions of atoms in the crystal.[78] The effect, originally viewed as a nuisance, is analogous to the photochromic effect,[79,80] in which the light striking the material can change the extent to which the medium absorbs light. The photochromic effect has been found to be very useful in the manufacture of certain types of sunglasses. A wide variety of materials are photorefractive; these include barium titanate (an insulator), gallium arsenide (a semiconductor) and 2-cyclooctylamino-5-nitropyridine doped with 7,7,8,8-tetracyanoquinodimethane. The requirement is that they form noncentrosymmetric crystals, that they contain defects, and that the crystal lattice can easily be distorted.

As the laser light passes through the photorefractive crystal, mobile electrons move away from the bright regions to darker areas. David M. Pepper, Jack Feinberg, and Nicolai V. Kukhtarev[78] wrote: "If light illuminates charges in one region of the crystal, they will diffuse away from that region and accumulate in the dark, in the way cockroaches scurry under furniture to avoid light. Each charge that moves inside the crystal leaves behind an immobile charge of the opposite sign. In the region between these positive and negative charges, the electric field is strongest, and the crystal lattice will distort the most. A beam of light that passes through this region of the crystal will experience a different refractive index from that of the unaffected regions." Mobile electrons are provided by crystal defects, which result in extra charge that responds to incident light. In an intense beam of light, these defects cause an electric field that distorts the crystal lattice. The speed of light will then be slower in some parts of the crystal and faster in others. As long as the crystal does not contain a center of symmetry, the change in refractive index is proportional to the strength of the electric field.

We noted earlier that when two light beams cross each other, photons in one beam do not affect those in the other beam. This is normally the case, but when the incident light beams are two intense laser beams, the first laser beam will alter the optical behavior of the material so that the second laser beam will behave differently from the first. When two laser beams of the same frequency intersect, they interfere and produce a pattern of light that varies sinusoidally through the crystal. If the crystal is photorefractive, the refractive index of the crystal will also be distorted in a sinusoidal manner (but with a 90° phase shift). As the beams pass through the crystal they may interfere with each other so that one beam, by constructive interference, gains intensity, while the other, by destructive interference, loses it. Effects such as these may provide important technology.

Summary

1. Crystals may show mechanical properties such as hardness, or cleavage, that provide information on their internal structure. Similarly a surface reaction to the environment is a result of the chemistry of the components on the crystal faces, and of the vapor pressure.
2. The quality of crystals, such as diamond, as gemstones, depends on their refractive indices and dispersion properties, that is, the variation of refractive index with wavelength.
3. A detailed optical examination of a crystal may reveal its symmetry.
    a. With crossed Nicol prisms it is possible to determine whether a crystal is isotropic or anisotropic.
    b. Refractive indices of isotropic crystals can be measured with ordinary light.
    c. The optical indicatrix can be used as a representation of the variation of refractive index with direction.
    d. Refractive indices of anisotropic crystals can be measured in polarized light.
    e. With convergent polarized light between crossed Nicol prisms, one can observe interference figures and determine the uniaxial or biaxial character of the crystal.
4. Light may be polarized by reflection or by passage through an anisotropic crystal.
5. Compounds that are optically active often form crystals with hemihedral faces.
6. If crystals possess physical properties such as cleavage or piezoelectric effects, information on their internal symmetry and/or structure may be obtained from measurements of these properties.

Glossary

Anisotropy: Variation of a physical property of a crystal with direction within that crystal.

Atomic and ionic refractivities: The contribution that each atom or ion makes to the total refractive index.

Biaxial crystals: Crystals that have two directions along which there is no double refraction (two optic axes). Such biaxial crystals are either orthorhombic, monoclinic, or triclinic, and have three principal refractive indices.

Birefringence: The difference between the refractive index for the extraordinary and ordinary rays ($n_E - n_O$) in a uniaxial or biaxial crystal. If a crystal, such as quartz, is positively birefringent, n_O is less than n_E, and the velocity of the ordinary ray is greater than that of the extraordinary ray. The reverse is true for a negatively birefringent crystal.

Brilliance: The extent to which a gem sparkles and returns light from within. It depends on the cut of the gem and its refractive indices.

Cleavage: The tendency of some crystals to break or be split readily in one or more definite directions to give a smooth surface, parallel to possible crystal faces. Cleavage indicates directions of weakest intermolecular forces in the crystal.

Critical angle: When light enters a medium of lower refractive index, if the angle of incidence exceeds the critical angle, light will not be refracted but will be reflected back into the initial medium, as if by a mirror. The critical angle is determined from the ratio of the refractive indices of the two media.

Crossed Nicol prisms: An arrangement of two materials (calcite polarizing prisms or Polaroid discs) such that the first, the polarizer, transmits plane–polarized light that passes through the second, the analyzer, only if oriented in specific ways.

Dextrorotation: The property of an optically active substance that results in the clockwise rotation of the plane of vibration of plane–polarized light during its transmission through the substance.

Dispersion: Variation of the velocity of light (and hence its refractive index) in a material (such as a crystal) as a function of the wavelength of the light. As a result of dispersion (for example, by a prism), white light is split (dispersed) into its component colors. The velocity of light in a medium usually increases smoothly as the wavelength increases, but if the incident radiation is strongly absorbed the curve becomes discontinuous at that point (anomalous dispersion).

Double refraction, see **Birefringence**

Enantiomorph: A molecule or crystal that is not superimposable on its mirror image (from the Greek: *enantio*, opposite). One of a pair of chiral objects related by mirror symmetry.

Extraordinary ray: A refracted ray in biaxial crystals which does not obey the ordinary laws of refraction because its velocity varies with direction through the crystal.

Fire: Dispersion of light to give various colors in a colorless transparent gem.

Hardness: The resistance of a material to scratching. It is measured on Mohs' scale (1–10). Talc (1), gypsum (2), calcite (3), fluorite (4), apatite (5), orthoclase (6), quartz (7), topaz (8), corundum (9), and diamond (10).

Hemihedry: Crystals with reduced symmetry, as opposed to holohedry (full symmetry of the class).

Holohedry: Crystals that display the full symmetry of the class.

Interference figure: Characteristic patterned images obtained when a crystal is viewed with convergent plane-polarized light by a microscope under special conditions. These images can be used to reveal the optical character of anisotropic crystals. The figures are different for uniaxial or biaxial crystals.

Isogyre: The black or gray areas of an interference figure (q.v.). They change position as the microscope stage is rotated.

Isotropy: Description of a property that is independent of the direction in which it is observed in a crystal.

Levorotation: The ability of an optically active substance to rotate plane–polarized light during its transmission through the substance such that the emergent beam has the plane of vibration rotated in a counterclockwise manner with respect to the incident beam.

Nonlinear optics: Effects resulting from the nonlinear portion of the equation connecting the polarization of a molecule with the applied field. Such effects are now observable as a result of the intense sources produced by laser technology.

Optic axis: A direction in a birefringent crystal along which the ordinary and extraordinary rays travel with the same velocity so that only one image is seen. Uniaxial crystals have one such axis; biaxial crystals have two.

Optical activity, optical rotatory power: The ability to rotate the plane of polarized light. This property is found for substances that do not have either a plane or center of symmetry. The light, viewed toward the source, may be rotated in a left-handed (levorotatory) or right-handed (dextrorotatory) direction. The degree of rotation depends on the substance, the wavelength of light, and the thickness of the crystal. Equal amounts of enantiomorphs rotate the plane of polarization the same amount but in opposite directions.

Optical indicatrix: A three-dimensional ellipsoid whose surface is defined by vectors from an origin. These vectors have lengths proportional to the refractive indices of a crystal in those directions.

Ordinary ray: A refracted ray that obeys the laws of refraction. It moves with equal velocity in all directions through a crystal.

Photorefractive effect: A change in refractive index in response to intense (laser) light.

Piezoelectric effect: The generation of a small potential difference across certain crystals when they are subjected to stress (direct effect), or the change in shape of a crystal that accompanies the application of a potential difference across a crystal (inverse effect). The piezoelectric effect is only shown by noncentrosymmetric crystals.

Plane-polarized light: Electromagnetic radiation in which the electric vectors of the waves are confined to a single plane.

Pleochroism: The variation of the color or intensity of color of a crystal as a function of direction of view through the crystal.

Polar axis: An axis in the crystal that exhibits different properties at its two ends (meaning that it has directionality, like an arrow). These properties include face development and charge accumulation, and may be used to define the directionality of the axis.

Polarization of atoms: The separation of charges in response to some stimulus.

Polarization of light: Polarization of light refers to the situation when the electric vector of electromagnetic radiation is no longer equally distributed throughout the plane normal to the direction of the propagation of light.

Pyroelectricity: The development of a small potential difference across certain crystals as a result of a change in temperature.

Refraction: A change in the direction of a light beam when it passes from one material to another in which it has a different velocity.

Schlieren effect: Regions in a fluid that have a density (and hence refractive index) that is different from that of the bulk. As a result, streaks are observed when light is viewed across the liquid.

Second harmonic generation: Frequency doubling as light passes through a noncentrosymmetric medium.

Snell's Law: For light passing between two transparent media, the ratio of the sines of the angle of incidence and the angle of refraction is a constant.

Total internal reflection: The reflection of light inside a material because of the incident angle of the light and the refractive index (see critical angle).

Triboluminescence: A flash of light produced as the result of friction between certain crystals, such as quartz. It is also called piezoluminescence, or fractoluminescence.

Uniaxial crystals: Crystals that are characterized by having one (and only one) direction along which there is no double refraction (one optic axis). These crystals are tetragonal, hexagonal, or rhombohedral, and have two principal refractive indices.

References

1. Kittel, C. *Introduction to Solid State Physics.*   6th edn. Wiley: New York (1986).
2. West, A. R. *Solid State Chemistry and Its Applications.* John Wiley: Chichester, New York, Brisbane, Toronto, Singapore (1987).
3. Nye, J. F. *Physical Properties of Crystals. Their Representation by Tensors and Matrices.* Paperback edn. of 1957 book, revised. Clarendon Press: Oxford (1985).
4. Azároff, L. V. *Introduction to Solids.* McGraw-Hill: New York, Toronto, London (1960).
5. Abrahams, S. C. Physical properties and atomic arrangement. In: *Crystallography in North America.* (**Eds.**, McLachlan, D., Jr., and Glusker, J. P.) Section E. Internal properties of matter. Chapter 1, pp. 291–293 American Crystallographic Association: New York. (1983).
6. Mohs, F. *Grundriss der Mineralogie, 1.* [Principles of mineralogy.] p. 374. Arnold: Dresden (1822).
7. Evans, R. C. *An Introduction to Crystal Chemistry.* 2nd edn. Cambridge University Press: Cambridge, UK (1966).

8. Wooster, N. The correlation of cleavage and structure. *Science Progress* **26**, 462–473 (1932).

9. Glusker, J. P., Patterson, A. L., Love, W. E., and Dornberg, M. L. X-ray crystal analysis of the substrates of aconitase. IV. The configuration of the naturally occurring isocitric acid as determined from potassium and rubidium salts of its lactone. *Acta Cryst.* **16**, 1102–1107 (1963).

10. Malus, E. Sur une propriété de la lumière réfléchie par les corps diaphanes. [Concerning a property of light reflected by transparent materials.] *Nouveau Bull. Scienc. Soc. Philomatique de Paris* (1807–1809), 266–269 (1808). *Mémoires de Physique et de Chimie de la Société d'Arcueil* **2**, 143–158 (1809).

11. Weber, H-J. Anisotropic bond polarizabilities in birefringent crystals. *Acta Cryst.* **A44**, 320–326 (1988).

12. Wahlstrom, E. E. *Optical Crystallography.* 5th edn. (1st edn., 1943). John Wiley: New York, Chichester, Brisbane, Toronto (1979).

13. Feynman, R. P., Leighton, R. B., and Sands, M. *The Feynman Lectures on Physics. Mainly Mechanics, Radiation and Heat.* Addison-Wesley: Reading, MA, Palo Alto, London (1963).

14. Reichert, J. D. Refraction. In: *Encyclopedia of Physics.* (**Eds.**, Lerner, R. G., and Trigg, G. L.) pp. 861–862. Addison-Wesley: Reading, MA (1981).

15. Wood, E. A. *Crystals and Light. An Introduction to Optical Crystallography.* Van Nostrand: New York (1977).

16. Kraus, E. H., and Slawson, C. B. *Gems and Gem Materials.* McGraw-Hill: New York (1947).

17. Shubnikov, A. V. *Principles of Optical Crystallography.* **English translation (from the Russian):** Consultants Bureau: New York (1960).

18. Bartholinus, E. *Experimenta crystalli islandici disdiaclastici quibus mira et insolita refractio detegitur.* [Experiments made on a crystal-like body sent from Iceland.] Daniel Paulli: Hafniae, Copenhagen (1669). *Phil. Trans. Roy. Soc. (London)* **5**, 2039–2048 (1670).

19. Wollaston, W. H. On the oblique refraction of Iceland crystal. *Phil. Trans. Roy. Soc. (London)* **92**, 381–386 (1802).

20. Hecht, E. *Optics.* 2nd edn. Addison-Wesley: Reading, MA, and other international cities (1987).

21. Biot, J. B. Sur un nouveau genre d'oscillation que les molécules de la lumière éprouvent en traversant certains cristaux. [On a new type of oscillation that molecules of light show on traversing certain crystals.] *Mémoires de la Classe des Sciences Mathématiques et Physiques de l'Institut Imperial de France (Paris).* I. Part I. **13**, 1–372 (1812).

22. Biot, J. B. Sur la découverte d'une propriété nouvelle dont jouissent les forces polarisantes de certains cristaux. [On the discovery of a new property that polarizing forces show in certain crystals.] *Mémoires de la Classe des Sciences Mathématiques et Physiques de l'Institut Impérial de France (Paris)* Part II, 19–30 (1812).

23. Brewster, D. On the affections of light transmitted through crystallized bodies. Part I. *Phil. Trans. Roy. Soc. (London)* pp. 187–218 (1812).

24. Brewster, D. On the laws which regulate the absorption of polarized light by doubly refracting crystals. *Phil. Trans.* **1**, 11–28 (1819).

25. Fletcher, L. The optical indicatrix and the transmission of light in crystals. *Min. Mag.* No. 44 (1891).

26. Dent Glasser, L. S. *Crystallography and its Applications.* Van Nostrand Reinhold: New York, Cincinnati, Toronto, London, Melbourne (1977).

27. Sundararajan, K. S. Optical studies on organic crystals. *Z. Krist.* **93**, 238–248 (1936).

28. Bragg, W. L. The refractive indices of calcite and aragonite. *Proc. Roy. Soc. (London)* **A105**, 370–386 (1924).

29. Rubin, B. H., Stallings, W. C., Glusker, J. P., Bayer, M. E., Janin, J., and Srere, P. A. Crystallographic studies of *E. coli* citrate synthase. *J. Biol. Chem.* **258**, 1297–1298 (1983).

30. Nicol, W. On a method of so far increasing the divergency of the two rays in calcareous spar that only one image may be seen at a time. *Edinburgh New Phil.* **6**, 83–84 (1829).

31. Land, E. H. Some aspects of the development of sheet polarizers. *J. Opt. Soc. America* **41**, 957–963 (1951).

32. Hartshorne, N. H., and Stuart, A. *Crystals and the Polarizing Microscope. A Handbook for Chemists and Others.* 2nd edn. Edward Arnold: London (1950).

33. Arago, F. Mémoire sur une modification remarquable qu'éprouvent les rayons lumineux dans leur passage à travers certains corps diaphanes et sur quelques autres nouveaux phénomènes d'optique. [Note on a remarkable phenomenon shown by light rays on their passage through certain transparent materials, and on other new optical phenomena.] *Mémoires de la Classe des Sciences Mathématiques et Physiques de l'Institut Impérial de France (Paris)* **12**, 93–134 (1811).

34. De Vries, A. Determination of the absolute configuration of α-quartz. *Nature (London)* **181**, 1193 (1958).

35. Herschel, J. F. W. On the rotation impressed by plates of rock crystal on the planes of polarisation of the rays of light, as connected with certain peculiarities in its crystallisation. *Trans. Camb. Phil. Soc.* **1**, 43–52 (1821).

36. Pasteur, L. Mémoire sur la relation qui peut exister entre la forme cristalline et la composition chimique, et sur la cause de la polarisation rotatoire. [Note on the relationship of crystalline form to chemical composition, and on the cause of rotatory polarization.] *Comptes Rendus, Acad. Sci. (Paris)* **26**, 535–538 (1848).

37. Pasteur, L. Recherches sur les relations qui peuvent exister entre la forme cristalline, la composition chimique, et le sens de la polarisation rotatoire. [Research on the relationships between crystalline form, chemical composition, and the sense of rotatory polarization.] *Annales de Chimie* **24**, 442–459 (1848).

38. Valasek, J. Piezoelectric and allied phenomena in Rochelle salt. *Phys. Rev.* **17**, 475–481 (1921).

39. Curie, J., and Curie, P. Développement, par pression, de l'électricité polaire dans les cristaux hémihèdres à faces inclinées. [Development of polar electricity in hemihedral crystals with inclined faces.] *Comptes Rendus, Acad. Sci. (Paris)* **91**, 294–295 (1880).

40. Curie, J., and Curie, P. Contractions et dilations produites par des tensions électriques dans les cristaux hémièdres à faces inclinés. [Contractions and expansions in hemihedral crystals with inclined faces produced by electrical forces.] *Comptes Rendus, Acad. Sci. (Paris)* **93**, 1137–1140 (1881).

41. Beevers, C. A., and Hughes, W. The crystal structure of Rochelle salt (sodium potassium tartrate tetrahydrate $NaKC_4H_4O_6 \cdot 4H_2O$). *Proc. Roy. Soc. (London)* **A177**, 251–259 (1941).

42. Mueller, H. Properties of Rochelle salt. *Phys. Rev.* **57**, 829–839 (1940).

43. Holden, A., and Singer, P. *Crystals and Crystal Growing*. Anchor Books, Doubleday: Garden City, NY (1960).

44. Bunn, C. *Crystals: Their Role in Nature and in Science*. Academic Press: New York, London (1964).

45. Cady, W. G. *Piezoelectricity. An Introduction to the Theory and Applications of Electromechanical Phenomena in Crystals*. McGraw-Hill: New York and London (1946). Revised edn. Dover Publications: New York (1964).

46. Walton, A. J. Triboluminescence. *Adv. Phys.* **26**, 887–948 (1977).

47. Sweeting, L. M., and Reingold, A. L. Crystal structure and triboluminescence. 1. 9-anthryl carbinols. *J. Phys. Chem.* **92**, 5648–5655 (1988).

48. Booth, C. F. The application and use of quartz crystals in telecommunication. *J. Inst. Elect. Eng.* **88**, 97–144 (1941).

49. Parrish, W. Role of X rays and crystallography in the manufacture of quartz oscillators in World War II. In: *Crystallography in North America*. (**Eds.**, McLachlan, D., Jr., and Glusker, J. P.) Section F. Applications to various sciences. Chapter 13. pp. 393–395. American Crystallographic Association: New York. (1983).

50. Mott, N. F., and Gurney, R. W. *Electronic Processes in Ionic Crystals*. Oxford University Press: Oxford (1940).

51. Megaw, H. D. *Ferroelectricity in Crystals*. Methuen: London (1957).

52. Thorp, J. H., and Buckley, H. E. Morphological and dielectric studies of some crystals of the Rochelle salt type. *Acta Cryst.* **2**, 333–337 (1949).

53. Brewster, D. Observations on the pyro-electricity of minerals. *Edinburgh J. Sci.* **1**, 208–215 (1824).

54. Aepinus, F. U. T. Mémoire concernant quelques nouvelles expériences électriques remarquables. [Note on several new remarkable electrical effects.] *Mémoires de l'Académie Royale des Sciences et Belles Lettres de Berlin*, **12**, 105–121 (1756).

55. Wilson, B. Experiments on the tourmaline. *Phil. Trans. Roy. Soc. (London)* **51**, 308–313 (1759).

56. Lang, S. B. *Sourcebook of Pyroelectricity*. Gordon and Breach: New York (1974).

57. Wooster, W. A. *A Text-book on Crystal Physics*. Cambridge University Press: London (1938).

58. Kundt, A. A. E. E. *Die neuere Entwicklung der Electricitäts-Lehre*. [Newer developments in electrical theory.] Otto Lange: Berlin (1891).

59. Patil, A. A., Curtin, D. Y. and Paul, I. C. Use of the pyroelectric effect to determine the absolute orientation of the polar axis in molecular crystals. *J. Amer. Chem. Soc.* **107**, 726–727 (1985).

60. Bhalla, A. S., Tongson, L. L. and Newnham, R. E. Relationship of crystallographic polarity to piezoelectric, pyroelectric and chemical etching effects in Li_2GeO_3 and $LiGaO_2$ single crystals. *J. Appl. Cryst.* **16**, 138–140 (1983).

61. Kerr, J. A new relation between electricity and light: dielectrified media birefringent. *Phil. Mag.* **50**, 337–348 (1875).

62. Kerr, J. A new relation between electricity and light: dielectrified media birefringent. (Second paper.) *Phil. Mag.* **50**, 446 (1875).

63. Pockels, F. Ueber die Aenderungen des optischen Verhaltens und die elastischen Deformationen dielektrischer Krystalle im elektrischen Felde. [Concerning the changes in the optical behavior and the elastic deformations of dielectric crystals in electric fields.] *Neues Jahrb. Mineral.* **7**, 201–231 (1891).

64. Pockels, F. Ueber die elastischen Deformationen piëzoelektrischer Krystalle im elektrischen Felde. [Concerning elastic deformations of piezoelectric crystals in electric fields.] *Neues Jahrb. Mineral.* **8**, 407–417 (1893).

65. Pockels, F. Ueber den Einfluss des elektrostatischen Feldes auf das optische Verhalten piëzoelektrischer Krystalle. [Concerning the influence of an electrostatic field on the optical behavior of piezoelectric crystals.] *Neues Jahrb. Mineral.* 241–255 (1894).

66. Franken, P. A., Hill, A. E., Peters, C. W., and Weinreich, G. Generation of optical harmonics. *Phys. Rev. Lett.* **7**, 118–119 (1961).

67. Tripathy, S., Cavicchi, E., Kumar, J., and Kumar, R. S. Organic materials for nonlinear optics. Part 1. *Chemtech* (October 1989) pp. 620–625 (1989).

68. Tripathy, S., Cavicchi, E., Kumar, J., and Kumar, R. S. Nonlinear optics and organic materials. Part 2. *Chemtech* (December 1989) pp. 747–752 (1989).

69. Kurtz, S. K., and Perry, T. T. A powder technique for the evaluation of nonlinear optical materials. *J. Appl. Phys.* **39**, 3798–3813 (1968).

70. Bloembergen, N. *Nonlinear Optics.* Benjamin: New York (1965).

71. Armstrong, J. A., Bloembergen, N., Ducuing, J., and Pershan, P. S. Interactions between light waves in a nonlinear dielectric. *Phys. Rev.* **127**, 1918–1939 (1962).

72. Greene, B. I., Orenstein, J., and Schmitt-Rink, S. All-optical nonlinearities in organics. *Science* **247**, 679–687 (1990).

73. Zyss, J. Nonlinear organic materials for integrated optics: a review. *J. Molec. Electronics* **1**, 25–95 (1985).

74. Williams, D. J. Organic polymeric and non-polymeric materials with large optical nonlinearities. *Angew. Chem., Int. Ed. Engl.* **23**, 690–703 (1984).

75. Dougherty, J. P., and Kurtz, S. K. A second harmonic analyzer for the detection of non-centrosymmetry. *J. Appl. Cryst.* **9**, 145–158 (1976).

76. Abrahams, S. C. Sensitive test for acentric point groups. *J. Appl. Cryst.* **5**, 143 (1972).

77. Ashkin, A., Boyd, G. D., Dziedzik, J. M., Smith, R. G., Ballman, A. A., Levinstein, J. J., and Nassau, K. Optically-induced refractive index inhomogeneities in $LiNbO_3$ and $LiTaO_3$. *Appl. Phys. Lett.* **9**, 72–74 (1966).

78. Pepper, D. M., Feinberg, J., and Kukhtarev, N. V. The photorefractive effect. *Sci. Amer.* **263** (4), 62–74 (1990).

79. ter Meer, E. Ueber Dinitroverbindungen der Fettreihe. [Concerning the dinitro compounds of lipids.] *Justus Liebigs Annalen der Chemie.* **181**, 1–22 (1876).
80. Dürr, H. Perspectives in photochromism: a novel system based on the 1,5-electrocyclization of hetero analogous pentadienyl anions. *Angew. Chem. Int. Ed., Engl.* **28**, 413–431 (1989).

6

Combining Waves to Obtain an Image

We have, so far in this book, described crystals, with their internal regularity and symmetry, and the manner by which they diffract X rays. Now we will show how the action of a lens can be simulated, as was diagrammed in Figure 1.2 (Chapter 1), so that the required magnified image of the arrangement of electron density around atoms in the crystal can be obtained.

In order to understand the procedures needed to obtain an image of a crystal on an atomic scale, it is best to use mathematical tools, such as the Fourier analysis, the Fourier synthesis and the Fourier transform. These greatly simplify our understanding of how to analyze a crystal diffraction pattern. In this Chapter we will describe the use of a Fourier synthesis to obtain an electron-density map, and the significance of a Fourier analysis in terms of the Bragg reflections measured in the diffraction experiment. In addition we will describe what a Fourier transform is, and its role as the "bridge" between the atomic arrangement in a crystal and its diffraction pattern.

6.1 Descriptions of waves

A periodic wave can be described by its amplitude, wavelength, and relative phase angle, as shown in Figure 3.1 (Chapter 3). The amplitude $[F(hkl)$ for a Bragg reflection with indices $hkl]$ is the height at the crest of a wave. It is equal to the square root of the intensity of the wave. The relative phase angle $[\alpha_{hkl}$ for a Bragg reflection with indices $hkl]$ is the position of the crest of the wave relative to some origin that has been chosen arbitrarily by the investigator. For convenience, the crystallographer usually represents a periodic wave by a vector, as shown in Figure 6.1(a). Thus the wave is represented by a line whose length is proportional to the amplitude, and whose orientation gives a measure

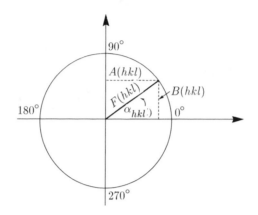

(a)

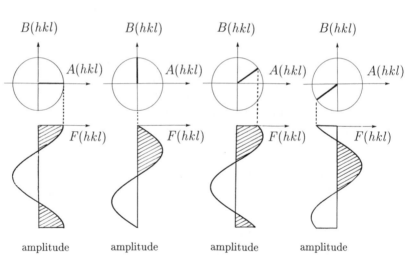

(b)

FIGURE 6.1. Geometric ways of depicting phase angles. (a) Representation of a wave by a vector **F(hkl)**, indicating relative phase angle α_{hkl} and amplitude, $|F(hkl)|$. The two components $A(hkl)$ and $B(hkl)$ of the wave are also shown. (b) The meaning of the phase angle in terms of the cosine component of the wave. Above is the vector diagram and below, moving vertically, is the wave. Note how the displacement of the wave at the origin is indicated by the value of $A(hkl)$. Analogous diagrams may be drawn for $B(hkl)$, represented by a wave that travels horizontally. The intensity at any point is the sum of the squares of the amplitudes of the A and B waves.

of the phase angle. It is usual to diagram the phase angle on a circle, with 360° representing one wavelength. By convention the phase angle is measured in a counterclockwise direction; if the relative phase angle is 0°, the line is horizontal, pointing to the right; if the phase angle is 90°, the line is vertical, pointing upwards [Figure 6.1(a)]. The line representing the vector of a Bragg reflection, therefore, has an angle α_{hkl} and a length $|F(hkl)|$. This vector can be considered to consist of two components: $A(hkl)$ parallel to the horizontal axis ($\alpha_{hkl} = 0°$), and $B(hkl)$ parallel to the vertical axis ($\alpha_{hkl} = 90°$), as shown in Figure 6.1(a).

Any periodic wave can be considered as a sum of cosine and sine waves [amplitudes $A(hkl)$ and $B(hkl)$, respectively]. The ratio of the amplitudes of the two waves gives a measure of the phase angle (Equation 6.1), and the sum of the squares of the amplitudes gives the intensity (Equation 6.2), which is the square of the amplitude.

$$\text{phase angle} = \tan \alpha_{hkl} = B(hkl)/A(hkl) \tag{6.1}$$

$$\text{intensity} = |A(hkl)|^2 + |B(hkl)|^2 = |F(hkl)|^2 \tag{6.2}$$

These two geometric equations are evident from an examination of Figure 6.1(a). The relationship between the vectorial representation and the actual wave (diagrammed, for convenience, as travelling vertically down the page) is shown in Figure 6.1(b). Note that pure cosine waves (relative phase 0°) and pure sine waves (relative phase 90°) of the same amplitude and frequency are related to each other merely by a change in the relative phase angle.

6.1.1 The meaning of the relative phase angle

The relative phase angle of a wave is a measure of a position of the crest of the wave under consideration with respect to some particular vantage point. This point may either be the crest of another wave traveling in the same direction, or the chosen origin of the unit cell. By convention, in X-ray crystallographic studies, the reference point for the measuring the relative phase of a Bragg reflection is generally the chosen origin of the unit cell, as shown in Figure 6.2.

The origin of the unit cell of the crystal is, however, merely a convenient geometrical construction that may be chosen in one of several ways (as was discussed in Chapter 4). This implies that there is no such thing as the absolute phase of a Bragg reflection. In practice, *the crystallographer makes some arbitrary choice of origin*, and the relative phase we use for a Bragg reflection is relative to this selected position for the origin. Such a choice will be described in more detail in Chapter 8, but we have already mentioned in Chapter 4 that the origin of a unit cell is best chosen with respect to any symmetry that is present in the crystal structure.

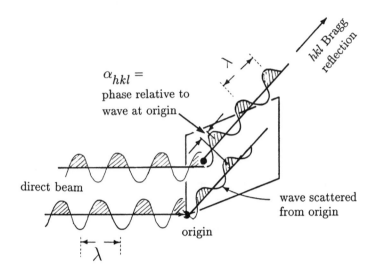

FIGURE 6.2. The meaning of the relative phase angle α_{hkl} of a Bragg reflection *hkl*. The relative phase is generally represented, as shown in Figure 6.1, as an angle (a fraction of a complete cycle, i.e., 360°). It is the phase of the wave of the *hkl* Bragg reflection (scattered by the atom represented by a black circle in the middle of the unit cell) relative to that of a similar wave traveling in the same direction, but scattered by an imaginary atom (also a black circle) at the origin of the unit cell. The relative phase of two waves is the difference between peaks (measured in a direction perpendicular to the direction of travel of the Bragg reflection, as shown in this Figure) to give α_{hkl}. Note that if the position of the atom in the middle of the unit cell had been selected as the origin of the unit cell, the Bragg reflection would have had a relative phase of 0°.

6.1.2 Graphical and algebraic representations of waves

When several periodic waves travel in the same direction they interfere with each other, and the overall effect is the wave obtained by summing these component periodic waves. There are several representations of waves that are convenient for mathematical summation, such as the geometrical diagrams in Figure 6.1, or the algebraic expressions in Figure 6.3, respectively (see Equation 6.3.1, for example). These methods of representing waves, equivalent to each other, are merely the ways that mathematicians, physicists and crystallographers have faced the problem of summing waves. This task is now greatly facilitated by the availability of computers.

The numerical summation of waves, illustrated in Figure 6.4, is quite straightforward. At each point along the horizontal axis, the displacement of each wave can be measured as the distance from this axis in a

The displacement of n cosine waves x_n can be represented as:

$$x_1 = c_1 \cos(\phi + \alpha_1) \tag{6.3.1}$$

$$x_2 = c_2 \cos(\phi + \alpha_2) \tag{6.3.2}$$

$$\cdots$$

$$x_n = c_n \cos(\phi + \alpha_n) \tag{6.3.3}$$

where c_n is the amplitude (the maximum displacement) of the wave, ϕ is proportional to the time since the traveling wave has passed the origin (a constant for a particular experiment), and α_n is the relative phase of the wave (see Figure 6.2). We assume coherent scattering so that the phase relationships are preserved and that the wavelengths of all waves are the same (monochromatic X rays).

When two waves are superimposed, the displacement x_r is the sum of the individual displacements:

$$x_r = x_1 + x_2 = c_1 \cos(\phi + \alpha_1) + c_2 \cos(\phi + \alpha_2) \tag{6.3.4}$$

$$x_r = (c_1 \cos \alpha_1 + c_2 \cos \alpha_2) \cos \phi - (c_1 \sin \alpha_1 + c_2 \sin \alpha_2) \sin \phi. \tag{6.3.5}$$

If we write the amplitude c_r and the phase α_r of the resulting waves such that:

$$c_r \cos \alpha_r = c_1 \cos \alpha_1 + c_2 \cos \alpha_2 = \sum_n c_n \cos \alpha_n \tag{6.3.6}$$

$$c_r \sin \alpha_r = c_1 \sin \alpha_1 + c_2 \sin \alpha_2 = \sum_n c_n \sin \alpha_n \tag{6.3.7}$$

then, substituting in Equation 6.3.5, we get:

$$x_r = c_r \cos \alpha_r \cos \phi - c_r \sin \alpha_r \sin \phi = c_r \cos(\phi + \alpha_r). \tag{6.3.8}$$

Therefore the wave obtained has the same frequency as the two original waves, and its phase is measured relative to the same origin as that of the original waves. This wave is expressed by the equation:

$$\tan \alpha_r = \frac{c_r \sin \alpha_r}{c_r \cos \alpha_r} = \frac{\sum_n c_n \sin \alpha_n}{\sum_n c_n \cos \alpha_n}. \tag{6.3.9}$$

The amplitude of the resultant wave c_r is:

$$c_r = [(c_r \cos \alpha_r)^2 + (c_r \sin \alpha_r)^2]^{\frac{1}{2}} = [(\sum_n c_n \cos \alpha_n)^2 + (\sum_n c_n \sin \alpha_n)^2]^{\frac{1}{2}}. \tag{6.3.10}$$

FIGURE 6.3. The mathematical (algebraic) method for summing waves.

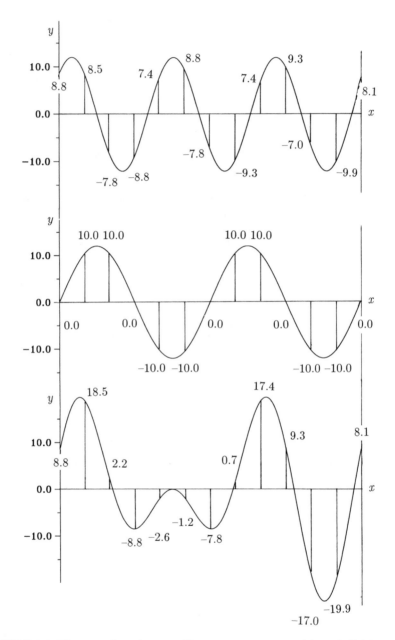

FIGURE 6.4. The summing of waves. Shown are two waves with periodicities 3 and 2 respectively, (3 0 0) and (2 0 0), and their sum. Values of the displacements of the waves, indicated by vertical lines, are summed to give the values shown.

direction parallel to the vertical axis. To sum two or more waves, the displacement values can be added point by point, and the shape of the composite wave can then be drawn. A vectorial example is shown in Figure 6.5.

6.2 The use of Fourier series

Any periodic function (such as the electron density in a crystal which repeats from unit cell to unit cell) can be represented as the sum of cosine (and sine) functions of appropriate amplitudes, phases, and periodicities (frequencies). This theorem[1] was introduced in 1807 by Baron Jean Baptiste Joseph Fourier (1768–1830), a French mathematician and physicist who pioneered, as a result of his interest in a mathematical theory of heat conduction, the representation of periodic functions by trigonometric series. Fourier showed that a continuous periodic function can be described in terms of the simpler component cosine (or sine) functions (a **Fourier series**). A **Fourier analysis** is the mathematical process of dissecting a periodic function into its simpler component cosine waves, thus showing how the periodic function might have been been put together. A simple

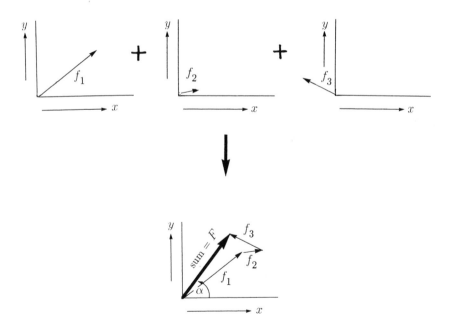

FIGURE 6.5. Summing waves by means of a vector diagram. Each wave is represented by a vector, as shown in Figure 6.1(a). The sum of these waves = vector **F** = the sum of the vectors $\mathbf{f}_1 + \mathbf{f}_2 + \mathbf{f}_3$. The phase angle of the sum is α.

example is illustrated in Figure 6.6. This type of analysis is also called a **harmonic** analysis and is used in acoustics to derive the fundamental and overtone (harmonic) components of a musical note.

A **Fourier synthesis**, illustrated in Figure 6.7, is the reverse of a Fourier analysis. A Fourier synthesis involves the summation of waves of known frequency, amplitude and phase in order to obtain a more complicated, but still periodic function. The relationship between a Fourier synthesis and a Fourier analysis is evident from the use of the same set of waves in both Figures 6.6 and 6.7. In a Fourier synthesis, if everything except the relative phases of the component waves are known, there will still be an almost infinite number of ways in which the waves can be combined.

6.3 Image formation by an optical microscope

Now we can consider how a microscope forms an image, and how this image formation can be simulated in a convenient way for use with data

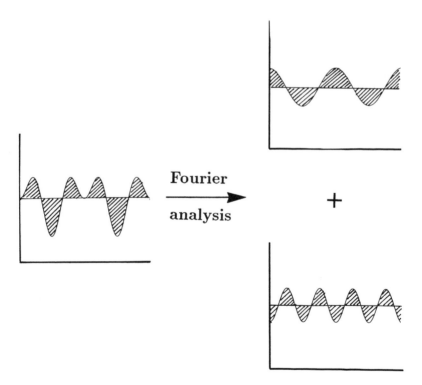

FIGURE 6.6. Fourier analysis. Take a regularly repeating function and break it down into its component simple cosine waves. Fourier analysis gives, for each component wave, its amplitude, its frequency, and its phase relative to a chosen origin.

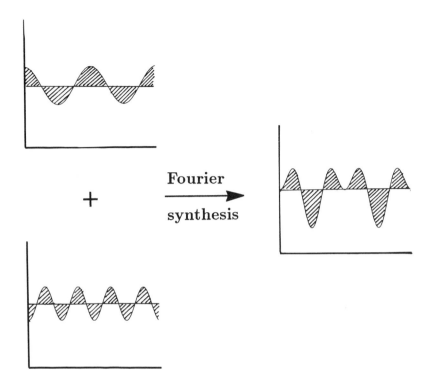

FIGURE 6.7. Fourier synthesis. Take a series of waves of different amplitudes, frequencies, and relative phases. Sum them. Because the waves are periodic, their sum is also periodic. Note the relative phase of each term.

from the X-ray diffraction pattern. Ernst Abbé at the University of Jena, Germany, worked in collaboration with Karl Zeiss on improvements to objective lenses in the optical microscope, and found that diffraction effects were important in optical microscopy.[2] These investigators observed that large apertures influenced the resolution obtained for the image. When an image of a periodic structure (a diffraction grating or a crystal structure) is formed in an optical microscope, the different diffracted beams are brought separately into focus on the back focal plane of the lens, as diagrammed in Figure 6.8. The reciprocal relationship between the spatial frequency (number of repeats per unit of pattern) and the order of a diffracted beam, that is, the number of wavelengths in the path difference, was described in Chapter 3. These waves then recombine in the image plane to give a reconstruction of the object being viewed. The more orders of diffraction that are combined in this way (for example, by having a larger aperture), the higher the resolution of the resulting image. For example, the resolution limit of half a wavelength,

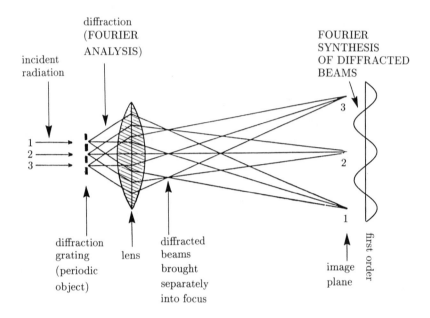

FIGURE 6.8. Imaging of a structure in the microscope. Diffracted beams are recombined in the image plane by a Fourier synthesis.

described in Chapter 1, is derived from the minimal requirement for at least the zero and first order diffracted beams for image formation. Ideally an infinite number of orders of diffraction are needed to form a perfect image, but this is not possible in practice, because of the limitations imposed by the wavelength of the radiation. Therefore the investigator has to make do with a less-than-infinite number of diffraction orders.

Albert B. Porter[3] continued the work of Ernst Abbé by analyzing the diffraction pattern of a finely ruled grating in terms of a Fourier series. First he considered the intensity of light passing through a diffraction grating as a step function. Light that interacts with the grating has zero amplitude where the grating is opaque (light cannot pass through it), and an appreciable amplitude where the grating is transparent (light can pass through it). Porter modeled these observations by a Fourier series and showed that each order of diffraction gives a cosine wave of illumination in the image plane. All waves, with differing amplitudes, frequencies and phases, are combined in a Fourier synthesis in the image plane of the optical microscope. *The frequencies of these waves to be summed in the image plane are their diffraction orders, and the amplitudes of the waves to be summed are the amplitudes of the diffracted beams* (the square roots of the measured intensities of the diffracted beams). Porter wrote of his analysis[3] "It thus appears that a diffraction grating performs a double process of harmonic analysis ⋯ it analyzes the distribution of wave-amplitude in its own plane, distributing the Fourier components of the amplitude curve in order among the successive spectra [Fourier analysis] ⋯ When a lens forms a real image of a grating, it does so by adding together in the focal plane the harmonic components of the diffracted light [Fourier synthesis] ⋯ "

In summary, in the action of an optical microscope,[3,4] as was shown in Figure 6.8, diffracted beams result from a Fourier analysis of the pattern of light passing through the object. The image of the object obtained by recombining these diffracted beams is a Fourier synthesis of the Fourier analysis of the object. Thus a Fourier analysis is involved in diffraction and a Fourier synthesis is involved in the formation of an image.

6.4 Formation of an image by X-ray diffraction

We will now consider how to simulate this method of image formation in the X-ray diffraction experiment where we have to use a mathematical replacement for the objective lens. The studies by Porter are of great importance because they show how the Bragg reflections give the amplitude components of a Fourier series representing the electron density in the crystal (**the electron-density map**). In effect, Fourier analysis takes place in the diffraction experiment, so that the scattering of X rays by the electron density in the crystal produces Bragg reflections, each with a different amplitude $|F(hkl)|$ and relative phase α_{hkl}.

6.4.1 Calculation of an electron-density map

The electron density in a crystal precisely fits the definition of a periodic function in which an exact repeat occurs at regularly fixed intervals in any direction (the crystal lattice translations). Therefore the electron density in a crystal with a periodicity d can be described by a Fourier synthesis in which each component cosine wave (which we will call an **electron-density wave**) has a periodicity (i.e., wavelength) d/n, and the amplitude of the nth-order Bragg reflection.

An electron-density map is calculated by a Fourier synthesis, as shown in Equation 6.3.

$$\rho(xyz) = \frac{1}{V} \sum_h \sum_k \sum_l |F(hkl)| \cos 2\pi(hx + ky + lz - \alpha_{hkl}). \quad (6.3)$$

This electron density $\rho(xyz)$, at a point x,y,z in the unit cell volume V, is expressed in electrons per cubic Å, and is highest near atomic centers. In the X-ray diffraction experiment the number of waves that must be summed in Equation 6.3 is large. The equation for the electron density at any point x,y,z in the unit cell then involves the **structure factor amplitude** $|F(hkl)|$, and the relative phase angle α_{hkl} of each Bragg reflection. The electron density can be plotted and contoured to give a representation of the object. In the summation of a series of waves, the contributions of the relative phase angles (Figure 3.3, Chapter 3) are as important as, and generally more important than, the amplitude of the waves.

If a single point within the unit cell is chosen with fractional coordinates x,y,z (distance $x\,a$ parallel to a, $y\,b$ parallel to b, and $z\,c$ parallel to c), the electron density $\rho(xyz)$ at that point is calculated by use of Equation 6.3. The right-hand side of this equation involves the summation of all measured Bragg reflections. If the intensities of 6,000 diffracted beams are measured, there will be 6,000 $|F(hkl)|$ values included in this summation for just one point. In practice, the electron density map is calculated by performing this summation at definite intervals of the asymmetric unit. We represent the electron density in electrons per Å^3, and the summation is divided by the unit-cell volume, V. An example is given in Figure 6.9.

An image of the object is obtained in this way by the summation of waves. The reader, however, needs to obtain a clear idea of the type of waves that are being summed. These are not X rays, but electron-density waves. Each Bragg reflection can be represented by an electron-density wave with a three-dimensional periodicity that is defined by its Miller indices hkl. For example, a Bragg reflection with indices 5 3 2 has a periodicity (frequency) of 5 along the a axial direction, 3 along the b axial direction, and 2 along the c axial direction, as shown in Figure 6.10. This should accentuate for the reader the fact that the waves added to obtain an electron density map (electron-density waves) are different from those (scattered waves of constant wavelength) summed in the analysis of diffraction effects. The scattered waves are electromagnetic radiation of known and constant wavelength. The electron-density waves have different periodicities (and hence wavelengths) depending on the order of diffraction they represent, as diagrammed in Figure 6.11.

6.4.2 Relationships of Bragg reflections to electron density

Each electron-density wave provides a component for summation to give the electron-density map, shown in Figure 6.11. *If the electron density of a crystal could be described precisely by a single cosine wave that repeats three times in the unit cell dimension d, then the electron density has a periodicity of d/3 and the diffraction pattern will have intensity only in the third order (only one diffracted beam, 3 0 0).*[5] This is the electron-density wave that is used in the summation that gives an electron-density map; if there is only one term because only one Bragg reflection is ob-

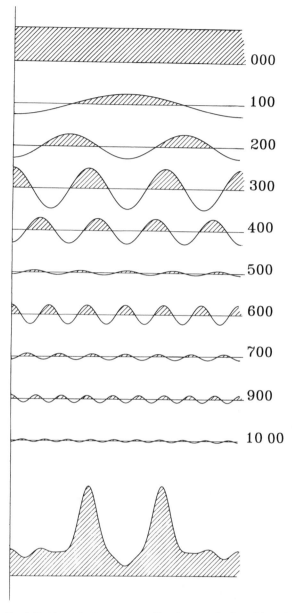

(a)

FIGURE 6.9. Contributions to terms in a Fourier synthesis. (a) Individual terms from 000 to 10 00 ($h = 10$, $k = 0$, $l = 0$) are represented with positive areas shaded. These ten electron-density waves combine to give the electron density shown at the bottom of the diagram. This electron density is dependent on the phases (+ = 0°, − = 180°) which are: 000 +; 100 −; 200 −; 300 +; 400 −; 500 −; 600 +; 700 −; 800 unobs'd; 900 +; 10 00 − (see Reference 6).

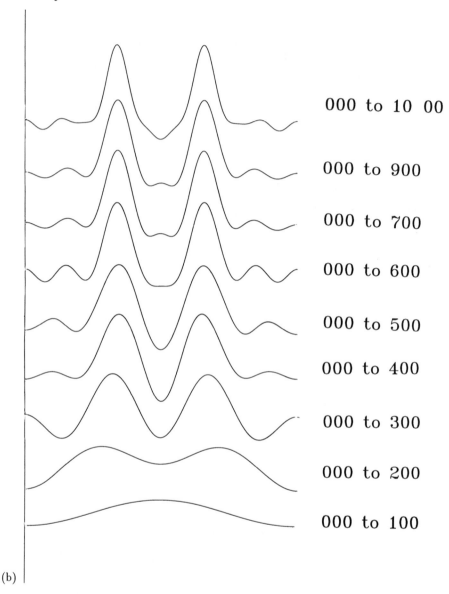

(b)

FIGURE 6.9 (cont'd). (b) Effects of successive addition of terms in (a). At the bottom of this diagram is shown the effect of adding the 000 and 100 terms [see Figure 6.9(a)]. As more terms are added the two peaks of electron density become more distinct (and the resolution of the image is increased). Note that, as a result of the various relative phases there is no negative electron density even though each of the electron-density waves (except 000) has equal positive and negative components. Note also that the 000 Bragg reflection adds a constant quantity to the electron-density map.

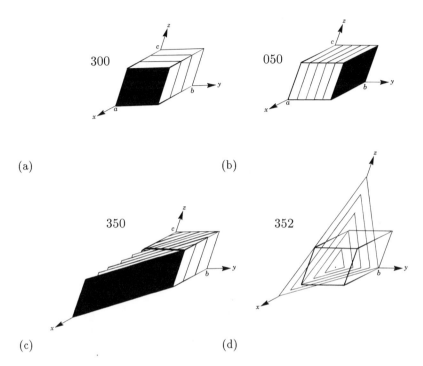

FIGURE 6.10. Periodicities of waves and their representation. Shown, for the same unit cell in each case, are various waves $F(hkl)$ with different periodicities. (a) 300, (b) 050, (c) 350, and (d) 352. Note that, for $F(352)$, the wave repeats 3 times in the a direction, 5 times in the b direction, and 2 times in the c direction.

served, the electron density will be a pure cosine wave, frequency 3 (see Figure 6.11). Bijvoet wrote:[5] "The single components have no physical reality and it is not therefore surprising that one half of such a wave represents a negative [electron density]. The negative density of one [electron-density] wave simply compensates a surplus of density produced by the other component [electron-density] waves at that particular place."

Conversely, Porter showed that, if only one order of the diffraction spectrum is observed (the Bragg reflection 3 0 0, for example), then the electron density that caused this diffraction can be represented by a single cosine wave with frequency $d/n = d/3$. Each electron-density component gives rise to one Bragg reflection the intensity of which corresponds to the square of the amplitude of this wave. The relationship between the diffracting intensity and the electron-density waves is well described in Reference 5. W. H. Bragg wrote[7] in 1915: "If we know the nature of

Electron-density waves hkl lie perpendicular to the sets of crystal lattice planes with Miller indices hkl. The wavelength of the electron-density wave is the spacing of hkl crystal lattice planes, i.e., d_{hkl}. The amplitude and the relative phase of the electron-density wave depends on the locations of the atoms in the unit cell. An example is given below for three electron-density waves:

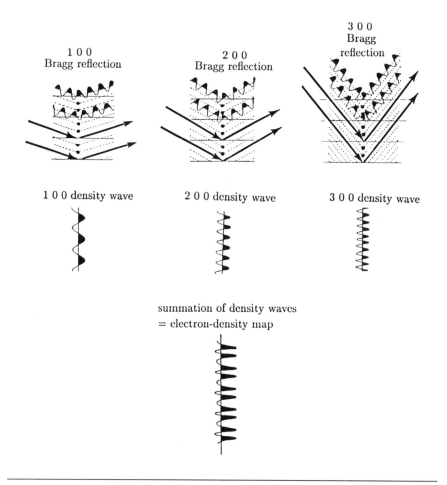

FIGURE 6.11. The meaning of electron-density waves from the Bragg reflections 100, 200, and 300, and their summation to give an electron-density map with peaks at atomic positions (PD means path difference). These electron-density waves have amplitudes and relative phases that depend on the atomic arrangement in the **a** direction. When the corresponding electron-density map is calculated, it should contain peaks in the positions corresponding to actual atomic positions.

the periodic variation of the density of the medium [the electron density in a crystal] we can analyze it by Fourier's method into a series of harmonic terms [amplitudes representing Fourier coefficients] $\cdots$ The series of spectra which we obtain for any given set of crystal planes may be considered as indicating the existence of separate harmonic terms." In this way the relationship between the order (hkl) of a Bragg reflection (see Figure 6.11) and its contribution to the electron density is established. These ideas were developed further by Paul S. Epstein and Paul Ehrenfest[8] in 1924, so that it became possible for William Duane[9] and Robert James Havighurst[10,11] to calculate the first one-dimensional electron density maps of sodium chloride.

6.5 Fourier transforms (between crystal and diffraction space)

We have shown that the electron density $\rho(xyz)$ (Equation 6.3) can be expressed as a Fourier series with the **structure factors** $F(hkl)$ as coefficients and with V as the unit-cell volume. In an analogous way, the structure factors can be expressed in terms of the electron density. The mathematical way of expressing these analogies involves **Fourier transforms**, also called Fourier inversions. There are many good sources for detailed information on Fourier transforms to which the reader is referred.[12-18] The expressions for Fourier transforms are given in Figure 6.12, and two examples are illustrated in Figure 6.13(a) and (b). A Fourier transform allows the crystallographer to flip between real space and diffraction (reciprocal) space. It is the mathematician's way of transforming Bragg reflection data to an electron-density map. The use of Fourier transforms is becoming more common in many experimental techniques used in chemistry and biochemistry, particularly in nuclear magnetic resonance and infrared and mass spectrometry.

The Fourier transform equations show that the electron density is the Fourier transform of the structure factor and the structure factor is the Fourier transform of the electron density. Examples are worked out in Figures 6.14 and 6.15. If the electron density can be expressed as the sum of cosine waves, then its Fourier transform corresponds to the sum of the Fourier transforms of the individual cosine waves (Figure 6.16). The "inversion theorem" states that the Fourier transform of the Fourier transform of an object is the original object, hence the opposite signs in Equations 6.12.1 and 6.12.2. This theorem provides the possibility of using a mathematical expression to go back and forth between reciprocal space (structure factors) and real space (electron density), so that the phrase "and vice versa" is applicable here.

Not only can one go back and forth when computing Fourier transforms, but one can also isolate and separate functions that are products of each other. The Fourier transform of the convolution of two functions is the product of their individual Fourier transforms (see Figure 2.17, Chapter 2, for information on convolutions). In other words, the diffrac-

In the pair of equations

$$f(x) = \int_{-\infty}^{\infty} e^{2\pi i x y} g(y) dy \qquad (6.12.1)$$

and

$$g(y) = \int_{-\infty}^{\infty} e^{-2\pi i x y} f(x) dx \qquad (6.12.2)$$

$g(y)$ *is the Fourier transform of* $f(x)$ *and* $f(x)$ *is the inverse transform (because of the negative sign) of* $g(y)$, *and* $i = \sqrt{-1}$. *In a similar way, the electron density* $\rho(xyz)$ *is the Fourier transform of the array of structure factors* $F(hkl)$ *and the array of structure factors is the inverse Fourier transform of the electron density.* V *is the unit-cell volume.*

Similarly:

$$\text{structure factor} = F(hkl) = \int_{V} \rho(xyz) e^{i\phi} dV \qquad (6.12.3)$$

and

$$\text{electron density} = \rho(xyz) = (1/V) \sum_{hkl} F(hkl) e^{-i\phi} dV \qquad (6.12.4)$$

where

$$\phi = 2\pi(hx + ky + lz) \qquad (6.12.5)$$

STRUCTURE FACTOR $\overset{\text{FT}}{\Longleftrightarrow}$ ELECTRON DENSITY

FIGURE 6.12. Mathematical expressions for Fourier transforms. These show that one can use a Fourier transform to convert structure factors (with phases) to electron density and electron density to structure factors with phases.

tion pattern of a crystal is the diffraction pattern of the contents of one unit cell (the molecular transform) multiplied by the Fourier transform of the crystal lattice (the reciprocal lattice). The diffraction pattern of two objects, for example, will look like the diffraction pattern of one object, such as the slit described earlier, multiplied by the transform of two points, which is a series of fringes. In this way we get the diffraction pattern of two slits such as that shown in Figure 3.7(h) and (i) (Chapter 3). Overall schemes are provided of the relationship of the diffraction pattern to the crystal structure (Figure 6.17) and of the method of crystal structure analysis (Figure 6.18).

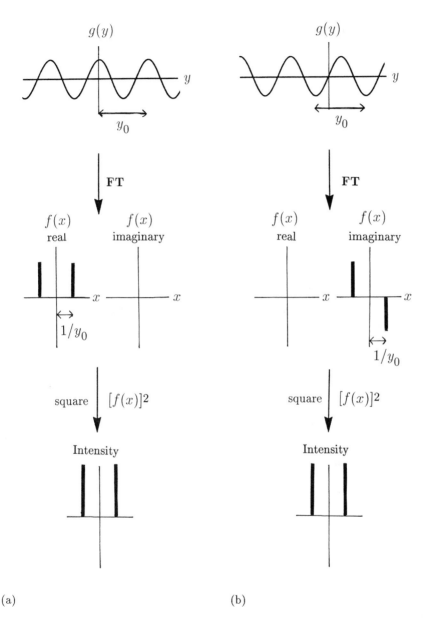

(a)

(b)

FIGURE 6.13. Graphical examples of the Fourier transforms of (a) a cosine and (b) a sine function. Note that the Fourier transform contains information on phase, but that this information is lost when intensities (which involve the square of the displacement) are measured. The designation "real" and "imaginary" derives from the presence of $i = \sqrt{-1}$ in the Fourier transform Equation 6.14.1.

The Fourier transform is

$$f(x) = \int g(y)[\cos(xy) + i\sin(xy)]dy. \qquad (6.14.1)$$

In this Figure we have chosen a centrosymmetric function, $g(y)$, that is a cosine function with a periodicity of 1. The integration has been approximated by a summation with small increments in y in order to demonstrate how a Fourier transform can be calculated.

$$f(x) \approx \Sigma g(y)\cos(xy). \qquad (6.14.2)$$

Note that only when $x = 1$ does the summation give a nonzero value. The reader can try this for other $g(y)$ functions, such as a cosine wave with a different periodicity form than used above, or a sine wave (see Figure 6.13).

		$x = 0$			$x=1$			$x = 2$		
y	$g(y)$	xy	$\cos(xy)$	$g(y)\times$ $\cos(xy)$	xy	$\cos(xy)$	$g(y)\times$ $\cos(xy)$	xy	$\cos(xy)$	$g(y)\times$ $\cos(xy)$
0.00	+1.00	0.00	1.00	+1.00	0.00	+1.00	+1.00	0.00	+1.00	+1.00
0.05	+0.95	0.00	1.00	+0.95	0.05	+0.95	+0.90	0.10	+0.81	+0.77
0.10	+0.81	0.00	1.00	+0.81	0.10	+0.81	+0.65	0.20	+0.31	+0.25
0.15	+0.59	0.00	1.00	+0.59	0.15	+0.59	+0.35	0.30	−0.31	−0.18
0.20	+0.31	0.00	1.00	+0.31	0.20	+0.31	+0.10	0.40	−0.81	−0.25
0.25	+0.00	0.00	1.00	+0.00	0.25	+0.00	+0.00	0.50	−1.00	+0.00
0.30	−0.31	0.00	1.00	−0.31	0.30	−0.31	+0.10	0.60	−0.81	+0.25
0.35	−0.59	0.00	1.00	−0.59	0.35	−0.59	+0.35	0.70	−0.31	+0.18
0.40	−0.81	0.00	1.00	−0.81	0.40	−0.81	+0.65	0.80	+0.31	−0.25
0.45	−0.95	0.00	1.00	−0.95	0.45	−0.95	+0.90	0.90	+0.81	−0.77
0.50	−1.00	0.00	1.00	−1.00	0.50	−1.00	+1.00	1.00	+1.00	−1.00
0.55	−0.95	0.00	1.00	−0.95	0.55	−0.95	+0.90	1.10	+0.81	−0.77
0.60	−0.81	0.00	1.00	−0.81	0.60	−0.81	+0.65	1.20	+0.31	−0.25
0.65	−0.59	0.00	1.00	−0.59	0.65	−0.59	+0.35	1.30	−0.31	+0.18
0.70	−0.31	0.00	1.00	−0.31	0.70	−0.31	+0.10	1.40	−0.81	+0.25
0.75	+0.00	0.00	1.00	+0.00	0.75	+0.00	+0.00	1.50	−1.00	+0.00
0.80	+0.31	0.00	1.00	+0.31	0.80	+0.31	+0.10	1.60	−0.81	−0.25
0.85	+0.59	0.00	1.00	+0.59	0.85	+0.59	+0.35	1.70	−0.31	−0.18
0.90	+0.81	0.00	1.00	+0.81	0.90	+0.81	+0.65	1.80	+0.31	+0.25
0.95	+0.95	0.00	1.00	+0.95	0.95	+0.95	+0.90	1.90	+0.81	+0.77
		sum = 0.0			sum = 10.0			sum = 0.0		

FIGURE 6.14. A numerical example of a Fourier transform.

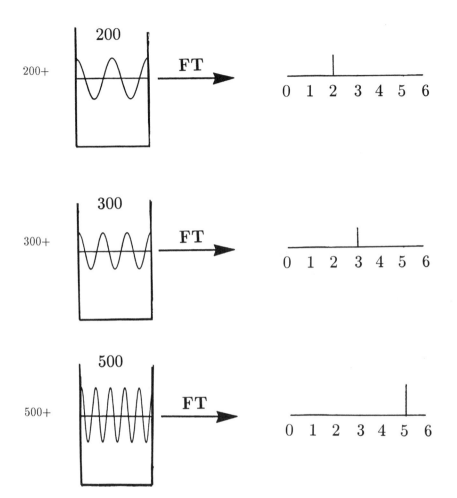

FIGURE 6.15. The Fourier transforms (**FT**s) of the 200, 300, and 500 electron-density waves. Shown on the left is one unit cell and an electron-density wave, and, on the right, its Fourier transform (Bragg reflection).

The Fourier transform thus provides the bridge between Bragg reflections and the electron-density map. The Bragg reflection, order n, contributes to the electron-density map (Equation 6.3) an electron-density wave with a periodicity d/n. For example, the Bragg reflection 300 represents an electron-density wave that repeats three times in the **a** direction of the unit cell and has an amplitude $|F(300)|$. If only the 300 Bragg reflection is observed, then the electron density repeats three times

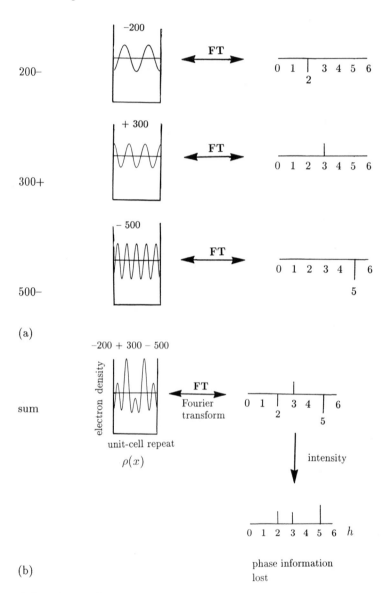

FIGURE 6.16. Summing Fourier transforms of the Fourier components of an electron-density map to get a representation of a diffraction pattern. (a) The Fourier transforms of three electron-density waves (200, 300, and 500). (b) The sum of the electron-density waves in (a) (with relative phases 200, $\alpha_{hkl} = 180°$, 300, $\alpha_{hkl} = 0°$, and 500, $\alpha_{hkl} = 180°$) give a diffraction pattern with intensity at 200, 300, and 500, but no phase information. The sign of the results of the Fourier transformation are lost when the values of $|F(hkl)|^2$ are calculated.

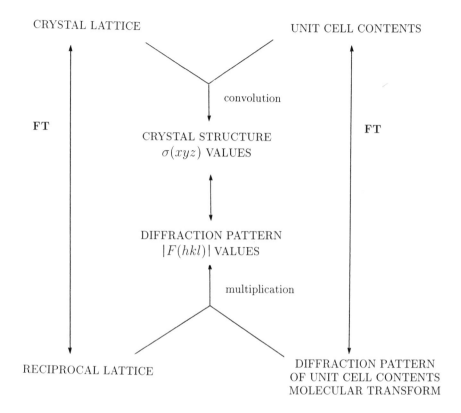

FIGURE 6.17. Scheme of the relationship between a crystal structure and its diffraction pattern in terms of Fourier transforms (FT), convolutions, and multiplications.

in the unit-cell repeat (equal to a). If the 200, 300 and 500 Bragg reflections are all observed, as shown in Figure 6.16, the electron density must equal the sum of the three cosine waves (electron-density waves). When all such density waves are summed with appropriate relative phases, the electron density in the crystal is obtained. This was shown in Figure 6.9 in which contains ten different electron-density waves and their sum are illustrated. Note that each electron-density wave has positive and negative areas that add or subtract from the sum. Therefore, when we measure the order (hkl) and amplitude of each diffracted beam, we obtain almost all the information we need to calculate the electron-density map. Only the relative phases of each term in the Fourier series are missing,

(a)   *Grow a crystal. This has a regular arrangement of atoms, e.g., at x = 0.3517 in a unit cell repeat of 8.88 Å (Reference 6).*

(b)   *The diffraction pattern of the crystal give the following structure amplitudes:*

1 0 0	52	6 0 0	45
2 0 0	57	7 0 0	14
3 0 0	96	8 0 0	unobs'd
4 0 0	59	9 0 0	17
5 0 0	10	10 0 0	6

(c)   *The relative phases of each electron-density wave (+ = 0°, − = 180°) are found. These phases are calculated from the deduced atomic arrangement.*

1 0 0	−52	6 0 0	+45
2 0 0	−57	7 0 0	−14
3 0 0	+96	8 0 0	unobs'd
4 0 0	−59	9 0 0	+17
5 0 0	−10	10 0 0	−6

(d)   *Calculate the electron-density map (see Figure 6.9.)*

(e)   *Put atoms at peaks in this map.*

The structure obtained should correspond to the real structure shown in (a) above.

FIGURE 6.18. Overall scheme of a crystal structure determination in one dimension. (a) Atomic structure, (b) Bragg reflections that are measured, (c) phases assigned to give electron-density waves with the correct phases, $\alpha(hkl)$, and amplitudes, $|F(hkl)|$, (d) the summation of density waves to give an electron-density map, and (e) this electron-density map has peaks at atomic positions (compare with the situation in (a), the true atomic arrangement).

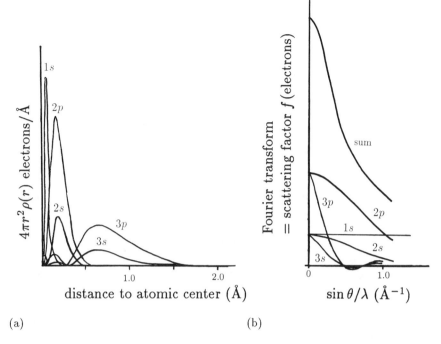

FIGURE 6.19. Fourier transforms of atomic orbitals give atomic scattering factors. In (a) is shown the theoretical electron density of various atomic orbitals, and in (b) is shown the Fourier transform of these.

because intensities rather than amplitudes are, in fact, measured. Methods of compensating for this deficiency will be described in Chapter 8.

If the electron density is known correctly, then structure factors and their relative phases can be computed by Fourier transform techniques. The calculation of X-ray scattering factors from the computed orbital electron densities as a function of distance from the nucleus, shown in Figure 6.19, provides an example of this. In a crystal structure analysis it is possible, from the measured diffraction pattern (structure factors and their phases) to compute the Fourier transform and thereby obtain an image of the entire crystal structure. In practice, only the contents of one unit cell are computed because the reciprocal lattice is the Fourier transform of the direct lattice and vice versa, so that the two transforms can be multiplied (Figure 6.17).

6.6 Modifications of the diffraction pattern

The diffraction pattern is affected by various factors that depend on crystal quality and experimental conditions. These will be examined in turn.

6.6.1 Kinematical and dynamical diffraction

Most crystals are composed of **mosaic blocks** about 10^{-5} cm in diameter (10^3 Å), each such block slightly misoriented with respect to its neighbors. The more mosaic the crystal, the more the blocks are misaligned with respect to each other, deviations as large as several minutes of an arc being found. Therefore a crystal containing mosaic blocks is considered not "perfect," since the internal periodicity is not exact. If a monochromatic beam of X rays were to impinge on a "perfect" crystal, only a few diffracted beams, having the correct relationship of d, λ, and θ implied by the Bragg Equation 3.2 (Chapter 3), would be observed. If the crystal is rotated, other diffracted beams will be observed. In each case the Bragg equation must be satisfied. In practice, however, X-ray beams are never truly monochromatic and crystals have "mosaic spread." As a result, Bragg reflections occur over a small range of θ rather than at just one specific value of θ.

The simplest type of crystal diffraction, and that considered here, is called **kinematical diffraction**. In such diffraction the X-ray beam, once diffracted, is not further modified by additional diffraction in its passage through the crystal. The phase differences between radiation scattered at different points in the crystal depend only on differences in the path lengths of the incident and diffracted waves. The summation of these waves with the appropriate amplitudes and relative phases determines the intensities.

Kinematical theory, however, fails to account correctly for multiple scattering events in the crystal. These cause a discrepancy between the measured intensity and that calculated by kinematical theory. This discrepancy is called **extinction** and is addressed by an alternate theory, dynamical theory. When **dynamical diffraction** occurs, it is necessary to take into account the dynamical interaction of the incident and various scattered beams as repeated scattering occurs within the crystal. This effect is significant for perfect crystals and, with electron diffraction, is common for all crystals. The incident beam is modified as it passes through a single block of a "perfect" crystal, that is, one in which the unit cells and their contents are all in perfect register. The intensity of a beam diffracted by a well-developed (perfect) crystal is proportional to the structure factor amplitude (magnitude) $|F|$, rather than to its square, $|F^2|$, as usually found in crystals for which only kinematical diffraction applies. In a perfect crystal, part of the incident beam may be diffracted twice so that it returns to its original direction. It is, however, not now perfectly in phase with the incident (primary) beam, because its path length differs. This serves to reduce the intensity of the beam (primary extinction). When the crystal is mosaic, part of the incident beam will be diffracted by one mosaic block, and this part of the beam will therefore not be available for diffraction by a nearby block, which is accurately aligned with respect to the first. The second block, therefore, receives incident radiation with a slightly diminished intensity, and so the diffracted (secondary) beam is reduced in intensity accordingly (secondary extinction).[19] The effect of secondary extinction is for $|F|_o$, the observed structure amplitude, to be systematically smaller than $|F|_c$, the calculated structure amplitude. Crystals may be made less perfect by dipping them in liquid nitrogen, a treatment that increases the mosaicity.

The effect of dynamical diffraction on X-ray diffraction intensities has been demonstrated by the work of Paul P. Ewald, Charles Galton Darwin and Max von Laue.[20-23] They showed that the kinematical theory of X-ray diffraction was an oversimplified description of what is found to occur experimentally. A "perfect" or "ideal" crystal will diffract less radiation in a given direction than will a crystal that has a distinct mosaic spread. Most crystals are neither perfect nor totally mosaic but are generally nearer the latter. Diamonds, because of their hardness and general lack of flaws, provide crystals that are nearly perfect, so that the intensities of the X-ray beams diffracted from diamonds have been the subject of many experimental studies.[24] For example, the intensity for the 111 Bragg reflection of diamond (using Cu $K\alpha$ radiation) can be fifty times larger for a mosaic crystal than for a perfect (essentially non-mosaic) diamond of the same size; the diffraction from mosaic crystals is "kinematic" diffraction, whereas the diffraction from the perfect crystal is essentially "dynamical." Nearly all studies described in this book involve kinematical diffraction.

Dynamical theory also can be used to explain **double reflections**. The diminution of some intensities was noted by Ernst Wagner[25] in 1920, and the enhancement of others (from the German, *Umweganregung*, excitation by a roundabout path) was observed by Moritz Renninger[26] in 1937. The effect is now often referred to as the "Renninger effect" or "multiple diffraction," and it occurs when several sets of atomic planes, with different values of *hkl* are simultaneously brought into a position to diffract X rays; that is, when two or more reciprocal lattice points are simultaneously on the surface of the Ewald sphere. By the name *n* beam diffraction this effect is contrasted to the simple Bragg reflection, which is called a "two-beam reflection," because only the incident beam and one diffracted (reflected) beam are involved. In multiple diffraction (*n*-beam diffraction) the diffracted beams interact with each other, causing an increase or decrease in the intensity over that which would be measured if only simple Bragg-type diffraction were to occur. This is a significant and common cause of errors in intensity data measurements. The double reflection appears in the position in reciprocal space expected for a normal Bragg reflection, but is sharper in appearance than is an ordinary diffracted beam. The effect may cause an ambiguity in the space group determination if a systematically absent Bragg reflection appears to have intensity as a result of the double reflection. Specific double reflections, however, can be eliminated in most instances by reorienting the crystal.

6.6.2 Effects of the atomic arrangement: the structure factor

The major determinants of the intensity of a Bragg reflection are the number of electrons in the atoms and the extent to which X rays scattered by various atoms reinforce those X rays scattered by other atoms in the crystal structure. Sometimes, as was mentioned for sodium and potassium chlorides in Chapter 1, certain diffracted beams have no intensity because the waves scattered by two different atoms in the repeat unit interfere destructively. For the same crystal *other* diffracted beams (with different values of *hkl*) will be reinforced by scattering from these two atoms, as indicated by histograms in Figure 6.20. For some Bragg reflections, the cations and anions scatter constructively, giving appreciable values to $| F(hkl)|$, while, for other Bragg reflections the scattered waves interfere destructively giving low values for $| F(hkl)|$.

If there is only one atom in the unit cell, then its scattering is affected only by the scattering in the next unit cell, and therefore the entire structure will behave as a simple three-dimensional diffraction grating. It is then convenient to consider that this atom lies at the origin of the unit cell. The scattering is now only from the origin. We have defined the relative phase angle as the difference in the phases of a wave that has been scattered by an (imaginary) atom at the origin and that scattered by a chosen atom. In the example described here, the relative phase angle for each Bragg reflection will be zero. But it is equally reasonable to place this imaginary atom at an arbitrary position x,y,z in the unit cell. Then the intensities of each diffracted beam will be unchanged, but the relative phase angles will change. This is another way of saying that the relative phase angle is referred to an arbitrary origin, which may be changed if necessary. It reinforces the fact that intensities are experimental quantities, while the relative phase angles are quantities that have to be derived after an origin has been selected.

The structure factor $F(hkl)$ is the Fourier transform of the unit cell contents sampled at reciprocal lattice points, hkl. The structure factor amplitude (magnitude) $|F|$ is the ratio of the amplitude of the radiation scattered in a particular direction by the contents of one unit cell to that scattered by a single electron at the origin of the unit cell under the same conditions (see Chapter 3). The first report of the structure factor expression was given by Arnold Sommerfeld[27] at a Solvay Conference. The structure factor F has both a magnitude $|F(hkl)|$ and a phase α_{hkl}, relative to the origin of the unit cell (see Figure 6.21). As will be described in Chapter 7, $|F|$ is derived experimentally from the square root of the intensity of a Bragg reflection. The measured "observed" structure factor amplitudes are denoted by $|F(hkl)|_o$; those calculated for a proposed crystal structure (a model) are designated $|F(hkl)|_c$.

The equations for calculating $|F(hkl)|$ including its phase angle α_{hkl}, when the crystal structure is known are given in Figure 6.21. These show that once x_j, y_j, and z_j are derived for each atom in the unit cell, the

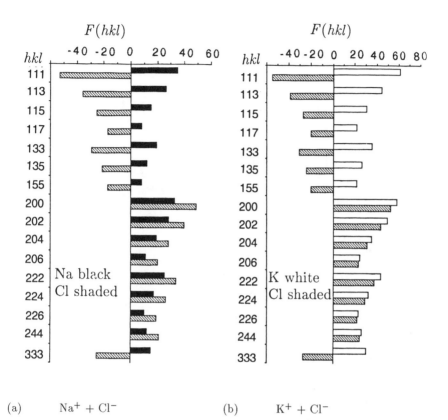

(a) $Na^+ + Cl^-$ (b) $K^+ + Cl^-$

FIGURE 6.20. Histograms of the calculation of structure factors for the alkali halides. Individual components (K^+, Na^+, Cl^-) and their sums which give the total structure factors are shown. Indices of Bragg reflections (111 to 333) are indicated beside each diagram and are the same for each. (a) Na contribution black, Cl contribution shaded, (b) K contribution white, Cl contribution shaded, and (c) NaCl black and KCl white. Note that some Bragg reflections are intense because the scattering from one type of atom reinforces that from another, while for other Bragg reflections the scattering from one type of atom interferes destructively with that from the other atom (since K^+ and Cl^- contain the same number of electrons). Note the different behavior of the Bragg reflections with *hkl* all odd (the difference of the structure factors) and all even (the sum of the structure factors) (see Figure 6.21).

information needed to compute F_{hkl} and its components, $|F(hkl)|$ and $(\alpha_{hkl})_c$, is available. The computation stresses the important fact that every atom contributes to every measured Bragg reflection. As mentioned earlier, a shift in the origin selected in the unit cell can change

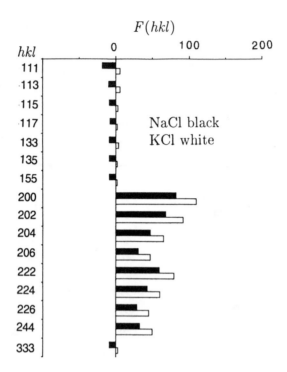

FIGURE 6.20 (cont'd). (c) Structure factors for NaCl (black) and KCl (white).

the relative phase angle, but not the magnitude of $|F(hkl)|$. The above computations show that the position of this origin has no effect on the diffraction pattern, a comforting finding, in view of the remarks at the beginning of this Chapter.

The calculation of $|F(hkl)|$, shown in Figure 6.21, is broken into two parts, one, $|A(hkl)|$, involving a cosine function and the other, $|B(hkl)|$, involving a sine function.[27] For each atom the scattering factor f_j at the value of $\sin\theta/\lambda$ appropriate for the diffracted beam, hkl, is multiplied by a cosine or sine function that contains h,k,l and x,y,z. This is done for each atom in the structure and the values are all summed to give $|A(hkl)|$, and $|B(hkl)|$. This entire computation is then repeated for every Bragg reflection. In practice, assumed phase angles can be applied to measured $|F(hkl)|$ values, and these quantities can then be used in the computation of an electron-density map. The coordinates of the centers of peaks in this map give the coordinates of atoms in the structure in three dimensions, usually as fractions of the three unit cell edges.

The equation for a structure factor is:

$$F(hkl) = | F(hkl) | e^{i\alpha_{hkl}} = A(hkl) + iB(hkl). \qquad (6.21.1)$$

The equation for a phase angle α_{hkl} is:

$$\alpha_{hkl} = \tan^{-1}[B(hkl)/A(hkl)], \qquad (6.21.2)$$

where the amplitude of a Bragg reflection is:

$$| F(hkl) | = [(A(hkl)^2 + B(hkl)^2)]^{\frac{1}{2}}. \qquad (6.21.3)$$

To compute $F(hkl)$ when atomic coordinates x_j, y_j, z_j for each atom j are known:

$$A(hkl) = \sum_j f_j \cos 2\pi(hx_j + ky_j + lz_j) \qquad (6.21.4)$$

$$B(hkl) = \sum_j f_j \sin 2\pi(hx_j + ky_j + lz_j). \qquad (6.21.5)$$

Sodium chloride as an example. There are 4 sodium and 4 chloride ions per unit cell, described by the following eight positions of x, y, z:

Na	$0, 0, 0$	$\frac{1}{2}, \frac{1}{2}, 0$; $\quad \frac{1}{2}, 0, \frac{1}{2}$;	$0, \frac{1}{2}, \frac{1}{2}$;
Cl	$\frac{1}{2}, \frac{1}{2}, \frac{1}{2}$;	$0, 0, \frac{1}{2}$; $\quad 0, \frac{1}{2}, 0$;	$\frac{1}{2}, 0, 0$.

The crystal structure is centrosymmetric, and therefore $B(hkl) = 0$. Only when $h + k, k + l, l + h$ are all even or all odd does $A(hkl)$ have a nonzero value (a requirement of the space group $Fm3m$).

$$F(hkl) = A(hkl) = f_{Na} \cos 2\pi(0) + f_{Na} \cos 2\pi(h/2 + k/2) +$$

$$f_{Na} \cos 2\pi(h/2 + l/2) + f_{Na} \cos 2\pi(k/2 + l/2) + f_{Cl} \cos 2\pi(h/2 + k/2 + l/2) +$$

$$f_{Cl} \cos 2\pi(l/2) + f_{Cl} \cos 2\pi(k/2) + f_{Cl} \cos 2\pi(h/2). \qquad (6.21.6)$$

Values calculated are, for example:

hkl	$F(hkl)_c$	systematic description	$|F(hkl)|_c$	$| F(hkl) |_o$
111	$4(f_{Na} - f_{Cl})$	h, k, l all odd	19.0	20.6
222	$4(f_{Na} + f_{Cl})$	h, k, l all even	53.3	52.9
333	$4(f_{Na} - f_{Cl})$	h, k, l all odd	7.3	8.1
120	0	h odd, k even, l even	0.0	0.0

FIGURE 6.21. Equations for calculating structure factors.

An example is provided by a comparison of the diffraction patterns of the isostructural chlorides of sodium and potassium (see Figure 6.20). It is noted that alternate rows of diffraction spots are very faint in the potassium chloride diffraction pattern, unlike the situation for sodium chloride. This alternating pattern of intensity is due to the fact that potassium and chloride ions are isoelectronic (with 18 electrons), and therefore have approximately identical powers to scatter X rays. On the other hand, the difference in scattering power between a sodium ion (10 electrons) and a chloride ion (18 electrons) is appreciable. Therefore those diffraction spots in which scattering from the metal ion interferes with scattering from the chloride ion will have a measurable intensity for diffraction by crystals of sodium chloride but almost no intensity for diffraction by crystals of potassium chloride.

In summary: it is shown that *each atom contributes intensity to each Bragg reflection* by an amount that depends on its position in the unit cell, its scattering power, and the scattering angle. Because there are many more Bragg reflections that can be measured than parameters (x,y,z) to be determined, the structure solution is overdetermined; the data-to-parameter ratio is generally around 10. The structure determination is treated by techniques appropriate to such a case, for example, the probability methods to be discussed in Chapter 8 and the least squares methods to be described in Chapter 10. Comparisons of the measured value of $|F(hkl)|_o$ with $|F(hkl)|_c$ for sodium chloride (space group $Fm2m$, cubic, $a = 5.6402$ Å) give some indication of how well the proposed model, with ions (Na^+ with $f = 10.0$ and Cl^- with $f = 18.0$) at fractional positions x_j, y_j, z_j in the unit cell, fits the observed data. The results show that the systematic absence 1 2 0 (because h, k and l are not all even or all odd) is accounted for by the model (see Equation 6.21.6).

6.6.3 Effects of atomic displacements and vibrations

A precise register of atomic position from unit cell to unit cell is not found in practice, and deviations from this exact register increase as a function of temperature because, at increased temperatures, atomic vibrations are larger and atomic **disorder** may be increased. Henry Gwyn Jeffrey Moseley and Charles Galton Darwin wrote[28] in 1913: "In the simple case of rocksalt the intensity falls off for the higher orders much more rapidly than theoretical considerations would suggest. This is doubtless due to the effect of the temperature oscillations of the atoms in the crystal." *During the experiment to obtain X-ray diffraction, the average of the contents of all the unit cells in the crystal is obtained* (all 10^{15-19} of them). The X-ray diffraction experiment is the equivalent of an instant snapshot of vibrating molecules, each with atoms in different unit cells displaced in random ways from their average positions. The molecules are apparently "frozen in time" because the frequencies of X rays are four orders of magnitude greater than the frequency of atomic vibrations. This means that if the molecule vibrates much, or is disordered in some way or another so that atoms lie in slightly different locations from unit cell to unit cell, the overall snapshot becomes fuzzy so that the apparent size of the individual atoms increases. It is difficult to differentiate between the two types of displacement (vibration and disorder) and therefore the term "displacement parameters" is now preferred[29] over the older term "temperature factor."

We stressed earlier that a larger size in real space means a smaller size in reciprocal space. Therefore, as the atoms appear to become broader and fuzzier, the diffraction pattern decreases in extent and becomes narrower. In other words, for reasons illustrated in Figure 3.12 (Chapter 3), the wider an atom as a result of increased vibration, the more out of phase are the rays scattered from various parts of the atom. *At higher scattering angles the intensity falls off more than calculated for a point atom*, as shown in this Figure, and repeated in Figure 6.22. The fall-off can be approximated by an exponential factor, $\exp(-B_j \sin^2 \theta / \lambda^2)$, applied to the scattering factors, where B_j, the displacement parameter for an atom, may be considered to be related to the mean-square amplitude $\overline{(u^2)}$ of atomic vibration by

$$B_j = 8\pi^2 \overline{u^2}. \tag{6.4}$$

So, a good approximation to the scattering factor for an atom j is

$$f_j = f_{j_o} e^{-B_j (\sin^2 \theta / \lambda^2)}, \tag{6.5}$$

where f_{j_o} is the scattering factor for a stationary atom. In minerals and some salts, B_j values are low, in the range of 1 to 3 Å², corresponding to 0.01 – 0.04 Å² in $\overline{u^2}$ or 0.1 to 0.2 Å in $\sqrt{\overline{u^2}}$. In organic compounds B_j values range from 2 to 5 or 6 Å². In crystals of large molecules or in crystals near their melting point, B_j values may rise to 30 to 40 Å² (see Figure 6.22). As noted above, however, B_j may contain input from disorder in atomic positions as well as from vibrations.

The atomic environment within the crystal is usually far from isotropic, and the next simplest model of atomic motion (after the isotropic model just described) is one in which the atomic motion is represented by the axes of an ellipsoid; this means that the displacements have to be described by six parameters (three to define the lengths of three mutually perpendicular axes describing the displacements in these directions, and three to define the orientation of these ellipsoidal axes relative to the crystal axes), rather than just one parameter, as in the isotropic case. Atomic displacement parameters, and their relationship to thermal vibrations and spatial disorder in crystals are covered in more detail in Chapter 13.

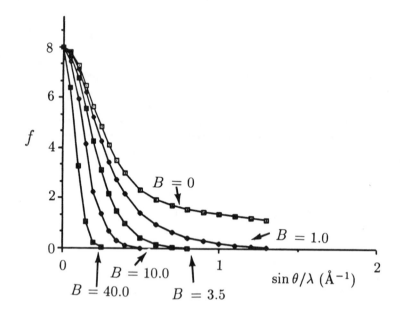

FIGURE 6.22. Isotropic displacement factors for oxygen, atomic number 8. Note that the larger the value of B, the greater the falloff in intensity as a function of $\sin\theta/\lambda$, but that the values of f when $\sin\theta = 0$ is always 8.0.

6.6.4 Modification by anomalous scattering

The extent that X rays are absorbed by an atom normally increases slowly as the wavelength of the X rays increases. Discontinuities in the amount of absorption are found at certain wavelengths; the absorption drops suddenly and then starts to rise again, as shown in Figure 6.23. These discontinuities are called **absorption edges** and occur at wavelengths that represent the energy necessary to excite a bound electron to a vacant higher energy level or to eject it altogether. At wavelengths just below an absorption edge, that is, at energies higher than the absorption edge, the scattering of X rays becomes complex. This means that the scattering factor f_j must be replaced by the "complex" value

$$f_j + \Delta f_j' + i\Delta f_j'', \tag{6.6}$$

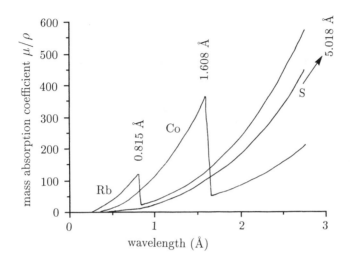

FIGURE 6.23. Absorption curves of cobalt, sulfur and rubidium as a function of the wavelength of the radiation. Note how the absorption by the element rises, falls abruptly at an absorption edge, and slowly rises again as the wavelength is increased.

where $i = \sqrt{-1}$ and $\Delta f_j{}'$ and $\Delta f_j{}''$ vary with the wavelength of the incident radiation. If an atom absorbs X rays strongly, there will be a phase change (related to $\Delta f_i' + i\Delta f_i''$) for the X rays scattered by the absorbing atom relative to the phases of the X rays scattered by other atoms. This is called **anomalous scattering**. The result *for a crystal in a noncentrosymmetric space group* is a change in the intensity of a Bragg reflection, *hkl*, and a different change for the *–h,–k,–l* Bragg reflection. Thus, while Friedel's Law[30] states that $|F|^2$ values of centrosymmetrically related Bragg reflections (*h, k, l* and *–h, –k, –l*) are equal, even for a noncentrosymmetric structure, this law does not apply to a noncentrosymmetric crystal if there is an anomalous scatterer in the structure.

In 1951, Johannes Martin Bijvoet[31] used such differences in intensity, resulting from anomalous scattering by an atom in a noncentrosymmetric crystal, to determine the chirality (absolute configuration) of the tartrate ion. Details of this method, which has been used extensively for finding the absolute configurations of natural products and for determining macromolecular structures, are given in Chapter 14.

Summary

1. The amplitudes of diffracted beams are summed with a periodicity inversely proportional to their order (and with a relative phase, which needs to be determined) to give an electron density map.

2. Fourier series are used in crystal structure analysis in several ways. An electron-density map is a Fourier synthesis with measured values of $F(hkl)$ and derived values of phase angles α_{hkl}. A Fourier analysis is the breakdown to component waves, as in the diffraction experiment. Fourier transform theory allows us to travel computationally between real space, $\rho(xyz)$, and reciprocal space, $F(hkl)$.

3. The molecular transform, computed as the structure factor $F(hkl)$ for each value of hkl, defines the relationship of the intensity of each Bragg reflection to the atomic arrangement causing the diffraction effect.

4. The intensities of diffracted beams are modified by crystal quality, disorder, temperature, and any anomalous scattering caused by the proximity of the incident X-ray wavelength to the absorption edge of an atom in the crystal.

Glossary

Absorption edges: The absorption of X rays by an atom increases as the wavelength increases. Discontinuities in the plot of absorption versus wavelength, called absorption edges, occur at energies (inversely related to wavelength) at which the incident X rays excite an electron in the atom to a higher orbital or else eject it altogether. At these absorption edges the absorption drops abruptly and then starts to rise again.

Anomalous scattering: A resonant interaction of X rays with the lowest energy levels of the crystal, that is, the atomic core levels. This interaction results in a change in both the magnitude and the phase of the scattering from those atoms whose core energy levels are near the X-ray energy. As a result the usual atomic scattering factor must be replaced by one that is wavelength-dependent:

$$f_j \text{ is replaced by } f_j + \Delta f_j' + i\Delta f_j'' \tag{6.6}$$

where $\Delta f_j'$ and $\Delta f_j''$ vary with the wavelength, being high near an absorption edge. This anomalous scattering is caused by high absorption by the medium at the absorption edge (q.v.). As a result, in crystals that lack symmetry elements of the second kind, Bragg reflections with indices hkl and $\overline{hkl}$ have different intensities (contrary to Friedel's Law). These differences in intensity may be used to determine the absolute configuration of the crystal.

Disorder (atomic and molecular): Lack of regularity. In crystal structures it implies that there is not exact register of the contents of one unit cell with those from all others. The atoms or molecules in the crystal structure pack randomly (nonperiodically) in alternative ways in different unit cells. Such disorder may cause some diffuse scattering around intense Bragg reflections. See also the definition in Chapter 2.

Double reflection: X rays diffracted by one set of crystal lattice planes may have sufficient intensity to be diffracted again by a second set of planes that are, by chance or design, in exactly the correct orientation. The twice-diffracted beam emerges in a direction that corresponds to a third set of planes with Miller indices equal to the sum of the indices of the two planes causing the double reflection. This double reflection, travelling in the same direction as a singly diffracted beam, will enhance or weaken the intensity of the latter.

Dynamical diffraction: Diffraction in which there are interactions between the incident and scattered beams ("reflection of the reflected beam"). This is particularly noticeable with perfect crystals and with electron diffraction.

Electron-density map: A contour representation of electron density in a crystal structure. Peaks appear at atomic positions. The map is computed by a Fourier synthesis, that is, the summation of waves of known amplitude, periodicity, and relative phase. The electron density is expressed in electrons per cubic Å.

Electron-density wave: A term in the Fourier summation of waves of different amplitudes and frequencies to give the electron density in the crystal. Each electron-density wave represents the contribution of a single structure factor $F(hkl)$ to the total electron density map.

Extinction: An effect of dynamical diffraction whereby the incident beam is weakened as it passes through the crystal. If the crystal is perfect there may be multiple reflection of the incident beam, which then is out of phase with the main beam and therefore reduces its intensity (primary extinction). If the crystal is mosaic, one block may diffract the beam, and is not then available to a second block similarly aligned (secondary extinction). Both effects result in a diminution of the intensities so that the most intense Bragg reflections are systematically smaller than those calculated from the crystal structure. The effect can be reduced by dipping the crystal in liquid nitrogen, thereby increasing its mosaicity.

Fourier analysis: The breaking down of a periodic function into its component cosine and sine waves (harmonics) with different amplitudes and frequencies.

Fourier series: A function $f(t)$ that is periodic with period T so that $f(t+T) = f(t)$, it may be represented by a Fourier series, which is an infinite series of the form

$$f(t) = a_0/2 + a_1 \cos 2\pi(t/T) + a_2 \cos 2\pi(2t/T) + \ldots\ldots$$

$$+ b_1 \sin 2\pi(t/T) + b_2 \sin 2\pi(2t/T) + \ldots..$$

$$= a_0/2 + \sum_{n=1}^{\infty} a_n \cos 2\pi(nt/T) + \sum_{n=1}^{\infty} \sin 2\pi(nt/T)$$

$$= \sum_{n=-\infty}^{\infty} c_n e^{2\pi i(nt/T)} \tag{6.7}$$

The Fourier theorem states that any periodic function may be resolved into cosine and sine terms involving known constants. Since a crystal has a periodically repeating internal structure, this can be represented, in a mathematically useful way, by a three-dimensional Fourier series, to give a three-dimensional Fourier or electron density map. In X-ray diffraction studies the magnitudes of the coefficients may be derived from the intensities of the Bragg reflections; the periodicities of the terms in the Fourier series are derived from the Miller indices h, k, l of the Bragg reflection, but the relative phases (which determine the ratio of sine to cosine expressions, b_1/a_1, b_2/a_2, etc.) are rarely determined experimentally.

Fourier synthesis: The summation of sine and cosine waves to give a periodic function; an example is the computation of an electron density map from waves of known phase, frequency, and amplitude $|F|$. See also the definition in Chapter 1.

Fourier transform (Fourier inversion): A mathematical procedure used in crystallography to interrelate the electron density and the structure factors. In the pair of equations

$$f(x) = \int_{-\infty}^{\infty} e^{2\pi i x y} g(y)\, dy \tag{6.8}$$

and

$$g(y) = \int_{-\infty}^{\infty} e^{-2\pi i x y} f(x)\, dx \tag{6.9}$$

$g(y)$ is the Fourier transform of $f(x)$ and $f(x)$ is the inverse transform (because of the negative sign) of $g(y)$. In a similar way the electron density is the Fourier transform of the array of structure factors, and the array of structure factors is the inverse Fourier transform of the electron density. Analogous equations, sometimes with a summation replacing the integral, relate these two quantities. If values for structure factors (including their relative phases) are known, the electron-density map can be calculated, and, if the electron density is known, the structure factors can be calculated.

$$\text{Note that } e^{ix} = \cos x + i \sin x \tag{6.10}$$

Harmonic: A sinusoidal component of a complex wave form with a frequency that is an exact multiple of the basic repetition frequency (fundamental). In acoustics harmonics are often called overtones.

Kinematical diffraction: Diffraction theory in which it is assumed that the incident beam only undergoes simple diffraction on its passage through the crystal. No further diffraction occurs that would change the beam direction after the first diffraction event. This type of diffraction is assumed in most crystal structure determinations by X-ray diffraction. Kinematical theory is well applicable to highly imperfect crystals made up of small mosaic blocks.

Mosaic blocks (mosaic spread): Tiny blocks within a crystal structure that are slightly misoriented with respect to each other. As a result of such mosaic spread, Bragg reflections have a finite width. Extinction is weaker in a mosaic crystal than in a perfect crystal, and therefore the intensities can be predicted by the rules of kinematical diffraction.

Structure factor: The structure factor $F(hkl)$ is the value, at the reciprocal lattice point hkl, of the Fourier transform of the electron density in the unit cell. When appropriately scaled, it is the coefficient, with indices hkl, of the Fourier series that gives the electron density in the crystal. It has both a magnitude (the structure amplitude) and a phase relative to the chosen origin of the unit cell.

Structure factor amplitude: The magnitude of a structure factor $|F|$ is the ratio of the amplitude of the radiation scattered in a particular direction by the contents of one unit cell to that scattered by a point electron in the same direction under the same conditions. Its relative value can be obtained from the intensity of a Bragg reflection.

References

1. Fourier, J. B. J. *Théorie Analytique de la Chaleur.* [Analytical theory of heat.] **English translation**: Freeman, A. (1872). Firmin Didot: Paris (1822).

2. Abbé, E. Beiträge zur Theorie des Mikroskops und der mikroskopischen Wahrnehmung. I. Die Konstruction von Mikroskopen auf Grund der Theorie; II. Die dioptrischen Bedingungen der Leistung des Mikroskops; III. Die physikalischen Bedingungen für die Abbildung feiner Structuren; IV. Das optische Vermögen des Mikroskops. [Contributions to the theory of the microscope and microscopic observations. I. Construction of a microscope on the basis of this theory. II. The dioptric conditions for the working of a microscope. III. Physical conditions for the imaging of small structures. IV. The optical properties of the microscope.] *Archiv für Mikroskopische Anatomie* **9**, 413–468 (1873).

3. Porter, A. B. On the diffraction theory of microscopic vision. *Phil. Mag.* **11**, 154–166 (1906).

4. Rayleigh, Lord. On the theory of optical images, with special reference to the microscope. Phil. Mag. 42, 167–195 (1896).

5. Bijvoet, J. M., Kolkmeyer, N. H., and McGillavry, C. H. *X-ray Analysis of Crystals.* Interscience: New York; Butterworths: London (1951). Originally *Röntgenanalyse van Kristallen.* [X-ray Analysis of Crystals.] 2nd edn., revised. D. B. Centen: Amsterdam (1948).

6. Trotter, J. The crystal structure of some anthracene derivatives. II. 9:10-Dibromoanthracene. *Acta Cryst.* **11**, 803–807 (1958).

7. Bragg W. H. X rays and crystal structure. (Bakerian lecture.) *Phil. Trans. Roy. Soc. (London)* **A215**, 253–274 (1915).

8. Epstein, P. S., and Ehrenfest, P. The quantum theory of the Fraunhofer diffraction. *Proc. Nat. Acad. Sci. (USA)* **10**, 133–139 (1924).

9. Duane, W. The calculation of the X-ray diffracting power at points in a crystal. *Proc. Nat. Acad. Sci. (USA)* **11**, 489–493 (1925).

10. Havighurst, R. J. The distribution of diffracting power in sodium chloride. *Proc. Nat. Acad. Sci. (USA)* **11**, 502–507 (1925).

11. Havighurst, R. J. The distribution of diffracting power in certain crystals. *Proc. Nat. Acad. Sci. (USA)* **11**, 507–512 (1925).

12. Waser, J. Pictorial representation of the Fourier method of X-ray crystallography. *J. Chem. Educ.* **45**, 446–451 (1968).

13. Glasser, L. Fourier transforms for chemists. Part I. Introduction to the Fourier transform. *J. Chem. Educ.* **64**, A228–A233 (1987).

14. Glasser, L. Fourier transforms for chemists. Part II. Fourier transforms in chemistry and spectroscopy. *J. Chem. Educ.* **64**, A260–A266 (1987).

15. Carslaw, H. S. *Introduction to the Theory of Fourier's Series and Integrals.* 3rd edn., revised. Dover Publications: New York (1930).

16. Wayne, R. P. Fourier transformed. *Chem. Britain.* **23**, 440–446 (1987).

17. Bracewell, R. N. The Fourier transform. *Sci. Amer.* **260(6)**, 86–95 (1989).
18. E. G. Steward. Fourier Optics: An Introduction. 2nd edn. Ellis Horwood: Chichester (1987).
19. Zachariasen, W. H. The secondary extinction effect. *Acta Cryst.* **16**, 1139–1144 (1963).
20. Ewald, P. P. Introduction to the dynamical theory of X-ray diffraction. *Acta Cryst.* **A25**, 103–108 (1969).
21. Darwin, C. G. The theory of X-ray reflexion. *Phil. Mag.* **27**, 315–333, 675–690 (1914).
22. Laue, M. von. Die dynamische Theorie der Röntgenstrahl-Interferenzen in neuer Form. [A new version of the dynamic theory of X-ray interference.] *Ergebnisse der Exacten Naturwissenschaften* **10**, 133–158 (1931).
23. Cowley, J. M. *Diffraction Physics.* 2nd edn. North-Holland: Amsterdam, Oxford (1981).
24. Renninger, M. X-ray measurements of diamonds. *Physik. Z.* **36**, 834–837 (1935).
25. Wagner, E. Über Spektraluntersuchungen an Röntgenstrahlen. [Spectroscopic observations of X rays.] *Physik. Z.* **21**, 621–626 (1920).
26. Renninger, M. "Umweganregung" eine bisher unbeachtete Wechselwirkungserscheinung bei Raumgitterinterferenzen. ["Detour-excitation" a hitherto unobserved interaction in space-lattice interference.] *Z. Physik* **106**, 141–176 (1937).
27. Sommerfeld, A. Remarks in: *La Structure de la Matière. Rapports et Discussions du Conseil de Physique tenu à Bruxelles du 27 au 31 Octobre 1913 sous les auspices de l'Institut International de Physique Solvay.* [The structure of matter.] p. 131. (Publication delayed because of the Great War). Gauthier-Villars: Paris (1921). (Reference supplied by C. A. Taylor.)
28. Moseley, H. G. J., and Darwin, C. G. The reflexion of X rays. *Phil. Mag.* **26**, 210 (1913).
29. Dunitz, J. D., Schomaker, V., and Trueblood, K. N. Interpretation of atomic displacement parameters from diffraction studies of crystals. *J. Phys. Chem.* **92**, 856–867 (1988).
30. Friedel, G. Sur les symétries cristallines que peut révéler la diffraction des rayons Röntgen. [Concerning symmetries of crystals that can be revealed by X-ray diffraction.] *Comptes Rendus, Acad. Sci. (Paris)* **157**, 1533–1536 (1913).
31. Bijvoet, J. M., Peerdeman, A. F., and van Bommel, A. J. Determination of the absolute configuration of optically active compounds by means of X rays. *Nature (London)* **168**, 271–272 (1951).

Measurement of Structure Amplitudes

A description of the first X-ray diffraction experiment by von Laue and his co-workers was written by W. L. Bragg in 1913 and reads as follows:[1] "Herren Friedrich, Knipping and Laue have lately published a paper entitled 'Interference Phenomena with Röntgen Rays,'[2] the experiments which form the subject of the paper being carried out in the following way. A very narrow pencil of rays from an X-ray bulb is isolated by a series of lead screens pierced with fine holes. In the path of this beam is set a small slip of crystal, and a photographic plate is placed a few centimetres behind the crystal at right angles to the beam. When the plate is developed, there appears on it, as well as the intense spot caused by the undeviated X rays, a series of fainter spots forming an intricate geometrical pattern."

The apparatus we use today for measuring the intensities of Bragg reflections consists of these same three components:

1. a source of X rays,
2. a crystal appropriately placed in the X-ray beam, and
3. a system to detect the spatial direction and intensity of each Bragg reflection.

While the methods for acquiring of X-ray diffraction data have not changed since the first diffraction studies by von Laue, improvements in their components have made them more efficient and "user friendly," so that crystal structures of much larger and more complicated molecules can now be determined. For example, during the years since 1912, X-ray sources have evolved from sealed tubes that produce only a nominal flux of radiation to synchrotron sources that produce radiation that is several orders of magnitude more intense. The first structural studies were done on crystals of simple salts containing only a few atoms in each unit cell and therefore relatively small numbers of Bragg reflections were available for measurement. Now we can study complicated macromolecules and determine atomic parameters for thousands of atoms in the asymmetric unit.

In this Chapter we will describe the components of the equipment used for the measurement of diffraction data, the Bragg reflections (Figures 7.1 and 7.2). Following this is a description of the conversion of the measured data into intensities, $I(hkl)$, and, from these, to values of the structure amplitudes, $|F(hkl)|$. The sophisticated instrumentation now available for measurement of Bragg reflections, when used with care and wisdom, can yield very precise intensity data. There are several general texts from which the reader may obtain more detailed information than given here, if experimental studies are planned.[3-14]

7.1 The intensity formula for diffracted X rays

The quantities we want to measure are the intensities of the Bragg reflections and their θ_{hkl} values. When a reciprocal lattice point passes through the Ewald sphere (see Figure 3.17, Chapter 3) and satisfies the condition

$$\lambda = 2d_{hkl} \sin\theta_{hkl}, \tag{7.1}$$

(which is Bragg's Law), a diffracted beam is obtained. Its intensity and direction of travel are directly amenable to measurement, as will be shown in this Chapter.

The measured intensity, $I(hkl)$, of a diffracted X-ray beam can be calculated using the formula of Charles Galton Darwin for a crystal rotating with a uniform angular velocity, ω, through a reflecting position:[15,16]

$$I(hkl) \propto I_o \frac{\lambda^3}{\omega} \frac{V_x \cdot L \cdot p \cdot A}{V^2} \mid F(hkl) \mid^2 . \tag{7.2}$$

This equation shows that the structure factor amplitude, $|F(hkl)|$, is a function of the intensity, $I(hkl)$, of a diffracted beam. Of the various quantities listed in Equation 7.2, the intensity of the incoming incident X-ray beam, I_0, and the wavelength of the radiation, λ, are selected by the experimenter when a system for generating X rays is chosen. The unit cell volume V influences the intensity of the diffracted beam in an inverse-square manner. Thus the greater the number of unit cells in the crystal, the more intense the Bragg reflection. Equation 7.2 also indicates that, as V_x, the volume of the crystal in the incident X-ray beam, increases, so do the intensities of the diffracted beams. There are, however, practical limits to the crystal size, because the crystal needs to be uniformly bathed in the X-ray beam during the experiment, so that the diameter of the incident X-ray beam limits the size of the crystal to a maximum of about 0.6 mm. The Lorentz factor L takes into account the relative time each reflection is in the diffracting position. It is a geometric correction, calculated as a function of the scattering angle 2θ, and it is

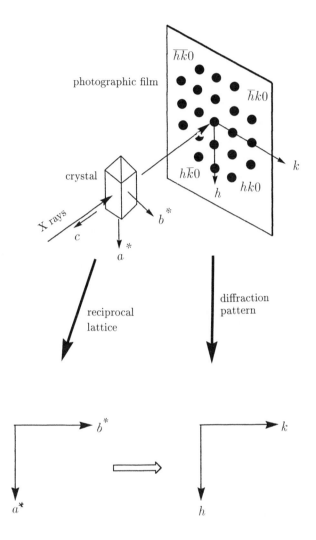

FIGURE 7.1. The relative orientations of the reciprocal lattice of a crystal (expressed as a^* and b^*), and its indexed X-ray diffraction pattern (expressed as h and k). In the diffraction pattern the intensities of the diffracted beams (I) (the blackness of spots on X-ray film, for example) and the directions of travel $(\sin\theta)$ (positions of spots on the X-ray film) are measured. Note the relationship of a^* to h, and b^* to k. From the positions of spots on the photographic film it is possible to deduce the dimensions of the reciprocal lattice, hence of the crystal lattice, hence the indices hkl of each Bragg reflection.

FIGURE 7.2 (a) A comparison of the various methods used to measure X-ray diffraction patterns.

Method (camera)	Crystal mount	Radiation	Detector	Detector motion
Rotation/ oscillation	rotates/ oscillates	monochromatic	film	stationary
Weissenberg	oscillation	monochromatic	film	translational motion as crystal oscillates
Precession	oscillation	monochromatic	film	precesses about axis
Laue	stationary	polychromatic	film	stationary
Diffractometer	moves to computed position on circle	monochromatic	scintillation counter	moves to computed position on circle
Area detector	stationary	monochromatic	multiwire proportional counter	stationary

FIGURE 7.2 (cont'd). (b) A comparison of information obtained by these methods.

Rotation/oscillation	Many spots on film. Rotation range controls number of spots.
Weissenberg	Distorted, but interpretable view of the reciprocal lattice.
Precession	Undistorted view of reciprocal lattice. Good for unit cell dimensions and checking crystal quality.
Laue	Large number of Bragg reflections with one crystal setting. More difficult to index.
Diffractometer	Can scan peak and measure intensity profiles. Precise data, measured sequentially.
Area detector	Can measure large numbers of data at the same time. Computer indexing of Bragg reflections. Reasonably precise data.

essentially independent of the source of radiation, the detection system, and the crystal. The polarization factor p accounts for the polarization of the X-ray beam after diffraction. Like the Lorentz factor, the polarization factor is a function of the scattering angle. The absorption factor A accounts for the absorption of each diffracted beam as it travels through the crystal. Such absorption implies that energy has been transferred to the crystal by the X rays passing through it. Therefore a correction for absorption is a function of the wavelength of the radiation, the atomic contents of the unit cell, and the path length of each diffracted beam through the crystal. All of these factors contribute to and modify the intensity of a Bragg reflection. Ultimately, the accuracy of $|F(hkl)|$ is dependent upon the precision with which we can measure $I(hkl)$, the intensity of the diffracted beam.

Therefore the initially measured intensity of a Bragg reflection must be modified by taking into consideration:

1. any problems in the shape of the Bragg reflection peak,
2. any intensity of the background on either side of the Bragg reflection,
3. effects of the polarization of X rays on scattering,
4. effects of the varying times that the crystal was in position to cause different Bragg reflections (Lorentz correction),
5. effects due to absorption of X rays by the crystal,
6. effects due to rediffraction of X rays within the crystal (extinction and other corrections), and
7. corrections necessary to take into account the method used for measuring the intensity, such as the geometry of the detecting device.

The number of Bragg reflections N that can be measured to any particular interplanar spacing, d_{min}, depends upon the volume of the unit cell.

$$N = \frac{4\pi}{3} \frac{V}{n} \left(\frac{1}{d_{min}}\right)^3. \tag{7.3}$$

At the limit of possible Bragg reflections at a wavelength λ, this equals:

$$N = \frac{4\pi}{3} \frac{V}{n} \left(\frac{8}{\lambda^3}\right). \tag{7.4}$$

In these Equations, V is the volume of the unit cell, and n is an integer ($n = 1$ for a primitive unit cell, 2 for a body-centered or single-face-centered unit cell, and 4 if it is fully face-centered). The value of $d_{min} = \lambda/(2\sin\theta_{max})$ in Equation 7.3 defines the resolution of the diffraction data and cannot be less than $\lambda/2$, because $\sin\theta$ cannot be greater than 1.0 (see Equation 7.4). Shorter-wavelength radiation therefore allows a greater number of Bragg reflections to be measured and at a higher resolution, provided the crystal quality is good enough for all possible data to be experimentally measurable (see Figure 7.3).

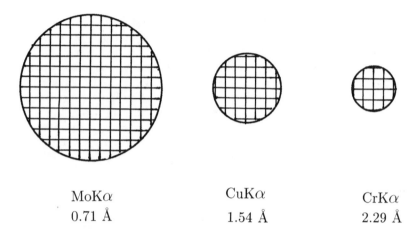

MoKα   CuKα   CrKα
0.71 Å   1.54 Å   2.29 Å

FIGURE 7.3. The effects of different wavelengths. The reciprocal lattice of a crystal is shown as a grid, with a circle indicating the limit of measurement ($\sin\theta = 1$). Note how, the longer the wavelength, the fewer the number of Bragg reflections amenable to measurement.

7.2 Sources of X rays

X rays, with wavelengths of 0.1 – 100 Å, are produced when fast-moving electrons are suddenly decelerated and their kinetic energy of motion is converted directly or indirectly to radiation. The equipment generally used to produce X rays consists of a cathode that produces an electron beam and a stationary metal anode that is the target for this beam and from which X rays are emitted. These components are enclosed in a glass tube under a vacuum. Large amounts of heat accompany the production of X rays, and therefore the X-ray target must be cooled by a stream of water circulating in a jacket enclosing it.

Two types of radiation are produced when the anode target is bombarded by fast-moving electrons in this way. A continuum of radiation, referred to as **white radiation**, is produced by simple collisions between electrons and target. In addition to simple energy exchange on collision, metal atoms in the target may become ionized by the loss of an inner-shell electron. An outer-shell electron then moves to an inner shell, and radiation of a wavelength specific to the target material is emitted [Figure 7.4(a)]. These X rays produced are called **characteristic X rays** because their nature is dependent on the atomic character of the material used to make the target. The characteristic radiation that is generally used for X-ray diffraction studies is from either a copper or a molybdenum metal target. Molybdenum radiation has a shorter wavelength and higher penetration through a crystal, and it is less readily absorbed. In practice, the characteristic X radiation is not of a single wavelength but

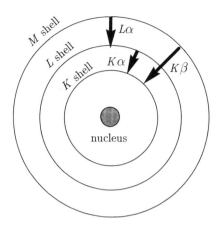

(a)

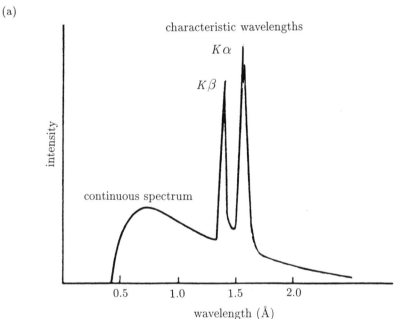

(b)

FIGURE 7.4. The X radiation emitted when an electron falls from an outer to an inner level. (a) Electronic energy levels involved. $K\alpha = L \to K = 2p \to 1s$. $K\beta = M \to K = 3p \to 1s$. Edge $= K \to \infty = 1s \to \infty$. (b) Characteristic X-ray spectrum for copper radiation. Wavelengths for copper and molybdenum radiation are: $CuK\alpha_1 = CuKL_{III} = 1.54056$ Å, $CuK\alpha_2 = CuKL_{II} = 1.54439$ Å, $CuK\beta_{1,2} = CuKM_{II}$ and $CuKM_{III} = 1.39222$ Å, the K edge is at 1.38 Å; $MoK\alpha_1 = MoKL_{III} = 0.70930$ Å, $MoK\alpha_2 = MoKL_{II} = 0.71359$ Å, $MoK\beta_1 = MoKM_{III} = 0.63229$ Å. The molybdenum spectrum is analogous to that of copper [shown in (b)].

contains several sharp lines, as shown in Figure 7.4(b) which illustrates the characteristic spectrum from a copper target. All but one of these spectral lines need to be filtered out in order to obtain a monochromatic beam. Usually $K\alpha$ radiation is selected. In addition, the white radiation, described above, must also be eliminated or minimized.

There are three accessories used to produce monochromatic radiation: metal foil **filters**, crystal **monochromators**, and **focusing mirrors**. An element with atomic number Z can be used as a selective filter for radiation produced by an element of atomic number $Z+1$. For example, a nickel ($Z=28$) absorption filter, may be used to cut out the Cu $K\beta$ ($Z=29$ for Cu) radiation, leaving only Cu $K\alpha$ radiation.[17,18] Not all white radiation, however, is eliminated by this method. Alternatively a single-crystal monochromator may be used. An intense Bragg reflection from the monochromator crystal is used as the incident beam for X-ray diffraction studies.[19,20] Focusing mirrors,[21] designed to produce a beam that is not only monochromatic but also convergent, may be used. In this case the incident beam is doubly deflected by two perpendicular mirrors.

The undeflected direct X-ray beam, which is very intense compared to any diffracted beams, is intercepted by a beam stop after passing through the crystal. Otherwise the direct beam could, by overloading the capabilities of the detection system, adversely affect the precision of the detection of the diffraction pattern, which is relatively much less intense than the direct beam. The beam stop is a small cuplike structure, lined with lead and designed so that if the X-ray beam is reflected by the internal metal surface it still cannot escape the cup. This use of a beam stop is analogous to the way that the moon obscures the sun in a solar eclipse, making the corona visible. The X-ray beam used for diffraction studies is also carefully collimated. A collimator is a hollow, straight tube, generally of internal diameter 0.2 to 1.0 mm, designed for producing a narrow, nearly parallel beam of radiation so that all points on the crystal may be completely bathed in X rays of a uniform intensity. If the collimator bore is too wide, the amount of background radiation that reaches the detector may be high enough to decrease the precision of measurement of a diffraction peak. If the collimator bore is too narrow the crystal may not be completely irradiated in the X-ray beam.

Intense sources of radiation (I_o in Equation 7.2) give better diffraction data because the peak-to-background ratio is increased. For the **rotating anode generator**,[22,23] which produces more intense radiation than normal X-ray tubes, it has been found to be advantageous to rotate the anode at high speed and to direct the electron beam to the outer edge of this rotating target so that the heat given off during the generation of X rays can be dissipated more readily. **Synchrotron radiation**[24,25] is far more intense than the radiation from a rotating anode. It is emitted when very high-energy electrons travelling at nearly the speed of light in an electron storage ring are decelerated. The X rays so produced consist of pulses (on a nanosecond scale) with a very high intensity and a high degree of polarization. The radiation is produced in the direction of travel of the electrons (tangential to the curve of the ring) and is characterized by a continuous spectral distribution (i.e., it is polychromatic). It can, however, be "tuned" to a selected wavelength by the use of a monochromator crystal.

7.2.1 Neutron diffraction

Crystals also diffract neutrons, and this fact is useful to the crystallographer because it is the nuclei of the atoms that scatter neutrons rather than the electrons (which are the scatterers in X-ray diffraction). Among the difficulties encountered if one wishes to use neutron diffraction, however, are a need for bigger crystals and the uncertainty of availability of time at a nuclear reactor where neutron flux is available. The results of such **neutron diffraction** studies[26] are particularly valuable for precise location of hydrogen atoms, for the differentiation of atoms of nearly the same atomic number, and for distinguishing isotopes.

7.2.2 Devices for detection of X rays

There are several types of detection devices that are used to record the intensities of Bragg reflections. Films and counters are commonly used detection systems for recording X-ray diffraction data. Both are highly sensitive to X rays and can provide a precise measure of the intensities of the diffracted beams (see Figure 7.5 for a diagram of the geometry of the detection system). If we know where on a film each Bragg reflection is recorded, then it is possible to determine $\sin \theta / \lambda$ and hence h,k,l. If we know the darkness of the spot on the film it is possible to integrate the peak and determine the intensity of the Bragg reflection.

The oldest and simplest recording device is photographic film. This is the detector that Wilhelm Konrad Röntgen used when he took photographs of the bones in his hands in order to demonstrate the great penetration of X rays. X rays, like visible light, interact with the silver halide contained in the emulsion of photographic film. When the film is developed, black metallic silver is deposited at the positions at which the diffracted rays made contact with the film. The optical density or the darkness of each spot is proportional to the intensity of a reflection. In visual estimations, a comparison is made of the intensity of each spot on a piece of photographic film on which one Bragg reflection has been recorded (in different positions on the film) for different measured lengths of time. Since the intensity is proportional to the time of exposure, a standard series of diffraction spots of known intensity is obtained. With experience it is possible to interpolate intensity measurements with fair precision, and obtain reasonably good experimental data.[27,28] More precise measurements of the intensity are made, however, by use of a film-

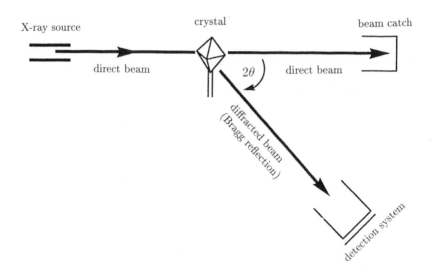

FIGURE 7.5. Geometry of the detection system.

scanning **photometer** (a microdensitometer)[29] which records the optical density at regularly spaced grid points on the film. Intensity measurements are obtained from the fraction of light transmitted by the film compared with the intensity of a source of light of constant intensity.

Electronic detectors incorporate the sensitivity of diffractometry with the efficiency of film methods. Electronic detectors can record the intensities of a large number of Bragg reflections simultaneously, in a manner analogous to that of X-ray film. There are basically two types of electronic detectors: multiwire proportional counters, and scintillation screens with television-type readout. Both detectors function by converting the incident energy to an electric charge and recording the charge and its position. There are two distinct advantages of electronic detectors over film. Electronic detectors are more sensitive to radiation. Consequently, the diffraction data can be collected more precisely and quickly, and a large number of intensity measurements can be made in less time, so crystals that decay or are sensitive to radiation can be studied more efficiently. The second advantage of an electronic device is that the intensity information is saved in a computer readable form. This eliminates the need for a densitometer and results in a tremendous saving of time.

Scintillation counters, **Geiger counters** (named for Hans Geiger),[30,31] and **proportional counters** are also used to detect X rays. Counters directly measure the intensity of an X-ray beam, as opposed to photometric devices that measure the intensity of the effect of an X-ray beam on photographic film. The scintillation counter makes use of the fact that X rays cause certain substances to emit photons of visible light by fluorescence. The X-ray detector used is often a sodium iodide crystal, activated by a small amount of thallous ion. The intensity of the emitted light is proportional to the intensity of the incident X-ray beam and is measured by a photomultiplier tube. Geiger counters and proportional counters are used less frequently because they are not as efficient as scintillation counters. A Geiger counter is a gas discharge tube that can electronically detect individual X-ray photons by the production of ions in the gas and an amplification of their effects. The proportional counter is similar to a Geiger counter and produces a measurable amplified voltage pulse of height proportional to the energy of the photon; it gives a linear response at high counting rates.

The **area detector**[32-39] is an electronic device for measuring many diffracted intensities at one time. It is an electronic substitute for film, and is now used, where possible, for crystals of biological macromolecules. It is a **position-sensitive detector**, and is coupled to an electronic device for recording the data in computer-readable form. The data so recorded include the intensity of a Bragg reflection (diffracted beam) and its precise direction (as a location on the detector). Both types of information are needed for each Bragg reflection so that $I(hkl)$, and $\sin\theta/\lambda$ can be determined.

Three types of area detectors are presently used in macromolecular crystallographic research: (1) multiwire proportional counters, (2) television area detectors, and (3) imaging plates. A multiwire proportional counter consists of an anode between two cathodes; these consist of arrays of parallel wires and are arranged perpendicular to each other. The chamber is filled with a gas, consisting partly of xenon which is ionized by the incident X rays, an effect recorded by the detecting device. Television area detectors contain a fluorescent phosphor that produces visible light when hit by an X-ray beam. After intensification, the photons are detected by a television photocathode.

The newest technological advance in X-ray detection is the **imaging plate**,[40,41] which is a storage phosphor. A latent image is produced on the plate when X rays hit it, and this image is then exposed to laser light. Light of a different wavelength is emitted and is converted by a photomultiplier into an electrical signal. The imaging plate can store information for a considerable time period, and can then be used again for another diffraction experiment. This detection system is being used in many macromolecular (and some small-molecule) crystallographic laboratories.

7.3 Preparing a crystal for diffraction studies

The results of an X-ray diffraction experiment are only as good as the quality of the crystal and the intensity data that result from it. It cannot be emphasized strongly enough that the time and care invested in obtaining a good crystal and setting up a careful experiment for collecting data will result in a more precise set of data. Some crystals ready for mounting are shown in Figure 7.6.

The polarizing microscope is a valuable tool that may aid in the choice of a good single crystal for diffraction studies, provided the crystal is not opaque. A crystal is examined with a microscope under crossed polarizers to check for imperfections such as cracks or voids, and to make sure there is complete extinction of light at certain angles of rotation of the stage (see Chapter 5). If the crystal is not single, it will cause problems when diffraction data are measured; the diffraction pattern will be a superposition of two or more patterns in differing orientations, and difficult to interpret.

Crystals need to be mounted in such a way that they can be manipulated in various devices used for intensity measurement. Two methods are commonly used to mount crystals (see Figure 7.7). Crystals that are not volatile or sensitive to the environment are glued onto a thin glass fiber with an epoxy glue. The fiber, via a brass pin, is inserted into the

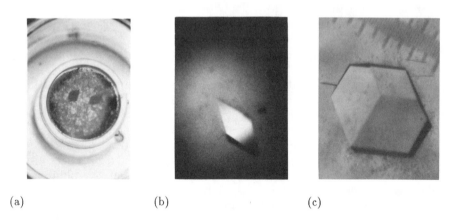

(a)   (b)   (c)

FIGURE 7.6. Crystals ready for mounting for X-ray diffraction studies. (a) A typical standing-drop crystallization experiment for a chemically modified horse hemoglobin. A sample of about 50 μl of protein solution was used. A precipitate typically formed before the single crystals grew. The crystals in the center of the photograph are approximately 0.6 × 0.5 × 0.5 mm in size. These are also used in Figures 7.7 and 7.12. (Courtesy J. J. Stezowski). (b) Crystals of lac repressor protein. One of these crystals was used for Figure 7.14. (c) A crystal of D-xylose isomerase. (Courtesy H. L. Carrell).

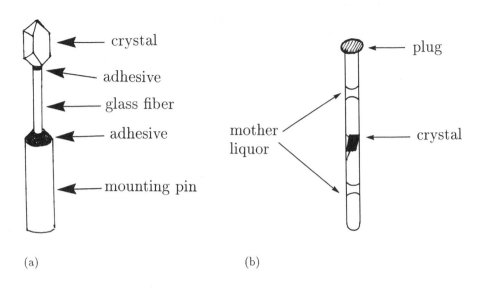

(a)

(b)

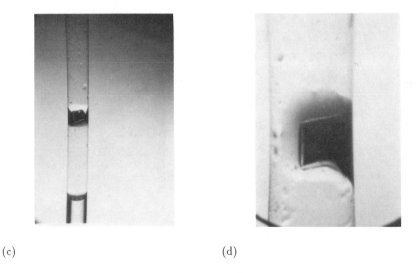

(c)

(d)

FIGURE 7.7. Methods of mounting crystals. (a) A crystal mounted in a glass fiber, as used for a small-molecule crystal that does not decompose on exposure to air. (b) Diagram of the mounting of a crystal in a capillary tube. (c) and (d). A crystal of a chemically modified horse hemoglobin enclosed with mother liquor in a thin walled glass capillary. (Courtesy J. J. Stezowski).

goniometer head (Figure 7.8). This device holds the crystal in place on a camera or diffractometer and allows it to be oriented in the X-ray beam by means of translational and angular motions. When crystals are volatile or sensitive to the environment, they are mounted inside a thin-walled glass capillary tube. A good diffraction pattern from protein crystals requires the mounting of a crystal in contact with its mother liquor inside a capillary along with a drop or two of the mother liquor so that an equilibrium atmosphere is maintained.[42,43] The capillary is then mounted on the goniometer head. Alternatively, a flow cell,[44] in which liquid with a controlled composition is allowed to flow over a crystal, can be used to maintain the required crystal environment.

The crystal, mounted on a goniometer head, is placed in the appropriate portion of the instrument that will be used to measure intensities. Then the crystal is viewed through the eyepiece of a telescope mounted

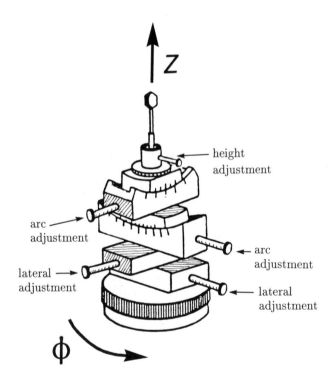

FIGURE 7.8. A goniometer head for orienting and centering a crystal in an X-ray camera or diffractometer. The arcs and lateral adjustments provide the means for the crystallographer to orient the crystal as needed. Note the directions of ϕ and Z. These define the goniometer head orientation.

on the instrument and the arcs and translations of the goniometer head are adjusted so that the crystal will be centered in the X-ray beam at all angles of study. It is essential that the crystal be securely attached to the mounting device of the measuring instrument so that no slippage can occur during the data collection. Ideally, the crystal should be small enough and, if possible, equidimensional, in order to be bathed totally in the X-ray beam and yet large enough to give good diffraction.[45] For macromolecules, a fine beam is needed to resolve the diffracted beams. As a rule of thumb, the crystal should contain material corresponding to 10^{18} to 10^{19} electrons. Very large crystals of small molecules can be cut to a smaller size with a thin razor blade, possibly along a cleavage plane. This is a fairly routine procedure, although it is rarely done for macromolecular crystals.[46]

7.3.1 Preliminary measurements of diffraction

When a stationary crystal is exposed to a beam of X rays, only those Bragg reflections with the relationship among θ, d, and λ defined by Bragg's Law are detectable. Since d, the lattice spacing, is a constant for a particular crystal, the only variables are λ and θ. As a result, for a stationary crystal with a small unit cell, very few Bragg reflections are found. Two methods are employed to overcome this paucity of Bragg reflections. Either a wide range of wavelength is used (polychromatic radiation, as in the Laue method) or, at one wavelength of the incident X rays, the crystal is rotated or oscillated about an axis to give a range of crystal orientations suitable for diffraction (as in the oscillation, rotation, Weissenberg, and precession methods). These two types of techniques are illustrated in Figure 7.9 by use of the Ewald sphere (described in Figure 3.17, Chapter 3). Precise intensity measurements are made when the X-ray detection system is in a precisely aligned orientation to receive the Bragg reflection. The crystal must be well centered and completely bathed in the incident X-ray beam. From the directions of the diffracted beams (the Bragg reflections) it is possible to assign a reciprocal lattice, hence a crystal lattice and, therefore, the Miller indices hkl of the Bragg reflection measured.

X-ray diffraction data are measured with X-ray cameras and photographic film, or **diffractometers** with counter methods of detection. The overall geometry of the crystal, X-ray beam, and detector, is shown in Figure 7.10. An advantage of the film method is that many Bragg reflections may be measured over the same time period. Also, the method can be used for very large structures with many Bragg reflections; the film is, in effect, a position-sensitive detector. Both the intensities and directions of spots on an X-ray film are measured. Diffractometers, on the other hand, measure the intensities by an automated process that is serial; one Bragg reflection is measured and then the diffractometer settings are changed for measurement of the next Bragg reflection. The

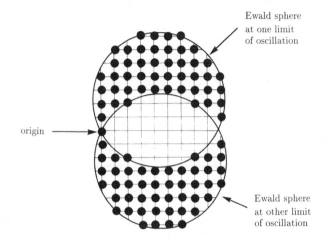

origin

Ewald sphere
at one limit
of oscillation

Ewald sphere
at other limit
of oscillation

(a)

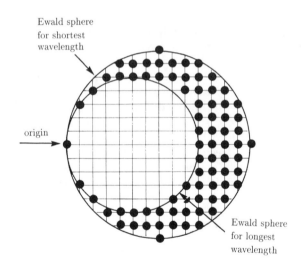

Ewald sphere
for shortest
wavelength

origin

Ewald sphere
for longest
wavelength

(b)

FIGURE 7.9. Use of the Ewald sphere to predict which Bragg reflections will be observed in a particular X-ray diffraction experiment (see Figure 3.17, Chapter 3). (a) Variation in the orientation of the crystal (by oscillation or rotation about ϕ). The two limits of oscillation are shown. (b) Variation in the wavelength of the radiation used, as in a Laue photograph, with a stationary crystal. The limits for two wavelengths are shown by the two circles. In both cases all Bragg reflections in the shaded area will be observed (where the surface of the Ewald sphere intercepts reciprocal lattice points).

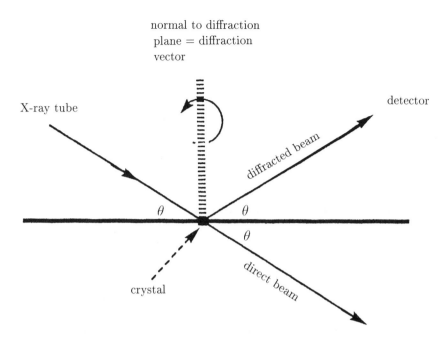

FIGURE 7.10. Diagram of the relative arrangement of the X-ray source, the detector, and the crystal.

location of the detector is adjusted so that it is in the correct orientation to intercept the required Bragg reflection.

Instruments used to record X-ray diffraction exist with varying degrees of automation. The **automatic diffractometer** is the instrument of choice for current small-molecule crystal structure determination. The automated instrument can move the crystal (on the goniometer head) and the detector through angular ranges under computer control to the required setting angles that orient the crystal in such a way that Bragg diffraction can occur and be measured by the detector. Diffractometer methods are less useful than film methods if the number of Bragg reflections is very large, or if the crystal deteriorates rapidly in the X-ray beam. For more details on the operation of these instruments the reader should consult texts, such as that by George H. Stout and Lyle H. Jensen.[7]

An X-ray camera is a device for holding X-ray film in an appropriate manner to intercept a diffracted beam. The various cameras commonly used are listed in Figure 7.2. The crystal quality (and, hence, its ability to generate accurate diffraction data) is also investigated by photographic techniques. Depending upon the type of experiment and the information that is required, different geometries are used. For example, some film holders are flat, and some are curved around the axis of rotation of the mounted crystal (see, for example, Figure 7.11); some cameras have an oscillating motion, and some precess about an axis. It is usual practice to employ a lead screen, with an appropriate hole or slit in it, to select the portion of the diffraction pattern that is of interest, since X rays will not pass through the lead screen. This screen may later be moved to select another set of data. Screens of this sort are of particular use in Weissenberg and **precession cameras** where they serve to separate out two-dimensional portions of the reciprocal lattice.

For proteins, on the other hand, the current state of the art is to use an electronic variation of film methods. Such large structures have specific experimental requirements as a result of the greater size of the unit cell. The number of reciprocal lattice points is increased for X rays of a selected wavelength; therefore, more diffracted Bragg reflections are observed. These are closer together (recall the inverse relationship between real unit cell spacing and the spacing in reciprocal space) and are also considerably weaker than for smaller molecules (there are fewer unit cells per crystal, and the atoms in the unit cell are generally very weak scatterers — carbon, nitrogen, oxygen and hydrogen atoms). The problem of weaker intensities can be counteracted, in part, by increasing the intensity of the incident beam through the use of rotating anodes or synchrotron radiation sources. If the unit cell is very large, the Bragg reflections may be so close together that they overlap, making it very difficult to obtain precise intensity measurements. In addition, the effects of radiation damage may have to be taken into account. The use of longer wavelength radiation, increasing the crystal-to-film distance, and sophisticated analyses of the shapes of the measured Bragg reflections (peak profiles), when these are scanned instrumentally, are typical ways of overcoming these experimental problems.

7.3.2 Oscillation, rotation, and Weissenberg methods

Preliminary photographs of a crystal may be taken in order to check for a cracked or twinned crystal, or for thermal diffuse scattering or super-lattice formation. Some of the instruments for doing this are those that were historically used for data collection. It is now debated whether it is necessary to take such preliminary photographs, because the more sophisticated data-collecting devices, together with a high-speed computer, can provide much of the same information. The reader, however, may encounter these other methods, which are briefly described here. More details can be obtained from the listed references.

A rotation photograph is obtained by rotating a crystal continuously (through 360°) about a fixed axis called the ϕ-axis (phi-axis) or spindle axis of the instrument on which the goniometer head is fastened.[47–49] This ϕ axis is perpendicular to the incident X-ray beam. Since a row of lattice points in a crystal lies perpendicular to a plane of reciprocal

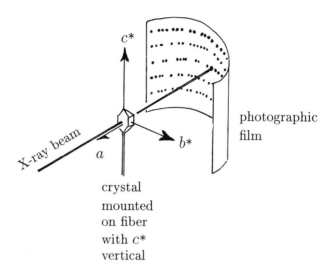

(a)

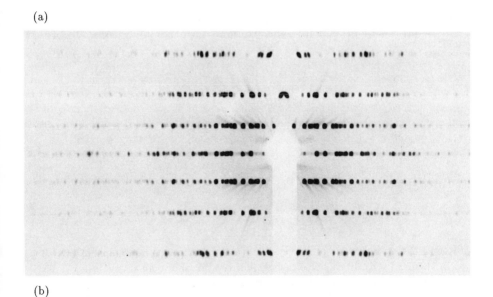

(b)

FIGURE 7.11. An X-ray diffraction photograph obtained by oscillating a crystal about a vertical axis. (a) Arrangement of camera and crystal, and (b) the resulting diffraction pattern. Note that the a axis is perpendicular to the b^* and c^* axes. With one crystal lattice axis vertical, layer lines (each with a constant value of l in a given layer) are produced.

lattice points, when the ϕ-axis of rotation is coincident with a principal axis in the crystal lattice, diffraction spots are arranged in a series of straight lines, called **layer lines**, that are perpendicular to the axis of rotation, shown in Figure 7.11. They are detected on film that is wrapped as a cylinder around the crystal. The number of Bragg reflections in any one layer line, however, is often very large, and therefore different spots may overlap each other on the film; these may be hard to separate analytically in order to obtain accurate intensities. Oscillation and Weissenberg cameras were designed to rectify this problem of overlap.

An oscillation camera[50-55] is similar to a rotation camera, but the crystal is only rotated back and forth about the spindle axis (ϕ) through a narrow predetermined angular range (see Figure 7.11). As a result, there are fewer Bragg reflections (spots) on each layer line than on a rotation photograph. The Bragg reflections which occur during this rotation may be determined by the use of the Ewald sphere and the orientation of the reciprocal lattice with respect to the two limits of oscillation [Figure 7.9(a)]. This method has proved to be very useful for protein crystals. Very small oscillation ranges, often less than 1°, are used, thereby necessitating many films for a complete data set. An example is shown in Figure 7.12.

In the Weissenberg method[4,13,56-59] the crystal is also oscillated about an axis. The camera has, in addition, a metal screen with a slit in it that selects out a single layer line and excludes all other layer lines. In order

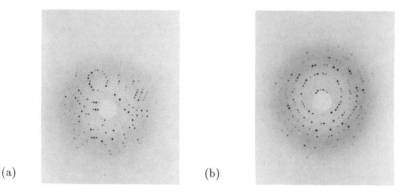

(a)   (b)

FIGURE 7.12. Orienting a protein crystal. (a) An oscillation photograph ($\Delta\phi = 1.0°$) for a chemically modified horse hemoglobin crystal, shown mounted in a capillary in Figure 7.7. The diffraction data were measured with Cu $K\alpha$ monochromatized radiation on a MARRESEARCH image plate detector. The crystal was mounted in a "random orientation." The diffuse darkening (ring) is scattering from the mother liquor. (b) A second oscillation photograph ($\Delta\phi = 1.0°$) for the same crystal in a different orientation. Note the circular pattern displayed by the diffracted X rays. This indicates that a potential unit-cell axis is nearly parallel to the incident X-ray beam. (Courtesy J. J. Stezowski).

to separate the various spots on this layer line so that they can be readily indexed and measured, the oscillation of the crystal is coupled to a back-and-forth linear motion of the camera. As a result, the position of a layer line on film, for a particular instant during the crystal oscillation, will vary as a function of the angle through which the crystal has been rotated. The one-dimensional layer line of an oscillation or rotation photograph is, therefore, in a Weissenberg photograph, converted to a two-dimensional pattern. The angle of rotation of the crystal about the ϕ-axis is the second dimension on the film. This method of separating the diffraction spots in a given layer line makes the indexing and interpretation of the photograph much simpler. The Weissenberg photograph is thus a distorted, but readily interpretable image of one layer (plane) of the reciprocal lattice.

7.3.3 Precession methods

The precession camera,[4,60-62] designed by Martin J. Buerger, records an undistorted, magnified image of a portion of the reciprocal lattice. The resulting X-ray diffraction photographs, usually of one plane through the reciprocal lattice, are simple to analyze, and therefore precession cameras are a basic tool for the preliminary examination of crystals. This method is especially useful if the unit cell is triclinic and provides conceptual problems because none of the axes are orthogonal to each other in either the crystal or the reciprocal lattices. The precession camera is a very useful tool for determining the space group, symmetry and quality of a crystal before intensity measurement is started.

Bragg reflections corresponding to a selected plane of the reciprocal lattice can be photographed by orienting the crystal in such a way that an axis perpendicular to this reciprocal lattice plane is inclined by an angle, μ, to the incident beam (Figure 7.13), and is then made to precess about the beam. A portion of the reciprocal lattice passes through the sphere of reflection and the resulting Bragg reflections are recorded on the film. An annular screen is used to isolate a single plane of the reciprocal lattice. The photograph resulting from this complicated set of motions is simple to interpret, and the indices (h,k,l) of the diffraction spots may be found readily by inspection. Examples are given in Figure 7.14.

7.3.4 Use of low temperatures

Most routine structure analyses are performed on data collected at room temperatures and pressures.[63] Low temperatures[64-74] can be achieved with either a liquid nitrogen ($T_{min} \approx 100$ K) or a liquid helium ($T_{min} \approx 10$ K) cryostat, both of which cool the crystal by passing a very cold gas stream over it. Lower temperatures reduce thermal motion, so that more intense Bragg reflection data are found at the higher scattering angles than if the temperature has not been lowered. Such low temperatures also serve to maintain crystal stability.

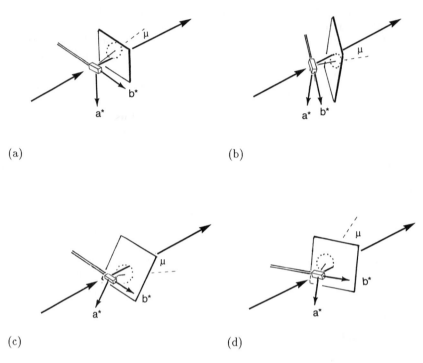

(a)

(b)

(c)

(d)

FIGURE 7.13. The geometry of a precession camera. Four stages, (a) to (d), in the formation of a precession photograph. The direct beam is inclined at an angle μ (the precession angle) to the normal to the film (center of circle of diffraction spots). A screen selects one ring of spots (one layer, $l = 0$, for example). The Bragg reflections, as a result of this screen, hit the film in a circle, as shown. For different portions of the precession cycle, these circles of spots appear in different areas of the film. During the complete precession cycle, all portions of the center of the film are exposed to Bragg reflections. Note that the $a^* b^*$ plane (perpendicular to **c**) is always parallel to the film.

Unfortunately most low temperature devices cause several problems, such as mechanical restrictions on measurement, particularly at high scattering angles, or temperature instability. It is important that the temperature be kept within a small range (± 1 K) during the data collection and that steps be taken to prevent the crystal from receiving a coat of ice from moisture in the environment. For low-temperature work, instead of using adhesives to mount the crystal on the metal pin, a silicone high vacuum grease, which hardens in the cold stream, is used. This method also obviates the need for corrections for absorption by the glass capillary in which the crystal is mounted. Macromolecules may also be studied at low temperatures, preferably by mounting them on a fiber rather than in a capillary tube;[75] at low temperatures protein crystals lose water more slowly than at room temperature and therefore are more stable during the diffraction experiment.

7.3.5 Powder diffraction

Powders generally contain a large number of randomly oriented **microcrystals**, each of which diffracts, yielding a continuous ring of Bragg reflections instead of individual spots. **Powder diffraction** patterns[76-82] are recorded on film or with two-circle diffractometers. A powder diffraction pattern of a partially ground up crystalline specimen is shown in Figure 7.15(a). While the diffraction pattern consists of circles, representing crystallites in all possible orientations, there is some evidence of crystallinity, indicated by the "grittiness" of the rings. When the crystals are fully ground, the diffraction pattern shown in Figure 7.15(b) is obtained. The circles can be thought of as the result of rotating an X-ray diffraction photograph, such as that in Figure 1.6 (Chapter 1). Since the diffraction pattern is spherically symmetrical, it is usual to simply show just a small slice from the center to the edge. Values of $\sin\theta/\lambda$ and intensities can then be measured.

The most common use for powder diffraction is to identify the components of samples[83] by what are colorfully described as "fingerprinting techniques." For example, the composition of material taken from the interior of an industrial chimney can be determined by matching the observed powder diffraction pattern with those of likely (and known) components. The entire identification process is now automated to use a data base, the Powder Diffraction File,[83] described in Chapter 16. This contains precise measurements of $\sin\theta/\lambda$ and intensities for a large number of powdered materials.

Currently, techniques for analyzing powder diffraction that involve the analytical dissection of measured Bragg reflection shapes[84-88] can be used to derive a data set similar to that for a single crystal. As a result, crystal structures may be determined, and in some cases, refined anisotropically using powder diffraction data for X rays or neutrons. Powder diffraction methods can be used at a wide variety of temperatures and pressures and are well suited for studies of phase transformations.

7.3.6 Fiber diffraction

Fiber diffraction is a technique used to determine the structures of molecules that are oriented to form fibers by virtue of a parallel assembly of molecules. Certain materials, such as cellulose, keratin and fibroin (the silk protein) occur naturally in this form. Some polymers can be drawn out to form fibers in which the same type of orientation occurs. This was the technique used by Rosalind Franklin[89] to study DNA. More recently,

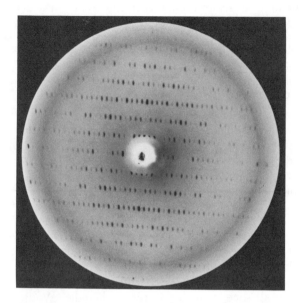

(a)

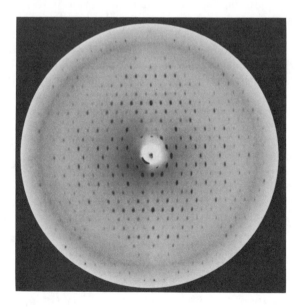

(b)

FIGURE 7.14. Two precession photographs of a protein crystal (lac repressor). (a) $0kl$ photograph. (b) $hk0$ photograph. The $hk0$ photograph was also shown in Figure 3.15 (Chapter 3).

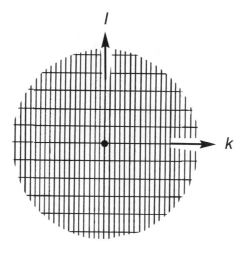

(c)

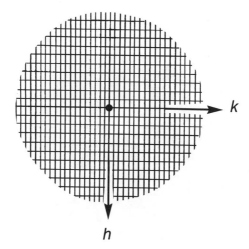

(d)

FIGURE 7.14 (cont'd). The reciprocal lattices of the two photographs shown in Figure 7.14 (a) and (b). (c) 0*kl* photograph. (d) *hk*0 photograph. The reciprocal lattice is drawn to scale as a grid for the central portion of each photograph.

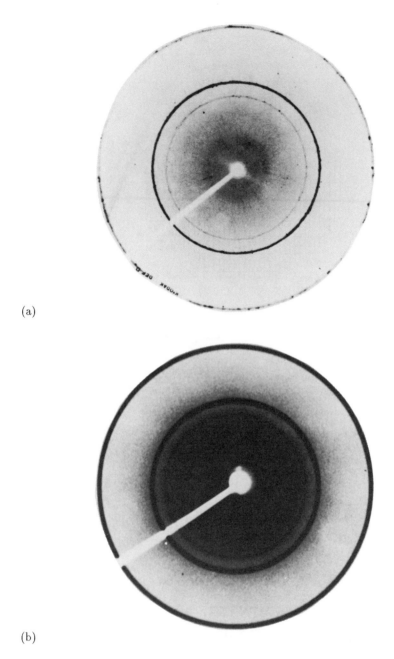

(a)

(b)

FIGURE 7.15. Diffraction patterns of crystals that have been (a) partly ground up and (b) totally ground up to a powder. Note that the rings in (a) show some spots which are evidence of crystallinity.

structures of large macromolecular assemblies, such as viruses, as well as multienzyme complexes, have been studied successfully.[90,91]

7.3.7 Small-angle scattering

Small-angle scattering is the diffraction of X rays, neutrons, or electrons at diffraction angles smaller than about 1°. Because of the reciprocal relationship between the dimensions of scattering objects and the spacings in the diffraction pattern, the objects studied at such low scattering angles have dimensions in the range of hundreds of Å units. The information obtained by the small-angle scattering method is the overall shape of the molecule and its approximate dimensions.[92-96]

7.4 Small-molecule studies by automated diffractometry

X-ray diffraction data for small structures are normally measured by use of an automatic diffractometer, diagrammed in Figure 7.16. This is, as the name implies, a device that systematically (and sequentially) measures the intensities of each diffracted beam.[97-108] The X-ray source is usually in a fixed position; the collimator and the monochromator are attached to this source. For simplification of design, the detector can move only in one plane, and the crystal is rotated about two axes to bring the Bragg reflection into the plane of the detector. The detection device is rotated in the **diffraction plane**. This plane (also called the equatorial plane) is usually parallel to the base of the instrument. It contains the X-ray source (more precisely, the focal spot of the X-ray beam), the crystal, and the detector, as shown in Figure 7.10. The angle defined by source–crystal–detector is $(180° - 2\theta)$ and the diffraction vector is the bisector of this angle (also in the diffraction plane). The detector is moved to the correct values of 2θ to intercept the diffracted beam and scan it. The crystal, bathed in the incident X-ray beam, is rotated in such a way that its center is fixed in space, always in the diffraction plane.

The diffractometer has, as shown in Figure 7.16, four mechanical circles, each identified by the angle it measures. These correspond to the three **Eulerian angles**, ω, χ and ϕ which define three-dimensional space, plus 2θ. The three circles (ϕ about the spindle axis of the goniometer head, χ that moves the goniometer head as a whole, and ω coincident with the plane of the 2θ circle) serve to orient the crystal so that a Bragg reflection to be measured will lie in the diffraction plane. In a common way of measuring data, called the $2\theta : \theta$ or bisecting mode, the motion of 2θ and ω are coupled in a 2:1 relationship. Alternatively an ω scan technique may be used. The **integrated intensity** of each Bragg reflection is then measured with an electronic recording device, such as a scintillation counter. A fifth angle ψ (psi), also known as the **azimuth angle**,[108] is a measure of the rotation of the sample about the normal to the diffracting plane of interest (the diffraction vector). This type of rot-

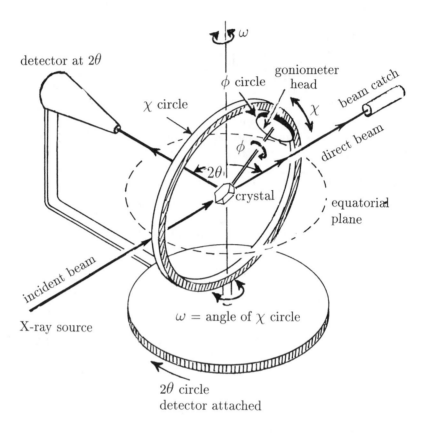

FIGURE 7.16. Diagram of an automatic diffractometer. The crystal is mounted on the goniometer head, which is attached, in turn, to the χ circle. The spindle axis of the goniometer head is ϕ. The angle χ is the angle between the ϕ axis of the goniometer head and the base of the diffractometer. The χ circle can be rotated about the ω axis, where ω is the angle between the diffraction vector and the plane of the χ circle. The detector is moved on the 2θ circle, where 2θ is the angle between the incident and diffracted beams.

ation can only be achieved by a combination of χ, ϕ, and ω changes, since there is not a specific circle for it. In practice, ψ-scans are useful for making empirical absorption measurements.

The mutual orientations of the crystal and of the detector with respect to the source of radiation are determined experimentally from some initial diffraction data, usually derived from photographic data or random search techniques, under computer control, for a few selected reflections. An **orientation matrix** is found which describes the orientation of the crystal reciprocal lattice to the coordinate frame of the instrument (Figure 7.17). The orientation matrix allows the detector and the crystal-orienting circles to be positioned so that the intensity of a particular Bragg reflection can be measured. A study of the profiles of some diffraction peaks as the crystal is rotated about the ω axis will give some measure of crystal quality.[109-115] The peak shape should be symmetrical and sharp; shoulders, split peaks, or very broad peaks are found if the crystal is of poor quality, such as a cracked crystal that consists of two portions almost aligned with respect to each other (Figure 7.18). It is essential to ensure that the entire diffraction peak is scanned, that no part is missed, and that portions of adjacent diffraction peaks are not included in the measurement. This implies that diffraction peaks must be well resolved.

To measure intensity data requires scanning the peak (which has an intensity P) and measuring background intensities (b_1 and b_2) on either side of the peak beyond where it is judged to have fallen to zero. These locations are generally described by the scan-range parameters. The background measurements are scanned for a fraction of the time needed to measure the peak P. The peak intensity minus that of the background gives a good measure of the net intensity,

$$I = P - B = P - 0.5(b_1 + b_2) \tag{7.5}$$

The e.s.d of the intensity $\sigma(I)$ can be derived by counting statistics from the expression:

$$\sigma^2_{\text{peak}}(I) = P + 0.25(b_1 + b_2) \tag{7.6}$$

and combined with an instrumental uncertainty. The quality of diffractometer data is usually monitored by repeatedly remeasuring a small number of Bragg reflections at regular intervals. In this way any crystal slippage, **radiation damage** and instrumental instability can be monitored and, often, corrected. **Equivalent reflections** are also measured to monitor the internal consistency of the data.

A general practice is to classify those Bragg reflections that have measured intensities less than an arbitrary multiple (usually two or three) of their estimated standard deviations as **unobserved reflections**. The term unobserved[116-118] is an unfortunate one since it also includes some weak reflections whose intensities have been measured. Reflections that are classified as weak or unobserved cannot simply be discarded from the data set; they are needed for statistical analysis purposes and may contain relevant information about the structure.[116-118]

Multiple reflections, described in Chapter 3, can cause enhancement in the intensities of some reflections and diminution in intensities of others. They are most clearly seen when reflections expected to be systematically absent have weak intensity. Generally only a few Bragg reflections are significantly affected, but there are various methods that can be used

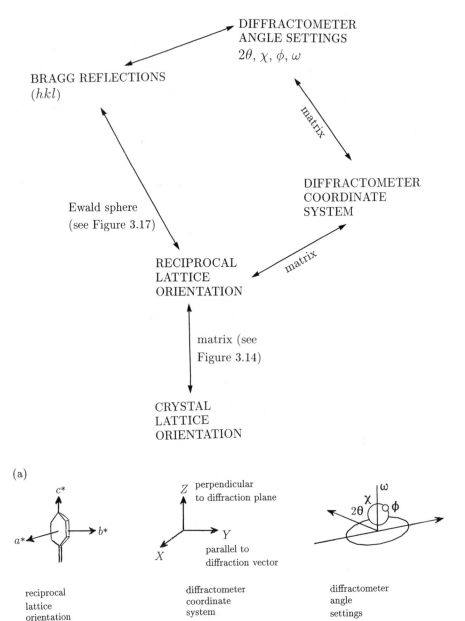

FIGURE 7.17. The orientation matrix. (a) The overall interconnections via matrices. (b) Details of the various components of the orientation matrix.

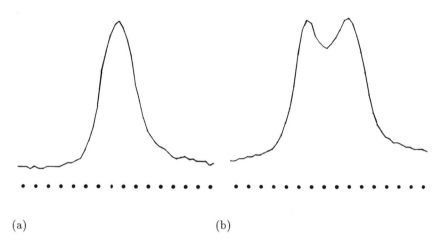

(a)   (b)

FIGURE 7.18. (a) A good and (b) a bad peak profile.

to reduce this effect. For example, a major crystal lattice vector should not be parallel to the ϕ axis. Equivalent reflections should be measured, their intensities compared, and any discrepancies checked by doing a ψ scan around the diffraction vector, as one would for absorption corrections. Some asymmetry in the profiles of the peaks of multiple reflections may also be detected.

Many different kinds of errors can affect the measurement of X-ray diffraction data.[119-124] Incorrect assignment of the space group, badly applied absorption corrections, insufficient data (possibly because the unit cell size was not correctly determined), and the misinterpretation of the Miller indices of Bragg reflections may give rise to data sets unsuitable for a structure determination.[7] In other words, if the experiment is not done properly, the structure will not be determined with the necessary precision, or, perhaps, even at all.

7.5 Diffraction by proteins and other macromolecules

The measurement of diffraction data from crystalline macromolecules presents additional problems. In the first place, the intensity of the diffraction is related to the size of the unit cell. Crystals with large unit cells diffract less strongly than do crystals with small unit cells. This is because there are fewer unit cells per unit volume for a macromolecular crystal. As a result, there is a need for very sensitive detection devices that can measure the intensities of weak Bragg reflections with high precision. A second related complication that arises with large unit cells is that the number of Bragg reflections is increased, and therefore the dimensions of the reciprocal lattice are decreased causing a related decrease in the angular spacings between the Bragg reflections. This means that the Bragg reflections are closer together than they would be if the unit cell were smaller. Although the structures of many proteins with intermediate-sized unit cells have been determined using diffractometers to measure the diffraction data, this method is not very efficient because only one Bragg reflection is recorded at a time. A more efficient procedure is to measure a large number of Bragg reflections at the same time, and this is now the general practice for crystalline macromolecules, by use of area detectors[32–39] and/or imaging plates.[40,41]

7.5.1 Oscillation methods for macromolecules

When the reciprocal lattice is densely populated (because the unit-cell lengths of biological macromolecules are large),[125] small angular movements of the crystal in the X-ray beam will bring a large number of lattice points into the diffracting condition (i.e., several reciprocal lattice points touch the surface of the Ewald sphere at the same time during the oscillation). The Bragg reflections that result are recorded on some type of detection device, together with their positions on the detection device.

In practice, three-dimensional diffraction data are measured as the crystal is oscillated over a small angular range about some arbitrary axis. The optimum oscillation range corresponds to the angle through which the maximum number of Bragg reflections will pass through the Ewald sphere and be recorded at unique positions on the detection device, without any overlap of these positions. If we define the point that the direct beam is recorded on the detection device as the origin of the diffraction pattern, then the coordinates X and Y of the Bragg diffraction spots can be measured in units (centimeters) relative to this origin (the direct beam spot). In addition, the angle about the rotation axis ϕ is noted. The difficulty we are now confronted with is determining the relationship between a Bragg diffraction spot measured as X, Y and ϕ to its reciprocal lattice index (h,k,l). The methodology for performing this is similar to that used for data measurement with a diffractometer (see Figure 7.17). It involves a knowledge of the orientation of the reciprocal lattice of the crystal with respect to a set of fixed orthogonal axes. With this information we can, by means of an orientation matrix, relate each reciprocal lattice point with an experimental observation in terms of X, Y, and ϕ.

7.5.2 Laue methods for proteins

A **Laue photograph**[126–137] is produced by irradiating a stationary crystal with a beam of X rays that has a wide range of wavelengths ("white" radiation). It differs from all of the other methods for collecting diffraction data in that the crystal is stationary throughout the experiment. Diffraction is therefore dependent on the multiwavelength feature of the incident beam. The Laue method is particularly useful for demonstrating the diffraction symmetry of a crystal. With the current accessibility of synchrotron radiation, Laue methods are often used for macromolecular studies. It is possible to collect large numbers of Bragg reflections in a very short period of time on a single photograph, such as the Laue photograph shown in Figure 7.19.[130]

7.6 Converting measured sin θ values to unit cell dimensions

The important preliminary data for a crystal, listed in the introduction to any crystallography article after the compound name, formula, and formula weight, are the unit cell dimensions and space group. These

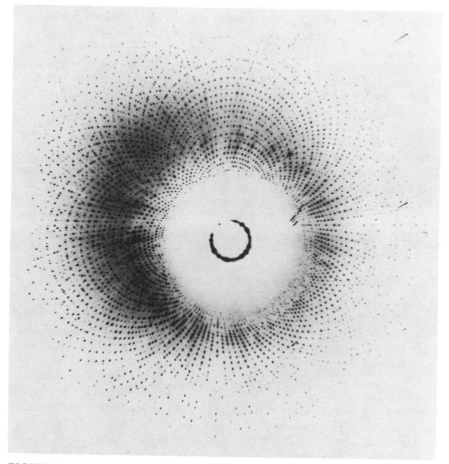

FIGURE 7.19. Laue photograph of lysozyme, 100 msec photograph, with synchrotron radiation 0.2 – 2.0 Å wavelength, taken at CHESS (Cornell High Energy Synchrotron Source). (Courtesy Edwin Westbrook).

provide the framework for the atomic positional parameters derived from the intensity measurements. Together with the density of the crystal, they give information on the weight contents of the unit cell. The unit cell dimensions are obtained by measuring the directions of Bragg reflections, represented on film by the distances between diffraction spots; the space group is obtained by examining the symmetry (or lack of it) among classes of Bragg reflections and the presence of systematic absences among them.

7.6.1 Unit cell dimensions

Unit cell dimensions are obtained from measurements of 2θ values of several Bragg reflections for which the indices h, k, and l are known. Values of 2θ are measured as accurately as possible, and, since the wavelength λ, of the radiation used is known, a value of d_{hkl} may be found by Bragg's Law, Equation 3.2 (Chapter 3). The value of d_{hkl} is related to the unit cell dimensions and, if 2θ values are measured for several reflections, values of the unit cell dimensions may be derived. The selected group of reflections chosen to do these calculations should contain a distribution of Miller indices (h,k,l) and they should have relatively high 2θ values; if possible it is best to measure 2θ at high enough values that the $K\alpha_1$ peaks are resolved from $K\alpha_2$ peaks (see Figure 7.3).

These unit-cell dimensions are listed in all published reports (see Figure 7.20) as: $a = \cdots$, $b = \cdots$, $c = \cdots$, $\alpha = \cdots$, $\beta = \cdots$, and $\gamma = \cdots$ (α, β, and γ are not listed if they are 90°). Also listed in the structure report are the estimated standard deviations of each of these unit cell dimensions, given with respect to the last digit of the measurement; for example, 15.63(2) Å means 15.63±0.02 Å. Following these the unit cell volume V (in Å^3), Z, the number of formula units per unit cell, and the observed and calculated densities (D_m and D_x, respectively) are listed.

7.6.2 Space group and crystal symmetry

The internal symmetry of the crystal is revealed in the symmetry of the Bragg reflection intensities, as discussed in Chapter 4. The crystal system is derived by examining the symmetry among various classes of reflections. Key patterns in the diffraction intensities indicate the presence of specific symmetry operations and lead to a determination of the space group. The translational component of the symmetry elements, as in glide planes or screw axes, causes selective and predictable destructive interference to occur. These are the systematic absences that characterize these symmetry elements, described in Chapter 4.

Unfortunately it is not always possible to characterize a space group unequivocally by systematic absences; two or more space groups may have the same systematic absences, leading to space group ambiguities. Often these are cases where one space group is centrosymmetric and the other is noncentrosymmetric. Alternative methods of symmetry assign-

(a) General information on the crystal

How it was grown, the faces that it developed, and its dimensions (in mm)
$C_{14}H_{10}\cdot C_{14}H_4O_6$ [chemical formula]
$M_r = 446.42$ [formula weight]
Monoclinic [crystal system]
$P2_1/a$ [space group]
$a = 17.572(10)$ Å. [unit-cell dimensions with e.s.d.s (in parentheses)]
$b = 7.727(4)$ Å.
$c = 7.398(4)$ Å.
$\beta = 101.90(4)°$ [α, β, γ only listed if not 90° by space-group symmetry]
$V = 982.9(9)$ Å³ [unit-cell volume]
Number and 2θ ranges of Bragg reflections used to measure unit-cell dimensions
$Z = 2$ [number of asymmetric units in the unit cell.]
$D_x = 1.508$ Mg m^{-3} [density calculated from V, formula weight and Z]
D_m [measured density]
λ Mo K$\alpha = 0.71069$ Å [radiation used and value used for the wavelength]
$\mu = 0.100$ mm^{-1} [linear absorption coefficient]
$T = 293$ K [temperature of study]

(b) Data measurement

1. Instrument used to measure the diffraction pattern.
2. Method used to measure the intensities of the Bragg reflections.
3. Whether a monochromator was used or not.
4. Ranges of h, k, and l studied.
5. Range of θ for measured data.
6. Number of data measured, number of independent data (not equivalent by symmetry).
7. Number of Bragg reflections considered "observed."
8. Criteria for "observed" data.
9. Internal agreement of two measurements of the same Bragg reflection or its equivalent.
10. Information on checks for intensity decay during the data collection.
11. Was an absorption correction applied? If yes, how was it determined? If not, why not?
12. Range of transmission factors in 11.
13. Checks that Lorentz and polarization factors were applied.

FIGURE 7.20. Crystal data for the molecular complex of anthracene with 1,8:4,5-naphthalenetetracarboxylic dianhydride.[138] [Remarks in brackets].

ment must be employed in these cases, such as testing, as described in Chapter 5, for effects in noncentrosymmetric crystals, for example second harmonic generation, or the failure of Friedel's Law if an anomalous scatterer is present. An optically "pure" chiral molecule cannot crystallize in a space group containing any inversion center, mirror or glide plane. Therefore, while space groups that contain these symmetry elements are possible for racemic mixtures of chiral molecules, they are not possible for a chiral molecule. This information may help settle a space group ambiguity.

7.6.3 Crystal density and unit cell contents

The weight of the unit cell contents can be calculated if the unit cell dimensions and the density of the crystal are known. The results may indicate the presence of solvent of crystallization in the crystal; it may also indicate that the crystal cannot possibly contain what it was expected to. Therefore, crystal density measurements are essential in the early stages of a crystal structure analysis.

The density of a crystal is given by:

$$D_x = \frac{M_c \times n}{V \times N_{\text{Avog.}}} = \frac{M_c \times n}{0.6022V} \tag{7.7}$$

where M_c is the sum of the atomic weights of all atoms in the asymmetric unit, V is the volume of the unit cell in Å^3, and n is the number of asymmetric units per unit cell. For most organic compounds containing oxygen but no metals, the densities of crystals lie near 1.3 g cm^{-3}. The computational use of an approximate volume of 18 Å^3 per atom, excluding hydrogen, is an indication of whether or not the unit cell volume is reasonable.[139]

The density of a crystal can be measured by flotation methods, which involves finding a homogeneous solution of two liquids in which the crystal is insoluble and in which it neither floats nor sinks, implying that the mixture of liquids has the same density as that of the crystal. The density of the solution can then be found by weighing a measured volume of this mixture. For example, a reasonable mixture for use for an organic compound can be made from heptane (density 0.684 g cm^{-3}) and carbon tetrachloride (density 1.589 g cm^{-3}). For macromolecules it is more convenient to use a vertical column with a density gradient,[140,141] calibrating the column with crystals of known density and then measuring where along the column the crystal equilibrates. From this measured density, M_c may be calculated by Equation 7.7. Most macromolecules crystallize with about 50% or more solvent of crystallization. If an approximate value for the molecular weight is known, then the number of subunits per cubic Å may be found. For a list of crystal data normally reported in a crystal structure determination, see Figure 7.20.

7.7 Converting $I(hkl)$ to $|F(hkl)|^2$

For each Bragg reflection, the **raw data** normally consist of the Miller indices (h,k,l), the integrated intensity $I(hkl)$, and its standard deviation $[\sigma(I)]$. In Equation 7.2 (earlier), the relationship between the measured intensity $I(hkl)$ and the required structure factor amplitude $|F(hkl)|$ is shown. This conversion of $I(hkl)$ to $|F(hkl)|$ involves the application of corrections for X-ray background intensity, Lorentz and polarization factors, absorption effects, and radiation damage. This process is known as **data reduction**.[142] The corrections for photographic and diffractometer data are slightly different, but the principles behind the application of these corrections are the same for both.

7.7.1 Lorentz and polarization corrections

The Lorentz and polarization corrections,[143-146] often called Lp, are geometrical corrections made necessary by the nature of the X-ray experiment. The **Lorentz factor** takes into account the different lengths of time that the various Bragg reflections are in the diffracting position. This correction factor differs for each type of detector geometry. For example, the Lorentz correction for a standard four-circle diffractometer is:

$$L = \frac{1}{\sin 2\theta}. \tag{7.8}$$

The partial polarization of X rays on scattering causes a reduction in their intensity that varies with 2θ. The correction for this, the **polarization factor**, is a function of the Bragg angle. For a diffractometer it is:

$$p = \frac{(1 + \cos^2 2\theta)}{2}. \tag{7.9}$$

Because the X-ray beam from a monochromator has already undergone diffraction from a monochromator crystal and hence some polarization, the beam diffracted from a monochromator (the subsequent incident beam) has a more complicated polarization correction.[146]

7.7.2 Absorption and extinction

As an X-ray beam passes through a crystal, its intensity is reduced as a result of its absorption by the material in the crystal. The extent of this intensity reduction[147-155] depends on the distance the diffracted beam must travel through the crystal, the nature of the atoms in the crystal, and the wavelength of the incident X-ray beam. It is described by the transmission factor (T, less than 1.0, also represented, causing great confusion, as A). The **absorption correction** (Abs or A^*, greater than 1.0) is the reciprocal of the transmission factor and is the amount by which the measured intensity should be artificially boosted in order to correct for absorption. The equation representing the reduction in intensity as the radiation passes through the crystal is:

$$\text{fraction of radiation transmitted} = T = \frac{1}{Abs} = \frac{I}{I_o} = e^{-\mu t}, \qquad (7.10)$$

where I is the intensity of a beam that is measured after passage through a thickness t (in cm) of an absorbing crystal, I_o is the intensity of the incident beam, and μ the total linear absorption coefficient for the primary beam (with units of cm^{-1}). Thus, we measure I and multiply it by $e^{\mu t}$. The value of μ, which can be calculated from published data,[154] is a function of wavelength and atomic number; heavy metals, if present, will greatly increase μ. Absorption corrections are necessary if μt is greater than 0.5 and should also be considered if the crystal is not equidimensional. If an absorption correction is not made when μt is high, the results of the structure determination may contain some unusual features that should be regarded with suspicion.

The absorption correction may be determined either by analytical or by experimental procedures. The analytical methods require a careful measurement of the crystal size and shape and an estimation of the path length for each Bragg reflection as it passes through the crystal. If absorption is high (for large crystals or for crystals containing heavy metals), it is important to use a crystal that is as equidimensional as possible. Sometimes it is possible to grind a crystal to a spherical shape in order to minimize absorption effects. This may be done by propelling the crystal with a gentle stream of air around a circular track lined with emery cloth. For crystals which need to be mounted in a capillary, experimental absorption corrections are the only option. Another method of applying an absorption correction is to use an experimental procedure, an **azimuthal scan** (ψ-scan). This is a scan of diffraction data measured as the crystal rotates about the diffraction vector. The intensities can be corrected for absorption as a function of the orientation of the crystal when the reflection under consideration was measured. Experimental methods such as the azimuthal scan are used for correcting for X-ray absorption by a crystal in a capillary tube.

Some other effects may need to be taken into account. Extinction corrections are related to mosaic spread[156,157] as described in Chapter 3. Corrections for this effect are generally made after refinement, and involve finding a factor needed to adjust the most intense Bragg reflections so that they agree with those calculated (which are initially larger). Sometimes, however, these corrections are not physically reasonable and care should be taken in their interpretation.[158] **Thermal diffuse scattering**,[159-161] which results from cooperative thermal motion of atoms in the crystal, is inelastic and diffuse, but may slightly enhance the intensity of a Bragg reflection. Such effects must be removed if exceptionally precise data are needed. Lowering the temperature of the crystal will decrease the effect.

7.7.3 Radiation damage

The exposure of a crystal to an X-ray beam is often harmful, and leads to radiation damage of the crystal. It will deteriorate, possibly as a result of formation of free radicals, loss of solvent of crystallization, heating of the crystals, or a variety of other factors. Radiation-induced damage causes Bragg reflections to change intensity as a function of time. If the damage results in the formation of a powder on the surface of the crystal by, for example, loss of solvent of crystallization, then intensity loss will occur because the effective size of the diffracting crystal is decreased. In most other cases radiation damage involves movements of atoms so that each Bragg reflection is changed in a specific way, depending on the crystal structure. This second type of damage is harder to correct for. Radiation damage may be detected by monitoring a set of reference reflections that are measured at regular intervals throughout the data collection. If the change in the intensities of the reference reflections as a function of time can be fit to a simple mathematical function describing a decay curve, this correction can be applied to the experimental data so that the effects of radiation damage are apparently reduced.

7.7.4 Estimating errors in measured intensities

There are two sources of error that have to be taken into account when assessing the experimental error in a measured intensity. There are errors that arise from the random fluctuations in the detection system; these errors follow a **Poisson distribution** and are proportional to the square root of the measured value (the count, hence the term "counting statistics").[162] The total error in an intensity,[145,158,163] represented by the estimated standard deviation (e.s.d.) $\sigma(I)$, is approximated by counting statistics, σ_{peak}^2, and the second source of error, the instrumental uncertainty, $K_{instrum}$. This relationship is represented by the equation:

$$\sigma(I) = \sqrt{\sigma_{peak}^2 + K_{instrum}}, \tag{7.11}$$

where σ_{peak} takes into account the errors in peak and background measurements, and $K_{instrum}$ is derived from measurements on a few Bragg reflections (standards) that are scanned periodically to monitor the reproducibility of the data as the experiment progresses. Since

$$I = k \, | \, F \, |^2, \tag{7.12}$$

where k is a constant; it follows, upon differentiation, that:

$$\frac{dI}{dF} = 2kF \text{ and therefore } \frac{dI}{I} = \frac{2kF \cdot dF}{kF^2} = \frac{2dF}{F} \tag{7.13}$$

so that

$$\frac{\sigma(I)}{I} = 2\frac{\sigma(F)}{F}, \tag{7.14}$$

that is, the relative error in F is half the relative error in I. It can also be shown that:

$$[\sigma(F)] = [\sigma(I)]/(2kF_o). \tag{7.15}$$

At this point we now have a useful data set, generally as a file in a computer-accessible storage device, which contains h, k, and l and the observed structure factor magnitude $|F|$. Many computer program systems place additional information for each reflection in the Bragg reflection data file. Typically, among these are whether the reflection is above or below a threshold value (observed or unobserved), the standard deviation $\sigma(F)$, $\sin\theta/\lambda$ values of the Bragg reflection, and atomic scattering factors, f_j, for each atomic type j in the crystal (Figure 7.21). Other information in the computer data file include unit cell dimensions with their estimated standard deviations, and the space-group symmetry of the crystal under study.

7.7.5 Absolute scaling of structure amplitudes

Structure amplitudes are on an **absolute scale** when they are expressed relative to the amplitude of scattering by a single electron under the same conditions. A scale factor is required to convert structure amplitudes derived from experimental X-ray diffraction intensities to such absolute values. In order to derive this scale factor, the average intensity from a crystal, as a function of scattering angle, is compared with the theoretical values to be expected for a random arrangement of the same atoms in the same unit cell. The plot, suggested by Arthur J. C. Wilson and known as a **Wilson plot**,[164] provides both the absolute scale factor K and an average temperature factor B for the crystal under study.

$$\ln\left(\Sigma \mid F(hkl)\mid^2_{\text{meas}} / \Sigma f^2\right) = \ln K - 2B\sin^2\theta/\lambda^2. \tag{7.16}$$

since

$$K \mid F(hkl)\mid^2_{\text{meas}} \approx \Sigma f^2 e^{-2B\sin^2\theta/\lambda^2} \approx \mid F(hkl)\mid^2. \tag{7.17}$$

Two Wilson plots are shown in Figure 7.22, one for a small molecule and the other for a macromolecule. In these plots the average values of $|F(hkl)|^2$ are compared in narrow ranges of $\sin^2\theta/\lambda^2$ to the values that can be calculated for a random arrangement of the same atoms in the same unit cell (that is, Σf_j^2). A plot of the natural logarithm of the ratio as a function of $\sin^2\theta/\lambda^2$ leads to the scale factor K that relates all structure factors to their absolute values, and an overall temperature factor B for the crystal structure. These values are generally used in the initial stages of a structure refinement, as will be described in more detail in Chapter 10.

h	k	l	F	$\sigma(F)$
0	2	0	66.31	0.48
1	2	0	5.78	1.57
0	3	0	4.97	2.07
2	0	0	88.77	0.62
2	1	0	60.49	0.56
1	3	0	4.86	2.57
2	2	0	16.91	0.66
0	4	0	85.59	0.65
0	0	1	8.11	0.91
0	1	1	21.93	0.84
1	4	0	5.97	2.41
2	3	0	210.19	1.16
1	0	1	37.83	0.65

(a)

h	k	l	I	$\sigma(I)$
0	0	10	0.9889E+04	0.1526E+03
0	0	12	0.3360E+05	0.5139E+03
0	0	14	0.9628E+04	0.1842E+03
0	0	24	0.1103E+06	0.1229E+04
0	0	26	0.4178E+05	0.7479E+03
0	0	28	0.7633E+05	0.6942E+03
0	0	30	0.1403E+04	0.5412E+02
0	0	32	0.2672E+05	0.3312E+03

(b)

FIGURE 7.21. Examples of experimental output. (a) Listed are diffraction data [as $F(hkl)$] for sodium citrate ($Na_3C_6O_7H_5 \cdot 5.5H_2O$) measured on an Enraf-Nonius FAST system with $MoK\alpha$ radiation from a rotating anode. There were a total of 3,067 Bragg reflections in this data set. (b) Listed are intensity data [as $I(hkl)$] for the protein D-xylose isomerase, measured on a Nicolet/Xentronics area detector with $CuK\alpha$ radiation from a Rigaku rotating anode. There was a total of 39,515 Bragg reflections in this data set. The e.s.d. values for $F(hkl)$ and $I(hkl)$ are listed as $\sigma(F)$ and $\sigma(I)$, respectively. (Courtesy H. L. Carrell.)

Summary

1. Measurements of X-ray diffraction requires a source of X rays, a crystal, and a detector of X radiation.
2. The crystal should be single, of good diffracting quality, and approximately equidimensional. It is mounted on a glass fiber or, if unstable, in a glass capillary.
3. The collimated beams of the X rays generally used are characteristic radiation [Cu $K\alpha$ (wavelength 1.5418 Å) or Mo $K\alpha$ (wavelength 0.7107Å)]. High-intensity radiation is obtained with rotating anode generators or by use of synchrotron radiation. Neutrons of similar wavelengths may also be used.
4. The diffracted beams are detected and their intensities measured by use of a scintillation counter or by use of photographic film; the latter is subsequently photometered. An area detector is used for many macromolecular studies.

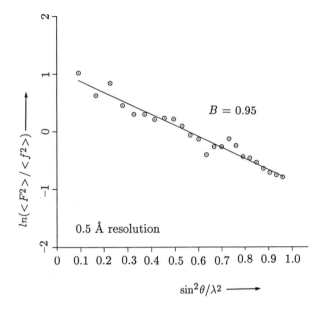

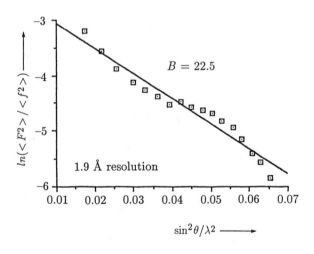

FIGURE 7.22. Two Wilson plots. (a) Diffraction data for a small structure, sodium citrate. (b) Diffraction data for macromolecule, D-xylose isomerase. Note that all the data for the macromolecule in (b) are within the value of $\sin\theta/\lambda$ for the first point for the small structure in (a). (Courtesy H. L. Carrell)

5. The Laue method involves a stationary crystal and polychromatic ("white") X rays. In the other camera methods, monochromatic radiation is used. In these cases the crystal may be oscillated over a small angular range (oscillation method) or rotated 360° about an axis (rotation method). The layer lines so formed may be selected individually. In the Weissenberg method, the oscillation of the crystal is coupled with a movement of the photographic film. The Buerger precession method, by a more complex motion of the instrument, produces an undistorted and magnified picture of the reciprocal lattice.

6. Diffractometer methods, usually automated, involve driving the detector and rotating the crystal to computed positions, scanning the Bragg reflection in steps and recording the X-ray intensity at each of these steps.

7. Unit cell dimensions are measured from the directions, 2θ, of diffracted beams and an application of Bragg's law, $\lambda = 2d\sin\theta$. Space groups are derived from systematic absences in the Bragg reflections.

8. Measurement of the density of a crystal permits calculation of the weight of the contents of the unit cell.

9. X-ray data collection at low temperatures increases the precision of the analysis.

10. The data obtained, called the "raw data," are corrected for background intensity, crystal decay, Lorentz and polarization factors, and absorption effects.

11. The results are values of $|F(hkl)|$ and its estimated standard deviation for each Bragg reflection (h,k,l).

12. Structure amplitudes are put on an absolute scale by a Wilson plot, which gives an overall scale factor and temperature factor.

Glossary

Absolute scale: When the structure factors are on an absolute scale, they represent the scattering of X rays by the total contents of the unit cell relative to the scattering of X rays by a single electron under the same condition.

Absorption correction: The linear absorption coefficient μ of a crystal is a function of its atomic composition and of the wavelength of the X rays. The ratio of the intensities of X rays entering and leaving a crystal of thickness t is $\exp(\mu t)$.

Area detector: A position-sensitive electronic device for measuring the intensities of a large number of diffraction data at one time in two dimensions. It gives information on both the direction and intensity of each diffracted beam. By analogy, it might be called electronic film.

Automated diffractometer: A computer-controlled instrument that automatically measures and records the intensities of Bragg reflections. The required mutual orientations of the crystal and the detector with respect to the X-ray source are computed from initial data on 20–30 selected Bragg reflections. These orientations are then achieved by computer-directed commands to electromechanical devices that position the crystal orienter and X-ray detector at the desired angular settings.

Azimuth angle: The azimuth angle is the angle measured in a plane perpendicular to a selected axial direction. For example, if a line joining the center of the earth to the North Pole is perpendicular to the horizon, then the azimuthal angle is the longitude (as measured with respect to a zero value at Greenwich, UK). In diffraction, the azimuth angle is a measure of the rotation of the crystal about the diffraction vector (the normal to the diffracting plane of interest).

Azimuthal scan: A scan of diffraction data measured as the crystal rotates about the diffraction vector. This is used to make an empirical absorption correction and to remove possible errors due to double reflections.

Characteristic X rays: X rays of a specific wavelength, characteristic of the chemical identity of the metal target, that are emitted from a target material that has been bombarded by fast electrons. When an electron that has been displaced from an inner shell of an atom in the target material is replaced by another electron from its outer shell, the energy lost by the second electron is emitted as an X ray characteristic of the target material.

Data reduction: Conversion of intensities, $I(hkl)$, to structure amplitudes, $|F(hkl)|$, by application of various correction factors including Lorentz factors, polarization factors, and absorption factors.

Diffraction plane: The plane containing the incident beam, crystal, and diffracted beam.

Diffractometer: An instrument for measuring the intensities and directions of the Bragg reflections comprising the X-ray diffraction pattern of a crystal.

Equivalent reflections: For a complete set of intensity data there are eight measurements for each h,k,l, corresponding to combinations of positive and negative values of each. Some of these, equivalent by the symmetry of the crystal, have (within experimental error) identical intensities. For high-symmetry crystals, other Bragg reflections may also be equivalent (e.g., hkl, klh, and lhk for cubic crystals).

Eulerian angle: The three successive angles of rotation needed to transform one set of Cartesian coordinates into another.

Filter: A material that absorbs some of the X radiation passing through it, with high absorptivity at some wavelengths and low at others. A narrow wavelength range of radiation can be selected by choosing appropriate filters with different wavelength absorptivities. The use of filters reduces the intensity of the X-ray beam.

Focusing mirror system: Two bent metal mirrors that deflect the X-ray beam and produce a small intense beam with a narrow angular divergence, uniform beam profile and a low background intensity. They are useful for experiments involving crystals with large unit cells.

Geiger counter: A gas discharge tube that can detect individual X-ray photons by amplification of their effects by producing ions in the gas. The amplified effect can then be detected electronically.

Goniometer head: A device for orienting and aligning a crystal for use in an instrument such as a diffractometer or X-ray camera. The orientation is done by moveable arcs and the alignment by translations.

Imaging plate: A surface that behaves like photographic film and can be used to store X-ray intensities as latent images in the form of color centers. The stored image is scanned by laser light. The plate is erasable and can be used many times. It is more sensitive than photographic film and useful because intensities can be retrieved electronically.

Integrated intensity: The total intensity measured at the detector as a reflection is scanned.

Laue photograph: A diffraction photograph produced with "white X rays" (i.e., those with a wide range of wavelengths).

Layer lines: When a crystal is rotated or oscillated about a principal axis, the diffraction spots on a cylindrical film surrounding the crystal are arranged in a series of straight lines called layer lines. These layer lines are perpendicular to the axis of rotation.

Lorentz factor: A factor that is used to correct diffraction intensities that allows for the varying time that different Bragg reflections (as reciprocal lattice points with finite sizes) take to pass through the sphere of reflection (Ewald sphere). The value of this correction factor depends on the scattering angle and the geometry of the measurement of the Bragg reflection.

Microcrystals: Crystals that are visible only through a microscope.

Monochromator: A means of producing radiation of a single wavelength. This is generally done by selecting a Bragg reflection from a crystal such as graphite, and using this as the incident beam.

Neutron diffraction: Neutrons of wavelengths near 1 Å can be used for diffraction by crystals. Neutrons are scattered by atomic nuclei. Neutron scattering factors are not a regular function of the atomic number of the scattering atom and therefore give information that often complements that from X-ray diffraction. The location of hydrogen atoms by neutron diffraction is much more precise since the scattering is higher relative to other atoms than is the case for X rays.

Orientation matrix: A matrix that relates the vectors of the reciprocal lattice of a crystal to the orientations of the circles of a diffractometer. It provides a connection between the orientation of the diffractometer circles and the production of a Bragg reflection so that the indices hkl of the Bragg reflection can be found from the orientation of the chosen unit cell of the crystal.

Photometer: An instrument for measuring the intensity of radiation, such as the intensity of light passing through any selected position on an X-ray diffraction photograph.

Poisson distribution: A distribution of measurements applied to rare events, in which the number of such events depends only on the length of the time interval. The mean and variance of the distribution are equal so that the e.s.d. is proportional to the square root of the measured value. The distribution is named for the French mathematician Siméon D. Poisson.

Polarization factor: A factor that allows for the variation in intensity of scattering as a result of polarization of the beam.

Position-sensitive detector: A detector that registers both the arrival of incident radiation intensity and also the position at which it was absorbed by the detector. This gives two-dimensional data.

Powder diffraction: Diffraction by a crystalline powder (a mass of randomly oriented microcrystals) consisting of lines or rings rather than separate diffraction spots. The diffraction pattern obtained is that expected for a set of randomly oriented crystals.

Precession camera: A specific camera used for recording the X-ray diffraction pattern of a crystal. The result is a photograph that is an undistorted magnified image of a given layer of the reciprocal lattice. The necessary camera and crystal motion involve the precession of one crystal axis about the direction of the direct beam. The film is continuously maintained in the plane perpendicular to this precessing axis. The photograph resulting from this complicated set of motions is simple to interpret.

Proportional counter: This radiation detector that produces a measurable amplified voltage pulse of height proportional to the energy of photons hitting it; it gives a linear response at high counting rates.

Radiation damage: Damage caused by radiation. Since a crystal is constantly irradiated by X-rays during the diffraction experiment, such damage can be an important source of error. As a result of such damage, molecules in the crystal may move, be ionized, form free radicals, or interact with other species in the crystal. Sets of three or more Bragg reflections are measured at regular intervals during intensity data measurement by a diffractometer in order to monitor any radiation damage.

Raw data: Data when they are first measured, before correction and other factors are applied to them.

Rotating anode generator: A tube in which an electron beam hits a wheel of target material rotating at high speeds. The anode moves while the X-ray beam remains fixed so that the heat generated during X-ray production is spread over a larger area than in a conventional sealed X-ray tube.

Scintillation counter: A device for measuring the intensity of an X-ray beam. X rays cause certain substances to emit visible light by fluorescence. The intensity of this visible light, which can be measured, is proportional to the intensity of the incident X-ray beam.

Synchrotron radiation: Radiation emitted by very high-energy electrons, such as those in an electron storage ring, when their path is bent by a magnetic field. This radiation is characterized by a continuous spectral distribution (which can, however, be 'tuned' by appropriate selection), a very high intensity, a pulsed-time structure, and a high degree of polarization.

Thermal diffuse scattering: Diffuse scattering results from a departure from a regular periodic character of a crystal lattice. It is evident as diffuse spots or blurs around normal diffraction spots. If it is a temperature-dependent effect, it is called thermal diffuse scattering.

Unobserved reflections: Bragg reflections that are too weak to be measured by the apparatus in use. The term is also used for Bragg reflections for which the intensity I is less than $n\sigma(I)$, where n is chosen (usually 2–3).

White radiation: Any radiation, such as X rays or sunlight, with a continuum of wavelengths.

Wilson plot: A plot that gives the factor necessary to place the observed intensities on an absolute scale and that also gives the average displacement factor (temperature factor) of the crystal. It is obtained by plotting the logarithm of the average ratio of observed Bragg reflections in successive ranges of $\sin^2 \theta/\lambda^2$ against the theoretical values in those same ranges for stationary atoms (with the same total number of electrons) arranged randomly in the same unit cell.

References

1. Bragg, W. L. The diffraction of short electromagnetic waves by a crystal. *Proc. Camb. Phil. Soc.* **17**, 43–57 (1913).
2. Friedrich, W., Knipping, P., and Laue, M. Interferenz-Erscheinungen bei Röntgenstrahlen. [Interference phenomena with X rays.] *Sitzungsberichte der mathematisch-physikalischen Klasse der Königlichen Bayerischen Akademie der Wissenschaften zu München*, pp. 303–322 (1912). **English translation**: Stezowski, J. J. In: *Structural Crystallography in Chemistry and Biology.* (**Ed.**, Glusker, J. P.) pp. 23–39. Hutchinson & Ross: Stroudsburg, PA (1981).
3. Henry, N. F. M., Lipson, H., and Wooster, W. A. *The Interpretation of X-ray Diffraction Photographs.* Macmillan: New York (1960).
4. Buerger, M. J. *X-ray Crystallography. An Introduction to the Investigation of Crystals by their Diffraction of Monochromatic X-radiation.* J. Wiley: New York (1942). Reprinted. Krieger: Melbourne, FL (1980).
5. Blundell, T. L., and Johnson, L. N. *Protein Crystallography.* Academic Press: New York, London, San Francisco (1976).
6. Ladd, M. F. C., and Palmer, R. A. *Structure Determination by X-ray Crystallography.* 2nd edn. Plenum: New York, London (1985).
7. Stout, G. H., and Jensen, L. H. *X-ray Structure Determination. A Practical Guide.* Macmillan: London (1968). 2nd edn. John Wiley: New York, Chichester, Brisbane, Toronto, Singapore (1989).
8. Glusker, J. P., and Trueblood, K. N. *Crystal Structure Analysis. A Primer.* 2nd edn. Oxford University Press: New York, Oxford (1985).
9. Jeffery, J. W. *Methods in X-ray Crystallography.* Academic Press: London, New York (1971).
10. Alexander, L. E. *X-ray Diffraction Methods in Polymer Science.* Wiley-Interscience: New York, London, Sydney, Toronto (1969)
11. Nuffield, E. W. *X-ray Diffraction Methods.* Wiley: New York, London, Sydney (1966).
12. Cullity, B. D. *Elements of X-ray Diffraction.* Addison-Wesley: Reading, MA (1956). 2nd edn. (1978).
13. Buerger, M. J. *Crystal-Structure Analysis.* John Wiley: New York (1960).
14. Warren, B. E. *X-ray Diffraction.* Addison-Wesley: Reading, PA, Menlo Park, CA, London, Ontario (1969); Dover Publications: New York (1990).
15. Darwin, C. G. The theory of X-ray reflexion. *Phil. Mag.* **27**, 315–333, 675–690 (1914).
16. Darwin, C. G. The reflexion of X rays from imperfect crystals. *Phil. Mag.* **43**, 800–829 (1922).
17. Soules, J. A., Gordon, W. L., and Shaw, C. H. Design of differential x-ray filters for low-intensity scattering experiments. *Rev. Sci. Instrum.* **27**, 12–14 (1956).
18. Nelmes, R. J. Removal of the white radiation background from data collected with β-filtered MoKα and AgKα X rays. *Acta Cryst.* **A31**, 273–279 (1975).
19. Fankuchen, I. A condensing monochromator for x-rays. *Nature (London)* **139**, 193–194 (1937).

20. Mathieson, A. McL. The relation of wavelength dispersion and scan mode in X-ray diffractometry with a monochromator. *J. Appl. Cryst.* **18**, 506–508 (1985).

21. Phillips, W. C., and Rayment, I. A systematic method for aligning double-focusing mirrors. *Methods in Enzymology.* **114**, 316–329 (1985).

22. Müller, A. A spinning target X-ray generator and its input limit. *Proc. Roy. Soc. (London)* **A125**, 507–516 (1929).

23. Taylor, A. A 5kW. crystallographic X-ray tube with a rotating anode. *J. Sci. Instruments* **26**, 225–229 (1949).

24. Usha, R., Johnson, J. E., Moras, D., Thierry J. C., Fourme, R., and Kahn, R. Macromolecular crystallography with synchrotron radiation: collection and processing of data from crystals with a very large unit cell. *J. Appl. Cryst.* **17**, 147–153 (1984).

25. Helliwell, J. R. *Macromolecular Crystallography with Synchrotron Radiation.* Cambridge University Press: Cambridge, UK (1992).

26. Bacon, G. E. *Neutron Diffraction.* 2nd edn. Oxford University Press: Oxford (1962). 3rd edn. (1975).

27. Brentano, J. C. M. The quantitative evaluation of photographic line patterns. *J. Opt. Soc. Amer.* **35**, 382–389 (1945).

28. Mees, C. E. K., and James, T. H. *The Theory of the Photographic Process.* 3rd edn. Macmillan: New York (1966).

29. Brown, J. R., Moneypenny, H. K., and Wakelin, R. J. A servo-controlled micro densitometer for x-ray diffraction photographs. *J. Sci. Instruments* **32**, 55–59 (1955).

30. Geiger, H., and Marsden, E. The laws of deflexion of alpha particles through large angles. *Phil. Mag.* **25**, 604–623 (1913).

31. Cochran, W. A Geiger-counter technique for the measurement of integrated reflexion intensity. *Acta Cryst.* **3**, 268–278 (1950).

32. Hamlin, R. Multiwire proportional counters for use in area X-ray diffractometers. *Trans. Amer. Cryst. Assn.* **18**, 95–123 (1982).

33. Hamlin, R. Glossary of frequently used detector terms. *Trans. Amer. Cryst. Assn.* **18**, 1–13 (1982).

34. Xuong, N. H., Nielsen, C., Hamlin, R., and Anderson, D. Strategy for data collection from protein crystals using a multiwire counter area detector diffractometer. *J. Appl. Cryst.* **18**, 342–350 (1985).

35. Arndt, U. W. X-ray position-sensitive detectors. *J. Appl. Cryst.* **19**, 145–163 (1986).

36. Messerschmidt, A., and Pflugrath, J. W. Crystal orientation and X-ray pattern prediction routines for area-detector diffractometer systems in macromolecular crystallography. *J. Appl. Cryst.* **20**, 306–315 (1987).

37. Blum, M., Metcalf, P., Harrison, S. C., and Wiley, D. C. A system for collection and on-line integration of X-ray diffraction data from a multiwire area detector. *J. Appl. Cryst.* **20**, 235–242 (1987).

38. Howard, A. J., Gilliland, G. L., Finzel, B. C., and Poulos, T. L. The use of an imaging proportional counter in macromolecular crystallography. *J. Appl. Cryst.* **20**, 383–387 (1987).

39. Kabsch, W. Evaluation of single-crystal X-ray diffraction data from a position-sensitive detector. *J. Appl. Cryst.* **21**, 916–924 (1988).

40. Tanaka, I., Yao, M., Suzuki, M., Hikichi, K., Matsumoto, T., Kozasa, M., and Katayama, C. An automatic diffraction data collection system with an imaging plate. *J. Appl. Cryst.* **23**, 334–339 (1990).

41. Sato, M., Yamamoto, M., Imada, K., Katsube, Y., Tanaka, N., and Higashi, T. A high-speed data-collection system for large-unit-cell crystals using an imaging plate as a detector. *J. Appl. Cryst.* **25**, 348–357 (1992).

42. Bernal, J. D., and Crowfoot, D. X-ray photographs of crystalline pepsin. *Nature (London)* **133**, 794–795 (1934).

43. King, M. V. An efficient method for mounting wet protein crystals for X-ray studies. *Acta Cryst.* **7**, 601–602 (1954).

44. Wyckoff, H. W., Doscher, M., Tsernoglou, D., Inagami, T., Johnson, L. N., Hardman, K. D., Allewell, N. N., Kelly, D. M., and Richards, F. M. Design of a diffractometer and flow cell system for X-ray analysis of crystalline proteins with applications to the crystal chemistry of ribonuclease-S. *J. Molec. Biol.* **27**, 563–578 (1967).

45. Bond, W. L. *Crystal Technology.* John Wiley: New York, London, Sydney, Toronto (1976).

46. Wood, E. A. *Crystal Orientation Manual.* Columbia University Press: New York, London (1963).

47. Bernal, J. D. On the interpretation of X-ray single-crystal rotation photographs. *Proc. Roy. Soc. (London)* **A113**, 117–160 (1926).

48. Arndt, U. W., and Wonacott, A. J. *The Rotation Method in Crystallography.* North-Holland: Amsterdam (1977).

49. Kabsch, W. Automatic indexing of rotation diffraction patterns. *J. Appl. Cryst.* **21**, 67–71 (1988).

50. Bernal, J. D. A universal X-ray photogoniometer. *J. Sci. Instruments* **4**, 273–284 (1927).

51. Greenhough, T. J., Helliwell, J. R., and Rule, S. A. Oscillation camera data processing. 3. General diffraction spot size, shape and energy profile: formalism for polychromatic diffraction experiments with monochromatized synchrotron X-radiation from a singly bent triangular monochromator. *J. Appl. Cryst.* **16**, 242–250 (1983).

52. Rossmann, M. G., and Erickson, J. W. Oscillation photography of radiation-sensitive crystals using a synchrotron source. *J. Appl. Cryst.* **16**, 629–636 (1983).

53. Sarma, R., McKeever, B., Gallo, R., and Scuderi, J. A new method for determination of the crystal setting matrix for interpreting oscillation photographs. *J. Appl. Cryst.* **19**, 482–484 (1986).

54. Vriend, G., and Rossmann, M. G. Determination of the orientation of a randomly placed crystal from a single oscillation photograph. *J. Appl. Cryst.* **20**, 338–343 (1987).

55. Greenhough, T. J., and Suddath, F. L. Oscillation camera data processing. 4. Results and recommendations for the processing of synchrotron radiation data in macromolecular crystallography. *J. Appl. Cryst.* **19**, 400–409 (1986).

56. Weissenberg, K. Ein neues Röntgengoniometer. [A new X-ray camera.] *Z. Physik* **23**, 229–238 (1924).

57. Böhm, J. Das Weissenbergsche Röntgengoniometer. [The Weissenberg camera.] *Z. Physik* **39**, 557–561 (1926).

58. Buerger, M. J. The Weissenberg reciprocal lattice projection and the technique of interpreting Weissenberg photographs. *Z. Krist.* **88**, 356–380 (1934).

59. Sakabe, N. A focusing Weissenberg camera with multi-layer-line screens for macromolecular crystallography. *J. Appl. Cryst.* **16**, 542–547 (1983).

60. Buerger, M. J. *The Photography of the Reciprocal Lattice.* (ASXRED Monograph Number 1) (1944)

61. Buerger, M. J. *The Precession Method in X-ray Crystallography.* Wiley: New York (1964).

62. Staudenmann, J.-L., Horning, R. D., and Knox, R. D. Buerger precession camera and overall characterization of thin films and flat-plate crystals. *J. Appl. Cryst.* **20**, 210–221 (1987).

63. Block, S., and Piermarini, G. J. High pressure crystallography. In: *Crystallography in North America.* (**Eds.**, McLachlan, D., Jr., and Glusker, J. P.) Section D. Apparatus and methods. Ch. 14. pp. 265–267. American Crystallographic Association: New York (1983).

64. Douzou, P. *Cryobiochemistry: An Introduction.* Academic Press: London, New York (1977).

65. Rudman, R. Low-temperature X-ray diffraction. In: *Crystallography in North America.* (**Eds.**, McLachlan, D., Jr. and Glusker, J. P.) Section D. Apparatus and methods. Ch. 11. pp. 250–253. American Crystallographic Association: New York (1983).

66. Rudman, R. *Low-Temperature X-ray Diffraction: Apparatus and Techniques.* Plenum: New York (1976).

67. Machin, K. J., Begg, G. S., and Isaacs, N. W. A low-temperature cooler for protein crystallography. *J. Appl. Cryst.* **17**, 358–359 (1984).

68. Hajdu, J., McLaughlin, P. J., Helliwell, J. R., Sheldon, J., and Thompson, A. W. Universal crystal cooling device for precession cameras, rotation cameras and diffractometer. *J. Appl. Cryst.* **18**, 528–532 (1985).

69. Cosier, J., and Glazer, A. M. A nitrogen-gas-stream cryostat for general X-ray diffraction studies. *J. Appl. Cryst.* **19**, 105–107 (1986).

70. Henriksen, K., Larsen, F. K., and Rasmussen S. E. Mounting a 10 K cooling device without rotating seals on a four-circle diffractometer. *J. Appl. Cryst.* **19**, 390–394 (1986).

71. Abrahams, S. C., Collin, R. L., Lipscomb, W. N., and Reed, T. B. Further technologies in single-crystal X-ray diffraction studies at low temperatures. *Rev. Sci. Instrum.* **21**, 396–397 (1950).

72. Kaufman, H. S., and Fankuchen, I. A low-temperature single crystal X-ray diffraction technique. *Rev. Sci. Instruments* **20**, 733–734 (1949).

73. Dulmage, W. J., and Lipscomb, W. N. The crystal structure of hydrogen cyanide, HCN. *Acta Cryst.* **4**, 330–334 (1951).

74. Williams, P. S. The cooling of crystals for X-ray scattering measurements. *Rev. Sci. Instruments* **4**, 334–336 (1933).

75. Tilton, R. F. Greatly reduced radiation damage in ribonuclease crystals mounted on glass fibers. *J. Appl. Cryst.* **20**, 130–132 (1987).

76. Hanawalt, J. D. History of the Powder Diffraction File (PDF). In: *Crystallography in North America.* (**Eds.**, McLachlan, D. Jr. and Glusker, J. P.) Section D. Apparatus and methods. Ch. 2. pp. 215–219. American Crystallographic Association: New York (1983).

77. Parrish, W. History of the X-ray Powder Method in the U.S.A. In: *Crystallography in North America.* (**Eds.**, McLachlan, D. Jr. and Glusker, J. P.) Section D. Apparatus and methods. Ch. 1. pp. 201–214. American Crystallographic Association: New York (1983).

78. Brentano, J. C. M. Parafocusing properties of the microcrystalline powder layers in X-ray diffraction applied to the design of X-ray goniometers. *J. Appl. Phys.* **17**, 420–434 (1946).

79. Debye, P., and Scherrer, P. Interferenzen an regellos orientierten Teilchen in Röntgenlicht. [X-ray interference by disordered particles.] *Physik. Z.* **17**, 277–283 (1916).

80. Hull, A. W. The crystal structure of iron. *Phys. Rev.* **9**, 84–87 (1917).

81. Hull, A. A new method of X-ray analysis. *Phys. Rev.* **10**, 661–696 (1917).

82. Young, R. A. Application of the Rietveld method for structure refinement with powder diffraction data. *Adv. X-ray Anal.* **24**, 1–23 (1981).

83. Hanawalt, J. D., Rinn, H. W., and Frevel, L. K. Chemical analysis by X-ray diffraction, classification and use by X-ray diffraction patterns. *Ind. Eng. Chem., Anal. Ed.* **10**, 457–512 (1938).

84. Rietveld, H. M. A profile refinement method for nuclear and magnetic structures. *J. Appl. Cryst.* **2**, 65–71 (1969).

85. Will, G., Parrish, W., and Huang, T. C. Crystal-structure refinement by profile fitting and least-squares analysis of powder diffractometer data. *J. Appl. Cryst.* **16**, 611–622 (1983).

86. Hill, R. J., and Madsen, I. C. The effect of profile step width on the determination of crystal structure parameters and estimated standard deviations by X-ray Rietveld analysis. *J. Appl. Cryst.* **19**, 10–18 (1986).

87. Thompson, P., and Wood, I. G. X-ray Rietveld refinement using Debye-Scherrer geometry. *J. Appl. Cryst.* **16**, 458–472 (1983).

88. Hill, R. J., and Howard, C. J. Quantitative phase analysis from neutron powder diffraction data using the Rietveld method. *J. Appl. Cryst.* **20**, 467–474 (1987).

89. Franklin, R. E., and Gosling, R. G. Molecular configuration in sodium thymonucleate. *Nature (London)* **171**, 740–741 (1953).

90. Stokes, A. R. The theory of X-ray fibre diagrams. *Prog. Biophys.* **5**, 140–167 (1955).

91. Namba, K., Pattanayek, R., and Stubbs, G. Visualization of protein-nucleic acid interactions in a virus. Refined structure of intact tobacco mosaic virus at 2.9 Å resolution by X-ray fiber diffraction. *J. Molec. Biol.* **208**, 307–325 (1989).

92. Guinier, A. *X-ray Diffraction in Crystals, Imperfect Crystals, and Amorphous Bodies.* W. H. Freeman: San Francisco and London (1963).

93. Schmidt, P. W. (**Ed.**) *Proceedings of the Symposium on Small-angle Scattering at the University of Missouri, Columbia, Missouri, March 15, 1983. Trans. Amer. Cryst. Assn.* **19** (1983).

94. Schmidt, P. W. Small-angle X-ray scattering. In: *Crystallography in North America.* (**Eds.** McLachlan, D., Jr. and Glusker, J. P.) Section D. Apparatus and methods. Ch. 13. pp. 257–264. American Crystallographic Association: New York (1983).

95. Sjoberg, B., and Osterberg, R. Small-angle X-ray scattering of chain molecules in a hydrodynamic field. The internal rigidity of double-stranded DNA. *J. Appl. Cryst.* **16**, 349–353 (1983).

96. Luzzati, V., and Taupin, D. Accuracy and resolution in small-angle crystallographic analysis. A comparison with solution scattering studies. *J. Appl. Cryst.* **19**, 51–60 (1986).

97. Arndt, U. W., and Willis, B. T. M. *Single Crystal Diffractometry.* Cambridge University Press: Cambridge, UK (1966).

98. Furnas, T. C., Jr. *Single Crystal Orienter Instruction Manual.* General Electric Company: Milwaukee (1957).

99. Arndt, U. W., and Phillips, D. C. On the determination of crystal and counter settings for a single-crystal X-ray diffractometer. *Acta Cryst.* **10**, 508–510 (1957).

100. Robinson, W. T. Diffractometer data collection. In: *Crystallographic Computing 4. Techniques and New Technologies.* (**Eds.**, Isaacs, N. W. and Taylor, M. R.) International Union of Crystallography/ Oxford University Press: New York (1988).

101. Busing, W. R., and Levy, H. A. Angle calculations for 3- and 4-circle X-ray and neutron diffractometers. *Acta Cryst.* **22**, 457–464 (1967).

102. North, A. C. T., Phillips, D. C., and Mathews, F. S. A semi-empirical method of absorption correction. *Acta Cryst.* **A24**, 351–359 (1968).

103. Sands, D. E. *Vectors and Tensors in Crystallography.* Addison-Wesley: Reading, MA (1982).

104. Clegg, W. Enhancements of the 'auto-indexing' method for cell determination in four-circle diffractometry. *J. Appl. Cryst.* **17**, 334–336 (1984).

105. Clegg, W. Diffractometer control techniques. In: *Methods and Applications in Crystallographic Computing.* (**Eds.**, Hall, S. R., and Ashida, T.) pp. 19–29. Oxford University Press: Oxford (1984).

106. Thomas, D. J. Modern equations of diffractometry. Diffraction geometry. *Acta Cryst.* **A48**, 134–158 (1992).

107. Thomas, D. J. Modern equations of diffractometry. Goniometry. *Acta Cryst.* **A46**, 321–343 (1990).

108. Schwarzenbach, D., and Flack, H. D. On the definition and practical use of crystal-based azimuthal angles. *J. Appl. Cryst.* **22**, 601–605 (1989).

109. Destro, R., and Marsh, R. E. Scan-truncation corrections in single-crystal diffractometry: an empirical method. *Acta Cryst.* **A43**, 711–718 (1987).

110. Mathieson, A. McL. Anatomy of a Bragg reflexion and an improved prescription for integrated intensity. *Acta Cryst.* **A38**, 378–387 (1982).

111. Clegg, W. Intensity measurements by diffractometry. In: *Methods and Applications in Crystallographic Computing.* (**Eds.**, Hall, S. R., and Ashida, T.) pp. 30–40. Oxford University Press: Oxford (1984).

112. Gabe, E. J. Diffractometer control with minicomputers. In: *Computing in Crystallography.* (**Eds.**, Diamond, R., Ramaseshan, S., and Venkatsesan, K.) pp. 1.01–1.31. Indian Academy of Sciences: Bangalore (1980).

113. Berger, H. Systematic errors in precision lattice-parameter determination of single crystals caused by asymmetric line profiles. *J. Appl. Cryst.* **19**, 34–38 (1986).

114. Birknes, B., and Hansen, L. K. A method for collection of step-scan data on a CAD 4 diffractometer; homogeneity of the monochromated X-ray beam. *J. Appl. Cryst.* **16**, 11–13 (1983).

115. Lehmann, M. S., and Larsen, F. K. A method for location of the peaks in step-scan-measured Bragg reflexions. *Acta Cryst.* **A30**, 580–584 (1974).

116. Seiler, P., Schweizer, W. B., and Dunitz, J. D. Parameter refinement for tetrafluoroterephthalonitrile at 98K: making the best of a bad job. *Acta Cryst.* **B40**, 319–327 (1984).

117. Hirshfeld, F. L., and Rabinovich, D. Treating weak reflections in least-squares calculations. *Acta Cryst.* **A29**, 510–513 (1973).

118. Arnberg, L., Hovmöller, S., and Westman, S. On the significance of 'non-significant' reflexions. *Acta Cryst.* **A35**, 497–499 (1979).

119. Gabe, E. J. Reducing random errors in intensity data collection. In: *Crystallographic Computing 3: Data Collection, Structure Determination, Proteins and Databases.* (**Eds.**, Sheldrick, G. M., Krüger, C., and Goddard, R.) pp. 3–17. Clarendon Press: Oxford (1985).

120. Flack, H. D. Avoidance, detection and correction of systematic errors in intensity data. In: *Crystallographic Computing 3: Data Collection, Structure Determination, Proteins and Databases.* (**Eds.**, Sheldrick, G. M., Krüger, C., and Goddard, R.) pp. 3–17. Clarendon Press: Oxford (1985).

121. Clegg, W. Data collection: avoiding blunders. In: *Crystallographic Computing 3: Data Collection, Structure Determination, Proteins and Databases.* (**Eds.**, Sheldrick, G. M., Krüger, C., and Goddard, R.) pp. 3–17. Clarendon Press: Oxford (1985).

122. Flack, H. D. Correcting intensity data for systematic effects. In: *Methods and Applications in Crystallographic Computing.* (**Eds.**, Hall, S. R., and Ashida, T.) pp. 41–55. Oxford University Press: Oxford (1984).

123. Mighell, A. D., Rodgers, J. R., and Karen, V. L. Protein symmetry: metric and crystal (a precautionary note). *J. Appl. Cryst.* **26**, 68–70 (1993).

124. Young, R. A. Background factors and technique design. *Trans. Amer. Cryst. Assn.*, **1**, 42–66 (1965).

125. Arndt, U. W., Champness, J. N., Phizackerly, R. P., and Wonacott, A. I. A single-crystal oscillation camera for large unit cells. *Acta Cryst.* **6**, 457–463 (1973).

126. Wyckoff, R. W. G. The crystal structure of some carbonates of the calcite group. *Amer. J. Science* **50**, 317–360 (1920).

127. Jacobson, R. A. An orientation-matrix approach to Laue indexing. *J. Appl. Cryst.* **19**, 283–286 (1986).

128. Fewster, P. F. Laue orientation and interpretation by microcomputer. *J. Appl. Cryst.* **17**, 265–268 (1984).

129. Hajdu, J., Acharya, K. R., Stuart, D. I., McLaughlin, P. J., Barford, D., Oikonomakos, N. G., Klein, H., and Johnson, L. N. Catalysis in the crystal: synchrotron radiation studies with glycogen phosphorylase *b*. *EMBO J.* **6**, 539–546 (1987).

130. Hajdu, J., Acharya, K. R., Stuart, D. I., Barford, D., and Johnson, L. N. Catalysis in enzyme crystals. *Trends Biochem. Sci.* **13**, 104–109 (1988).

131. Helliwell, J. R., Habash, J., Cruickshank, D. W. J., Harding, M. M., Greenhough, T. J., Campbell, J. W., Clifton, I. J., Elder, M., Machin, P. A., Papiz, M. Z., and Zurek, S. The recording and analysis of synchrotron X-radiation Laue diffraction photographs. *J. Appl. Cryst.* **22**, 483–497 (1989).

132. Cruickshank, D. W. J., Helliwell, J. R., and Moffat, K. Angular distribution of reflections in Laue diffraction. *Acta Cryst.* **A47**, 352–373 (1991).

133. Cruickshank, D. W. J., Helliwell, J. R., and Moffat, K. Multiplicity distribution of reflections in Laue diffraction. *Acta Cryst.* **A43**, 656–674 (1987).

134. Shrive, A. K., Clifton, I. J., Hajdu, J., and Greenhough, T. J. Laue film integration and deconvolution of spatially overlapping reflections. *J. Appl. Cryst.* **23**, 169–174 (1990).

135. Bartunik, H. D., Bartsch, H. H., and Qichen, H. Accuracy in Laue X ray diffraction analysis of protein structures. *Acta Cryst.* **A48**, 180–188 (1992).

136. Hajdu, J., and Johnson, L. N. Progress with Laue diffraction studies on proteins and virus crystals. *Biochemistry* **29**, 1669–1678 (1990).

137. Helliwell, J. R., Ealick, S., Doing, P., Irving, T., and Szebenyi, M. Towards the measurement of ideal data for macromolecular crystallography using synchrotron sources such as the ESRF. *Acta Cryst.* **D49**, 120–128 (1993).

138. Hoier, H., Zacharias, D. E., Carrell, H. L., and Glusker, J. P. Structure of the molecular complex of anthracene with 1,8:4,5-naphthalenetetracarboxylic dianhydride. *Acta Cryst.* **C49**, 523–526 (1993).

139. Clegg, W. Measuring and collecting single crystal data. *Notes: International Summerschool on Crystallography and its Teaching. September 15–24, 1988.* pp. 1–15. Tianjin Normal University: Tianjin, China (1988).

140. Low, B. W., and Richards, F. M. The use of the gradient tube for the determination of crystal densities. *J. Amer. Chem. Soc.* **74**, 1660–1666 (1952).

141. Matthews, B. W. Solvent content of protein crystals. *J. Molec. Biol.* **33**, 491–497 (1968).

142. Blessing, R. H. Data reduction and error analysis for accurate single crystal diffraction intensities. *Crystallography Reviews* **1**, 3–58 (1987).

143. Mark, H., and Szilard, L. Die Polarisierung von Röntgenstrahlen durch Reflexion an Kristallen. [The polarization of X-rays on reflection by crystals.] *Z. Physik.* **35**, 743–747 (1926).

144. Kirkpatrick, P. The polarization factor in X-ray reflection. *Phys. Rev.* **29**, 632–636 (1927).

145. von Laue, M. Lorentz-Faktor und Intensitätsverteilung in Debije-Scherrer-Ringen. [The Lorentz factor and the intensity distribution in Debye-Scherrer powder diffraction rings.] *Z. Krist.* **64**, 115–142 (1926).

146. Azaroff, L.V. Polarization corrections for crystal-monochromatized X-radiation. *Acta Cryst.* **8**, 701-704 (1955).

147. Busing, W. R., and Levy, H. A. High-speed computation of the absorption correction for single-crystal diffraction measurement. *Acta Cryst.* **10**, 180–182 (1957).

148. Bond, W. L. Equi-inclination Weissenberg intensity correction factors for absorption in spheres and cylinders, and for crystal monochromatized radiation. *Acta Cryst.* **12**, 375–381 (1959).

149. Walker, N., and Stuart, D. An empirical method for correcting diffractometer data for absorption effects. *Acta Cryst.* **A39**, 158–166 (1983).

150. DeTitta, G. T. ABSORB: An absorption correction program for crystals enclosed in capillaries with trapped mother liquor. *J. Appl. Cryst.* **18**, 75–79 (1985).

151. Sears, V. F. Absorption factor for cylindrical samples. *J. Appl. Cryst.* **17**, 226–230 (1984).

152. De Meulenær, J., and Tompa, H. The absorption correction in crystal structure analysis. *Acta Cryst.* **19**, 1014–1018 (1965).

153. Chidambaram, R. Computing in crystallography. Absorption corrections for single crystal X-ray and neutron data. In: *Computing in Crystallography.* (**Eds.,** Diamond, R., Ramaseshan, S., and Venkatesan, K.) pp. 2.01–2.20. Indian Academy of Sciences: Bangalore (1980).

154. H. Lipson. Absorption corrections. In: *International Tables for X-ray Crystallography.* Volume II. *Mathematical Tables. Section 5. Physics of Diffraction Methods.* (**Eds.,** Kasper, J. S., and Lonsdale, K.) pp. 291–312. Kynoch Press: Birmingham (1959).

155. Koch, B., and MacGillavry, C. H. Absorption. In: *International Tables for X-ray Crystallography.* Volume III. *Physical and Chemical Tables. Section 3. Measurements and Interpretation of Intensities.* (**Eds.,** MacGillavry, C. H., and Rieck, G. D.) pp. 157–192. Kynoch Press: Birmingham (1962).

156. Zachariasen, W. H. A general theory of X-ray diffraction in crystals. *Acta Cryst.* **23**, 558–564 (1967).

157. Becker, P. J., and Coppens, P. Extinction within the limit of validity of the Darwin transfer equations. III. Non-spherical crystals and anisotropy of extinction. *Acta Cryst.* **A31**, 417–425 (1975).

158. Seiler, P. Measurement of accurate Bragg intensities. In: *Accurate Molecular Structures. Their Determination and Importance.* (**Eds.**, Domenicano, A. and Hargittai, I.) Ch. 7, 170–198. Oxford University Press: Oxford (1992).

159. Matsubara, E., and Georgopoulos, P. Diffuse scattering measurements with synchrotron radiation: instrumentation and techniques. *J. Appl. Cryst.* **18**, 377–383 (1985).

160. Welberry, T. R. Routine recording of diffuse scattering from disordered molecular crystals. *J. Appl. Cryst.* **16**, 192–197 (1983).

161. Butler, B. D., and Welberry, T. R. Calculation of diffuse scattering from simulated disordered crystals: a comparison with optical transforms. *J. Appl. Cryst.* **25**, 391–399 (1992).

162. Poisson, S. D. Recherches sur la probabilité des jugements en matière criminelle et en matière civile, précédés des règles générales du calcul des probabilités. [Research on the probability of decisions on criminal and civil matters, preceded by general rules for calculating probabilities.] p. 206. Bachelier: Paris (1837).

163. Rees, B. Assessment of accuracy. *Isr. J. Chem.* **16**, 180–186 (1977).

164. Wilson, A. J. C. Determination of absolute from relative X-ray intensity data. *Nature (London)* **150**, 152 (1942).

8

Estimation of Relative Phase Angles

This Chapter is concerned with methods for obtaining the relative phase angles for each Bragg reflection so that the *correct electron-density map* can be calculated and, from it, the *correct molecular structure determined*. When scattered light is recombined by a lens, as described in Chapters 3 and 6, the relationships between the phases of the various diffracted beams are preserved. In X-ray diffraction experiments, however, only the intensities of the Bragg reflections are measured, and information on the relative phases is lost. An attempt is made to remedy this situation by deriving relative phases by one of the methods to be described in this Chapter. Then Equation 6.3 (Chapter 6) is used to obtain the electron-density map. Peaks in this map represent atomic positions.

When a diffraction grating, such as a crystal, interacts with X rays, the electron density that causes this diffraction can be described by a Fourier series, as discussed in Chapter 6. The diffraction experiment effects a Fourier analysis, breaking down the Fourier series (of the electron density) into its components, that is, the diffracted beams with amplitudes, $|F(hkl)|$. The relative phases $\alpha(hkl)$ are, however, lost in the process in all usual diffraction experiments. This loss of the phase information needed for the computation of an electron-density map is referred to as the phase problem. The aim of X-ray diffraction studies is to reverse this process, that is, to find the true relative phase and hence the true three-dimensional electron density. This is done by a Fourier synthesis of the components, but it is now necessary to know both the actual amplitude $|F(hkl)|$ and the relative phase, $\alpha(hkl)$, in order to calculate a correct electron-density map (see Figure 8.1). *We must be able to reconstruct the electron-density distribution in a systematic way by approximating, as far as possible, a correct (but so far unknown) set of phases!* In this way the crystallographer, aided by a computer, acts as a lens for X rays.

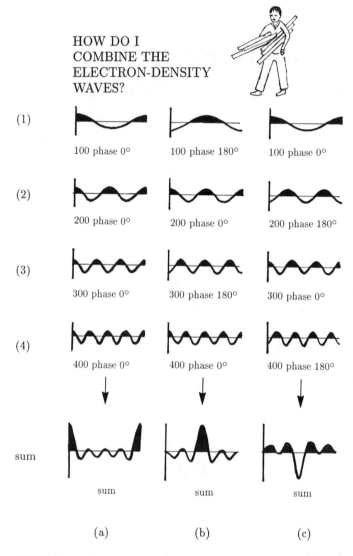

HOW DO I COMBINE THE ELECTRON-DENSITY WAVES?

(1) 100 phase 0°   100 phase 180°   100 phase 0°

(2) 200 phase 0°   200 phase 0°   200 phase 180°

(3) 300 phase 0°   300 phase 180°   300 phase 0°

(4) 400 phase 0°   400 phase 0°   400 phase 180°

sum

sum   sum   sum

(a)   (b)   (c)

FIGURE 8.1. Four waves with the same amplitude and periodicity are combined in three different ways [(a), (b), and (c)] as a result of different relative phase angles. In each case the result of the Fourier synthesis (addition of waves) is different. Shown at the top of this Figure is a crystallographer with information on amplitudes and periodicities of the electron-density waves to be summed (on cardboard strips), but no information on relative phases (how to align the cardboard strips).

The relative phase angle of a scattered wave is defined[1] as the difference between the crest of the diffracted wave and the crest of a wave scattered in the same direction by an imaginary point electron at the origin of the unit cell (as illustrated in Figure 6.2, Chapter 6). It was also pointed out in Chapter 2 that there may be several ways that a unit cell may be chosen in the crystal lattice. As a result, the origin could lie at one of a variety of possible sites. Therefore, whenever the phase angle of a Bragg reflection is discussed, it is necessary to know how the origin of the unit cell has been defined by the crystallographer.

The importance of relative phase angles to the process of obtaining the correct electron density map is stressed in Figure 8.1, where it is shown that different sets of relative phases (for the same sets of waves with known amplitudes and periodicities) can give completely different sums (electron-density maps). The impact of the phase problem on the calculation of an electron-density map can be illustrated for the simplest case, a **centrosymmetric crystal structure** in which the atomic arrangement contains a center of symmetry at the origin, so that each $|F(hkl)|$ may have only one of the two possible phase angles, $0°$ or $180°$ (see Figure 8.2) for the electron-density wave to be symmetrical about the origin along the unit cell. If there are N structure factors with a measurable intensity, there are 2^N possible electron-density maps that can be calculated. If N is as small as 15, the number of such maps is $2^{15} = 32{,}768$. Only *one* of these maps represents the *true* electron-density.

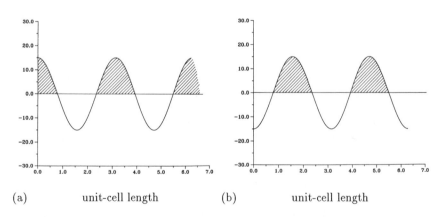

(a)   unit-cell length   (b)   unit-cell length

FIGURE 8.2. Two possibilities for the 200 electron-density wave in one unit-cell length (horizontal) of a centrosymmetric crystal. The structure factor magnitudes have relative phase angles of (a) $0°$ or (b) $180°$ (π radians), implying two possible values for each structure factor, $|F(hkl)| \cos 0° = + |F(hkl)|$ and $|F(hkl)| \cos 180° = - |F(hkl)|$. Relative phases of $| F(hkl) |$ are denoted $+$ or $-$ ($0°$ or $180°$, respectively). If the crystal structure is noncentrosymmetric, there are no such constraints on relative phase angles (any value between $0°$ and $360°$).

The phase problem is more extensive for noncentrosymmetric structures, because then most of the structure factors have phase angles with *any* value between 0° and 360°. The number of relative phase combinations is, then, *infinite*. If, however, the relative phase angle is in the correct quadrant (0 to 90°, 90 to 180°, etc.), the map is generally interpretable. Therefore, as an approximation, the number of possibilities is reduced from infinity to 4^N. Of these, only *one* is totally correct, several are essentially correct, while the majority are incorrect.

8.1 Determination of a phase angle

There are three ways commonly used to deduce the relative phase angles of a Bragg reflection, *hkl*, and one additional method, presently being developed.

1. **Constraints of direct methods.** The phase angles can be estimated by statistical methods that are based on the concept that the electron density is never negative, and that it consists of isolated, sharp peaks at atomic positions. The statistical methods for combining electron-density waves subject to these conditions are called **direct methods**. They make it possible to derive phases for a set of structure factors when only information on the magnitudes of $|F(hkl)|$ is available. At present this is the method of choice for small molecules.

2. **Interpretation of interatomic vectors.** Use of known atomic positions for an initial "trial structure" (a preliminary postulated model of the atomic structure) can be made, by application of Equations 6.21.4 and 6.21.5 (Chapter 6), to give **calculated phase angles**. Methods for obtaining such a trial structure include **Patterson** and **heavy-atom** methods. Such methods are particularly useful for determining the crystal structures of compounds that contain heavy atoms (e.g., metal complexes) or that have considerable symmetry (e.g., large aromatic molecules in which the molecular formula includes a series of fused hexagons). The Patterson map also contains information on the orientation of molecules, and this may also aid in the derivation of a trial structure.

3. **Comparison of isostructural crystals.** Phases can be estimated by comparing intensities of isomorphous (isostructural) crystals that differ only in the identity of one atom. This **isomorphous replacement method** is the method of choice for macromolecular (protein and nucleic acid) phase determination. In this case the isomorphism is generally between the crystalline macromolecule and its heavy-atom derivative obtained by replacing some of the solvent in the crystal by a compound containing a heavy atom.

4. **Measurement of multiple Bragg diffraction.** Some sets of phases may be derived from experimental measurements of ψ-scan profiles of reflections exhibiting the effects of **multiple Bragg diffraction.** Groups of relative phases have absolute values that can be detected by this method which is currently in the developmental stage, and is only applicable to those sets of Bragg reflections that display the effect.

8.2 Direct methods of relative phase determination

How is it possible to derive phase information when only structure amplitudes have been measured? An answer can be found in what are called direct methods of structure determination. By these methods the crystallographer estimates the relative phase angles *directly* from the values of $|F(hkl)|$ (the experimental data). An electron-density map is calculated with the phases so derived, and the atomic arrangement is searched for in the map that results. This is why the method is titled "direct." Other methods of relative phase determination rely on the computation of phase angles *after* the atoms in a trial structure have been found, and therefore they may be considered "indirect methods." Thus, the argument that phase information is lost in the diffraction process is not totally correct. The phase problem therefore lies in finding methods for extracting the *correct phase information* from the experimental data.

The basis of direct methods[2-6] is that there are two imposed conditions that restrict the relative values of the phase angles. These conditions that restrict the possible relative phases of Bragg reflections seem quite obvious to our chemical intuition, and are diagrammed in Figure 8.3. First, the electron-density function calculated with deduced phases should never be negative. An electron has a finite probability of being found at any position (x,y,z) in the unit cell, in which case the electron density $\rho(xyz)$ has a positive value; if, however, this probability is zero, then $\rho(xyz)$ is zero, but it is never negative (see Figure 6.9, Chapter 6). Therefore, in the Fourier summation, relative phases that reinforce negative electron density are unlikely to be correct. Second, the electron-density maps should have high values at and near atomic positions and have nearly zero values everywhere else. This is surmised from the expected shape of the electron density of an atom or ion (see Figure 6.19, Chapter 6).

Possible phase angles are constrained by these two conditions (Figure 8.3), so that relative phase determination hinges on the mathematical expressions for Fourier series. In the total Fourier synthesis involving $F(hkl)$ with the correct value of $\alpha(hkl)$, these two conditions should apply. This can be appreciated by an examination of Figure 6.9 (Chapter 6), where the negative features of all the electron-density waves have disappeared in the final summation. Exceptions occur in neutron scattering where

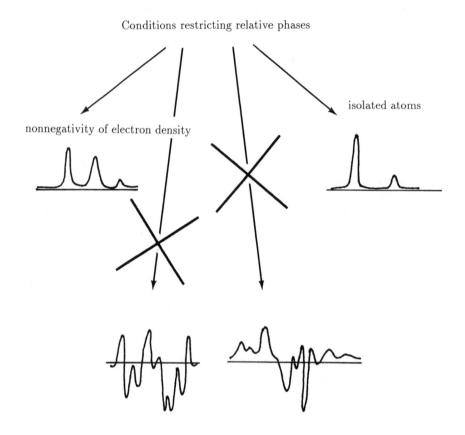

FIGURE 8.3. Two conditions restricting relative phases. Left hand side: nonnegativity of the electron-density map. Right hand side: isolated atoms.

some scattering factors are negative, and when anomalous scattering occurs in a noncentrosymmetric structure (where the scattering factor becomes complex). The possibility that there is such a relationship between measured structure amplitudes and the correct relative phase angles was first explored by Heinrich Ott,[7] Kedareswar Banerjee,[8] and Melvin Avrami.[9]

8.2.1 Normalized structure factors (E values)

In direct methods, the measured structure factors are modified so that the maximum information on atomic position can be extracted from them. Other effects, such as the falloff of intensity at high scattering angles due to atomic size and atomic vibrations (see Figures 3.12 and 3.13, Chapter 3) are eliminated, to be considered when the structure has been determined. The structure-factor data are therefore converted to those expected for a structure composed of point atoms at rest (not vibrating). The scattering factor expression for real atoms (described in Chapter 3) involves the scattering of an atom with a finite size (represented by f_{jo}), and an exponential factor representing atomic vibrations and disorder. The expression is:

$$f_j = f_{j_o} e^{-B_j(\sin^2 \theta / \lambda^2)}. \tag{8.1}$$

In direct methods it is usual to replace this expression by one for that of point atoms, so that X-ray scattering is now essentially independent of $\sin \theta / \lambda$. This is done by dividing $F(hkl)$ by a function of f_j that eliminates any fall-off of intensity as a function of $\sin \theta / \lambda$ (see Figure 8.4). The resulting **normalized structure factor**, $|E(hkl)|$ is:

$$| E(hkl) | = \frac{| F(hkl) |}{\left[\epsilon \sum_j f_j^2(hkl) \right]^{\frac{1}{2}}} \tag{8.2}$$

where f_j was defined in Equation 8.1, and the integer ϵ the **epsilon factor** is a factor that is generally unity, but is needed to account that certain classes of Bragg reflections, such as those with one zero index, that have, as a group, an average intensity higher than that for the rest of the Bragg reflections.[10-13] Values for ϵ for each type of Bragg reflection

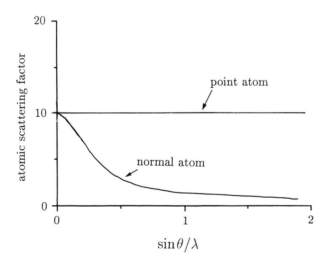

FIGURE 8.4. Scattering curves for regular atoms and stationary point atoms. Point atoms are considered to have no width and no vibrational amplitude. Compare this Figure with Figure 3.13, Chapter 3.

can be found in *International Tables*.[12] Some examples are give in Table 8.1. The resulting normalized structure factors (E values) have an average value (over the entire data set) for their square equal to unity ($< E_{hkl}^2 > = 1$), hence the name "normalized."

An advantage of using $|E(hkl)|$ rather than $|F(hkl)|$ values is that peaks in the resulting **E map** [equivalent to an electron-density map using $E(hkl)$ in place of $F(hkl)$] are sharp and heightened, so that they are easier to identify. Also, since all classes of Bragg reflections are normalized to the same value, it is possible to compare sets of reflections. High $|E(hkl)|$ values imply that most atoms scatter in phase for that particular Bragg reflection. As a result, there is a high probability that atoms lie somewhere on the most positive area of the electron density waves corresponding to that Bragg reflection, as shown for the 3 0 0 electron-density wave in Figure 6.9(a) (Chapter 6).

It is somewhat surprising that, in spite of what has been said in earlier chapters about the loss of phase information in the X-ray diffraction experiment, the intensity distribution contains information on whether the structure is centrosymmetric or noncentrosymmetric. This information is important when there is a space group ambiguity in which one possibility is centrosymmetric, while the other is not (see Chapter 4). Arthur J. C. Wilson[10,14,15] noted that, while the intensities of Bragg reflections on the average depend only on the atomic contents of the unit cell, the **distribution of intensities** is different depending on whether the structure is centrosymmetric or noncentrosymmetric. The intensities

TABLE 8.1. Values of the ϵ factor for three crystal systems.

	Symmetry	*hkl*	*0kl*	*h0l*	*hk0*	*h00*	*0k0*	*00l*
triclinic	1	1	1	1	1	1	1	1
	$\bar{1}$	1	1	1	1	1	1	1
monoclinic	2	1	1	1	1	1	2	1
	m	1	1	2	1	2	1	2
	2/*m*	1	1	2	1	2	2	2
orthorhombic	222	1	1	1	1	2	2	2
	*mm*2	1	2	2	1	2	2	4
	mmm	1	2	2	2	4	4	4

Note that ϵ is a factor needed to account for the enhancement of intensities of certain classes of Bragg reflections with at least one index zero, i.e., certain classes of Bragg reflections that have an average intensity higher than the average.

from a noncentrosymmetric crystal are generally nearer the mean than are those from a centrosymmetric crystal, i.e., centrosymmetric crystals give a relatively higher number of weak Bragg reflections than do non-centrosymmetric crystals. This difference is usually measured in terms of E values. For example, as shown in Table 8.2 the average value of $|E|$ is $(2/\pi)^{1/2} = 0.798$ if the crystal structure is centrosymmetric, but larger, $[(\pi/4)^{1/2} = 0.886]$, if the crystal structure is noncentrosymmetric (see Figure 8.5).[16] When intensity data have been converted to $|E|$ values, one of the first things that the crystallographer does is to check the average value of $|E|$ to determine whether the crystal structure is more likely to be centrosymmetric (overall mean $E \approx 0.9$), or noncentrosymmetric (overall mean $E \approx 0.8$) [see Table 8.2(a)]. This method generally works fairly well.[3] A note of caution, however, is appropriate! The E statistics must be viewed as a useful but not conclusive criterion for whether the crystal structure is centrosymmetric or not, because crystals do not consist of randomly distributed atoms, as assumed in this statistical analysis. It may be necessary to wait until the refinement stage has been nearly completed before a final decision can confidently be made.

TABLE 8.2. A comparison of theoretical E-value statistics.

(a) Average value	*noncentrosymmetric*	*centrosymmetric*	*hypercentric*
$< E >$	0.886	0.798	0.718
$< E^2 >$	1.000	1.000	1.000
$< E^2 - 1 >$	0.736	0.968	1.145
$< (E^2 - 1)^2 >$	1.0	2.0	
$< (E^2 - 1)^3 >$	2.0	3.0	

(b) Fraction of E values greater than a certain value.

E greater than	% if noncentro	% if centro	% if hypercentric
0.20	96	84	73
1.00	36.8	31.7	23
2.00	1.8	4.6	6.2
3.00	0	0.3	1.0

$< >$ *is the average over all Bragg reflections hkl. A hypercentric distribution implies centrosymmetric molecules in general positions of a centrosymmetric space group.*

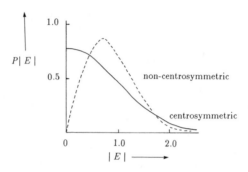

(a)

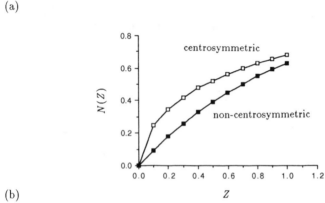

(b)

FIGURE 8.5. (a) The probability $P(E)$ of $|E|$ having certain values depends on whether the structure is centrosymmetric (solid line) or noncentrosymmetric (broken line). Thus, if $|E|$ values are known, a plot of the number with different ranges of $|E|$ will indicate the centrosymmetry (or lack of it) in the crystal structure. (b) The cumulative distribution curves for centrosymmetric and noncentrosymmetric crystals. $N(Z)$ is the fraction of Bragg reflections with intensities (or $|E|^2$ values) less than or equal to Z times the mean intensity.[17]

8.2.2 Development of phase information

The methods used to ensure that the Fourier summation does not give a negative electron-density map are mathematical in nature. David Harker and John Kasper[18,19] in 1948 used the **inequality** relationships of Augustin Louis Cauchy, Hermann Amandus Schwarz, and Victor Buniakowsky [Buniakovski][20-23] (generalized to the **Cauchy–Schwarz inequality**) to derive relationships between the structure factors (the **Harker–Kasper inequalities**). These were used by David Harker, John Kasper, and Charlys Lucht[24] to determine the structure of decaborane, $B_{10}H_{14}$, which was unknown at that time. For this study they used **unitary structure factors**, that is, structure factors (ranging in absolute magnitude from 0 to 1) with the effects of atomic size removed:

$$U(hkl) = F(hkl)_{\text{point atom}}/F(000)$$

$$= |F(hkl)| / (\sum_j f_j(hkl) \exp(-B \sin^2 \theta \lambda^2)). \qquad (8.3)$$

When $|U(hkl)|$ is unity, all atoms scatter in phase, although such high values are rarely found. For example, by an application of inequality relationships, they found:[24,25]

$$U^2(hkl) - \frac{1}{2} \leq \frac{1}{2}[U(2h, 2k, 2l)] \qquad (8.4)$$

so that if $|U^2(hkl)|$ is greater than 0.7, the left-hand side of Equation 8.4 is zero or positive. As a result, $U(2h,2k,2l)$ must be positive (phase $0°$, see Figure 8.2). Normalized $|E(hkl)|$ structure factors, rather than unitary structure factors $|U(hkl)|$, are now generally used, but the principles of phase determination are similar.

Certain linear combinations of relative phases are uniquely determined by the crystal structure and are independent of the choice of origin. These relative phase combinations are of great importance in direct methods, and they are called **structure invariants.** They generally involve sets of three reflections, often called "triplets;" these are $h_1 k_1 l_1$, $h_2 k_2 l_2$, $h_3 k_3 l_3$ such that $h_1 + h_2 + h_3 = k_1 + k_2 + k_3 = l_1 + l_2 + l_3 = 0$. The notation that we use here is: $\mathbf{h} = h_1 k_1 l_1$ or the indices of one Bragg reflection, $\mathbf{h'} = h_2 k_2 l_2$, the indices of a second Bragg reflection, and $- (\mathbf{h} + \mathbf{h'}) = - (h_1+h_2), - (k_1+k_2), - (l_1+l_2)$, the indices of a third Bragg reflection, whose indices are equal to the negative of the sum of the first two (so that the sum of all three is zero). For example, the Bragg reflections $2\bar{1}3$, $42\bar{5}$ and $\bar{6}12$ provide such a triplet. There also exist many more linear combinations of relative phases whose values remain unchanged when the location of the origin is changed provided these changes are made subject to specific space-group symmetry constraints. These are called **structure seminvariants.**

There is a high probability that if phase information is available on two of the component Bragg reflections ($\mathbf{h}$ and $\mathbf{h'}$, or $2\bar{1}3$ and $42\bar{5}$) the phase of the third ($\mathbf{h} + \mathbf{h'}$, or $61\bar{2}$) can be derived. For most space groups, the phases of three Bragg reflections may be chosen arbitrarily (according to certain rules, see Figure 8.6), thereby defining the origin of the unit cell. It is necessary to define where the origin of the unit cell lies because, although this will not affect $|F(hkl)|$ values, it will affect relative phase values. As shown in Figure 8.6, if h, k, and l are all even, the choice of origin is not important, but for all other cases it is important. In centrosymmetric cases three Bragg reflections are chosen with h, k, and l not all even, and not linearly dependent (see Figure 8.6). Then if h is even, k is odd, and l is odd (*even, odd, odd*), the other two Bragg reflections chosen cannot be in this class. For noncentrosymmetric

| Origin: X | 0, | $\frac{1}{2}$, | 0, | 0, | 0, | $\frac{1}{2}$, | $\frac{1}{2}$, | $\frac{1}{2}$, |
| Y | 0, | 0, | $\frac{1}{2}$, | 0, | $\frac{1}{2}$, | 0, | $\frac{1}{2}$, | $\frac{1}{2}$, |
Z	0.	0.	0.	$\frac{1}{2}$.	$\frac{1}{2}$.	$\frac{1}{2}$.	0.	$\frac{1}{2}$.
$h\,E,\,k\,E,\,l\,E$	+	+	+	+	+	+	+	+
$h\,O,\,k\,E,\,l\,E$	+	−	+	+	+	−	−	−
$h\,E,\,k\,O,\,l\,E$	+	+	−	+	−	+	−	−
$h\,E,\,k\,E,\,l\,O$	+	+	+	−	−	−	+	−
$h\,E,\,k\,O,\,l\,O$	+	+	−	−	+	−	−	+
$h\,O,\,k\,E,\,l\,O$	+	−	+	−	−	+	−	+
$h\,O,\,k\,O,\,l\,E$	+	−	−	+	−	−	+	+
$h\,O,\,k\,O,\,l\,O$	+	−	−	−	+	+	+	−

FIGURE 8.6. Effects of changes of the unit-cell origin on the signs of $\alpha(hkl)$ for a centrosymmetric structure. Origin locations (X, Y, Z) are listed. The change in relative phase angle is shown for various types of Bragg reflections (E = even, O = odd). In this Figure, + means no change in relative phase (sign); − means a change in relative phase (sign). Note that when h, k, and l are all even, the choice of origin is not important.

space groups, an enantiomorph must also be specified by an additional choice of phase. Further development of phases by the methods just described can lead to relative phases for large numbers of Bragg reflections, as will be shown in detail later.

Work on direct methods was continued independently by Jerome Karle and Herbert Hauptman,[5,26–30] Joseph Gillis,[31,32] William H. Zachariasen,[33,34] David Sayre,[35] William Cochran,[36] and Isabella Karle.[37] Phase relationships were clearly found for the crystal structure of decaborane (which is centrosymmetric), but were not so clear for some other structures. Gillis[31,32] showed, however, that often certain inequalities were nearly satisfied, and this observation led to subsequent investigations of the **probability relationships** among the structure factors. Karle and Hauptman[27] went on to show how the use of inequalities can restrict the range of phase angles for noncentrosymmetric structures. All of these studies led to the direct methods now used routinely in small-molecule crystallographic laboratories.

For a centrosymmetric structure, using a triplet of Bragg reflections [(**h**, **h′**, − (**h** + **h′**)] each with high E values (implying intense Bragg reflections), the product of the signs of the three relative phase angles has a high probability of being positive [since in a centrosymmetric structure the signs of (− **h** − **h′**) and (**h** + **h′**) are equal]. Otherwise the constraints on the structure factors will not hold, and the calculated electron-density map may contain several deeply negative areas. The relationship so derived is shown in Equation 8.5. It is called the Σ_2 formula:

$$\Sigma_2 \text{ formula}: \quad S(\mathbf{h}) \times S(\mathbf{h'}) \times S(\mathbf{h + h'}) \approx +, \qquad (8.5)$$

where $S(\mathbf{h})$ means the sign of the centrosymmetric Bragg reflection hkl, and $\approx$ means "is probably equal to." This relationship is called the "triple-product sign relationship." In practice, since Bragg reflections with the highest $E(hkl)$ values contribute most to the subsequent E map, it is usual to restrict a consideration of sign relationships to the 10% of the highest E values. There is no information in Equation 8.5 on the individual relative phases, only on their product, which is independent of the choice of origin.

Equation 8.5 is at the heart of direct methods procedures. Some examples of its use are given in Figures 8.7 and 8.8. For example, if Bragg reflections $2\bar{1}3$, $42\bar{5}$, and $61\bar{2}$ are all intense and if it is already known (by some means or other) that the Bragg reflections $2\bar{1}3$ and $42\bar{5}$ have phases of $0°$ (signs +), then $61\bar{2}$ probably also has a phase angle of $0°$ (Equation 8.6a). If $2\bar{1}3$ has a phase angle of $180°$ (sign −) and $42\bar{5}$ has a phase angle of $0°$ (sign +), then $61\bar{2}$ would probably (but not definitely) have a phase angle of $180°$ (sign −) (Equation 8.6b).

$$\{S(2\bar{1}3) = +\} \times \{S(42\bar{5}) = +\} \times \{S(61\bar{2}) = +\} \approx + \qquad (8.6a)$$

$$\{S(2\bar{1}3) = -\} \times \{S(42\bar{5}) = +\} \times \{S(61\bar{2}) = -\} \approx + \qquad (8.6b)$$

Note that h, k and l may each be positive or negative; the relationships of the phases to that of $F(hkl)$ when the signs of h, k, or l are changed are tabulated for each space group in *International Tables, Volume 1*.[38]

Statistical methods are used to estimate the probability of each of the relationship [shown in Equation 8.5], and if the probability relationship is high, it is accepted as true. In the centrosymmetric case, the probability that a triple product (involving **h**, **h′** and **h** − **h′**) is positive is:

$$P[\mathbf{h}, \mathbf{h'}, \mathbf{h} - \mathbf{h'}]_+ = 0.5 + 0.5 \tanh\left[(N^{-\frac{1}{2}}) \mid E(\mathbf{h})E(\mathbf{h'})E(\mathbf{h} - \mathbf{h'}) \mid\right] \qquad (8.7)$$

where N is the number of atoms in the unit cell. The probability that $E(\mathbf{h})$ is positive is

$$P[E(\mathbf{h})]_+ = 0.5 + 0.5 \tanh\left[(N^{-\frac{1}{2}}) \mid E(\mathbf{h}) \mid \sum_{\mathbf{h'}} E(\mathbf{h'})E(\mathbf{h} - \mathbf{h'})\right]. \qquad (8.8)$$

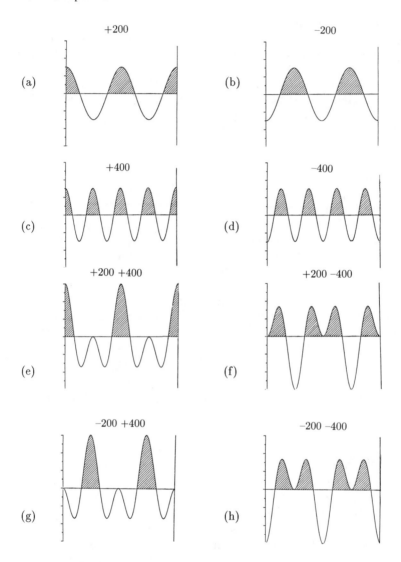

FIGURE 8.7. Direct sign determination by analytical methods, dependent on the least negative electron-density map. In general, for centrosymmetric structures, if $F(hkl)$ is large, whatever its sign and $F(2h, 2k, 2l)$ is also large, then the latter is probably positive with a phase of $0°$. Possible situations are (a) and (b) for $F(200)$, and (c) and (d) for $F(400)$. (a), (b), (c) and (d) are consistent only if 400 is positive. The effects of different signs on the summed electron density is shown in (e), (f), (g), and (h). In (e) and (g), where 400 is positive, the electron density is less negative than in (f) and (h). Therefore it is likely that $F(400)$ is positive. Nothing is learned in this example about the sign of $F(200)$.

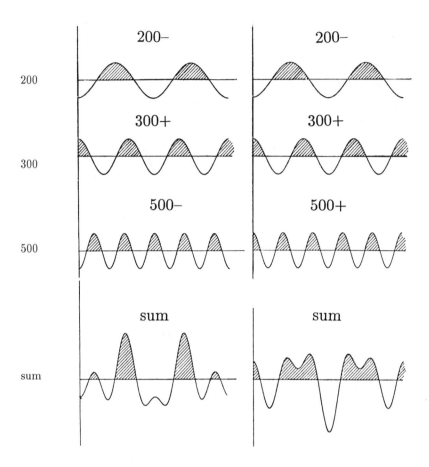

FIGURE 8.8. Direct phase determination for a centrosymmetric crystal structure. Two possible summations of waves are shown (read vertically). If | $F(200)$ |, | $F(300)$ |, and | $F(500)$ | are all large, they will make a significant contribution to the electron-density map. If it is known that | $F(200)$ | has a negative sign (relative phase 180°) and | $F(300)$ | has a positive sign (relative phase 0°), then it is most likely, by summing terms, that | $F(500)$ | is negative (relative phase 180°), since this gives a sum with the less negative electron-density map (compare the summations on the left and on the right). The sum (electron-density map) on the left is better because its background is less negative.

In the determination of the crystal structure of hexamethylbenzene (Figure 1.9, Chapter 1), the Bragg reflections $7\bar{3}0$, 340 and $4\bar{7}0$ form a triplet. It was found that by placing atoms at the intersections of high positive values for the electron-density waves (plots called structure factor graphs), Kathleen Lonsdale (in 1929) was able to solve the crystal structure, shown for the $7\bar{3}0$ and 340 electron-density waves in Figure 8.9.[39]

The **symbolic addition method**[33,34,37,40] has proved to be a very useful and much-used procedure in direct methods of structure analysis. Origin-defining phases are selected. Additional phases are initially represented by symbols *a*, *b*, etc. to give a basis set. More phase angles, expressed in terms of those in the chosen basis set, are derived by use of phase relations, such as Equation 8.5. The result for a centrosymmetric structure, will be sets of relative phases that are designated each +, −, *a*, *b*, or *c*, etc. Finally, the correct signs for the total set (i.e., signs for *a*, *b*, *c*, etc.) may become evident from obvious relationships, for example, those that contain a^2 (which must be positive), or may be checked by calculating the two possible electron density maps, one with *a* = + and

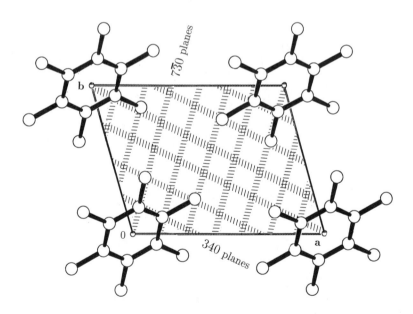

FIGURE 8.9. Hexamethylbenzene structure (Ref. 39), and the electron-density waves 340 and $7\bar{3}0$ (hatched). The actual crystal structure is shown (cf., Figure 1.9, Chapter 1). Note that all atoms lie on the positive intersections of these electron-density waves.

the other with $a = -$, and distinguishing, by use of chemical information, which is the correct map. Several direct-methods computer programs[41-47] provide an automated way of using triple-phase invariants. The most useful relationship for the solution of centrosymmetric structures is the Σ_2 formula (a variant of Equation 8.5):

$$S(h) = \sum_{h'} S(h') \times S(h - h').\tag{8.9}$$

The derivation of phases for noncentrosymmetric structures is more difficult, because the values are not restricted to 0° or 180°. In addition to the origin-defining phases, an enantiomorph-defining phase must be selected. Usually phases are derived by the **tangent formula**:

$$\tan[\alpha(h)] \approx \frac{\sum_{h'} |E(h')E(h - h')| \sin[\alpha(h') + \alpha(h - h')]}{\sum_{h'} |E(h')E(h - h')| \cos[\alpha(h') + \alpha(h - h')]},\tag{8.10}$$

where all available terms are summed. This Equation is used to develop additional phases and to refine them. The estimated phases $\alpha(h)$ are thus the average taken over all the triplet reflections involving h'. The main contributions to this Equation will be those with high $|E|$ values. If most of the resulting phase angles found by this method are within 45° of the true values, the resulting electron-density map will generally reveal a recognizable portion of the structure. For noncentrosymmetric structures, the formulæ used in computer programs are general versions of those for centrosymmetric structures, for example:

$$\alpha(h) \approx \langle \alpha(h') + \alpha(h - h') \rangle_{h'}.\tag{8.11}$$

Most direct methods phasing procedures follow the sequence listed in Figure 8.10, illustrated by an example in Figure 8.11. The relative phases are derived by application of some of the equations listed above, an E map is calculated, coordinates are determined for peaks in this map (on the assumption that they represent atomic positions), and the resulting trial structure is examined for chemical reasonableness. Sometimes there is more than one set of phases that must be tested in this way because the results of direct methods are not conclusive for this crystal structure. Only the results of refinement (see Chapter 10) will indicate if the correct structure has been found.

Despite the success of direct methods, there are still certain structures that are not readily solved, if at all, by these methods. This may be due to a breakdown in the certainty with which the triplet relationships are developed, or to a breakdown of the assumption that the atomic position vectors form a set of random variables, so that the probability functions that are derived may no longer be strictly valid. Some new developments for dealing with these problems have focused on improving the techniques for calculating the triplets[48] and including in the direct methods procedure as much structural information as possible, such as

Procedures in direct methods.

1. *Normalized structure factors $|E|$ are calculated from the observed magnitudes $|F|$ of the structure factors. Only high $|E|$ values (e.g., $|E| > 1.5$) are used, because they will be the main contributors to the E map.*

2. *Phases are assigned to a limited number of selected reflections. These are the starting set and include those that define the location of the origin (origin-defining phases) and those that define which enantiomorph of the crystal structure, if any, is chosen (enantiomorph-defining phases).*

3. *Sets of three Bragg reflections are selected with indices that satisfy the triple-product sign relationship Σ_2 formula:*

$$\alpha(h) \approx \left\langle \alpha(h') + \alpha(h - h') \right\rangle_{h'}. \qquad (8.11)$$

   where $\alpha(\mathbf{h})$ means the relative phase angle of the centrosymmetric Bragg reflection hkl, and $\approx$ means "is probably equal to." In other words: h,k,l (or $\mathbf{h}$); h', k', l' (or $\mathbf{h}'$); $h + h'$, $k + k'$, $l + l'$ (or $\mathbf{h} + \mathbf{h}'$), are found. This is called the Σ_2 (sigma-2) listing and is prepared by considering each $|E(hkl)| > 1.5$ and searching for all possible interactions of $\mathbf{h}$ with $\mathbf{h}'$ and $\mathbf{h} + \mathbf{h}'$. Some Bragg reflections have many interactions, while others have only a few.

4. *Probabilities are calculated that each triple product is positive and that each E value is positive. The tangent formula:*

$$\tan[\alpha(h)] \approx \frac{\sum_{h'} |E(h')E(h - h')| \sin[\alpha_h + \alpha_{(h-h')}]}{\sum_{h'} |E(h')E(h - h')| \cos[\alpha_h + \alpha_{(h-h')}]}. \qquad (8.10)$$

   is used for the refinement of noncentrosymmetric structures, to give a consistent set of relative phases.

5. *The most probably correct (best) phase set is used to calculate an E map. This map is calculated the same way as an electron-density map, but uses $|E(hkl)|$ instead of $|F(hkl)|$ as the Fourier coefficient.*

6. *The area around each peak in the E map is searched for evidence of atomic connectivity and hence molecular fragments. The coordinates of peaks selected from this map represent atomic positions and are used as an initial trial structure.*

FIGURE 8.10. Procedures in direct methods.

Development of phases

0kl values for a small-molecule crystal structure are listed for E values greater than 1.5. 0kl data were chosen in order to simplify the description of the method. In practice all E(hkl) greater than 1.5 are used. The crystal is orthorhombic, unit-cell dimensions $a = 13.52$, $b = 26.58$, $c = 7.26$ Å, space group $P2_1 2_1 2_1$.

(a) *E values from observed Bragg reflection intensities (see* Chapter 7).

hkl	E	hkl	E
0 18 5	3.90	0 20 4	3.38
0 2 1	2.93	0 13 8	2.76
0 7 3	2.21	0 23 4	2.21
0 7 1	2.07	0 22 3	2.03
0 16 5	1.99	0 20 3	1.91
0 14 6	1.85	0 14 1	1.83
0 32 1	1.82	0 8 6	1.67
0 5 1	1.62		

(b) *Choice of origin. Listed are h, k, l, $\alpha(hkl)$.*

0 7 1	90°	0 16 5	180°

Note that even though the crystal structure is centrosymmetric in projection (view down **a**), *in this space group the origin is, by convention, chosen so that the relative phase angle $\alpha(0kl)$ equals 0, 90, 180 or 270°.*

(c) *Relationships between phases when the signs of h, k, and/or l are changed. (From International Tables)*[38] *are:*

0kl	hkl	$\overline{h}\overline{k}\overline{l}$	$\overline{h}kl$	$h\overline{k}l$	$hk\overline{l}$	possible phases
k even, l even	+	−	−	−	−	0, or 180°
k even, l odd	+	−	−	$\pi-$	$\pi-$	0, or 180°
k odd, l even	+	−	$\pi-$	$\pi-$	−	90, or 270°
k odd, l odd	+	−	$\pi-$	−	$\pi-$	90, or 270°

These changes in sign must be taken into consideration when relationships between the phases of Bragg reflections are being developed. Note, however, that values of the relative phases for the 0kl Bragg reflections are limited to 0, 90, 180, and 270°.

(d) Σ_2 *relationships*

$$\alpha(0\,2\,1) = -\alpha(0\,18\,5) + \alpha(0\,20\,4) + 180° \qquad (A)$$
$$\alpha(0\,2\,1) = -\alpha(0\,14\,6) + \alpha(0\,16\,5) + 180° \qquad (B)$$
$$\alpha(0\,2\,1) = -\alpha(0\,20\,4) + \alpha(0\,22\,3) + 180° \qquad (C)$$
$$\alpha(0\,7\,1) = -\alpha(0\,16\,5) + \alpha(0\,23\,4) \qquad (D)$$
$$\alpha(0\,14\,6) = -\alpha(0\,18\,5) + \alpha(0\,32\,1) + 180° \qquad (E)$$
$$\alpha(0\,23\,4) = \alpha(0\,5\,1) + \alpha(0\,18\,5) \qquad (F)$$

(e) *From Equation D,* $\alpha(0\,23\,4) = 270°$.

(f) *Symbolic phases may be chosen:*

$$\alpha(0\,5\,1) = a$$

$$\alpha(0\,18\,5) = b$$

$$\alpha(0\,2\,1) = c.$$

(g) *From Equation F,* $a + b = 270°$; *therefore* $b = 270° - a$.
(h) *From Equation A,* $\alpha(0\,20\,4) = c + b - 180° = c - a + 90°$.
(i) *From Equation C,* $\alpha(0\,22\,3) = c - c + a - 90° - 180° = a - 270°$.
(j) *From Equation B,* $\alpha(0\,14\,6) = 360° - c = -c$.
(k) *From Equation E,* $\alpha(0\,32\,1) = 360° - c + b - 180° = -c - a + 90°$

(l) *In this way relative phases for 10 out of 15 of the highest E values are derived (with respect to a and c which are not yet known). Both a and c can either equal 0, 90, 180, or 270°. Now only 16 electron-density maps need be calculated, instead of 4^8 (= 65,536) electron-density maps.*

(m) *The relative probabilities of Equations A to F are:*

A	6.8	C	3.5	E	2.3
B	1.9	D	1.6	F	2.5

(n) *Working with more E values (for all hkl Bragg reflections with E greater than 1.5), and accounting for the probability of each relationship, it is found that $a = 270°$ and $c = 180°$. At this stage an electron-density map can be calculated.*

FIGURE 8.11. Development of phases.

known stereochemistry, known atomic positions, or both known position and orientation of some of the structure. Other methods have focused on the use of higher-order structure invariants ("quartets" and "quintets").[49] These are sets of four (or five) Bragg reflections with indices that have a zero sum. A so-called negative quartet has a phase sum that is probably near 180° rather than 0°. Newer methods, such as these, have greatly enhanced the power of direct methods in structure determination.

8.3 Patterson methods

A second method of relative phase determination is to find a preliminary atomic arrangement (trial structure), calculate the relative phase angles of all Bragg reflections, obtain the electron-density map, and then check if the trial structure has been chosen correctly. The Patterson method has proved to be a very powerful and successful way of determining an initial trial structure. The method can also be used for determining the locations of heavy atoms in protein derivative crystal structures.

The **Patterson function** is a map that indicates all the possible relationships (vectors) between atoms in a crystal structure. It was introduced by A. Lindo Patterson[50-52] in 1934, inspired by earlier work on radial distribution functions in liquids and powders.[53-54] In crystals the directionality as well as the lengths of vectors between atoms (atomic distances) can be deduced. By contrast, in liquids and powders the geometric information that can be obtained is limited to interatomic distances, because in these the molecules are randomly oriented. While the use of the Patterson function revolutionized the determination of crystal structures of small molecules in the 1930s to 1950s, direct methods are now the most widely used methods for obtaining structures of small organic molecules. The Patterson function, however, continues to play an essential part in the determination of crystal structures of inorganic compounds and macromolecules. It is also very useful when the structure of a small molecule proves difficult to solve by direct methods.

The Patterson map, commonly designated $P(uvw)$, is a Fourier synthesis that uses the indices, h,k,l, and the square of the structure factor amplitude, $|F(hkl)|^2$, of each diffracted beam. It is usual to describe the Patterson map in vector space defined by u, v, and w, rather than x,y,z as used in electron-density maps.

$$P(uvw) = \frac{1}{V} \sum_{h,k,l} \sum \sum |F(hkl)|^2 \cos 2\pi(hu + kv + lw). \qquad (8.12)$$

This Equation 8.12 has the same form as the equation for electron density (Equation 6.3, Chapter 6), but note that *there is no phase angle in the expression!* The coefficients of the Patterson function are the observed intensities, after some geometric corrections involved with the data collection process are made. Because the Patterson function uses coefficients that are the squares of the structure factor expression, only the cosine terms of the structure factor are conserved and no phases are needed. This is an extremely useful feature since all the quantities required are readily accessible to direct measurement in the X-ray diffraction experiment, and the Patterson function can then be calculated (by a Fourier synthesis) without any ambiguity. The Patterson function can also be described as the Fourier transform of $F(hkl)^2$. There is only one Patterson map for any one crystal structure. In rare circumstances it is possible to have two different interpretations (two unique sets of atomic coordinates) for the same Patterson map, that is, **homometric structures**, but these have only been observed in heavy-atom positions in a crystal structure that is no longer homometric when lighter atoms are taken into account.[55,56]

   If the electron density is known, the Patterson function at any point u,v,w may be calculated by multiplying the electron density at a point x,y,z with that at $x+u$, $y+v$, $z+w$, doing this for all values of x, y, and z and summing the products:

$$P(uvw) = \int_V \rho(xyz)\, \rho(x+u, y+v, z+w)\, dV. \tag{8.13}$$

In other words, the Patterson function is the convolution of the electron density at all points x, y, z in the unit cell with the electron density at points $x+u$, $y+v$, $z+w$ (see Figure 2.17, Chapter 2, for the definition of a convolution). A peak at u,v,w in the Patterson map represents a vector from the origin of the Patterson function to the point u,v,w. This means that if any two atoms in the unit cell are separated by a vector u,v,w, then there will be a peak in the Patterson map at u,v,w. This peak represents the vector between two atoms, one at x_1,y_1,z_1 and one at x_1+u, y_1+v, z_1+w and tells us what u,v,w are but not what x_1,y_1,z_1 are (see Figure 8.12). The important feature is that, since there is a peak at this location in the Patterson map, then, within the actual arrangement of atoms in the crystal structure, there are *at least two atoms separated by this particular vector*. The height of a peak in a Patterson map is approximately proportional to the products of the atomic numbers, $Z_i Z_j$ of the atoms at the two ends of the vector. The Patterson map has a very high peak at the origin representing a vector between each atom and itself ($u = v = w = 0$). For a centrosymmetric structure with atoms at x, y, z, and $-x$, $-y$, $-z$, the Patterson map, shown in Figure 8.13, is obtained. If the unit cell contains a small number of relatively heavy atoms, the vectors between them will dominate the vector map. Examples of the Patterson maps of potassium dihydrogen phosphate,[57] copper sulfate,[58] and hexachlorobenzene[59] are shown in Figures 8.14 to 8.16, respectively.

   In order to enhance peaks in a Patterson map, it is common to use $|E|^2$ values as coefficients, corresponding to point atoms, and compute a **sharpened Patterson function.** This can be done by dividing the

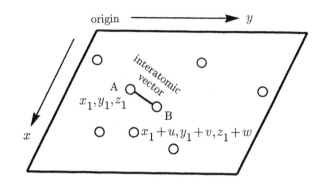

(a)

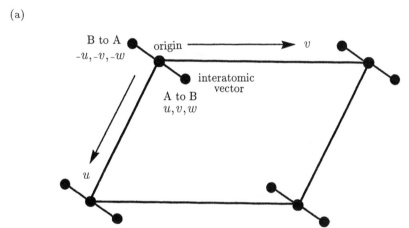

(b)

FIGURE 8.12. The Patterson map represents all interatomic vectors, moved so that one end of each vector is at the origin of the map. (a) An interatomic vector, between two atoms A and B is represented in the Patterson map by a vector with one end at the origin of the Patterson map. (b) In a centrosymmetric triclinic crystal structure with one heavy atom per asymmetric unit, if the Patterson map has a peak at u, v, then the heavy atom lies at $x = u/2$, $y = v/2$. If u and v are measured, then x and y will be found.

values of $|F(hkl)|^2$ by an exponential function, for example, $\exp(-2B\sin^2\theta/\lambda^2)$, where the value of B is obtained from a Wilson plot (see Figure 7.22, Chapter 7). It is now more common to use values of $|E^2-1|$ as coefficients. These will give the vector map with the origin peak

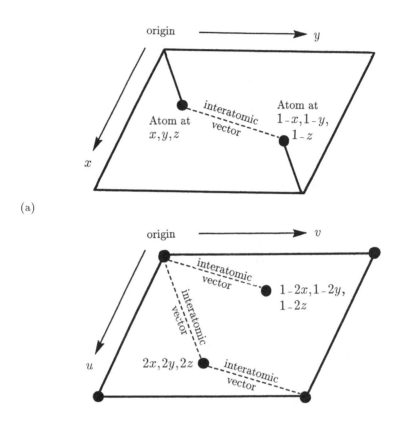

(a)

(b)

FIGURE 8.13. Interatomic vectors in a centrosymmetric structure. (a) The atomic arrangement and (b) its Patterson map.

(which normally dominates a Patterson map) removed. The interatomic vectors then appear as sharper peaks, as expected for point atoms, and interpretation of the Patterson map is thereby simplified. In favorable cases the vectors in the Patterson map can be analyzed to give a model for all or part of the structure that then can be used to develop a set of phases for the diffraction data.[56]

Horace Freeland Judson, in *The Eighth Day of Creation*,[60] likened a Patterson map to the impressions of one hundred strangers at a cocktail party. While there are only one hundred invitations, there are approximately five thousand $[\approx (100 \times 99)/2]$ introductions to be made between

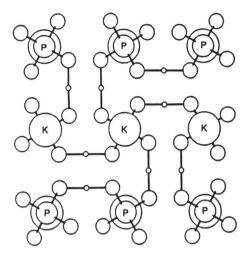

(a) crystal structure

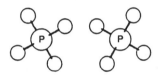

(b)

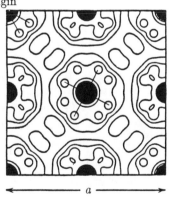

(c) Patterson map

FIGURE 8.14. (a) and (b) Crystal structure of potassium dihydrogen phosphate, and (c) the Patterson map The potassium ions and phosphorus atoms lie over each other in this projection. Note in (b) that there are two orientations of the phosphate groups, so that there are eight rather than four P–O vectors around the origin (Ref. 57).

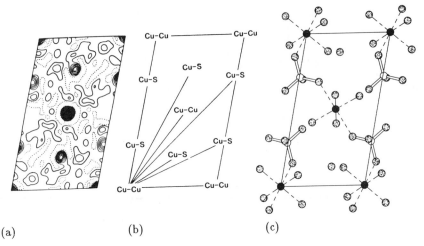

FIGURE 8.15. (a) Copper sulfate pentahydrate Patterson map with (b) Cu–Cu and Cu–S vectors shown, and (c) the crystal structure (Ref. 58). Cu black circles, S open circles, O stippled circles, H atoms omitted for clarity.

guests. The concept of the interatomic vectors and their directionality is likened to the guests having their shoes nailed to the ground and having to twist in different directions and extend their arms by differing lengths, with differing strengths of grip, in order to greet everyone. These hand-shakes (with lengths and directions) are equivalent to the vectors between atoms in the Patterson map.

A Patterson map is centrosymmetric, regardless of whether the space group is centrosymmetric or noncentrosymmetric. Any pair of atoms yield two vectors, one at u,v,w, and one at $-u, -v, -w$, that is, a vector from A to B or from B to A. Therefore the Patterson map frequently has symmetry that is higher than that of the original crystal space group. For example, the Patterson maps of the two triclinic space groups both exhibit $\bar{1}$ symmetry, those of the monoclinic space groups have $2/m$ symmetry, and those of all orthorhombic space groups have mmm symmetry. Also, any symmetry elements in the crystal space group containing translational elements such as glide planes or screw axes are replaced in the Patterson vector set by a mirror plane or simple rotation axes, respectively. If the space group is $P2_1/c$, the Patterson vector symmetry is $P2/m$, while if the space group is $C2/c$, the Patterson vector set corresponds to $C2/m$.

With a knowledge of the space–group symmetry, strong vectors can be analyzed to give the fractional atomic coordinates of those atoms in the structure that have the highest atomic numbers. If there is only one such heavy atom in the asymmetric unit, the interpretation of the Patterson map is simplified because the map is dominated by heavy-atom–heavy

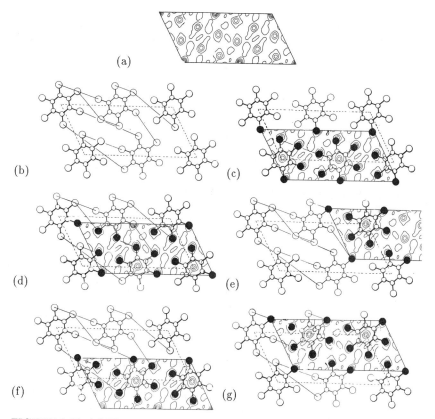

FIGURE 8.16. (a) The Patterson map and (b) the crystal structure of hexachloroben-zene (Ref. 59). (c) through (g) show views of the crystal structure superimposed on the Patterson map in order to show where the Patterson peaks come from.

-atom vectors. The next highest vector peaks are heavy-atom–light-atom peaks. Thus, most of the vectors, which are light-atom–light-atom vec-tors, may be neglected. For example, in the space group $P2_1$, the general atomic positions for a crystal structure containing one heavy atom per asymmetric unit are:

$$x,\ y,\ z \text{ and } -x,\ \tfrac{1}{2}+y,\ -z.$$

The vectors between symmetry-related atoms can be found by subtract-ing the coordinates of one atom from that of another atom at a symmetry-related site. In this particular case these vectors lie at:

$$u = 2x,\ v = \tfrac{1}{2},\ w = 2z.$$

Apart from the high peak at the origin (the sum of the vectors from each atom to itself), there should be one high peak in the Patterson map at $v = \frac{1}{2}$, and the position of this peak will give values for x and z of the heavy atom in the unit cell (with the screw axes at $x = 0$, $z = 0$). The y coordinate of one atom is arbitrary in this particular space group. This is evident in the general positions of the space group (listed above) since if there is an atom at y there is also one at $\frac{1}{2} + y$, but no information on where y is. The y value of a second atom is necessarily not arbitrary, but is relative to that of the first atom. On the other hand, if the space group were $P\bar{1}$ with heavy atoms at:

$$x, \ y, \ z \text{ and } -x,-y,-z,$$

then, if there is an atom at x there is also one at $-x$, thereby fixing the origin halfway between them. The high peak in the Patterson map will lie at:

$$u = 2x, \ v = 2y \text{ and } w = 2z,$$

so that the atomic coordinates (x, y, and z) can be found.

8.3.1 Harker sections

The symmetry of the Patterson function is the same as the Laue symmetry of the crystal. The Patterson function for space groups that have symmetry operations with translational components (screw axes and glide planes) has an added property that is very useful for the determination of the coordinates of heavy atoms. Specific peaks, first described by David Harker,[61] are associated with the vectors between atoms related by these symmetry operators. These peaks are found along "lines" or "sections" (Figure 8.17). For example, in the space group $P2_12_12_1$ there are atoms at

$$x, \ y, \ z;$$
$$\tfrac{1}{2} - x, \ - y, \ \tfrac{1}{2} + z;$$
$$\tfrac{1}{2} + x, \ \tfrac{1}{2} - y, \ - z;$$
$$-x, \ \tfrac{1}{2} + y, \ \tfrac{1}{2} -z.$$

The interatomic vector peaks between them occur at 0,0,0 (a vector from each atom to itself), and:

$$\tfrac{1}{2} - 2x, \ - 2y, \ \tfrac{1}{2};$$
$$\tfrac{1}{2}, \ \tfrac{1}{2} - 2y, \ - 2z;$$
$$- 2x, \ \tfrac{1}{2}, \ \tfrac{1}{2} - 2z.$$

Suppose there is one heavy atom in the asymmetric unit. Then in the Patterson map at $u = \frac{1}{2}$ (a Harker section) there should be a peak at:

$$v = \tfrac{1}{2} - 2y \text{ and } w = -2z.$$

Similarly on the Harker section $v = \frac{1}{2}$ of the Patterson map, there are peaks at:

$$u = -2x,$$
$$w = \tfrac{1}{2} - 2z,$$

and similarly for the (shown in Figure 8.17) Harker section $w = \frac{1}{2}$ there should be two peaks with positions related to those on the other two Harker sections. In other words, there are, on the various sections of the Patterson map, pairs of peaks that lead to a determination of x, y, and z as shown in the example in Figure 8.17. Methods like these, using Harker sections or lines if they pertain to the space group of the crystal structure, are used routinely in the determination of the positions of heavy atoms in proteins.

In theory, the Patterson map contains all the information necessary to determine the complete crystal structure and thereby solve the phase

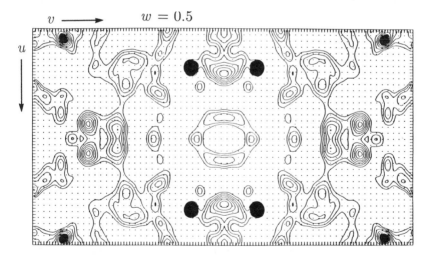

FIGURE 8.17. Harker section, $w = \frac{1}{2}$, of the Patterson map of methylcobalamin,[62] space group $P2_12_12_1$. Hint: the cobalt (heavy) atom lies at $x = 0.340$, $y = 0.294$, $z = 0.387$. The highest peaks on this map appear at $u = \pm 0.18$, and $v = \pm 0.41$, as expected. ($1/2 - 2x \approx 1.50 - 0.68 = 0.82 \approx -0.18$; $-2y = 0.59 \approx -0.41$.) Note the equivalency of positions u, $1+u$, etc. due to the periodicity of the unit cell.

problem. In practice, however, it is not easy to obtain the actual atomic arrangement from a Patterson map unless there is one (or are a few) heavy atom(s) in the crystal structure. There are N^2 interatomic vectors in the Patterson map of a unit cell containing N atoms. Of these N^2 vectors, N lie at the origin because N atoms are zero distance from themselves. As N increases to the value found for a protein, 10^3 for example, the number of vectors increases to 10^6. As a result, the vector density in the unit cell makes the resolution of peaks very difficult or impossible. If the unit cell, however, contains no more than about 100 light atoms together with one or more heavy atoms (large atomic number), then the vectors between the heavy atoms and the rest of the structure will dominate the Patterson map and the determination of the structure will be much simplified.

8.3.2 Heavy-atom method

When one atom in the asymmetric unit has a sufficiently high atomic number compared to that of any other atom in the structure, then the scattering of the heavy atom will dominate the X-ray diffraction pattern and hence the Patterson map. This will generally make it simple to determine the position of the heavy atom in the unit cell.

If this atom is sufficiently heavy, it will also dominate the relative phase angles of the Bragg reflections, as shown in Figure 8.18. An electron-density map calculated with observed structure amplitudes, and phases calculated from the positions of the heavy atoms (see Figure 6.21, Chapter 6), will contain a high peak at the position of the heavy atom.

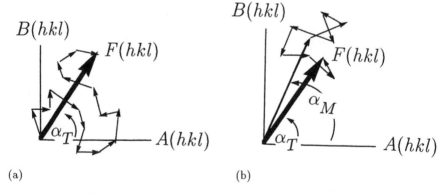

(a)   (b)

FIGURE 8.18. Relative phase angles of small-molecule structures composed of (a) many light atoms, and (b) one heavy atom and many light atoms. The heavy arrows show the total vector **F(hkl)**. Structure-factor components of individual atoms are indicated by lighter arrows. Note that in (b) the relative phase angle of the total structure (α_T) lies near to the relative phase angle of the heavy atom (α_M). In a macromolecule there are too many light atoms for this to be true.

From this map, in addition, peaks representing the lighter atoms in the structure can then be located. This was the method used[63] to determine the then-unknown structure of vitamin B_{12}. The heavy atom (cobalt, atomic number $Z = 27$) was located from the Harker sections of the Patterson map (see Figure 8.17 for the Harker section of another B_{12} derivative). Then a three-dimensional electron-density map was calculated with phase angles computed only for a cobalt atom but using the measured structure amplitudes. Peaks around the cobalt atom in this map revealed atomic positions for carbon and nitrogen atoms in what is now called a corrin ring. Those peaks near the cobalt atom were selected in order to extend the trial structure. Then another three-dimensional electron-density was calculated. The process was repeated until the entire molecular structure was found. This method has been used successfully in many structure determinations. In fact, until direct methods became simple to use, it was common to modify a compound with a heavy-atom-containing group, such as a *p*-bromophenyl group, so that the heavy-atom method could be used for the structure determination.

8.3.3 Superposition and molecular replacement methods

Sometimes clues as to the orientation of molecules in a unit cell are seen in the Bragg reflections. For example, when molecules lie parallel to low-order lattice planes, the Bragg reflections corresponding to these planes are sometimes extremely intense. For example,[64] in the crystal structure of a methylated anthracene derivative, molecules lay in planes perpendicular to the *b* axis, separated by one quarter of the unit cell ($14.387/4 = 3.85$ Å), the 0 4 0 Bragg reflection was by far the most intense in the data set. The molecules lie in planes with *y* values near $\frac{1}{8}$, $\frac{3}{8}$, $\frac{5}{8}$, and $\frac{7}{8}$. Within these planes the orientations of the six-membered rings could also be found from an inspection of the Patterson map. Thus the Patterson map may give general information on packing, even if the precise locations of atoms are more difficult to find.

The Patterson map will also give information on the orientation of a molecule of known structure in the unit cell. An example is given in Figure 8.19 of the vectors around the origin of a Patterson map for benzene. Note that the peaks indicate the orientation of the sixmembered ring. Several systematic methods that make use of Patterson maps have been devised for locating atoms in the unit cell of a crystal. These involve transcribing a Patterson map upon itself with different relative origins (**a minimum function** or **a vector superposition map**), by rotating and/or translating the map if it is suspected that there are two or more identical groups in the crystal (a **rotation function** or a **translation function**), or by comparing the map with a vector map calculated for known molecular fragments of the molecule in the crystal under study (see Figure 8.20). These methods greatly simplify the interpretation of the Patterson function, and can be used for fairly complicated molecules

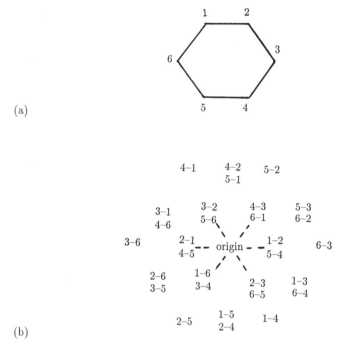

(a)

(b)

FIGURE 8.19. Self vectors A–B between atoms A and B in a benzene molecule. (a) The atomic numbering used. (b) Self vectors corresponding to the structure in (a). Note that the arrangement of the six nearest peaks around the origin (indicated by - - - - -) gives the orientation of the molecule. The origin of the Patterson map is marked as such.

that have no heavy atoms in them. More detailed discussion of these methods is found elsewhere.[65,66]

In a minimum-function map,[67,68] the origin of the Patterson map is put in turn on each of the known symmetry-related positions of a heavy atom that has already been located from a Patterson map. On each superposition of the origin of the Patterson map onto the various symmetry-related heavy-atom positions, the lowest value at each superimposed grid point in the pairs of maps is recorded. This **superposition process** is repeated until the structure is revealed. In this way the lighter atoms can be located. The method is an alternative to the heavy-atom method just described and has proved useful in many cases.

The **molecular replacement method** is often used for protein structure determination. It involves determining the orientation and the position of the known structure (or a portion of the structure) with respect to the crystallographic axes; three angles describe the orientation

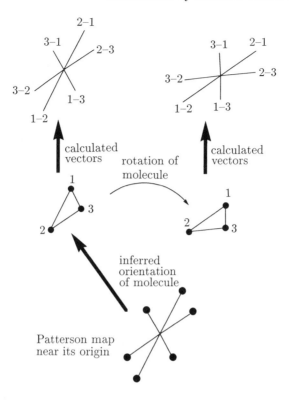

FIGURE 8.20. Use of a rotation function to determine the orientation of a known component of a structure. Peaks around the origin of the Patterson map are shown at the bottom of the Figure. Expected peak positions, calculated from the known atomic arrangement (the model), are shown at the top. The model (center) must be **rotated** until observed and expected vector maps agree.

and three vectors describe the translations. The basic ideas[69,70] were described by Michael G. Rossmann and David M. Blow in 1962. The method[71-75] is similar to the superposition method, just described.

The Patterson function of a crystal consists of two parts: self vectors, which are a set of vectors within the particular molecule, and cross vectors, which are a set of vectors from atoms in one molecule in the crystal to atoms in another molecule. Suppose that a vector map is calculated for a molecule or part of a molecule whose structure is known.[76] Then the set of self vectors in this vector map will be identical to those in the Patterson map computed for the crystal, but will probably be rotated by some set of angles. If the two Patterson functions are superimposed, there would be no particular agreement except when the two sets of self-vectors have the same orientation. If one of these Patterson maps is rotated on the second, keeping the locations of the origins of the two maps at the same point, we will find one (or more) angles at which the two Patterson maps show considerable agreement (peak overlap). Thus, for the rotation function, the Patterson map is systematically laid down upon itself in all possible orientations. Maxima in a function describing the extent of overlap will reveal the relative orientations of molecules in the unit cell. The rotation function is thus a computational tool used to assess the agreement or degree of coincidence of two Patterson functions, one from a model and the other from the diffraction pattern. The method is diagrammed in Figure 8.20, and an example is given in Figure 8.21.

Once the orientation of the molecule is determined, it is necessary to position the fragment with respect to the crystallographic axes.[69,76,77] A translation function[78-80] was developed that essentially takes the molecule whose structure and orientation are now known and calculates the degree of overlap or coincidence for the two maps (the experimental map and a copy, moved in direction (but not orientation) to different grid positions with respect to the first. Computer program systems[81] exist that do rotation-translation searches for macromolecules, although the translation problem (diagrammed for a simple case in Figure 8.22) still seems to be difficult to solve analytically for macromolecules, and must be used with considerable care.

Many protein crystals, and a significant number of small-molecule crystals, exist with more than one molecule per asymmetric unit. If the different molecules in the asymmetric unit have the same three-dimensional structures, they are related to each other by some general transformation which is generally a **noncrystallographic symmetry** operation. This additional symmetry, which may be as simple as a twofold rotation, is not part of the symmetry defined by the space group. Noncrystallographic symmetry can, however, be extremely valuable when it is used to improve a preliminary set of phases, provided the transformation can be accurately defined.[82-87] Suppose there are two molecules in the asymmetric unit that are related by a known transformation. Then the electron density associated with one molecule should be very similar to the electron density of the second molecule. Any observed differences in the electron density in the regions of the two molecules may be a result of poor phases. If the electron density associated with the two molecules is averaged, a more accurate set of phases can be calculated by back transforming the averaged electron density. This is known as **real-space averaging**. In a study of the crystalline disc protein of the tobacco mosaic virus[87] it was found that there were 17 copies of the coat protein in the asymmetric unit and that these copies were related by a definable transformation. By averaging the 17 copies of electron density associated with the coat protein, and then calculating the relative phases (by a Fourier transform), the structure determination was greatly facilitated. It is also possible for a compound to crystallize in different forms with different packing (polymorphism), and if a transformation can be

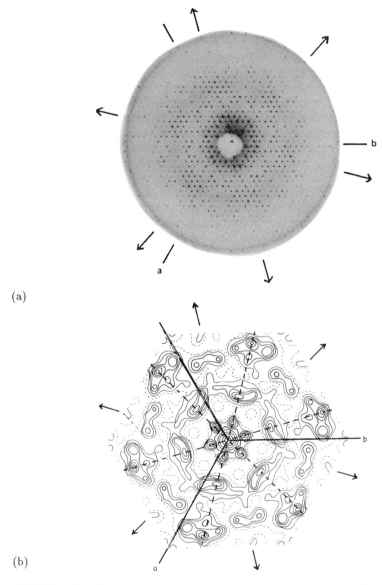

(a)

(b)

FIGURE 8.21. The use of a rotation function. Shown in (a) is the diffraction pattern of the 2Zn insulin hexamer (Ref. 86). Monomers form dimers by way of a noncrystallographic twofold axis. Dimers form hexamers by way of a crystallographic threefold axis perpendicular to the twofold axis. The twofold axes (indicated by spikes in the diffraction pattern) are highlighted in (a) through (d) by arrows. (b) The corresponding Patterson function [calculated from the diffraction data in (a)]. Harker planes between equivalent atoms in the two molecules are indicated by - - - - - .

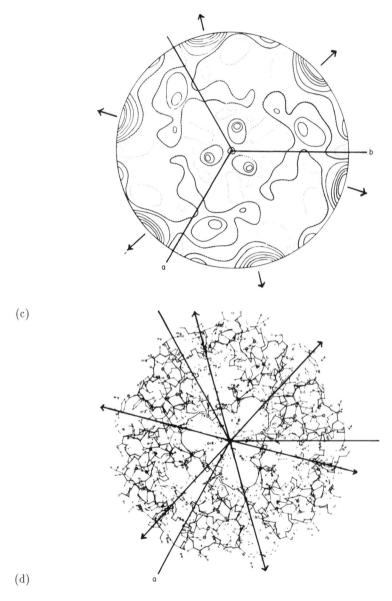

(c)

(d)

FIGURE 8.21 (cont'd). (c) The rotation function for crystalline insulin.[86] Peaks (highlighted by arrows) correctly indicate the direction of the local twofold symmetry axes. These twofold axes were also indicated in the Patterson function in (b). The actual crystal structure, in the same orientation as in (a) to (c), is shown in (d), with the unit-cell axes **a** and **b**, and the local twofold axes indicated by arrows. (From Ref. 65. Courtesy the authors and Academic Press.)

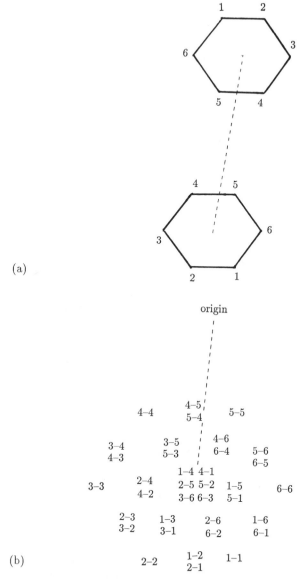

FIGURE 8.22. A set of translation vectors between two benzene molecules. (a) The two molecules with atomic numbering used, and (b) their cross vectors with the atoms involved in each vector so marked. Compare these with the self vectors in Figure 8.19. The origin of the vector (Patterson) map is labelled as such. The highest peak in the vector map (indicated by - - - - -) represents six coincident vectors between molecules, and corresponds to the distance between the centers of the two rings (also indicated by - - - - -).

defined that relates the molecules in the two crystal forms, the electron density associated with the molecules can generally be averaged and used in a similar way to improve the phases.

8.4 Isomorphous replacement method

The method of isomorphous replacement is the primary method used to determine the relative phases of protein crystal structures. The phenomenon of isomorphism was first described by Mitscherlich[88,89] in 1819, and is described in Chapter 2. Isomorphous crystals have, by definition, almost identical structures, but with one or more atoms replaced by chemically similar ones (with different X-ray scattering power). The method by which relative phases are determined for a pair of isomorphous crystals depends on a knowledge of the intensity differences between the data sets for the two isomorphous crystals and the location of the varied atom, a quantity that is available from an analysis of the Patterson map or difference map.

8.4.1 Isomorphous replacement for small molecules

The method of isomorphous replacement is rarely used for small molecules, in part because small unit cells are seldom exactly isomorphous, the change in the identity of one atom causing a significant change in unit-cell dimensions.[90] The use of this method from small molecules, however, illustrates the steps in the procedure.

If two centrosymmetric crystal structures are isomorphous, the arrangement of atoms is the same in both and only one atom (sometimes more than one) has a different atomic number in the two structures. The differences in the intensities for the Bragg reflections, therefore, result only from the differences in the scattering powers of the two atoms, M_1 and M_2, that can replace each other. The contribution to the structure factors made by the rest of the structure, F_R, is the same for both crystal structures. If the structure amplitudes are F_1 and F_2 for a given Bragg reflection in the two structures, then the calculated difference is illustrated by the use of vectors as:

$$\mathbf{F}_1 - \mathbf{F}_2 = (\mathbf{F}_{M1} + \mathbf{F}_R) - (\mathbf{F}_{M2} + \mathbf{F}_R) = (\mathbf{F}_{M1} - \mathbf{F}_{M2}) \qquad (8.14)$$

If the replaceable atoms are heavy, they can be located by a Patterson map. Then, since $\mathbf{F}_{M1}$ and $\mathbf{F}_{M2}$ are calculated for identical positions of M_1 and M_2, their relative phases are identical. The value of $\mathbf{F}_{M1} - \mathbf{F}_{M2}$ is compared with that of $|F_1|$ and $|F_2|$ obtained from experimental data. This process is shown in Figure 8.23.

An excellent example of a series of isomorphous compounds is that of the alums, described in Chapter 2. The space group of the alums, which are cubic, is $Pa3$ with four $R_1R_3(SO_4)_2 \cdot 12H_2O$ per unit cell. The potas-

	$\mathbf{F}_{M1}$ $-\mathbf{F}_{M2}$	$\mid F_1 \mid$	$\mid F_2 \mid$	$\mid F_2 \mid >$ $\mid F_1 \mid ?$	Possibilities	Deductions on sign of F_1 and F_2
(a)	+3	9	12	yes	$(+3)+(+9)=(+12)$	+ +
(b)	−3	9	6	no	$(-3)+(+9)=(+6)$	+ +
(c)	−3	9	12	yes	$(-3)+(-9)=(-12)$	− −
(d)	+3	9	6	no	$(+3)+(-9)=(-6)$	− −

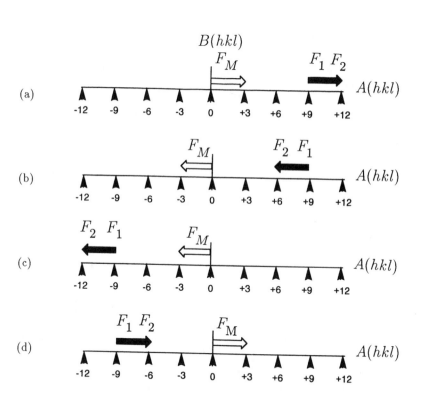

FIGURE 8.23. Calculation of phase angles for a centrosymmetric crystal by the method of isomorphous replacement. Two isomorphous crystals have structure amplitudes $\mid F_1 \mid$ and $\mid F_2 \mid$. The replaceable atom M has calculated structure factors $\mathbf{M} = \mathbf{M}_1 - \mathbf{M}_2$. From these it is possible to deduce relative phases (signs) for F_1 and F_2. In each case the vector from F_1 to F_2 must be the same as the vector $\mathbf{F}_{M1} - \mathbf{F}_{M2}$.

sium ions ($R_1 = $ K) and the aluminum atoms ($R_3 = $ Al) are located at positions strictly defined by the space group. The positions of the sulfur atoms are defined only by an x coordinate, while water molecules occupy general positions, defined by x, y, and z coordinates. The first attempts to use isomorphous replacement to study their structure were made by James M. Cork[91] in 1927, several years before the Patterson function was derived. Cork measured the intensities of the diffraction patterns of several alums, differing only in the identity of the monovalent cation. The metal ion and sulfur positions were found by him by a comparison of diffraction pattern intensities. Unfortunately the water molecules and some of the oxygen atoms were incorrectly located by this method, but Henry Lipson and C. Arnold Beevers[92] later corrected the structure. They studied three different alums — so-called potash alum $KAl(SO_4)_2 \cdot 12H_2O$, chrome alum $KCr(SO_4)_2 \cdot 12H_2O$ and selenium alum $KAl(SeO_4)_2 \cdot 12H_2O$. From their measured differences in intensities it was possible to derive the phases for all Bragg reflections, calculate the electron-density map, and locate all the oxygen and water atoms in the crystal structure (Figure 8.24).

Many other isomorphous pairs of crystals have been studied, including copper sulfate and copper selenate[93] and the phthalocyanines.[94-97] The phthalocyanines are crystalline macrocyclic organic pigments from which it is possible to remove or replace the metal. The structure of platinum phthalocyanine was solved by direct application of the heavy-atom method; this method was used because the platinum atom has such a high atomic number that its X-ray scattering dominates the scattering of the rest of the structure. An electron-density map phased on platinum atom positions was found to reveal additional peaks that represented the lighter atoms of the phthalocyanine molecule. The structures of a second series of eight phthalocyanines, isomorphous with each other, but not with the platinum compound, were then determined by use of the method of isomorphous replacement.

When crystals are noncentrosymmetric, the situation becomes more complicated because there are two possible phases angles that are derived for each reflection for data from one isomorphous pair. Therefore, in order to derive the phase angle, more than one isomorphous pair is required. This problem was first tackled by Johannes M. Bijvoet, Cornelis Bokhoven and Jean C. Schoone, who solved the noncentrosymmetric isomorphous structures of strychnine sulfate and selenate.[98,99]

8.4.2 Isomorphous replacement for macromolecules

Macromolecular crystal structure analyses are carried out by different methods than those used for small molecule crystal structure analyses. The method employed for macromolecules involves isomorphous replacement of light atoms (such as solvent molecules) by "heavy atoms" (elements with a high atomic number). Macromolecules contain large num-

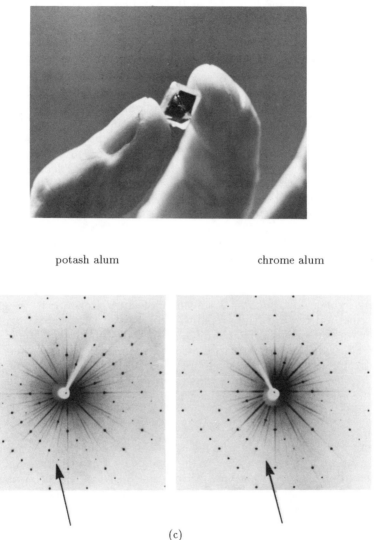

potash alum   chrome alum

(a)

(b)   (c)

FIGURE 8.24. A view of (a) potash alum (transparent) growing on a single crystal of isomorphous chrome alum (purple interior). Diffraction photographs of (b) potash alum and (c) chrome alum. The radial streaks are from white radiation and can be ignored. Note some differences in intensity. The arrows indicate two differences that are visible to the naked eye. (Photographs courtesy Maria Flocco and Henry Katz.)

bers of atoms, and the number of Bragg reflection intensities measured is also much higher for macromolecules than for small molecules. In spite of this, the statistical methods ("direct methods") used for small molecules cannot generally be used for macromolecular structures because the intensity data are not measured to atomic resolution (often far from it). In addition, the phase problem cannot generally be solved by generating a trial structure by Patterson methods because there are too many peaks in the Patterson map of a macromolecular structure (N^2).

Most of the structures of large biological molecules are currently determined by the method of isomorphous replacement.[100-107] Proteins and nucleic acids crystallize with a large amount (30–90%) of water in the unit cell, so that there are aqueous channels in the crystals. These channels provide routes along which solutions of compounds containing heavy atoms may diffuse and interact with side chains on the surface of the protein. If the heavy atoms attach to the macromolecule in well defined positions that are the same from unit cell to unit cell, intensity differences due to the addition of the heavy atom will be observed in the X-ray diffraction pattern.[108,109] The method of isomorphous replacement may then be used to derive phases for the "native" macromolecule (no heavy-atom binding). The isomorphism between the "native" (normal) macromolecular structure and crystals soaked (leached) with appropriate "heavy-atom" vehicles (salts or reactants) is the basis of this method of relative phase determination. The metal is presumed to have replaced some light group, such as a solvent molecule. In order to be effective, this method requires the preparation of several heavy-atom derivatives, each with metal attached to different sites on the protein.

The value of the structure factor $\mathbf{F}_P$ depends on the positions and scattering powers of each atom in the unit cell. If the identity of an atom is changed or a new atom is added, and it does not cause any disruption in the crystal structure, modified $\mathbf{F}_{PH}$ values will be obtained by vector addition as follows:

$$\mathbf{F}_{PH} = \mathbf{F}_P + \mathbf{F}_H, \tag{8.15}$$

where P represents the original isomorph (one isomorph or a protein, for example) and H represents the changed atom (a change in identity or addition of a heavy atom, for example). PH is the other isomorph, that is, the heavy-atom derivative of the protein. Each of these structure factors has a phase and an amplitude. If the magnitudes of $|F_P|$ and $|F_{PH}|$ have been measured, and if, from an analysis of a Patterson map it has been possible to locate the variable atom or the heavy atom, then $\mathbf{F}_H$, which includes in this case a phase angle, can be calculated. Equation 8.15 may then be written:

$$\mathbf{F}_{PH} - \mathbf{F}_P = \mathbf{F}_H. \tag{8.16}$$

The Patterson function used[65] has $||F_{PH}|-|F_P||^2$ as coefficients in Equation 8.12. The main peaks in such a map (see Figure 8.25) represent

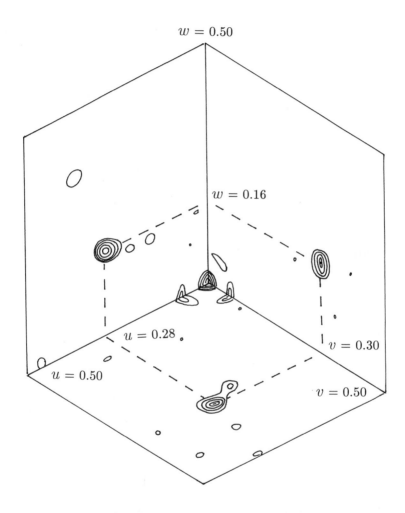

FIGURE 8.25. A difference Patterson map. The macromolecule crystallizes in the space group $I222$.[110] Atoms lie at $(0,0,0;$ and $\frac{1}{2}, \frac{1}{2}, \frac{1}{2})+ (x, y, z; -x, -y, z; x, -y, -z; -x, y, -z$. Peaks are seen in the Harker sections at $u = 2x, v = 2y, w = 0$, at $u = 2x, v = 0, w = 2z$, and at $u = 0, v = 2y, w = 2z$ as marked. The heavy atom (uranium) lies at $x = 0.14, y = 0.35, z = 0.42$.

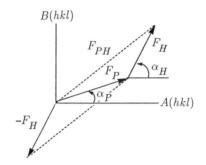

(a)

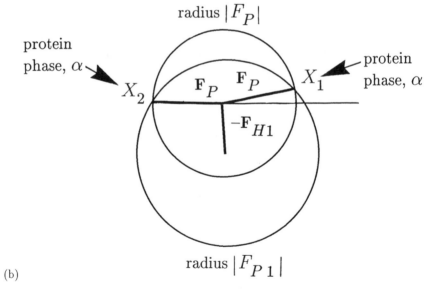

(b)

FIGURE 8.26. Phase angles by isomorphous replacement. P = protein, H = heavy atom, PH, $P1$, $P2$ = heavy-atom derivatives of the protein. The center of each circle is set to the end of the vector $-\mathbf{F}_H$ and the radius is set to $\mid F_{P1} \mid$ or $\mid F_{P2} \mid$. The arrows point to selected phase angles. How well the circles intersect at one place gives a measure of the precision of the relative phase angle $\alpha(hkl)$. (a) The geometry used. (b) The case for a single derivative. The circles radii $\mid F_P \mid$ and $\mid F_{P1} \mid$ intersect at X_1 and X_2, giving two possible values for $\alpha(hkl)$.

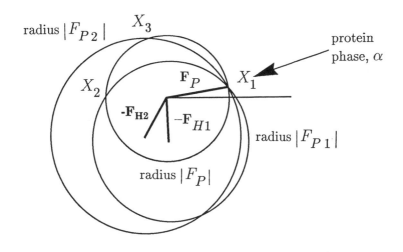

(c)

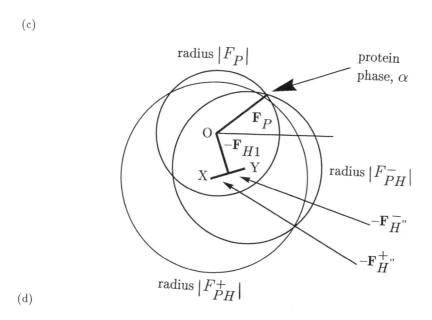

(d)

FIGURE 8.26 (cont'd). Phases by isomorphous replacement (c) for two derivatives. The circles radii $|F_P|$, $|F_{P1}|$, and $|F_{P2}|$, intersect at X_1 only, giving only one value of $\alpha(hkl)$. (d) Anomalous dispersion (to be described in more detail in Chapter 14) may also be used for relative phase determination. The calculated relative phase angle of the heavy atom has perpendicular vectors representing anomalous dispersion components. The three resulting circles intersect at one point. Measurements on one heavy-atom derivative may be sufficient to give good phase information.

vectors between the heavy atoms, provided these heavy atoms have a high occupancy in their sites. If, however, the occupancy is low, it may be difficult to locate the heavy atom because of noise in the map resulting from the approximations inherent in the equations used, and any errors in the measured data. Since $\mathbf{F}_H$, $|F_P|$, and $|F_{PH}|$ are known, it is possible to start deriving the phase angle $\alpha(hkl)$. If there is only one heavy-atom derivative, there is an ambiguity, as shown in Figure 8.26. At least two heavy-atom derivatives are needed to resolve this ambiguity so that a value for the phase angle $\alpha(hkl)$ can be found for a biological macromolecule (which necessarily crystallizes in a noncentrosymmetric space group).

The effect of the introduction of heavy atoms on the scattering by proteins is illustrated by the early work on the oxygen-carrying protein hemoglobin. Austen F. Riggs[111] in 1952 found that the heavy-atom compound *para*-mercuribenzoate would react with hemoglobin without affecting its sigmoid oxygen equilibrium curve. In other words, the interaction of a heavy-atom containing compound had not appreciably disturbed the arrangement of atoms in the crystal structure of hemoglobin. Perutz[112] wrote about hemoglobin that "··· the vast majority of the scattering contributions of the light atoms cancel out by interference, while the 80 electrons of a mercury atom would scatter in phase." This realization enabled Riggs to compute an uninterpretable electron-density map, in projection, for hemoglobin.[111] He continued: "Then Dorothy Hodgkin came over from Oxford and drew my attention to the paper by Bokhoven, Schoone, and Bijvoet [on strychnine sulfate], who had pointed out that the ambiguity left in general phases by a single isomorphous replacement could be resolved by isomorphous replacement with a second heavy atom occupying a position different from that of the first."[98,99] He then applied this same method of relative phase determination to hemoglobin and this work led to the structure of the protein.[103,113–116]

The construction, shown in Figure 8.26, can also be assessed for phase probabilities. If the circles for several isomorphs all intersect at exactly the same point, the probability is high that the phase is well determined. The overall probabilities are represented by a "figure of merit." The lack of closure is an indication of the error in the phase angle. The electron-density map and its interpretation can be further improved by comparing the images formed by calculated and observed data and analyzing any differences. This is generally done by studying a difference map.

This method of isomorphous replacement (Figure 8.27), together with anomalous dispersion data collection (see Chapter 14) is, to date, the principal method that has been successful for phase determination of macromolecules.[114,117,118] Unfortunately, it is common to find that, although a heavy-atom solution has been soaked into a protein crystal, no regular (ordered) substitution has occurred, and solutions of other heavy-atom compounds must be tried.

Isomorphous replacement for a protein.

---

1. *Measure the unit-cell dimensions and X-ray diffraction intensity data for crystals of the native protein.*

2. *Prepare heavy-atom derivatives of the protein with heavy atoms in different positions in the unit cell. Measure an intensity data set for each heavy-atom derivative.*

3. *Find the locations of the heavy atom(s) from a difference Patterson map.*

4. *Refine the heavy-atom positions by difference Fourier maps.*

5. *Estimate the phases for each hkl of the native protein (parent crystal).*

6. *Calculate the multiple-isomorphous replacement (MIR) map.*

FIGURE 8.27. Isomorphous replacement for a protein.

8.5 Multiple Bragg diffraction

Phase information can also be obtained from an effect that is found when precise intensity measurements are made, but that has generally not been used until recently. It has, so far, only been used in specific cases for a few Bragg reflections with specialized equipment, but shows promise for the future. When a crystal is oriented in the X-ray beam so that two reciprocal lattice points lie simultaneously in the diffracting position, double reflections are formed,[119-122] as shown in Figure 8.28. If two reciprocal lattice points ($h_1k_1l_1$ and $h_2k_2l_2$) lie simultaneously on the Ewald sphere, there are two interfering beams diffracted in the direction normally expected for $h_1k_1l_1$. They are the normal $h_1k_1l_1$ Bragg reflection and the one resulting from successive Bragg reflections at $h_2k_2l_2$ and $h_1 - h_2$, $k_1 - k_2$, and $l_1 - l_2$. In other words, the $h_2k_2l_2$ Bragg reflection has acted as the incident beam for the $h_1 - h_2$, $k_1 - k_2$, $l_1 - l_2$ Bragg reflection, and the Bragg reflection $h_1k_1l_1$ contains an additional intensity component (positive or negative) from this second diffraction event. This situation is analogous to optical holography with one beam ($h_1k_1l_1$ for example) acting as a phase reference for the second ($h_2k_2l_2$). Similarly, in Fourier transform infrared spectroscopy, the incident beam is split into two parts, and one of these undergoes a known phase shift while the other goes through the sample. The two beams are then recombined before being recorded on the detector. The amplitude of the resultant wave contains information on the phase difference of these two component waves.

In the *Aufhellung* effect,[119] there is a depletion of intensity of $I(h_1 k_1 l_1)$. In the *Umweganregung effect*,[120] the intensity of $I(h_1 k_1 l_1)$ is increased at the expense of the intensity of of $I(h_2 k_2 l_2)$, i.e., the diffracted beam acts as an incident beam for further diffraction. When multiple diffraction occurs, an asymmetry is noted in the shape of a Bragg reflection peak. The shape of this asymmetry depends on the value of the phase invariant. The effect is a physical manifestation of the direct method–triple product relationship. Since this method gives a measure of the value of this relationship, it provides an experimental method for direct phase determination. The fact that there is phase information in multiple X-ray diffraction has long been appreciated.[123–131] William N. Lipscomb,[132] in 1949, made an attempt to determine phase information from "umweg" reflections, that is, double reflections (see Chapter 6), but came to the conclusion at that time that it would be very hard to do.

In the multiple Bragg diffraction experiment the phase angle α_{sum} is measured. It is expressed as:

$$\alpha_{\text{sum}} = \alpha_{(-h_1,-k_1,-l_1)} + \alpha_{(-h_2,-k_2,-l_2)} + \alpha_{(h_1-h_2,k_1-k_2,l_1-l_2)}, \quad (8.17)$$

which is like the structure-invariant triple-phase relationships discussed earlier in the direct methods section, except that the probability sign is now an equality. This reinforces the earlier statement that groups of phases (triplets, for example) have unique values for their sums as a group for a given crystal structure. The value of α_{sum} may be obtained by a ψ-scan experiment through a few minutes of arc. This is difficult to do with a normal single crystal diffractometer, and therefore a special six-circle diffractometer had to be built.[133] The asymmetry of the multiply diffracted beam in the ψ-scan is then recorded. It is also necessary to note whether the multiple diffraction occurs at the incoming position of the second reciprocal lattice point (initially outside the Ewald sphere), or at the outgoing position (initially inside the Ewald sphere). The latter confers an additional negative sign on the product of the measured phases and the result gives a measure of the quadrant in which α_{sum} may be found (see Figure 8.29). This method has been used successfully to obtain phase information.[133–143] Even in the case of a small protein, myoglobin,[137] synchrotron radiation could be used to determine triplet phases to within 45°.

8.6 Effects of choices of unit cell origin

As we stressed earlier, relative phases are a function of the choice of unit cell origin.[144] A Bragg reflection does not have an absolute phase angle, but only one relative to the chosen origin. If another origin were chosen, the intensity of the Bragg reflection would remain the same, but the relative phase angle would be different.

When "direct methods" are used, origin-defining phases are arbitrarily chosen according to rules developed to define the unit-cell origin. The probabilities of the phases of various structure invariants (triplets, which are origin-independent) are assessed. The E map is then calculated with respect to the chosen unit-cell origin.

From the Patterson and heavy-atom methods, once the direction of the first vector to be selected has been assigned, the origin of the unit

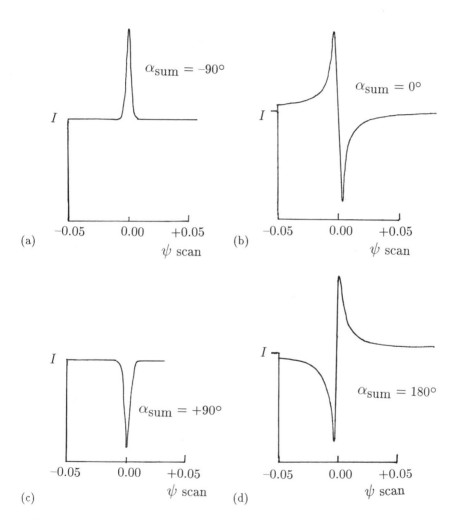

FIGURE 8.28. Multiple Bragg reflections. Shown are four idealized ψ scans (in degrees), and the values of α_{sum} derived from the peak profile. (a) $-90°$, (b) $0°$, (c) $+90°$, and (d) $180°$. Values intermediate between these relative phase angles these can be measured for noncentrosymmetric crystals.

Monoclinic crystals of $Cs_{10}Ga_6Se_{14}$ *have unit-cell dimensions* $a = 18.23$, $b = 12.89$, $c = 9.67$ Å, $\beta = 108.2°$. *The space group is* $C2/m$.

(a) *Three beam diffraction measured:* $S_P = S_L S_R$.
     S_P = *relative phase of the triplet consisting of* $\mathbf{h}_1$ *(primary beam),* $\mathbf{h}_2$ *(secondary beam), and* $\mathbf{h}_1 -\mathbf{h}_2$ *(coupling beam).*
     S_L = *relative phase from peak profile (see* Figure 8.28).
     S_R = *sign depending on whether the secondary beam reciprocal lattice point is entering* (+1) *or leaving* (−1) *the Ewald sphere during rotation.*

(b) *Measurements:*

primary	S_R	secondary	coupling	S_L	$S_L S_R$
$\overline{3}11$	out−	$\overline{4}24$	$1\overline{1}3$	+	−
	in+	$\overline{2}0\overline{2}$	$\overline{1}13$	+	+
	out−	$2\overline{2}2$	$\overline{5}3\overline{1}$	−	+
	in+	$2\overline{2}\overline{2}$	$\overline{1}\overline{1}3$	−	−

(c) *International Tables relationships*[38] *for this space group are:* $\alpha(hkl) = \alpha(\overline{hkl}) = \alpha(h\overline{k}l) = \alpha(\overline{h}k\overline{l})$.

(d) *To illustrate how the method works we give the answers. The correct relative phases are:*
$$3 1 \overline{1} - \qquad 1 1 \overline{3} - \qquad 2 2 2 -$$
$$4 2 \overline{4} - \qquad 2 0 2 + \qquad 5 3 1 +$$

(e) *This gives*

primary	secondary	coupling	$S_L S_R$
−	−	−	−
−	+	−	+
−	−	+	+
−	−	−	−

In agreement with the results in (b) *above.*

FIGURE 8.29. Multiple Bragg diffraction.

---

cell has been arbitrarily chosen. This unit-cell origin is maintained as the structure is found, and all phase angles are relative to that of an imaginary beam diffracted at this origin. The isomorphous replacement method requires a knowledge of the location of the replaceable atom, and

this is generally found by Patterson methods for which the origin definition has just been described. If more than one isomorphous replacement
is used (with a different position for the replaceable atom), care must
be taken to ensure that all replaceable atoms are located with respect
to the same origin. Once the position of an atom is determined from
a Patterson map, the location of the unit cell origin is implied in the
coordinates.

Finally, the determination of phases of structure invariants (groups
of phases that have combined values independent of the choice of origin)
by ψ-scans of multiple reflections is a recent technique. A comparison of
the usefulness in a general laboratory of the various methods is given in
Table 8.3. The result of all these methods consist of the relative phase
angles of the reflections, h,k,l, which is all the information needed in
order to calculate an electron density map.

Summary

The phase angles α_{hkl} for each Bragg reflection hkl is dependent on the arrangement
of atoms in the unit cell and the choice of origin of this unit cell. Values may be
derived in several general ways:

1. "Direct methods" of phase determination are based on the premise that the
   electron-density map should not have negative areas, and should have peaks only
   at discrete locations (atomic positions). Phase relationships between groups of
   three Bragg reflections ("triplets") are developed to give relative phase angles for
   which an electron-density map (or more than one map if there is an ambiguity
   in phase determination) may be calculated. The structure is generally revealed
   in the resulting electron-density map.

2. The Patterson map gives all interatomic vectors in the crystal structure. This
   is hard to interpret unless the crystal contains only a few atoms of high atomic
   number that dominate the X-ray scattering. A "trial structure" (model) derived from an analysis of the vectors in the Patterson map, is used to calculate
   phase angles that are then used to compute an electron-density map. Rotation
   and translation functions may aid in the analysis of the Patterson function of a
   macromolecule. In the heavy-atom method, the calculated relative phase angles
   of the heavy atom are used in the calculation of a three-dimensional electron-
   density map. Some images of other atoms usually appear in such maps and the
   positions of these peaks can be used to calculate better relative phase angles.
   The complete structure can be developed in this way.

3. If isomorphous crystals are available, differences in the intensities in the diffraction patterns of the two isomorphs lead to phase information. This is the method
   of choice for finding phase angles of macromolecules. Heavy-atom salts are diffused into crystals. Intensity differences between the native macromolecule and
   its heavy-atom derivative are measured and used to determine relative phases.

4. When a ψ scan is done through certain multiple reflections, it is possible to
   determine the sum of the phases of the three reflections involved.

TABLE 8.3. Phase-determining methods.

Method	Notes	Use
Direct methods	*Aim for no negative areas in electron-density map. Probabilities of phases analyzed.*	*Used for small molecules.*
Patterson methods	*Map of interatomic vectors. Analyses complicated unless few atoms or heavy atom.*	*Used to locate heavy atom in isomorphous replacement for macromolecules. Can also be used for small molecules even if they contain no heavy atom.*
Isomorphous replacement	*Intensity differences for isomorphous crystals. Best if several derivatives are made. Replaced atom located by Patterson methods.*	*General method for macromolecules.*
Molecular replacement	*Known part of structure positioned in unit cell.*	*Used when molecule large. Eliminates need for heavy-atom derivatives.*
Noncrystallographic symmetry	*Used when several copies of subunit in asymmetric unit.*	*Very useful for large molecules especially viruses.*
Anomalous scattering	*Needs good data. Used as an addition to isomorphous replacement. Requires anomalous scattering.*	*Good with a tunable X-ray source, such as synchrotron radiation.*
Multiple Bragg reflections	*Needs very special and precise equipment.*	*In developmental stages.*

Glossary

Calculated phase angle: The phase angle α calculated from the atomic positional and displacement parameters for a model structure.

$$\alpha = \tan^{-1}[\,B(hkl)/A(hkl)\,],$$
$$\text{where } A(hkl) = \Sigma f \cos 2\pi(hx + ky + lz),$$
$$B(hkl) = \Sigma f \sin 2\pi(hx + ky + lz),$$
$$\text{and } |\,F(hkl)\,|^2 = [A(hkl)]^2 + [B(hkl)]^2.$$

Cauchy–Schwarz inequality : This inequality (q.v.) is used in direct methods of phase determination. Cauchy's inequality states:

$$\left|\sum_{j=1}^{N} a_j b_j\right|^2 \leq \left(\left|\sum_{j=1}^{N} a_j\right|^2\right)\left(\left|\sum_{j=1}^{N} b_j\right|^2\right)$$

where a_j and b_j are real integers, not proportional to each other. Schwarz's inequality, first put forward by Buniakowsky, states:

$$\left|\int fg\,d\tau\right|^2 \leq \left(|\int f\,|^2\,d\tau)(|\int g\,|^2\,d\tau\right).$$

This corresponds to Cauchy's inequality, but is applied to integrals. These equations were used in the derivation of direct methods.

Centrosymmetric crystal structure: A crystal structure whose space group, and therefore arrangement of atoms, contains a center of symmetry. When the origin of the unit cell is at a center of symmetry, the relative phase angle for each Bragg reflection is either $0°$ or $180°$ in the absence of X-ray anomalous dispersion.

Direct methods, direct phase determination: A method of deriving relative phases of diffracted beams by consideration of relationships among the Miller indices and among the structure factor amplitudes of the stronger Bragg reflections. These relationships come from the conditions that the structure is composed of atoms and that the electron-density map must be positive or zero everywhere. Only certain values for the phases are consistent with these conditions.

Distribution of intensities: The overall variation in the intensities of the Bragg reflections obtained on X-ray diffraction of a crystal. Intensities from a non-centrosymmetric crystal tend to be clustered more tightly around the mean than do those from a centrosymmetric one. This information forms the basis for one test for the presence or absence of a center of symmetry in the crystal.

E map: A Fourier map (equivalent to an electron-density map) with phases derived by "direct methods" and structure amplitudes $|\,F(hkl)\,|$ replaced by normalized structure amplitudes $|\,E(hkl)\,|$. Since the $|\,E\,|$ values correspond to sharpened atoms, the peaks on the resulting Fourier map are higher and sharper, and therefore easier to identify, than those in an electron-density map calculated with $|\,F\,|$ values.

Epsilon factor, ϵ: This is a weighting factor used in the calculation of normalized structure factors. It takes into account the fact that, depending on which of the 32 crystal classes the crystal belongs to, there will be certain groups of reflections that have an average intensity greater than that for the general reflections. The ϵ factor is used to eliminate these differences.

Harker–Kasper inequalities: Inequalities among structure factors, dependent on the space group, that lead to equations that allow the determination of the relative phases of certain intense Bragg reflections.

Heavy-atom derivative of a protein: The product of soaking a solution of the salt of a metal of high atomic number into a crystal of a protein. If the heavy-atom derivative is to be of use in structure determination, the heavy atom must be substituted in only one or two ordered positions per asymmetric unit. Then the method of isomorphous replacement can be used to determine the relative phase angles of the Bragg reflections.

Heavy-atom method: Relative phases calculated for a heavy atom in a location determined from a Patterson map are used to calculate an approximate electron-density map. Further portions of the molecular structure may be identified in this map and used to calculate better relative phases, and therefore a more realistic electron-density map results. Several cycles of this process may be necessary in order to determine the entire crystal structure.

Homometric structure: A structure with a uniquely different arrangement of atoms from another, but having the same sets of interatomic vectors, and hence the same Patterson map. Examples to date consist of homometric heavy-atom positions in crystal structures that also contain lighter atoms. The total crystal structure is not, however, homometric.

Inequality: A mathematical statement that the value of one expression is greater than or less than (but not equal to) the value of another.

Isomorphous replacement method: A method for deriving relative phases by comparing the intensities of corresponding Bragg reflections from two or more isomorphous crystals. If the locations in the unit cell of atoms that vary between each isomorph have been located, for instance from a Patterson map, then the relative phase of each Bragg reflection can be assessed if a sufficient number of isomorphs is studied (at least 1 for a centrosymmetric crystal, at least 2 for a noncentrosymmetric crystal).

Minimum function: Analysis of the Patterson map by setting its origin in turn on the known positions of certain atoms, and then recording the minimum value for all of these superpositions. The resulting three-dimensional map should contain an indication of the atomic arrangement.

Molecular replacement method: The use of rotation and translation functions (q.v.), of noncrystallographic symmetry (q.v.), or of structural information from related structures, to determine a protein crystal structure.

Multiple Bragg diffraction: Further diffraction of a Bragg reflection by a second set of lattice planes. This occurs when two reciprocal lattice planes lie simultaneously on the surface of the Ewald sphere. It affects the intensity of the Bragg reflection and a detailed analysis of the effect can lead to some phase information.

Noncrystallographic symmetry: Symmetry within an asymmetric unit in a crystal structure that is not accounted for by any space group symmetry. For example, one asymmetric unit of a crystalline protein may contain a dimer whose two subunits have identical molecular structures, but, since they are not related by crystallographic symmetry, may have different environments. This noncrystallographic symmetry can be used as an aid in structure determination.

Normalized structure factor: The ratio of the value of the structure amplitude $| F |$ to its root-mean-square expectation value. This ratio is denoted $E(hkl)$.

$$| E(hkl) | = | F(hkl) | / (\epsilon \sum_j f_j^2(hkl))^{\frac{1}{2}} \tag{8.2}$$

The epsilon factor ϵ (q.v.) takes into account the crystal class and the deviation of the average intensities of certain groups of Bragg reflections from those of the general Bragg reflections.

Patterson function: A Fourier summation that has $| F |^2$ as coefficients and all phases zero. Ideally the positions of maxima in this map represent the end points of vectors between atoms, all referred to a common origin.

$$P(uvw) = \frac{1}{V} \sum_{h,k,l} \sum \sum | F(hkl) |^2 \cos(hu + kv + lw). \tag{8.12}$$

Patterson methods: Methods for analyzing the Patterson map to obtain the final structure.

Probability relationships: In crystallographic use this term refers to equations that express the probability that a relative phase angle will have a certain value. Such equations are the basis of phase determination by direct methods.

Real-space averaging: A computational method for improvement of phases, when there are two or more identical chemical units in the crystallographic asymmetric unit. The electron densities of these identical units are averaged. Then a new set of phases is computed by Fourier transformation of the averaged structure, and with these a new map is synthesized with the observed $| F |$ values. By iteration of this procedure, the electron density is improved.

Rotation function: The rotation function gives a measure of the of the degree of correspondence between a vector set calculated for a known structure and the Patterson function of that crystal, as a function of the rotation of one with respect to the other about the origin. Peaks in the rotation function define a likely orientation of the known fragment of the structure. All vectors are calculated for a known portion of a crystal structure, and the resulting set of vectors is rotated about its origin until it best matches the orientation of vectors in the Patterson map.

Sharpened Patterson function: A Patterson map computed with values of $| F |^2$ enhanced as $\sin \theta / \lambda$ increases. As a result the map contains sharper peaks and is easier to interpret. The Patterson map with $| E^2 - 1 |$ as coefficients is commonly used for small molecules and gives a sharpened map with the origin peak removed.

Structure invariant: A linear combination of the phases of a particular set of Bragg reflections such that the combination is independent of the choice of origin of the unit cell.

Structure seminvariant: A linear combination of the relative phases of a particular set of Bragg reflections. The value of this combination remains unchanged when the location of the origin is changed, provided this change is made subject to space-group symmetry constraints.

Superposition method: Analysis of the Patterson map by setting the origin of the Patterson map in turn on the positions of certain atoms whose positions may already be known, and then recording those areas of the superposed maps in which peaks appear that are derived from both maps. As a result, it may be possible to derive the atomic arrangement.

Symbolic addition method: The use of symbols (a, b, etc.) in the derivation of relative phases from relationships between them. As the analysis by direct methods proceeds, values for these symbols may become evident. Otherwise, electron-density maps with all possible values for the undetermined symbols must be computed, and one hopes to find the one that contains the structure.

Tangent formula: A formula of use in direct methods for developing additional relative phases and refining them.

Translation function: A function that can be calculated in order to determine (with respect to the unit cell axes) how a molecule, for which the orientation has been found (see **Rotation function**), is positioned with respect to the origin of the unit cell. This function is important in protein crystallography.

Unitary structure factor: The ratio of the structure amplitude $F(hkl)$ to its maximum possible value for point atoms at rest.

$$U(hkl) = \mid F(hkl) \mid / \left(\sum_j f_j(hkl) \exp(-B \sin^2 \theta / \lambda^2) \right) \tag{8.3}$$

This is now generally replaced by a normalized structure factor (q.v.).

Vector superposition map: See Superposition method.

References

1. Nyburg, S. C. *X-ray Analysis of Organic Structures.* Academic Press: New York (1961).

2. Hauptman, H. A. The phase problem in X-ray crystallography. *Physics Today* **November**, 24–29 (1989).

3. Woolfson, M. M. *Direct Methods in Crystallography.* Oxford University Press: Oxford (1961).

4. Ladd, M. F. C., and Palmer, R. A. (**Eds.**) *Theory and Practice of Direct Methods in Crystallography.* Plenum: New York, London (1980).

5. Hauptman, H., and Karle, J. *Solution of the Phase Problem. I. The Centrosymmetric Crystal.* American Crystallographic Association Monograph No. 3: New York (1953).

6. Woolfson, M. M. Direct methods — from birth to maturity. *Acta Cryst.* **A43**, 593–612 (1987).

7. Ott, H. Zur Methodik der Strukturanalyses. [On structure analysis methods.] *Z. Krist.* **66**, 136–153 (1928).

8. Banerjee, K. Determinations of the Fourier terms in complete crystal analysis. *Proc. Roy. Soc.* (*London*) **A141**, 188–193 (1933).

9. Avrami, M. Direct determination of crystal structure from X-ray data. *Phys. Rev.* **54**, 300–303 (1938).

10. Wilson, A. J. C. The probability distribution of X-ray intensities. III. Effects of symmetry elements on zones and rows. *Acta Cryst.* **3**, 258–261 (1950).

11. Stout, G. H., and Jensen, L. H. *X-ray Structure Determination. A Practical Guide.* Macmillan: New York (1968); 2nd edn. John Wiley: New York, Chichester, Brisbane, Toronto, Singapore (1989).

12. Karle, J. Direct methods for structure determination: origin specification, normalized structure factors, formulas, and the symbolic-addition procedure for phase determination. In: *International Tables for X-ray Crystallography.* Volume IV. *Revised and Supplementary Tables. Section 6.* (**Eds.**, Ibers, J. A., and Hamilton, W. C.) pp. 339–358. Kynoch Press: Birmingham (1974).

13. Karle, J., and Hauptman, H. A theory of phase determination for the four types of non-centrosymmetric space groups $1P222$, $2P22$, $3P_12$, $3P_22$. *Acta Cryst.* **9**, 635–651 (1956).

14. Wilson, A. J. C. The probability distribution of X-ray intensities. *Acta Cryst.* **2**, 318–321 (1949).

15. Wilson, A. J. C. (**Ed.**) *Structure and Statistics in Crystallography.* Adenine Press: Guilderland, NY (1985).

16. Dunitz, J. D. *X-ray Analysis and the Structure of Organic Molecules.* Cornell University Press: London, Ithaca, NY (1979).

17. Howells, E. R., Phillips, D. C., and Rogers, D. The probability distribution of X-ray intensities. II. Experimental investigations and the X-ray detection of centres of symmetry. *Acta Cryst.* **3**, 210–214 (1950).

18. Harker, D., and Kasper, J. S. Phases of Fourier coefficients directly from crystal diffraction data. *Acta Cryst.* **1**, 70–75 (1948).

19. Harker, D., and Kasper, J. S. Phases of Fourier coefficients directly from crystal diffraction data. *J. Chem. Phys.* **15**, 882–884 (1947).

20. Cauchy, A. L. *Cours d'Analyse de l'École Royale Polytechnique. Ire partie. Analyse Algébrique.* [Lectures on Analysis from the Royal Polytechnic School. Part I. Algebraic Analysis.] Paris (1821). (*Œuvres complètes*, IIe série, III, p. 373.)

21. Schwarz, H. A. Über ein die Flächen kleinsten Flächeninhalts betreffendes Problem der Variationsrechnung. [Concerning the calculus of variation problem of the surface area of least volume.] *Acta Soc. Scient. Fenn.* **15**, 315–362 (1885). [*Werke*, **I**, 224–269 (p. 251).]

22. Buniakowsky, V. Sur quelques inégalités concernant les intégrales ordinaires et les intégrales aux différences finies. [On some inequalities regarding the common integrals and integrals of finite differences.] *Mémoires de l'Academie des Impériale Sciences de St-Pétersbourg.* Series VII. No. 9. pp. 1–18 (1859).

23. Hardy, G. H., Littlewood, J. E., and Pólya, G. *Inequalities.* 2nd edn. Cambridge University Press: Cambridge, UK (1952).

24. Kasper, J. S., Lucht, C. M., and Harker, D. The crystal structure of decaborane, $B_{10}H_{14}$. *Acta Cryst.* **3**, 436–455 (1950).

25. Woolfson, M. M. Some thoughts on Harker–Kasper inequalities. *Acta Cryst.* **A44**, 222–225 (1988).

26. Karle, J., and Hauptman, H. The probability distribution of the magnitude of a structure factor. I. The centrosymmetric crystal. II. The noncentrosymmetric crystal. *Acta Cryst.* **6**, 131–135 (1953).

27. Karle, J., and Hauptman, H. The phases and magnitudes of the structure factors. *Acta Cryst.* **3**, 181–187 (1950).

28. Karle, J., and Hauptman, H. A unified program for phase determination, type 1P. *Acta Cryst.* **12**, 404–410 (1959).

29. Hauptman, H., and Karle, J. Solution of the phase problem for space group $P\bar{1}$. *Acta Cryst.* **7**, 369–374 (1954).

30. Hauptman, H. A new method in the probabilistic theory of the structure invariants. *Acta Cryst.* **A31**, 680–687 (1975).

31. Gillis, J. Structure-factor relations and phase determination. *Acta Cryst.* **1**, 76–80 (1948).

32. Gillis, J. The application of the Harker–Kasper method of phase determination. *Acta Cryst.* **1**, 174–179 (1948).

33. Zachariasen, W. H. A new analytical method for solving complex crystal structures. *Acta Cryst.* **5**, 68–73 (1952).

34. Lavine, L. R. Corrections to Grison's paper on the Harker-Kasper inequalities and to Zachariasen's paper on the 'statistical method.' *Acta Cryst.* **5**, 846–847 (1952).

35. Sayre, D. The squaring method: a new method for phase determination. *Acta Cryst.* **5**, 60–65 (1952).

36. Cochran, W. A relation between the signs of structure factors. *Acta Cryst.* **5**, 65–67 (1952).

37. Karle, J., and Karle, I. L. The symbolic addition procedure for phase determination for centrosymmetric and noncentrosymmetric crystals. *Acta Cryst.* **21**, 849–859 (1966).

38. *International Tables for X-ray Crystallography, Vol. 1. Symmetry Groups,* (**Eds.,** Henry, N. F. M., and Lonsdale, K.) Kynoch Press: Birmingham (1952).

39. Lonsdale, K. The structure of the benzene ring in $C_6(CH_3)_6$. *Proc. Roy. Soc. (London)*, **A123**, 494–515 (1929).

40. Karle, I. L., and Karle, J. An application of the symbolic addition method to the structure of L-arginine dihydrate. *Acta Cryst.* **17**, 835–841 (1964).

41. Stewart, J. M., Kundell, F. A., and Baldwin, J. C. *The XRAY 70 System.* College Park, MD: Computer Science Center, University of Maryland (1970).

42. Sheldrick, G. M. *SHELX-76 Program for Crystal Structure Determination.* University of Cambridge: Cambridge, England (1976).

43. Germain, G., and Woolfson, M. M. On the application of phase relationships to complex structures. *Acta Cryst.* **B24**, 91–96 (1968).

44. Germain, G., Main, P., and Woolfson, M. M. The application of phase relationships to complex structures. III. The optimum use of phase relationships. *Acta Cryst.* **A27**, 368–376 (1971).

45. Main, P., Woolfson, M. M., Lessinger, L., Germain, G., and Declerq, J. P. *MUL-TAN 74: A System of Computer Programs for the Automatic Solution of Crystal Structures from X-ray Diffraction Data.* University of York and University of Louvain (1974).

46. Sheldrick, G. M. Phase annealing in *SHELX*-90: direct methods for larger structures. *Acta Cryst.* **A46**, 467–473 (1990).

47. Gilmore, C. J. *MITHRIL* – an integrated direct-methods computer program. *J. Appl. Cryst.* **17**, 42–46 (1984).

48. White, P. S., and Woolfson, M. M. The application of phase relationships to complex structures. VII. Magic integers. *Acta Cryst.* **A31**, 367–372 (1975).

49. Hauptman, H. On the identity and estimation of those cosine invariants, $\cos(\psi_1 + \psi_2 + \psi_3 + \psi_4)$, which are probably negative. *Acta Cryst.* **A30**, 472–476 (1974).

50. Patterson, A. L. A direct method for the determination of the components of interatomic distances in crystals. *Z. Krist.* **A90**, 517–542 (1935).

51. Patterson, A. L. A Fourier series method for the determination of the components of interatomic distances in crystals. *Phys. Rev.* **46**, 372–376 (1934).

52. Patterson, A. L. An alternative interpretation for vector maps. *Acta Cryst.* **2**, 339–340 (1949).

53. Zernike, F., and Prins, J. A. Die Beugung von Röntgenstrahlen in Flüssigkeiten als Effekt der Molekülanordnung. [The diffraction of X rays in liquids as an effect of the molecular arrangement.] *Z. Physik* **41**, 184–194 (1927).

54. Warren, B. E., and Gingrich, N. S. Fourier integral analysis of X-ray powder patterns. *Phys. Rev.* **46**, 368–372 (1934).

55. Pauling, L., and Shappell, M. D. The crystal structure of bixbyite and the C-modification of the sesquioxides. *Z. Krist.* **75**, 128–142 (1930).

56. Glusker, J. P., Patterson, B. K., and Rossi, M. (**Eds.**) *Patterson and Pattersons. Fifty Years of the Patterson Function.* International Union of Crystallography Crystallographic Symposia. Oxford University Press: Oxford (1987).

57. West, J. A quantitative X-ray analysis of the structure of potassium dihydrogen phosphate. *Z. Krist.* **74**, 306–332 (1930).

58. Beevers, C. A., and Lipson, H. The crystal structure of copper sulphate pentahydrate, $CuSO_4 \cdot 5H_2O$. *Proc. Roy. Soc. (London)* **A146**, 570–582 (1934).

59. Lonsdale, K. An X-ray analysis of the structure of hexachlorobenzene using the Fourier method. *Proc. Roy. Soc. (London)* **A133**, 536–552 (1931).

60. Judson, H. F. *The Eighth Day of Creation. The Makers of the Revolution in Biology.* Simon & Schuster: New York (1979).

61. Harker, D. The application of the three-dimensional Patterson method and the crystal structures of proustite, Ag_3AsS_3, and pyrargyrite, Ag_3SbS_3. *J. Chem. Phys.* **4**, 381–390 (1936).

62. Rossi, M., Glusker, J. P., Randaccio, L., Summers, M. F., Toscano, P. J., and Marzilli, L. G. The structure of a B_{12} coenzyme: methylcobalamin studies by X-ray and NMR methods. *J. Amer. Chem. Soc.* **107**, 1729–1738 (1985).

63. Hodgkin, D. C., Pickworth, J., Robertson, J. H., Trueblood, K. N., Prosen, R. J., and White, J. G. The crystal structure of the hexacarboxylic acid derived from B_{12} and the molecular structure of the vitamin. *Nature (London)* **176**, 325–328 (1955).

64. Chomyn, A., Glusker, J. P., Berman, H. M. and Carrell, H. L. The crystal structure of 10-chloromethyl-2,3,9-trimethylanthracene. *Acta Cryst.* **B28**, 3512–3517 (1972).

65. Blundell, T. L., and Johnson, L. N. *Protein Crystallography.* Academic Press: New York, London, San Francisco (1976).

66. Diamond, R. A note on the rotational superposition problem. *Acta Cryst.* **A44**, 211–216 (1988).

67. Beevers, C. A., and Robertson, J. H. Interpretation of the Patterson synthesis. *Acta Cryst.* **3**, 164 (1950).

68. Nordman, C. E., and Nakatsu, K. Interpretation of the Patterson function of crystals containing an unknown molecular fragment. The structure of an *Alstonia* alkaloid. *J. Amer. Chem. Soc.* **85**, 353–354 (1963).

69. Rossmann, M. G., and Blow, D. M. The detection of sub-units within the crystallographic asymmetric unit. *Acta Cryst.* **15**, 24–31 (1962).

70. Rossmann, M. G. (Ed.) *The Molecular Replacement Method.* Gordon and Breach: New York (1972).

71. Fitzgerald, P. M. D. Molecular replacement. In: *Crystallographic Computing 5. From Chemistry to Biology.* (**Eds.**, Moras, D., Podjarny, A. D., and Thierry, J. C.) International Union of Crystallography/Oxford University Press: Oxford (1991).

72. Arnold, E., and Rossmann, M. G. The use of molecular-replacement phases for the refinement of the human rhinovirus 14 structure. *Acta Cryst.* **A44**, 270–282 (1988).

73. Rossmann, M. G., McKenna, R., Tong, L., Xia, D., Dai, J-L., Wu, H., Choi, H-K., and Lynch, R. E. Molecular replacement real-space averaging. *J. Appl. Cryst.* **25**, 166–180 (1992).

74. Brünger, A. T. Solution of a Fab (26-10)/digoxin complex by generalized molecular replacement. *Acta Cryst.* **A47**, 195–204 (1991).

75. Tollin, P., Main, P., and Rossmann, M. G. The symmetry of the rotation function. *Acta Cryst.* **20**, 404–407 (1966).

76. Egert, E., and Sheldrick, G. M. Search for a fragment of known geometry by integrated Patterson and direct methods. *Acta Cryst.* **A41**, 262–268 (1985).

77. Beurskens, P. T., Gould, R. O., Slot, H. J. B., and Bosman, W. P. Translation functions for the positioning of a well oriented molecular fragment. *Z. Krist.* **179**, 127–159 (1987).

78. Crowther, R. A. and Blow, D. M. A method of positioning a known molecule in an unknown structure. *Acta Cryst.* **23**, 544–548 (1967).

79. Pavelčík, F. A comment on the asymmetric part of the translation function. *Acta Cryst.* **A47**, 292–293 (1991).

80. Fujinaga, M., and Read, R. J. Experiences with a new translation-function program. *J. Appl. Cryst.* **20**, 517–521 (1987).

81. Fitzgerald, P. M. D. *MERLOT*, an integrated package of computer programs for the determination of crystal structures by molecular replacement. *J. Appl. Cryst.* **21**, 273–278 (1988).

82. Rossmann, M. G., Arnold, E., Erickson, J. W., Frankenberger, E. A., Griffith, J. P., Hecht, H-J., Johnson, J. E., Kamer, G., Luo, M., Mosser, A. G., Rueckert, R. R., Sherry, B., and Vriend, G. Structure of a human common cold virus and functional relationship to other picornaviruses. *Nature (London)* **317**, 145–153 (1985).

83. Tong, L., and Rossmann, M. Patterson-map interpretation with noncrystallographic symmetry. *J. Appl. Cryst.* **26**, 15–21 (1993).

84. Rossmann, M. G., and Blow, D. M. Determination of phases by the conditions of non-crystallographic symmetry. *Acta Cryst.* **16**, 39–45, (1963).

85. Jones, E. Y., Walker, N. P. C., and Stuart, D. I. Methodology employed for the structure determination of tumor necrosis factor, a case of high non-crystallographic symmetry. *Acta Cryst.* **A47**, 753–770 (1991).

86. Dodson, E., Harding, M. M., Hodgkin, D. C., and Rossmann, M. G. The crystal structure of insulin. III. Evidence for a 2-fold axis in rhombohedral zinc insulin. *J. Molec. Biol.* **16**, 227–241 (1966).

87. Champness, J. N., Bloomer, A. C., Bricogne, G., Butler, P. J. G. and Klug, A. The structure of the protein disk of tobacco mosaic virus to 5 Å resolution. *Nature (London)* **259**, 20–24 (1976).

88. Mitscherlich, E. Über die Kristallisation der Salze in denen das Metall der Basis mit zwei Proportionen Sauerstoff verbunden ist. [On the crystallization of salts in which the base metal is bound to two proportions of oxygen.] *Abhandlungen der Königlichen Akademie der Wissenschaften in Berlin.* **1820**, 427–437 (1818–1819).

89. Mitscherlich, E. Sur la relation qui existe entre la forme cristalline et les proportions chimiques. [On the relationship between crystalline form and chemical composition.] *Ann. Chem. Phys.* **14**, 172–190 (1820). (Read at the Academy of Sciences of Berlin, December 9, 1819).

90. Glusker, J. P., van der Helm, D., Love, W. E., Dornberg, M. L., Minkin, J. A., Johnson, C. K., and Patterson, A. L. X-ray analysis of the substrates of aconitase. VI. The structures of sodium and lithium dihydrogen citrates. *Acta Cryst.* **19**, 561–572 (1965).

91. Cork, J. M. Crystal structures of the alums. *Phil. Mag.* **4**, 688–698 (1927).

92. Lipson, H., and Beevers, C. A. The crystal structure of some of the alums. *Proc. Roy. Soc. (London)* **A148**, 664–680 (1935).

93. Beevers, C. A., and Lipson, H. Crystal structure of copper sulphate pentahydrate, $CuSO_4 \cdot 5H_2O$. *Proc. Roy. Soc. (London)* **A146**, 570–582 (1934).

94. Robertson, J. M. An X-ray study of the structure of the phthalocyanines, Part I. The metal-free, nickel, copper and platinum compounds. *J. Chem. Soc. (London)* 615–621 (1934).

95. Robertson, J. M. An X-ray study of phthalocyanines. Part II. Quantitative structure determination of the metal-free compound. *J. Chem. Soc.* 1195–1209 (1936).

96. Robertson, J. M., and Woodward, I. An X-ray study of the phthalocyanines, Part III. Quantitative structure determination of nickel phthalocyanine. *J. Chem. Soc.* 219–230 (1937).

97. Robertson, J. M., and Woodward, I. An X-ray study of the phthalocyanines, Part IV. Direct quantitative analysis of the platinum compound. *J. Chem. Soc.* 36–48 (1940).

98. Bokhoven, C., Schoone, J. C., and Bijvoet, J. M. The crystal structure of strychnine sulfate and selenate. III. [001] projection. *Proc. Koninklijke Nederlandse Akademie van Wetenschappen.* [*Proc. Roy. Soc.* (*Amsterdam*)] **52**, 120–121 (1949).

99. Bokhoven, C., Schoone, J. C., and Bijvoet, J. M. The Fourier synthesis of the crystal structure of strychnine sulfate pentahydrate. *Acta Cryst.* **4**, 275–280 (1951).

100. Blow, D. M., and Rossmann, M. G. The single isomorphous replacement method. *Acta Cryst.* **14**, 1195–1202 (1961).

101. Crick, F. H. C., and Magdoff, B. S. The theory of the method of isomorphous replacement for protein crystals. I. *Acta Cryst.* **9**, 901–908 (1956).

102. Harker, D. The determination of the phases of the structure factors of non-centrosymmetric crystals by the method of double isomorphous replacement. *Acta Cryst.* **9**, 1–9 (1956).

103. Perutz, M. F. Isomorphous replacement and phase determination in non-centrosymmetric space groups. *Acta Cryst.* **9**, 867–873 (1956).

104. Bijvoet, J. M. Structure of optically active compounds in the solid state. *Nature* (*London*) **173**, 888–891 (1954).

105. Kendrew, J. C., Bodo, G., Dintzis, H. M., Parrish, R. G., and Wyckoff, H. A three-dimensional model of the myoglobin molecule obtained by X-ray analysis. *Nature* (*London*) **181**, 662–666 (1958).

106. Blow, D. M., and Crick, F. H. C. The treatment of errors in the isomorphous replacement method. *Acta Cryst.* **12**, 794–802 (1959).

107. Robertson, J. M. Vector maps and heavy atoms in crystal analysis and the insulin structure. *Nature* (*London*) **143**, 75–76 (1939).

108. King, M. V. Referred to by Wood, E. A. In: The development of X-ray diffraction in the U.S.A. From the beginning of World War II to 1961. In: *Fifty Years of X-ray Diffraction.* (**Ed.**, Ewald, P. P.) N. V. A. Oosthoek: Utrecht (1962).

109. Lebioda, L., and Zhang, E. Soaking of crystals for macromolecular crystallography in a capillary. *J. Appl. Cryst.* **25**, 323–324 (1992).

110. Carrell, H. L., Rubin, B. H., Hurley, T. J., and Glusker, J. P. X-ray crystal structure of D-xylose isomerase at 4-Å resolution *J. Biol. Chem.* **259**, 3230–3236 (1984).

111. Riggs, A. F. Sulfhydryl groups and the interaction between the hemes in hemoglobin. *J. Gen. Physiol.* **36**, 1–16 (1952).

112. Perutz, M. Early Days of Protein Crystallography. *Methods in Enzymology* **114A**, 3–18 (1985).

113. Green, D. W., Ingram, V. M., and Perutz, M. F. The structure of hæmoglobin IV. Sign determination by the isomorphous replacement method. *Proc. Roy. Soc.* (*London*) **A225**, 287–307 (1954).

114. Perutz, M. F., Rossmann, M. G., Cullis, A. F., Muirhead, H., Will, G., and North, A. C. T. Structure of hæmoglobin. A three-dimensional Fourier synthesis at 5.5-Å resolution, obtained by X-ray analysis. *Nature* (*London*) **185**, 416–422 (1960).

115. Blow, D. M. The structure of hæmoglobin. VII. Determination of the phase angles in the noncentrosymmetric (100) zone. *Proc. Roy. Soc.* (*London*) **A247**, 302–336 (1958).

116. Bijvoet, J. M. Phase determination in direct Fourier-synthesis of crystal structures. *Proc. Koninklijke Nederlandse Akademie van Wetenschappen.* [*Proc. Roy. Soc.* (*Amsterdam*)] **52**, 313–314 (1949).

117. Bodo, G., Dintzis, H. M., Kendrew, J. C., and Wyckoff, H. W. Crystal structure of myoglobin. V. Low-resolution three-dimensional Fourier synthesis of spermwhale myoglobin crystals. *Proc. Roy. Soc.* (*London*) **A253**, 70–102 (1959).

118. Dickerson, R. E., Kendrew, J. C., and Strandberg, B. E. The crystal structure of myoglobin: Phase determination to a resolution of 2 Å by the method of isomorphous replacement. *Acta Cryst.* **14**, 1188–1195 (1961).

119. Wagner, E. Über Spektraluntersuchungen an Röntgenstrahlen. [Spectroscopic observations of X rays.] *Physik. Z.* **21**, 621–626 (1920).

120. Renninger, M. "Umweganregung" eine bisher unbeachtete Wechselwirkungserscheinung bei Raumgitterinterferenzen. ["Detour-excitation" by a hitherto unobserved interaction effect in space-lattice interference.] *Z. Physik.* **106**, 141–176 (1937).

121. Ewald, P. P., and Heno, Y. X-ray diffraction in the case of three strong rays. I. Crystal composed of non-absorbing point atoms. *Acta Cryst.* **A24**, 5–15 (1968).

122. Chang, S.-L. *Multiple Diffraction of X-rays in Crystals.* Springer-Verlag: Berlin, Heidelberg, New York, Tokyo (1984).

123. Cole, H., Chambers, F. W., and Dunn, H. M. Simultaneous diffraction: indexing umweganregung peaks in simple cases. *Acta Cryst.* **15**, 138–144 (1962).

124. Colella, R. Multiple diffraction of X-rays and the phase problem. Computational procedures and comparison with experiment. *Acta Cryst.* **A30**, 413–423 (1974).

125. Post, B. Comments on *Multiple diffraction of X-rays and the phase problem* by R. Colella. *Acta Cryst.* **A31**, 153–155 (1975).

126. Colella, R. Reply to Post's comments on my paper 'Multiple diffraction of X-rays and the phase problem. Computational procedures and comparison with experiment.' *Acta Cryst.* **A31**, 155 (1975).

127. Shen, W., and Colella, R. Solution of phase problem for crystallography at a wavelength of 3.5 Å. *Nature* (*London*) **329**, 232–233 (1987).

128. Miyake, S., and Kambe, K. A dynamical theory of the simultaneous reflection by two lattice planes. II. The effect of the phase factor of the structure amplitude. *Acta Cryst.* **7**, 220 (1954).

129. Post, B. Solution of the X-ray 'phase problem.' *Phys. Rev. Lett.* **39**, 760–763 (1977).

130. Juretschke, H. J. Invariant-phase information of X-ray structure factors in the two-beam Bragg intensity near a three-beam point. *Phys. Rev. Lett.* **48**, 1487–1489 (1982).

131. Burbank, R. D. Intrinsic and systematic multiple diffraction. *Acta Cryst.* **19**, 957–962 (1965).

132. Lipscomb, W. N. Relative phases of diffraction maxima by multiple reflection. *Acta Cryst.* **2**, 193–194 (1949).

133. Hümmer, K., and Billy, H. Experimental determination of triplet phases and enantiomorphs. In: *Crystallographic Computing 3: Data Collection, Structure Determination, Proteins, and Databases.* (**Eds.**, Sheldrick, G. M., Krüger, C., and Goddard, R.) pp. 33–42 Clarendon Press: Oxford (1985).

134. Colella, R. Multiple diffraction of X-rays and the phase problem. In: *P. P. Ewald and his Dynamical Theory of X-ray Diffraction. A memorial Volume for Paul P. Ewald. 23 January 1888 — 22 August 1985.* (**Eds.**, Cruickshank, D. W. J., Juretschke, H. J., and Kato, N.) International Union of Crystallography/Oxford University Press: Oxford (1992).

135. Chang, S-L. Solution of the X-ray phase problem using multiple diffraction – a review. *Cryst. Rev.* **1**, 87–187 (1987).

136. Billy, H., Burzlaff, H., and Hümmer, K. Experimental determination of triplet phases for a non-centrosymmetric structure: L-asparagine. *Acta Cryst.* **A40**, C-409 (1984).

137. Hümmer, K., Schwegle, W., and Weckert, E. A feasibility study of experimental triplet-phase determination in small proteins. *Acta Cryst.* **A47**, 60–62 (1991).

138. Post, B. The experimental determination of the phases of X-ray reflections. *Acta Cryst.* **A39**, 711–718 (1983).

139. Shen, Q., and Colella, R. The current status of phase determination by means of multiple Bragg diffraction. *Acta Cryst.* **A42**, 533–538 (1986).

140. Han, F-S., and Chang, S-L. A novel method for indexing multiple diffraction peaks. *J. Appl. Cryst.* **15**, 570–571 (1982).

141. Han, F-S., and Chang, S-L. Determination of a centrosymmetric crystal structure using experimentally determined phases with the direct method. *Acta Cryst.* **A39**, 98–101 (1983).

142. Hart, M., and Lang, A. R. Direct determination of X-ray reflection phase relationships through simultaneous reflection. *Phys. Rev. Lett.* **7**, 120–121 (1961).

143. Hümmer, K., and Billy, H. W. Experimental determination of triplet phase and enantiomorphs of non-centrosymmetric structures. I. Theoretical considerations. *Acta Cryst.* **A42**, 127–133 (1986).

144. Burzlaff, H., and Zimmerman, H. On the choice of origin in the description of space groups. *Z. Krist.* **153**, 151–179 (1980).

CHAPTER
9

Electron-Density Maps

When considering the results of diffraction experiments, it is important to keep in mind exactly which component of the crystal structure caused the diffraction effect, since this will be what we "view" in the subsequent Fourier series calculated from the diffraction data. X rays are scattered by electrons, and the result of an X-ray diffraction study is an electron-density map, that is, a representation of the variation in the local concentration of electrons (mostly core electrons in the atoms, together with some valence electrons) throughout the crystal, expressed as the number of electrons per unit volume. With respect to X-ray diffraction, Reginald William James[1] wrote "It is true that it is often convenient to regard the unit cell as built up of discrete atoms, but the correct idea, and for our present purpose the essential one, is that of a continuous distribution of diffracting matter having maxima in the regions occupied by atoms." In an analogous way, since neutrons are scattered by atomic nuclei, the result of a neutron diffraction analysis is the corresponding image of the positions of the atomic nuclei in the crystal structure.

The Fourier method of calculating an electron-density map was introduced by William Henry Bragg[2] in 1915. The equations he used had been published by Albert Porter[3] in 1906 who showed that the coefficients in the Fourier expression are the amplitudes of the diffracted beams, as described in Chapter 6. Initially such calculations were done along specific directions (as lines) through the crystal. For example, such one-dimensional Fourier series were calculated by William Duane[4] and Robert James Havighurst[5] in the 1920s for some compounds with the rocksalt structure. Sodium chloride, sodium fluoride, and lithium fluoride were selected for such calculations. George Shearer[6] computed one-dimensional Fourier series for long-chain ketones. The first two-dimensional electron-density map was calculated for the mineral diopside by William Lawrence

Bragg following advice to do so from his father, W. H. Bragg.[7] The calculations involved were many and tedious, and therefore Henry Lipson, C. Arnold Beevers, A. Lindo Patterson, and George Tunell provided more convenient methods of computing the electron-density functions in the days before high-speed computers were available.[8,9] Currently, the computation of an electron-density map is simple and fast because of the efficiency of available computers.

9.1 Calculations of electron-density maps

The **electron density** in a crystal is periodic from unit cell to unit cell. Therefore it can be represented as a Fourier series (as discussed in Chapter 6). The coefficients of this Fourier series are the amplitudes of the Bragg reflections; the periodicities ($h,k,$ and l) are the indices of each Bragg reflection. *Only the relative phase angles $\alpha(hkl)$ are still needed, and once these have been estimated* (see Chapter 8), *all of the information for calculating the electron density becomes available.*

The formula for the Fourier summation used to calculate an electron-density map is:

$$\rho(xyz) = \left(\frac{1}{V}\right) \sum_h \sum_k \sum_l F(hkl)\, e^{-2\pi i(hx+ky+lz)}, \qquad (9.1)$$

where x, y and z are the coordinates of each point in the unit cell at which the electron density is calculated. In this equation $F(hkl)$ is the structure factor and V is the unit-cell volume. The summation is over all values of h, k and l.

The meaning of Equation 9.1 in terms of the summation of waves was illustrated in Figure 6.9 (Chapter 6). Each value of $|F(hkl)|$ contributes to the total electron density a wave of this specific amplitude, and with a periodicity h, k and l in three dimensions, and a phase angle $\alpha(hkl)$. A series of two-dimensional terms (E values) that contribute to the electron-density map are shown are shown in Figure 9.1. Thus, if an intense Bragg reflection is measured, its electron-density wave will dominate the electron-density map, as shown in Figure 9.2 for the structure of a crystalline portion of DNA in which the 10 0 0 Bragg reflection is intense in a unit cell 34 Å long. As a result, the regularity of the electron density in the x direction is preserved in the map, although the phase of this Bragg reflection with respect to an origin chosen by the crystallographer is not known.

Equation 9.1 may also be written, in a more convenient form, as

$$\rho(xyz) = \left(\frac{1}{V}\right) \sum_h \sum_k \sum_l |F(hkl)| \cos 2\pi[hx + ky + lz - \alpha(hkl)], \qquad (9.2)$$

where the coefficient, $|F(hkl)|$, may be modified to give specific kinds of maps, listed in Table 9.1. In an electron-density map, this coefficient,

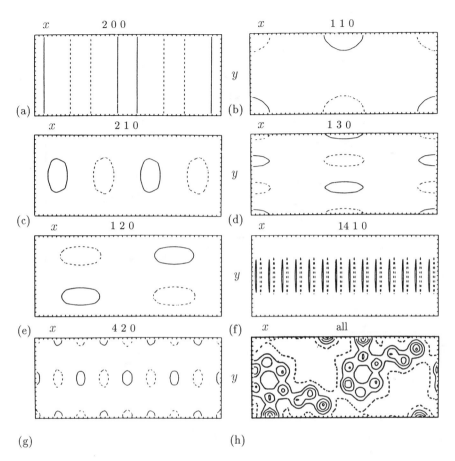

FIGURE 9.1. (a) Two-dimensional numerical output of a calculated three-dimensional map, contoured and showing, (a) to (g), the contributions of various terms to this map. (a) 2 0 0 $F_c = 19.1$; (b) 1 1 0 $F_c = -4.8$ (too weak to contribute); (c) 2 1 0 $F_c = -45.5$; (d) 1 3 0 $F_c = -29.0$; (e) 1 2 0 $F_c = 65.4$; (f) 14 1 0 $F_c = -11.6$; (g) 4 2 0 $F_c = -35.5$; (h) all terms. Note that y extends from 0.00 to 0.50 (down the page), while x extends the full length of the unit cell (0.00 to 1.00) (across the page).

$|F(hkl)|$, may be either the observed, $|F(hkl)|_o$, or calculated, $|F(hkl)|_c$, structure factor; in a difference electron-density map, the coefficient is the difference between these two quantities,

$$|F(hkl)|_o - |F(hkl)|_c.$$

The resulting electron-density map gives a representation of that part of the electron density that has not been correctly accounted for in the

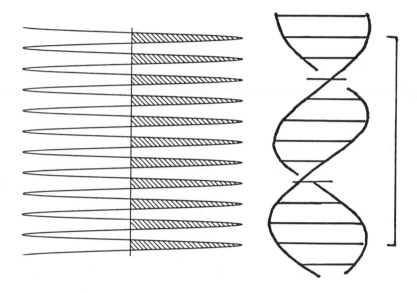

(a)   electron density →   (b)

FIGURE 9.2. The relationship between a high-amplitude electron-density wave (left, traveling vertically) and periodicity in the structure (right). The repeat unit is indicated at the far right. The 10 0 0 electron-density wave is intense (see Figure 3.16, Chapter 3) for DNA-like structures. The repeat unit is 34 Å long and the distance between nucleic acid bases is 3.4 Å.

atomic model that was used to calculate $F(hkl)_c$, i.e., the manner by which the data [measured as $F(hkl)_o$] differ from those found in the model [in $F(hkl)_c$]. The reader must, however, be warned that any of these maps are biased towards the model because it was the model, rather than the true structure, that was used to calculate the relative phase angles.

Three general ways in which electron-density maps are used in structure determination will be considered in turn.

1. **The F(obs) map:** this Fourier summation is calculated by use of Equation 9.1.1 in Table 9.1. It can be used to determine the arrangement of the atoms in the crystal structure. The relative phase angles $\alpha(hkl)$ used to calculate this map have been derived by one of the methods described in Chapter 8. The coefficients are generally either experimentally determined structure amplitudes, $|F(hkl)|_o$, structure amplitudes calculated from a model, $|F(hkl)|_c$, or normalized structure factor amplitudes $|E(hkl)|$.

2. **Difference electron-density maps:** these maps are calculated with Fourier coefficients $| \, |F(hkl)|_o - |F(hkl)|_c \, |$, as shown in Equation 9.1.4, or with variations on this as shown, for example, in Equation 9.1.5. Such maps are primarily used to refine a trial structure, to find a part of the structure that may not yet have been identified or located, to identify errors in a postulated structure, or to refine the positional and displacement parameters of a model structure. A difference map is very useful for analyses of the crystal structures of small molecules. It is also very useful in studies of the structures of crystalline macromolecules, since it can be used to find the location of substrate or inhibitor molecules that have been soaked into a crystal once the macromolecular structure is known. A formula like that in Equation 9.1.5 is then used. When a structure determination is complete, it is usual to compute a difference electron-density map to check that the map is flat, and approximately zero at all points.

3. **Deformation density maps:** these types of maps are calculated when intensity data have been measured to very high resolution (high values of $\sin \theta / \lambda$). Peaks and valleys in such maps indicate the deformation of the true electron density from the model in which an ellipsoidal peak is placed at each atomic position. The picture of such deformations provided by these maps may contain some evidence of bonding electrons or of lone-pair electrons, that is, the true electron density at very high resolution. The map, is, however, really only the difference between the model and the map from the diffraction data.

The electron density in a crystal, $\rho(xyz)$, is a continuous function, and it can be evaluated at any point x,y,z in the unit cell by use of the Fourier series in Equations 9.1 and 9.2. It is convenient (because of the amount of computing that would otherwise be required) to confine the calculation of electron density to points on a regularly spaced three-dimensional grid, as shown in Figure 9.3, rather than try to express the entire continuous three-dimensional electron-density function. The electron-density map resulting from such a calculation consists of numbers, one at each of a series of grid points. In order to reproduce the electron density properly, these grid points should sample the unit cell at intervals of approximately one third of the resolution of the diffraction data.[10] They are therefore typically 0.3 Å apart in three dimensions for the crystal structures of small molecules where the resolution is 0.8 Å.

Extensive use of three-dimensional electron-density or difference electron-density maps in crystallography became possible only with the advent of high-speed computers. The magnitude of the problem can be illustrated[11] for a compound crystallizing in an orthorhombic unit cell with dimensions $a = 11.98$, $b = 15.82$, $c = 11.49$ Å, $Z = 4$, for which 4397 (independent) Bragg reflections were measured at a resolution of

---

TABLE 9.1 The types of coefficients used in Fourier series (Equation 9.2)

---

1. F_o **map**

$$\rho_o(xyz) = \frac{1}{V}\sum_h\sum_k\sum_l |F_o(hkl)| \cos[2\pi(hx+ky+lz)$$

$$-\alpha_c(hkl)] \qquad (9.1.1)$$

2. F_c **map**

$$\rho_c(xyz) = \frac{1}{V}\sum_h\sum_k\sum_l |F_c(hkl)| \cos[2\pi(hx+ky+lz)$$

$$-\alpha_c(hkl)] \qquad (9.1.2)$$

3. E **map**

$$\rho_{\text{point atoms}}(xyz) = \frac{1}{V}\sum_h\sum_k\sum_l |E_o(hkl)| \cos[2\pi(hx+ky+lz)$$

$$-\alpha_c(hkl)] \qquad (9.1.3)$$

4. **Difference map** $(F_o - F_c)$

$$\rho_o - \rho_c = \frac{1}{V}\sum_h\sum_k\sum_l \Big||F_o(hkl)| - |F_c(hkl)|\Big| \cos[2\pi(hx+ky+lz)$$

$$-\alpha_c(hkl)] \qquad (9.1.4)$$

5. $2F_o - F_c$ **map for proteins**
   (P = protein, PL = protein-ligand complex.)

$$2\rho_{PL} - \rho_P = \frac{1}{V}\sum_h\sum_k\sum_l \Big|2|F_{PL}(hkl)| - |F_P(hkl)|\Big| \cos[2\pi(hx+ky+lz)$$

$$-\alpha_P(hkl)] \qquad (9.1.5)$$

0.77 Å. An electron-density map, sampled at a grid separation of 0.3 Å in each direction means that, with these unit cell dimensions and four asymmetric units in the unit cell, there will be $76{,}602/4 = 19{,}151$ grid points (a quarter of the unit cell). The value of $\rho(xyz)$ must be computed at each of these 19,151 positions. This means that it is necessary to sum 4397 terms (the number of Bragg reflections) at each of 19,151 points (the number of grid points in one quarter of the unit cell) to compute *one* three-dimensional Fourier map, *a total of* 84,226,947 *calculations*. For such a large calculation the use of a high-speed computer is essential

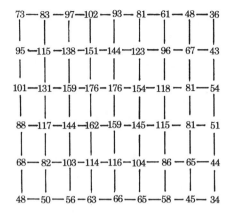

(a)

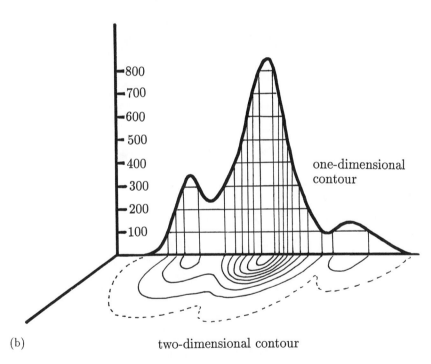

(b)   two-dimensional contour

FIGURE 9.3. Numerical values for the calculated electron density (a) at grid points, and (b) a two-dimensional plot, showing how contours are drawn in two dimensions. The level of contours (in electrons per cubic Å) can be calculated if the volume of each three-dimensional grid block is known in Å^3, and the absolute scale is know for the electron density (from the Wilson plot initially, and then from the subsequent least-squares refinement).

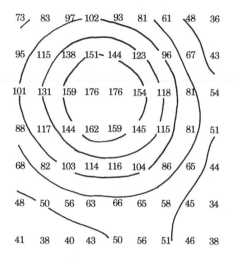

(c)

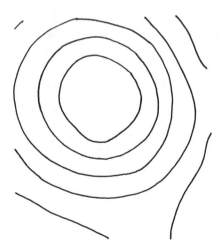

(d)

FIGURE 9.3 (cont'd). (c) Numerical values at grid points from (a) are contoured at regular intervals (50, 75, 100, 125, 150 here). (d) General contours minus the grid points and numerical values. This is what is now usually shown in published crystal-structure reports.

Various methods have been derived for decreasing the number of calculations needed, such as the **fast Fourier transform**,[12-16] but the equations used still require present-day computer power in order to be practicable.

9.1.1 Contouring electron-density maps

Since the electron density is calculated only at grid points, it is useful to draw contours at arbitrary values of electron density so that the locations and shapes of peaks can be evaluated. Contouring, illustrated in Figure 9.3, is the drawing of lines connecting points of equal value (generally electron density) and is done for successive plane sections, usually, but not always, parallel to planes defined by two of the three crystallographic axes. This is analogous to the contouring for altitude on geographic maps found in atlases. Contouring can be done manually (as in Figure 9.3), but more frequently it is now done by computer (as in Figure 9.4). Contoured sections are then superimposed at appropriate distances apart to obtain a three-dimensional image of the contents of the map.

The availability of high-performance computer graphics systems has made it possible for contoured electron-density maps to be displayed on a graphics screen. The maps so calculated[17] may have peaks that look like chicken wire, as shown in Figure 9.5. A stick model of the molecular fragment can be fit, again by computer graphics technology, into this electron density. The whole map with the molecule positioned in it can then be rotated and viewed from all directions in order to check the correctness of the fit. The coordinates of the atomic positions in the trial structure with respect to the axes of the electron-density map are then automatically determined and recorded.

9.1.2 Peak-search routines

There exist automated methods for searching the numerical output of the calculation of an electron-density map to find regions that are rich in electrons. The positional coordinates of the true maxima of such peaks are determined by evaluating the numerical value of $\rho(xyz)$ at each grid point in the electron-density map, and selecting the highest values, that is, those at the top of the peak, in a stepwise manner. The electron density is interpolated between the grid points assuming an ellipsoidal electron-density distribution about an atom. The coordinates of the peak center are considered to represent atomic positions. They may be further refined (see Chapter 10), so that they can be used for model building, or for calculation of the molecular geometry (see Chapter 11).

9.1.3 F(obs) electron-density maps

The simplest approach to structure completion involves calculating a Fourier series using the observed structural amplitudes and the most reliable set of phases. This is commonly referred to as the $F_o(hkl)$ map.

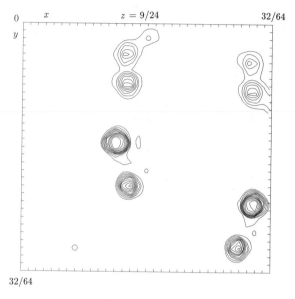

(a)

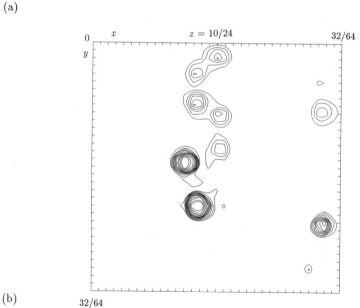

(b)

FIGURE 9.4. Sections of a three-dimensional electron-density map (an orthorhombic crystal, $a = 22.9$, $b = 22.1$, $c = 7.42$ Å). Each section is at a constant value of z. The contouring grid was $a/64$, $b/64$, $c/24$ Å. Of the other two axes x is across the page from 0.00 to 0.50, and y is down the page from 0.00 to 0.50. (a) $z = 9/24 = 0.375$, and (b) $z = 10/24 = 0.417$.

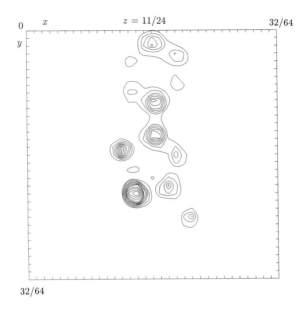

(c)

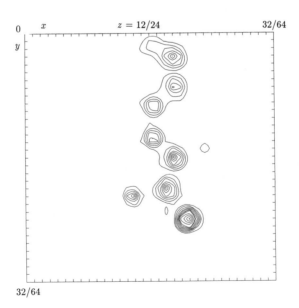

(d)

FIGURE 9.4 (cont'd). Successive sections a three-dimensional electron-density map. (c) $z = 11/24 = 0.458$, and (d) $z = 12/24 = 0.500$.

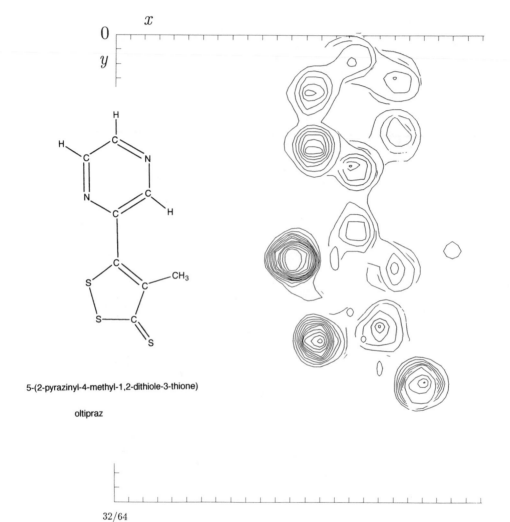

5-(2-pyrazinyl-4-methyl-1,2-dithiole-3-thione)

oltipraz

32/64

(e)

FIGURE 9.4 (cont'd). A composite of the maps in (a) to (d) to give the molecular structure. Note that the methyl group lies on a section not drawn here.

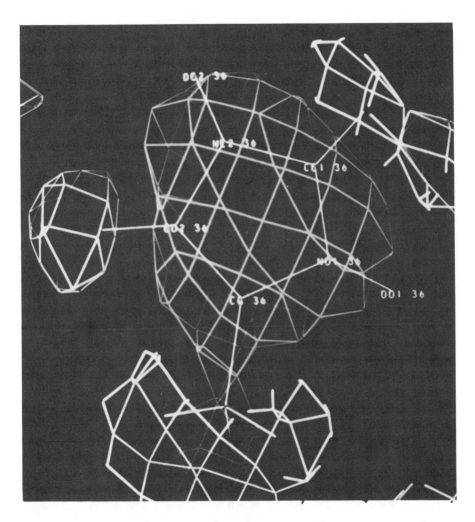

FIGURE 9.5. Nuclear density on a computer graphics screen. Shown is a histidine residue in a neutron-diffraction study. (Courtesy Benno P. Schoenborn.)

The coordinates of atomic positions can be located and a molecular model derived, either by contouring the map, or by use of a peak search routine. An example[18] is provided in Figure 9.6. In the most favorable cases this map will also give an indication of the locations of atoms that were not included in the phasing model. Such a map can also be used to improve the accuracy of a preliminary model by adjusting the model to best fit the electron density.

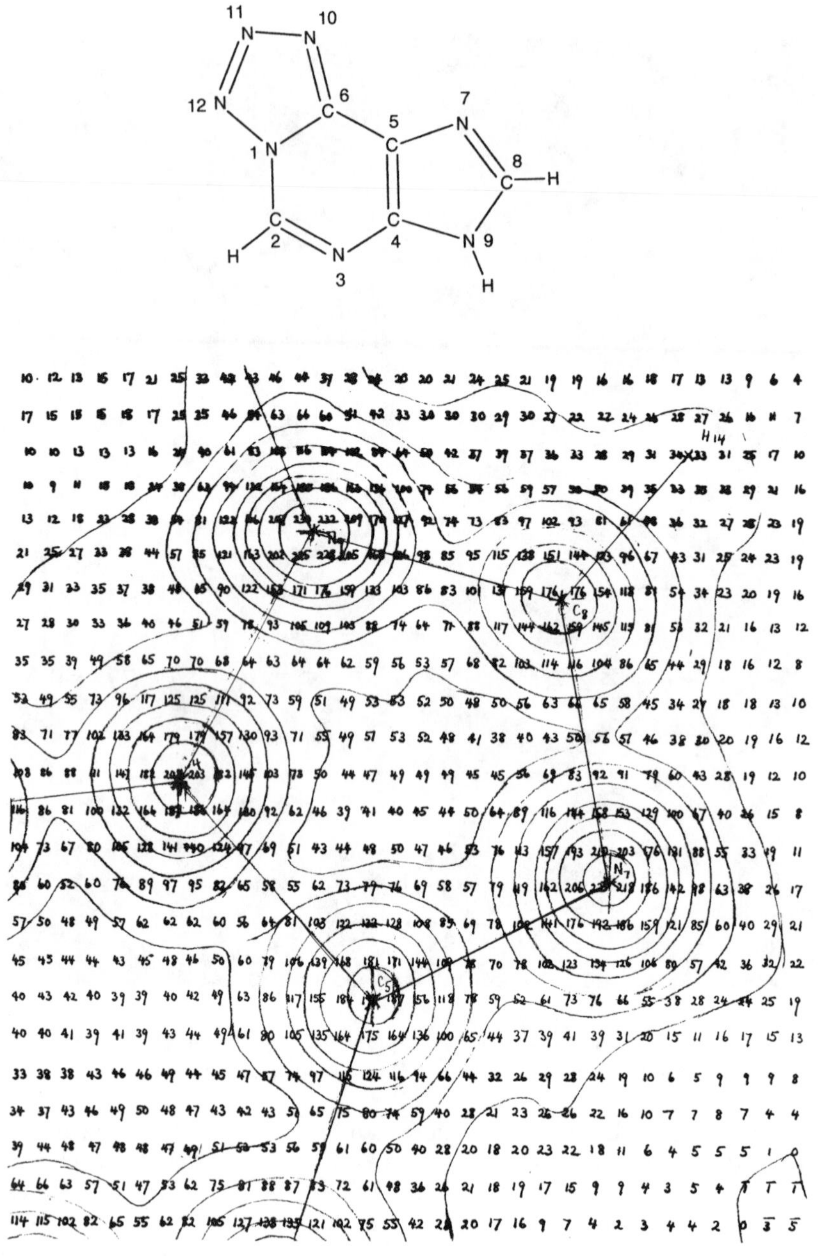

(a)

(b)

FIGURE 9.6. A tricyclic compound with one very short axis. (a) Chemical formula. (b) A portion of the electron density (view down 3.78 Å).

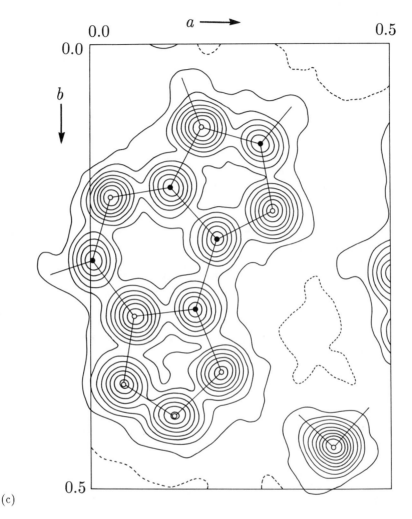

FIGURE 9.6 (cont'd). The fully contoured electron-density map. Carbon atoms represented by filled circles, nitrogen and oxygen atoms by open circles. Contour interval 1 electron/Å^2, zero contour dotted.

Variations in $F(obs)$ maps are possible. The phases from direct methods can be combined with $|E(hkl)|$ values to calculate an E map, that is, an analogue of a normal electron-density map (which has $F(hkl)$ values as coefficients). The ideal E map represents the electron density of a structure composed of point atoms, rather than the "smeared" atoms found if values of $F(hkl)$ are used. As a result, peaks are higher and sharper in an E map than in an F_o map.

Ideally, a map calculated with experimental coefficients $|F_o(hkl)|$, and phases $\alpha(hkl)_c$ should accurately reproduce the electron density of the crystal to the precision of the experiment. In practice, there are experimental errors in the measured values of $F(hkl)$ and errors in the model, such as the assumption of spherical atoms in the calculation of the atomic scattering factors, and limits in the extent to which disorder and thermal motion of atoms can be modeled.

9.1.4 Difference electron-density maps

When a partial model for a crystal structure has been obtained, it is usual to modify the coefficients $F(hkl)$ by subtracting the contributions of the "known" atoms in the crystal structure. This gives an electron-density map in which the electron density of "known" atoms has been removed. Peaks corresponding to any missing atoms are therefore made more prominent than they were in the $F(obs)$ map, and thus more easily recognizable. Disorder of atoms or groups may also be revealed in difference electron-density maps. Errors in relative phase angles (calculated only for the known part of the structure) may be severe when part of the structure is either missing or incorrect and may affect the usefulness of a difference map.

Peaks occur in a difference map in positions in the unit cell where the model did not include enough electron density; valleys appear in places where the model contained too much electron density. This information may be used to obtain more precise atomic positions, atomic displacement parameters, or atomic numbers. For example, in the last category, the identities of atoms (carbon or nitrogen) in a tricyclic molecule[18] were established by setting all atoms to one type (carbon in this case) in the structure factor calculation. A difference map was calculated with the calculated phases and examined for excess electron density at atomic positions (Table 9.2). It was found to be possible to distinguish between nitrogen (seven electrons) and carbon (six electrons), even though these atoms are adjacent in the Periodic Table.

TABLE 9.2. Assignment of atom types in a ring system (see Figure 9.6).[18]

Atom		ΔB	new	Atom		ΔB	new	Atom		ΔB	new
1	C	−0.5	N	5	C	−0.1	*C	9	C	−0.4	*N
2	C	+0.4	C	6	C	−0.2	*C	10	C	−0.2	*N
3	C	−0.6	N	7	C	−0.4	*N	11	C	−0.5	N
4	C	0.0	C	8	C	+0.5	C	12	C	−0.5	*N

* implies identity known. ΔB means $B - B_{av}$

The most valuable Bragg reflections for use in difference maps are those for which the calculated structure amplitudes are much larger than those observed. The model has contributed too much to these Bragg reflections, and a difference map with $-|F|_c$ as coefficients, i.e., $|F|_o - |F|_c$, with $|F|_o$ set to zero for these specific Bragg reflections, will indicate how to correct the model. The calculated relative phase angle is probably approximately correct.

Difference maps (difference electron-density maps) can be used to complete and refine a model of the atomic positions in the crystal. For example, such maps are used for finding the last few atoms in a trial structure,[19-22] or hydrogen atoms in small molecules (Figure 9.7). Because of their low electron density, hydrogen atoms make a relatively minor contribution to the structure factors. Consequently, it may be difficult to determine their atomic coordinates from electron-density maps calculated using the F_o synthesis. On the other hand, difference electron-density maps often give good peaks at hydrogen atom positions if all other atoms (and their displacement parameters) have been correctly accounted for and subtracted out. It is a common practice to refine a partial model, perhaps with anisotropic temperature factors, before attempting to locate hydrogen atoms in difference electron-density maps.

Difference electron-density maps can be used to refine atomic parameters by what is called a **differential synthesis**. If an atom is near but not at its correct position, there will be a gradient in the difference electron density map at the assumed atomic position. From this gradient the necessary correction to the atomic position can be estimated, as shown in Figure 9.8(a). Convergence to the true atomic position will be obtained if the assumed position (in F_c) lies within an atomic radius (in the true electron-density map). If the deviation is greater than an atomic radius, negative electron density will be centered at the wrong position, and a corresponding positive peak will appear at the correct position.

The original method for refining an atomic position, developed for F_{obs} maps, was to analyze the slope and curvature of the electron density at the input position of an atom. Andrew D. Booth[23] showed how this could be done. The slope and curvature of the electron density correspond, respectively, to the first and second derivatives of the electron density with respect to the atomic positions. Expressing the slope and curvature at the atomic center by a truncated **Taylor's series**,[24] the correction to the atomic positions was derived:

$$\text{shift} = -\frac{\text{slope at atomic position}}{\text{curvature at atomic position}} \tag{9.3}$$

A cyclic procedure employing structure factor calculations, followed by difference Fourier maps, is usually employed to locate all or nearly all nonhydrogen atoms, before the model is subjected to refinement by least-squares techniques (Chapter 10).

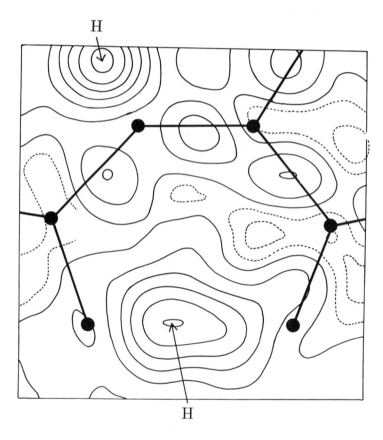

(a)

(b)

FIGURE 9.7. Hydrogen atoms in a section of a three-dimensional difference electron-density map (Ref. 25). The contour interval is 0.1 electrons /Å^3 with negative contours indicated by broken lines. The resolution is approximately 0.36 Å. Note the difference density in the centers of the bonds as well as at the locations of hydrogen atoms.

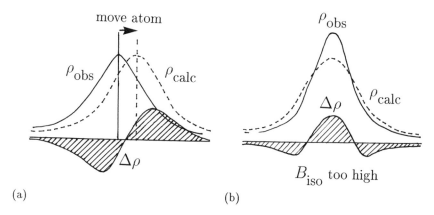

FIGURE 9.8. Use of difference maps to estimate (a) errors in atomic positions and (b) errors in thermal parameters or atomic occupancies.

Difference maps also allow adjustments to be made to displacement parameters. If the displacement parameter of an atom in the model is too large, it will be spread over more space than necessary, and its peak height will be lower than it should be. As a result, there will be a positive peak at the atomic center in the difference map; if the displacement factor is too small, a negative valley will appear in that position [Figure 9.8(b)]. The final difference Fourier map is generally not completely flat because it contains indications of both errors in the data ($|F_o|$) and inadequacies in the model ($|F_c|$), including, of course, the relative phase angle, $\alpha(hkl)_c$.

Another type of difference map that is used for macromolecular crystal structures, is the **omit map**. Part of the structure in a defined area of the unit cell is omitted from the phasing and then F_{obs} and difference maps are checked to determine if the portion that was omitted appears (at least to some extent) in the resulting electron-density map. In this way, for example, the continuity of electron density in areas selected for the backbone of a protein can be ascertained.[26,27]

9.2 The resolution of an electron-density map

The ability to distinguish two close objects as separate entities rather than as a single, blurred object is a measure of the resolution of the image under study. Many crystals, because of poor quality or disorder in atomic positions, do not scatter to the limit of the resolution imposed by the wavelength of the X rays used. In X-ray crystallography the value conventionally used for the resolution is $\lambda/2\sin\theta_{max}$, the maximum value of $\lambda/2\sin\theta$ in the measured data set. This is equal, by Bragg's Law (Equation 3.2, Chapter 3) to the interplanar spacing d_{min} As shown in Figure 9.9, the value of θ_{max} is measured by how far out from the

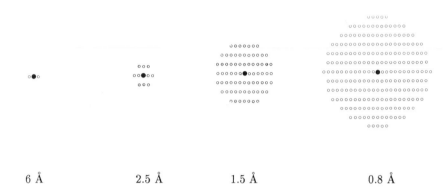

6 Å   2.5 Å   1.5 Å   0.8 Å

FIGURE 9.9. The resolution of a protein crystal structure in terms of the data measured. The higher the value of θ to which the data are measured, the higher the resolution of the data.

center of an X-ray diffraction photograph the diffraction data are visible. The variation in resolution of a crystal structure is shown in Figure 9.10. Most X-ray structures of small molecules are determined with Cu $K\alpha$ radiation, $\lambda = 1.54$ Å, and the maximum resolution that can be obtained with that radiation (because 2θ cannot be greater than $180°$) is $\lambda/2\sin\theta = 0.77$ Å. If Mo $K\alpha$ radiation is used, $\lambda = 0.71$ Å, then the maximum resolution would be $\lambda/2\sin\theta = 0.35$ Å. A limit to how high a resolution may be obtained experimentally is also imposed by the scattering power of the atoms present and the quality of the crystal, since disorder or high thermal motion will reduce θ_{max}. It is essential, when inspecting an electron density map, to know the resolution of the map.

At high resolution the scattering angles 2θ extend to large values where the interplanar spacings d_{hkl} are small. When data are measured to interplanar spacings corresponding to a resolution of 0.77 Å (normal for small molecules), each atom (except, possibly, a hydrogen atom) is distinct. Most small-molecule crystal structures are studied at this resolution or better. Macromolecules seldom diffract to a resolution better than 1.5 Å. An example of the appearance of an electron-density map of a protein at two different resolutions is shown in Figure 9.11. It is clear from this Figure that, at high resolution, the atomic positions in the molecular line diagram provide an excellent fit to the electron-density map. At low resolution, however, the electron density is more diffuse, and, while the line diagram fits reasonably well, it might have been more difficult to find the exact atomic arrangement.

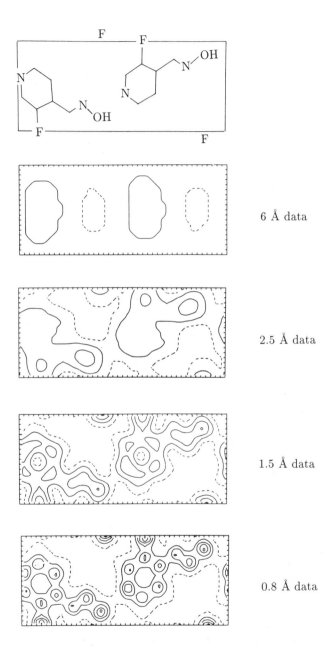

6 Å data

2.5 Å data

1.5 Å data

0.8 Å data

FIGURE 9.10. Different resolutions of the same structure. These diagrams were obtained by only including terms that diffract to the quoted resolution.

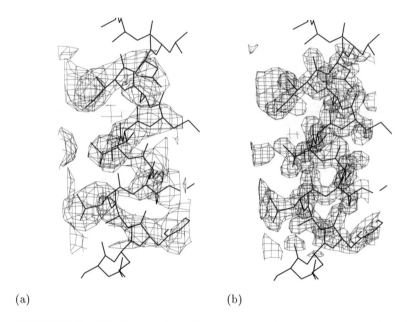

(a)   (b)

FIGURE 9.11. Fit of a protein molecule to a graphics system output (D-xylose iso-merase). (a) 4 Å resolution, and (b) 1.7 Å resolution. Note that the fit of the model, drawn by lines, is better at the higher resolution. (Courtesy H. L. Carrell).

When the intensities of Bragg reflections are measured only up to a cer-tain value of 2θ, but are of measurable intensity beyond this scattering angle, the electron density will show effects from the truncation of the data. The scattering factors have not, at the limit of the data, fallen to zero and therefore there will be **series termination errors** (also called termination-of-series errors) that appear as waves of positive and nega-tive electron density around each atom.[28,29] This problem is particularly noticeable in the regions around heavy atoms in electron-density maps, as diagrammed in Figure 9.12.

When small crystal structures are studied, all Bragg reflection data are used, and relative phase angles are derived by one of the methods described in Chapter 8, and electron-density maps are calculated to the maximum possible resolution that the wavelength of the X rays permit. On the other hand, because isomorphous replacement methods are used to obtain relative phase angles for macromolecular structures, it is usual to calculate electron-density maps at low resolution initially, and to in-crease the resolution as more phases from isomorphous replacement data become available. Traditionally the structure determination is divided into three resolution shells that correspond to the minima of the radial distribution of intensities.[30]

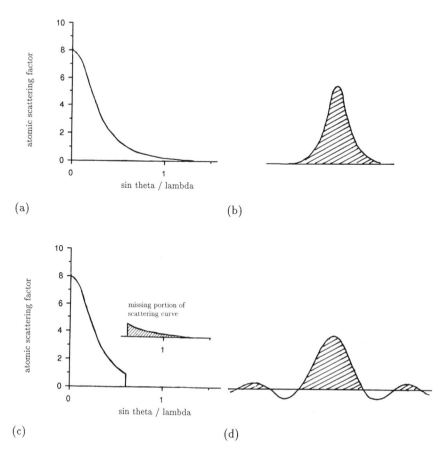

FIGURE 9.12. Series-termination errors. (a) A normal atomic scattering factor curve and (b) the atomic peak obtained by Fourier transformation. (c) A truncated atomic scattering factor curve, such as that used for data that are measured to a lower $\sin\theta/\lambda$ value than advisable. The missing portion of the scattering curve is indicated. (d) The atomic peak obtained by Fourier transformation. Note the ripples caused by loss of the missing portion of the atomic scattering curve.

The radial distribution of intensity as a function of scattering angle shows features that can be correlated with what is seen at each resolution level of a protein. The first minimum of this function corresponds to intensity data with d_{hkl} spacings greater than 6 Å. At this resolution the overall shape of the molecule can be assessed, and the boundary between protein and solvent may possibly be seen. At this low resolution, however, the electron density is averaged over a volume greater than that of an individual atom, so that detailed information cannot be obtained. One low-resolution feature is the α helix (see Chapter 12) which appears as a high-density cylinder with a diameter of approximately 5 Å. A complete tracing of the polypeptide chain is normally not possible at this resolution. The resolution limit of 4.5 Å is generally avoided in the calculation of protein electron density because this resolution is near a maximum in the radial distribution function of Bragg reflection intensities. Truncation of the data at this resolution may introduce unwanted noise (such as termination-of-series errors) in the electron density map. At a resolution between 3.5 and 2.8 Å it is usually possible to trace the polypeptide chain correctly, provided the amino acid sequence is known.[31] A large number of macromolecular structures are determined at this resolution because the resultant electron-density map is usually of sufficient quality to allow one to position most of the side-chain atoms and fit the complete linear sequence in three dimensions.

High resolution electron-density maps for a macromolecule usually imply data with spacings less than 2 Å. At this resolution the chain tracing of the backbone, $-NH-C_\alpha-CO-$, may be unambiguous in places because it is generally possible to see the electron density associated with the carbonyl oxygen of the main chain and that will fix the orientation and position of the peptide unit. The identities of amino acid sidechain residues can often be assigned because of their characteristic shapes in the electron-density map. Unfortunately, some regions of the polypeptide backbones of proteins have been erroneously interpreted, particularly in loop regions. At 2.9 Å resolution, individual atoms are not resolved and errors in the phase angles produce ambiguities in the maps.[32] In essence the problem is analogous to that found when looking at an object through a low-power microscope; sometimes it is difficult to tell what is joined to what. At higher power this problem diminishes as does the problem of locating the backbone and side-chain atoms in proteins when higher-resolution X-ray diffraction data are obtainable.

There are many cases in crystallography where not all the necessary Bragg reflection data are available. This is usually a consequence of poor crystal quality, wavelength restrictions, or problems in obtaining phases. What is generally done with data that cannot be measured is to set their amplitudes to zero. This may be satisfactory in early stages but it causes problems when refining structure, such as the "termination-of-series effects" just described. Entropy maximization methods are now being used in an attempt to obtain more unbiased information from measured data.[33-38] The aim of entropy maximization is *to prevent the introduction of extra pattern that may result if one ignores unmeasured data.* Lack of pattern, which is what is aimed at for unmeasured data, can be interpreted as "chaos" or "maximum entropy." Thus, the method is an attempt to eliminate the introduction of the ordered (false) information, thereby introducing additional false, unjustified pattern into the known data. Rather, the unmeasured data are constrained to provide the least regular (most chaotic) perturbations to the known data. The procedures employed in entropy maximization are highly complex and involve a way of inventing amplitudes and phases for missing data using the available data. The result is a less biased map than that obtained by the other currently used methods.

9.3 The interpretation of electron-density maps

The locations of peaks in small-molecule electron-density maps lead directly to a model of the atomic arrangement. This is then interpreted in terms of molecular structure, as will be described in Chapter 11. The interpretation of the electron-density maps of larger molecules may encounter one or more problems, some of which will be described. Such maps are rarely at atomic resolution. As a result, interpretation is somewhat more subjective than is the interpretation of the electron-density maps of small molecules.

9.3.1 Building a model of a protein structure

We will now describe some details of the interpretation of the electron-density map that has been calculated. There are two major methods used in protein crystallography to build a model of the molecule from the various features found in an electron-density map. These methods differ from those used for small-molecules because the number of atoms is so large, and because individual atoms are not resolved in most protein crystal structures. A scheme, which is a continuation of Figure 8.10 (Chapter 8), is given in Figure 9.13.

The first method for building a model from a protein electron-density map, developed by Frederic Richards,[39,40] was an "optical comparator," conceived independently of the constructions of nineteenth century magicians.[41] It is commonly referred to as "Fred's Folly" (or the "Richards box"). A half-silvered mirror is used to provide a superposition image of the model, built from Watson–Kendrew brass-wire models,[42] onto the electron density map. The electron density is displayed as contours on large transparent plastic sheets, to a scale of, for example, 0.5 cm to the Å. Each section of electron density is inserted in sequence with the appropriate spacing. The electron-density map and a model are appropriately scaled so that they can be optically superpositioned. The model of the protein molecule can then be adjusted to fit this three-dimensional electron-density map. When building or improving a model of a macromolecule, the brass pieces are manipulated by rotation about single bonds while, at the same time, the superposition onto the three-dimensional map is observed in the mirror. Once the three-dimensional structure has been built, it is necessary to record the positions of all the atoms as required for structural analysis and refinement. Several methods have been developed for measuring the atomic positions, all of which are time consuming, labor intensive, and not very precise.

1. *The multiple-isomorphous replacement (MIR) map has been calculated. Polyala-nine is fit to the map on a graphics system. Side chains can also be looked for if the sequence is known.*

2. *Symmetry averaging and solvent flattening to obtain more interpretable maps.*

3. *Refine by least-squares techniques or X-PLOR at end of least-squares refinement. Use simulated annealing because more ability to move than in least squares. Raise temperature, therefore can sample different conformations.*

4. *If model still in doubt use OMIT maps and difference maps and rebuild.*

5. *Cycle the refinement until it converges.*

6. *Soak in substrates or other interesting molecules and look at difference maps. These difference maps have slightly wrong phases.*

7. *Check agreement of observed and calculated structure factors.*

FIGURE 9.13. Building a model of a protein from an electron-density map.

High-performance computer graphics systems and their associated software have greatly improved methods of model building. Protein structures can be built more quickly and efficiently by computer techniques. Several computer programs have been developed over the years that display the electron density and allow one to construct a model that will fit it. Although the features of model-building programs vary, the basic ideas behind them are the same. Computer programs have been written to provide an extensively used means of model building.[17] The three-dimensional electron-density map is displayed on a computer graphics terminal as contours as a netlike structure. A model can then be fit to this electron density map, as was done with the Richards box. The use of interactive computer graphics for fitting and building a molecular model has several advantages over mechanical modeling methods. Among these are the fact that computer models can be built faster, more easily, and generally more accurately than mechanical ones. Further, large mechanical models are difficult to manipulate, while computer modeling allows one to view the fit of the model dynamically by rotation of the image and recording the location of each atom at the same time. There are, however, some disadvantages of the computer-generated model. The display is two dimensional, and the model can not be handled directly. In addition, the amount of information that can be displayed at one time is limited. Over the past few years, however, it has been shown that the advantages of computer modeling far outweigh its disadvantages, and it has become the approach of choice.

9.4 Improving electron-density maps

For macromolecules it may be difficult to determine the overall conformation of the protein from the electron-density map, but some interesting methods are in use for aiding in the interpretation of this map.

9.4.1 Modifying a macromolecular electron-density map

Density modification is a procedure that is used to improve protein structures, If phase information is poor, it still is possible to calculate an electron-density map, modify it in some way by use of chemical or crystallographic information, and then to calculate the Fourier transform of this modified map (to give new structure factors and phases), and recompute the electron-density map with what are hopefully improved phases.[43-45] The new electron density map should be easier to interpret than the first.

The method of density modification by **solvent flattening** was originally introduced to help "correct" the phases of partially known structures.[46] Two assumptions are made when this method is used. The first is that the electron density in solvent areas is fairly smooth and of constant value, and the second is that there is an absolute value below which the electron density may not fall. Inside a protein the atoms are generally bonded and lie approximately 1.2–1.5 Å apart, while the solvent consists mainly of hydrogen-bonded water molecules, at distances of about 2.7 Å apart. Therefore the electron density within a protein is higher than in the solvent and it is possible to determine from the electron-density map, as shown in Figure 9.14, an "envelope" defining the approximate boundary of the molecule.[47] Since both the size of the protein and the percentage of the unit cell that is solvent are known (from the molecular weight, the crystal density and its unit-cell dimensions), the volume within the molecular envelope and the average electron density in the solvent area can be calculated. All electron density inside this envelope is considered to contain protein molecule; all electron density outside the envelope is set to the average value for water background. A new set of phases can be determined by calculation of the Fourier transform of this "solvent flattened" map.[47-53]

The second type of electron-density modification is noncrystallographic symmetry averaging. If there is more than one copy of the molecule in the asymmetric unit, and the relationship between them is known (for example, from the rotation function), constraints can be placed on the electron density map to make the two molecules look approximately the same. This information can then be used to improve the electron density, and hence the relative phases of the structure, and

FIGURE 9.14. Protein electron-density map showing several sections. The overall shape of the molecule can be seen in this set of sections of the electron-density map.

hence the electron-density map. For example, the orientation of a twofold rotation axis relating two molecules in the asymmetric unit can probably be determined from the rotation function, described in Chapter 8. This symmetry will also be evident in the diffraction pattern, as shown in Figure 8.21 (Chapter 8).[65,83] If the two-fold axis can be located, then the electron density for the two symmetry-related portions of the molecule can be averaged. A Fourier transform will then give better relative phases, and the subsequent electron-density map may be easier to interpret than was the previous map.

9.4.2 Refinement of an electron-density map

Atomic coordinates of atoms in the model of a protein structure can be refined to more precise values that give a better fit to the experimental data. The so-called real-space refinement procedure optimizes the fit of a macromolecular model to the electron-density map. A method for doing this was developed for proteins by Robert Diamond,[56,57] with the object of finding the best fit of the protein model to the electron density map. The gradient of the electron density at each atomic position is calculated and the extent to which atoms must be moved so as to best fit the density is estimated (see Equation 9.3). This method of structure refinement is generally replaced by reciprocal-space refinement of the type to be described in Chapter 10,[58–60] although both methods are also used together.[61]

9.5 Ligand binding in protein crystal structures

Difference electron-density maps are of great use in protein crystallography for locating the binding of small molecules such as substrates and inhibitors,[62] and examining the effects of changes in pH. Since the protein crystal is approximately half water (generally in the range 27–65%),[54,63] there is plenty of space for such molecules to be soaked into the crystal. A complete set of diffraction data for the protein under these different conditions are then measured. A difference Fourier map using the already determined relative phases (as in Equation 9.1.5, Table 9.1) will produce an electron density map of the type shown in Figure 9.15 which reveals the locations of any changes that occur when substrate, inhibitor, or drug binds when a mutation occurs. In this way, information on the mechanism of action of the macromolecule, if it is a catalyst, can be obtained.

When a substrate or inhibitor binds to a protein, it displaces water.[54] As a result, electron density for ordered water is replaced by electron density for part of the ligand molecule. This means that there may be no appreciable peak in the difference map. In addition, because of somewhat incorrect phases, the substrate or inhibitor will appear in a difference map with reduced electron density, usually about half that of a well-phased map (half-weight ΔF map). Consequently, the practice has sometimes been to multiply coefficients. Often a map that combines the features of a difference map $F_{PH} - F_P$, enhanced by a factor of two, and of the native protein map F_P, is used. The coefficients of the Fourier synthesis are then:

$$[(n + 1)|F_{PH}| - n|F_P|], = \left(|F_{PH}|\right) + \left(n||F_{PH}| - |F_P||\right)$$

where $n = 1$ or 2 is usual, but may be as large as 5. By this method the macromolecular difference electron-density map is enhanced to emphasize those features of greatest interest. Thus, $2F_{PH} - F_P$ (F_{obs} + half-weight ΔF map) is used rather than $F_{PH} - F_P$ (half-weight ΔF map). Note that in all such cases, it is assumed that the phases of the protein-ligand crystal structure are the same as those (determined as described in Chapter 8) of the native structure (protein only).

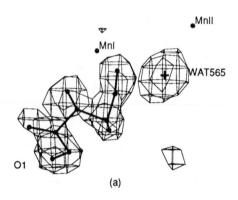

(a)

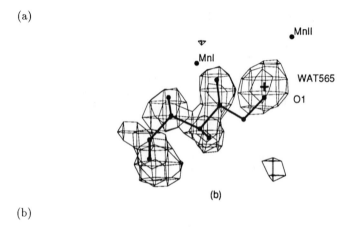

(b)

FIGURE 9.15. A substrate binding to a protein shown by difference maps. The fit to the electron density map is better in (a) than in (b).

9.6 Deformation density maps

As the resolution of the Bragg reflection data is improved, it becomes possible to obtain information on the more minute details of electron density in a molecule. At high enough resolution information can be obtained on the redistribution of electron density (deformation density) around atoms when they combine to form a molecule. Electrons in molecules may form bonds or exist as lone pairs, thereby distorting the electron density around each atom and requiring a more complicated function to describe this overall electron density than normally used, in which it is treated as if it were spherically symmetrical (deformed to an ellipsoid in order to account for anisotropic displacements). This assumption is inherent in the use of spherically-symmetrical scattering factors[64] although the electron density around an isolated atom may only be very approximately spherical. The extent to which this assumption is not true can be shown by high-resolution electron-density maps. Some hint of the inadequacy of the model is shown by the peaks in the centers of bonds that appear in the difference map shown in Figure 9.7. In the **deformation density map**[65-80] the difference between the total observed electron density and the calculated **promolecule density**,[81] obtained by superposition of the calculated electron densities of spherical atoms, is computed, as shown in Figure 9.16. This map gives a measure of the redistribution of the electron density of spherically symmetrical atoms when chemical bonding occurs. The deformation density map is, of course, affected by the precision of the data,[82] the resolution of the diffraction experiment, the treatment of data, the refinement procedures, and the correctness of the relative phases used to calculate it.

Deformation density maps can be obtained from high-order X-ray or neutron diffraction analyses. X-ray and neutron diffraction analyses of crystal structures provide different types of chemical information. X-ray studies give the electron-density distribution throughout the unit cell. By contrast, the neutron diffraction pattern gives locations of the nuclei of atoms in the crystal structure, that can also be approximated by X-ray studies if Bragg reflections are measured to very high $\sin \theta / \lambda$ values. The differences between the results of X-ray and neutron studies can be exploited to give information on perturbations of the electron density of isolated atoms on chemical bonding (deformation density). There are two ways to obtain precise atomic positions and a good deformation density map: one is by using only **high-order data** (subscript ho) in the refinement of the structural parameters (X_{ho}) and called the $X-X_{ho}$ method (which gives an **X–X map**); the other is by using neutron diffraction (N) for the refinement of the data and is called the X–N method (which gives an **X–N map**).

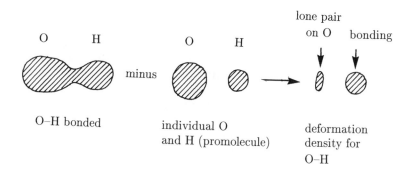

FIGURE 9.16. Principles of deformation density maps.

In the carbonyl C=O group, the presence of two lone pair electrons on the oxygen atom and high bonding density in the C=O double bond will cause the electron density around the oxygen atom to be nonspherical. The centroid of the electron density of the oxygen atom may not, then, correspond to the nuclear position found by a neutron diffraction analysis. Another example is provided by bonds involving a hydrogen atom, which has only one electron. The centroid of the electron distribution of the hydrogen atom is found in electron-density maps to be significantly displaced toward the atom that is bonded to the hydrogen atom. This means that bond lengths involving hydrogen atoms will be shorter in X-ray studies than in neutron diffraction studies. Hydrogen atom positions based on low-temperature high-order X-ray data, however, are close to those obtained from neutron diffraction data, although the corresponding e.s.d. values of the positional parameters are larger in the X-ray study.[55]

This asymmetrical nature of the valence electron density around atoms can be accounted for in the molecular model by the introduction of more parameters. These parameters are refined in order to obtain the best possible fit between calculated and experimental Bragg reflection data. This type of analysis gives a better approximation to the true phases in noncentrosymmetric structures than does a model that only contains ellipsoidal electron density around atoms. If the structural parameters take into account vibrational smearing, that is the inclusion of vibrational parameters, the resulting density is known as dynamic deformation density; if the refinement is with atoms corrected to point atoms, it is called static deformation density. When more parameters are added, however, one has to be sure that the parameter-to-data ratio is good, 1 to 10, for example. Therefore many more Bragg reflection data, to higher $\sin\theta/\lambda$ values (at least to 1.1 Å^{-1}), are needed. The resulting deformation density map will show a balance of the effects of the valence electrons and of any imprecisions in the experimental data.

A general equation can be derived that describes the variation in direction of the valence electron density about the nucleus. The distortion from sphericity caused by valence electrons and lone-pair electrons is approximated by this equation, which includes a population parameter, a radial size function, and a spherical harmonic function, equivalent to various lobes (multipoles). In the analysis the core electron density of each atom is assigned a fixed quantity. For example, carbon has 2 core electrons and 4 valence electrons. Hydrogen has no core electrons but 1 valence electron. Experimental X-ray diffraction data are used to derive the parameters that correspond to this function. The model is now more complicated, but gives a better representation of the true electron density (or so we would like to think). This method is useful for showing lone-pair directionalities, and bent bonds in strained molecules. Since a larger number of diffraction data are included, the geometry of the molecular structure is probably better determined.

An excellent example in which the information obtained from X–X$_{ho}$ studies of deformation density maps has been compared with that from theoretical calculations is given by tetrafluoroterephthalonitrile.[84–87] This compound is ideal for these studies because it contains no hydrogen atoms, and all the atoms have similar scattering power (6 to 9 electrons for C, N, and F). The molecule has high symmetry which simplifies the number of parameters to be refined. A very precise set of X-ray Bragg reflections were measured at a low temperature (98 K) and to $\sin\theta/\lambda = 1.15$ Å^{-1}.[73] Least-squares refinement based on the high-order data ($\sin\theta/\lambda > 0.85$ Å) and removal of the effects of atomic motion resulted in a static deformation density map. The main featured of the dynamic and static deformation density maps are shown in Figure 9.17. Note in Figure 9.17(b) and (c) that there is a large peak of electron density along the C $\equiv$ N triple bond and very weak density along the C–F bond. Also, in the static deformation density map, there is some evidence of lone-pair electrons on the nitrogen and fluorine atoms. For comparison, a deformation density map from theoretical calculations[87] (by means of a local-density-functional approximation), in Figure 9.17(d), shows good agreement with the static deformation density map in Figure 9.17(c). This implies that, if a high-quality experimental X-ray diffraction data set is refined, the atomic parameters obtained can be used for reliable and precise deformation density studies.

9.7 The importance of having the correct phase information

At this point it must be clear to the reader that the *electron density maps or other Fourier maps obtained are only as good as the relative phases used in their calculation.* The structure amplitudes are important in this calculation, but if the relative phases are not approximately correct, the crystallographer may be in the situation of refining an incorrect structure, often to a quite reasonable result, but to a local minimum in deviations of observed and calculated structure amplitudes. Calculated relative phases

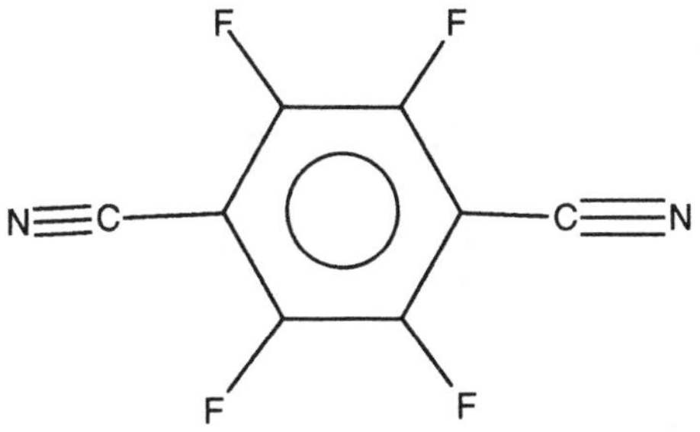

FIGURE 9.17. Examples of deformation density maps. (a) Chemical formula of tetrafluoroterephthalonitrrile.

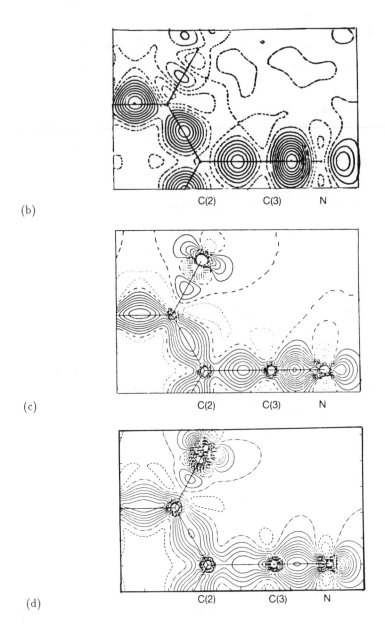

(b)

(c)

(d)

FIGURE 9.17 (cont'd).   Examples of deformation density maps.   (b) Dynamic deformation-density map in which motion of the atoms has not been corrected for.[86] (c) Static deformation density map in which motion of atoms has been corrected for.[85] (d) Theoretical deformation density map.[87] (Reproduced from Seiler, Schweizer, and Dunitz (Ref. 86), Hirshfeld (Refs. 76, 85), and Delley (Ref. 87), with permission.)

have a bias towards the atomic positions that were input into their computation. In centrosymmetric structures, where the relative phase angles are either 0° or 180°, if most of the trial structure is correct, the electron-density map using relative phases from this trial structure may contain an excellent representation of the true crystal structure. But if the crystal structure is noncentrosymmetric, the relative phases may have any value from 0° to 360°. If the model is not correct, it may be difficult, if not impossible, to find the correct structure from the calculated electron-density map. Unusual features in a crystal structure, such as nonbonded atoms that approach each other too closely, or unusually long bonds, may require one to return to the determination of the relative phases (see Chapter 8) and check that there was not a better set of relative phases that could have been used. The heights of peaks in the centers of bonds in a deformation density map appear to depend on the nature of the atoms at each and of the bond.[88,89] Thus, as shown for tetrafluoroterephthalonitrile, the deformation density is greater in C–C than in C–F bonds. A similar situation is found in hydrogen peroxide, where there is apparent negative bonding density in the O–O bond.[90] Electronegative elements such as N, O, and F have lone pair electrons which cause an increase in electron density in that region of the atom. This means that the differences between free atoms and bonded atoms in the region of a chemical bond are positive for carbon, near zero for nitrogen and negative for O and F, while, in the lone-pair region they are positive for N, O and F. Deformation-density maps can, because they represent molecules in a crystalline environment, be used in conjunction with *ab initio* molecular orbital calculations (on isolated molecules) to study the effects of intermolecular interactions on the molecule.[89]

Deformation density can also be fit by Equation 9.4. The shape of the valence shell electron density is represented by a parameter κ which is a scale factor for a radial function.[91–96] An expanded spherical harmonic function Y_{lm} provides the angular aspects of the model:

$$\rho_{\text{atom}} = \rho_c + P_v \rho_v(\kappa r) + \sum_{lmn} P_{lmn} R_n(r) Y_{lm}(\theta, \phi) \tag{9.4}$$

where ρ_c is the core density and is fixed, ρ_v is the valence density and varies radially by means of the parameter κ. P_v and P_{lmn} are population coefficients ($l = 0$ monopole, $l = 1$ dipole, $l = 2$ quadrupole, $l = 3$ octopole) and m ranges from $-l$ to $+l$. The term Y_{lm} is a spherical harmonic shape function, which, together with the radial size functions $R_n(r)$ is determined from tables. The parameters that are refined are P_v, P_{lmn}, and κ. In this way an analytical description of the deformation density can be obtained. Such studies can give useful chemical insight. For example, in a study of ammonium hexaaquachromium(II) disulfate (an ammonium Tutton salt)[97] it was shown that a large Jahn-Teller distortion existed. The expected hole in the $3d$ density was identified in these maps. Those interested further in the subject are urged to read the various listed articles.

Summary

1. The electron density in a crystal may be calculated by Equations 9.1 or 9.2.
2. An electron-density map will have a resolution that depends on the limit of the value of $\sin\theta/\lambda$ to which the Bragg reflections have been measured.
3. Electron-density maps are used to develop a trial structure.
4. Difference electron-density maps are used to determine if that trial structure needs to have electron density added or subtracted from it in various areas of the structure. They are particularly used for locating hydrogen atoms, which have small scattering effects because of their low atomic number.
5. Deformation density maps give some indications of the deformation from sphericity (or ellipticity) of the electrons of the atoms in the model as a result of chemical bonding or the existence of lone-pair electrons. They should, however, be interpreted with caution, especially with respect to the resolution of the electron-density map obtained from them. The maximum value of $\sin\theta/\lambda$ should be much higher than normally used, generally requiring short-wavelength X rays and low temperatures of measurement.
6. Each of these maps is only as good as the set of relative phases (see Chapter 8) used to calculate it.

Glossary

Deformation density: The difference between the electron density in a molecule, with all its distortions as a result of bonding, and the promolecule density, obtained by forming a molecule with spherical electron density around each atom (free atoms). This map contains effects caused both by the errors in the relative phases of Bragg reflections, experimental errors in the data, and inadequacies in the representations of the scattering factors of free atoms.

Density modification: A computational method for improvement of phases. The electron density is modified and a new set of phases is then determined by Fourier transformation, so that an improved electron-density map can be calculated.

Difference map: A Fourier synthesis (or "map") for which the Fourier coefficients are the difference between the observed and calculated structure factor amplitudes. Such a map, normally calculated during the refinement of a structure, will have peaks where not enough electron density was included in the trial structure and troughs where too much was included. It has proved to be an exceedingly valuable tool both for locating missing atoms and for correcting the positions of those already present.

Differential synthesis: A method of refining parameters of an atom from a mathematical consideration of the slope and curvature of the difference map (q.v.) in the region of the atom.

Electron density: The concentration of electrons per unit volume, expressed as electrons per Å^3.

Fast Fourier transform: An algorithm for the calculation of the Fourier transform in a rapid manner.

High-order data: Bragg reflections with values of $\sin\theta/\lambda$ greater than a value at which scattering due to valence electrons is judged to be small (around 0.8 Å^{-1}).

Omit map: A difference map in which part of the structure in a specific area of the unit cell is omitted from the phasing calculation. The resulting electron-density map is then examined to check if the proposed structure can still be recognized in that area.

Promolecule density: The electron density of spherically symmetrical free atoms with no effects of chemical bonding or other factors that distort the electron density.

Series-termination errors: Errors that result from a limitation in the number of terms in a Fourier series. Ideally an infinite number of data is required to calculate a Fourier series. In practice, the number of data depends on the resolution (reciprocal radius or $\sin\theta/\lambda$) to which the data have been measured. Because of truncation of the Fourier series at the highest value of $\sin\theta/\lambda$ of the data, peaks in the resulting Fourier syntheses are surrounded by a series of ripples. These are especially noticeable around a heavy atom because its scattering factor is still appreciable at the highest values of $\sin\theta/\lambda$ measured. Difference maps (q.v.) can be used to obviate most of the effects of series-termination errors.

Solvent flattening: A density modification technique useful when a unit cell contains a high proportion of solvent, as is found in macromolecular structures, which may be 50% or more solvent. An "envelope" defining the approximate boundary of the molecule is determined from the electron-density map, and all electron density outside this envelope is set to the average value. The usefulness of the method increases as the fraction of solvent in the unit cell increases.

Taylor's series: A convergent power series used to calculate the value of a function of x at values of x near x_0, a fixed number, by expanding the function in terms of powers of $(x - x_0) = \Delta x$, a small number.

$$f(x) = f(x_0) + \Delta x \frac{\partial f(x_0)}{\partial x} + (\Delta x)^2 \frac{\partial^2 f(x_0)}{2\partial^2 x} + \cdots .$$

Termination-of-series errors: See **Series-termination errors.**

X–N map: The difference between the experimental electron-density map and that calculated with X-ray scattering factors for spherical atoms and positional and displacement parameters derived from a neutron diffraction experiment.

X–X map: The difference between the experimental electron-density map and that calculated with X-ray scattering factors for spherical atoms and positional and displacement parameters derived from high-order X-ray diffraction data.

References

1. James, R. W. *Optical Principles of the Diffraction of X-rays.* Bell: London (1954).

2. Bragg, W. H. IX Bakerian lecture. X-rays and crystal structures. *Trans. Roy. Soc. (London)* **A215**, 253–274 (1915).

3. Porter, A. B. On the diffraction theory of microscopic vision. *Phil. Mag.* **11**, 154–166 (1906).

4. Duane, W. The calculation of the X-ray diffracting power at points in a crystal. *Proc. Natl. Acad. Sci. USA* **11**, 489–493 (1925).

5. Havighurst, R. J. The mercurous halides. *J. Amer. Chem. Soc.* **48**, 2113–2125 (1926).

6. Shearer, G. An X-ray investigation of certain organic esters and other long-chain compounds. *J. Chem. Soc.* **123**, 3152–3156 (1923).

7. Warren, B., and Bragg, W. L. The structure of diopside, $CaMg(SiO_3)_2$. *Z. Krist.* **69**, 168–193 (1928).

8. Lipson. H., and Beevers, C. A. An improved numerical method of two-dimensional Fourier synthesis for crystals. *Proc. Phys. Soc.* **48**, 772–780 (1936).

9. Patterson, A. L., and Tunell, G. A method for the summation of the Fourier series used in the X-ray analysis of crystal structures. *Amer. Mineralogist* **27**, 655–679 (1942).

10. Shannon, C. E. Communication in the presence of noise. *Proc. Inst. Radio Eng. (New York)* **37**, 10–21 (1949).

11. Stezowski, J. J. Chemical structural properties of tetracycline derivatives. 1. Molecular structure and conformation of the free base derivatives. *J. Amer. Chem. Soc.* **98**, 6012–6018 (1976).

12. Danielson, G. C., and Lanczos, C. Some improvements in practical Fourier analysis and their application to X-ray scattering from liquids. *J. Franklin Inst.* **233**, 365–380, 435–452 (1942).

13. Cooley, J. W., and Tukey, J. W. An algorithm for machine calculation of complex Fourier series. *Math. Computation* **19**, 297–301 (1965).

14. Ten Eyck, L. F. Crystallographic fast Fourier transforms. *Acta Cryst.* **A29**, 183–191 (1973).

15. Ten Eyck, L. F. Fast Fourier transform calculation in electron density maps. *Methods in Enzymology. Diffraction Methods for Biological Macromolecules. Part B.* (**Eds.**, Wyckoff, H. W., Hirs, C. H. W., and Timasheff, S. N.) **115** 324–337 (1985).

16. Brigham, E. O. *The Fast Fourier Transform.* Prentice-Hall: Englewood, NJ (1974).

17. Jones, T. A. A graphics model building and refinement system for macromolecules. *J. Appl. Cryst.* **11**, 268–272 (1978).

18. Glusker, J. P., van der Helm, D., Love, W. E., Minkin, J. A., and Patterson, A. L. The molecular structure of an azidopurine. *Acta Cryst.* **B24**, 359–366 (1968).

19. Cochran, W. The structure of pyrimidines and purines. V. The electron distribution in adenine hydrochloride. *Acta Cryst.* **4**, 81–92 (1951).

20. Cruickshank, D. W. J. The accuracy of atomic co-ordinates derived from least-squares or Fourier methods. *Acta Cryst.* **2**, 154–157 (1949).

21. Cruickshank, D. W. J. The accuracy of electron density maps in X-ray analysis with special reference to dibenzyl. *Acta Cryst.* **2**, 65–82 (1949).

22. Cochran, W. Some properties of the $(F_o - F_c)$-synthesis. *Acta Cryst.* **4**, 408–411 (1951).

23. Booth, A. D. Application of the method of steepest descents to X-ray structure analysis. *Nature (London)* **160**, 196 (1947).

24. Taylor, B. *Methodus Incrementorum Directa et Inversa. Proposition 7.* [Direct and indirect methods of incrementation.] Pearsniansis: Londini (1715)

25. Glusker, J. P., Orehowsky, W. Jr., Casciato, C. A., and Carrell, H. L. X-ray crystal analysis of the substrates of aconitase. X. The structure of dipotassium *cis*-aconitate. *Acta Cryst.* **B28**, 419–425 (1972).

26. Bhat, T. N., and Cohen, G. H. *OMITMAP*: an electron density map suitable for the examination of errors in a macromolecular model. *J. Appl. Cryst.* **17**, 244–248 (1984).

27. Bhat, T. N. Calculation of an OMIT map. *J. Appl. Cryst.* **21**, 279–281 (1988).

28. Sayre, D. Emerging new methodologies in the imaging of nanostructures. In: *Patterson and Pattersons. Proceedings of a Symposium held at the Fox Chase Cancer Center, Philadelphia, PA, USA, November 13–15, 1984.* pp. 68–85. (**Eds.**, Glusker, J. P., Patterson, B. K., and Rossi, M.) Oxford University Press: Oxford (1987).

29. Bricogne, G. Geometric sources of redundancy in intensity data and their use for phase determination. *Acta Cryst.* **A30**, 395–405 (1974).

30. Richardson, J. S., and Richardson, D. C. Interpretation of electron density maps. *Methods in Enzymology* **115**, 189–206 (1985).

31. Stenkamp, R. E., Sieker, L. C., Jensen, L. H., and McQueen, J. E. Jr. Structure of methemerythrin at 2.8-Å resolution: computer graphics fit of an averaged electron-density map. *Biochemistry* **17**, 2499–2504 (1978).

32. Brändén, C-I. and Jones, T. A. Between objectivity and subjectivity. *Nature (London)* **343**, 687–689 (1990).

33. Jaynes, E. T. Prior probabilities. *I. E. E. E. Transactions on Systems Science and Cybernetics* **4**, 227–241 (1968).

34. Collins, D. M. Electron density images from imperfect data by iterative entropy maximization. *Nature (London)* **298**, 49–51 (1982).

35. Collins, D. M. Extrapolative filtering. I. Maximization of resolution for one-dimensional positive density functions. *Acta Cryst.* **A34**, 533–541 (1978).

36. Bricogne, G. A Bayesian statistical theory of the phase problem. I. A multichannel maximum-entropy formalism for constructing generalized joint probability distributions of structure factors. *Acta Cryst.* **A44**, 517–545 (1988).

37. Bricogne, G. Direct phase determination by entropy maximization and likelihood ranking: status report and perspectives. *Acta Cryst.* **D49**, 37–60 (1993).

38. Prince, E. Construction of maximum-entropy maps, and their use in phase determination and extension. *Acta Cryst.* **D49**, 61–65 (1993).

39. Richards, F. M. The matching of physical models to three-dimensional electron-density maps: a simple optical device. *J. Molec. Biol.* **37**, 225–230 (1968).

40. Richards, F. M. Optical matching of physical models and electron density maps: early developments. *Methods in Enzymology* **115**, 145–154 (1985).

41. Hopkins, A. A. *Magic: Stage Illusions and Scientific Diversions.* Chapter II. Benjamin Blom: New York (1897).

42. Kendrew, J. C., Dickerson, R. E., Strandberg, B. E., Hart, R. G., and Davies, D. R. Structure of myoglobin. A three-dimensional Fourier synthesis at 2 Å resolution. *Nature (London)* **185**, 422–427 (1960).

43. Raghavan, N. V., and Tulinsky, A. The structure of α-chymotrypsin. II. Fourier phase refinement and extension of the dimeric structure at 1.8 Å resolution by density modification. *Acta Cryst.* **B35**, 1776–1785 (1979).

44. Tulinsky, A. Phase refinement/extension by density modification. *Methods in Enzymology* **115**, 77–89 (1985).

45. Schevitz, R. W., Podjarny, A. D., Zwick, M., Hughes, J. J., and Sigler, P. B. Improving and extending the phases of medium- and low-resolution macromolecular structure factors by density modification. *Acta Cryst.* **A37**, 669–677 (1981).

46. Hoppe, W., and Gassmann, J. Phase correction, a new method to solve partially known structures. *Acta Cryst.* **B24**, 97–107 (1968).

47. Wang, B-C. Resolution of phase ambiguity in macromolecular crystallography. *Methods in Enzymology.* **115**, 90–112 (1985).

48. Wang, B-C., Yoo, C. S. and Sax, M. Crystal structure of Bence Jones protein Rhe (3 Å) and its unique domain-domain association. *J. Molec. Biol.* **129**, 657–674 (1979).

49. Fraser, R. D. B., MacRae, T. P., and Suzuki, E. An improved method for calculating the contribution of solvent to the X-ray diffraction pattern of biological molecules. *J. Appl Cryst.* **11**, 693–694 (1978).

50. Chambers, J. L., and Stroud, R. M. Difference Fourier refinement of the structure of DIP-trypsin at 1·5 Å with a minicomputer technique. *Acta Cryst.* **B33**, 1824–1837 (1977).

51. Teeter, M. M. Water structure of a hydrophobic protein at atomic resolution: pentagon rings of water molecules in crystals of crambin. *Proc. Natl. Acad. Sci. USA* **81**, 6014–6018 (1984).

52. Watenpaugh, K. D., Sieker, L. C., Herriott, J. R., and Jensen, L. H. The structure of a non-heme iron protein: rubredoxin at 1.5 Å resolution. *Cold Spring Harbor Symposia in Quantitative Biology* **36**, 359–367 (1971).

53. Subbiah, S. Low-resolution real-space envelopes: improvements to the condensing protocol approach and a new method to fix the sign of such an envelope. *Acta Cryst.* **D49**, 108–119 (1993).

54. Blundell, T. L., and Johnson, L. N. *Protein Crystallography.* Academic Press: New York, London, San Francisco (1976).

55. Dodson, E., Harding, M. M., Hodgkin, D. C., and Rossmann, M. G. The crystal structure of insulin. III. Evidence for a 2-fold axis in rhombohedral zinc insulin. *J. Molec. Biol.* **16**, 227–241 (1966).

56. Diamond, R. A real-space refinement procedure for proteins. *Acta Cryst.* **A27**, 436–452 (1971).

57. Diamond, R. Real-space refinement of the structure of hen egg-white lysozyme. *J. Molec. Biol.* **82**, 371–391 (1974).

58. Deisenhofer, J., Remington, S. J., and Steigemann, W. Experience with various techniques for the refinement of protein structures. *Methods in Enzymology* **115**, 303–323 (1985).

59. Hendrickson, W. A. Stereochemically restrained refinement of macromolecular structures. *Methods in Enzymology* **115**, 252–270 (1985).

60. Jensen, L. H. Overview of refinement in macromolecular structure analysis. *Methods in Enzymology* **115**, 227–234 (1985).

61. Cowtan, K. D., and Main, P. Improvement of macromolecular electron-density maps by the simultaneous application of real and reciprocal space constraints. *Acta Cryst.* **D49**, 148–157 (1993).

62. Stryer, L., Kendrew, J. C., and Watson, H. C. The mode of attachment of the azide ion to sperm whale myoglobin. *J. Molec. Biol.* **8**, 96–104 (1964).

63. Matthews, B. W. Solvent content of protein crystals. *J. Molec. Biol.* **33**, 491–497 (1968).

64. *International Tables for X-ray Crystallography, Vol. III. Physical and Chemical Tables.* (**Eds.**, MacGillavry, C. H., and Rieck, G. D.) Atomic scattering factors. pp. 201–245. Kynoch Press: Birmingham, New York (1962).

65. Coppens, P., and Hall, M. B. (**Eds.**) *Electron Distributions and the Chemical Bond.* Plenum: New York, London (1982).

66. Bats, J. W., and Fuess, H. Charge density distribution in thiosulfates: $Na_2S_2O_3$ and $MgS_2O_3 \cdot 6H_2O$. *Acta Cryst.* **B42**, 26–32 (1986).

67. Swaminathan, S., Craven, B. M., Spackman, M. A., and Stewart, R. F. Theoretical and experimental studies of the charge density in urea. *Acta Cryst.* **B40**, 398–404 (1984).

68. Stevens, E. D., DeLucia, M. L., and Coppens, P. Experimental observation of the effect of crystal field splitting on the electron density of iron pyrite. *Inorg. Chem.* **19**, 813–820 (1980).

69. Coppens, P. Can we see electrons? *J. Chem. Educ.* **61**, 761–765 (1984).

70. Stevens, E. D., and Coppens, P. Refinement of metal d-orbital occupancies from X-ray diffraction data. *Acta Cryst.* **A35**, 536–539 (1979).

71. Stevens, E. D. Electronic structure of metalloporphyrins. 1. Experimental electron density distribution of ($meso$-tetraphenylporphinato)cobalt(II). *J. Amer. Chem. Soc.* **103**, 5087–5095 (1981).

72. Massa, L., Goldberg, M., Frishberg, C., Boehme, R. F. and LaPlaca, S. J. Wave functions derived by quantum modeling of the electron density from coherent X-ray diffraction: beryllium metal. *Phys. Rev. Lett.* **55**, 622–625 (1985).

73. Massa, L. J., Boehme, R. F., and La Placa, S. J. X-ray imaging of quantum electron structure. In: *Patterson and Pattersons. Proceedings of a Symposium held at the Fox Chase Cancer Center, Philadelphia, PA, USA, November 13–15, 1984.* pp. 427–449. (**Eds.**, Glusker, J. P., Patterson, B. K., and Rossi, M.) Oxford University Press: Oxford (1987).

74. Hansen, N. K., and Coppens, P. Testing aspherical atom refinements on small-molecule data sets. *Acta Cryst.* **A34**, 909–921 (1978).

75. Stewart, R. F. Electron population analysis with rigid pseudoatoms. *Acta Cryst.* **A32**, 565–574 (1976).
76. Hirshfeld, F. L. Difference densities by least-squares refinement: fumaramic acid. *Acta Cryst.* **B27**, 767–781 (1971).
77. Hirshfeld, F. L., and Havel, M. Difference densities by least-squares refinement. II. Tetracyanocyclobutane. *Acta Cryst.* **B31**, 162–172 (1975).
78. Smith, V. H., Jr. Concepts of charge density analysis: the theoretical approach. In *Electron Distributions and the Chemical Bond.* (**Eds.**, Coppens, P., and Hall, M. B.) pp. 3–59. Plenum: New York, London (1982).
79. Coppens, P. Concepts of charge density analysis: the experimental approach. In: *Electron Distributions and the Chemical Bond.* (**Eds.**, Coppens, P., and Hall, M. B.) pp. 61–92. Plenum: New York, London (1982).
80. Stevens, E. D. Analyses of electronic structure from electron density distributions of transition metal complexes. In: *Electron Distributions and the Chemical Bond.* (**Eds.**, Coppens, P., and Hall, M. B.) pp. 331–349. Plenum: New York, London (1982).
81. Hirshfeld, F. L. Bonded-atom fragments for describing molecular charge densities. *Theor. Chim. Acta* **44**, 129–138 (1977).
82. Becker, P. J., Coppens, P., and Hirshfeld, F. L. Recommendations of the *ad hoc* Committee on Criteria for Publication of Charge Density Studies. *J. Appl. Cryst.* **17**, 369 (1984).
83. Hope, H., and Ottersen, T. Accurate determination of hydrogen positions from X-ray data. I. The structure of *s*-diformohydrazide at 85 K. *Acta Cryst.* **B34**, 3623–3626 (1978).
84. Dunitz, J. D., Schweizer, W. B., and Seiler, P. X-ray study of the deformation density in tetrafluoroterephthalodinitrile: weak bonding density in the C,F–bond. *Helv. Chim. Acta* **66**, 123–133 (1983).
85. Hirshfeld, F. L. The static deformation density of tetrafluoroterephthalonitrile (TFT) from the Zürich X-ray data at 98 K. *Acta Cryst.* **B40**, 484–492 (1984).
86. Seiler, P., Schweizer, W. B., and Dunitz, J. D. Parameter refinement for tetrafluoroterephthalonitrile at 98 K: making the best of a bad job. *Acta Cryst.* **B40**, 319–327 (1984).
87. Delley, B. Calculated electron distribution for tetrafluoroterephthalonitrile (TFT). *Chem. Phys.* **110**, 329–338 (1986).
88. Dunitz, J., and Seiler, P. The absence of bonding electron density in certain covalent bonds as revealed by X-ray analysis. *J. Amer. Chem. Soc.* **105**, 7056–7058 (1983).
89. Angermund, K., Claus, K. H., Goddard, R., and Krüger, C. High-resolution X-ray crystallography — an experimental method for the description of chemical bonds. *Angew. Chem., Intl. Edn. Engl.* **24**, 237–247 (1985).
90. Savariault, J. M., and Lehmann, M. S. Experimental determination of the deformation electron density in hydrogen peroxide by combination of X-ray and neutron diffraction measurements. *J. Amer. Chem. Soc.* **102**, 1298–1303 (1980).

91. Stewart, R. F. Valence structure from X-ray diffraction data: physical properties. *J. Chem. Phys.* **57**, 1664–1668 (1972).

92. Su, Z., and Coppens, P. On the mapping of electrostatic properties from the multipole description of the charge density. *Acta Cryst.* **A48**, 188–197 (1992).

93. Hansen, N. K., and Coppens, P. Testing aspherical atom refinements on small-molecule data sets. *Acta Cryst.* **B34**, 909–921 (1978).

94. Klein, C. L., Stevens, E. D., Zacharias, D. E., and Glusker, J. P. 7,12-dimethylbenz[*a*]anthracene: refined structure, electron density distribution and *endo*-peroxide structure. *Carcinogenesis* **8**, 5–18 (1987).

95. Destro, R., Bianchi, R., and Morosi, G. Electrostatic properties of L-alanine from X-ray diffraction data. *J. Phys. Chem.* **93**, 4447–4457 (1989).

96. Klooster, W. T., and Craven, B. M. Electrostatic potential for $O \cdots H \cdots O$ in tetragonal ammonium dihydrogen-phosphate. *Acta Cryst.* **C47**, 2196–2198 (1991).

97. Figgis, B. N., Kucharski, E. S., and Reynolds, P. A. Charge density in $(NH_4Cr(SO_4)_4 \cdot 6H_2O$ at 84 K: a Jahn-Teller distorted complex. *Acta Cryst.* **B46**, 577–586 (1990).

10

Least-Squares Refinement of the Structure

Preliminary three-dimensional atomic coordinates of atoms in crystal structures are usually derived from electron-density maps by fitting atoms to individual peaks in the map. The chemically reasonable arrangement of atoms so obtained is, however, not very precise. The observed structure amplitudes and their relative phase angles, needed to calculate the electron-density map, each contain errors and these may cause a misinterpretation of the computed electron-density map. Even with the best electron-density maps, the precisions of the atomic coordinates of a preliminary structure are likely to be no better than several hundredths of an Å. In order to understand the chemistry one needs to know the atomic positions more precisely so that better values of bond lengths and bond angles will be available. The process of obtaining atomic parameters that are more precise than those obtained from an initial model, referred to as **refinement of the crystal structure**, is an essential part of any crystal structure analysis.

Changes in parameters are found that increase the agreement between measured data and those calculated from the revised model that results from these changes. The positions x, y, and z and the atomic displacement (thermal) parameters B derived for each atom in a preliminary crystal structure are adjusted so as to improve the agreement between the observed structure amplitudes, $|F(hkl)|_o$, and those calculated from the determined model $|F(hkl)|_c$. The progress in improving this agreement is usually monitored by a residual index known as the **R value**, defined as:

$$R = \sum_{\text{all } hkl} \left| \, |F(hkl)|_o - |F(hkl)|_c \, \right| \bigg/ \sum_{\text{all } hkl} |F(hkl)|_o. \qquad (10.1)$$

The R value is calculated as a measure (although, as we shall see, not a rigorous measure) of the precision of the results of the refinement. When atoms are placed at random positions in a unit cell, the theoretical values of R are 0.83 for a centrosymmetric structure and 0.59 for a noncentrosymmetric structure. Therefore, if the R value is this high, the derived structure is not likely to be correct. Much lower R values are needed.

10.1 Precision and accuracy

This chapter is concerned with the precision of the atomic arrangement that we derive from a crystal structure analysis. Therefore it is necessary at this point to pause and discuss errors and estimated standard deviations, and the methods used to assign a confidence or "weight" to a measurement. **Precision**, which should not be confused with accuracy, is a measure of how closely a series of measurements of the same quantity agree with each other.[1-3] **Accuracy** is a measure of how close a measurement is to the *true value*. If rulers have become distorted with time, measurements made with them may be highly precise but are unlikely to be accurate. The true electron density in a crystal is not yet attainable by theory or experiment. As detection and theoretical methods improve, we hope that the data on the geometry of molecular structures, measured in various ways, will converge and approach those of the true molecular structure.

Since we are not able to find the "true" value for any parameter, we often "make do" with the average of all of the experimental data measured for that parameter, and consider this as "the most probable value." Measured values, which necessarily contain experimental errors, should lie in a random manner on either side of this most probable value as expressed by the normal or **Gaussian distribution**. This distribution is a bell-shaped curve[3] that represents the number of measurements N that have a specific value x (which deviates from the mean or most probable value x_0 by an amount $x - x_0$, representative of the error). Obviously the smaller the value of $x - x_0$, the higher the probability that the quantity being measured lies near the most likely value x_0, which is at the top of the peak. A plot of N against x, shown in Figure 10.1, is called a Gaussian distribution or error curve, expressed mathematically as:

$$N = \frac{h}{\sqrt{\pi}} e^{-h^2(x-x_0)^2}, \tag{10.2}$$

where $x = x_0$ at the peak of the plot and therefore x_0 is the mean value of the Gaussian distribution. The plot is symmetrical about this mean value x_0. Its **variance** (another estimate of the error), $\sigma^2 = 1/(2h^2)$, describes the spread (width) for a given Gaussian distribution. The square root of the variance, σ, is called the standard deviation (or root mean square error). The smaller the value of σ, the sharper the peak in the Gaussian distribution.

When measurements lie on such a Gaussian error curve, they can be analyzed to assess the probability that one observation is *significantly*

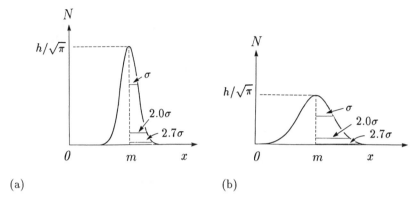

(a)   (b)

FIGURE 10.1. A Gaussian distribution function. In (a) is shown a more precise measurement, compared with that in (b). Consequently the estimated standard deviations of the distribution in (a) are smaller than for the distribution in (b). Values of σ, 2.0σ, and 2.7σ are shown for each distribution.

different from another. In other words, the Gaussian distribution is a means for finding out whether the difference between two measurements is significant or merely a result of random errors in the two measurements. The question raised is, if further sufficiently precise measurements were to be made, what is the probability that they would substantiate this difference?

In order to address this problem the concept of an **estimated standard deviation** (abbreviated to e.s.d.), which is related to σ described previously, was introduced.[1-3] It is defined by an equation analogous to Equation 10.2 and differs from the standard deviation only in that, rather than calculating the difference between a measurement and the true value, the mean value has to be used, rather than the (unknown) true value. For a large number of measurements, the standard deviation and the e.s.d are approximately equal (unless there is a systematic error), and it is common to use the symbol σ interchangeably for these two measures of the error.

A consequence of the Gaussian error distribution shown in Figure 10.1 and defined in Equation 10.2 is that there is a 99% chance that a given measurement will differ by less than 2.7 e.s.d. from the value of highest probability (i.e., the mean). This implies that if two measurements differ by 2.7 e.s.d., the probability is 99% that they represent two distinct values (95% probability if the difference is 2.0 e.s.d.). The magnitude of the e.s.d. is usually listed in parentheses after the refined parameter. For example, 1.542(7) Å means 1.542 Å with an e.s.d. of 0.007 Å. Thus, at the 99% probability level a measurement of 1.556(9) Å is not significantly different from one of 1.542(7) Å.

The more precisely a measurement is made, the smaller is the e.s.d. of that measurement and the larger the weight or confidence that should be assigned to it. As a first approximation, the **weight of a measurement** can be taken as the reciprocal of the square of its e.s.d. Another type of weighting function also takes systematic errors, such as instrumental errors, into account by use of a small constant k,

$$w(hkl) = \frac{1}{\sigma^2[F_o(hkl)]} \quad \text{or} \quad \frac{1}{\sigma^2[F_o(hkl)] + kF_o(hkl)^2}. \tag{10.3}$$

Weighting schemes, or how much confidence is placed on each of the measurements made, are the subject of much discussion in crystal structure analyses, and a variety of weighting schemes are used, according to the preference of the user. The nature of this weighting scheme used should be listed when the crystal structure is reported.

10.2 The method of least squares

A very powerful procedure for obtaining the most reliable possible information from a set of experimental observations is the **method of least squares**. This method, described by Adrien-Marie Legendre[4–6] in 1806 and Carl Friedrich Gauss,[7] relies on the existence of many more experimental observations than parameters to be determined and the assumption that the experimental errors in the data follow a Gaussian distribution. Legendre wrote[4] (translated): "By this method, a kind of equilibrium is established among the errors which, since it prevents the extremes from dominating, is appropriate for revealing the state of the system which most nearly approaches the truth."

Before describing this method, let us examine the probability that a measurement will have a particular value (Equation 10.2 with $\sigma = 1/h\sqrt{2}$). The probability that a measurement x_i will lie within an interval dx of the mean value x_0 and have a spread (standard deviation) σ is

$$\text{Single measurement}: P_i = \frac{1}{\sigma\sqrt{2\pi}}e^{-(x_i-x_0)^2/2\sigma^2}dx, \tag{10.4}$$

and the probability that a whole set of N measurements simultaneously lies within this interval dx is the product P of the separate probabilities P_i of the individual measurements,

$$\text{Multiple measurements}: P = \prod_{i=1}^{N} P_i = k_1 e^{-\left(\sum_{i=1}^{N}(x_i-x_0)^2/k_2\right)}, \tag{10.5}$$

where k_1 and k_2 are constants from Equation 10.4 in simpler form. The principle of maximum likelihood then states that the best value for x_0 is the one that maximizes the value of the probability of occurrence of that value P. The value of P is a maximum when the exponents $\sum_i(x_i - x_0)^2$, are minimized (that is, maximizing the probability in Equation 10.5). Ideally, if there were no errors, this term would be zero and x_i would be the value of x_o that is being sought. In practice, since there are errors in measurements, we must use the principle of maximum likelihood. This leads to the principle of least squares,[8-14] which states:

The most probable value of a quantity is obtained from a set of measurements by choosing the value that minimizes the sum of the squares of the deviations of all the measurements from that chosen value (see Equation 10.5). The simple example of a fitting of points to a straight line is given in Figure 10.2.

Consider a set of n measurements, $F(i)_o$, where $i = 1$ to n, and where the subscript o indicates an observed value. The most probable value of this function, $F(i)_c$, where the subscript c indicates a calculated value, is that which minimizes the quantity D, where

$$D = \sum_{i=1}^{n} w_i[F(i)_o - F(i)_c]^2. \tag{10.6a}$$

In this equation n is the number of measurements, w_i the weight of each measurement, and, as in Figure 10.3,

$$a_{i1}x_1 + a_{i2}x_2 + \cdots + a_{im}x_m = F(i)_c, \tag{10.6b}$$

where m is the number of unknowns, x_1 to x_m.

The method of least squares, described in Figures 10.2 to 10.9, allows us to minimize the difference between a function we observe, $F(i)_o$, and one that we can calculate (from known atomic position, see Figure 6.21, Chapter 6), $F(i)_c$. We want to find a set of values of x_i such that when the function $F(i)_c$ is calculated and subtracted from the observed values of $F(i)_o$, the sum of the squares of these differences will be a minimum. This method was introduced into X-ray crystallography by Edward W. Hughes[15,16] in order to obtain the best possible coordinates for a crystal structure, that of melamine.

It is important for the reader to understand that in a least-squares refinement of a crystal structure it is *the shifts in parameters* that are calculated *in order to improve the structure, not the parameters themselves*. The preliminary parameters that are shifted to more appropriate values come from the trial structures (see Chapters 8 and 9).

10.2.1 Least-squares method for linear equations

First we will show how the least-squares method can work for a simple **linear equation**, that is, one in which $F(i)_c$ is a function of x but no higher powers of x. A typical problem might consist of n observations leading to linear equations (called **observational equations**) with m

parameters (variables), x_1, x_2, $\cdots$, x_m (the quantities we want to determine from the experiment) and m independent constants, a_1, a_2, $\cdots$, a_m [known numbers that describe the relationship of $F(i)_c$, to any x_i, $i = 1$ to m].

To calculate the least-squares line through a series of points.

$$y = A_o x + B_o$$

where A_o and B_o are the best values for describing the relationship of x to y.

The error ϵ_i of any measurement i is:

$$\epsilon_i = A_o x_i + B_o - y_i.$$

The principle of least squares requires that the sum of the squares of the errors:

$$\sum_i \epsilon_i^2 = \sum_i (A_o x_i + B_o - y_i)^2$$

be minimized for the N observations from $i = 1$ to N.

In order to minimize ϵ^2 it is necessary to partially differentiate it with respect to A_o and B_o in turn and then equate the derivatives to zero.

$$\epsilon^2 = (A_o x_i + B_o - y_i)^2 =$$

$$A_o^2 x_i^2 + B_o^2 + y_i^2 + 2A_o x_i B_o - 2A_o x_i y_i - 2B_o y_i.$$

$$\frac{\partial \epsilon^2}{\partial A_o} = 2A_o x_i^2 + 2B_o x_i - 2x_i y_i = 0$$

$$\frac{\partial \epsilon^2}{\partial B_o} = 2A_o x_i + 2B_o - 2y_i = 0$$

For the many measurements of x_i and y_i:

$$A_o \sum_i x_i^2 + B_o \sum_i x_i = \sum_i x_i y_i$$

$$A_o \sum_i x_i + B_o \sum_i 1 = \sum_i y_i$$

(where $\sum_i 1 = N$). Weights may be added if desired.

These equations are simultaneous and may be solved as such. They imply that the best values of A_o and B_o can be obtained from $\sum_i x_i^2$, $\sum_i x_i$, $\sum_i x_i y_i$, and $\sum_i y_i^2$.

FIGURE 10.2. Finding the best straight line through a series of points.

$$a_1 x_1 + a_2 x_2 + \cdots + a_m x_m = F(i)_c. \qquad (10.7)$$

If we merely measure the function $F(i)_o$ at m different points, in the situation where m (the number of variables) now equals n the number of observations), we can solve the m simultaneous equations that result and therefore the parameters x_1, x_2, $\cdots$, x_m can be determined directly. The minimum number of measurements that allow us to determine m values of x_i ($i = 1$ to m) have been made in this example. Because there are necessarily imprecisions and errors in each of the m measurements, however, the values we obtain for x_i may deviate considerably

---

1. *A set of observations $F(1)_o, F(2)_o, F(3)_o, \cdots, F(i)_o$ is made. We then want to determine the best values for the parameters $x_1, x_2, \cdots, x_m$. $F(i)_o$ is related to $x_1, x_2, \cdots, x_m$, where $i = 1$ to n, by the constants $a_{i1}, a_{i2}, \cdots, a_{im}$, which are known (from the preliminary structure chosen for this case).*

2. *The preliminary structure is expressed as follows:*

$$a_{11} x_1 + a_{12} x_2 + \cdots + a_{1m} x_m = F(1)_c$$

$$a_{21} x_1 + a_{22} x_2 + \cdots + a_{2m} x_m = F(2)_c$$

$$a_{i1} x_1 + a_{i2} x_2 + \cdots + a_{im} x_m = F(i)_c$$

3. *The observational equations, one for each observation, are obtained by replacing the exact $F(i)_c$ by the observed $F(i)_o$, which contain random errors.*

$$a_{11} x_1 + a_{12} x_2 + \cdots + a_{1m} x_m = F(1)_o$$

$$a_{21} x_1 + a_{22} x_2 + \cdots + a_{2m} x_m = F(2)_o$$

$$a_{i1} x_1 + a_{i2} x_2 + \cdots + a_{im} x_m = F(i)_o$$

4. *Our ability to solve these equations depends on the number of variables m and the number of observations n.*

   a. *If $n < m$ there is no unique solution.*

   b. *If $n = m$ the equations can be solved directly, assuming that there are no errors in any $F(i)_o$, and that the equations are linearly independent.*

   c. *If $n > m$ the problem is overdetermined and amenable to the least squares procedure.*

FIGURE 10.3. Setting up the observational equations: a simple example. Here we have the observed values $F(i)_o$ (a structure amplitude, for example), and we define them by suitable equations (shown here) in terms of the parameters in which we are interested (x,y,z of the atoms, for example).

1.   *The errors in any measurement (assuming a correct model) are:*

$$\epsilon_i = F(i)_o - F(i)_c = F(i)_o - (a_{i1}x_1 + a_{i2}x_2 + \cdots + a_{im}x_m).$$

2.   *Set*

$$D = \sum_{i=1}^{n} w_i \epsilon_i^2 = \sum_{i=1}^{n} w_i[F(i)_o - (a_{i1}x_1 + a_{i2}x_2 + \cdots + a_{im}x_m)]^2,$$

where w_i is the weight of each observation and is inversely proportional to the variance, σ^2, of $F(i)_o$.

3.   *The value of D is minimized. This is done by differentiating D with respect to each x_j and setting the derivative equal to zero.*

$$\frac{1}{2}\frac{\partial D}{\partial x_j} = \sum_{i=1}^{n} w_i a_{ij}[-F(i)_o + a_{i1}x_1 + a_{i2}x_2 + \cdots + a_{im}x_m] = 0.$$

FIGURE 10.4. Differentiating so that it will be possible to minimize the difference between observed and calculated values of $F(i)$. The strategy is to adjust parameters $(x_1, x_2,$ etc.) until a minimum is reached in the comparison of $F(i)_c$, the calculated structure factor, and $F(i)_o$, the observed structure factor. The calculated values of $F(i)$ are obtained by use of equations in Figure 10.3, together with approximate parameters for the atomic arrangement (see Chapters 8 and 9).

from accurate values. We can increase the precision of the determination of x_1, x_2, $\cdots$, x_m. This can be done by increasing the number of measurements so that the number of observations is very much greater than the number of parameters. Then the "best possible" set of values of x_1, x_2, $\cdots$, x_m is obtained such that the sum of the squares of the differences between observed and calculated values of $F(i)$ is minimized. Thus, in classical least-squares problems there are many more observations than variables. The observational equations (Figure 10.3) are obtained by replacing the exact $F(i)_c$ in Equation 10.7 by the measured value, $F(i)_o$. The minimization in the least-squares method is done by differentiating D in Figure 10.4 with respect to each of the parameters in turn and then equating each value of ϵ_i to zero. This gives m equations of m unknowns, the **normal equations**, much less than the number n of observational equations. There is now the same number of parameters to be determined as there are equations, and so a direct solution is possible.

If the function is linear with respect to the variables x_1, x_2, $\cdots$, x_m, then solution of the least-squares problem is straightforward without any prior knowledge about the parameter values. If, however, the functional

1. *The partial derivatives listed at the bottom of* Figure 10.4, *which are equated to zero in order to minimize the sum of the squares, may be rearranged to give the m normal equations for each x_j, where $j = 1$ to m.*

$$\sum_{i=1}^{n} w_i a_{i1} a_{i1} x_1 + \sum_{i=1}^{n} w_i a_{i1} a_{i2} x_2 + \cdots \sum_{i=1}^{n} w_i a_{i1} a_{im} x_m = \sum_{i=1}^{n} w_i F(i)_o a_{i1}$$

$$\sum_{i=1}^{n} w_i a_{i2} a_{i1} x_1 + \sum_{i=1}^{n} w_i a_{i2} a_{i2} x_2 + \cdots \sum_{i=1}^{n} w_i a_{i2} a_{im} x_m = \sum_{i=1}^{n} w_i F(i)_o a_{i2}$$

$$\vdots \qquad \vdots \qquad \vdots \qquad \vdots$$

$$\sum_{i=1}^{n} w_i a_{im} a_{i1} x_1 + \sum_{i=1}^{n} w_i a_{im} a_{i2} x_2 + \cdots \sum_{i=1}^{n} w_i a_{im} a_{im} x_m = \sum_{i=1}^{n} w_i F(i)_o a_{im}.$$

2. *There are now as many equations as there are parameters to be determined. This allows us to solve for $x_1, x_2, \cdots, x_m$ directly.*
3. *This direct solution requires matrix inversion.*

FIGURE 10.5. Setting up the **"normal"** equations of least squares. These are the equations derived in Figure 10.4.

form of the observational equations is not linear, the normal equations are more complicated, and for these problems there may be no direct least-squares solution.

10.2.2 The least-squares method for nonlinear equations

Most problems in crystallography[17-23] are not linear; therefore, it is not possible to solve for the parameters directly. For example, the structure factor $F(hkl)$ is not a linear function of x, y, z, and B, because it contains cosine functions (Figure 6.21, Chapter 6). If, however, we have an initial set of parameters that are significantly better than a random guess, we can use the method of least squares to improve the parameters. The use of a Taylor's series[24] allows certain functions, such as cosines and sines, to be expanded as a convergent power series so that the higher powers of x, the parameter, make successively smaller and smaller contributions to the total function. If $f(x)$ is a differentiable function for $x = x_0 + \Delta x$, then the Taylor's series expression is:

$$f(x) = f(x_0) + \Delta x \frac{\partial f(x_0)}{\partial x} + (\Delta x)^2 \frac{\partial^2 f(x_0)}{2 \partial^2 x} + \cdots. \qquad (10.8a)$$

1. *The set of simultaneous equations:*

$$a_{11}x_1 + a_{12}x_2 + a_{13}x_3 = b_1$$

$$a_{21}x_1 + a_{22}x_2 + a_{23}x_3 = b_2$$

$$a_{31}x_1 + a_{32}x_2 + a_{33}x_3 = b_3$$

   may be rewritten as:

$$\begin{pmatrix} a_{11} & a_{21} & a_{31} \\ a_{12} & a_{22} & a_{32} \\ a_{13} & a_{23} & a_{33} \end{pmatrix} \begin{pmatrix} x_1 \\ x_2 \\ x_3 \end{pmatrix} = \begin{pmatrix} b_1 \\ b_2 \\ b_3 \end{pmatrix}.$$

2. *This may be further shortened to:*

$$[\mathbf{A}]\mathbf{x} = \mathbf{b}.$$

3. *The solution is:*

$$\mathbf{x} = [\mathbf{A}]^{-1}\mathbf{b}.$$

FIGURE 10.6. Solving a set of simultaneous equations. There are now, as shown in Figure 10.5, as many equations as there are variables. These simultaneous equations are solved (by computer) by matrix techniques, as shown here.

If Δx is small and if the series is a convergent power series, this expression may be truncated, and only the first two terms in the approximation need be obtained,

$$f(x) \approx f(x_0) + \Delta x \frac{\partial f(x_0)}{\partial x}, \qquad (10.8b)$$

and the result is sufficiently precise to be helpful. This can be applied to the structure factors as follows:

$$F(x_n) = F(x_0) + (\Delta x_i)\frac{\partial F_c}{\partial x_i} + (\Delta x_i)^2 \frac{\partial^2 F_c}{2\partial^2 x_i} + \cdots \qquad (10.8c)$$

where x_i refer to the different parameters being refined (e.g., positional or the atomic displacement parameters). The underlying assumption is that, although the function itself is nonlinear, the incremental shifts Δx are so small that higher-order terms (Δx^2, Δx^3, etc.) may be ignored; the equation is then considered to be approximately linear. Because *the Taylor's series has been truncated, the refinement will require succes-*sive iterations (called "cycles of refinement") until there is no significant

1. *The observational equations*, step 3, Figure 10.3, *can be written more suc-cinctly in matrix notation as:*

$$\begin{pmatrix} a_{11} & a_{12} & a_{13} & \cdots & a_{1m} \\ a_{21} & a_{22} & a_{23} & \cdots & a_{2m} \\ \cdot & \cdot & \cdot & & \cdot \\ \cdot & \cdot & \cdot & & \cdot \\ \cdot & \cdot & \cdot & & \cdot \\ a_{i1} & a_{i2} & \cdot & \cdot & a_{im} \end{pmatrix} \begin{pmatrix} x_1 \\ x_2 \\ x_3 \\ \cdot \\ \cdot \\ x_m \end{pmatrix} = \begin{pmatrix} F(1)_o \\ F(2)_o \\ F(3)_o \\ \cdot \\ \cdot \\ F(i)_o \end{pmatrix}.$$

2. *These can be further abbreviated in matrix notation as*

$$[\mathbf{A}]x = F.$$

3. *Similarly,* **normal equations** *can be written more succinctly in matrix notation as:*

$$\begin{pmatrix} a_{11} & \cdots & a_{i1} \\ a_{12} & \cdots & a_{i2} \\ \cdot & \cdot & \cdot \\ \cdot & \cdot & \cdot \\ \cdot & \cdot & \cdot \\ a_{1m} & \cdot & a_{im} \end{pmatrix} \begin{pmatrix} a_{11} & \cdots & a_{1m} \\ a_{21} & \cdots & a_{2m} \\ \cdot & \cdot & \cdot \\ \cdot & \cdot & \cdot \\ \cdot & \cdot & \cdot \\ a_{i1} & \cdot & a_{im} \end{pmatrix} \begin{pmatrix} x_1 \\ \cdot \\ x_m \end{pmatrix} = \begin{pmatrix} a_{11} & \cdots & a_{i1} \\ a_{12} & \cdots & a_{i2} \\ \cdot & \cdot & \cdot \\ \cdot & \cdot & \cdot \\ \cdot & \cdot & \cdot \\ a_{1m} & \cdot & a_{im} \end{pmatrix} \begin{pmatrix} F(1)_o \\ \cdot \\ F(i)_o \end{pmatrix}.$$

4. *This can be abbreviated or written even more compactly in matrix notation as*

$$([\mathbf{A}]^{\mathrm{T}}[\mathbf{A}])x = [\mathbf{A}]^{\mathrm{T}}F,$$

where $[\mathbf{A}]^{\mathrm{T}}$ *is the transpose of* $[\mathbf{A}]$.

5. *Rearrangement of this gives the solution to the least-squares equation*

$$x = ([\mathbf{A}]^{\mathrm{T}}[\mathbf{A}])^{-1}[\mathbf{A}]^{\mathrm{T}}F.$$

6. Item 4 *may be represented as:*

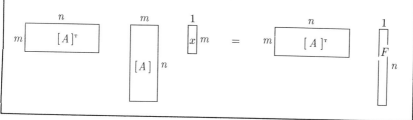

FIGURE 10.7. Describing the above equations (Figures 10.3 to 10.6) in matrix nota-tion.

Nonlinear equations are solved in a similar fashion to linear equations except that the function is expanded using a Taylor series, which is then truncated. Then the shifts in parameters that will give a minimum value of D (see Figure 10.4) are calculated. These shifts will not be exact, because the formula was only approximate. Therefore several cycles of refinement, each cycle with smaller shifts, are necessary.

1. *As before we want D to be a minimum*

$$\frac{1}{2}\frac{\partial D}{\partial x_m} = \sum_{i=1}^{n} w_i[F(i)_o - F(i)_c]\frac{\partial F_c}{\partial x_m} = 0.$$

2. *This time we expand the function F_c using a Taylor series with the higher order terms truncated.*

$$F_c = f(a) + (x - a)\frac{\partial f(a)}{\partial x}$$

3. *Written more fully, with Δx_i as the shifts in parameters.*

$$\sum_{i=1}^{n} w_i \left(F(i)_o - F(i)_c(x_{1o}, x_{2o}, \cdots, x_{mo}) - \frac{\partial F(i)_c}{\partial x_1}\Delta x_1 - \cdots - \frac{\partial F(i)_c}{\partial x_m}\Delta x_m \right) \frac{\partial F(i)_c}{\partial x_j}$$

4. *Minimization is achieved as before, by taking the derivative with respect to each parameter x_m and equating it to zero. This leads to the m normal equations.*

$$\sum_{i=1}^{m} w_i \frac{\partial F(i)_c}{\partial x_1}\frac{\partial F(i)_c}{\partial x_1}\Delta x_1 + \sum_{i=1}^{m} w_i \frac{\partial F(i)_c}{\partial x_1}\frac{\partial F(i)_c}{\partial x_2}\Delta x_2 + \cdots \sum_{i=1}^{m} w_i \frac{\partial F(i)_c}{\partial x_1}\frac{\partial F(i)_c}{\partial x_i}\Delta x_i$$

$$= \sum_{i=1}^{m} w_i \Delta F_i \frac{\partial F(i)_c}{\partial x_m}$$

5. *As in Figure 10.6 this can be written more compactly in matrix notation as*

$$[\mathbf{A}]\,\Delta\mathbf{x} = \mathbf{b}$$

    where

$$a_{ij} = \sum_{i=1}^{m} w_i \frac{\partial F(i)_c}{\partial x_m}\frac{\partial F(i)_c}{\partial x_j}$$

$$b_i = \sum_{i=1}^{m} w_i \Delta F_i \frac{\partial F(i)_c}{\partial x_m}.$$

6. *The solution of the problem is:*

$$\Delta\mathbf{x} = ([\mathbf{A}]^{\mathbf{T}}[\mathbf{A}])^{-1}[\mathbf{A}]^{\mathbf{T}}\mathbf{b}.$$

FIGURE 10.8. Least squares for nonlinear equations. The relationship between structure factors and the x, y, z of atoms is nonlinear because it involves cosine functions.

1. *Determine an initial value for x_{j_o} (the model)* (Chapters 8 and 9).
2. *Calculate $D = [F(i)_o - F(i)_c]^2$* (Figure 10.4).
3. *Expand $F(i)_c$ as a truncated Taylor series* (Figure 10.8).
4. *Calculate the derivatives of D* (Figure 10.4).
5. *Set up the normal equations* (Figure 10.5).
6. *Solve for the shifts Δx_j* (Figure 10.8).
7. *Calculate a better estimate of $x_j = x_{j_o} + \Delta x_j$.*
8. *Go back to* step 3.
9. *Iterate until $(shift/\sigma)$ is much less than 1.0 for each parameter being refined.*

FIGURE 10.9. Solving for shifts in x_j in nonlinear problems by least-squares methods.

change in Δx between two successive cycles. The course of least-squares refinement for nonlinear equations is shown in Figures 10.8 to 10.10. The function most commonly minimized in a crystallographic least-squares refinement is

$$D = \sum_{h,k,l} w_{hkl} (F_o - F_c)^2, \qquad (10.9)$$

where the sum is over the set of crystallographically independent reflections F_o and w_{hkl} are weights (cf., Equation 10.6a). The refinement is considered completed when changes in individual parameters are no longer significant, i.e., a fraction of the e.s.d. for each parameter.

10.2.3 Constraints and/or restraints

The ability to determine atomic parameters by least-squares methods depends greatly on the **overdeterminancy** of the problem, that is, the ratio of the number of independent observations to the number of variable parameters. This overdeterminancy ratio is important in crystallographic refinements. Generally, at least 5 to 10 Bragg reflection intensity measurements are desired to determine each parameter (x, y, z, U or B, etc.). Each atom normally has three positional parameters and either one isotropic temperature factor or six anisotropic temperature factors. In addition, it is also possible to refine a factor that defines the extent to which an atom fills a given site in the crystal structure (an **occupancy factor**). Thus, if there are n atoms in the asymmetric unit, if just the atomic positions are to be refined, there are $3n + 2$ parameters ($3n$ positional parameters, plus a scale and overall temperature factor). For a refinement in which each atom vibrates isotropically, $4n + 1$ parameters need to be refined. To refine anisotropically vibrating atoms, $9n + 1$ parameters are required. For small organic molecules, such as simple sugars, or individual amino acids, the ratio of the number of observations to parameters is generally about 10:1. These refinements lead to a

1. Select the parameters $x_1, x_2, \cdots, x_m$ that are to be refined. These may be positional parameters, displacement parameters, occupancy factors. In this simplified example we consider only the positional parameters x_n, y_n, z_n for the nth atom. Determine an initial value for each parameter x_{jo} from the atomic arrangement at this stage and calculate structure factors.

$$F_c(hkl) = \sum_{n=1}^{n \text{ atoms}} f_n \exp 2\pi i (hx_i + ky_i + lz_i).$$

2. Calculate $\Delta \mid F(hkl) \mid = \mid F(hkl)_o \mid - \mid F(hkl)_c \mid$ and $D = \Sigma w [\Delta F(hkl)]^2$.

3. In order to expand by a Taylor's series (Equation 10.8a), calculate the derivatives $\frac{\partial F(hkl)_c}{\partial x_{jo}}$, remembering that the structure factors

$$\mid F(hkl) \mid = \sqrt{A(hkl)^2 + B(hkl)^2},$$

$$\text{where } A(hkl) = \sum_{n=1}^{n \text{ atoms}} f_n \cos 2\pi (hx_n + ky_n + lz_n) = \mid F(hkl) \mid \cos \alpha_{hkl}$$

$$\text{and } B(hkl) = \sum_{n=1}^{n \text{ atoms}} f_n \sin 2\pi (hx_n + ky_n + lz_n). = \mid F(hkl) \mid \sin \alpha_{hkl}$$

In these equations f_n is the scattering factor, including displacement parameters, for the nth atom at position x_n, y_n, z_n. Therefore:

$$\frac{\partial \mid F_c \mid}{\partial x_j} = \frac{\partial A}{\partial x_j} \cos \alpha + \frac{\partial B}{\partial x_j} \sin \alpha$$

$$\frac{\partial A}{\partial x_j} = -\sum_n 2\pi h_x f_n \sin 2\pi (hx_n + ky_n + lz_n)$$

$$\frac{\partial B}{\partial x_j} = \sum_n 2\pi h_x f_n \cos 2\pi (hx_n + ky_n + lz_n),$$

$$\text{where } h_x = \frac{\partial (hx + ky + lz)}{\partial x_j}, \quad x_j = x, y, \text{ or } z.$$

4. Set up the normal equations.
5. Solve for the shifts in parameters Δx_j.
6. Calculate a better estimate of $x_j = x_{jo} + \Delta x_j$.
7. Go back to step 2.
8. Iterate until $(shift/\sigma)$ is much less than 1.0 for each parameter being refined.

FIGURE 10.10. The least-squares equations for structure factors.

high precision in atomic parameters. In general, however, for many large structures, such extensive data are not all always available.

Macromolecular crystals are generally rather flexible and have a high solvent content. The resulting larger thermal vibrations and disorder for the atoms causes a weaker diffraction intensity with increasing scattering angle. As a result, the ratio of the number of observations to the number of parameters is decreased. Two ways can be considered for improving this ratio. One is to reduce the number of parameters by the use of **constraints**, and the other is to increase the number of observations by adding **restraints**. To differentiate between these, think of a dog constrained in a small pen, so that he cannot move, or restrained by a leash, so that the extent of his movement is controlled.

Prior stereochemical knowledge is at the heart of either of these computational approaches.[25-34] Much is known about the geometry of the components from which the molecules are built. A constraint will reduce the number of formal parameters by transforming individual atomic coordinates to other parameters related to groups of atoms. For example, a phenyl group with 6 atoms has $3 \times 6 = 18$ positional parameters. If the entire phenyl group were to be refined with the constraints of a rigid body, then the number of parameters would be reduced from 24 to 6 (3 rotational and 3 translational degrees of freedom). Thus, when constraints are used, some bond lengths, bond angles, and planes are forced to retain ideal values. This greatly reduces the number of parameters, but it must be remembered that addition of constraints has changed the least-squares model that is being fit.

The use of restraints requires the addition of more observations, such as known bond distances and bond angles, to a set of observational equations. For a phenyl group (just described) the number of parameters is not reduced (as it is with constraints), but observations are added that describe the relationship of the atoms in the phenyl group with respect to bond lengths, angles, and planarity of the ring. In most cases when restraints are used, they are introduced not as specific values for certain features, but rather as a distribution of values centered about average values, so that variability from these average values is allowed in the refinement. The model proposed for the structure is therefore made to match the expected distribution for the stereochemical data by a relative weighting of these stereochemical observations. In other words, restraints restrict such a model to a realistic range of possibilities rather than confining it rigidly to certain specific values in the use of constraints.

Both methods have the very important function of increasing the rate of convergence of the least-squares refinement to the final values. The only condition in applying constraints or restraints is that the weights of all the individual terms be on a common, relative scale. One way of achieving this is to impose constraint conditions that interrelate parameters by use of a **Lagrange multiplier**, λ (see Figure 10.11), as

---

1. *If the parameters* $x_1, x_2, \cdots, x_m$ *that are solutions of the normal equations*

$$a_{11}x_1 + a_{12}x_2 + \cdots + a_{1m}x_m = b_1$$

$$a_{m1}x_1 + a_{m2}x_2 + \cdots + a_{mm}x_m = b_m$$

must also satisfy the additional conditions

$$c_{11}x_1 + c_{12}x_2 + \cdots + c_{1m}x_m = h_1$$

$$c_{r1}x_1 + c_{r2}x_2 + \cdots + c_{rm}x_m = h_r$$

then we can augment the normal equations with Lagrangian multipliers $\lambda_1, \lambda_2, \cdots, \lambda_r$.

$$a_{11}x_1 + a_{12}x_2 + \cdots + a_{1m}x_m = c_{11}\lambda_1 + \cdots c_{r1}\lambda_r = b_1$$

$$a_{m1}x_1 + a_{m2}x_2 + \cdots + a_{mm}x_m = c_{1m}\lambda_1 + \cdots c_{rm}\lambda_r = b_m$$

$$c_{11}x_1 + c_{12}x_2 + \cdots + c_{1m}x_m = h_1$$

$$c_{r1}x_1 + c_{r2}x_2 + \cdots + c_{rm}x_m = h_r.$$

2. *Then:*

$$\frac{1}{2}\frac{\partial D}{\partial x_m} = \sum_{n=1}^{m} \lambda_1 w_n (f_n^o - f_n^c)\frac{\partial f_c}{\partial x_m} + \sum_{l=j}^{k} \lambda_2 w_j (g_j^o - g_j^c)\frac{\partial g_c}{\partial x_m} = 0.$$

FIGURE 10.11. Use of Lagrange multipliers. These are necessary when there are extra conditions (constraints) to be satisfied in the solution of the least-squares problem.

suggested by Hughes when he introduced the method of least squares to crystallographic refinement.[15] Minimization with respect to each variable parameter is then achieved as before.

A simple example of the use of a Lagrange multiplier (constraint) can be seen in the calculation of the equation of a plane:

$$lx_i + my_i + nz_i - p = 0, \tag{10.10}$$

where p equals the length of the normal from the origin to the plane, and l, m, and n are the cosines of the angles this perpendicular makes with the positive x, y, and z axes. The aim is to minimize the difference between p and $lx_i + my_i + nz_i$ for all atoms. The parameters l, m, n and p, however, are not independent; they are subject to the constraint that the sum of the squares of the direction cosines should be unity ($l^2 + m^2 + n^2 = 1$). This must be taken into account in the calculation, and is done by use of a Lagrange multiplier.[35,36] For details on how to do the calculation the reader should study Reference 35.

10.3 Refinement by simulated annealing

The least-squares method is a very powerful tool, provided the model is sufficiently close to the true structure. If the initial model is basically correct, the shifts in parameters indicated by the least-squares refinement will drive the energy of the structure to a **global minimum**. Unfortunately, if the model is not quite correct, the least-squares refinement will produce a structure that is trapped in a **local minimum** of energy which is not the true structure (see Figure 10.12). This problem manifests itself by monitoring the R value. Often there is a hint of trouble in that the R value is higher than expected and will not decrease to acceptable values. Several crystal structures have been reported with this type of problem, but they are generally corrected in the subsequent literature.

This problem occurs quite often with macromolecules because there is a low data-to-parameter ratio and there are so many atoms to position. It is generally not possible to get out of this false minimum without considerable revision of the model being refined. The strength of the standard least squares refinement is that small adjustments can be made to the model so that it best agrees with the diffraction data. However, conventional least squares type refinements are not sufficiently robust to generate large changes and can only make small changes to a structure that is essentially correct. When the model is sufficiently poor or regions of the model are built improperly, then least squares has little chance of

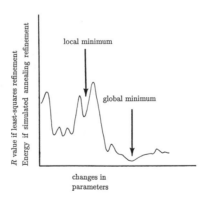

FIGURE 10.12. Diagram of a local and global minimum during refinement. Not that the ordinates are the R vale in the least-squares refinement, or the energy in the simulated annealing refinement. This emphasizes that the R value is not the only indicator of the correctness of a structure that can be used.

moving the model correctly. In mathematical terms one says that such a model is outside the radius of convergence.

An alternative tactic, called **simulated annealing**, has been shown to have a larger effective radius of convergence and therefore is an extremely powerful tool for refinement. Simulated annealing is a nonlinear approach that allows the model to explore a large range of conformations and find models that are closer to the global minimum[37-40] It is an optimization technique that can overcome barriers and reach a lower minimum than those obtained by conventional least squares. Thus, simulated annealing is a process whereby a molecule is, in a sense, "heated" giving the atoms a large amount of freedom to move and adopt alternate conformations. It is analogous to the physical process of heating a solid to a temperature where all the particles of the solid are randomly arranged as in a liquid. By slowly cooling the liquid, the molecules are able to rearrange themselves into the lowest energy ground state of the solid. In a similar way the heating allows the molecule to take on a large number of conformations which may deviate significantly from the starting model. The rearrangement if done by allowing the coordinates and velocities of the atoms to change according to Newton's Laws of Motion. When the structure has been built improperly or the chain has been traced incorrectly the simulated annealing approach may be able to sort out the problem and position the atoms correctly with respect to the diffraction data. In those instances where the model has been built correctly and only small shifts in the atomic coordinates are needed to improve the fit of the model to the diffraction data the conventional least squares refinement process is quite sufficient.

10.4 Estimated standard deviations

Values of the e.s.d.s of parameters can be obtained, as shown in Figure 10.13, in the least-squares refinement from values of the diagonals of the inverse matrix. Similarly, any **correlations between parameters**, such as is often found to occur between occupancy and atomic displacement parameters, can be identified and taken into account in the description of the resulting molecular structure. The e.s.d.s for the refined parameters can then be used to calculate e.s.d.s of derived parameters, such as distances, angles, and torsion angles.[36]

Summary

The refinement of the parameters of the atomic arrangement that best fits the experimental Bragg intensity data is generally done by the method of least squares. This involves setting up observational equations, differentiating them in order to obtain a minimum value for the error, setting up normal equations, and solving the set of simultaneous equations so obtained. Matrix techniques are used, and they give not only the shift in each parameter that is refined but also its estimated standard deviation

1. *The normal equations*

$$c_{11}x_1 + c_{12}x_2 + \cdots + c_{1n}x_n = h_1$$

$$c_{n1}x_1 + c_{n2}x_2 + \cdots + c_{nn}x_n = h_n$$

can be inverted to give

$$c_{11}^{-1}d_1 + c_{12}^{-1}d_2 + \cdots + c_{1n}^{-1}d_n = x_1$$

$$c_{n1}^{-1}d_1 + c_{n2}^{-1}d_2 + \cdots + c_{nn}^{-1}d_n = x_n.$$

2. *Let us assume that*
   a. *All x_i are small.*
   b. *The weights have been chosen correctly.*
   c. *The errors are not correlated.*

   Then we find for the variances that

   $$\sigma^2(x_i) = kc_{ii}^{-1}$$

   the diagonal terms of the least-squares matrix and

   $$\text{covariance}(x_i, x_j) = kc_{ij}^{-1},$$

   where

   $$k = \sum_{i=1}^{n} w_i [\,|\,F(i)_o\,| - |\,F(i)_c\,|\,]^2/(n - s)$$

   for the case of n values of $|\,F_o\,|$ and s parameters.

FIGURE 10.13. Estimated standard deviations of parameters are derived as part of the output of the least-squares refinement.

(e.s.d.). Because of the use of Taylor's series as a representation of the cosine, sine, and other functions, it is usually necessary to perform several cycles of refinement. For macromolecules it is often necessary to use methods that take into account the very small observation-to-parameter ratio and use optimization techniques such as "simulated annealing."

Glossary

Accuracy: Deviation of a result obtained by a particular method from the value accepted as true (cf., **Precision**).

Constraints: Limits on the changes that may occur in parameters in a least squares refinement. This may be done by reducing the number of formal parameters in the least squares equations. For example, individual atomic coordinates may be transformed to other parameters, fewer in number, that describe the position and orientation of a whole group of atoms. A benzene ring is an example. It can be defined as a hexagon with fixed geometry and only orientational flexibility.

Correlation between parameters: A correlation is a measure of the extent to which two mathematical variables are dependent on each other. In the least-squares refinement of a crystal structure, parameters related by symmetry are completely correlated, and temperature factors and occupancy factors are often highly correlated.

Estimated standard deviation (e.s.d.): A measure of the precision of a quantity. If the distribution of errors is normal, then there is a 99% chance that a given measurement will differ by less than 2.7 e.s.d. from the mean value.

Gaussian distribution: A symmetrical bell-shaped curve described by the equation $y = A\exp(-x^2/B)$. The value of x is the deviation of a variable from its mean value. The variance of such measurements (the square of the e.s.d.) is $B/2$. In many kinds of experiments, repeated measurements follow such a Gaussian or normal error distribution.

Global minimum: The lowest value of some quantity (such as energy) that can be obtained from all possible values of the parameters describing the system. The parameters of this minimum should correspond to the actual structure. In X-ray crystallography it is the R value (q.v.) that is used as a measure of the quantity to be minimized, while in simulated annealing (q.v.) the energy is used.

Lagrange multipliers: Terms in a method used to find the maximum or minimum of a function that is subject to constraints. The function is written so that

$$F(x_i, y_i, \lambda) = f(x_i, y_i) + \lambda g(x_i, y_i), \tag{10.11}$$

where $f(x_i, y_i)$ is the function whose maximum and minimum are to be found and $\lambda g(x, y)$ is the constraint function. In order to solve this, the simultaneous solution of the equations, the partial derivative of F with respect to x, y, and λ, are set equal to zero. The result contains those points that will give the maximum or minimum values of the function.

Linear equation: An equation is linear if a change in one input variable gives a response that is proportional to that variable.

Local minimum: A minimum in some quantity that is obtained by small shifts in parameters. This minimum does not necessarily correspond to the global minimum (q.v.) which may require more extensive parameter shifts.

Method of least squares: A statistical method for obtaining the best fit of a large number of observations to a given equation. This is done by minimizing the sum of the squares of the deviations of the experimentally observed values from their respective calculated ones. The individual terms in the sum are usually weighted to take into account their relative precisions. In crystal structure analyses, atomic coordinates and other parameters may be fitted in this way to the measured structure factors. Ideally, there should be at least 10 such measurements for each parameter to be determined. In a similar way the least-squares criterion can be applied to the computation of a plane through a group of atoms and to many other problems.

Normal equations: Any set of simultaneous equations involving experimental unknowns and derived from a larger number of observational equations, used in the course of a least-squares adjustment of observations. The number of normal equations is equal to the number of parameters to be determined.

Observational equation: An equation expressing a measured value as some function of one or more unknown quantities. Observational equations are reduced to normal equations during the course of a least-squares refinement.

Occupancy factor: A parameter that defines the partial occupancy of a given site by a particular atom. It is most frequently used to describe disorder in a portion of a molecule, or for describing nonstoichiometric situations, such as when a solvent molecule is being lost to the atmosphere.

Overdeterminancy: The extent to which the number of measurements is greater than the number of unknowns.

Precision: A measure of the experimental uncertainty in a measured quantity, an indication of its reproducibility (cf., **Accuracy**).

R value or R factor: An index that gives a crude measure of the correctness of a structure and the quality of the data. It is defined as

$$R = \sum_{\text{all } hkl} \left| \, | \, F(hkl) \, |_o - | \, F(hkl) \, |_c \, \right| \Big/ \sum | \, F(hkl) \, |_o, \qquad (10.1)$$

and values of 0.06 to 0.02 for "observed data" are considered good for present-day structure determinations. Some partially incorrect structures, however, have had R values below 0.10, and many basically correct but imprecise structures have higher R values.

Refinement of a crystal structure: A process of improving the parameters of an approximate (trial) structure until the best fit of calculated structure factor amplitudes to those observed is obtained. The process usually requires many successive stages.

Restraints: Limits on the possible values that parameters may have. For example, additional observations, such as known bond distances and angles, can be added to the least squares equations, and these must hold true for the results of the least-squares calculation.

Simulated annealing: Annealing is a method used to make steel or glass more soft and less brittle, and involves heating and then cooling. This process is simulated in crystallographic refinements by adjusting the parameters of a macromolecule to simulate "heating" of the molecule, and then "cooling" it so as to minimize the energy. In this way a global minimum may be more readily obtained than with other methods of refinement, such as least-squares methods where a local minimum (q.v.) may be the end point of the refinement.

Variance: The mean square deviation of a frequency distribution from its arithmetic mean. The variance is the square of the estimated standard deviation σ. For a random variable x, the variance is $\sum[(x_i - x_m)^2/n]$ where x_m is the mean value of x, and n is the number of measurements.

Weight of a measurement: A number assigned to express the relative precision of each measurement. In least-squares refinement the weight should be proportional to the reciprocal of the square of the estimated standard deviation of the measurement. Other weighting schemes frequently are used.

References

1. Young, H. D. *Statistical Treatment of Experimental Data*. McGraw-Hill: New York, San Francisco, Toronto, London (1964).
2. Margenau, H., and Murphy, G. M. *The Mathematics of Physics and Chemistry*. van Nostrand: Princeton, Toronto, New York, London (1950).
3. Wilson, E. B. *An Introduction to Scientific Research*. McGraw-Hill: New York (1952).
4. Legendre, A. M. Appendix. Sur la méthode des moindres quarrés. In: *Nouvelles Méthodes pour la Détermination des Orbites des Comètes*. pp. 72–75 Courcier: Paris (1805). **English translation**: Harvey, G. On the method of minimum squares, employed in the reduction of experiments, being a translation of the appendix to an essay of Legendre's entitled "Nouvelles Méthodes pour la Détermination des Orbites des Comètes," with remarks. *Edinburgh Philosophical Journal* **7**, 292–301 (1822).
5. Legendre, A. M. *Nouvelles méthodes pour la détermination des orbites des comètes. Un supplément contenant divers perfectionnements de ces méthodes et leur application aux deux comètes de 1805*. [New methods for the determination of the orbits of comets. A supplement containing various improvements in these methods and their application to two comets in 1805.] Courcier: Paris (1806).
6. Stigler, S. M. *The History of Statistics. The Measurement of Uncertainty before 1900*. Belkap Press of Harvard University Press: Cambridge, MA, and London (1986).
7. Gauss, C. F. *Theoria Motus Corporum Cœlestium in Sectionibus Conicus Solum Ambientum*. Perthes et Besser: Hamburg (1809). **English translation:** Davis, C. H. *The Theory of the Motion of Heavenly Bodies Moving about the Sun in Conic Sections*. Little, Brown: Boston (1857). Reprinted. Dover: New York (1963). **French translation:** Bertrand, J. *Méthode des Moindres Carrés. Mémoires sur la Combination des Observations*. Mallet-Bachelier: Paris (1855).
8. Whittaker, E. T., and Robinson, G. *The Calculus of Observations*. Ch. 9. (1st edn. 1924.) 4th edn. Blackie: Glasgow (1958) .
9. DuMond, J. W. M., and Cohen, E. R. Least squares adjustment of the atomic constants 1952. *Rev. Mod. Phys.* **25**, 691–708 (1953).
10. Pepinsky, R., Robertson, J. M., and Speakman, J. C. (**Eds.**) *Computing Methods and the Phase Problem in X-ray Crystal Analysis*. Pergamon Press: New York, Oxford, London, Paris (1961).

11. Stout, G. H., and Jensen, L. H. *X-ray Structure Determination.* Macmillan: London (1968). 2nd edn. John Wiley: New York (1989).

12. Prince, E. *Mathematical Techniques in Crystallography and Materials Science.* Springer-Verlag: New York, Heidelberg, Berlin (1982).

13. Jensen, L. H. Overview of refinement in macromolecular structure analysis. *Methods in Enzymology* **115**, 227–234 (1985).

14. Hestenes, M. R., and Stiefel, E. Methods of conjugate gradients for solving linear systems. *J. Res. Natl. Bur. Stand.* **49**, 409–436 (1952).

15. Hughes, E. W. The crystal structure of melamine. *J. Amer. Chem. Soc.* **63**, 1737–1752 (1941).

16. Hughes, E. W., and Lipscomb, W. N. The crystal structure of methylammonium chloride. *J. Amer. Chem. Soc.* **68**, 1970–1975 (1946).

17. Rollett, J. S. Structure factor routines. In: *Computing Methods in Crystallography.* (**Ed.**, Rollett, J. S.) Ch. 5, pp. 38–46. Pergamon Press: Oxford (1965).

18. Rollett, J. S. Basic processes involved in least-squares and Fourier refinement. In: *Computational Crystallography.* (**Ed.**, Sayre, D.). pp. 338–353. Clarendon Press: Oxford (1982).

19. Hendrickson, W. A. Refinement of protein structures. In: *Proceedings of the Daresbury Study Weekend (15–16 November, 1980) — Refinement of Protein Structures.* (**Eds.**, Machin, P. A., Campbell, J. W. and Elder, M.) pp. 1–8. SRC Daresbury Laboratory: Warrington, U.K. (1981).

20. Sussman, J. L. The use of CORELS for the refinement of biological macromolecules. In: *Proceedings of the Daresbury Study Weekend (15–16 November, 1980) — Refinement of Protein Structures,* (**Eds.**, Machin, P. A., Campbell, J. W. and Elder, M.) pp. 13–23. SRC Daresbury Laboratory: Warrington, U.K. (1981).

21. Hendrickson, W. A., and Konnert, J. H. Computing in crystallography. Incorporation of stereochemical information into crystallographic refinement. In: *Computing in Crystallography.* (**Eds.**, Diamond, R., Ramaseshan, S. and Venkatsesan, K.) pp. 13.01–13.25. Indian Academy of Science: Bangalore (1980).

22. Jack, A., and Levitt, M. Refinement of large structures by simultaneous minimization of energy and R factor. *Acta Cryst.* **A34**, 931–935 (1978).

23. Jones, P. G. Crystal structure determination: a critical view. *Chem. Soc. Rev.* **13**, 157–172 (1984).

24. Taylor, B. *Methodus Incrementorum Directa et Inversa. Proposition 7.* [Direct and indirect methods of incrementation.] Pearsniansis: Londini (1715)

25. Watenpaugh, K. D., Sieker, L. C., Herriott, J. R., and Jensen, L. H. Refinement of the model of a protein: rubredoxin at 1.5 Å resolution. *Acta Cryst.* **B29**, 943–956 (1973).

26. Pawley, G. S. Constrained refinements in crystallography. *Adv. Struct. Res. Diff. Methods* **4**, 1–64 (1972).

27. Waser, J. Least-squares refinement with subsidiary conditions. *Acta Cryst.* **16**, 1091–1094 (1963).

28. Scheringer, C. Least-squares refinement with the minimum number of parameters for structures containing rigid-body groups of atoms. *Acta Cryst.* **16**, 546–550 (1963).

29. Hendrickson, W. A. Stereochemically restrained refinement of macromolecular structures. *Methods in Enzymology* **115**, 252–262 (1985).

30. Hendrickson, W. A. and Konnert, J. H. A restrained-parameter thermal-factor refinement procedure. *Acta Cryst.* **A36**, 344–350 (1980).

31. Konnert, J. H. A restrained-parameter structure-factor least-squares refinement procedure for large asymmetric units. *Acta Cryst.* **A32**, 614–617 (1976).

32. Sussman, J. L., and Podjarny, A. D. The use of a constrained-restrained least-squares procedure for the low-resolution refinement of a macromolecule, yeast tRNA$_f^{Met}$. *Acta Cryst.* **B39**, 495–505 (1983).

33. Herzberg, O., and Sussman J. L. Protein model building by the use of a constrained-restrained least-squares procedure. *J. Appl. Cryst.* **16**, 144–150 (1983).

34. Cheng, X., and Schoenborn, B. P. Repulsive restraints for hydrogen bonding in least-squares refinement of protein crystals. A neutron diffraction study of myoglobin crystals. *Acta Cryst.* **A47**, 314–317 (1991).

35. Schomaker, V., Waser, J., Marsh, R. E. and Bergman, G. To fit a plane or a line to a set of points by least squares. *Acta Cryst.* **12**, 600–604 (1959).

36. *International Tables for X-ray Crystallography. II. Mathematical Tables.* (**Eds.**, Kasper, J., and Lonsdale, K.) International Union of Crystallography. Kynoch Press: Birmingham (1959).

37. Brünger, A. T., Kuriyan, J., and Karplus, M. Crystallographic R factor refinement by molecular dynamics. *Science* **235**, 458–460 (1987).

38. Fujinaga, M., Gros, P., and van Gunsteren, W. F. Testing the method of crystallographic refinement using molecular dynamics. *J. Appl. Cryst.* **22**, 1–8 (1989).

39. Brünger, A. T. Simulated annealing in crystallography. *Annu. Rev. Phys. Chem.* **42**, 197–223 (1991).

40. Brünger, A. T., Krukowski, A., and Erickson, J. W. Slow-cooling protocols for crystallographic refinement by simulated annealing. *Acta Cryst.* **A46**, 585–593 (1990).

CHAPTER

11

Interpreting x, y, and z (Atomic Coordinates)

Now that the positions of atoms in the crystal structure are known, they can be analyzed to give the molecular geometries of its components. The resulting detailed molecular structures are important findings in chemistry and biochemistry that have assisted in the development of many theories. In this Chapter we discuss how to calculate molecular geometry from the information obtained by the methods of structure determination described in Chapters 8 through 10. The meaning of the tables of values of x, y, and z are described in detail, and the reader can repeat some of the numerical calculations, if so inclined. We also aim in this Chapter to present a concept of precision resulting from the diffraction experiment, and what a crystal structure can reveal. The e.s.d. values (estimated standard deviations) of atomic parameters from the least-squares refinement are also listed in publications of crystal structures, and they can be used to estimate errors in derived geometric parameters. This makes it possible to assess the precision with which these parameters were measured.

11.1 How to calculate molecular geometry

The relevant **atomic coordinates** for the calculation of molecular geometry consist of the values of x, y, and z, obtained as precisely as possible, for all atoms. From values of x, y, and z we can derive geometrical information such as bond distances, bond angles, torsion angles, and the mean planes through groups of atoms. In this way the geometry of the molecule or ion is found.

The reader should appreciate that because atoms are always vibrating, the atomic positions found in an X-ray diffraction experiment represent the average positions of the atoms during vibration. The **atomic parameters** include atomic displacement parameters (described in Chapter 13) which give some measure of the amplitude of this

413

vibration and of any disorder in atomic positions from unit cell to unit cell.

11.1.1 Listings of x, y, and z for a crystal structure

Three atomic coordinates, x, y, and z, describe the position of an atom in the asymmetric unit of the crystal structure. They are each measured as a fraction of the distance along the appropriate unit cell edge and therefore are also called **fractional coordinates**. The value listed for x, for example, is a fractional coordinate relative to the length of the a dimension, and is measured parallel to this unit cell edge. This is shown in Figure 11.1 which is a diagram illustrating the meaning of x, y, and z coordinates for three atoms in a unit cell. One unit cell repeat, whatever its length, corresponds to 1.000 in fractional coordinates, so that if the value of x is 0.3245, the atom lies 32.45% along the a spacing, measured in a direction parallel to the a axis. The y and z coordinates are defined in an analogous manner.

Because atomic arrangements in crystal structures are periodic from unit cell to unit cell, one part of a molecule may lie in one unit cell, and another part may lie in an adjacent unit cell. If there is an atom at x, because of this translational symmetry there is another at $1 + x$, and another at $n + x$, where n is any integer. Values of x, y, and z that are reported usually correspond to a complete and distinct molecule so that, for convenience, some atomic coordinates may have negative values, and other atomic coordinates may have values greater than 1.0000. The symmetry of the space group and the identical contents of adjacent unit cells can lead to a diagram of the complete crystal structure and an analysis not only of the molecular structure, but also of its surroundings.

A simple method for obtaining a view of a crystal structure, given the atomic coordinates and unit cell dimensions, is shown in Figure 11.2. If two edges of a unit cell and their enclosed angle are drawn to scale, as shown in this Figure, a ruler, placed as shown, can be used to divide this scale drawing into a 5 × 5 or 10 × 10 grid. Then the atomic coordinates can be plotted as for a normal graph. This provides a useful "arm-chair" method for perusing crystal structures of interest in the literature in the absence of an immediately available computer graphics system. For such a plot, it is then possible to measure distances between pairs of atoms from the difference of atomic coordinates perpendicular to the plane of the paper, together with the known unit cell dimension involved. If the third axis is not perpendicular to the plane of the paper (for example, in a triclinic structure) such distances will be very approximate unless the appropriate projection angle is used.

Typical sets of atomic coordinates are given in Tables 11.1 and 11.2 for the alkali halides and copper sulfate pentahydrate, respectively. Fractional atomic coordinates are listed separately from unit cell dimensions

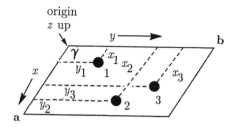

(a)

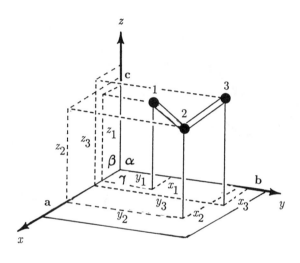

(b)

FIGURE 11.1. Atomic coordinates. (a) Two-dimensional diagram with measures of x and y for three atoms (1, 2, and 3 at x_1, y_1, x_2, y_2 and x_3, y_3 respectively), and (b) a three-dimensional representation of the same three atoms. Note that x is measured parallel to the a axis, y parallel to the b axis, and z parallel to the c axis.

because they are derived from different measurements, and their errors are also estimated separately. *Unit cell dimensions* are derived from 2θ-values, that is, the directions and indices of the Bragg reflections, whereas *fractional atomic coordinates* are derived from the measured intensities of the entire set of Bragg reflections (see Figure 11.3). There are advantages in having unit cell dimensions and atomic coordinates in two separate lists. For example, the isostructural character of the the crystal structures of sodium and potassium chlorides is stressed, as shown in Table 11.1. These two crystal structures have identical fractional coordinates for the various ions and differ only in their unit-cell dimensions. The crystal structure of cesium chloride is quite different from those of the other two alkali halides.

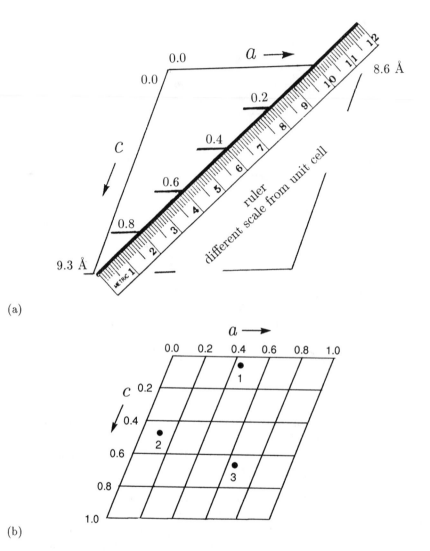

(a)

(b)

FIGURE 11.2. A convenient way to draw the contents of a unit cell, given the atomic coordinates, is shown here. (a) The outline of the unit cell is drawn to scale in two dimensions as shown. It is then divided into one tenths in each dimension, by means of a ruler (any scale, inches, centimeters), inclined as shown, so that each side can be divided into ten parts. (b) The result is a grid on which the positions of atoms can be plotted, as shown. In the third dimension, if the third unit-cell axis is perpendicular to the plane of the paper, Pythagoras' theorem can be used to measure interatomic distances; if it is not perpendicular, only an approximate estimate can be made.

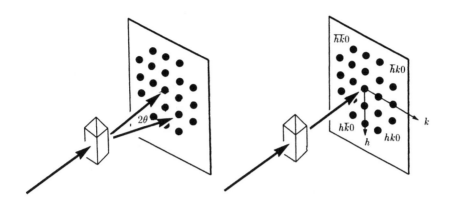

(a) $\sin\theta/\lambda$ gives
   unit-cell dimensions

(b) *hkl* values give periodicities
   of electron-density waves.

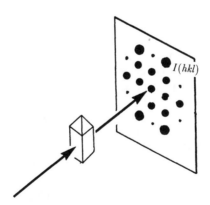

(c) $I(hkl)$ gives amplitudes of
   electron-density waves, and hence x, y, and z
   of the atomic arrangement in the crystal.

FIGURE 11.3. The sources from the Bragg reflections in the X-ray diffraction pattern
of (a) unit-cell dimensions (from $\sin\theta/\lambda$ values), (b) periodicities of electron-density
waves (from *hkl* values), and (c) atomic coordinates x, y, and z (from intensities of
the Bragg reflections).

TABLE 11.1. Unit cell dimensions and atomic coordinates for (a) sodium, (b) potassium, and (c) cesium chlorides (Refs. 6 and 7). $Fm3m$ is face centered, with atoms at $(0,0,0;\ 0,\frac{1}{2},\frac{1}{2};\ \frac{1}{2},0,\frac{1}{2};\ \frac{1}{2},\frac{1}{2},0)\ \pm x,y,z$ for NaCl and KCl. $Pm3m$ is primitive with atoms at x,y,z for CsCl.

(a) NaCl. *Space group $Fm3m$ (face-centered).* $a = 5.640$ Å, Na$^+$ $\cdots$ Cl$^-$ $= 2.820$ Å.

	x	y	z
Na	0.00	0.00	0.00
Cl	0.50	0.00	0.00

(b) KCl. *Space group $Fm3m$ (face-centered).* $a = 6.291$ Å, K$^+$ $\cdots$ Cl$^-$ $= 3.146$ Å.

	x	y	z
K	0.00	0.00	0.00
Cl	0.50	0.00	0.00

(c) CsCl. *Space group $Pm3m$ (primitive).* $a = 4.121$ Å, Cs$^+$ $\cdots$ Cl$^-$ $= 4.874$ Å.

	x	y	z
Cs	0.00	0.00	0.00
Cl	0.50	0.50	0.50

Since, however, the unit cell dimensions are different for these two salts, the distances between anions and cations are not the same in the two crystal structures. Sometimes the atomic coordinates are converted to **Cartesian** or **orthogonal coordinates**, relative to three mutually perpendicular axes, with values in Å. The way to do this is shown in Figure 11.4. The conversion is useful for simple calculations of interatomic distances but, if such coordinates are reported, then information about the crystal packing is lost. Therefore crystal structures are reported in terms of the unit-cell geometry rather than in orthogonal geometry.

The bond length between two atoms is a function not only of the identities of the two atoms, but also of strain, the extent to which electrons are shared, and any other significant interactions with other atoms. For example, the C–C bond length measured in many organic compounds is, as shown in Table 11.3, not a constant value. The length of a C–C single bond in diamond[1] is 1.54 Å, but most so-called single C–C bonds in organic compounds are somewhat shorter than that. Some representative examples of bond lengths are given in Table 11.3.

Much of our current knowledge of bonding geometry of atoms has been derived from crystal structure determinations. Atomic coordinate data that have been published are included in databases, as will be described in more detail in Chapter 16. For organic compounds the Cambridge Structural Database[2–4] can be accessed by computer. Similar data for inorganic compounds are found in the Inorganic Crystal Structure Database.[5] Comparisons with other experimental methods, such as

TABLE 11.2. Unit cell dimensions and atomic coordinates for copper sulfate penta-hydrate, $CuSO_4 \cdot 5H_2O$ (Refs. 8 and 9).

Unit-cell dimensions:
$a = 6.141$ Å; $b = 10.736$ Å; $c = 5.986$ Å;
$\alpha = 82.27°$; $\beta = 107.43°$; $\gamma = 102.67°$.

Atom	x	y	z	Atom	x	y	z
Cu	0.000	0.000	0.000	H5a	0.103	0.859	0.750
S1	0.011	0.287	0.624	H5b	0.281	0.989	0.771
O1	0.093	0.848	0.325	H6a	0.300	0.202	0.067
O2	0.757	0.682	0.203	H6b	0.335	0.128	0.322
O3	0.141	0.627	0.365	H7a	0.320	0.378	0.342
O4	0.045	0.300	0.384	H7b	0.397	0.606	0.575
O5	0.817	0.074	0.154	H8a	0.147	0.616	0.841
O6	0.290	0.118	0.149	H8b	0.195	0.600	0.112
O7	0.465	0.406	0.299	H9a	0.410	0.196	0.697
O8	0.756	0.416	0.019	H9b	0.601	0.132	0.670
O9	0.434	0.125	0.630				

electron diffraction, generally confirm the information obtained by X-ray diffraction studies. For example, the well-known tetrahedral geometry of a carbon atom covalently linked to four other atoms was first observed experimentally in the crystal structure determination for diamond,[1] as described in Chapter 1 (Figure 1.7). In the case of biological macromolecules, crystal structure analyses have greatly extended our understanding of how these molecules are folded, the nature of the contacts between different parts of the molecule, and the dimensions and stereochemistries of their active sites.

11.1.2 Atomic connectivity

The first piece of information obtained from a structure determination is the connectivity of atoms. This leads to the chemical formula, if previously unknown. Penicillin and vitamin B_{12} (see Figures 1.12 and 1.13, Chapter 1) are examples of compounds for which X-ray diffraction analyses established the atomic connectivity and, as a result, the chemical formulæ and subsequent geometries of these molecules. Other examples include the boron hydrides, which were found to have unexpected arrangements of atoms. For many of these compounds the atomic connectivity was not known before the X-ray diffraction analyses.

TABLE 11.3. Selected bond distances, in Å, with e.s.d. values in parentheses.

Note: The number of atoms bonded to a given atom is designated, where needed, in parentheses ( ). The coordination numbers of the metals ions are designated by brackets [ ]. C* denotes the atoms in the bond listed in this Table. (From Ref. 3).

Carbon–carbon bonds

Csp^3-Csp^3	C–C overall	1.53(2)
	$C_3-C^*-C^*-C_3$	1.59(3)
	$C_2-C^*-C^*-C_2$	1.54(1)
	$C-CH_2^*-CH_2^*-C$	1.52(1)
Csp^3-Csp^2	$C^*-C^*=C$	1.51(2)
Csp^3-Csp^1	$C^*-C^* \equiv C, C^*-C^* \equiv N$	1.47(1)
$Csp^2=Csp^2$	C=C	1.32(1)
C=C aromatic	C⋯C	1.38(1)
Csp^1-Csp^1	C≡C	1.18(1)

Carbon–halogen bonds

C–F	alkyl/aryl C–F	1.40(2)/1.36(4)
Cl–C	alkyl/aryl C–Cl	1.79(1)/1.74(1)
C–Br	alkyl/aryl C–Br	1.97(3)/1.90(1)
C–I	alkyl/aryl C–I	2.16(2)/2.10(2)

Carbon–nitrogen bonds

$Csp^3-N(3)$	$C-Nsp^3$	1.47(1)
$C_{Arom}-N(3)$	$C_{Arom}-NO_2$	1.47(1)
$C_{Arom}=N(2)$	C=N=C	1.34(1)
CO–NH	peptides	1.33(1)
$Csp^1 \equiv N$	C=C–C≡N	1.14(1)

Carbon–oxygen bonds

$Csp^3-O(2)$	alcohols, ethers	1.43(1)
	oxiranes (epoxides)	1.45(1)
	carboxylic acids	1.31(2)
	esters	1.34(1)
$C_{Arom}-O(2)$	phenols	1.36(2)
	alkyl/aryl ethers	1.37(1)
$Csp^2=O(1)$	aldehydes	1.19(1)
	ketones	1.21(1)
	carboxylates	1.25(2)

TABLE 11.3 (cont'd).

Other bonds to carbon

Csp^3–S(2)	thioethers	1.82(2)
Csp^3–Se	selenium compounds	1.97(3)

Other bonds to nitrogen

N(2)=N(2)	C_{Arom}–N=N–C_{Arom}	1.26(2)
N(3)–O(1)	pyridine *N*-oxides	1.30(2)
N(3)=O(1)	nitro groups	1.22(1)

Other bonds to oxygen

O(2)–O(2)	peroxides	1.47(1)
O(2)–P(4)	phosphodiester	1.61(1)
O(1)=P(4)	phosphodiester	1.48(1)
O(1)–S(4)	C–SO_2–C	1.44(1)
O(1)=S(3)	C–S(=O)–C	1.50(1)

Cations-to-ammonia [Coordination number]

Cr(III)···N	2.07	[6]
Ru(II)···N	2.13	[6]
Co(III)···N	1.96	[6]
Ru(III)···N	2.11	[6]
Ni(II)···N	2.07	[4–6]
Rh(III)···N	2.12	[4]
Cu(II)···N	1.99	[4–6]

Cations-to-carboxylates [Coordination number]

Ti(IV)···O	2.00	[6]	Cu(II)···O	1.96,2.34	[6]
Cr(III)···O	1.96	[6]	Zn(II)···O	1.95	[4]
Mn(II)···O	2.16	[6]	Zn(II)···O	2.02	[5]
Fe(III)···O	2.02	[6]	Zn(II)···O	2.07	[6]
Co(II)···O	1.97	[4]	Ru(II)···O	2.09	[6]
Co(II)···O	2.09	[6]	Rh(III)···O	2.02	[6]
Ni(II)···O	2.06	[6]	Eu(III)···O	2.39	[9]
Cu(II)···O	1.95	[4]	Ir(III)···O	2.07	[6]
Cu(II)···O	1.99	[5]	U(VI)···O	2.36	[7]

Converting crystal coordinates to Cartesian coordinates

For a unit cell, dimensions a, b, c, α, β, γ, the equations relating the crystallographic coordinates x, y, z, to coordinates in an orthogonal system X, Y, Z, are:

$$X = ax + by \cos \gamma + cz \cos \beta$$

$$Y = by \sin \gamma + \frac{cz(\cos \alpha - \cos \beta \cos \gamma}{\sin \gamma}$$

$$Z = \frac{zV}{ab \sin \gamma}$$

where

$$V = (1 - \cos^2 \alpha - \cos^2 \beta - \cos^2 \gamma + 2 \cos \alpha \cos \beta \cos \gamma)^{1/2} abc$$

For example, for $CuSO_4 \cdot 5H_2O$ (Refs. 8 and 9):

$a = 6.141$ Å; $b = 10.736$ Å; $c = 5.986$ Å; $\alpha = 82.27°$; $\beta = 107.43°$; $\gamma = 102.67°$.

$\sin \alpha = 0.9909$	$\sin \beta = 0.9541$	$\sin \gamma = 0.9756$
$\cos \alpha = 0.1345$	$\cos \beta = -0.2995$	$\cos \gamma = -0.2193$

$$V = 0.9283abc \text{ Å}^3 = 366.36 \text{ Å}^3$$

$$X = 6.141x - 2.355y - 1.793z$$
$$Y = 10.474y + 0.422z$$
$$Z = 5.696z$$

Atom	x	y	z	Atom	X	Y	Z
Cu	0.000	0.000	0.000	Cu	0.00	0.00	0.00
Cu	1.000	0.000	0.000	Cu	6.14	0.00	0.00
Cu	0.000	1.000	0.000	Cu	-2.36	10.47	0.00
Cu	1.000	1.000	0.000	Cu	3.78	10.47	0.00
Cu	0.500	0.500	0.000	Cu	1.89	5.24	0.00
S1	0.011	0.287	0.624	S1	-1.73	3.28	3.55
O1	0.093	0.848	0.325	O1	-2.01	9.02	1.85
O2	0.757	0.682	0.203	O2	2.68	7.24	1.16
O3	0.141	0.627	0.365	O3	-1.26	6.72	2.08

FIGURE 11.4. From crystal coordinates to orthogonal (Cartesian) coordinates (Å).

Once the atomic coordinates have been determined, often before any refinement, it is usual to calculate interatomic distances to 3.5 or 4 Å, in order to check the connectivity of atoms, that is, to determine which atoms are bonded to which. This will show whether the chemical formula is correct. The connectivity also demonstrates whether or not the experimental atomic positions are for atoms connected in one molecule, or for atoms which are in different molecules (related by space-group symmetry). Finally, if the connectivity calculation includes longer distances, it will show how the molecules or ions pack with respect to each other in the crystal. Information on the different intermolecular interactions, such as hydrogen bonds present in the crystal, are found in this way. If any intermolecular distance is substantially less than the expected value (see Table 11.4, for example), implying that molecules approach each other too closely, the reported crystal structure may not be correct. The derived set of relative phases (see Chapter 8) should be scrutinized, as there may be another more suitable set. Any truly unusual geometrical features in a structure determination should be carefully analyzed before being accepted as experimental evidence.

11.1.3 Bond lengths and their calculation

A **bond length** is the distance between two atomic nuclei that are joined together by some type of electronic bonding. Because atoms undergo vibrational and other motion, reported bond lengths are the average or equilibrium bond lengths. If the temperature is lowered, vibrations are reduced in amplitude, and the atomic positions can be located more reliably. The equilibrium interatomic distance in bonded and nonbonded interactions is a result of attractive and repulsive interatomic forces. The major effect of temperature is to change the distances between molecules, rather than to affect bond lengths. Molecular geometry at low temperatures does not change appreciably from that at room temperature (if suitably corrected for thermal motion), but, at the lower temperatures, the atoms are more precisely located. Therefore distances at very low temperatures are usually the best to use, since they are more precise.

In order to compute interatomic distances, it is necessary to know x, y, and z values of each atom, the unit cell dimensions, and the space

---

TABLE 11.4. Van der Waals radii in Å, after Pauling (Ref. 10).

H	1.2	N	1.5	O	1.4
F	1.35	P	1.9	S	1.85
Cl	1.8	As	2.0	Se	2.0
Br	1.95	Sb	2.2	Te	2.2
I	2.15	methyl	2.0	aromatic	2.15

group of the crystal structure. The distance between two atoms in a triclinic crystal structure can be calculated by use of Equation 11.1. If two atoms at x_1, y_1, z_1 and x_2, y_2, z_2, lie in a unit cell that has edges with lengths a, b, and c and interaxial angles of α, β, and γ, the square of the distance r between these two atoms is:

$$r^2 = [(x_1 - x_2)a]^2 + [(y_1 - y_2)b]^2 + [(z_1 - z_2)c]^2$$
$$+ [2ab \cos \gamma (x_1 - x_2)(y_1 - y_2)] + [2ac \cos \beta (x_1 - x_2)(z_1 - z_2)]$$
$$+ [2bc \cos \alpha (y_1 - y_2)(z_1 - z_2)]. \tag{11.1}$$

The notation may be simplified by setting $\Delta x = x_1 - x_2$ so that it equals the difference in x values for the two atoms at the ends of the bond for which the distance is being calculated, and similar notations for y and z. This gives

$$r^2 = (a \, \Delta x)^2 + (b \, \Delta y)^2 + (c \, \Delta z)^2 + 2ab \cos \gamma \Delta x \Delta y$$
$$+ 2ac \cos \beta \, \Delta x \, \Delta z + 2bc \cos \alpha \, \Delta y \, \Delta z. \tag{11.2}$$

An example of the calculation of a bond length for a triclinic crystal $(CuSO_4 \cdot 5H_2O)$ is given in Figure 11.5.

In $CuSO_4 \cdot 5H_2O$ the $P\bar{1}$ space-group symmetry is x, y, z and $-x$, $-y$, $-z$. The distances in the sulfate ion are required, but it is necessary to have all atoms in one sulfate ion (x', y', z'). Unit-cell dimensions $a = 6.141$ Å; $b = 10.736$ Å; $c = 5.986$ Å; $\alpha = 82.27°$; $\beta = 107.43°$; $\gamma = 102.67°$ (Refs. 8 and 9). $\Delta x = x'_1 - x'_2$, etc.

Atom	x	y	z	Atom	x'	y'	z'
S1	0.011	0.287	0.624	S1	0.011	0.287	0.624
O1	0.093	0.848	0.325	O1	-0.093	0.152	0.675
O2	0.757	0.682	0.203	O2	0.243	0.318	0.797
O3	0.141	0.627	0.365	O3	-0.141	0.373	0.635
O4	0.045	0.300	0.384	O4	0.045	0.300	0.384

S1 *and* O4 *at* x, y, z; O1 *and* O3 *at* $-x$, $1-y$, $1-z$; *and* O2 *at* $1-x$, $1-y$, $1-z$.

---

bond	$\lvert \Delta x \rvert$	$\lvert \Delta y \rvert$	$\lvert \Delta z \rvert$	length (Å)
S1–O1	0.104	0.135	0.051	1.481
S1–O2	0.232	0.031	0.173	1.488
S1–O3	0.130	0.086	0.011	1.470
S1–O4	0.034	0.013	0.240	1.497

FIGURE 11.5. Some bond distances in a triclinic crystal.

Sometimes a molecule contains symmetry within it that is determined by the space group. As a result, only a fraction of the molecule lies in the asymmetric unit. For example, benzene[11,12] crystallizes in the space group *Pbca* with eight asymmetric units in the unit cell. Since there are only four benzene molecules per unit cell, it follows that the asymmetric unit is only half a molecule. The unit cell dimensions (at 15 K)[12] are $a = 7.360$, $b = 9.375$, $c = 6.703$ Å. The atomic coordinates are:

Atom	x	y	z
C1	−0.061	0.141	−0.006
C2	−0.140	0.044	0.127
C3	−0.078	−0.096	0.133
H1	−0.109	0.251	−0.011
H2	−0.249	0.077	0.226
H3	−0.138	−0.171	0.238

The eight equivalent positions in the space group *Pbca* are:

$$
\begin{array}{lll}
x & y & z \\
\tfrac{1}{2}-x & \bar{y} & \tfrac{1}{2}+z \\
\tfrac{1}{2}+x & \tfrac{1}{2}-y & \bar{z} \\
\bar{x} & \tfrac{1}{2}+y & \tfrac{1}{2}-z
\end{array}
\qquad
\begin{array}{lll}
\bar{x} & \bar{y} & \bar{z} \\
\tfrac{1}{2}+x & y & \tfrac{1}{2}-z \\
\tfrac{1}{2}-x & \tfrac{1}{2}+y & z \\
x & \tfrac{1}{2}-y & \tfrac{1}{2}+z
\end{array}
$$

When interatomic distances are calculated, it is found that coordinates x, y, z must be supplemented by those at $-x$, $-y$, $-z$ to give the entire molecule. The two halves are related by a center of symmetry at the origin (0.0, 0.0, 0.0) and give six carbon atoms and six hydrogen atoms (the complete benzene molecule) at:

Atom	x	y	z
C1	−0.061	0.141	−0.006
C2	−0.140	0.044	0.127
C3	−0.078	−0.096	0.133
H1	−0.109	0.251	−0.011
H2	−0.249	0.077	0.226
H3	−0.138	−0.171	0.238
C1′	0.061	−0.141	0.006
C2′	0.140	−0.044	−0.127
C3′	0.078	0.096	−0.133
H1′	0.109	−0.251	0.011
H2′	0.249	−0.077	−0.226
H3′	0.138	0.171	−0.238

From these coordinates the C$\doteq$C and C–H bond distances can be calculated. When the unit cell is orthorhombic (as it is for crystalline benzene),
or tetragonal or cubic, then $\alpha = \beta = \gamma = 90°$, and, as a result, Equation
11.1 can be simplified to:

$$r^2 = (a\Delta x)^2 + (b\Delta y)^2 + (c\Delta z)^2 \qquad (11.3)$$

Application of Equation 11.3 to the atomic coordinates just listed for benzene gives its molecular geometry, shown in Figure 11.6. Other molecules
lie at $x + 1$, y, z; $1 - x$, $-y$, $-z$; $x + 1$, $y + 1$, z; $1 - x$, $1 - y$, $-x$, etc.
Distances between nonbonded atoms may be calculated, again by appli-
cation of Equation 11.3. For example, in benzene, the coordinates of C1
and H3 in different molecules are:

	x	y	z
C1 at x, y, z	−0.061	0.141	−0.006
H3 at $-\frac{1}{2} - x$, $-y$, $z - \frac{1}{2}$	−0.362	0.171	−0.262

so that $\Delta x = (0.301 \times 7.360$ Å$)$, $\Delta y = (0.030 \times 9.375$ Å$)$, and $\Delta z = (0.256$
$\times 6.703$ Å$)$. By application of Equation 11.3, r, the distance between C1
and the selected H3 atom is 2.816 Å. Note that this distance appears
in the list below, but from H3 to C1, so that symmetry positions are
different for the target atom in the two cases.

(a) Intramolecular distances, i.e., bond lengths (Å).

C1–C2	1.400	C1–H1	1.091
C2–C3	1.390	C2–H2	1.086
C1–C3′	1.396	C3–H3	1.088

(b) Intermolecular distances (Å)

H1 ⋯ C3	$-x - \frac{1}{2}$, $\frac{1}{2} + y$, z	2.880
H2 ⋯ C1	$-x - \frac{1}{2}$, $-y$, $\frac{1}{2} + z$	3.059
H2 ⋯ C2	$-x - \frac{1}{2}$, $-y$, $\frac{1}{2} + z$	3.030
H2 ⋯ C2	$x - \frac{1}{2}$, y, $\frac{1}{2} - z$	3.025
H2 ⋯ C3	$-x - \frac{1}{2}$, $-y$, $\frac{1}{2} + z$	3.016
H2 ⋯ C3	$x - \frac{1}{2}$, y, $\frac{1}{2} - z$	3.062
H3 ⋯ C1	$-x - \frac{1}{2}$, $-y$, $\frac{1}{2} + z$	2.816
H3 ⋯ C1	$-x$, $y - \frac{1}{2}$, $\frac{1}{2} - z$	2.912

FIGURE 11.6. (a) Intramolecular and (b) intermolecular distances in Å in crystalline
benzene. In (b) the symmetry operation for each carbon atom is shown.

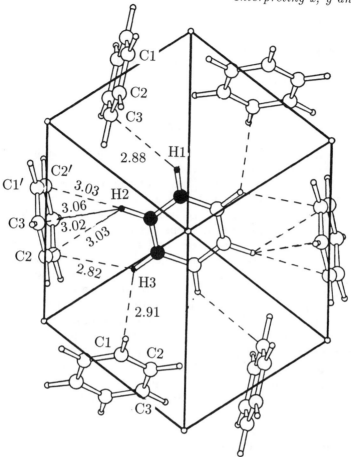

FIGURE 11.6 (cont'd). (c) View onto a benzene molecule in the crystal structure, showing the intermolecular interactions listed in (b). Carbon and hydrogen atoms drawn with filled circles are those in one asymmetric unit. Positions of all other atoms are generated from these by application of the symmetry operations of the space group.

The van der Waals radius of an atom[10,13] is half the distance that separates two contiguous but nonbonded atoms that have the same atomic number (see Table 11.4). The van der Waals radius of chlorine, for example, is found by determining the shortest distances between nonbonded chlorine atoms in crystal structures. An analogous definition can be applied to the van der Waals radius of a group. In this way it is found that the van der Waals radius of a methyl group is 2.0 Å, and that of a hydrogen atom is 1.2 Å. Therefore, the nonbonded H $\cdots$ H distance will be about 2.4 Å and methyl groups will pack with their centers about 4 Å apart (the nonbonded minimum C $\cdots$ C distance) (see Table 11.4).

11.1.4 Interbond angles and their calculation

The internuclear bond angle, A–B–C, is the angle between the bonds A–B and B–C formed by three atoms A, B and C, connected in that order. If the length of A–$B = l_1$, B–$C = l_2$ and $A\cdots C = l_3$, then the angle A–B–$C = \delta$ may be calculated from the formula

$$\cos \delta = \frac{l_1^2 + l_2^2 - l_3^2}{2l_1 l_2}. \tag{11.4}$$

The calculation of bond angles in the sulfate group in $CuSO_4 \cdot 5H_2O$ is shown in Figure 11.7. The sulfur atom lies at the center of a tetrahedron with four bonds directed to the corners of this tetrahedron, at angles near 109.5°. The situation is similar for a saturated (four-valent) carbon atom. For example, methane, CH_4, has H–C–H angles of 109.5°. This is the arrangement in space that keeps the four hydrogen atoms as far apart as possible. This equality of bond angles is strongly perturbed when one or more hydrogen atoms is replaced by another atom. It is found that a major determinant of the size of a bond angle is the size of the various ligands that are attached to a given atom. Variations in tetrahedral bond angles in compounds with longer alkyl chains can be explained by the effects of the van der Waals radii of atoms or groups of atoms. Repulsion between methyl or methylene groups (van der Waals radii 2.0 Å) will increase the C–C–C angle and necessarily mean that another angle (such as the H–C–H angle) will have to be decreased. For example, in n-pentane,[14] the C–C–C angle is 112° and the H–C–H angle is only 106°. If, however, the two hydrogen atoms were to be replaced by groups that are larger than the methyl groups, other geometrical relief of strain, such as by bond twisting (see Chapter 12), would be observed. Another feature shown to affect bond angles is the presence of a non-bonding lone pair of electrons in the valence shell; this pair takes up more space than does a bonding electron pair.[15] For example,[16] water

bond	length l_1, l_2 (Å)	O$\cdots$O	length l_3 (Å)	angle	angle δ (degrees)
S1–O1	1.481	O1–O2	2.409	O1–S1–O2	108.4
S1–O2	1.488	O1–O3	2.426	O1–S1–O3	110.6
S1–O3	1.470	O1–O4	2.414	O1–S1–O4	108.3
S1–O4	1.497	O2–O3	2.436	O2–S1–O3	110.9
		O2–O4	2.419	O2–S1–O4	108.3
		O3–O4	2.432	O3–S1–O4	110.1

FIGURE 11.7. Calculation of a bond angle. Shown are data for the angles in the sulfate ion in $CuSO_4 \cdot 5H_2O$, obtained by application of Equation 11.4.

and ammonia have bond angles smaller than 109.5°. Ammonia contains a nitrogen atom with a lone pair of electrons and H–N–H bond angles of 107°. In water, where the oxygen atom has two lone pairs of electrons, the H–O–H angle is 104.5°.

11.2 Limitations: precision and its assessment

The reader should be aware that there are cautions and limitations to be applied to the structural data derived by a crystal structure analysis. Interpretation of results must be made with an awareness of the precision of the coordinates being used, as discussed in Chapter 10. A description of terms used in statistical analyses of data and the resulting refined parameters is provided by Reference 17 which is recommended to the reader.

11.2.1. Estimated standard deviations of geometry

One way to assess the precision of a structure determination is to use the estimated standard deviation (e.s.d.) of the geometric quantity of interest. This is obtained from the least-squares refinement (see Chapter 10). The more precisely a measurement is made, the smaller is the e.s.d. of that measurement. The equations for calculating the e.s.d. values of a bond length and of a bond angle[18] are given in Figure 11.8.

In assessing precision, it is important to be able to decide whether two bonds that differ in length are really different, or whether the observed difference is merely that expected from the imprecision of the measurements (indicated by the e.s.d. of each atomic position). The significance of differences in bond lengths can be approached by probability theory. The probability that x does not differ from its mean value by more than $q\sigma = p$ is given for various values of q, in Table 11.5. For example, if two bond lengths differ by 0.04 Å, and the positional e.s.d. $= 0.005$ Å for both atoms, the e.s.d. of a bond is:

$$\sigma_{\text{bond}} = \sqrt{(0.005^2 + 0.005^2)} = 0.007.$$

The difference in bond length of 0.04 Å is 6 times the e.s.d., so that p in Table 11.5 is less than 0.001, and the two bond lengths are inferred to be significantly different.

E.s.d. values contain a measure of the random error in each atomic coordinate. They do not give any indication of systematic errors, such as wrong unit cell dimensions, incorrect scattering factors, incorrect atomic number for an atom, or a wrong space group choice. For a well-determined, high-resolution crystal structure determination, e.s.d.s in bond distances will be well under 0.01 Å and those in bond and torsion angles 0.1° or less. The e.s.d.s of individual atom coordinates in a given structure are approximately inversely proportional to the number of electrons (scattering power) in that atom. The influence of different

(a) e.s.d.s of bond lengths

For two atoms, A and B, σ_{xA} is the e.s.d. of x for atom A, and $\Delta x = x_B - x_A$. The required e.s.d. values are obtained by least-squares refinement (see Chapter 10). The e.s.d. of the bond distance between A and B is:

$$\sigma_r^2 = [(\sigma_{xA}^2 + \sigma_{xB}^2)(\frac{\Delta x + \Delta y \cos \gamma + \Delta z \cos \beta}{r})^2$$

$$[(\sigma_{yA}^2 + \sigma_{yB}^2)(\frac{\Delta y + \Delta x \cos \gamma + \Delta z \cos \alpha}{r})^2$$

$$[(\sigma_{zA}^2 + \sigma_{zB}^2)(\frac{\Delta z + \Delta x \cos \beta + \Delta y \cos \alpha}{r})^2]. \qquad (11.8.1)$$

(b) e.s.d.s of interbond angles

The e.s.d. of the angle between the bonds A–B and B–C is calculated by:

$$\sigma_\theta = \left[\frac{\sigma_A^2}{(AB)^2} + \frac{\sigma_B^2 (AC)^2}{(AB)^2 (BC)^2} + \frac{\sigma_C^2}{(BC)^2} \right]^{1/2} \qquad (11.8.2)$$

where σ_A, σ_B, and σ_C are the e.s.d.s of the positions of atoms A, B, and C. If the errors are not equal in all directions, the expression is more complicated.

FIGURE 11.8. Calculation of estimated standard deviations (e.s.d. values) for bond lengths and angles.

TABLE 11.5. Probabilities that two quantities differ from their mean values.

If two average values are $\langle x_1 \rangle$ and $\langle x_2 \rangle$, with e.s.d. values of σ_1 and σ_2, then

$$q = \frac{\langle x_1 \rangle - \langle x_2 \rangle}{(\sigma_1^2 + \sigma_2^2)^{\frac{1}{2}}}.$$

where q is a measure of the relationship of the difference ($\langle x_1 \rangle - \langle x_2 \rangle$) to the respective e.s.d.s. Shown below are probabilities p that $\langle x_1 \rangle$ and $\langle x_2 \rangle$ are the same for various values of the ratio q. If the probability p is greater than 0.05, the difference between $\langle x_1 \rangle$ and $\langle x_2 \rangle$ is not significant. If p lies between 0.01 and 0.05, the difference is possibly significant. If it is less than 0.01, this difference is probably significant.

p	q	p	q	p	q
0.1	1.65	0.05	1.96	0.01	2.58
0.001	3.29	0.0001	3.89		

atomic numbers can be dramatic when a crystal structure contains atoms with widely different atomic numbers. For example, in organometallic complexes the e.s.d.s in bond distances for metal–metal bonds, metal–carbon, or metal–oxygen bonds are frequently less than 0.001 Å. On the other hand, in the same structure, the e.s.d.s in the C–C or C–O bonds are typically between 0.005 and 0.015 Å, while those in C–H or O–H bonds can be on the order of 0.05 Å or more. Generally, e.s.d. values for hydrogen atoms are 8–10 times higher than those for carbon atoms in the same crystal structure. If more precise hydrogen atom locations are required, it is necessary to use neutron diffraction methods.

From the above description it follows that the reported R value, defined in Chapter 10, is not necessarily a good measure of the precision of bond distances. The e.s.d. of the measurement for a similar value depends on the atomic number of the heaviest atom and the resolution of the data set. Therefore, when comparing structure determinations with the same R value, if a crystal contains only carbon and hydrogen atoms, the e.s.d.s of the carbon atoms are much smaller than they would be if a heavier element, such as cobalt,[19] were present. The reason is that since $Z = 27$ for cobalt, while $Z = 6$ for carbon, their maximum contributions to intensities are in the ratio 729:36, or about 20:1.

11.2.2 Comparisons of similar geometries

Another way of estimating the precision of a crystal structure is to compare the geometries (especially distances and angles) of chemically known moieties, for example, phenyl rings. If there is more than one such group in the asymmetric unit, a comparison of their respective geometries, taking into account the e.s.d.s of the quantities found, may indicate the precision of the model; this must be done with care, however, because there may be a chemical reason why the two groups differ. If the errors are large, comparisons with known dimensions of such a group measured in other crystal structures may be helpful.

11.2.3 Comparison of X-ray and neutron diffraction results

When analyzing bonding geometry, the fact that X rays are diffracted by electrons must not be forgotten. Refined atomic coordinates give the center of the electron distribution around each atom. Often there is asymmetry in the electron distribution around an atom as, for example, for hydrogen and certain *sp*-hybridized atoms (such as carbon and nitrogen atoms in cyano groups). In these cases, as described in the section on deformation density in Chapter 9, the center of the electron distribution is displaced significantly from the nuclear center. Therefore, the bond lengths determined by X-ray diffraction will differ from the true values.

In addition, the determination of hydrogen atom positions by X-ray diffraction is imprecise because hydrogen has only one electron. If all of the atoms in the molecular structure are light (up to fluorine in atomic weight), a peak in the electron-density map corresponding to the position for a hydrogen atom can be found and its coordinates refined. Neutron diffraction studies, however, which give nuclear positions, provide a more nearly correct value for bond lengths involving hydrogen atoms.

11.2.4 Precision in macromolecular structures

The main concerns in estimating the precision of a small molecule crystal structure are how well the intensity data have been measured and to what extent the derived relative phases are correct. In addition, it is important to ascertain that the space group has been correctly assigned. In macromolecular crystal structures the quality of the relative phases are of highest importance, but the interpretability of the electron-density map must also be assessed. When the relative phases are poorly determined some regions of the molecule may be interpreted in an erroneous way. Such misinterpretations have mainly occurred in loop regions where the protein molecule is often decidedly floppy. Remember that, as shown in Figure 9.10 (Chapter 9), it is not possible to see individual atoms at, for example, 2.5 Å resolution, and the electron-density map is only as good as the relative phases that went into it. In essence the problem is analogous to that found when looking at an object through a low-power microscope; sometimes it is difficult to distinguish the various features of the object from each other. Atoms in the more flexible portions of protein molecules have higher displacement parameters[20] (B values), as will be described in Chapter 13.

11.3 Representations of structural results

Various methods can be used to display the results of a crystal structure determination. These include pictures of ball-and-stick models, space-filling models, and simple or sophisticated diagrams produced by computer graphics methods.[21-24] Stereoviews of molecules or of the arrangement of molecules in a crystal provide a simple and practical way of illustrating three-dimensional molecules. Some of these methods are illustrated throughout this book. The theoretical chemist may take issue with such representations of molecules, since atoms are better thought of as point nuclei surrounded by clouds of electron density defined by probability functions. For the metric description that follows, however, we generally use ball-and-stick models of molecules or crystal structures. They are simple, informative and easily comprehended, although incomplete because vibrational and other types of motion generally are ignored in these representations.

The reader should now appreciate that the use of fractional atomic coordinates to arrive at molecular geometry is straightforward. This is evident on inspection of crystal structure determinations in the scientific literature. The effects of thermal motion and slight disorder are illustrated by use of the computer program ORTEP.[24] The amplitudes of anisotropic vibration are indicated by an ellipsoid for each atom, and this ellipsoid is cut open to reveal the three principal axes. A typical ORTEP diagram[25] is shown in Figure 11.9. The appearance of an ORTEP diagram gives further information to a seasoned crystallographer on the precision or even the correctness of a crystal structure determination. Many other computer programs for a variety of graphics devices have been written in order to produce drawings of crystallographic results.[22,26-28] Very sophisticated diagrams, in color, with shading and highlighting can now be obtained.[29,30] Stereodiagrams can be readily viewed, even without a stereoviewer, if the centers of the two pictures, one for the left eye and the other for the right eye, are the correct distance apart (45–55 mm).

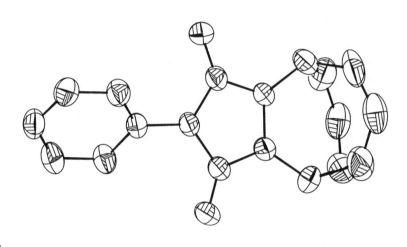

(a)

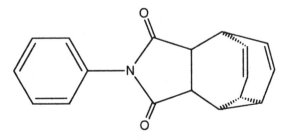

(b)

FIGURE 11.9. An ORTEP diagram of an organic molecule (Ref. 25). (a) The thermal-ellipsoid plot, and (b) the chemical formula. The shaded lines in (a) are drawn parallel to the principal axes and highlight the cutaway portion of the ellipsoid.

11.4 What can be learned from a crystal structure?

To assess what can be learned from a crystal structure determination the results of a few selected structure analyses will now be analyzed in terms of the molecular geometry and packing in the crystal.

11.4.1 Tin

The crystal structures of the chemical elements provide a simple starting point for illustrating what can be learned from listed atomic coordinates. Tin has both a nonmetallic and a metallic **allotrope** (polymorph).[31,32] The familiar stable allotrope at room temperature and above (18° to 212°) is metallic "white tin" (β-tin). It is tetragonal, space group $I4_1/amd$ with unit cell dimensions $a = b = 5.832$ Å, and $c = 3.182$ Å. Because of the symmetry of this space group, the e.s.d. values can only be reported for the unit-cell dimensions since the four tin atoms are fixed in position by symmetry and lie in the unit cell [Figure 11.10(a) (right)] at $(0,0,0; \frac{1}{2}, \frac{1}{2}, \frac{1}{2}) \pm 0, \frac{1}{4}, \frac{3}{8}$ with the origin of the unit cell on a center of symmetry. Note that $(0, \frac{1}{4}, \frac{3}{8},$ and $\frac{1}{2}$ are each exact numbers with no e.s.d. values).

In crystals of white tin the tin atoms are located in the unit cell at:

x	y	z
0.000	0.250	0.375
0.000	0.750	0.625
0.500	0.750	0.875
0.500	0.250	0.125

Each tin atom has six nearest neighbors. Four tin atoms lie 3.023 Å from one selected tin atom, and two tin atoms lie 3.182 Å (one unit cell dimension c) from it.

White tin transforms very slowly at temperatures below about 18°C (and down to –130°C) to grey tin (α-tin), which is nonmetallic.[33] This allotrope has a cubic structure, space group $Fd3m$, with eight atoms at $(0, 0, 0; \frac{1}{2}, \frac{1}{2}, 0; \frac{1}{2}, 0, \frac{1}{2}; 0, \frac{1}{2}, \frac{1}{2}) \pm \frac{1}{8}, \frac{1}{8}, \frac{1}{8}$ [Figure 11.10(a) (left)]. The unit cell dimension is $a = b = c = 6.489$ Å. Each tin atom in grey tin is surrounded by four others at distances of $\frac{1}{4}, \frac{1}{4}.\frac{1}{4}$ away, equal to:

$$(6.489/4) \times \sqrt{3} = 2.810 \, \text{Å}$$

Thus, both the coordination number and the Sn $\cdots$ Sn distances are decreased in the low-temperature form which is isostructural with diamond.

Tin has been used by mankind for centuries, but the transformation of metallic white tin to nonmetallic grey tin at low temperatures has had some devastating consequences. While sometimes the transition from crystalline β to α (white to grey) gives a crystal of the latter,[34] generally a powder is obtained. It has been reported that the tin buttons on the

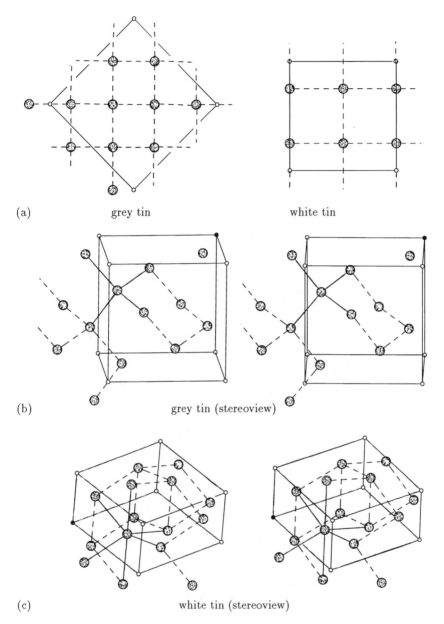

FIGURE 11.10. The two allotropes (polymorphs) of tin. Grey tin is the low-temperature form. (a) View down a unit cell axis of grey tin (left) and white tin (right). (b) Stereoview of grey tin and (c) stereoview of white tin. In (b) and (c) the surroundings of one tin atom are indicated by solid lines. Other Sn···Sn distances are shown by broken lines. Small circles indicate the corners of the unit cell.

uniforms of Napoleon's troops crumbled in the freezing Russian winter of 1812 as a result of such a conversion, and that this transformation to the nonmetallic allotrope contributed to their defeat.[35]

11.4.2 Boron hydrides

Boron, like carbon, can form a large number of compounds with hydrogen, referred to as the "boron hydrides." They were first prepared and their properties described by the German chemist Alfred Stock.[36] The simplest of this group of compounds is diborane, B_2H_6. Its formula appears to parallel that of ethane, C_2H_6, a saturated hydrocarbon formed by tetravalent carbon. Boron, however, contains one fewer electron than does carbon, and therefore there are not sufficient electrons to form a structure analogous to that of ethane with two electrons in each bond. The correct arrangement of atoms in diborane, predicted by Walther Dilthey[37] in 1921, was established by W. Charles Price[38,39] in 1947 from its infrared spectrum. Two of the hydrogen atoms form a bridge between the two boron atoms, as shown in Figure 11.11.

The three-dimensional structures of the higher hydrides of boron, established by X-ray crystallographic techniques, were unexpected. In 1948 John Kasper, Charlys Lucht, and David Harker[40,41] published the structure of decaborane, $B_{10}H_{14}$ (Figure 11.12) by a method newly derived by them (Harker–Kasper inequalities, see Chapter 8). This boron hydride has a basket-like arrangement of atoms, completely different from any structure predicted for a boron hydride at the time. The relationship between the structures of the boron hydrides and that of the various polymorphs of boron[42-48] (which exists as an icosahedral arrangement of boron atoms) is of interest since the geometry of each boron hydride can be thought of as being structurally related to that of a portion of an icosahedron.

William N. Lipscomb[49-55] and co-workers studied the structures of many other boron hydrides and their derivatives (Figure 11.13). These small neutral compounds are volatile and unstable. They often explode when in contact with air and have to be studied at low temperatures (well below $-100°C$). The problems of maintaining the low temperatures necessary to preserve crystals and of solving the structures have been

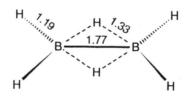

FIGURE 11.11. Chemical structure of diborane (Refs. 38 and 39).

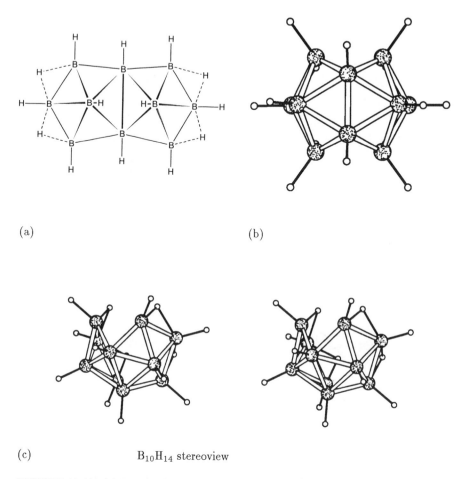

(a)

(b)

(c) $B_{10}H_{14}$ stereoview

FIGURE 11.12. (a) Formula and connectivity of $B_{10}H_{14}$, (b) view down a symmetry axis, and (c) a stereoview of the molecule.

vividly detailed in the descriptions of the structure determinations by Lipscomb and co-workers. From these it became possible to predict the three-dimensional structures of other boron hydrides, and to provide a structural basis for their reactivities.

In order to explain the hydrogen bridges in diborane and their apparent electron deficiencies, the concept of **three-center bonding,** in which one pair of electrons links three atoms, was introduced by Hugh C. Longuet-Higgins and Ronald P. Bell.[56,57] Lipscomb and co-workers[52] extended the concept to multicenter bonds in general.

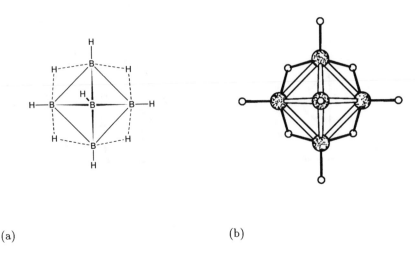

(a)

(b)

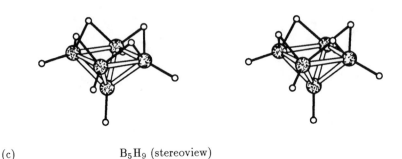

(c) B_5H_9 (stereoview)

FIGURE 11.13. (a) Formula and connectivity of B_5H_9. (b) View of B_5H_9 down a symmetry axis, and (c) a stereoview of the molecule.

11.4.3 Buckminsterfullerene

Diamond and graphite, two allotropes of carbon, were described in Chapter 1 (Figures 1.7 and 1.8). A new allotrope of carbon, C_{60}, with a geodesic dome-like structure, was reported[58,59] in 1985. This compound, with no hydrogen atoms, is called "buckminsterfullerene" in honor of Richard Buckminster Fuller, designer of the geodesic dome.[60] In this structure, which is like that of a soccer ball, there are 32 faces (12 five-sided and 20 six-sided) and 60 vertices (at which the carbon atom is found). Buckminsterfullerene is obtained from graphite at very high temperatures, such as those in an electric arc or from a laser beam, and also is present in soot. This molecule forms a close-packed cubic structure, unit cell dimensions $a = 14.17$ Å. The space group is $Pa\bar{3}$ and the molecules are disordered with respect to orientation.[61] Therefore it was necessary to break the spherical symmetry to obtain an ordered crystal structure,[62–64] and this was done by preparing a platinum derivative[65] $[(C_6H_5)_3P]_2Pt(\eta^2-C_{60})-(C_4H_8O)$, shown in Figure 11.14. Metal ions can occupy interstitial space between unsubstituted buckminsterfullerene molecules, and the structures of M_6C_{60} have been determined for M = potassium, rubidium or cesium. The cubic unit cells have dimensions $a = 11.39$, 11.59 and 11.79 Å, respectively, demonstrating the different sizes of the metal cations.[64] The crystal structures are body-centered cubic.

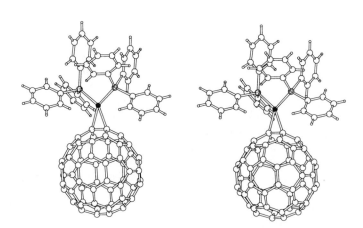

(a)

(b)   Pt derivative of buckminsterfullerene (stereoview)

FIGURE 11.14. A buckminsterfullerene derivative $[(C_6H_5)_3P]_2Pt(\eta^2-C_{60})\cdot(C_4H_8O)$. (a) Chemical formula (Ph = phenyl), and (b) a stereoview of the molecule. Pt black, P stippled, H small circles. (Coordinates courtesy J. C. Calabrese.)

11.4.4 Copper sulfate pentahydrate

The crystal structure of copper sulfate pentahydrate ($CuSO_4 \cdot 5H_2O$) (Figure 11.15), was determined in 1934 by C. Arnold Beevers and Henry S. Lipson.[8] The hydrogen atoms were located later by neutron diffraction.[9] Crystals of the compound are easily prepared by evaporation of water from a copper sulfate solution; they are a beautiful blue color and decompose to a pale blue powder upon heating. The sulfate anion is tetrahedral. The importance of hydrogen bonds in the overall stability of the crystal structure is evident. The copper is surrounded by six oxygen atoms in an approximately octahedral arrangement; four of these are water oxygen atoms, while the remaining two are oxygen atoms from sulfate anions. The fifth water molecule is held in a tetrahedral arrangement with four hydrogen bonds, to two water molecules (attached to the Cu^{2+}) and to two oxygen atoms from the sulfate anion (those not coordinated to the copper cation). Because the fifth water molecule is held by hydrogen bonds, which are weaker than coordination bonds to copper, this water molecule can be easily lost on heating. As a result, the crystal structure is destroyed and the blue crystals are reduced to a powder of composition $CuSO_4 \cdot 4H_2O$. As shown in Figure 11.16, the coordination bonds to the copper ion are not all equal in length, the result of Jahn–Teller distortion[66,67] caused by the d^9 valence structure of Cu^{2+}. This effect is also seen in the crystal structure of the trihydrate.[68]

11.4.5 Bond lengths in aromatic rings

The stronger a bond, the shorter its length.[69] For example, in diamond the C–C (sp^3–sp^3) bond is 1.54 Å in length, while an isolated, nonconjugated C=C (sp^2–sp^2) double bond is 1.34 Å in length (Table 11.3). In aromatic compounds, such as benzene, each C∹C bond is 1.38–1.40 Å, while a C≡C triple bond is 1.20 Å. Correlations of C–C bond lengths with their strengths have involved considerations of bond order[10] or π-bond order.[70] These are two quite different quantities. The **bond order** is the number of bonds used to describe a bond, 1.0 for a single bond, 2.0 for a double bond, and 3.0 for a triple bond. For intermediate bonds, a weighted average of the bond orders in all possible resonance hybrids is calculated. By this method the bond order of all C∹C bonds in benzene is 1.50. A correlation between bond character and C–C distance is shown in Figure 11.17(a).

The **π-bond order**, on the other hand, is derived from molecular orbital theory, and is defined as:

(number of electrons in bonding orbitals minus number in antibonding orbitals)/2

The π bond order is zero for a single bond, 0.667 for C∹C in benzene, 1.00 for a double bond, and 2.00 for a triple bond. This measure of bond order is derived from the weights of the atomic orbitals in each of the

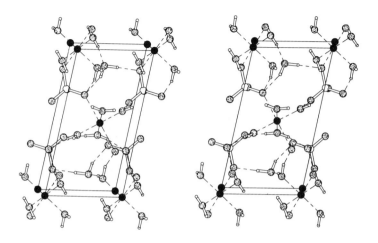

(a) $CuSO_4 \cdot 5H_2O$ (stereoview)

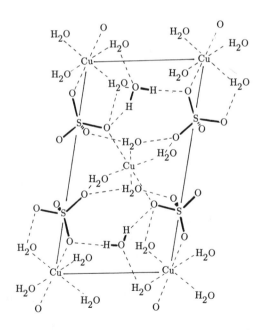

(b)

FIGURE 11.15. (a) Stereoview of the unit cell contents of $CuSO_4 \cdot 5H_2O$ (Refs. 8 and 9), and (b) identification of atoms on (a). The fifth water molecule is drawn in (b) with thick bonds, while the other four are simply labelled H_2O.

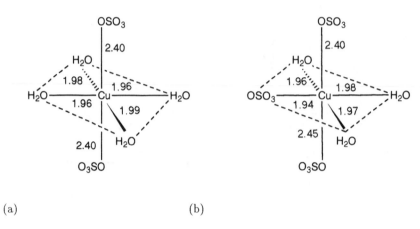

(a)   (b)

FIGURE 11.16. Jahn-Teller distortions in the bond lengths of oxygen to Cu(II) in (a) $CuSO_4 \cdot 5H_2O$. and (b) $CuSO_4 \cdot 3H_2O$. Note that one *trans* pair of bonds (drawn vertically) has longer bond lengths than do the four other Cu $\cdots$ O bonds.

molecular orbitals. The resulting plot of π-bond order versus C–C bond distance, shown in Figure 11.17(b), is approximately a straight line. Such π bond orders are useful for deriving free valences or unsaturation indices f_r for an atom r. The π-bonding power is the difference between the sum of the orders of bonds attached to that atom r and the maximum π-bond order that can be achieved. The equation is:

$$\text{free valence} = \text{maximum } \pi \text{ bond number} - \sum(\pi \text{ bond orders}). \quad (11.5)$$

(Values for the maximum π bond number are given as $\sqrt{3}$ for carbon, $\sqrt{2}$ for nitrogen, and 1 for oxygen; see Ref. 70). The unused π-bonding power obtained by use of Equation 11.5 makes atoms with high values of f_r very susceptible to attack, particularly by free radicals.[71]

The length of each C–C bond in a polycyclic aromatic hydrocarbon is near that expected from the various **resonance hybrids** that can be drawn, and the bond orders that can be calculated.[72] The main nonionized contributors for benzene[12] are available from neutron studies of crystalline deuterobenzene at 15 K (–258° C) and for crystalline naphthalene at 92 K (–181° C)[73] and are shown in Figure 11.18. The **delocalization of electrons** in benzene is shown by its equal bond lengths while, in naphthalene there is also evidence of some localization of double bonds.

An example of one of the problems that can result when interpreting X-ray diffraction results comes from "bond-stretch isomerism." The complex *mer*-$MoCl_2(PMe_2Ph)_3$ was observed to exist as blue

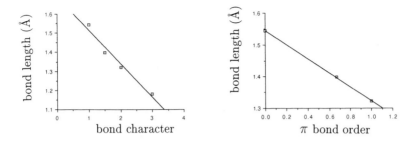

FIGURE 11.17. Plots of C–C bond lengths in Å versus (a) bond character and (b) π-bond order. Values in this plot are 1.54 Å (single bond), 1.40 Å (aromatic bond), 1.34 Å (double bond), and 1.18 Å (triple bond). From these graphs the approximate bond character or π-bond order may be obtained for a measured C–C bond length.

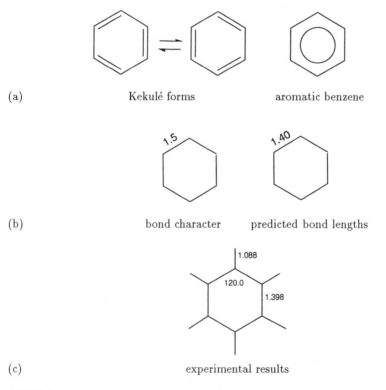

FIGURE 11.18. Resonance hybrids and bond lengths in aromatic compounds. For benzene, (a) Kekulé forms and a common mode of depicting the aromatic character of benzene, (b) bond character and predicted bond lengths, and (c) experimental bond lengths and angles.

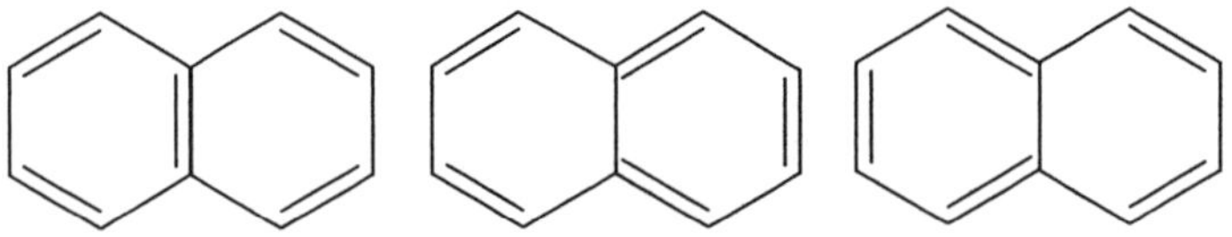

(d)   Kekulé forms

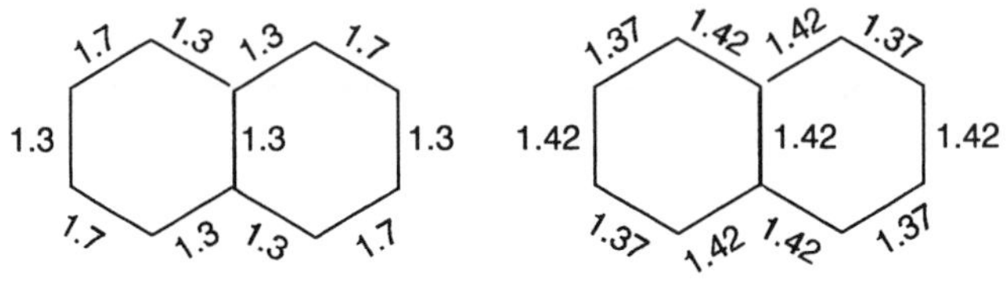

(e)   bond character   predicted bond lengths

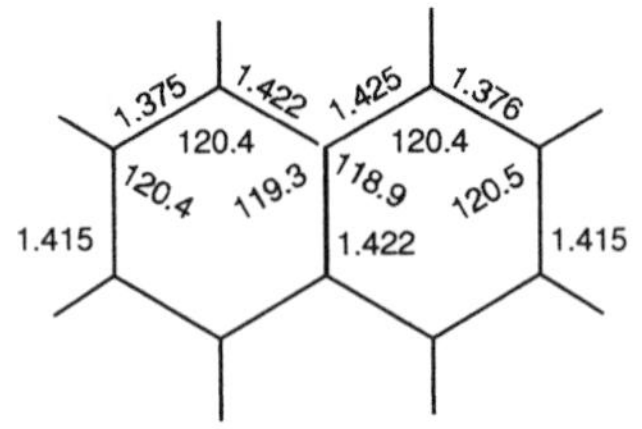

(f)   experimental results

FIGURE 11.18 (cont'd). Resonance hybrids of naphthalene. (d) Kekulé forms, (e) bond character and predicted bond lengths, and (f) experimental bond lengths and angles.

and green isomers.[74,75] X-ray diffraction studies, showed that the complexes *mer*-$MoCl_2(PMe_2Ph)_3$ and *mer*-$MoCl_2(PEt_2PH)_3$ (where Me = methyl, Et = ethyl, and Ph = phenyl) had the same coordination environment around the molybdenum atom but that the Mo=O bond lengths, normally 1.65–1.70 Å, differed significantly [1.803(11) Å in green *cis–mer*-$MoCl_2(PEt_2Ph)_3$ and 1.676(7) Å in blue *cis–mer*-$MoCl_2(PMe_2Ph)_3$]. The phenomenon was termed "bond-stretch" or "distortional isomerism." Eventually it was shown[75] by careful studies of difference electron-density maps that the crystals each contained two components, one with Mo=O (1.68 Å) and the other with Mo–Cl (2.45 Å). The X-ray experiment had given an average over all molecules in the crystal. Other examples of geometrical anomalies as a result of the formation of cocrystals have been reported.

11.4.6 Variation in bond angles

Electronic and steric effects can cause changes in interbond angles as well as in bond lengths. An example is provided by X-ray diffraction studies of derivatives of benzene. When one of the hydrogen atoms of benzene is replaced by a functional group X, the benzene ring is distorted from perfect D_{6h} symmetry (a perfect hexagon). The amount of this distortion, diagrammed in Figure 11.19, depends on the electronic properties of the substituent, X. In benzene, the internal ring angle α is 120° at the site of substitution, but ranges in values from 114° to 125° in monosubstituted benzene derivatives.[76-80] The value of α increases with the electron-withdrawing character of X.[81-83] Although the ring remains planar, the vicinal angles change much more than do the remote angles.

11.4.7 Tautomerism

Tautomerism is an easy and rapid equilibrium between two isomeric forms of a molecule that differ in the manner by which an atom, such as a hydrogen atom, is bonded. The term is usually applied to molecules in which the hydrogen atom location is changed between the two tautomers, as in keto-enol tautomerism. In the crystalline state, the tautomeric form of a molecule can be found directly from the locations of hydrogen atoms and from the bond lengths that indicate which are single and which are double bonds. An example is provided by the studies of the crystal structures of the bases in DNA. It was Jerry Donohue who, in 1953, pointed out[84,85] how the pairs of bases in DNA could form hydrogen bonds to each other, and thereby corrected a misconception about the relevant tautomeric forms of these bases in DNA. This was an important component of the resulting structure proposed for DNA by Watson and Crick.[85,86]

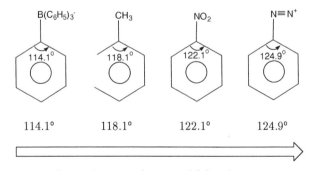

Approximate σ electron-withdrawing
power of substituent

FIGURE 11.19. Bond angle variations in monosubstituted benzene derivatives (Ref. 76). Note how the angle at the site of substitution varies with the electron-withdrawing power of the substituent.

Another example is provided by some 1-nitroacridines.[87] Amino-acridines, as shown in Figure 11.20, can exist with more than one tau-tomeric structure, that is, with hydrogen atoms located in different po-sitions. Thus, the amino tautomer may lose a hydrogen atom to some other functional group in the molecule, so that $-NH_2$ is replaced by $=NH$, resulting in the formation of an imino tautomer. Further, if an additional hydrogen atom is acquired to give a protonated acridine, the charge may, as shown in Figure 11.20, reside in the pyridine ring, or on the exocyclic nitrogen atom. This relationship between free base and monocation, il-lustrated in Figure 11.21, is termed amphoterism. Examples of these possibilities have been observed in the crystalline state.[87] It is found that all free bases that have 9-alkylamino substituents, but no 1-nitro groups (and therefore no steric problems) crystallize as amino tautomers. On

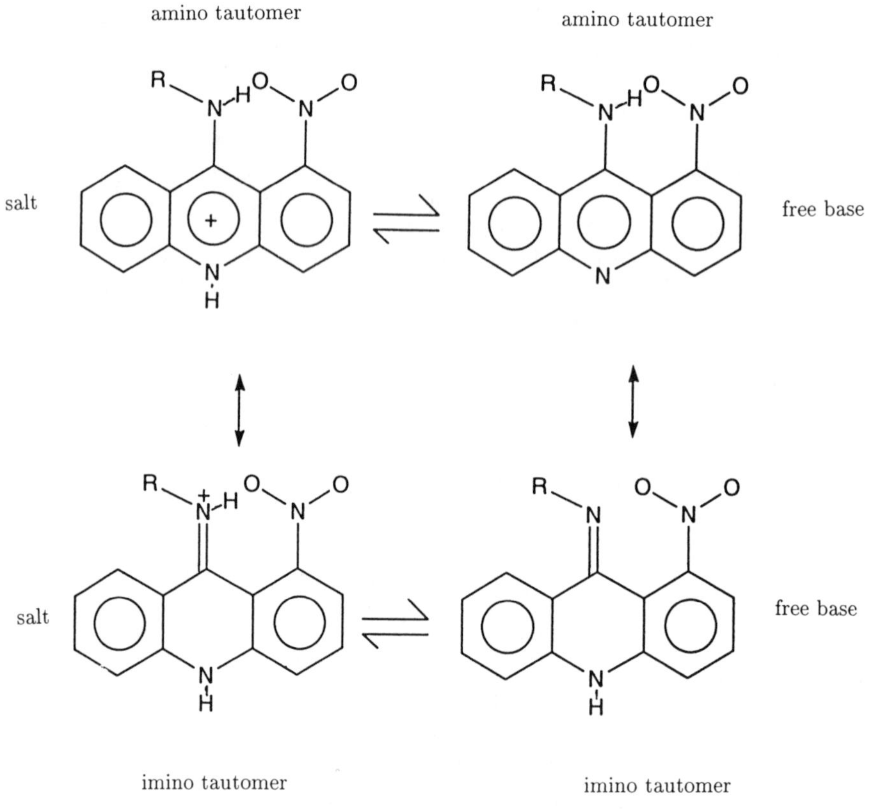

FIGURE 11.20. Tautomerism of 1-nitro-9-(alkylamino)acridines, revealed by X-ray studies (Ref. 87). Shown are chemical formulæ for both amino and imino tautomers of the nonprotonated and protonated forms. Crystal structures of all but salts of the imono tautomer have been reported (see Ref. 87).

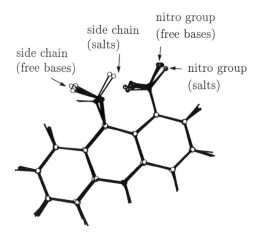

FIGURE 11.21. Various conformations of the side chains in 1-nitro-9-(alkylamino)acridines (Ref. 87). Note that the conformations are different for the free bases and for the salts.

the other hand, all free bases with both 9-alkylamino and 1-nitro substituents are found as imino tautomers. The alkyl group on N9 is essential for the imino tautomer to be observed. Furthermore, the various tautomeric and amphoteric states lead to different side-chain arrangements (conformations) in the 1-nitroacridines, as shown in Figure 11.21.

Summary

1. The data obtained from an X-ray crystal structure determination are the unit cell dimensions, the space group, the **atomic coordinates**, the atomic displacement parameters (to be described) and the atomic **occupancy factors** (which are generally unity).
2. Atomic connectivities and **bond lengths** can be calculated by use of Equations 11.1 and 11.2.
3. **Bond angles** (interbond angles) can be calculated by use of Equation 11.4.
4. Estimated standard deviations for atomic coordinates are derived from the least-squares refinement, as described in Chapter 10. These can be applied to the bond lengths and interbond angles.

Glossary

Allotrope: A polymorph of a chemical element.

Atomic coordinates: A set of numbers that defines the position of an atom with respect to a specified coordinate system. Atomic coordinates are generally expressed as the dimensionless quantities x, y, and z which are fractions of unit cell edges.

Atomic parameters: A set of numbers that specifies the position of an atom in the unit cell (atomic coordinates), the extent of its displacement about an equilibrium position (vibration), and an occupancy factor (generally 1.0). Three parameters define position, one parameter can be used to define isotropic displacements, or six to define anisotropic displacements and one parameter defines the site occupancy.

Bond length: The distance between two atomic nuclei that are joined together. This definition is generally applied to bonds that are primarily covalent.

Bond order: A theoretical index of the degree of bonding between two atoms relative to that for a normal single bond, that is, one localized electron pair. In valence bond theory it is expressed by the weighted average of the bond numbers between the respective atoms in the contributing structures. By this criterion each $C \cdots C$ bond in benzene has a bond order of 1.5.

Cartesian coordinate: The three-dimensional location of a point X, Y, and Z (in Å) measured parallel to three orthogonal (mutually perpendicular) axes with equal units along the axes. (Named after Descartes, the mathematician and philosopher.)

Delocalization of electrons: The π bonding in a conjugated system is not considered to consist of localized bonding, but to have the electrons delocalized over the entire system of double bonds. Each bond is assigned a "fractional double-bond character" or bond order. The result of this delocalization is some stabilization, expressed as "delocalization energy," the difference in energy between the delocalized system and a hypothetical one in which the chemical bonds are completely localized.

Fractional coordinates: Coordinates of atoms expressed as fractions of the unit cell lengths (see Atomic coordinates).

Orthogonal coordinates: (see **Cartesian coordinates**):

π-bond order: This measure of bond order is derived from the weights of the atomic orbitals in each of the occupied molecular orbitals. It is zero for a single bond, 0.667 for $C \cdots C$ in benzene, 1.00 for a double bond and 2.00 for a triple bond.

Resonance hybrids: Various possible chemical structures of molecules, each with identical atomic connectivity, but differing in the disposition of electrons. The wave function of the molecule is approximately represented by "mixing" the wave functions of the contributing structures. The energy calculated for such a mixture is lower, presumably because the representation is more nearly correct than it would be if formally represented by a single structure.

Tautomerism: Isomerism in which isomers, called tautomers, are readily interconvertible. Keto–enol tautomerism is an example

$$O = C - C - H \quad \leftrightarrow \quad H - O - C = C.$$

Three-center bonding: Bonding in which electrons occupy orbitals that encompass three or more atomic centers, as in diborane.

References

1. Bragg, W. H., and Bragg, W. L. The structure of the diamond. *Nature (London)* **91**, 557 (1913).

2. Allen, F. H., Kennard, O., and Taylor, R. Systematic analysis of structural data as a research technique in organic chemistry. *Acc. Chem. Res.* **16**, 146–153 (1983).

3. Allen, F. H., Kennard, O., Watson, D. G., Brammer, L., Orpen, A. G., and Taylor, R. Tables of bond lengths determined by X-ray and neutron diffraction. Part 1. Bond lengths in organic compounds. *J. Chem. Soc., Perkin Trans. II*, S1–S19 (1987).

4. Allen, F. H., Bellard, S., Brice, M. D., Cartwright, B. A., Doubleday, A., Higgs, H., Hummelink, T., Hummelink-Peters, B. G., Kennard, O., Motherwell, W. D. S., Rodgers, J. R., and Watson, D. G. The Cambridge Crystallographic Data Centre: computer-based search, retrieval, analysis and display of information. *Acta Cryst.* **B35**, 2331–2339 (1979).

5. Bergerhoff, G., Hundt, R., Sievers, R., and Brown, I. D. The Inorganic Crystal Structure Database. *J. Chem. Inf.* **23**, 66–69 (1983).

6. Bragg, W. L. The structure of some crystals as indicated by their diffraction of X rays. *Proc. Roy. Soc. (London)* **A89**, 248–277 (1913).

7. Broch, E., Oftedal, I., and Pabst, A. New determinations of the lattice constants of potassium fluoride, cesium chloride and barium fluoride. *Z. Physik. Chem.* **B3**, 209–214 (1929).

8. Beevers, C. A., and Lipson, H. Crystal structure of $CuSO_4$. $5H_2O$. *Proc. Roy. Soc. (London)* **A146**, 570–582 (1934).

9. Bacon, G. E., and Curry, N. A. The water molecules in $CuSO_4 \cdot 5H_2O$. *Proc. Roy. Soc. (London)* **A266**, 95–108 (1962).

10. Pauling, L. *The Nature of the Chemical Bond.* (3rd edn., 1973.) Cornell University Press: Ithaca, NY (1948).

11. Cox, E. G., and Smith, J. A. S. Crystal structure of benzene at $-3°C$. *Nature (London)* **173**, 75 (1954).

12. Jeffrey, G. A., Ruble, J. R., McMullan, R. K., and Pople, J. A. The crystal structure of deuterated benzene. *Proc. Roy. Soc. (London)* **A414**, 47–57 (1987).

13. Bondi, A. Van der Waals volumes and radii. *J. Phys. Chem.* **68**, 441–451 (1964).

14. Bonham, R. A., Bartell, L. S. and Kohl, D. A. The molecular structures of *n*-pentane, *n*-hexane and *n*-heptane. *J. Chem. Phys.* **81**, 4765–4769 (1959).

15. Gillespie, R. J. and Nyholm, R. S. Inorganic stereochemistry. *Quart. Rev. (Chem. Soc., London)* **11**, 339–380 (1957).

16. Herzberg, G. *Infrared and Raman Spectra of Polyatomic Molecules.* Van Nostrand: New York (1945).

17. Schwarzenbach, D., Abrahams, S. C., Flack, H. D., Gonschorek, W., Hahn, T., Huml, K., Marsh, R. E., Prince, E., Robertson, B. E., Rollett, J. S. and Wilson, A. J. C. Statistical descriptors in crystallography. Report of the International Union of Crystallography Subcommittee on Statistical Descriptors. *Acta Cryst.* **A45**, 63–75 (1989).

18. Cruickshank, D. W. J. Fourier synthesis and structure factors. In: *International Tables for X-ray Crystallography. Volume II. Mathematical Tables.* (**Eds.**, Kasper. J. S., and Lonsdale, K.) Section 6. pp. 317–340. Kynoch Press: Birmongham (1959).

19. Rossi, M., Glusker, J. P., Randaccio, L., Summers, M. F., Toscano, P. J., and Marzilli, L. G. The structure of a B_{12} coenzyme: methylcobalamin studies by X-ray and NMR methods. *J. Amer. Chem. Soc.* **107**, 1729–1738 (1985).

20. Sternberg, M. J. E., Grace, D. E. P., and Phillips, D. C. Dynamic information from protein crystallography. An analysis of temperature factors from refinement of the hen egg-white lysozyme structure. *J. Molec. Biol.* **130**, 231–253 (1979).

21. Motherwell, W. D. S., and Clegg, W. *PLUTO78. A program for drawing crystal and molecular structures.* Cambridge, UK (1978).

22. Carrell, H. L. Computer program VIEW. The Institute for Cancer Research, The Fox Chase Cancer Center, Philadelphia, PA (1976).

23. Smith, G. M., and Gund, P. Computer-generated space-filling molecular models. *J. Chem. Inf. Comput. Sci.* **18**, 207–210 (1978).

24. Johnson, C. K. ORTEP: A FORTRAN Thermal-Ellipsoid Plot Program for Crystal Structure Illustration. Report ORNL-3794. Oak Ridge National Laboratory: Oak Ridge, TN (1965).

25. Krow, G. R., Lee, Y. B., Szczepanski, S. W., Zacharias, D. E., and Bailey, D. B. Structural studies of 4-phenyl-2,4,6-triazatetracyclo[$6.3.2.0^{2,6}.0^{7,9}$]trideca-10,12-diene-3,5-dione, a dihydrodiazabullvalene. Preference for a bisected aminocyclopropane derivative having an unusual planar triazolidinedione ring. *J. Amer. Chem. Soc.* **109**, 5744–5749 (1987).

26. Keller, E. Some computer drawings of molecular and solid-state structures. *J. Appl. Cryst.* **22** 19–22 (1989).

27. Olson, A. J., and Goodsell, D. S. Visualizing biological molecules. *Scientific American* **267(5)**, 76–81 (1992).

28. Goodsell, D. S. A look inside the living cell. *American Scientist.* **80**, 457–465 (1992).

29. Dickerson, R. E., and Kopka, M. L. Rational design of DNA-binding drugs or How to read a helix. In: *Patterson and Pattersons. Fifty Years of the Patterson Function.* (**Eds.**, Glusker, J. P., Patterson, B. K. and Rossi, M.) Ch. 12, pp. 252–278. International Union of Crystallography/Oxford University Press: New York (1987).

30. Bash, P. A., Pattabiraman, N., Huang, C., Ferrin, T. E. and Langridge, R. Van der Waals surfaces in molecular modelling: implementation with real-time computer graphics. *Science* **222**, 1325–1327 (1983).

31. Donohue, J. *The Structure of the Elements.* Wiley: New York, 1974.

32. Deshpande, V. T., and Sirdeshmukh, D. B. Thermal expansion of tin in the β–γ transition region. *Acta Cryst.* **15**, 294–295 (1962).

33. Thewlis, J., and Davey, A. R. Thermal expansion of grey tin. *Nature (London)* **174**, 1011 (1954).

34. Kuo, K. and Burgers, W. G. An X-ray investigation of the white to gray transformation of tin. *Proc. Koninklijke Nederlandse Akademie van Wetenschappen* [*Proc. Roy. Acad.*] (*Amsterdam*) **B59**, 288–297 (1956).

35. Harris, J. Scourge of tin soldiers in cold climates. *New Scientist* **57**, 1525 (1986).

36. Stock, A. *Hydrides of Boron and Silicon.* Cornell University Press: Ithaca NY (1933).

37. Dilthey, W. Über die Konstitution des Wass7rs. [Concerning the structure of water.] *Z. Angew. Chemie* **34**, 596 (1921). [This really does have the correct structure of diborane drawn in it. JPG]

38. Price, W. C. The structure of diborane. *J. Chem. Phys.* **15**, 614 (1947).

39. Price, W. C. The absorption spectrum of diborane. *J. Chem. Phys.* **16**, 894–902 (1948).

40. Kasper, J. S., Lucht, C. M., and Harker, D. The crystal structure of decaborane, $B_{10}H_{14}$. *Acta Cryst.* **3**, 436–455 (1950).

41. Kasper, J. S., Lucht, C. M., and Harker, D. The structure of the decaborane molecule. *J. Amer. Chem. Soc.* **70**, 881 (1948).

42. Sullenger, D. B., and Kennard, C. H. L. Boron crystals. *Sci. Amer.* **215**, 96–107 (1966).

43. Hoard, J. L., Sullenger, D. B., Kennard, C. H. L., and Hughes, R. E. The structural analysis of β-rhombohedral boron. *J. Solid State Chem.* **1**, 268–277 (1970).

44. Hoard, J. L., Geller, S., and Hughes, R. E. On the structure of elementary boron. *J. Amer. Chem. Soc.* **73**, 1892–1893 (1951).

45. Johnson, R. D., Meijer, G., and Bethune, D. S. C_{60} has icosahedral symmetry. *J. Amer. Chem. Soc.* **112**, 8983–8984 (1990).

46. Hoard, J. L., Hughes, R. E., and Sands, D. E. The structure of tetragonal boron. *J. Amer. Chem. Soc.* **80**, 4507–4515 (1958).

47. Decker, B. F., and Kasper, J. S. The crystal structure of a simple rhombohedral form of boron. *Acta Cryst.* **12**, 503–506 (1959).

48. Sands, D. E., and Hoard, J. L. Rhombohedral elemental boron. *J. Amer. Chem. Soc.* **79**, 5582–5883 (1957).

49. Dulmage, W. J. and Lipscomb, W. N. The crystal and molecular structure of pentaborane. *Acta Cryst.* **5**, 260–264 (1952).

50. Hirshfeld, F. L., Eriks, K., Dickerson, R. E., Lippert, E. L. Jr. and Lipscomb, W. N. Molecular and crystal structure of B_6H_{10}. *J. Chem. Phys.* **28**, 56–61 (1958).

51. Lipscomb, W. N. Structures of the boron hydrides. *J. Chem. Phys.* **22**, 985–988 (1954).

52. Eberhardt, W. H., Crawford, B., Jr., and Lipscomb, W. N. The valence structure of the boron hydrides. *J. Chem. Phys.* **22**, 989–1001 (1954).

53. Lipscomb, W. N. Crystallographic aspects of an autobiography. In: *Crystallography in North America.* (**Eds.** McLachlan, D. Jr. and Glusker, J. P.) Chapter 18. pp. 97–101. American Crystallographic Association: New York. (1983).

54. Dickerson, R. E., Wheatley, P. J., Howell, P. A., Lipscomb, W. N., and Schaeffer, R. Boron arrangements in a B_9 hydride. *J. Chem. Phys.* **25**, 606–607 (1956).

55. Dickerson, R. E., Wheatley, P. J., Howell, P. A., and Lipscomb. W. N. Crystal and molecular structure of B_9H_{15}. *J. Chem. Phys.* **27**, 200–209 (1957).

56. Longuet-Higgins, H. C., and Bell, R. P. The structure of the B hydrides. *J. Chem. Soc.* 250–255 (1943).

57. Longuet-Higgins, H. C. The structure of some electron-deficient molecules. *J. Chem. Soc.* 139–143 (1946).

58. Kroto, H. W., Heath, J. R., O'Brien, S. C., Curl, R. F., and Smalley, R. E. C_{60}: buckminsterfullerene. *Nature (London)* **318**, 162–163 (1985).

59. Krätschmer, W., Lamb, L. D., Fostiropoulos, K., and Huffman, D. R. Solid C_{60}: a new form of carbon. *Nature (London)* **347**, 354–358 (1990).

60. Marks, R. W. (1960) *The Dymaxion World of Buckminster Fuller.* Reinhold: New York (1960).

61. Fischer, J. E., Heiney, P. A., and Smith, A, B. III. Solid-state chemistry of fullerene-based materials. *Acc. Chem. Res.* **25**, 112–118 (1992).

62. Hawkins, J. M., Meyer, A., Lewis, T. A., Loren, S., and Hollander, F. J. Crystal structure of osmylated C_{60}: conformation of the soccer ball framework. *Science* **252**, 312–313 (1991).

63. Hawkins, J. M. Osmylation of C_{60}: proof and characterization of the soccer-ball framework. *Acc. Chem. Res.* **25**, 150–156 (1992).

64. Fagan, P. J., Calabrese, J. C., and Malone, B. Metal complexes of buckminster-fullerene (C_{60}). *Acc. Chem. Res.* **25**, 134–142 (1992).

65. Fagan, P. J., Calabrese, J. C., and Malone, B. The chemical nature of buck-minsterfullerene (C_{60}) and the characterization of a platinum derivative. *Science* **252**, 1160–1161 (1991).

66. Jahn, H. A., and Teller, E. Stability of polyatomic molecules in degenerate electronic states. I. Orbital degeneracy. *Proc. Roy. Soc. (London)* **A161**, 220–235 (1937).

67. Orgel, L. E. *An Introduction to Transition-state Chemistry.* 2nd edn. Methuen: New York (1966).

68. Zahrobsky, R. F., and Baur, W. H. On the crystal chemistry of salt hydrates. V. The determination of the crystal structure of $CuSO_4\cdot3H_2O$ (bonattite). *Acta Cryst.* **B24**, 508–513 (1968).

69. Pauling, L. The dependence of bond energy on bond length. *J. Phys. Chem.* **58**, 662–666 (1954).

70. Pullman, B., and Pullman, A. *Quantum Biochemistry.* Interscience: New York, London (1963).

71. Kier, L. B. *Molecular Orbital Theory in Drug Research.* Academic Press: New York, London (1971).

72. Robertson, J. M. The measurement of bond lengths in conjugated molecules of carbon centres. *Proc. Roy. Soc. (London)* **A207**, 101–110 (1951).

73. Brock, C. P., and Dunitz, J. D. Temperature dependence of thermal motion in crystalline naphthalene. *Acta Cryst.* **B38**, 2218–2228 (1982).

74. Butcher, A. V., and Chatt, J. Complexes of tertiary phosphines and tertiary amines with molybdenum(IV). *J. Chem. Soc., A* 2562–2656 (1970).

75. Parker, G. Do bond-stretch isomers really exist? *Acc. Chem. Res.* **25**, 455–460 (1992).

76. Domenicano, A., Vaciago, A., and Coulson, C. A. Molecular geometry of substituted benzene derivatives. I. On the nature of ring deformations induced by substitution. *Acta Cryst.* **B31**, 221–234 (1975).

77. Domenicano, A., Vaciago, A., and Coulson, C. A. Molecular geometry of substituted benzene derivatives. II. A bond angle *versus* electronegativity correlation for the phenyl derivatives of second-row elements. *Acta Cryst.* **B31**, 1630–1641 (1975).

78. Domenicano, A., Mazzeo, P., and Vaciago, A. Substituent effects in the benzene series: a structural approach. *Tetrahedron Letters* 1029–1032 (1976).

79. Domenicano, A., and Hargittai, I. (**Eds.**) *Accurate Molecular Structures. Their Determination and Importance.* Oxford University Press: Oxford (1992).

80. Domenicano, A., and Hargittai, I. (**Eds.**) *Accurate Molecular Structures. Their Determination and Importance.* International Union of Crystallography/ Oxford University Press: Oxford (1992).

81. Hammett, L. P. *Physical Organic Chemistry.* McGraw-Hill: New York (1940).

82. Ehrenson, S., Brownlee, R. T. C., and Taft, R. W. Generalized treatment of substituent effects in the benzene series. A statistical analysis by the dual substituent parameter equation (1). *Progr. Phys. Org. Chem* **10**, 1–80 (1973).

83. Johnson, C. D. *The Hammett Equation.* Cambridge University Press: Cambridge (1973).

84. Judson, H. F. *The Eighth Day of Creation. The Makers of the Revolution in Biology.* Simon & Schuster: New York (1979).

85. Watson, J. D., and Crick, F. H. C. A structure for deoxyribose nucleic acid. *Nature (London)* **171**, 737–738 (1953).

86. Franklin, R. E., and Gosling, R. G. Molecular configuration in sodium thymonucleate. *Nature (London)* **171**, 740–741 (1953).

87. Stezowski, J. J., Kollat, P., Bogucka-Ledóchowska, M., and Glusker, J. P. Tautomerism and steric effects in 1-nitro-9-(alkylamino)acridines (ledakrin or nitracrine analogues): probing structure-activity relationships at the molecular level. *J. Amer. Chem. Soc.* **107**, 2067–2077 (1985).

CHAPTER

12

Conformation

What is the best way to describe the overall shape of a molecule? Typically, in a journal article describing a crystal structure, there are lists of bond distances, interbond angles, and torsion angles, which describe the three-dimensional structure. Of these parameters, the **torsion angles** are the most used in a description of the overall shape of a molecule.[1-3] In this Chapter we will describe some basic principles that have resulted from measurements of torsion angles. We aim to show how X-ray crystallographic results provide clear illustrations of molecular shapes.

Conformation is the word that is generally applied to the three-dimensional structure or shape of a molecule. Different conformations are structures differing as a result of rotation about single bonds. The word conformation should not be confused with "**configuration**." The conformation of a molecule can be changed without having to break any covalent bonds whereas, in order to change the configuration of a molecule, covalent bonds must be broken and remade. Configuration (as in "absolute configuration") is discussed in Chapter 14. We will concentrate in the present Chapter on conformation.

A molecule that contains several single bonds may be altered in shape by rotation about these single bonds. Such rotations occur frequently in the liquid or gaseous state, where intermolecular forces that keep the molecule fairly rigid in the crystal have been lost. Different rotational positions about the various bonds are represented by different torsion angles; these demonstrate the inherent flexibility of the molecule. Such rotations about bonds are evident in the different shapes of the same molecule in different crystal structures, or by disorder in the crystal.

A relatively stable molecular conformation corresponds to a minimum in the graph of potential energy versus those molecular parameters representing that conformation. There may be many of these minima, each of which corresponds to different combinations of rotations about the single bonds (different conformational isomers or conformers). The barriers to rotation give a measure of the **conformational flexibility** of the molecule. Provided the energy barrier between the various **conformers** is not too high, these different conformers may rapidly interconvert in solution or in the gas phase at room temperature. In the crystalline state, on the other hand, certain conformations of the molecules under study are "frozen out," and there may not be enough space in the crystal structure for any bond rotation to take place. In addition, the conformation found for a molecule in the crystalline state represents a compromise between the requirement for *the* lowest-energy conformation of the free molecule and the requirement for the best overall packing, with optimized intermolecular interactions, of this molecule in the crystal. As a result the conformation measured in the crystalline state may not be *the* lowest energy conformer, but it certainly is *a* low-energy conformer.

Generally, as we shall see in this Chapter, there is only one conformation of a molecule in any one crystal structure. One of the most common questions asked by solution chemists about the results of a crystal structure analysis is: how can one be sure that the solid-state conformation is the same as that observed in solution? The conformation found for a flexible molecule in the crystalline state is that of one of the various conformers found in solution. This has been verified by other physical methods such as nuclear magnetic resonance.[4-6] If, however, a molecule is found to have the same conformation in several different crystal structures, it is reasonable to assume that this conformation has a low (although not necessarily the lowest) energy. This assumption can often be tested by calculation (by *ab initio* molecular orbital calculations, for example) of the appropriate theoretical potential energy curve.

12.1 Torsion angles: What are they?

The concept of a torsion angle was introduced to describe steric relationships across single bonds and for describing ring geometry.[7,8] If one looks directly along a bond of interest in a ball-and-stick model of the molecule, the two atoms of the bond become apparently superposed as a point. Groups attached to the two atoms of the bond stick out radially, as shown in Figure 12.1. A graphical description of this is familiar to organic chemists as a **Newman projection**,[9,10] in which the atom nearer the viewer is designated by radii spaced at 120° angles (Figure 12.1). The atom further from the viewer is designated by a circle with equally spaced radial extensions. The bonds to the nearer atom are drawn so that they penetrate the circle, while bonds to the further atom do not. As a result, a Newman projection clearly differentiates between substituents on the near atom and those on the far atom. X-ray diffraction results can give precise values for the angles between these radii.

By contrast, a **Fischer projection** is drawn with the atom under consideration at the center of a cross. The upper and lower substituents, drawn with vertical bonds, lie below the plane of the paper [*A* and *D* in

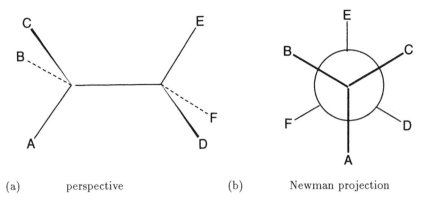

(a)   perspective   (b)   Newman projection

FIGURE 12.1. Ways of depicting molecular conformation. (a) Perspective view and (b) Newman projection.

Figure 12.2(a)] and the two substituents above this plane are drawn with horizontal bonds [B and C in Figure 12.2(a)]. These Fischer projections do not give information on the conformation, only on configuration,[11] as is evident in Figure 12.3.

An alternative way of analyzing molecular shape, besides drawing a Newman projection, is to analyze the amount of twist about the bond that is necessary to cause some or all of the bonds from the near (upper) and far (lower) atoms to be totally **eclipsed**. This leads us to a definition of torsion angle.

12.1.1 Calculation of planes through groups of atoms

Two substituted sp^2 carbon atoms, connected by a double bond, normally form a planar unit. Groups lying on the same side of the double bond are described as **cis** and those on opposite sides of the double bond as **trans** (Figure 12.4). The **best plane** through such a group of atoms[12-14] can readily be calculated, by use of least-squares methods, as described in Figure 12.5 (compare with Figure 10.2, Chapter 10).

12.1.2 Mathematical definition of a torsion angle

The mathematical way of describing conformation is by the calculation of the torsion angles about each bond; the input data are the atomic coordinates. A torsion is the extent of the twist of substituents about a bond. It is a way of looking along a bond and describing the disposition of the substituents attached to the atoms at both ends. It will indicate if the bonds to the substituent groups are eclipsed, **staggered**, or intermediate between these.

The method used to measure a torsion angle is to measure how much a bond has to be twisted to cause two substituents on it to be eclipsed

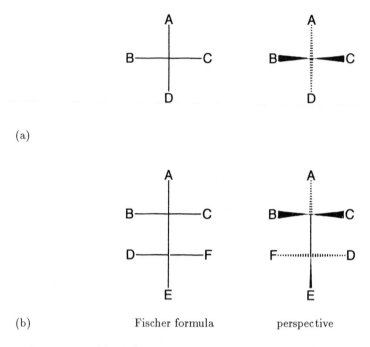

(a)

(b)   Fischer formula   perspective

FIGURE 12.2. (a) Fischer projection and perspective of a molecule containing only one asymmetric carbon atom. (b) Fischer projection and perspective of a molecule containing two bonded asymmetric carbon atoms.

(Figure 12.6). By definition, a positive torsion angle is the clockwise twist about the central of three connected bonds that is needed to make the near (upper) bond ($A-B$ in a series of four-bonded atoms $A-B-C-D$) eclipse the far (lower) bond ($C-D$); a negative torsion angle is a counterclockwise twist. It does not matter which way you look along the bond [that is, which is nearer to you ($A-B$) or ($D-C$)]; the value and sign of the torsion angle stay the same. Convince yourself with a ball-and-stick model that reversing the sequence of view from $A-B-C-D$ to $D-C-B-A$ does not change the twist necessary to make $A-B$ and $C-D$ lie on top of each other, as illustrated in Figure 12.6. There is no overall change in the shape of the molecule as you turn it over to look at it from the other end. On the other hand, when there are two enantiomers, the torsion angles of the mirror images are reversed in sign since the mirror operation also pertains to the torsion angles. This is also shown in Figure 12.7. Newman projections[9,10] illustrate this enantiomerism in an idealized way.

The calculation of torsion angles and their e.s.d. values (from the results of the refinement of parameters) is relatively simple (see Fig-

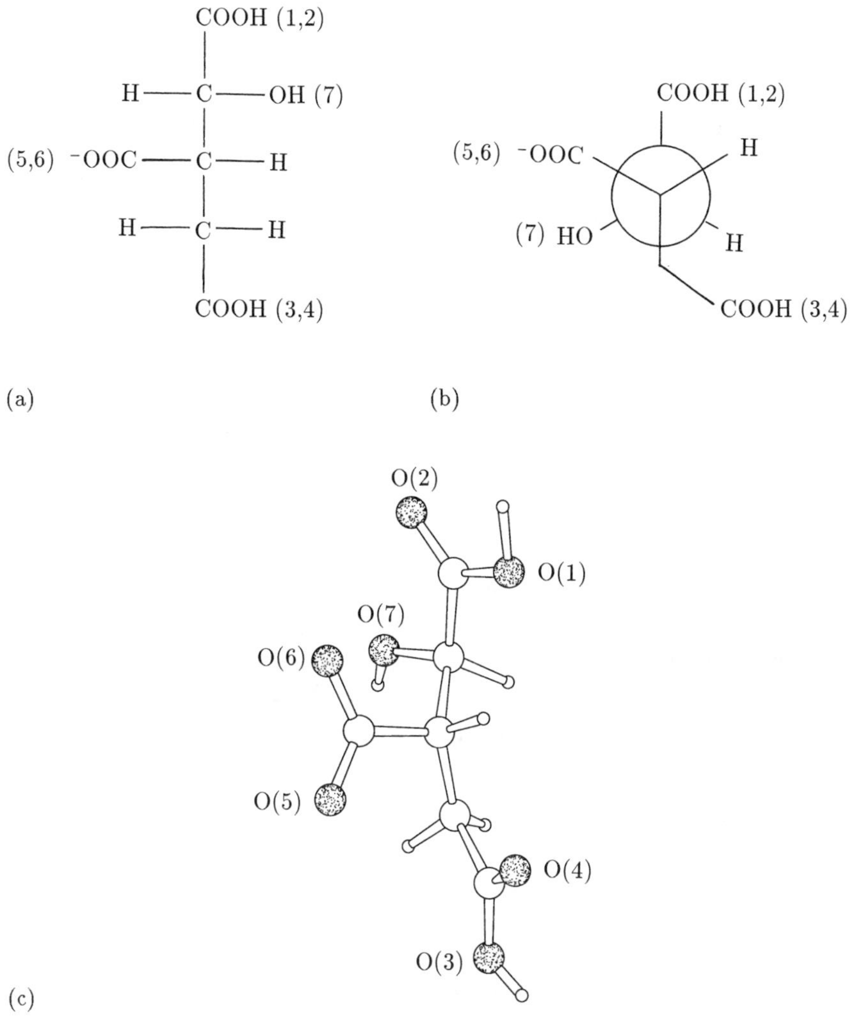

FIGURE 12.3. (a) Fischer and (b) Newman projections of the isocitrate ion in potassium isocitrate (Ref. 11), and (c) the appearance of the ion in the crystal structure. Note that the Fischer projection does not give an easily interpretable view of the molecule. For example, O(7) and the carboxyl group containing O(5) and O(6) are on the same side of the molecule in the Newman projection and in the crystal structure, but on opposite sides in the Fischer projection.

ure 12.8).[15,16] Their significance, however, may not spring out when the chemist or biochemist looks at lists of torsion angles. They need to be viewed in Newman-type projections and, when analyzed in this way, are among the interesting results that the X-ray crystallographer obtains.

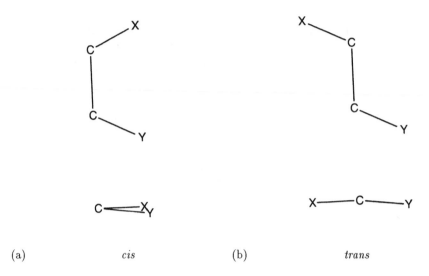

FIGURE 12.4. A planar group X–C–C–Y showing (a) *cis* and (b) *trans* isomers. The upper diagram is a view onto the plane of the four atoms, and the lower diagram is a view along the plane.

Calculation of a least-squares plane.

The least-squares best plane through a series of points expressed in Cartesian (orthogonal) coordinates relative to their center of mass is obtained by minimizing:

$$\sum_i w_i (lx_i + my_i + nz_i)^2$$

with respect to l, m, and n, subject to the condition that

$$l^2 + m^2 + n^2 = 1$$

(because they are direction cosines). w are the weights of the observations.

$$\delta = l^2 x^2 + m^2 y^2 + n^2 z^2 + 2lmxy + 2lnxz + 2mnyz$$

$$\frac{d\delta}{dl} = 2lx^2 + 2mxy + 2nxz = 0$$

$$\frac{d\delta}{dm} = 2my^2 + 2lxy + 2nyz = 0$$

$$\frac{d\delta}{dn} = 2nz^2 + 2lxz + 2myz = 0$$

This gives (if weights as set equal to 1.0 for simplification):

$$l \sum X^2 + m \sum XY + n \sum XZ = 0$$

$$l \sum XY + m \sum Y^2 + n \sum YZ = 0$$

$$l \sum XZ + m \sum YZ + n \sum Z^2 = 0$$

A principal axis transformation is carried out on the matrix:

$$\begin{pmatrix} \sum X^2 - \lambda & \sum XY & \sum XZ \\ \sum YX & \sum Y^2 - \lambda & YZ \\ \sum ZX & \sum ZY & \sum Z^2 - \lambda \end{pmatrix} = \begin{pmatrix} A^{11} - \lambda & A^{12} & A^{13} \\ A^{21} & A^{22} - \lambda & A^{23} \\ A^{31} & A^{32} & A^{33} - \lambda \end{pmatrix}$$

where coordinates have been transformed to an origin at the center of mass. This calculation is described in detail by Verner Schomaker, Jürg Waser, Richard E. Marsh and Gunnar Bergman (Ref. 12) with numerical assistance for the reader by David M. Blow (Ref. 13).
The equation for λ is

$$-\lambda^3 + \alpha\lambda^2 + \beta\lambda + \gamma$$

where (see Ref. 15):

$$\alpha = A^{11} + A^{22} + A^{33} \tag{12.5.1a}$$

$$\beta = (A^{12})^2 + (A^{13})^2 + (A^{23})^2 - [A^{11}A^{22} + A^{11}A^{33} + A^{22}A^{33}] \tag{12.5.1b}$$

$$\gamma = A^{11}A^{22}A^{33} + 2A^{12}A^{13}A^{23} - [A^{11}A^{23}A^{23} + A^{22}A^{13}A^{13}$$
$$+ A^{33}A^{12}A^{12}] \tag{12.5.1c}$$

Coordinates of four atoms with respect to their center of mass are:

X	Y	Z
−0.207	−1.887	+1.288
−0.553	−0.432	−0.564
−0.141	+0.961	−0.854
+0.900	+1.358	+0.131

All weights are set here to 1.00 for simplification.

$$\sum X^2 = 1.1785 \qquad \sum XY = 1.7162$$
$$\sum Y^2 = 6.5151 \qquad \sum XZ = 0.2836$$
$$\sum Z^2 = 2.7235 \qquad \sum YZ = -2.8297$$

Thus, from Equation 12.5.1, $\alpha = 10.4171$, $\beta = -17.5986$, $\gamma = 0.1746$

$$-\lambda^3 + 10.4171\lambda^2 - 17.5986\lambda + 0.1746 = 0 \tag{12.5.2}$$

There are three roots to this Equation 12.5.2. *The smallest is* $\lambda = 0.01$.

$$\begin{vmatrix} A^{22} - \lambda & A^{23} \\ A^{12} & A^{13} \end{vmatrix} = \begin{vmatrix} 6.51 & -2.83 \\ 1.72 & 0.28 \end{vmatrix}$$

$$\begin{vmatrix} A^{11} - \lambda & A^{13} \\ A^{12} & A^{23} \end{vmatrix} = \begin{vmatrix} 1.17 & 0.28 \\ 1.72 & -2.83 \end{vmatrix}$$

$$\begin{vmatrix} A^{12} & A^{11} - \lambda \\ A^{22} - \lambda & A^{12} \end{vmatrix} = \begin{vmatrix} 1.72 & 1.17 \\ 6.51 & 1.72 \end{vmatrix}$$

These three determinants are equal to 6.6904, −3.7927, *and* −4.6583, *respectively. Each must be divided by the square root of the sum of their squares to give* $l = 0.744$, $m = -0.422$, *and* $n = -0.517$. *Thus the equation of the plane is*

$$0.744X - 0.422Y - 0.517Z = 0.000$$

Deviations of the four atoms from this plane are, respectively, −0.025 Å, +0.062 Å, −0.068 Å, +0.030 Å.

FIGURE 12.5. Calculation of a least squares plane.

Another example is provided by a set of four connected carbon atoms in a chain, C–C–C–C. If the atoms are in a planar zigzag arrangement (staggered conformation) the C–C–C–C torsion angle is 180°. If this torsion angle is 60°, one end atom is twisted 60° out of the plane of the other three, as shown in Figure 12.9. Torsion angles measured from an X-ray crystal structure determination[11] are shown in Figure 12.10. Note the alternation of signs to the values of the torsion angles.

12.1.3 Definition of dihedral angles

If the orientation of two or more portions of a molecule with respect to each other are needed, one method used for this is to calculate the **dihedral angle** between two planes defined by appropriate sets of atoms. The dihedral angle between two planes is defined[17] as the angle between the normals to these planes. If a dihedral angle is calculated for the two planes in a group of four connected atoms (Figure 12.11) it is found that the torsion and dihedral angles add up to 180° (i.e., they are supplementary). There is considerable misuse in the literature of the term "dihedral angle" (used where the term "torsion angle" is intended). It is better just to list torsion angles for a description of the molecule.

12.1.4 Relative energies of different conformations

Flexible molecules with single bonds can readily interconvert into different conformers, provided there is sufficient energy. It is possible to calculate conformational energies and thus predict preferred conformations by

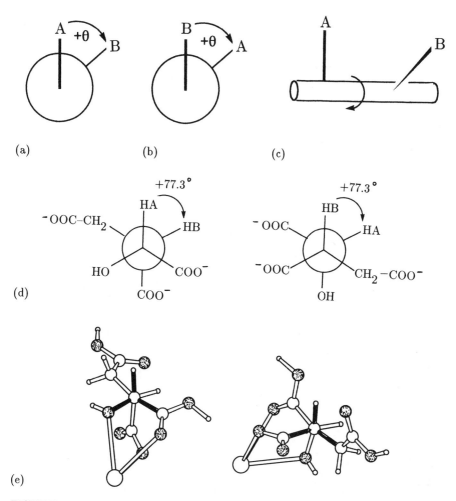

(a)   (b)   (c)

(d)

(e)

FIGURE 12.6. Definition of a torsion angle. In a molecule containing the string of atoms A–X–Y–B and viewed from X to Y (unlabelled), the central X–Y bond is twisted until A eclipses B. The positive sense is a clockwise rotation. (b) Reversal of the direction in which the molecule is viewed gives the same sign and magnitude to the torsion angle. (c) General view of the twisting operation. (d) and (e) The effect of turning the molecule over and viewing it from the other end. (d) The H–C–C–H torsion angle remains +77.3°. (e) The three-dimensional structure viewed in the same direction with the left-hand and right-hand diagrams (not a stereopair) showing the torsion angle viewed from the two ends of the bond and corresponding to the Newman projections in (a). Upper bonds (nearer the reader) are filled. The three upper bonds in the torsion angle are filled. A metal cation that coordinates to an α hydroxycarboxylate group is drawn with a larger circle.

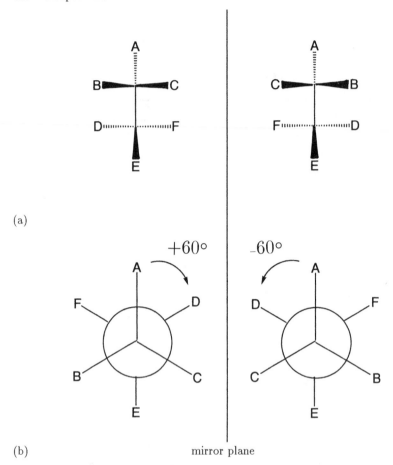

(a)

(b)   mirror plane

FIGURE 12.7. Fischer and Newman projections of enantiomers (mirror images of each other). One enantiomer is on the left and the other is on the right, divided by a vertical line (a mirror perpendicular to the plane of the page). (a) Fischer diagram, and (b) Newman projection.

theoretical methods. As with other processes that have energy barriers, rotations about bonds in flexible molecules are temperature-dependent; they are slower at low temperatures and faster at higher temperatures.

It is useful to keep in mind that different kinds of distortions in bonding geometry affect the internal energy (potential energy) of the molecule differently. As a rule of thumb, the energy necessary to distort bond distances is greater than that required to distort bond angles, and even less energy is necessary to change torsion angles.

One of the simplest organic molecules showing a variety of conformations is ethane. The free rotation of such a molecule (CH_3–CH_3) about

Calculation of a torsion angle.

For four linked atoms A–B–C–D, the torsion angle is the angle between A–B and C–D when viewed in projection down B–C [(Figure 12.6(a)].

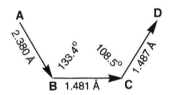

The atomic coordinates are:

Atom	x	y	z	Atom	X	Y	Z
A	0.000	0.000	1.000	A	−1.793	0.422	5.696
B	−0.093	0.152	0.675	B	−2.139	1.877	3.844
C	0.011	0.287	0.624	C	−1.727	3.270	3.554
D	0.243	0.318	0.797	D	−0.686	3.667	4.539

The torsion angle A–B–C–D may be determined in several ways:

(*a*) *The equation for calculating the torsion angle* ω *is given* (*see* Ref. 15) *by*

$$\cos \omega = (\mathbf{AB} \times \mathbf{BC}) \cdot (\mathbf{BC} \times \mathbf{CD}) \Big/ \left(AB \cdot BC \cdot BC \cdot CD \sin \theta_{\mathrm{ABC}} \sin \theta_{\mathrm{BCD}} \right) \quad (12.8.1)$$

where $\mathbf{AB}$ *is the vector between A and B, etc. Follow the computer program listed at the end of Jack Dunitz' book* (Ref. 15).

Interatomic distances are: A–B $= d_1 = 2.380$ Å, *B–C* $= d_2 = 1.481$ Å, *and C–D* $= d_3 = 1.487$ Å.

For vectors $\mathbf{R}$ *and* $\mathbf{S}$ (**AB**, **BC**, *and* **CD** *in* Equation 12.8.1, *for example*)

$$\mathbf{R}(x_1, y_1, z_1) \cdot \mathbf{S}(x_2, y_2, z_2) = x_1 x_2 + y_1 y_2 + z_1 z_2$$

$$\mathbf{R}(x_1, y_1, z_1) \times \mathbf{S}(x_2, y_2, z_2) = \begin{pmatrix} \mathbf{i}_1 & \mathbf{i}_2 & \mathbf{i}_3 \\ x_1 & y_1 & z_1 \\ x_2 & y_2 & z_2 \end{pmatrix}.$$

where $\mathbf{i}_1, \mathbf{i}_2, \mathbf{i}_3$ *are Cartesian basis vectors* (*that is* 1, 0, 0 *for* **u**, 0, 1, 0 *for* **v**, *and* 0, 0, 1 *for* **w**).

$(\mathbf{AB} \times \mathbf{BC})/AB{\cdot}BC = u_1,\ v_1,\ w_1,$ and $(\mathbf{BC} \times \mathbf{CD})/BC{\cdot}CD = u_2,\ v_2,\ w_2.$
Designating $\Delta x_{21} = x_2 - x_1/d_1$ to obtain $\mathbf{AB}/AB$, etc.

	Δx	Δy	Δz
Δ_{21}/d_1	-0.145	$+0.611$	-0.778
Δ_{32}/d_2	$+0.278$	$+0.941$	-0.196
Δ_{43}/d_3	$+0.700$	$+0.267$	$+0.662$

$$u_1 = \Delta y_{21}\Delta z_{32} - \Delta z_{21}\Delta y_{32} = +0.612$$
$$u_2 = \Delta z_{21}\Delta x_{32} - \Delta x_{21}\Delta z_{32} = -0.245$$
$$u_3 = \Delta x_{21}\Delta y_{32} - \Delta y_{21}\Delta x_{32} = -0.306$$
$$v_1 = \Delta y_{32}\Delta z_{43} - \Delta z_{32}\Delta y_{43} = +0.675$$
$$v_2 = \Delta z_{32}\Delta x_{43} - \Delta x_{32}\Delta z_{43} = -0.321$$
$$v_3 = \Delta x_{32}\Delta y_{43} - \Delta y_{32}\Delta x_{43} = -0.584$$
$$c_3 = (u_1v_1 + u_2v_2 + u_3v_3)/(s_1 \cdot s_2) = +0.672/s_4s_5$$
$$s_3 = (\Delta x_{21}v_1 + \Delta y_{21}v_2 + \Delta z_{21}v_3)/(s_1 \cdot s_2) = +0.161/s_4s_5$$

The answer is that

$$\omega = \tan^{-1}(s_3/c_3) = \tan^{-1}0.240 = 13.5° \tag{12.8.2}$$

(b) *The torsion angle is the angle between the two planes through A, B, C, and B, C, D, respectively. The equation of the plane through A, B, and C is:*

$$0.841X - 0.337Y - 0.422Z + 4.057 = 0,$$

and the equation of the plane through B, C, and D, is:

$$0.712X - 0.339Y - 0.615Z + 4.524 = 0.$$

The angle between the two planes

$$l_1 X + m_1 Y + n_1 Z + d_1 = 0 \tag{12.8.3a}$$
$$l_2 X + m_2 Y + n_2 Z + d_2 = 0 \tag{12.8.3b}$$

is $\cos^{-1}(l_1l_2 + m_1m_2 + n_1n_2)$. *Substituting in these* Equations (12.8.3) *it is found that* $(l_1l_2 + m_1m_2 + n_1n_2) = 0.97256 = \cos 13.5°.$

(c) *The torsion angle is the angle required to rotate A–B onto C–D viewing directly down B–C. The coordinates are expressed with respect to C at the origin. The coordinates are rotated so that the first two coordinates (X'' and (Y'') for B are also zero. Then the angle can be found by application of Equation 11.4 (Chapter 11).*

Atom	X	Y	Z
A	−0.066	−2.848	+2.142
B	−0.412	−1.393	+0.290
C	+0.000	+0.000	+0.000
D	+1.041	+0.397	+0.985

Rotation of axes with an equation chosen to make $Y' = 0$:

$$X' = X \sin \theta + Y \cos \theta = 0.2836X + 0.9589Y$$
$$Y' = -X \cos \theta + Y \sin \theta = -0.9589X + 0.2836Y$$

X'	Y'	Z'
−2.749	−0.744	+2.142
−1.452	+0.000	+0.290
+0.000	+0.000	+0.000
+0.676	−0.886	+0.985

Another rotation of axes with an equation chosen to make $X'' = 0$:

$$X'' = 0.1985X' + 0.9806Y'$$
$$Y'' = +0.9806X' - 0.1958Y'$$

X''	Y''	Z''
+1.562	−0.744	−3.116
+0.000	+0.000	−1.481
+0.000	+0.000	+0.000
+1.098	−0.886	+0.470

The resulting structure is compressed so that all $Z'' = 0.000$ and the angle between projections of A–B and C–D are calculated (see Equation 11.4, Chapter 11) as:

$$\cos \omega = \frac{r_{12}^2 + r_{34}^2 - r_{14}^2}{2r_{12}r_{34}} \tag{12.8.4}$$

$$\frac{(1.7304)^2 + (1.4108)^2 - (0.4849)^2}{2 \times 1.7304 \times 1.4108} = \frac{4.7496}{4.8825} = 0.9728 = \cos 13.4°$$

(d) *The torsion angle may be calculated from the distances.*

$$\cos \omega = P/Q \tag{12.8.5}$$

where

$$P = d_{12}^2(d_{23}^2 + d_{34}^2 - d_{24}^2) + d_{23}^2(-d_{23}^2 + d_{34}^2 + d_{24}^2)$$
$$+ d_{13}^2(d_{23}^2 - d_{34}^2 + d_{24}^2) - 2d_{23}^2 d_{14}^2$$
$$Q = \sqrt{[(d_{12} + d_{23} + d_{13})(d_{12} + d_{23} - d_{13})(d_{12} - d_{23} + d_{13})}$$
$$(-d_{12} + d_{23} + d_{13})(d_{23} + d_{34} + d_{24})(d_{23} + d_{34} - d_{24})$$
$$(d_{23} - d_{34} + d_{24})(-d_{23} + d_{34} + d_{24})$$

The interatomic distances in this example are: $d_{12} = 2.380$ Å, $d_{23} = 1.481$ Å, $d_{34} = 1.487$ Å, $d_{13} = 3.564$ Å, $d_{14} = 3.619$ Å, *and* $d_{24} = 2.408$ Å. *Application of* Equation 12.8.5 *gives* $P = 20.83$, $Q = 21.40$ *and* $\omega = 13.3°$.

FIGURE 12.8. Calculation of a torsion angle.

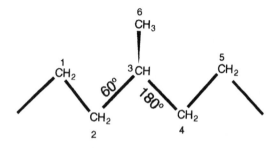

FIGURE 12.9. Torsion angles in a carbon chain. The torsion angle for C1–C2–C3–C6 is 60° with C6 out of the plane of the paper, while that for C1–C2–C3–C4 is 180° and all four atoms lie in the plane of the paper.

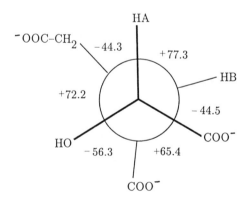

FIGURE 12.10. Torsion angles (in degrees) from X-ray diffraction data, drawn as for Newman projections (Ref. 11). Note the alternation of positive and negative signs. The two hydrogen atoms are named HA and HB to distinguish them.

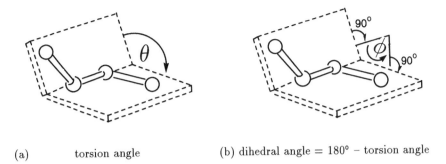

(a)   torsion angle   (b) dihedral angle = 180° – torsion angle

FIGURE 12.11. (a) Torsion angles versus (b) dihedral angles.

the C–C bond, and the sinusoidal potential energy barrier to such internal rotation, were postulated by Jacob D. Kemp and Kenneth S. Pitzer.[18] The energy profile of ethane [Figure 12.12(a)] has been well studied: there are three low-energy staggered forms of comparable energy, but separated by a rotational barrier of about 3 kcal/mol (approximately 13 kJ/mol). The rotation barrier for propane is similar (3.3 kcal/mol, approximately 14 kJ/mol).

The molecule *n*-butane C_4H_{10} has conformational flexibility,[19–23] as illustrated in the plots of potential energy versus the C–C–C–C torsion angle in Figure 12.12(b). This graph indicates that in the lowest energy form the methyl groups are completely staggered (180°, **antiperiplanar** or **trans**) while the highest-energy conformer has the methyl groups totally eclipsed (0°, **synperiplanar**).[7,24,25] There are also conformers in which the methyl groups are staggered (60° and 300°, low-energy conformations but less stable than the *antiperiplanar* (*trans*) conformation, **gauche** or **synclinal**) (see Figure 12.13). Note that the two *gauche* conformations (*gauche*+ and *gauche*−) are energetically equivalent and are actually mirror images of each other. These enantiomeric conformational isomers may interconvert simply by rotation about the central C–C bond. In addition there are higher-energy eclipsed (120° and 240°, **anticlinal**) conformations. All intermediate conformations are known as "skewed."

At room temperature, *n*-butane is a mixture of conformations; about 72% of the molecules are in the *antiperiplanar* conformation (see Figure 12.13). The energy difference between the more stable *anti* and the two enantiomeric *gauche* forms is 0.8 kcal/mol (approximately 3 kJ/mol); again the less stable forms are the eclipsed forms, of which there are two varieties. In one of these the two methyl groups are eclipsed, as are two pairs of hydrogen atoms. This is the highest-energy form of butane, about 4.5 kcal/mol (approximately 19 kJ/mol) above that of the *anti* conformation. In the second *anti* conformation the two methyl groups are eclipsed with hydrogen atoms, and two hydrogen atoms are eclipsed.

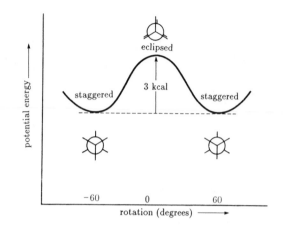

(a)

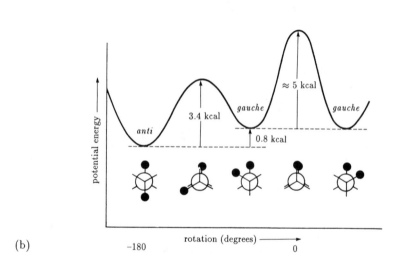

(b)

FIGURE 12.12. Potential energy versus conformation (rotation about the central C–C bond) for (a) ethane, CH_3–CH_3, and (b) n-butane, CH_3–CH_2–CH_2–CH_3.

This form is 3.4 kcal/mol higher (approximately 14 kJ/mol higher) in energy than the lowest-energy *anti* form.

12.1.5 Simple cyclic alkanes

The determination of structures of carbon-ring compounds of all sizes $[(CH_2)_n]$ has been tackled by X-ray crystallographers as well as other physical chemists. Cyclic alkanes with n greater than 5 must exist with some bonds in the *gauche* conformation (that is, with torsion angles of

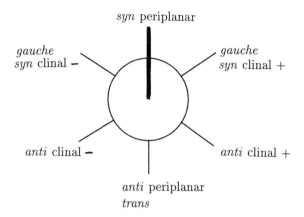

syn periplanar

gauche
syn clinal −

gauche
syn clinal +

anti clinal −

anti clinal +

anti periplanar
trans

FIGURE 12.13. Descriptions of various torsion angles.

± 60°); otherwise the two ends of the C–C chain would not join together. Small ring alkanes $(CH_2)_n$, with n less than six, exist in a rather
limited number of conformations because of the constraining geometric
factors. Larger rings have more degrees of freedom, and therefore can
exist in many conformations. The X-ray analyses of medium-sized ring
compounds[26] ($n = 6 – 12$) have, along with thermochemical data and
vibrational parameters, provided an experimental basis for a complete
conformational analysis of each structure.

Cyclobutane derivatives are usually, but not always, nonplanar. Cyclobutane itself exists as two butterflylike conformers that are easily interconvertible by inversion. Cyclopentane is a cyclic five-membered ring
structure that can exist as a set of half-chair (twist-boat) forms (C_2) and
a set of envelope (C_S) conformations. These are low-energy conformations and are readily interconverted by twists about bonds without any
bond angle changes, only changes in torsion angles. These interconversions are called **pseudorotations**.

Cyclohexane contains a six-membered cyclic arrangement of carbon
atoms. Each carbon atom has a tetrahedral arrangement of atoms surrounding it — two carbon atoms and two hydrogen atoms. Some possible conformations of saturated six-membered rings are shown in Figure
12.14.[27–29] The **chair** and **boat** forms of this compound were suggested
by Hermann Sachse.[30] Spectroscopic,[31,32] electron diffraction,[33] and X-ray
studies of hexachlorocyclohexane[34,35] and cyclohexane[36] showed that in
the chair form[37–40] all C–C bonds are staggered. Six of the C–H bonds are
parallel to the threefold axis of the molecule and are designated **axial**;[41]
the other six lie at tetrahedral angles to it and are designated **equatorial**
(Figure 1.10, Chapter 1).[42] In dimethylcyclohexanes the *equatorial* con-

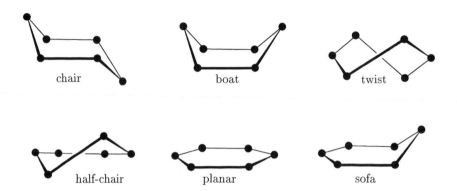

chair   boat   twist

half-chair   planar   sofa

FIGURE 12.14. Conformations of six-membered rings.

formation for the methyl group is preferred over the *axial* conformation by about 1.8 kcal/mole (approximately 8 kJ/mol).[42] The boat conformation of cyclohexane lies at an energy maximum because, in this conformation, two bonds are completely eclipsed. An energy-minimum species, although of higher energy than the chair form, is the twist-boat form. This can exist either in left- or right-handed forms. Conversions between the boat and twist-boat forms occur by pseudorotations. If the six-membered ring contains one double bond, bonds from atoms attached to that double bond will not lie far from the ring plane as they would have in cyclohexane, and are described as **pseudoaxial** and **pseudoequatorial** (see Figure 12.15).

Cycloheptane is found in two main groups of conformers: the chair form (including the half-chair) and the boat form (including the twist boat), with the chair form lower in energy by about 7 kcal/mol (ap-

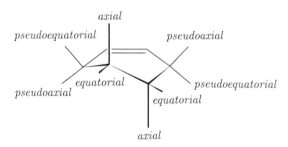

FIGURE 12.15. Dispositions of bonds in cyclohexene, a six-membered ring with one double bond.

proximately 29 kJ/mol). Once again, interconversions between chair and half-chair as well as boat and twist-boat are by pseudorotations. The lowest-energy conformer of cyclooctane is the boat–chair form. Multiring structures can be classified into four categories. The simplest are those with independent rings, connected either by a single bond or bridging atom. Such compounds, called *spiro* compounds, share only a single atom. When two rings are connected at adjacent atoms, they are called *fused ring* structures. Many molecules are found in this category. When rings are joined at nonadjacent atoms, the structures are called *bridged* rings. An excellent example of a fused carbon-ring system is provided by steroids,[43,44] which contain four fused rings: three cyclohexane rings (labeled A, B and C) and one cyclopentane ring (labeled D). Two of the ring junctions are always *trans* and the third either *trans* (cholestanes) or *cis* (coprostanes). The two faces of a steroid are designated as α and β (Figure 12.16). The atoms C18 and C19 point up on the β side of the steroid. This α, β- nomenclature is often also used for similarly shaped molecules, such as certain polycyclic aromatic hydrocarbons.

12.1.6 Steric strain

Adolph von Baeyer[45–47] was the first to note that the differences in ring stability of differently sized saturated hydrocarbon rings could be explained by the amount of **strain** caused by the deviation of the angles within the ring from the normally expected tetrahedral angle of 109°28′. This concept of strain, however, only pertains to small rings; larger rings containing six or more carbon atoms can take up nonplanar strainless conformations.

If a molecule is overcrowded with substituents, considerable steric strain may result. This strain can most efficiently by relieved by changes in torsion angles, since small displacements in bond angles, and bond lengths, require greater energy. Experimentally it is found that all three of these types of relief of strain may be operative. The effect of steric factors can be illustrated in the nonplanarity of the carcinogen 7,12-dimethylbenz[*a*]anthracene (Figure 12.17). This strain, a form of **steric hindrance**, is caused by the bulkiness of the methyl groups, one of which would "bump" into and interfere with a ring hydrogen atom if the molecule were flat.[48,49] In this strained hydrocarbon the overcrowding is relieved by increases in bond angles and by torsion about bonds in the region of overcrowding. All sp^2 carbon atoms in 7,12-dimethylbenz[*a*]anthracene have a planar arrangement of atoms around them, but torsion angles (expected to be zero degrees in a planar molecule) are found to be as high as 22°. Because chemists are accustomed to illustrations of molecules as two-dimensional diagrams, some are surprised to learn that methyl groups, or even hydrogen atoms, occupy so much space. According to Alexandr I. Kitaigorodsky,[50] the van der Waals radius for hydrogen is 1.12 Å, that for carbon 1.80 Å. Therefore

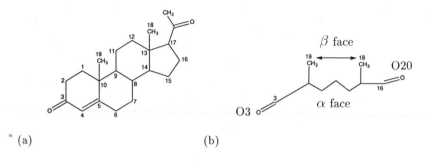

(a)   (b)

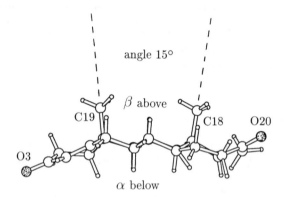

(c)

(d)

FIGURE 12.16. Designations of the two faces of a steroid (progesterone). (a) The chemical formula, and (b) a side view showing the α and β faces. (c) A side view of the molecule in the crystalline state, and (d) a view of the same from above with van der Waals radii on the hydrogen atoms on C18 and C19, to show how they block access to the β face. The β face is above the plane of the paper in (d) and the α face is below it.

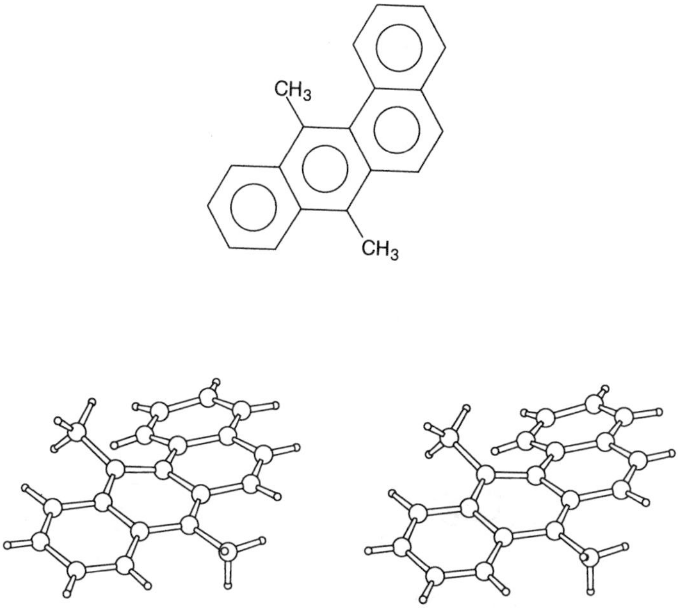

(a)

(b)   DMBA stereoview

FIGURE 12.17. (a) Chemical formula and (b) stereoview of 7,12-dimethylbenz[*a*]-anthracene (DMBA), showing the buckling of the molecule as a result of the presence of the 12-methyl group (Refs. 48 and 49).

two nonbonded hydrogen atoms cannot come closer to each other than 2.2–2.4 Å without considerable cost in energy.

Another molecule that is forced, for steric reasons, to be nonplanar is hexahelicene. This molecule, shown in Figure 12.18, is phenanthro-[3,4-*c*]phenanthrene.[51,52] It is composed of six benzene rings connected as if to a central hexagon, but the first and last added rings are not joined together. Since this arrangement of rings causes the molecule to be "overcrowded," it distorts into a helical shape in order to increase the distance between the first and last rings from a bonded value of 1.4 Å to a nonbonded value of 3.4 Å. This helicity results in optical isomers, one with a left-handed helix and the other with a right-handed helix. Hexahelicene was resolved into its two stereoisomers,[52] and it was shown, by X-ray techniques,[53,54] that the levorotatory isomer had, by chance, a left-handed helical configuration. This was done by preparing crystalline (−)-2-bromohexahelicene, studying the anomalous dispersion effect, and so determining its absolute configuration, converting the resolved bromo compound to hexahelicene, and determining the optical rotation of this

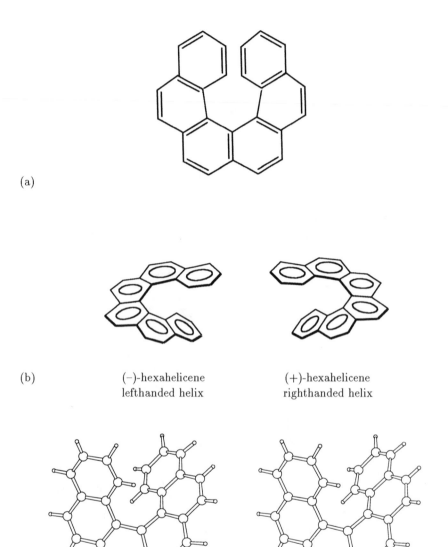

(a)

(b)   (−)-hexahelicene   (+)-hexahelicene
       lefthanded helix   righthanded helix

(c)   hexahelicene (stereoview)

FIGURE 12.18. Hexahelicene, showing the nonplanarity of the molecule (Refs. 53 and 54). (a) The chemical formula, (b) the conformations of left-and righthanded hexahelicene, and (c) a stereoview of (−)-hexahelicene.

product. Further details on absolute configuration are given in Chapter 14.

12.2 Macromolecular conformation

The term "macromolecule" refers to a variety of compounds, including biologically important compounds such as proteins, nucleic acids, and also many important manmade polymers. They may further aggregate into what are called macromolecular assemblies. The molecular shapes of these compounds, because of their large size, have greater variability and are somewhat different from those of smaller compounds. Their possible conformations are an important component of their biological and chemical function.

12.2.1 Proteins

Proteins perform a wide variety of functions in living systems, and the nature of their folding is very important in this context. Proteins may act as catalysts for a large number of reactions (enzymes), recognize foreign objects (antibodies), transport compounds (globins), and provide structural frameworks for living systems. These molecules are all simple, unbranched polymers composed of selections from the 20 naturally occurring amino acids (Figure 12.19). The amino acids are joined together in a topologically linear manner via a peptide linkage [Figure 12.20(a)]. The sequence determines the physical, chemical and biochemical properties of the protein. The α-carbon atom in these amino acids, except for glycine, is asymmetric. Only L-amino acids and glycine [Figure 12.20(b)] are found in naturally occurring proteins. As a result the polypeptide chain in a protein lacks symmetry. As the protein folds there is a possibility for crosslinking, provided by cysteine. Two cysteine residues, if nearby in the three-dimensional structure, can be oxidized to form a disulfide link, cystine (Figure 12.21), thereby joining together different portions of the polypeptide chain. Proteins may also bind metal ions and cofactor molecules as ligands. These also serve to control how the protein is folded. Additionally the protein may be glycosylated to give a glycoprotein.

The polypeptide chain folds in a systematic way to give a specific three-dimensional structure for the protein. Kaj U. Linderstrøm-Lang described the hierarchical organization of protein structure in terms of four levels, diagrammed in Figure 12.22.[55]

"Primary structure" refers to the amino acid sequence, the simplest level of structure. "Secondary structure" is any regular local structure of a segment of a polypeptide chain, such as a helical fold or an extended hydrogen-bonded pair of strands. "Tertiary structure" is the overall topology of the folded polypeptide chain. "Quaternary structure" describes the aggregation of folded polypeptides with each other by means of specific interactions, such as the aggregation of subunits to

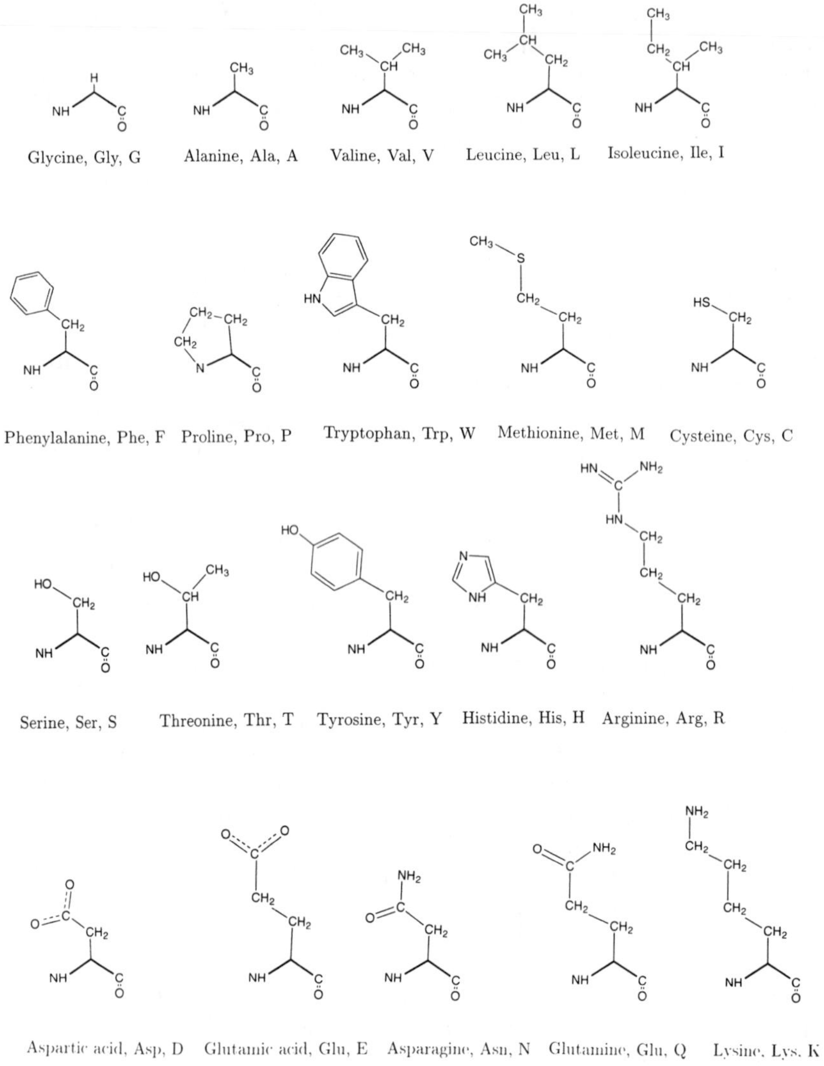

FIGURE 12.19. The twenty naturally occurring amino acids and their three- and one-letter designations.

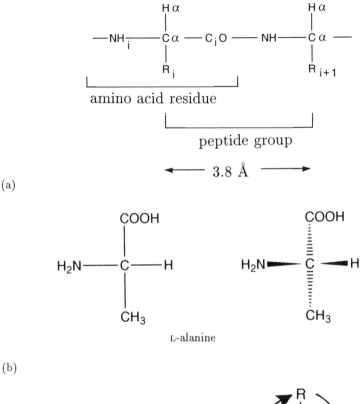

(a)

(b)

(c)

FIGURE 12.20. (a) Peptide linkages, (b) L-amino acids, and (c) an L amino acid in a peptide showing the mnemonic CORN, with the hydrogen atom H up toward the reader, CO down to the right, N down to the left, and the side chain R pointing to the top of the diagram, below the plane of the paper.

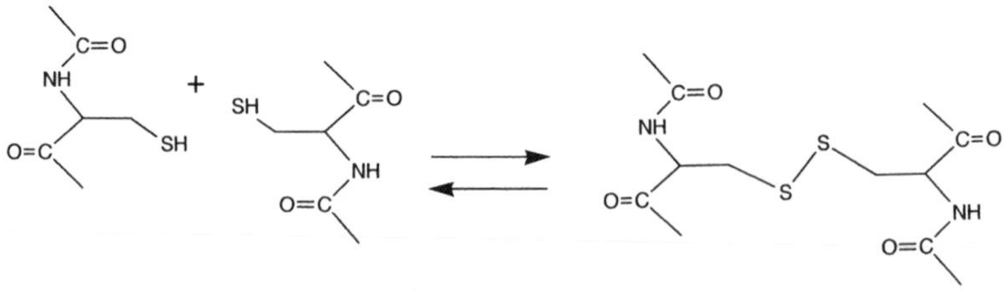

FIGURE 12.21. Conversion of cysteine to cystine, resulting in an S–S bond.

give a complete protein molecule.[56] Diagrams for the building of a model of quaternary structure are provided in Reference 57. The complexity of the three dimensional structure of a protein arises from the intrinsic ability of single covalent bonds to be rotated. The 20 different amino acid side chains have a variety of chemical properties that, when combined in a single molecule, a protein, provide the chemical diversity that is vital for its unique structure and function.

12.2.2 Conformations of polypeptides

The polypeptide backbone of a protein [Figure 12.20(a)] consists of a re-peated sequence of three atoms — the amide nitrogen (N_i), the α-carbon (C_α), and the carbonyl carbon (C_i), where i is the number of the residue starting from the amino end (remember the order of words in "amino acid"). The repeat distance for peptide units in a *trans* conformation is approximately 3.8 Å. The peptide group has a permanent dipole mo-ment with the negative charge on the carbonyl oxygen atom. The peptide bond is not chemically very reactive, and protons are lost or added only at extremes in *p*H.

Linus Pauling in 1933 first realized on the basis of resonance theory that the partial double-bond character in the C–N bond of the amide group in peptides, shown in Figure 12.23, would hinder rotation about the bond.[58] This is generally found to occur, and the peptide group is approximately planar, with bond lengths 1.33 Å between the carbonyl carbon and amide nitrogen, 1.45 Å between the α-carbon and amide nitrogen and 1.5 Å between the α-carbon and the carbonyl carbon (Figure 12.24).[59–61] This planar peptide group can exist in either the *cis* or the *trans* form. In practice the *trans* form is that most commonly found. A protein is essentially built up of planar, rigid peptide units.

Conventions have been established for the conformational analysis of protein structures. Three torsion angles, ϕ (phi), ψ (psi) and ω (omega), are used to describe the conformation of the main chain of a protein, as shown in Figure 12.25(a). The angle ω describes rotation about the peptide bond, $C_i(=O)$–N. Since this group is nearly planar (Figure 12.24),

PRIMARY (amino-acid sequence)

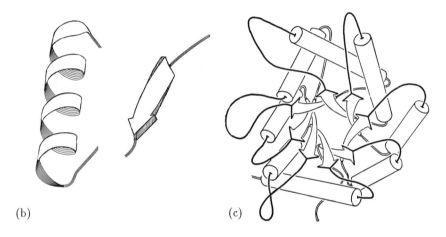

(a)

SECONDARY (α helices, β sheets)   TERTIARY (domains)

(b)   (c)

QUATERNARY (entire protein)

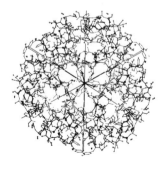

(d)

FIGURE 12.22. Levels of protein structure. (a) Primary structure, the amino-acid sequence. (b) Secondary structure, α helices, and β sheets. (c) Tertiary structure, domains. (d) Quaternary structure, the entire protein.

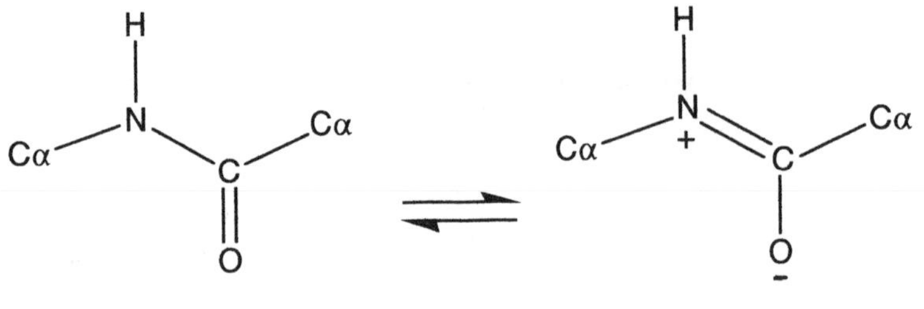

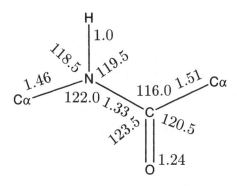

FIGURE 12.23. Planarity of the amide group. Two resonance forms of the peptide group are shown. The possibility of a double bond between C and N argues for restriction of rotation about this bond. The charged resonance hybrid gives rise to a dipole moment.

the torsion angles have values near 180° for a normal *trans* peptide or 0° for a *cis* peptide (see Figure 12.26). Small deviations from planarity, up to ±20° for ω, occur. Conceptually each peptide group can be thought of as lying on a flat rectangle, as shown in Figure 12.25(b), and it is necessary only to describe how one of these rectangles is related to the next. This means that while, in principle, there is the possibility for rotation about each of the three bonds that make up a peptide unit in the protein backbone, in practice this number is reduced to two (plus the possibility of *cis* or *trans* peptides), and these two angles describe how the peptide planes of adjacent residues lie with respect to each other. As shown in Figure 12.25(b), ϕ is the torsion angle about the RC_α–N bond and ψ is the torsion angle about the $(R)C_\alpha$–$C_i(=O)$ bond.

FIGURE 12.24. Approximate dimensions of the peptide group (in Å and degrees.)

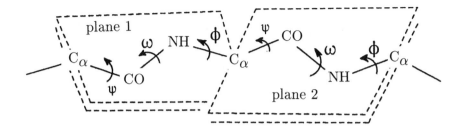

FIGURE 12.25. The three torsion angles that describe a peptide group, and the representation of the planarity of the peptide group.

When a protein crystal structure has been determined by fitting a model to the electron density map, the ϕ and ψ torsion angles are calculated for each amino acid residue in the model that has been fit to the map. These values give a good indication of the types of folding throughout the backbone of the protein. There are geometrical restraints on possible values of ϕ and ψ, so that most combinations are not possible. These restraints come about for steric reasons. For example, it is not possible for ϕ and ψ both to equal zero because these torsion angles would cause the carbonyl oxygen and the hydrogen atom on the nitrogen to overlap. Glycine, which has only a hydrogen atom as the side chain, has more conformational flexibility, and hence variability in its torsional parameters. The analysis is simplified by the construction of a conformational map, obtained by plotting ϕ against ψ for each amino acid to give a **Ramachandran plot**,[62,63] named after G. N. Ramachandran who, with V. Sasisekharan, first performed such an analysis. The Ramachandran plot in Figure 12.27 shows peptide torsion angles for D-xylose isomerase.[64]

Torsion angles, in addition, may be used to designate the conformation of the side chains. These are denoted by χ (χ^1 to χ^n working along the chain away from C_α). The steric interactions within the side chains in the *trans* form of the peptide bond ($\omega = 180°$) are much more favorable than those in the *cis* form ($\omega = 0°$), where there may also be steric interference with side chains from residues $i+2$. If the residue $i+1$ is proline,[65] however, the *cis* and *trans* forms (Figure 12.25) have similar energies. Proline[66] is the only amino acid taking part in a *cis* peptide that is normally encountered in proteins.

The long polypeptide chain of proteins can fold in a variety of ways depending on how hydrogen bonds are formed between backbone carbonyl and amino groups.[67-77] An example is given in Reference 78. A given polypeptide chain, however, if long enough, will fold in a characteristic way. In other words, information on how to fold is believed

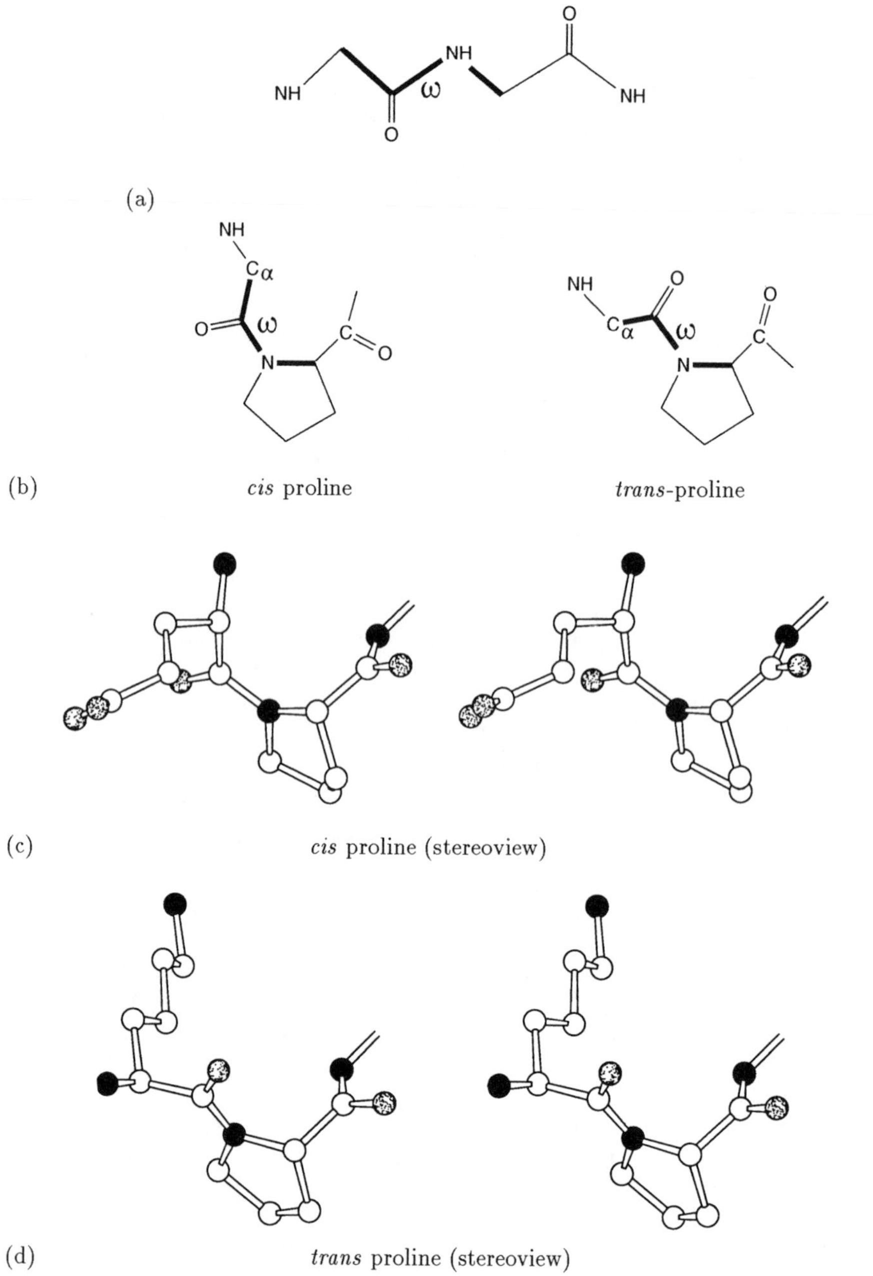

(a)

(b)   *cis* proline   *trans*-proline

(c)   *cis* proline (stereoview)

(d)   *trans* proline (stereoview)

FIGURE 12.26. *Cis* and *trans* proline. (a) Diagram of the torsion angle ω, (b) *cis* and *trans* prolines, (c) stereoviews of *cis* proline, and (d) *trans* proline.

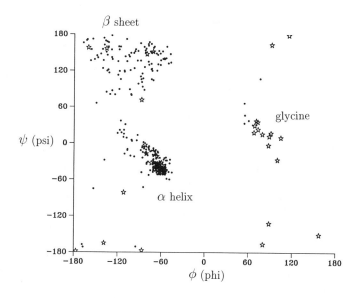

FIGURE 12.27. Ramachandran plot for D-xylose isomerase, a $(\beta\alpha)_8$ barrel. This graph contains a plot of ϕ versus ψ values for each peptide residue. Glycine residues are marked by asterisks. (Courtesy H. L. Carrell)

to be contained in the polypeptide itself. This was demonstrated by the reconstitution of ribonuclease after denaturation.[79]

During the folding of a water-soluble globular protein, the peptide group remains planar, hydrophobic amino acid residues are generally folded into the interior of the protein and the hydrophilic amino acid residues tend to lie on the surface of the protein. The result is a molecule with a hydrophobic core and a hydrophilic surface. The main chain of the protein, however, contains polar C=O and N–H groups, and these polarities must be neutralized in the hydrophobic core. Two types of secondary structure satisfy this requirement well — α **helices** and β **sheets**, and these are the commonly found motifs in proteins. It is important that, during protein folding, any hydrophobic patches on the surface of a protein be avoided because, as shown by the mutation of hemoglobin to sickle-cell hemoglobin, such hydrophobic areas tend to cause the protein molecules to adhere to each other, i.e., to aggregate. As a result, sickle-cell hemoglobin forms fibers.[80,81]

William Thomas Astbury found that fibrous proteins, such as keratin, could behave as if they had more than one structure.[82] Mammalian keratin, such as wool, normally gives a diffraction pattern referred to as the "α-keratin X-ray pattern" with repeat distances of 5.1 Å.[83–85] On stretching, however, a different diffraction pattern, the "β-keratin pattern," is obtained. He found that it is possible to reconvert β-keratin (stretched) to α-keratin (folded) when the tension is relaxed. These names, α and β, have been preserved in present-day nomenclature.

12.2.3 The α helix

The α helix is the most common conformational component of proteins, and a single helix may contain up to 40 amino acids. The polypeptide chain is wrapped in a right-handed spiral manner, held together by intra-chain hydrogen bonds (Figure 12.28). This helix has 3.6 residues per turn and a translation per residue of 1.5 Å (3.6 × 1.5 = 5.41 Å for one complete turn, cf., the 5.1 Å α repeat). Pauling predicted that this helical structure would be stable as a result of favorable hydrogen-bonding patterns. In 1948, he built models of polypeptide chains in helical arrangements.[86] After varying the pitch of the model helices, he realized that, if the helix had a nonintegral number of residues per turn, it was possible for hydrogen bonds to be formed between carbonyl and N–H groups on different turns of a single helical polypeptide.[87] These hydrogen bonds are formed between the carbonyl oxygen atom of one residue, i, and the amide nitrogen of residue, $i+4$ (see Figure 12.29). The closed loop so formed contains 13 atoms, including the hydrogen atom. All hydrogen bonds point in the same direction, nearly parallel to the helix axis, resulting in a dipole,[88,89] where the N terminus is positive and the C terminus is negative. This dipole moment may be considerable since the partial charges at the two ends are often well separated in space. The average length of a helix in reported protein crystal structures is about 12 residues or roughly three turns. The torsional angles of an α helix, $\phi = -60°$, $\psi = -50°$, are highly favorable and it is often considered the most natural conformation for a polypeptide. A left handed α helix is sterically possible; however, this conformation is not favorable energetically as the side chains are then too close to the backbone carbonyl group.

The periodicity of 3.6 residues per α-helical turn means that if the helix is viewed down its axis, the side chains stick out approximately every 100°. Therefore, if the helix is on the outside of the protein, the nature of the side chains must change from hydrophobic for the part that faces the interior of the protein to hydrophilic for the part of the protein that can interact with solvent. This can be represented by a **helical wheel**,[90] shown in Figure 12.29. It is generally found that one side of the wheel contains hydrophobic residues that pack into the interior of the protein, while the other contains hydrophilic residues that point toward the surface of the protein. Such a helix, with side chains changing from hydrophobic to hydrophilic and vice versa every three to four residues, is described as an **amphipathic helix**.

Pauling believed that hydrogen-bond energy is the dominant force for helix stability, but there are other factors making the helix a favorable structure. The backbone atoms are tightly packed together inside the helix and are in van der Waals contact; as a result there is no void

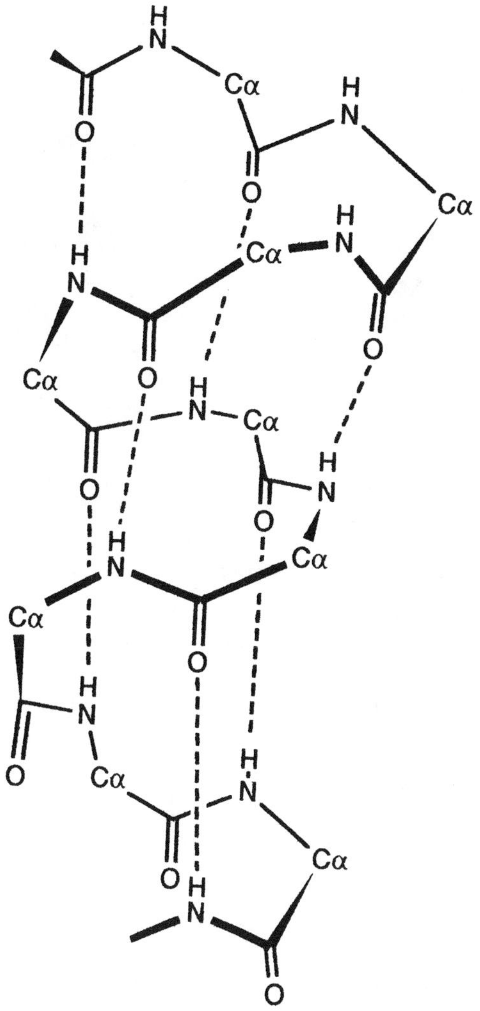

(a)

FIGURE 12.28. Views of an α helix. (a) General diagram of the backbone folding showing the importance of hydrogen bonding.

volume, and this probably contributes greatly to the high stability of the α helix. It has been observed that the carbonyl groups are tipped away from the helix axes, and that side-chain residues point out from the helix and toward its N terminus. As a result they do not interfere in any way with the helix formation. The exceptions are proline residues for which the side chain is bonded to the backbone nitrogen atom, preventing it from participating in hydrogen bonding and interfering in the packing.

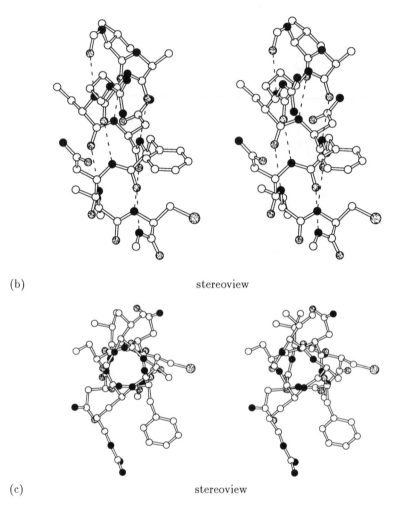

(b)   stereoview

(c)   stereoview

FIGURE 12.28 (cont'd). Stereoviews of a portion of an α helix in crambin (Ref. 91). (b) View perpendicular to the helix axis, and (c) view down the helix axis. Hydrogen atoms are not shown. Carbon atoms, open circles; nitrogen atoms, filled circles; oxygen atoms, stippled circles.

In the case of proline, the amide group is part of a five-membered ring and rotation about the C–N bond is not possible. Therefore, proline is considered a "helix breaker" and is rarely found in helices. If proline is found in an α helix, a bend or kink in the helix is usually observed at that point. Alanine, glutamic acid, leucine, and methionine are good helix formers, while proline, glycine, tyrosine, and serine are not.

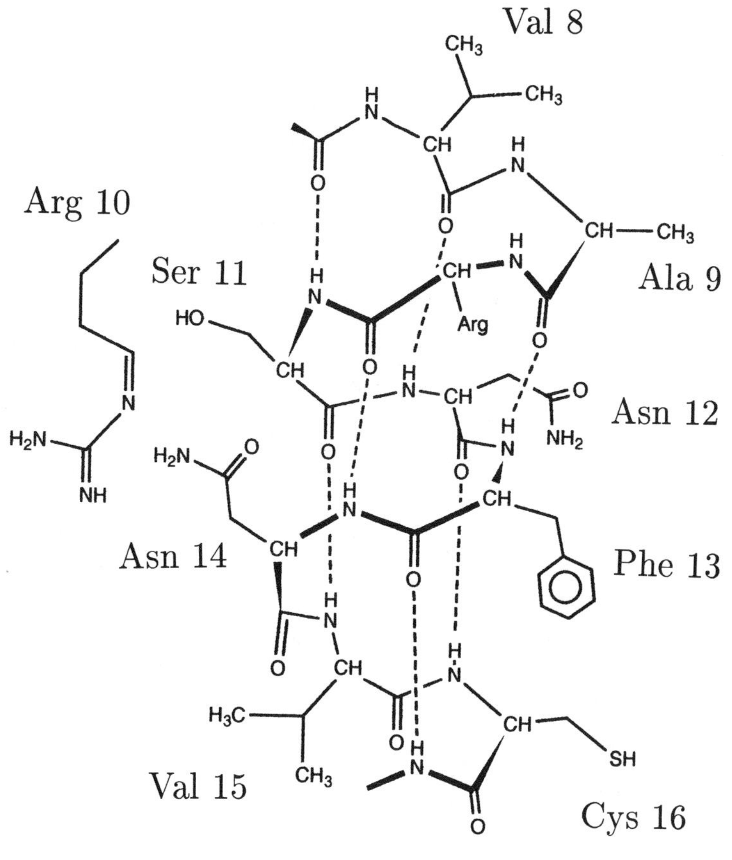

Val 8–Ala 9–Arg 10–Ser 11–Asn 12–Phe 13–Asn 14–Val 15–Cys 16

(d)

FIGURE 12.28 (cont'd). (d) The amino acid sequence of the portion of the protein crambin shown in (b) and (c)

The model for the α helix received strong support when Max Perutz,[92] working with hemoglobin and tilted fibers of keratin, observed a prominent meridional reflection at 1.5 Å. This had been previously overlooked, and corresponds to the translation per amino acid residue along the polypeptide chain. The observation of such a **meridional** Bragg reflection is a good indication of the presence of an α helix. The existence of α helices was finally verified in the crystal structure of myoglobin.[93]

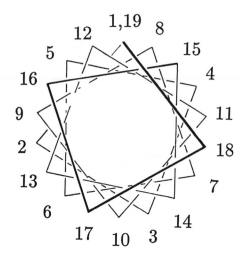

(a)

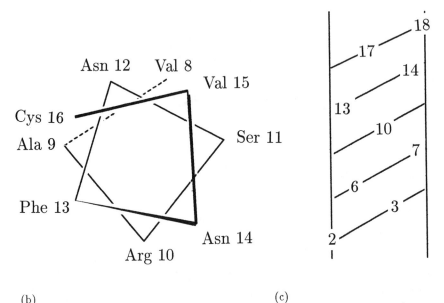

(b)   (c)

FIGURE 12.29. A helical wheel. (a) General view, and (b) and (c) views for the portion of crambin (Ref. 91) shown in Figure 12.27. (b) View down the helix axis, and (c) with the helix axis vertical.

A convention for describing helices involves listing the number of residues per turn and then denoting, by a subscript, the number of atoms necessary to close the hydrogen bonding (Figure 12.30). By this scheme the α helix would be called the 3.6_{13} helix. Another type of helix often found in protein structures is the 3_{10} helix, which has ϕ values of $-60°$ and ψ values of $-30°$.[94] It is less stable than the α helix and usually forms only a single turn. Frequently the carboxyl ends of α helices form 3_{10} helices. The hypothetical π helix (4.4 residues per turn)[95] would be described as 4.4_{16}, but does not have good packing energy, although the hydrogen bonding requirements are satisfied.

12.2.4 The β sheet

A second type of hydrogen bonding between different portions of a polypeptide chain, also first postulated by Pauling[96] and found later in many structures, is called the β sheet (Figure 12.31). In this structure, the polypeptide chains are nearly fully extended. The first β-sheet in a protein was observed in lysozyme[97] (hemoglobin and myoglobin are mainly α-helical). Individual strands of sheet aggregate side by side, forming hydrogen bonds between the carbonyl group oxygen atom of one

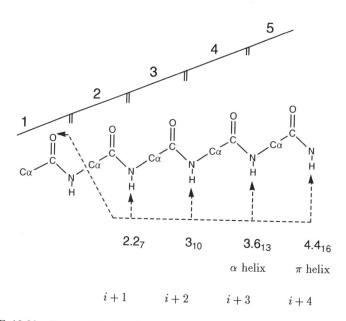

FIGURE 12.30. Types of helices in proteins, including the α helix. Shown is the backbone of a protein and the various types of hydrogen bonding to give a helical structure. The number indicates the number of amino-acid residues per turn and the subscript the number of atoms in the hydrogen-bonding scheme (count them).

chain and the amide-hydrogen atom of another chain.[98-101] The polypeptide chains in this side by side packing may be either parallel or antiparallel, as shown in Figure 12.31. When examined end on, the β strand appears pleated, meaning that successive C_α atoms are slightly above or below the sheet plane. The side chains then point in these directions as shown in Figure 12.31. In addition to the hydrogen bonds, the dipoles of the peptide bonds are alternated along the chain providing favorable conditions for interaction.

Parallel and antiparallel sheets have different patterns of hydrogen bonding. Antiparallel sheets have unevenly spaced hydrogen bonds per-

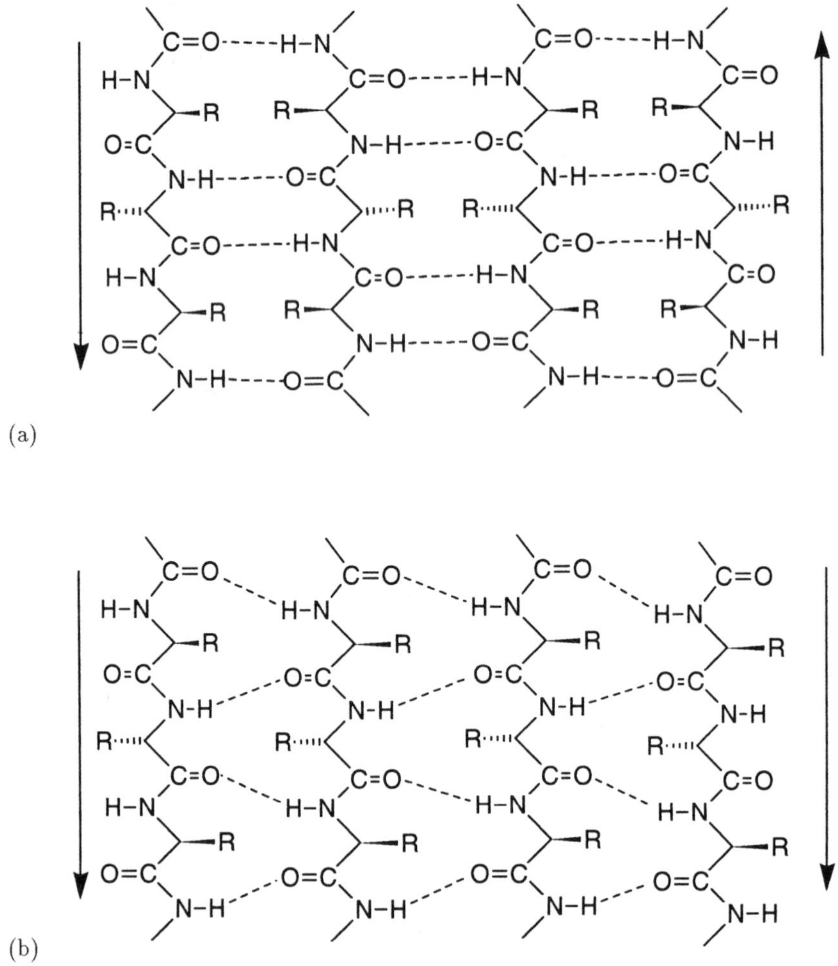

(a)

(b)

FIGURE 12.31. β-sheet. (a) Antiparallel and (b) parallel.

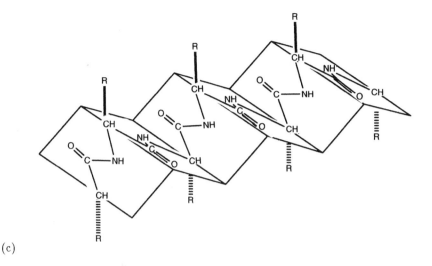

(c)

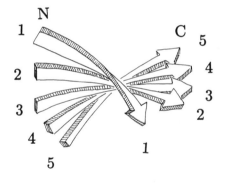

(d)

FIGURE 12.31 (cont'd). (c) View showing pleatedness of structure, and (d) view of β-sheet twist.

pendicular to the strands and the narrowly spaced bond pairs alternate with widely spaced pairs. Parallel strands, on the other hand, have hydrogen bonds that are more evenly spaced. The side chains are opposite one another on neighboring strands and extend above and below the plane of the sheet. Parallel and antiparallel sheets have slightly different torsion angles, although parallel sheets are more regular than antiparallel sheets; the torsion angles of parallel sheets cluster tightly together in a Ramachandran plot, while the torsion angles for antiparallel sheets are spread over an entire quadrant. Parallel sheets usually contain more than five strands, whereas two strands are quite sufficient for the formation of antiparallel sheets which can fold into a twisted ribbonlike structure. Often, β-sheets have a right-handed twist[102] that varies from 0 to 30°. Parallel strands tend to be buried in the interior of the protein, while antiparallel ones have one side exposed to solvent.

12.2.5 Connecting segments of polypeptide chains

The regions between the α helices and β sheets in a folded globular protein are loop regions.[103–107] These are the main sites of insertions and deletions of amino acids in like proteins from different species. Reverse turns (hairpins, loops) cause changes in the overall direction of the polypeptide chain. They were first noted by Cherayathumadom M. Venkatachalam,[103] and generally contain hydrophilic residues and occur at the surface of the protein. Turns can be identified in proteins by a distance criterion. If the α-carbon atoms of residues i and $i+3$ are less than 7 Å apart and the structure is not part of an α helix, that portion of the chain is considered to be a turn. There are three basic types of turns, which are defined by the dihedral angles of the $i+1$ and the $i+2$ residues (Figure 12.32). Type-1 turns have a characteristic pucker, and the α-carbon atoms do not lie in a plane. In Type-2 turns, all of the α carbon atoms lie in a plane. Frequently the residue at position $i+2$ is glycine; larger side chains at position $i+2$ would cause steric interference with the carbonyl oxygen atoms at position $i+1$. Type-3 turns are the same as 3_{10} helices. Turns are frequently found at the ends of α helices, or else joining β strands. George Rose has identified structures referred to as Ω **loops**.[108] These loops, which look like the Greek letter Ω, are compact structures but, unlike turns, have side chains in their interiors.

To simplify reported crystal structures, α helices are represented by cylinders and β strands by flat arrows, the arrow pointing to the C terminus (see Figure 12.33). The resulting topology diagrams, introduced by Jane Richardson[70,71] and extended by Arthur Lesk and Karl Hardman,[72] are useful for highlighting the similarities in folding of different proteins. A list of conformation angles for common structural motifs is given in Table 12.1.

12.2.6 Protein architecture

Certain assemblies of secondary structural elements have been observed a sufficient number of times in protein structures to merit classification. These common structural motifs are termed "supersecondary structures." Some examples of these structural motifs are shown in Figure 12.34. They are recognized as a level of organization higher than secondary structure, but do not constitute entire structural domains. For example, two α helices joined by a loop (the **helix–loop–helix motif**) can give a calcium-binding motif [Figure 12.34(a)], or a DNA-binding motif [(Figure 12.34(b)]. Two adjacent β strands joined by a loop give a β –β

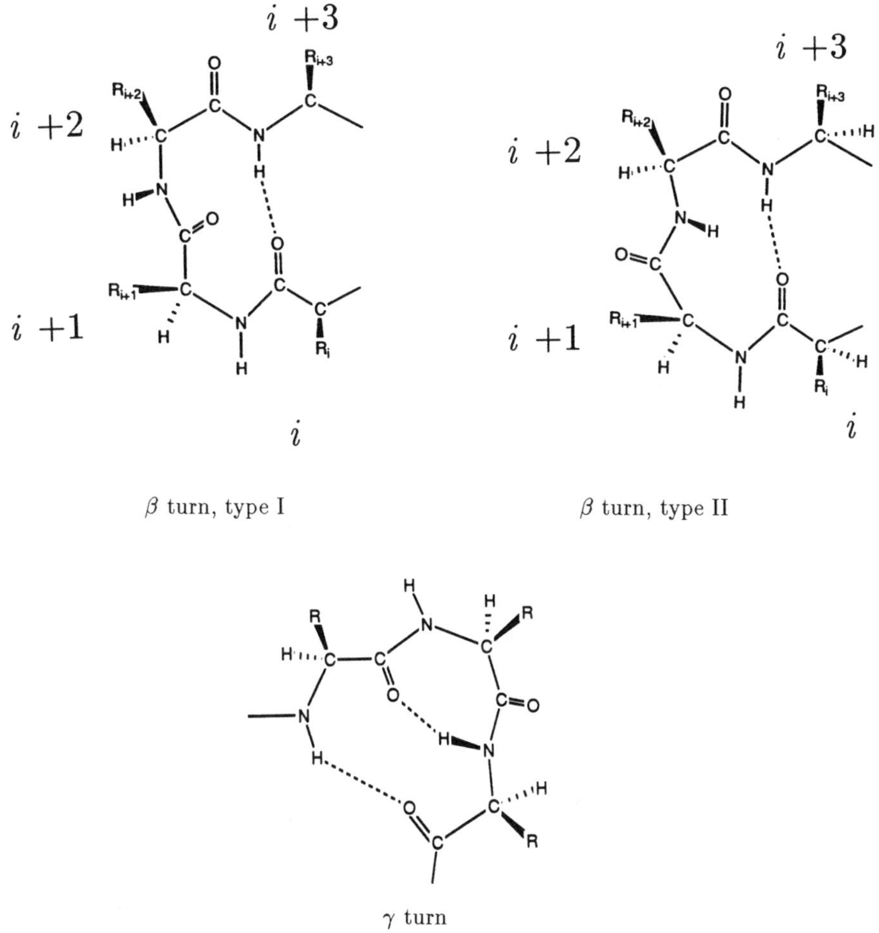

β turn, type I

β turn, type II

γ turn

FIGURE 12.32. Some turns commonly encountered in protein crystal structures.

unit or a β meander [Figure 12.35(a)]. The **Greek key motif** consists of four antiparallel β strands arranged in a pattern reminiscent of one found in ancient Greek friezes [Figure 12.35(b)]. Strands that are sequential in primary structure are also adjacent in the β sheet and are connected by relatively tight turns. The $\beta\alpha\beta$ **motif** [Figure 12.35(c)] is found in most proteins with parallel β sheets. It consists of two β strands, which are parallel but not necessarily adjacent, connected by an α helix such that the helix axis is parallel to the β strands.

As more and more protein structures are determined, it became apparent that it is not unusual for proteins that are dissimilar in function and sequence to have some common structural features. For example,

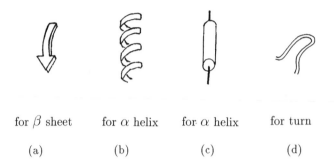

for β sheet   for α helix   for α helix   for turn

(a)   (b)   (c)   (d)

FIGURE 12.33. Representations of (a) a β sheet, (b) and (c) an α helix, and (d) a turn in protein crystal structures.

TABLE 12.1. Conformation angles for secondary structures.

Structure	ϕ	ψ	Structure	ϕ	ψ
α helix (3.6_{13})	−57	−48	β bend I, step 2	−70	−30
3_{10} helix (3.0_{10})	−76	−4	β bend I, step 3	−90	10
π helix (4.4_{16})	−57	−70	β bend II, step 2	−60	130
Parallel β sheet	−119	113	β bend II, step 3	80	0
Antiparallel β sheet	−139	135	γ turn, step 1	172	128
β bulge, step 1	−95	−65	γ turn, step 2	68	−61
β bulge, step 2	−130	150	γ turn, step 3	−131	162

the backbone chains of the first two proteins for which structures were determined by X-ray diffraction, myoglobin and hemoglobin, were found to be folded in a rather similar manner. As another example, the functionally dissimilar proteins, lysozyme and α-lactalbumin (although they have considerable sequence homology), are folded in similar ways.[109] Even more surprising was the discovery that superoxide dismutase and the immunoglobulin molecules share structural similarities, without having any functional or sequence relationships.[110]

Now that the crystal structures of a large number of proteins have been determined, it is possible to analyze the overall types of architecture, that is, the arrangement of motifs shown by such molecules (Figure 12.36). Proteins are found to be constructed of modular systems or **domains** (essentially the tertiary structure, mentioned earlier). These are portions of the polypeptide chain that can fold independently into a stable structure. A protein may be just one domain, or may be comprised

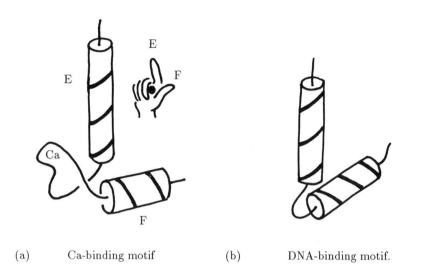

(a)   Ca-binding motif   (b)   DNA-binding motif.

FIGURE 12.34. Two helix–loop–helix motifs in protein structures. These bind (a) calcium ions, or (b) nucleic acids.

of many domains, which are, on the average, approximately 25 Å in diameter (100–150 amino acid residues).

Michael Levitt and Cyrus Chothia[111] constructed taxonomic classifications of protein structure and showed that secondary structural features within protein domains are organized in a limited number of ways. These modules can be divided into categories on the basis of the organization of the structural elements, the topology of their connections, and the number of major levels of backbone structure. The four main subcategories,[70,112] based upon the topology of their connections and the number of major levels of backbone structure, are α for α-helical proteins, β for proteins that are primarily β sheets, $\alpha + \beta$ for proteins with both α helices and β sheets, and α/β for those proteins that have alternating helix and sheet arrangements.

When more than 60% of the residues in a protein adopt a helical conformation, the protein is usually classified as α. The helices are usually in contact with one another. The simplest antiparallel α structure, a four-helix bundle, looks like a bundle of sticks. It usually consists of four α helices, aligned to form a hydrophobic core (Figure 12.37). The helices that are adjacent in the amino-acid sequence are also adjacent (and antiparallel) in the three-dimensional structure. The total structure, which is cylindrical, has a left-handed twist of approximately 15° between the axes of the helices. Hemerythrin is one example of a protein that adopts this topology. The "globin fold," so-named because it was found in myo-

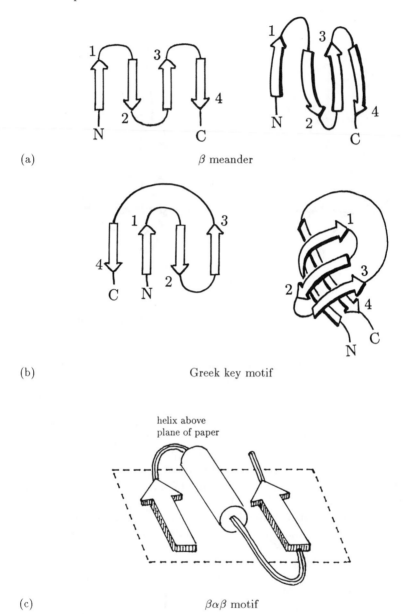

(a) β meander

(b)   Greek key motif

(c) $\beta\alpha\beta$ motif

FIGURE 12.35. Motifs in protein architecture. (a) The β meander and (b) the Greek key motif. On the left is shown a commonly used simplified diagram, while on the right is shown how the folding occurs in three dimensions. Note the three dimensionality of the Greek key motif. (c) The $\beta\alpha\beta$ motif with the helix above the plane (marked by broken lines).

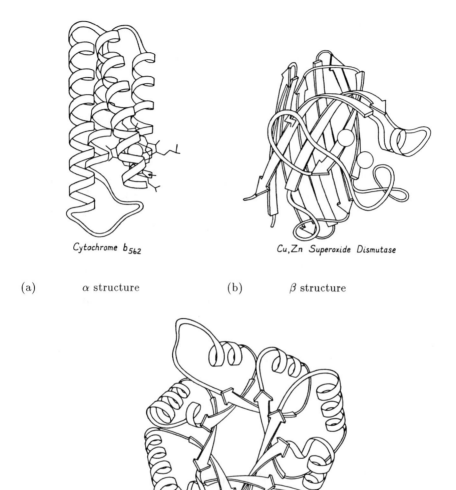

Cytochrome b_{562}

Cu,Zn Superoxide Dismutase

(a) α structure   (b) β structure

Triose Phosphate Isomerase

(c) α/β structure

FIGURE 12.36. Various types of protein tertiary structure. (a) An antiparallel α structure (cytochrome b_{562}) an up-and-down helix bundle (Ref. 113), (b) an antiparallel β structure (Cu,Zn superoxide dismutase) a Greek key β barrel (Ref. 114), and (c) and a parallel α/β structure (triose phosphate isomerase) a singly wound parallel β barrel (Ref. 115). (Courtesy Jane S. Richardson)

globin and hemoglobin, consists of eight α helices arranged to enfold the active site (the heme group). The angles between the helical axes are approximately 50°.

Proteins that have primarily a β sheet conformation are folded either as barrels or as layered sheets. The simplest antiparallel β barrel is analogous to the up-and-down helix bundles. Such barrel structures are made with either six or eight strands of β structure wrapped into a cylinder forming the surface of the barrel [Figure 12.38(a) and (d)]. Depending upon the topology of the strands, barrels are formed that can be described by a Greek key motif or the **jelly-roll motif** [Figure 12.38(b) and (e)].

Another major class of proteins, the $\alpha + \beta$, has approximately equal amounts of α-helical and β sheet structure. Often there is a cluster of helices at one or both ends of a single β sheet. β structure is generally found as antiparallel strands. One particular type of $\alpha + \beta$ structure is termed the **open-face sandwich** β structure.[71] The subtilisin inhibitor has a five-stranded β structure with two α helices protecting its open face.

Proteins that are classified as α/β usually have one major β sheet per domain. This is generally parallel sheet. The α helices occur in the loops connecting the strands and pack against the β sheet. The simplest arrangement is the singly wound parallel $(\beta\alpha)_8$ barrel consisting of eight parallel β strands forming a central β barrel with eight α helices surrounding the β strands [Figure 12.37(h)], first seen in triose phosphate isomerase. A large number of structures have been observed to have this particular type of protein folding conformation. β barrels are discussed

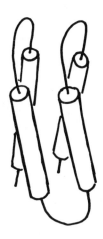

FIGURE 12.37. A four-helix bundle.

further in Chapter 16. An alternate form of the α/β proteins includes those that are doubly wound. In this conformation, the β sheet (4 to 10 strands) does not wrap around to form a barrel, but instead forms a single twisted β sheet. The sheet connections are again made by α helices.

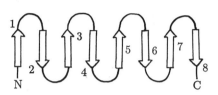

(a)   up-and-down barrel

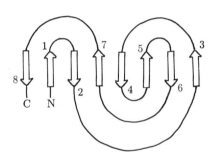

(b)   jelly-roll motif

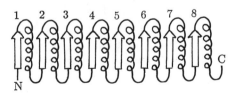

(c) $(\beta\alpha)_8$ barrel.

FIGURE 12.38. Motifs in protein architecture. (a) Up-and-down barrel, (b) jelly-roll motif, and (c) $(\beta\alpha)_8$ barrel. Compare these with (d), (e), and (f), which is what the folding looks like in three dimensions.

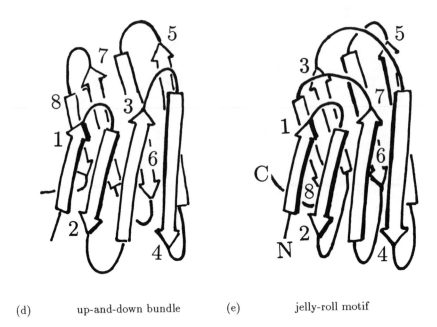

(d)   up-and-down bundle   (e)   jelly-roll motif

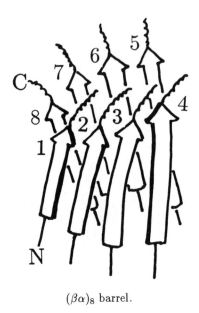

(f) $(\beta\alpha)_8$ barrel.

FIGURE 12.38 (cont'd). (d) Up-and-down barrel, (e) jelly-roll motif, and (f) $(\beta\alpha)_8$ barrel.

12.2.7 Interactions with water

Water has the significant property of being able to act both as a proton donor and as a proton acceptor.[116] Individual molecules that contain several polar groups capable of forming hydrogen bonds do not tend to form hydrogen bonds between each other in aqueous solution because these groups can so readily form hydrogen bonds to water. In nonpolar solvents, however, they tend to form hydrogen bonds to each other.

The manner by which solvent surrounds a protein provides useful information for a general understanding of protein structure. Protein–protein interactions are weak compared to protein–solvent and solvent–solvent interactions. This causes the hydrophobic effect, first noted by Isidor Traube,[117] who showed that long-chain hydrocarbons with a polar group at one end have, at low concentrations in water, a surface-to-bulk ratio that increases threefold for each additional CH_2 group. The important component is not the attraction of nonpolar groups for each other; it is the strong attractive forces between water molecules that determine these effects.[118,119] Thus the role of the hydrocarbon is to break some water–water interactions. Ionic salts and protein side chains can form strong bonds to water and compensate for the loss of water–water interactions; by contrast, hydrophobic compounds do not do this.

12.2.8 Nucleic acids

The diffraction pattern to be expected for a helical structure was worked out in a theoretical study by William Cochran, Francis H. C. Crick, and Vladimir Vand[120] using the α helix as a model. This work provided the basis for the interpretation of the diffraction patterns of proteins,[121,122] and also led, unexpectedly, to an understanding of nucleic acid structure. This culminated in the determination of the three-dimensional structure of DNA by James D. Watson and Frances H. C. Crick[123] from an X-ray diffraction photograph taken by Rosalind Franklin.[122,124]

Naturally occurring nucleotides and nucleic acids are composed of three units:[17,125]

1. A sugar, β D-ribose (in RNA) or β D-2'-deoxyribose (in DNA). This sugar is in the cyclic five-membered furanoside form.
2. A heterocyclic base, guanine, adenine, cytosine, uracil (in RNA), or thymine (in DNA). The sugar is substituted at C1' by the base, which is attached by a β glycosyl C1'–N link.
3. A phosphate group.

If the 3' or 5' hydroxyl of the sugar is phosphorylated, the combination of sugar base and phosphate is called a **nucleotide**; if the sugar is not phosphorylated and the molecule is composed only of base and sugar, it is called a **nucleoside**.

In nucleic acids, the main chain is referred to as a phosphodiester backbone, shown with numbering in Figure 12.39. Only C3′ and C4′ of the sugar are included in the backbone, which has the sequence:

$$\text{phosphodiester backbone}: \text{P} \rightarrow \text{O5}' \rightarrow \text{C5}' \rightarrow \text{C4}' \rightarrow \text{C3}' \rightarrow \text{O3}' \rightarrow \text{P}. \qquad (12.1)$$

Conformations of the furanose sugar are shown in Figure 12.40. The furanose ring is generally found in one of two conformations: an envelope (E) form with four atoms in a plane and the fifth out by as much as 0.5 Å, and a twist (T) form with two adjacent atoms displaced on opposite sides of the plane through the other three atoms.[126] Variations between these are common, and one conformation can readily, in the absence of steric constraints, be converted to another. If the displacement of an atom bound to the ring lies on the same side as the substituent C5′, it is called **endo**; if it lies on the other side, it is called **exo**.

In nucleosides and nucleotides, the base can adopt two main orientations about the glycosyl C1′–N link, called **anti** and **syn** and defined by the torsion angle χ, which is the O1′–C1′–N9–C4 torsion angle for

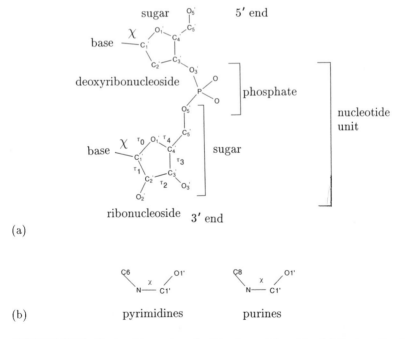

(a)

(b)   pyrimidines   purines

FIGURE 12.39. Nucleosides and nucleotides in nucleic acids. (a) Designation in DNA, and (b) pyrimidines and purines.

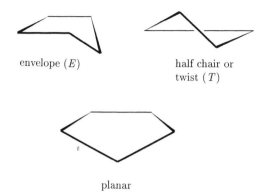

envelope (*E*)

half chair or
twist (*T*)

planar

(a)

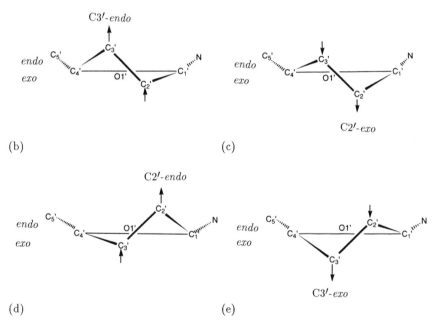

(b)

(c)

(d)

(e)

FIGURE 12.40. The conformations of furanose sugars. (a) Conformations of five-membered rings. When a line is drawn through C4′–O1′–C1′, the pucker is described by which atom in the five-membered ring is most out of this plane of the five-membered ring. (b) C3′-*endo* with C3′ most out of the plane on the C5′ side, (c) C2′-*exo* with C2′ most out of the plane on the opposite side from C5′, (d) C2′-*endo* with C2′ most out of the plane on the C5′ side, and (e) C3′-*exo* with C3′ most out of the plane on the opposite side from C5′.

purines and the O1′–C1′–N1–C2 torsion angle for pyrimidines (Figure 12.41). In the *anti* conformation the bulk of the heterocyclic base points away from the sugar, and in the *syn* conformation it points towards the sugar.

Polynucleotides fold into helices, right-handed (*A* and *B*)[123,127,128] or left-handed (*Z*).[129] Helical parameters are the pitch, which is the distance *P* vertically along the helix axis for one complete turn (360° rotation), the axial rise per residue *h* and the unit twist *t*. The numbering of atoms in the bases and the base pairing is illustrated in Figure 12.42. The overall shapes of B-, A-, and Z-DNA are diagrammed in Figure 12.43. Conformational variations in nucleic acids[130] have been defined rigorously for single and for successive base pairs. The reader should consult this article (Reference 130) for more information. For single base pairs, an axis of reference is taken as the line joining C6 of the pyrimidine and C8 of the purine. The perpendicular deviation of this axis from the helix axis is the **displacement**. The dihedral angle between pyrimidine and purine planes is called the **propeller twist**. The **inclination** is the angle between the plane of a base pair and a plane perpendicular to the

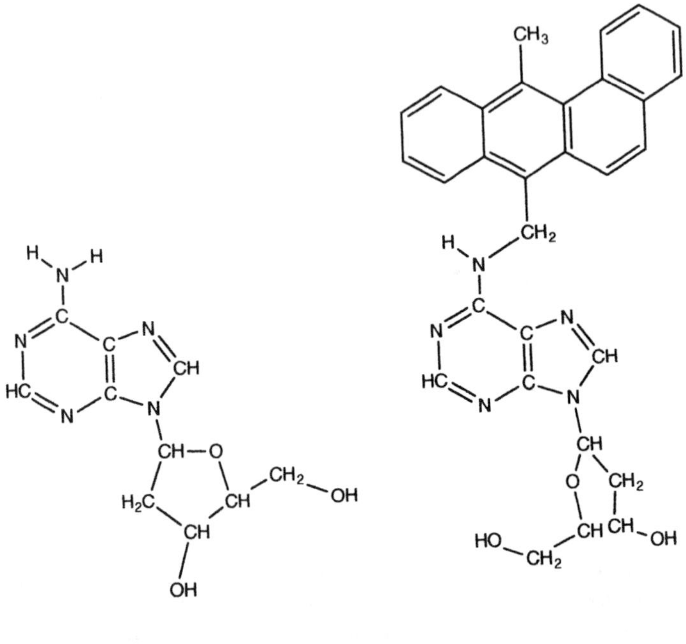

FIGURE 12.41. *Syn* and *anti* conformations of nucleosides and nucleotides.
(a) Chemical formula (left: deoxyadenosine, and right: *N*-alkylated deoxyadenosine).

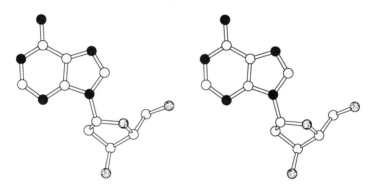

(b)   *anti* conformation (stereoview)

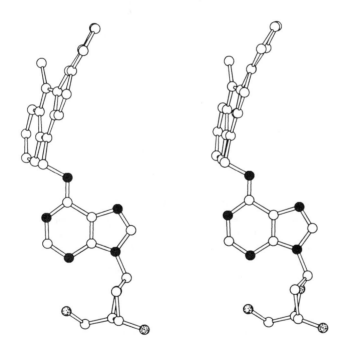

(c)   *syn* conformation (stereoview)

FIGURE 12.41 (cont'd). (b) *anti* conformation (stereoview), and (c) *syn* conformation (stereoview) of the compounds in (a).

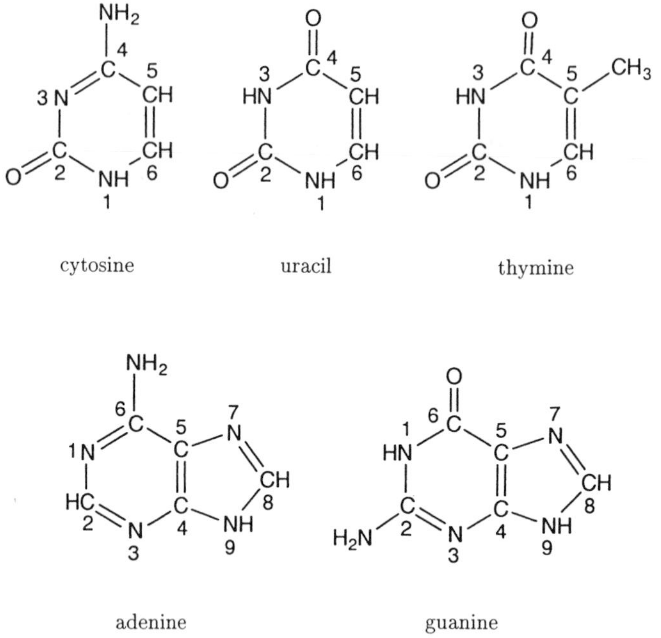

cytosine

uracil

thymine

adenine

guanine

(a)

FIGURE 12.42. Base pairs in nucleic acids. (a) Numbering of nucleic-acid bases.

helix axis. For successive base pairs, the helical twist is the orientation of the base pairs with respect to the helix axis. "Roll" and "tilt" define angles between successive base pairs in the directions of the short and long axes of the base pair, respectively. The distance between successive base pairs along the helix axis is called the "rise," while "shift" is the relative displacement of the midpoint of C6–C8 vectors of successive base pairs perpendicular to the long axis of the base pair and "slide" is their displacement parallel to this same axis. Propeller twist, roll, and tilt are illustrated in Figure 12.44. These descriptors[130] are very useful in nucleic-acid structural chemistry.

Summary

1. The conformation of a molecule may be described by its torsion angles. A torsion angle is measured by looking along a bond and describing the disposition of substituents attached to the atoms at the ends of the bond.

2. A positive torsion angle is the clockwise twist about the central of three connected bonds that is needed to make the near (upper) bond (*A–B* in a series of four-bonded atoms *A–B–C–D*) eclipse the far (lower) bond (*C-D*).

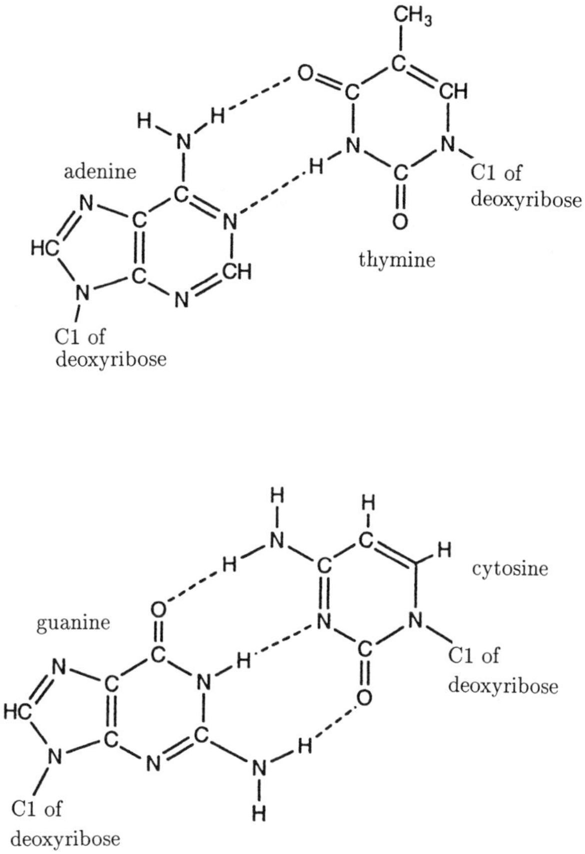

(b)

FIGURE 12.42 (cont'd). (b) Arrangement of atoms in base pairs in nucleic acids.

3. Steric strain will affect torsion angles.
4. Torsion angles are important in the description and characterization of portions of proteins and nucleic acids. A Ramachandran plot of the torsion angles of the peptide backbone provides a graphical description of the protein.
5. The α helix contains a helical folding of the polypeptide backbone such that a hydrogen bond is formed between the carbonyl group of residue i and the amide nitrogen atom of residue $i+4$.
6. A β sheet is formed when individual polypeptide strands line up side by side and form hydrogen bonds of the type C=O $\cdots$ H–N.
7. The manner in which a protein folds gives rise to four general subcategories: α, β, $\alpha + \beta$, and α/β.

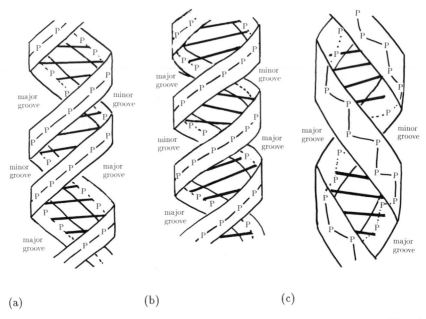

(a)   (b)   (c)

FIGURE 12.43. DNA shapes. (a) B-DNA, (b) A-DNA, and (c) Z-DNA. The phosphodiester backbone is drawn as a ribbon, and the bases are drawn as approximately horizontal lines.

8. Nucleic acids have a phosphodiester backbone that includes a ribose or deoxyribose group that is bound to a base (G, A, C, U, or T). The furanose sugar can adopt different conformations, and the link of the sugar to the base may have various torsion angles.

9. The major structural types of DNA are B, A, and Z.

Glossary

α **helix:** A mode of protein folding in which each carbonyl group accepts a hydrogen bond from the amide nitrogen atom of the amino acid four residues beyond (towards the C-terminus). This results in 3.6 amino acids per turn (360° rotation) and a translation of 1.5 Å per residue along the helix axis.

Amphipathic helix: An α helix that can be described as a cylinder with polar (hydrophilic) and nonpolar (hydrophobic) sides). Since there are 3.6 amino acid residues in one turn of an α helix, amino acid residues that are 7 apart will line up on one side of the helix.

Anti: Two groups on opposite sides of a reference plane. If they are on the same side the term is *syn* (q.v.).

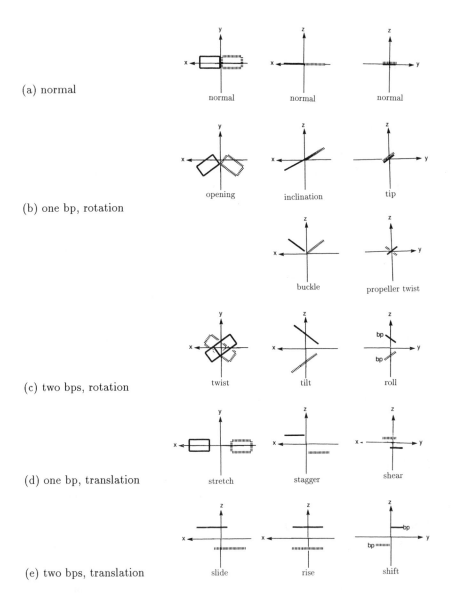

FIGURE 12.44. Descriptors of base and base-pair (bp) orientations. Shown at the top (a) are idealized base-pair arrangements viewed down the helix axis (z). Perturbations are illustrated for (b) rotation of one base pair, (c) rotation of two base pairs adjacent in the nucleic-acid sequence, (d) translation for one base pair, and (e) translation for two base pairs adjacent in the nucleic-acid sequence.

Anticlinal conformation (ac): Torsion angle within 30° of ±120°.

Antiperiplanar conformation (ap): Torsion angle within 30° of ±180°.

Axial bond (a): Bonds to a tetrahedral atom in a six-membered ring are termed *axial* if they make a small angle with the normal (perpendicular) to the plane containing the majority of the ring atoms (cf. *equatorial* bond).

Best plane: The plane through a group of atoms that best satisfies the least-squares criterion.

$\beta\alpha\beta$ motif: A commonly encountered motif found in protein structures. It consists of two β strands connected by an α helix. The crossover that this provides is right-handed.

β barrel: The folding of a polypeptide chain to form a barrel-shaped structure with eight β strands as the lining. Eight α helices lie outside this β sheet. Both the α helices and the β strands follow a right-handed spiral around the axis of the barrel. The amino acid sequence in such a protein is such that β sheet and α helix alternate to give $(\beta\alpha)_8$. This motif was first seen in triose phosphate isomerase, and has since been observed in many other protein structures.

β sheet: A mode of protein folding in which two polypeptide chains (β strands) lie side by side and either parallel or antiparallel with respect to the direction of the $NH-C_\alpha-CO$ group. Hydrogen bonds are formed from a carbonyl group of one chain to an amide nitrogen atom of the other chain and vice versa.

Boat conformation: For a six-membered ring with the atoms numbered sequentially 1 to 6, in the boat conformation atoms 1, 2, 4, and 5, but not 3 and 6, are coplanar; 3 and 6 are on the same side of this plane.

Chair conformation: For a six-membered ring with the atoms numbered sequentially 1 to 6, in the chair conformation atoms 1, 2, 4, and 5, but not 3 and 6, are coplanar; 3 and 6 are on opposite sides of this plane.

Cis: Two groups occupying adjacent positions on adjacent atoms or lying on the same side of a reference plane. The torsion angle is near 0° (see *syn*).

Configuration: The arrangement of atoms around one atom. Those arrangements that differ only as a result of rotation about one or more single bonds are not considered relevant.

Conformation: One of the various arrangements of molecules that occur by virtue of rotation about single bonds.

Conformational flexibility: The extent to which a molecule may take up different conformations.

Conformer: One of a set of possible conformations of a molecule.

Dihedral angle: A dihedral angle between two planes is the acute angle between the normals (perpendiculars) to these planes.

Displacement: A conformational description of base pairs in nucleic acids. An axis of reference is taken as the line joining C6 of a pyrimidine and C8 of a purine. The displacement is the perpendicular deviation of this axis from the helix axis of the nucleic acid.

Domain: A structurally independent region in a crystal or in a protein structure. It is described as a semiindependent unit that is important for folding stability, evolutionary recombination, or functional rigid-body motions.

Eclipsed conformation: If two atoms or groups attached at opposite ends of a bond appear one directly behind the other when the molecule is viewed along this bond, these atoms or groups are described as **eclipsed**, and that portion of the molecule is described as being in the **eclipsed conformation**.

Equatorial bond (e): Bonds to tetrahedral atoms in a six-membered ring are termed *equatorial* if they lie approximately in the best plane through the six atoms of the ring (see *axial* bond).

Endo-: If the displacement of an atom bound to a furanose (five-membered) ring lies on the same side as the substituent C5', it is called *endo*.

Exo-: If the displacement of an atom bound to a furanose (five-membered) ring lies on the opposite side to the substituent C5', it is called *exo*.

Fischer projection: A diagram that illustrates the spatial arrangement of bonds around an atom, usually carbon. In a compound with several connected asymmetric carbon atoms, the principal chain is drawn vertically (in a convex manner) with the lowest number chain member at the top. Atoms appearing above or below the atom of interest lie behind the plane of the paper. Those drawn to the left or right lie above the plane of the paper.

Gauche (g): Torsion angle within 30° of ±60°.

Greek key motif: A geometric motif found in Greek pottery that can be used to describe a certain type of folding pattern of a protein structure.

Helical wheel: A diagrammatic view directly down the helix axis of an α helix. It is drawn with amino acid residues projected at 100° from the previous residue (since there are 3.6 amino acid residues in a turn of 360°). This diagram may help identify α helices with hydrophobic and hydrophilic sides (see Amphipathic helix).

Helix-loop-helix motif: A structural motif found in some DNA-binding proteins. It consists of an α helix followed by a turn and then another α helix. This provides a finger-like projection that can fit in a groove of DNA.

Inclination (DNA): The angle between the plane containing a base pair in a nucleic acid structure and a plane perpendicular to the helix axis.

Jelly-roll motif: The folding of a protein in a manner reminiscent of the fold of a jelly roll.

Meridional: Pertaining to the great circle of s sphere that passes through its poles.

Newman projection: A projection of a molecule viewed down the bond between two atoms. A circle represents these atoms, the larger circle for the more distant atom. Lines from the center to the exterior of the circles represent bonds to the nearer atoms. Lines representing bonds to the further atom do not penetrate the inner circle. When two bonds are coincident, they are drawn at a small angle to each other.

Nucleoside: A component of a nucleic acid consisting of a base and a sugar.

Nucleotide: A component of a nucleic acid consisting of a base, a sugar and a phosphate group.

Omega loop: A loop with a narrow neck in the backbone of a protein structure that is shaped like a capital Greek Ω.

Open-face sandwich: In protein structures this term describes a single antiparallel β sheet that has a layer of α helices and loops on one side only.

Propeller twist (DNA): The dihedral angle between the planes of the pyrimidine and the purine in a base pair of a nucleic acid.

Pseudoaxial bond (a′), pseudoequatorial (e′): Bonds from atoms directly attached to the doubly bonded atoms in a monounsaturated six-membered ring are termed **pseudoequatorial** and **pseudoaxial** depending on whether the angles that they make with the best plane through the six ring atoms is near $0°$ or $90,°$ respectively.

Pseudorotation: The progression of one conformer of a five-membered ring to another conformer. In the case of cyclopentane there is no planar intermediate; all conformers have at least one carbon atom out of the plane of the other carbon atoms. The maximum pucker can, in this case, rotate with almost no potential energy barrier between conformers. Each of the multitude of possible conformers can be described in terms of the maximum pucker and the pseudorotation phase angle, that is, where the conformer lies on a pseudorotation cycle (with an arbitrarily chosen origin).

Ramachandran plot: A conformational map obtained by plotting the torsion angles about the $C\alpha$-N bond (ϕ) against the torsion angle about the C_{α}–C (carbonyl) bond (ψ) for each amino acid. This graph is named after Gopalasamudram Narayana Ramachandran.

Staggered conformation: If two atoms or groups attached at opposite ends of a bond appear at angles within $30°$ of $60°$ or $180°$ to one another when the molecule is viewed along this bond, these atoms or groups are described as **staggered**.

Strain: Strain is present in a chemical species if the bond lengths, bond angles, or dihedral angles are different from their typical values in related molecules or if the distances between nonbonded atoms are shorter than the sum of the van der Waals radii of the atoms.

Steric hindrance: Strain in a molecule caused by two atoms in that same molecule that would (under normal bonding conditions for each) approach closer than normal. To alleviate this closeness of approach the atoms readjust their positions slightly with resultant bond angle and torsion angle changes.

Syn: On the same side of a reference plane (cf. *anti* and *cis*).

Synclinal conformation (sc): Torsion angle within $30°$ of $\pm60°$.

Synperiplanar conformation (sp): Torsion angle within $30°$ of $0°$.

Torsion angle: (Sometimes called "conformational angle"). The torsion angle (or the angle of twist) about the bond B–C in a series of bonded atoms A–B–C–D is defined as the angle of rotation needed to make the projection of the line B–A coincide with the projection of the line C–D, when viewed along the B–C direction. The positive sense is clockwise. If the torsion angle is $180°$, the four atoms lie in a planar zigzag (Z shaped). If the torsion angle is $60°$, the bond to one end atom is twisted $60°$ out of the plane of the other two bonds. Enantiomers have torsion angles of equal absolute value but opposite sign.

Trans: Two groups lying on the opposite side of a reference plane. A prefix designating two groups directly across a central double bond from each other, that is, in the positions of the poles on a sphere.

Turn: A fold in the polypeptide backbone of a protein, not part of a helical structure, such that the α-carbon atoms of residues i and $i + 3$ are less than 7 Å apart.

References

1. Haworth, W. N. *The Constitution of Sugars*. Edward Arnold: London (1929).
2. Barton, D. H. R. The conformation of the steroid nucleus. *Experientia* **6**, 316–320 (1950).
3. Eliel, E. L. Conformational analysis – the last 25 years. *J. Chem. Educ.* **52**, 762–767 (1975).
4. Karplus, M. Contact electron-spin coupling of nuclear magnetic moments. *J. Chem. Phys.* **30**, 11–15 (1959).
5. Barfield, M., and Smith, W. B. Internal H–C–C angle dependence of vicinal ^{1}H–^{1}H coupling constants. *J. Amer. Chem. Soc.* **114**, 1574–1581 (1992).
6. Zacharias, D. E., Glusker, J. P., Fu, P. P., and Harvey, R. G. Molecular structures of the dihydrodiols and diol epoxides of carcinogenic polycyclic aromatic hydrocarbons. X-ray crystallographic and NMR analysis. *J. Amer. Chem. Soc.* **101**, 4043–4051 (1979).
7. Klyne, W., and Prelog, V. Description of steric relationships across single bonds. *Experientia* **16**, 521–568 (1960).
8. Bassindale, A. *The Third Dimension in Organic Chemistry*. Wiley: Chichester, New York, Brisbane, Toronto, Singapore (1984).
9. Newman, M. S. A useful notation for visualizing certain stereospecific reactions. *Record Chem. Progr., Kresge-Hooker Sci. Libr.* **13**, 111–116 (1952).
10. Newman, M. S. A notation for the study of certain stereochemical problems. *J. Chem. Educ.* **32**, 344–347 (1955).
11. van der Helm, D., Glusker, J. P., Johnson, C. K., Minkin, J. A., Burow, N. E., and Patterson, A. L. X-ray crystal analysis of the substrates of aconitase. VIII. The structure and absolute configuration of potassium dihydrogen isocitrate isolated from *Bryophyllum calycinum*. *Acta Cryst.* **B24**, 578–592 (1968).
12. Schomaker, V., Waser, J., Marsh, R. E., and Bergman, G. To fit a plane or a line to a set of points by least squares. *Acta Cryst.* **12**, 600–604 (1959).
13. Blow, D. M. To fit a plane to a set of points by least squares. *Acta Cryst.* **13**, 168 (1960).
14. Scheringer, C. A method of fitting a plane to a set of points by least squares. *Acta Cryst.* **B27**, 1470–1477 (1971).
15. Dunitz, J. D. *X-ray Analysis and the Structure of Organic Molecules*. Cornell University Press: London, Ithaca, NY, (1979).
16. Stanford, R. H., Jr., and Waser, J. The standard deviation of the torsion angle. *Acta Cryst.* **A28**, 213–215 (1972).
17. Saenger, W. *Principles of Nucleic Acid Structure*. Springer-Verlag: New York, Berlin, Heidelberg, Tokyo (1983).
18. Kemp, J. D., and Pitzer, K. S. Hindered rotation of the ethyl groups in ethane. *J. Chem. Phys.* **4**, 749 (1936).

19. Bonham, R. A., and Bartell, L. S. The molecular and rotational isomerization of *n*-butane. *J. Amer. Chem. Soc.* **81**, 3491–3496 (1959).

20. Bonham, R. A., Bartell, L. S., and Kohl, D. A. The molecular structures of *n*-pentane, *n*-hexane and *n*-heptane. *J. Chem. Phys.* **81**, 4765–4769 (1959).

21. Jorgenson, W. L., and Allen, L. C. Charge-density analysis of rotational barriers. *J. Amer. Chem. Soc.* **93**, 567–574 (1971).

22. Kingsbury, C. A. Conformations of substituted ethanes. *J. Chem. Educ.* **56**, 431–437 (1979).

23. Schrumpf, G. The conformational equilibrium of the lower alkanes. *Angew. Chem., Int. Edn. Engl.* **21**, 146 (1982).

24. Cross, L. C., and Klyne, W. IUPAC Commission of Nomenclature of Organic Chemistry. Rules for the nomenclature of organic chemistry. Section E: stereochemistry (recommendations 1974). *Pure Appl. Chem.* **45**, 11–30 (1976).

25. IUPAC tentative rules for the nomenclature of organic chemistry. Section E. Fundamental stereochemistry. *J. Org. Chem.* **35**, 2849–2867 (1970).

26. Dunitz, J. D., and Waser, J. Geometric constraints in six- or eight-membered rings. *J. Amer. Chem. Soc.* **94**, 5645–5660 (1972).

27. Hassel, O. Cykloheksanproblemet. *Tidsskrift for Kjemi, Bergvesen og Metallurgi* **3**, 32–34 (1943). **English translation.** Hedberg, K. The cyclohexane problem. In: *Topics in Stereochemistry.* (**Eds.**, Allinger, N. L., and Eliel, E. L.) Vol. 6, pp. 11–17. Wiley: New York (1971).

28. Dunitz, J. D. Approximate relationships between conformational parameters in 5- and 6-membered rings. *Tetrahedron* **28**, 5459–5467 (1972).

29. Dunitz, J. D. Conformations of medium rings. In: *Perspectives in Structural Chemistry.* (**Eds.**, Dunitz, J. D., and Ibers, J. A.) Vol. 2. pp. 1–70. Wiley: New York (1968).

30. Sachse, H. Über die geometrischen Isomerien der Hexamethylenderivate. [Concerning the geometrical isomers of hexamethylene derivatives.] *Ber. Deutsch. Chem. Gesell.* **23**, 1363–1370 (1890).

31. Kohlrausch, K. W. F., and Stockmair, W. Studien zum RAMAN-Effekt. Mitteilung LIV. Cyclohexylderivate und die Symmetrie des Cyclohexans und Dioxans. [The Raman effect. LIV. Cyclohexyl derivatives and the symmetry of cyclohexane and dioxane.] *Z. Phys. Chem.* **B31**, 382–401 (1936).

32. Rasmussen, R. S. The infra-red absorption spectrum and configuration of cyclohexane. *J. Chem. Phys.* **11**, 249–252 (1943).

33. Hassel, O., and Viervoll, H. Electron diffraction investigation of molecular structures. II. Results obtained by the rotating sector method. *Acta Chem. Scand.* **1**, 149–168 (1947).

34. Hendricks, S. B., and Bilicke, C. The space-group and molecular symmetry of β benzene hexabromide and hexachloride. *J. Amer. Chem. Soc.* **48**, 3007–3015 (1926).

35. Dickinson, R. G., and Bilicke, C. The crystal structures of beta benzene hexabromide and hexachloride. *J. Amer. Chem. Soc.* **50**, 764–770 (1928).

36. Kahn, R., Fourme, R., Andre, D., and Renaud, M. Crystal structures of cyclohexane I and II. *Acta Cryst.* **B29**, 131–138 (1973).

37. Bastiansen, O., and Hassel, O. Structure of the so-called *3cis*-decalin. *Nature (London)* **157**, 765 (1946).

38. Bastiansen, O., Ellerson, O., and Hassel, O. Electron diffraction investigation of α, β, γ, δ and ϵ benzene hexachloride. *Acta Chem. Scand.* **3**, 918–925 (1949).

39. Orloff, H. D. The stereoisomerism of cyclohexane derivatives. *Chem. Rev.* **54**, 347–447 (1954).

40. Dunitz, J. D. The two forms of cyclohexane. *J. Chem. Educ.* **47**, 488–490 (1970).

41. Barton, D. H. R., Hassel, O., Pitzer, K. S., and Prelog, V. Nomenclature of cyclohexane bonds. *Nature (London)* **172**, 1096–1097 (1953); *Science* **119**, 49 (1953).

42. Pitzer, K. S., and Beckett, C. W. Tautomerism in cyclohexane derivatives: reassignment of configuration of the 1,3-dimethylcyclohexanes. *J. Amer. Chem. Soc.* **69**, 977–978 (1947).

43. Duax, W. L., and Norton, D. A. (**Eds.**) *Atlas of Steroid Structure.* Vol. 1. Plenum Press: New York, Washington, London (1975).

44. Griffin, J. F., Duax, W. L., and Weeks, C. M. (**Eds.**) *Atlas of Steroid Structure.* Vol. 2. Plenum: New York, Washington, London (1984).

45. Baeyer, A. Über Polyacetylenverbindungen. [Concerning polyacetylene compounds.] *Ber. Deutsch. Chem. Gesell.* **18**, 2269–2281 (1885). **English translation**: In: Leicester, H. M., and Klickstein, H. S. *A Source Book in Chemistry 1400–1900.* p. 465. McGraw-Hill: New York, Toronto, London (1952).

46. Meyer, V. Ergebnisse und Ziele der stereochemischen Forschung. [Results and objectives of stereochemical research.] *Ber. Deutsch. Chem. Gesell.* **23**, 567–630 (1890).

47. Mohr, E. Die Baeyersche Spannungstheorie und die Struktur des Diamanten. [Baeyer's strain theory and the structure of diamond.] *Chemisches Zentralblatt* **2**, 1065 (1915).

48. Sayre, D., and Friedlander, P. H. Crystal structure of 9:10-dimethyl-1:2-benzanthracene. *Nature (London)* **187**, 139–140 (1960).

49. Klein, C. L., Stevens, E. D., Zacharias, D. E., and Glusker, J. P. 7,12-dimethylbenz[a]anthracene: refined structure, electron density distribution, and endoperoxide structure. *Carcinogenesis* **8**, 5–18 (1987).

50. Kitaigorodsky, A. I. *Molecular Crystals and Molecules.* Academic Press: New York (1973).

51. Newman, M. S., Lutz, W. B., and Lednicer, D. A new reagent for resolution by complex formation; the resolution of phenanthro[3,4-c]phenanthrene. *J. Amer. Chem. Soc.* **77**, 3420–3421 (1955).

52. Newman, M. S., and Lednicer, D. The synthesis and resolution of hexahelicene. *J. Amer. Chem. Soc.* **78**, 4765–4770 (1956).

53. Mackay, I. R., Robertson, J. M., and Sime, J. G. The crystal structure of hexahelicene. *J. Chem. Soc., Chem. Commun.* 1470–1471 (1969).

54. Lightner, D. A., Hefelfinger, D. T., Powers, T. W., Frank, G. W., and Trueblood, K. N. Hexahelicene: the absolute configuration. *J. Amer. Chem. Soc.* **94**, 3492–3497 (1972).

55. Linderstrøm-Lang, K. U. *Proteins and Enzymes*. Lane Medical Lectures, No. VI, p. 58. Stanford University Press: Stanford, CA (1952).

56. Bernal, J. D. Structure arrangements of macromolecules. *Disc. Faraday Soc.* **25**, 7–18 (1958).

57. Smith, J. M. A., Stansfield, R. F. D., Ford, G. C., White, J. L., and Harrison, P. M. A molecular model for the quaternary structure of ferritin. *J. Chem. Educ.* **65**, 1083–1084 (1988).

58. Pauling, L., and Sherman, J. The nature of the chemical bond. VI. The calculation from thermochemical data of the energy of resonance of molecules among several electronic structures. *J. Chem. Phys.* **1**, 606–617 (1933).

59. Corey, R. B., and Pauling, L. Fundamental dimensions of polypeptide chains. *Proc. Roy. Soc. (London)* **B141**, 10–20 (1953).

60. Momany, F. A., McGuire, R. F., Burgess, A. W., and Scheraga, H. A. Energy parameters in polypeptides. VII. Geometric parameters, partial atomic charges, nonbonded interactions, hydrogen bond interactions, and intrinsic torsional potentials for the naturally occurring amino acids. *J. Phys. Chem.* **79**, 2361–2381 (1975).

61. Ashida, T., Tsunogae, Y., Tanaka, I., and Yamane, T. Peptide chain structure parameters, bond angles and conformational angles from the Cambridge Structural Database. *Acta Cryst.* **B43**, 212–218 (1987).

62. Ramachandran, G. N., and Sasisekharan, V. Stereochemistry of polypeptide chain configurations. *J. Molec. Biol.* **7**, 95–99 (1963).

63. Ramachandran G. N., and Sasisekharan, V. (1968) Conformation of polypeptides and proteins. *Adv. Protein Chem.* **23**, 283–437 (1968).

64. Carrell, H. L., Glusker, J. P., Burger, V., Manfre, F., Tritsch, D., and Biellmann, J-F. X-ray analysis of D-xylose isomerase at 1.9 Å: native enzyme in complex with substrate and with a mechanism-designed inactivator. *Proc. Natl. Acad. Sci. USA* **86**, 4440–4444 (1989).

65. Levitt, M. Effect of proline residues on protein folding. *J. Molec. Biol.* **145**, 251–263 (1981).

66. Brandts, J. F., Halvorson, H. R., and Brennan, M. Consideration of the possibility that the slow step in protein denaturation reactions is due to cis-trans isomerism of proline residues. *Biochemistry* **14**, 4953–4963 (1975).

67. Nemethy, G., and Scheraga, H. A. Protein folding. *Quart. Rev. Biophys.* **10**, 239–352 (1977).

68. Rossmann, M. G., and Argos, P. The taxonomy of protein structure. *J. Molec. Biol.* **109**, 99–129 (1977).

69. Rossmann, M. G., and Argos, P. Protein folding. *Annu. Rev. Biochem.* **50**, 497–532 (1981).

70. Richardson, J. S. The anatomy and taxonomy of protein structure. *Adv. Protein Chem.* **34**, 167–339 (1981).

71. Richardson, J. S. Describing patterns of protein tertiary structure. *Methods in Enzymology* **115**, 349–358 (1985).

72. Lesk, A. M., and Hardman, K. D. Computer-generated pictures of proteins. *Methods in Enzymology* **115**, 381–390 (1985).

73. Karle, I. L. X-ray analysis: conformation of peptides in the crystalline state. In: *The Peptides.* (**Eds.**, Gross, E., and Meienhofer, J.) Vol. 4., pp. 1–54. Academic Press: New York (1981).

74. IUPAC-IUB Commission on Biochemical Nomenclature. *Biochemistry* **9**, 3471–3479 (1970).

75. Cheng, X., and Schoenborn, B. P. Repulsive restraints for hydrogen bonding in least-squares refinement of protein crystals. A neutron diffraction study of myoglobin crystals. *Acta Cryst.* **A47**, 314–317 (1991).

76. Vriend, G., and Sander, C. Quality control of protein models: directional atomic contact analysis. *J. Appl. Cryst.* **26**, 47–60 (1993).

77. Richards, F. M. The protein folding problem. *Scientific American* **264(1)**, 54–63 (1991).

78. Baker, E. N., Blundell, T. L., Cutfield, J. F., Cutfield, S. M., Dodson, E. I., Dodson, G. G., Hodgkin, D. M. C., Hubbard, R. E., Isaacs, N. W., Reynolds, C. D., Sakabe, K., Sakabe, N., and Vijayan, N. M. The structure of 2Zn pig insulin crystals at 1.5 Å resolution. *Phil. Trans. Roy. Soc. (London)* **B319**, 369–456 (1988).

79. Anfinsen, C. B., Haber, E., Sela, M., and White, F. H., Jr. The kinetics of formation of native ribonuclease during oxidation of the reduced polypeptide chain. *Proc. Natl. Acad. Sci. USA* **47**, 1309–1314 (1961).

80. Magdoff-Fairchild, B., Swerdlow, P. H., and Bertles, J. F. Intermolecular organization of deoxygenated sickle haemoglobin determined by X-ray diffraction. *Nature (London)* **239**, 217–219 (1972).

81. Finch, J. T., Perutz, M. F., Bertles, J. F., and Döbler, J. Structure of sickled erythrocytes and of sickle-cell hemoglobin fibers. *Proc. Natl. Acad. Sci. USA* **70**, 718–722 (1973).

82. Astbury, W. T., and Woods, H. J. The X-ray interpretation of the structure and elastic properties of hair keratin. *Nature (London)* **126**, 913–914 (1930).

83. Astbury, W. T., and Street, A. X-ray studies of the structure of hair, wool and related fibers. *Phil. Trans. Roy. Soc. (London)* **A230**, 75–101 (1931).

84. Astbury, W. T. *Fundamentals of Fibre Structure.* Oxford University Press: Oxford (1933).

85. Astbury, W. T., and Sisson, W. A. X-ray studies of the structures of hair, wool and related fibers. III. The configuration of the keratin molecule and its orientation in the biological cell. *Proc. Roy. Soc. (London)* **A150**, 333–351 (1935).

86. Hussain, F. Pauling and the alpha helix. *Trends Biochem. Sci.* **1**, N37–N38 (1976).

87. Pauling, L., Corey, R. B., and Branson, H. R. The structure of proteins: two hydrogen-bonded helical configurations of the polypeptide chain. *Proc. Natl. Acad. Sci. USA* **37**, 205–211 (1951).

88. Hol, W. G. J., van Duijnen, P. T., and Berendsen, H. J. C. The α-helix dipole and the properties of proteins. *Nature (London)* **273**, 443–446 (1978).

89. Fairman, R., Shoemaker, K. R., York, G. J., Stewart, J. M., and Baldwin, R. L. Further studies of the helix dipole model: effects of a free α-NH_3^+ or α-COO^- group on helix stability. *Proteins: Structure, Function, and Genetics* **5**, 1–7 (1989).

90. Schiffer, M., and Edmundson, A. B. Use of helical wheels to represent the structures of proteins and to identify segments with helical potential. *Biophys. J.* **7**, 121–135 (1967).

91. Hendrickson, W. A., and Teeter, M. Structure of the hydrophobic protein crambin determined directly from the anomalous scattering of sulphur. *Nature (London)* **290**, 107–113 (1981).

92. Perutz, M. F. New X-ray evidence on the configuration of polypeptide chains. Polypeptide chains in poly-γ-benzyl-L-glutamate, keratin and hemoglobin. *Nature (London)* **167**, 1053–1054 (1951).

93. Kendrew, J. C., Dickerson, R. E., Strandberg, B. E., Hart, R. G., and Davies, D. R. Structure of myoglobin. *Nature (London)* **185**, 422–427 (1960).

94. Donohue, J. Hydrogen-bonded helical configurations of the polypeptide chain. *Proc. Natl. Acad. Sci. USA* **39**, 470–478 (1953).

95. Low, B. M., and Baybutt, R. B. The π-helix — a hydrogen bonded configuration of the polypeptide chain. *J. Amer. Chem. Soc.* **74**, 5806–5807 (1952).

96. Pauling, L., and Corey, R. B. The pleated sheet, a new layer configuration of polypeptide chains. *Proc. Natl. Acad. Sci. USA* **37**, 251–256 (1951).

97. Blake, C. C. F., Koenig, D. F., Mair, G. A., North, A. C. T., Phillips, D. C., and Sarma, V. R. Structure of hen egg-white lysozyme: a three-dimensional Fourier synthesis at 2 Å resolution. *Nature (London)* **206**, 757–761 (1965).

98. Richardson, J. S. β sheet topology and the relatedness of proteins. *Nature (London)* **268**, 495–500 (1977).

99. Lifson, S., and Sander, C. Antiparallel and parallel β strands differ in amino acid residue preferences. *Nature (London)* **282**, 109–111 (1979).

100. Lifson, S., and Sander, C. Specific recognition in the tertiary structure of β sheets of proteins. *J. Molec. Biol.* **139**, 627–639 (1980).

101. Karle, I. L., Karle, J., Mastropaolo, D., Camerman, A., and Camerman, N. [Leu[5]]enkephalin: four cocrystallizing conformers with extended backbone that form an antiparallel β sheet. *Acta Cryst.* **B39**, 625–637 (1983).

102. Chothia, C. Conformation of twisted β pleated sheets in proteins. *J. Molec. Biol.* **75**, 295–302 (1973).

103. Venkatachalam, C. M. Stereochemical criteria for polypeptides and proteins. V. Conformation of a system of three linked peptide units. *Biopolymers* **6**, 1425–1436 (1968).

104. Lewis, P. N., Momany, F. A., and Scheraga, H. A. Chain reversals in proteins. *Biochim. Biophys. Acta* **303**, 221–229 (1973).

105. Smith, J. A., and Pease, L. G. Reverse turns in peptides and proteins. *CRC Crit. Rev. Biochem.* **8**, 315–399 (1980).

106. Wilmot, C. M., and Thornton, J. M. Analysis and prediction of the different types of β turn in proteins. *J. Molec. Biol.* **203**, 221–232 (1988).

107. Richardson, J. S., Getzoff, E. D., and Richardson, D. C. The β bulge: a common small unit of non-repetitive protein structure. *Proc. Natl. Acad. Sci. USA* **75**, 2574–2578 (1978).

108. Leszczynski, J., and Rose, G. Loops in globular proteins: a novel category of secondary structure. *Science* **234**, 849–855 (1986).

109. Brew, K., Vanaman, T. C., and Hill, R. C. Comparison of the amino acid sequence of bovine α-lactalbumin and hens eggwhite lysozyme. *J. Biol. Chem.* **242**, 3747–3749 (1967).

110. Richardson, J. S., Richardson, D. C., Thomas, K. A., Silverton, E. W., and Davies, D. R. Similarity of three-dimensional structure between the immunoglobulin domain and the copper, zinc superoxide dismutase subunit. *J. Molec. Biol.* **102**, 221–235 (1976).

111. Levitt, M., and Chothia, C. Structural patterns in globular proteins. *Nature (London)* **261**, 552–558 (1976).

112. Branden, C., and Tooze, J. *Introduction to Protein Structures.* Garland: New York, London (1991).

113. Czerwinski, E. W., Mathews, F. S., Hollenberg, P., Drickamer, K., and Hager, L. P. Crystallographic study of cytochrome b_{562} from *E. coli*. *J. Molec. Biol.* **71**, 819–821 (1972).

114. Richardson, J. S., Thomas, K. A., Rubin, B. H., and Richardson, D. C. Crystal structure of bovine Cu, Zn superoxide dismutase at 3 Å resolution: chain tracing and metal ligands. *Proc. Natl. Acad. Sci. USA* **72**, 1349–1353 (1975).

115. Banner, D. W., Bloomer, A. C., Petsko, G. A., Phillips, D. C., Pegron, C. I., Wilson, I. A., Corran, P. H., Furth, A. J., Milman, J. D., Offord, R. E., Priddle, J. D., and Waley, S. G. Structure of chicken muscle triose phosphate isomerase determined crystallographically at 2.5 Å resolution using amino acid sequence data. *Nature (London)* **255**, 609–614 (1975).

116. Eisenberg, D., and Kauzmann, W. *The Structure and Properties of Water.* Oxford University Press: New York, Oxford (1969).

117. Traube, J. [I.] Ueber die Capillaritätsconstanten organischer Stoffe in wässerigen Lösungen. [On the capillary constants of organic compounds in aqueous solutions.] *Justus Liebigs Annalen der Chemie* **265**, 27–55 (1891).

118. Kauzmann, W. Some factors in interpretation of protein denaturation. *Adv. Protein Chem.* **14**, 1–63 (1959).

119. Tanford, C. *The Hydrophobic Effect: Formation of Micelles and Biological Membranes.* Wiley: New York, London, Sydney, Toronto (1973).

120. Cochran, W., Crick, F. H. C., and Vand, V. The structure of synthetic polypeptides. I. The transform of atoms on a helix. *Acta Cryst.* **5**, 581–586 (1952).

121. Stokes, A. R. The theory of X-ray fibre diagrams. *Prog. Biophys.* **5**, 140–167 (1955).

122. Franklin, R. E., and Gosling, R. G. Molecular configuration in sodium thymonucleate. *Nature (London)* **171**, 740–741 (1953).

123. Watson, J. D., and Crick, F. H. C. A structure for deoxyribose nucleic acid. *Nature (London)* **171**, 737–738 (1953).

124. Judson, H. F. *The Eighth Day of Creation. The Makers of the Revolution in Biology.* Simon & Schuster: New York (1979).

125. Dickerson, R. E. DNA structures form A to Z. *Methods in Enzymology* **211**, 67–111 (1992).

126. Altona, C., Geise, H. J., and Romers, C. Conformation of non-aromatic ring compounds–XXV. Geometry and conformation of ring D in some steroids from X-ray structure determinations. *Tetrahedron* **24**, 13–32 (1968).

127. Wing, R. M., Drew, H. R., Takano, T., Broka, C., Tanake, S., Itakura, K., and Dickerson, R. E. Crystal structure analysis of a complete turn of B-DNA. *Nature* (*London*) **287**, 755–758 (1980).

128. Dickerson, R. E., Drew, H. R., Conner, B. N., Wing, R. M., Fratini, A. V., and Kopka, M. L. The anatomy of *A-*, *B-*, and *Z-*DNA. *Science* **216**, 475–485 (1982).

129. Wang, A. H.-J., Quigley, G. C., Kolpak, F. J., Crawford, J. L., van Boom, J. H., van der Marel, G., and Rich, A. Molecular structure of a left-handed double helical DNA fragment at atomic resolution. *Nature* (*London*) **282**, 680–686 (1979).

130. Dickerson, R. E., Bansal, M., Calladine, C. R., Diekmann, S., Hunter, W. N., Kennard, O., Lavery, R., Nelson, H. C. M., Olson, W. K., Saenger, W., Shakked, Z., Sklenar, H., Soumpasis, D. M., Tung, C-S., von Kitzing, E., Wang, A. H. -J., and Zhurkin, V. B. Definitions and nomenclature of nucleic acid structure parameters. *J. Molec. Biol.* **205**, 787–791 (1989).

Atomic and Molecular Displacements

Atoms are in motion in the crystalline state and even at absolute zero. The extent of their motion increases as the temperature is raised. But the amount and directions of their motion also depend on the nature of their immediate atomic environment in the crystal and the space available for such motions. Leopold Ruzicka once remarked to Jack Dunitz[1] that "a crystal is a chemical cemetery." In his view a crystal consists of "long rows of molecules, fixed in a rigid geometrical arrangement, lifeless compared with the molecular mazurkas that can be imagined to occur in solution." This view of "lifeless" atoms is, however, incorrect; atoms and molecules are in motion in the crystal, a fact that was recognized by Peter J. W. Debye,[2] soon after the first X-ray diffraction experiment. He predicted that diffraction intensities at higher 2θ-values would be decreased if the temperature were higher because atomic motion endows a larger "apparent size" to these atoms, as shown in Figure 13.1. This implies that a measure of the falloff in the intensity of the X-ray diffraction pattern as a function of scattering angle will provide information on the extent of molecular motion in crystals.

This predicted intensity falloff was observed experimentally by W. H. Bragg.[3] He studied rock salt at two temperatures, 15 and 37°C, and found that the intensities of Bragg reflections, particularly those at high 2θ-values, were smaller at the higher temperature. This effect was described in Chapter 7, where it was shown that the falloff in intensity as a function of scattering angle (2θ) is routinely used to obtain an average B value for the entire crystal structure by way of a Wilson plot.[4]

Atomic vibrations are displacements from equilibrium positions, with frequencies of such vibration typically of the order of 10^{13} per second. The frequencies of X rays (velocity of light/wavelength of X rays) are much faster, of the order of $(3 \times 10^{10}$ cm/sec$)/(1.5 \times 10^{-8}$ cm$) = 2 \times 10^{18}$ per second. As a result, the atom may vibrate and be "viewed" by X rays

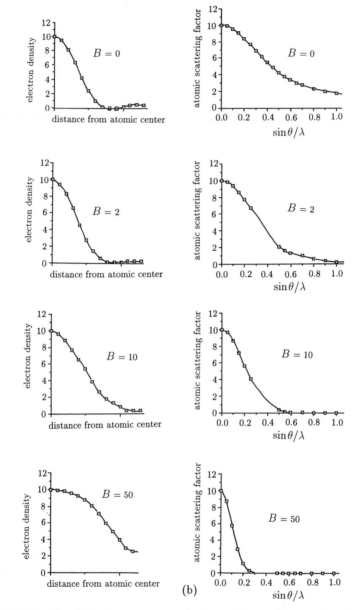

FIGURE 13.1. The effect of apparent atomic size on the atomic scattering factors. (a) (left) Falloff in intensity (represented by the atomic scattering factor) with increasing temperature (larger *B* values), and (b) (right) the corresponding apparent atomic size. As one proceeds vertically down the page in this Figure, the displacement parameters increase (implying a higher temperature), and disorder is increased. Electron-density scale arbitrarily scaled to 10 at the nucleus.

as apparently stationary, but displaced in a random manner from its average location to some other location along one of its possible vibration pathways. Bertram Terence M. Willis and Alan W. Pryor write[5] "An X-ray crystallographer wishing to determine the positions occupied by the atoms in a crystal would find his task easier if only the atoms kept still. ··· A portrait photographer trying to take a high-definition photograph of a fidgeting subject faces similar difficulties."

Because the diffraction experiment involves the average of a very large number of unit cells (of the order of 10^{18} in a crystal used for X-ray diffraction analysis), minor static displacements of atoms closely simulate the effects of vibrations on the scattering power of the "average" atom. In addition, if an atom moves from one disordered position to another, it will be "frozen in time" during the X-ray diffraction experiment. This means that atomic motion and spatial disorder are difficult to separate from each other by simple experimental measurements of intensity falloff as a function of $\sin\theta/\lambda$. For this reason, **atomic displacement parameter** is considered a more suitable term[6] than the terms that have been used historically, such as **temperature factor, thermal parameter**, or **vibration parameter** for each of the correction factors included in the structure factor equation. A displacement parameter may be isotropic (with equal displacements in all directions) or anisotropic (with different values in different directions in the crystal).

The displacements of atoms may take several forms (Figure 13.2). These have been described by Jack Dunitz, Verner Schomaker, and Kenneth N. Trueblood as follows: "The perfectly ordered crystal would have every atom firmly fixed to its own perfectly defined site in each unit cell for the entire period of observation."[6] There are, however, various types of disorder from unit cell to unit cell. "If the atom jumps to a different site, that is one kind of disorder [a mixture of static and dynamic disorder]; if it moves to and fro, that is another kind of disorder [dynamic disorder]; if it is forever in one site in a certain unit cell and in a different site in another cell, that is still another kind [static disorder]."[6] Each of these types of vibrations, displacements, and disorder has somewhat similar effects on the intensities of Bragg reflections; the effect they have in common is that they *reduce these intensities by an amount that increases with increasing scattering angle*, 2θ, as shown in Figure 13.1.

Now that fairly precise measures of electron density can be made, atomic displacement parameters can be refined so that the best possible fit to the experimental electron-density profiles of each atom is obtained. This is done by the introduction of additional atomic parameters, one parameter if the displacements are isotropic, six if they are anisotropic. When this least-squares refinement of displacement parameters is completed, the crystallographer is then left with the problem of explaining the atomic displacement parameters so obtained in terms of *vibration, static disorder, dynamic disorder, or a combination of these.*

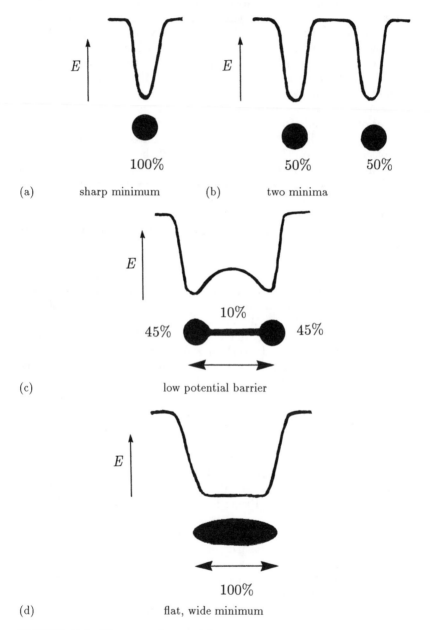

FIGURE 13.2. Diagrams of potential energy curves for an atomic position showing (a) perfect order, (b) static order, (c) dynamic disorder, and (d) a mixture of static and dynamic disorder of the position of one atom. The percentage occupancy of each site is listed. Vertical axes = energy, horizontal axes = position.

The purpose of this Chapter is to acquaint the chemist and biochemist with some of the possibilities of studying details of molecular motion in crystals by analyses of X-ray diffraction data. Dunitz, Schomaker, and Trueblood remarked:[6] "Nowadays, with improvements in the precision and accuracy of intensity measurement, it has become possible to detect quite subtle effects arising from the motion of molecules, and portions of molecules, in crystals."

13.1 Vibration effects in crystals

Atomic vibrations and disorder cause an apparent broadening of the electron density profiles of atoms. This is represented by a **probability density function (pdf)**, which is a combination of the scattering factors and the displacement parameters of the atoms. It is defined as the probability of finding an atom in a small defined volume when the atom in question is displaced from its rest position. It is diagrammed in Figure 13.3. The Fourier transform of this broadened electron-density profile (see Figure 13.3) yields the "temperature factor" (which, as we have just described, may or may not be a function of the actual temperature) multiplied by the scattering factor of the atom of interest. As the extent of atomic displacements increases, the profile of each atom in the electron-density map becomes more smeared out and the X-ray scattering power for each atom is reduced at high $\sin\theta/\lambda$ values. The number of Bragg reflections that

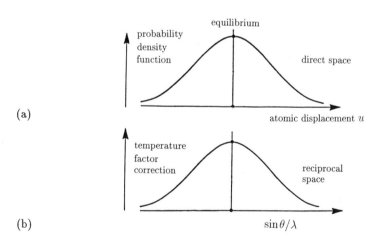

FIGURE 13.3. (a) The probability density function as a function of atomic displacement, and (b) its Fourier transform, the displacement parameter (temperature factor) measured in terms of $\sin\theta/\lambda$.

can be measured is then reduced, because, at higher scattering angles (2θ), the X-ray scattering is negligible as a result of destructive interference in the disordered structure (see Figure 13.4). This means that, at lower temperatures, the resolution of the diffraction data is increased, with concomitant higher precision in the crystal structure determination.

13.1.1 The Debye–Waller equation

The falloff in the intensity of Bragg reflections as a function of temperature, I_T, may be expressed, in the simplest case (that of cubic crystals composed of only one type of atom) by the exponential function

$$I_T = I_0 e^{-2M}. \tag{13.1}$$

where Equation 13.1 is called the Debye–Waller equation, first proposed by Peter Debye[7–9] and later modified by Ivar Waller.[10] In Equation 13.1, I_0 is the intensity that should be obtained if the atom were "stationary." I_T is the intensity that is actually measured,[7–11] (reduced from I_0 by

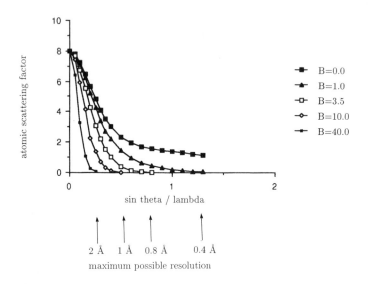

FIGURE 13.4. Effect of different atomic displacement parameters on the resolution of a crystal structure determination. As the Debye–Waller (displacement) factor B is increased, the scattering factors fall to zero at smaller and smaller scattering angles θ, thereby limiting the resolution (as shown). Thus, if the overall value of B is greater than 40 Å^2 in any crystal, the resolution of the structure determination cannot be better than 2 Å, because beyond this limit the intensities of Bragg reflections are generally too weak to measure.

an exponential factor $-2M$). The term M is related to $\langle u^2 \rangle$, the mean-square displacement of each atom from its average position, in the following way:

$$M = 8\pi^2 \langle u^2 \rangle (\sin^2 \theta / \lambda^2). \tag{13.2}$$

As a result, Equation 13.1 may be rewritten for a scattering factor f_T at temperature T (without the factor of 2, because the intensity is related to f^2) as

$$f_T = f_o e^{-M} = f_o e^{-B \sin^2 \theta / \lambda^2}, \tag{13.3}$$

where:

$$B = 8\pi^2 \langle u^2 \rangle = 8\pi^2 U. \tag{13.4}$$

This is the form used for calculating structure factors. One of the terms in Equation 13.4, B or U, is generally used as the displacement parameter by crystallographers, and some form of this parameter is refined in the least-squares procedure.

Dunitz[12] wrote of these equations: "Debye's paper, published only a few months after the discovery of X-ray diffraction by crystals, is remarkable for the physical intuition it showed at a time when almost nothing was known about the structure of solids at the atomic level." Ewald described how:[13] "The temperature displacements of the atoms in a lattice are of the order of magnitude of the atomic distances $\cdots$ The result is a factor of exponential form whose exponent contains besides the temperature the order of interference only [h,k,l, hence $\sin^2 \theta / \lambda^2$]." The importance of Debye's work, as stressed by Ewald,[14] was in "paving the way for the first immediate experimental proof of the existence of zero-point energy, and therewith of the quantum statistical foundation of Planck's theory of black-body radiation."

The energetics of such atomic motion can be investigated. If the probability density function is a Gaussian function, the potential energy in which the atom vibrates will be isotropic and harmonic and will have a normal **Boltzmann distribution** over energy levels. This potential energy will have the form:

$$V(x) = \kappa x^2 / 2, \tag{13.5}$$

where κ is a force constant and x represents the displacement of the atom from its equilibrium position. The probability of finding an atom at a position x is then:

$$\text{probability} = k_B \exp(-\kappa x^2 / 2kT), \tag{13.6}$$

where k_B is Boltzmann's constant, and T is the absolute temperature. This equation is also a Gaussian distribution. When the temperature is low and the force constant is large, anisotropic displacements will have small amplitudes and the Gaussian approximation will be a good one.

13.1.2 Disorder in crystals

Atomic disorder rather than atomic vibration also constitutes a displacement phenomenon. A common manifestation of the possibility of atomic disorder in crystals is the presence of two or more conformers that differ only modestly in, for example, the orientations of side chains or the conformations of ring structures.[15-18] Such flexibility may be observed either with two ordered conformers in the asymmetric unit, or as a disordered portion of the molecule. The latter possibility means that in some unit cells the molecule is in one conformation, while in other unit cells the molecule is in an alternate conformation (Figure 13.5), and the resulting electron density will be an average of the situation in all the unit cells throughout the crystal. Such disorder causes a breakdown in the regular periodicity in the packing in the crystal, and may be so severe that the crystals do not diffract well. This is equivalent to a very high B value in Figure 13.4.

The method used to distinguish between atomic vibration and disorder is to study the effect of changing the temperature on the displacement parameters.[19] If the displacements are mainly due to atomic vibrations, they will decrease as the temperature is lowered. If, however, there is disorder, meaning that there is more than one site in the unit cell, separated by energy barriers, then the magnitudes of the atomic displacement parameters would not be expected to be decreased significantly as the crystal is cooled. An example is provided by crystalline ferrocene,[1,20,21] which has a phase transition at 164 K. Above this temperature the displacement parameters of the ring atoms remain constant as the temperature is varied because there is disorder involving two or more ring orientations. Below the transition temperature, values of the displacement parameters depend on the temperature,[20-22] indicating that there is extensive atomic motion.

Disorder may strongly influence the precision with which *all* atomic positions in that crystal can be measured, and it is often difficult to obtain a model that accounts adequately for the disorder. Consequently diffraction effects resulting from disorder may be compensated for by (erroneous) shifts in other parameters that involve ordered atoms. Bond distances may appear unusually long or short and bond angles may be atypical. These effects almost certainly result from inadequacies in the model. If the disorder occurs in only a small fraction of the content of the asymmetric unit, it is presumed that the conformation of the (well-ordered) remainder of the asymmetric unit (molecule) is driving the crystal packing rather than vice versa. Disorder is often observed in solvent of crystallization, since there may be space in the crystal structure for the solvent molecules to organize in one of several manners.[23]

Different molecular components of a crystal structure may have very different atomic displacement parameters such that the relative influences of these different components on the diffraction pattern will also differ.

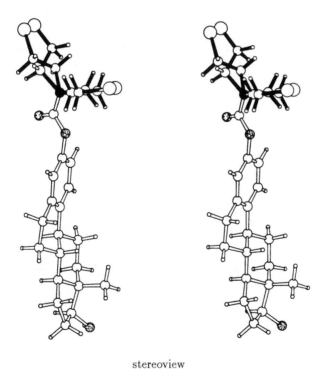

(a)

(b)   stereoview

FIGURE 13.5. Disorder in the side chains of estramustine (Ref. 18). (a) Chemical formula and (b) stereoview of estramustine with bonds in black in the disordered area of the crystal structure (two possible arrangements of the $N(CH_2CH_2Cl)_2$ group).

An example (Figure 13.6) is provided by the crystal structure of a β-cyclo-dextrin complex of *N*-acetylphenylalanine methyl ester.[24] X-ray diffraction data were measured for the same single crystal at four temperatures (297, 220, 160, and 110 K). The structure is an inclusion complex with the ester held loosely within its framework. Therefore it is not surprising to find that the host (β-cyclodextrin) molecules have lower displacement parameters than do the guest (*N*-acetylphenylalanine methyl ester) molecules. At room temperature the contribution of the guest molecules to the diffraction intensities is very small because their displacement parameters are so high, while at 110 K the contribution to the diffracted intensities of these same guest molecules becomes significant. Plots at two temperatures of the averaged X-ray scattering curves of the two components as a function of data set resolution, $\sin\theta/\lambda$ (Figure 13.6), provides an indication of the relative scattering power of these two different components.

13.2 Representations of displacement parameters

The simplest assumption to make about atomic displacements is that they are the same in all directions, that is, isotropic. Then only a single term is needed to describe them. The exponential factor is $\exp(-B_{iso}\sin^2\theta/\lambda^2)$, where B_{iso} is the isotropic displacement parameter,

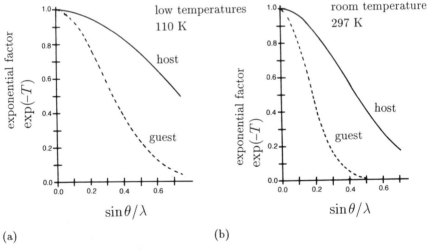

(a)   (b)

FIGURE 13.6. Scattering curves for the host and guest molecules in a β-cyclodextrin (host) – 2*N*-acetylphenylalanine methyl ester (guest) complex. Data for the host (β cyclodextrin) are indicated by solid lines, and for the guest (the methyl ester) by broken lines. The guest is somewhat disordered in the channel provided by the more ordered structure of the crystallized host (Ref. 24). (a) Low and (b) high temperatures.

and B_{iso} equals $8\pi^2 \langle u^2 \rangle$ (*cf.* Equation 13.4), and $\langle u^2 \rangle$ is the mean square displacement of the atom from its equilibrium position.

$$B_{iso} = 8\pi^2 U_{iso} = 8\pi^2 \langle u^2 \rangle \simeq 79 \langle u^2 \rangle. \tag{13.7}$$

Since $U_{iso} = \langle u^2 \rangle$, it is the best form of the displacement parameter to list (see Table 13.1) because it represents a physical feature of the structure, the mean-square amplitude of displacement.

The magnitudes of these displacements may interest the reader. If B_{iso}, found from a Wilson plot (Figure 7.14, Chapter 7), is 4 Å^2, then $U = \langle u^2 \rangle$ is about 0.05 Å^2 and the root-mean-square amplitude $\sqrt{\langle u^2 \rangle}$ is approximately 0.22 Å. Experimental values of $\langle u^2 \rangle$, and $\sqrt{\langle u^2 \rangle}$ for organic molecules are generally 0.02–0.07 Å^2 or 0.1–0.3 Å^2, respectively ($B =$ 2–6 Å^2). In minerals, which are harder materials, B may be below 1 Å^2, while values of B for larger molecules, such as proteins, are generally much higher.[25]

Atoms in crystals seldom have isotropic environments, and a better approximation (but still an approximation) is to describe the atomic motion in terms of an ellipsoid, with larger amplitudes of vibration in some directions than in others. Six parameters, the anisotropic vibration or displacement parameters, are introduced for each atom. Three of these parameters per atom give the orientations of the principal axes of the ellipsoid with respect to the unit cell axes. One of these principal axes is the direction of maximum displacement and the other two are perpendicular to this and also to each other. The other three parameters per atom represent the amounts of displacement along these three ellipsoidal axes. Some equations used to express anisotropic displacement parameters, which may be reported as B_{ij}, U_{ij}, or β_{ij}, are listed in Table 13.1. Most crystal structure determinations of all but the largest molecules include anisotropic temperature parameters for all atoms, except hydrogen, in the least-squares refinement. Usually, for brevity, the **equivalent isotropic displacement factor** U_{eq}, is published.[26,27] This is expressed as:

$$U_{eq} = \frac{1}{3} \sum_i \sum_j U_{ij} a_i^* a_j^* \mathbf{a}_i \cdot \mathbf{a}_j$$
$$= \frac{1}{(24\pi^2)} \sum_i \sum_j B_{ij} a_i^* a_j^* \mathbf{a}_i \cdot \mathbf{a}_j \tag{13.8}$$

13.2.1 ORTEP diagrams

Visual representations of the extent to which an atom is displaced in various directions are obtained from "thermal" ellipsoid plots of displacement parameters; these are commonly drawn by use of the program ORTEP

TABLE 13.1. Equations used for displacement parameters.

Isotropic displacements:

$$\exp\left(-B\frac{\sin^2\theta}{\lambda^2}\right) \tag{13.1.1}$$

$$\exp\left(-8\pi^2 U\frac{\sin^2\theta}{\lambda^2}\right) \tag{13.1.2}$$

Anisotropic displacements:

$$\exp\left[-(h^2\beta_{11} + k^2\beta_{22} + l^2\beta_{33} + 2hk\beta_{12} + 2hl\beta_{13} + 2kl\beta_{23})\right] \tag{13.1.3}$$

$$\exp\left[-\frac{1}{4}(h^2 B_{11}(a^*)^2 + k^2 B_{22}(b^*)^2 + l^2 B_{33}(c^*)^2 + 2hk B_{12}(a^*)(b^*)\right.$$
$$\left. + 2hl B_{13}(a^*)(c^*) + 2kl B_{23}(b^*)(c^*))\right] \tag{13.1.4}$$

$$\exp\left[-2\pi^2(h^2 U_{11}(a^*)^2 + k^2 U_{22}(b^*)^2 + l^2 U_{33}(c^*)^2 + 2hk U_{12}(a^*)(b^*)\right.$$
$$\left. + 2hl U_{13}(a^*)(c^*) + 2kl U_{23}(b^*)(c^*))\right] \tag{13.1.5}$$

(*Oak Ridge Thermal-Ellipsoid Plot*), written by Carroll K. Johnson.[28] An example is given in Figure 13.7. The lengths of the principal axes of the ellipsoids in such diagrams are proportional to the root-mean-square (r.m.s.) displacements of atoms in these directions.[29-31] The size of an ellipsoid is determined by the probability that the atom lies inside it. A preselected value of the probability that the atom lies within a surface, e.g. 50%, may arbitrarily be chosen. An octant of the ellipsoid may be cut out in order to enhance the impression of a three-dimensional image, and indicate the relative lengths of the principal axes. The larger an ellipsoid, the larger the anisotropic displacement parameters that it represents and the more the atom is vibrating or is otherwise displaced. The most precise crystal structure determinations, usually carried out at low temperatures, show very small ellipsoids. Disorder is represented in such ORTEP plots by apparently elongated ellipsoids, covering the alternate positions of an atom (see Figure 13.8). The value of ORTEP plots for the representation of anisotropic displacement parameters remains, although newer programs[32,33] are now available that provide for studies of differences in anisotropic displacements (which can be negative), and that allow for real-time graphics display of results.

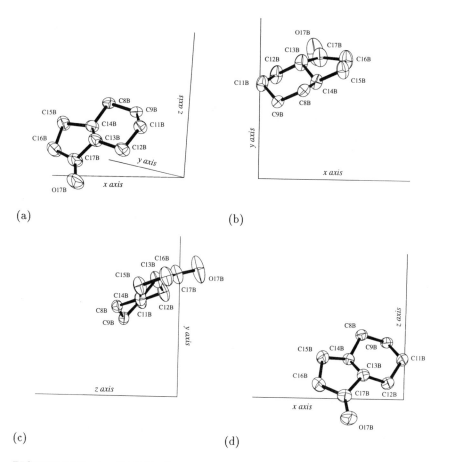

FIGURE 13.7. An ORTEP plot showing the anisotropy of O17B. The lengths l of the principal axes and their direction cosines are listed for this atom. Note in (c) how the longest principal axis of the thermal ellipsoid of O17B lies approximately in the direction of the y axis, with a direction cosine ($\cos \theta_2$) near −1.0.

	principal axes	direction cosines		
	l	$\cos \theta_1$	$\cos \theta_2$	$\cos \theta_3$
1	0.49	0.22	−0.94	−0.26
2	0.29	0.97	0.18	0.18
3	0.20	−0.12	−0.29	0.95

(a) General view, (b) xy plane, (c) yz plane, and (d) xz plane. (Courtesy Jonah Erlebacher.)

(e) *Directions of principal axes (compare with the ellipsoids illustrated in* (b), (c), *and* (d).

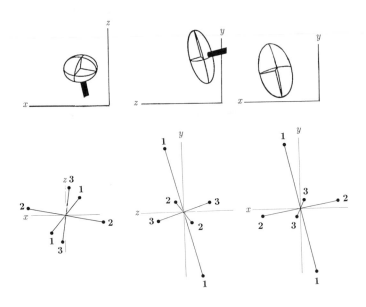

(f) *The method for obtaining the principal axes of a thermal ellipsoid is described with a numerical example by Donald Sands.*[34] *If the aniostropic temperature factors are β^{ij} in a monoclinic unit cell with dimensions a, b, c, α, β, γ, the equation used is:*

$$\left| \begin{pmatrix} \beta^{11} & \beta^{12} & \beta^{13} \\ \beta^{21} & \beta^{22} & \beta^{23} \\ \beta^{31} & \beta^{32} & \beta^{33} \end{pmatrix} \begin{pmatrix} a^2 & 0 & ac\cos\beta \\ 0 & b^2 & 0 \\ ac\cos\beta & 0 & c^2 \end{pmatrix} - \lambda \begin{pmatrix} 1 & 0 & 0 \\ 0 & 1 & 0 \\ 0 & 0 & 1 \end{pmatrix} \right| = 0.$$

The determinant is expanded to give the cubic equation:

$$A - B\lambda + C\lambda^2 - \lambda^3 = 0$$

which has three roots (see Figure 12.5, Chapter 12). *The reader should consult* Ref. 34 *for further information).*

FIGURE 13.7 (cont'd). (e) An enlarged view of a thermal ellipsoid, and (f) calculation of principal axes.

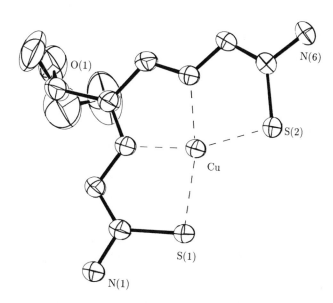

(a)

(b)

FIGURE 13.8. (a) Formula of copper KTS [3-ethoxy-2-oxobutyraldehyde bis (thiosemicarbazonato) copper II] (Ref. 17). (b) and (c). Two views of the thermal ellipsoids. Note the high thermal motion in the side chain, but not in the ring system.

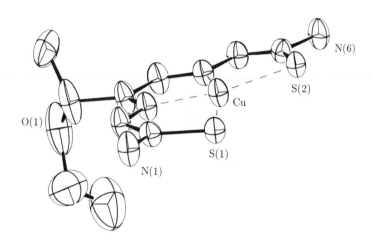

FIGURE 13.8 (cont'd). (c) Side view of the copper complex (Ref. 17).

13.2.2 Potential functions: harmonic and anharmonic

The vibrations of molecules in crystals can be divided into two groups.[5] The **external or lattice modes of vibration** are oscillations in which the entire (rigid) molecule moves, and, in so doing, involves neighboring molecules. The **internal or molecular modes of vibration** are the stretching and bending of bonds, that is, intramolecular vibrations. It is generally found that a low-frequency external mode (also called a soft mode) will contribute more to the molecular motion than will a high-frequency internal mode.

When deviations from a (parabolic) harmonic potential energy function are significant, as, for example, when rotating groups show large displacements in specific directions, this energy function is better represented by a more complicated (anharmonic) function,[35–37] (see Figure 13.9). Anharmonicity or the curvilinear motion of atoms make it necessary to introduce additional parameters (called higher cumulants) into the exponential expression for anisotropic displacement parameters used in the least-squares refinement.[38,39] These higher cumulants include, in addition to the six anisotropic parameters describing a quadratic potential, 10 cubic terms, 15 quartic ones, and so forth. Alternatively a three-dimensional Gram–Charlier series[40] (which is a Taylor's series-like expansion) may be used.[38] An analysis of these higher-order terms, if mathematically justified, can often lead to a better description of the probability density function and will therefore give a better picture of the complicated atomic motion.[38]

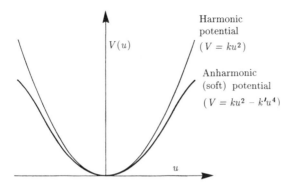

FIGURE 13.9. Harmonic and anharmonic potential energy functions. The harmonic function $V(u)$ is a simple parabola, but the anharmonic potential falls off at high displacements u.

13.2.3 Atomic displacement parameters: what to believe

It is important that highly anisotropic temperature parameters should make physical sense, given the molecular structure and the packing environment. For example, a large flat molecule would be expected to show most motion perpendicular to its molecular plane. Anisotropic displacement parameters are the parameters most susceptible (in the least-squares refinement) to the effects of errors in the data set and also to correlations between different parameters. For example, there is a high correlation between the scale factor and the displacement parameters, since both reduce the magnitudes of the structure factor, although in different ways; the reduction involves a constant term for the scale factor, but an exponential term for displacement parameters. Atomic occupancy factors (population parameters), in a similar way, are usually strongly correlated with displacement parameters. It is often difficult to determine the extent to which the observed effect involves each of the parameters. Systematic errors in the data set can also affect anisotropic displacement parameters (see Chapter 7). For example, if the absorption corrections applied to data from crystals with unequal dimensions, such as those with a needle-like morphology, are inadequate, then systematic errors will be introduced by the crystal shape into the anisotropic displacement parameters to compensate (incorrectly) for these effects.

Disorder in a crystal structure is frequently revealed by the shapes of the thermal ellipsoids obtained from the least-squares refinement of the anisotropic displacement parameters. An example is provided by the crystal structure determination of potassium dihydrogen isocitrate.[15] One carboxyl oxygen atom is very anisotropic as a result of two possible hydrogen bonding schemes in which it can take part (Figure 13.10).

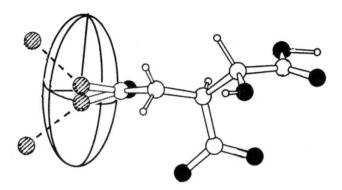

(a)

(b)

FIGURE 13.10. Alternate hydrogen bonding schemes in potassium isocitrate (Ref. 15), demonstrated by a highly anisotropic "thermal ellipsoid." (a) Chemical formula, and (b) results of X-ray study (ordered oxygen atoms black). Two possible locations of an oxygen atom (shaded) give two modes of hydrogen bonding. Thus, the crystal structure is disordered. Some oxygen atoms form hydrogen bonds in one way, others hydrogen bonding in the other manner.

The crystal structure found represents an average of these two possibilities. If the anisotropic displacement parameters do not correspond to ellipsoids but to other quadratic surfaces that are not everywhere positive, the atomic displacement parameters may lose their physical significance (they become **nonpositive definite**).

Some criteria have been suggested in order to decide when a thermal motion analysis can be expected to yield meaningful results.[6,41] Two of them, the rigid-bond test and the rigid-molecule test, investigate whether or not bonds or portions of molecules are considered to move in a physically reasonable way. They will be described in turn.

13.2.4 The rigid-bond test

The rigid-bond test explores whether two bonded atoms have mean-square displacements that are nearly the same along the direction of the bond, so that the actual bond length is not changed during vibration. The test was introduced by Fred Hirshfeld[42] and is based on the observation that bond-stretching vibrations are generally much smaller than other types of vibration, such as bond bending or torsional vibrations. As a result, a bond length should not change appreciably even if the two atoms composing it are vibrating; the two atoms should move in synchrony along the direction of the bond even if they do not move in synchrony in other directions. For example, if two atoms, A and B, form a bond, their displacement parameters (obtained from the least squares refinement (Chapter 10) may be examined as follows:

$$\Delta_{A-B} = (\text{displacement}_{A \to B}) - (\text{displacement}_{B \to A}) = 0$$
$$= \sqrt{\langle u^2 \rangle}_{A \to B} - \sqrt{\langle u^2 \rangle}_{B \to A} = 0, \tag{13.9}$$

where Δ_{A-B} is the difference in the mean square displacement amplitude of atom A in the direction of atom B, ($\text{mean square displacement}_{A \to B}$), and the mean square displacement amplitude of atom B in the direction of atom A, ($\text{mean square displacement}_{B \to A}$). If the difference in Equation 13.9 is not comparable to a value smaller than the e.s.d. values of $\langle u^2 \rangle$, the relevance of the reported anisotropic displacement parameters as indicators of thermal motion should be investigated further.

13.2.5 The rigid-molecule test

Sometimes when crystal structures are refined anisotropically it becomes clear that certain groups, phenyl groups, for example, are vibrating as rigid bodies. Indications of rigid segments within a molecule are provided by an examination of Δ_{A-B} values (as defined in Equation 13.9). In rigid segments, *all* of the Δ_{A-B} values for *all* intramolecular distances should be near zero; this means not just bonded atoms but other atoms in the same segment. If this is so, the motion of the atoms in this segment of the molecule is correlated, and the segment is behaving as a rigid body. Δ_{A-B} values for intramolecular distances other than those in the rigid segment of the molecule may be large, however, indicating that the rest of the molecule is moving relative to the rigid segment. If all the values of Δ_{A-B} are large (compared with their variances), it is not appropriate to consider any part of the molecule to be behaving as a rigid body.

13.2.6 Nonrigid molecules

The internal motion of molecules in crystals is, however, not always confined to rigid-body motion or uncorrelated motion of segments of the structure. Often the overall motion is the result of a combination of motions, for example torsional librations about bonds and out-of-plane bending of the same segment. These motions may or may not be correlated and each of these effects must be taken into account in the thermal motion analysis technique.[43,44] The molecule can be considered to be composed of coupled rigid-body segments (the segmented-body model),[45] or the effects of any internal librations can be evaluated.[43]

An example of an analysis of the motion in a nonrigid molecule is provided by the structures of two crystalline modifications of triphenyl phosphine oxide, one monoclinic and one orthorhombic. These were each determined at two temperatures, 100 and 150 K.[46] The internal molecular motion is similar at both temperatures but, as expected, is of higher amplitude at the higher temperature. The molecules did not, however, behave as rigid bodies but displayed internal torsions of the three phenyl rings about their C–P bonds. In each case, one phenyl ring displayed a mean-square libration of considerably greater amplitude than the other two. This finding was consistent with features of the distribution of triphenylphosphine oxide conformations in 62 crystal structures.[47] The effect is illustrated in Figure 13.11. Initially the molecule is like a propellor with three O–P–C–C torsion angles of 40°. One of these torsion angles increases while the other two decrease by about half the amount. Then the torsion angles adjust to give a transition state structure for the rotations with torsion angles of 90°, 10°, and –10°.

13.2.7 Thermal diffuse scattering

There are several experimental clues to the presence of motion or disorder in crystals. The falloff in intensity as a function of scattering angle may be much greater than expected, indicating motion and/or disorder. In addition to this information, the character of the diffraction spots on photographs may give a hint of disorder. Thermal diffuse scattering is evidenced by additional spots or areas of intensity, aptly described as "diffuse," that appear around normal Bragg reflections on X-ray photographs, even when the incident radiation is truly monochromatic.[48] These effects were described by William A. Wooster as follows:[49] "Experimental crystallographers have generally regarded this diffuse scattering as an unwanted effect which produced a background darkening of their X-ray photographs or an equivalent effect in their diffractometers ..." This thermal diffuse scattering, however, originally considered a nuisance to the experimenter, can now be analyzed to give information on the elastic properties of crystals, and the force constants between their constituent atoms.

Normal Bragg reflections are produced as a result of the interaction of X rays with the electrons of atoms that are arranged on a regular periodic lattice. Diffuse scattering is due to crystal lattices that depart from a regular periodic character.[50–52] When atoms vibrate in a crystal they must affect their neighbors and, because of their sizes, are more likely

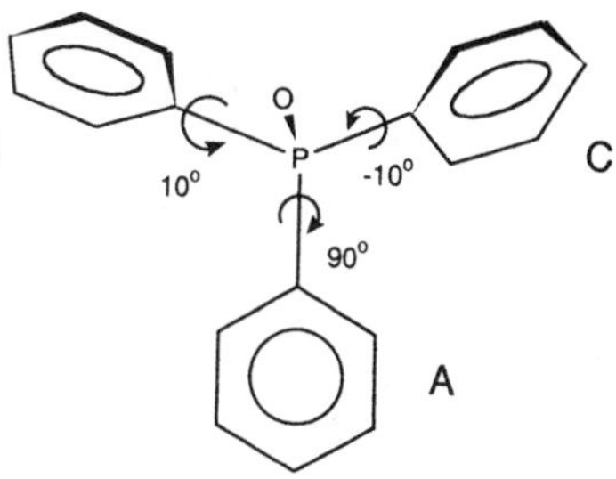

(a)

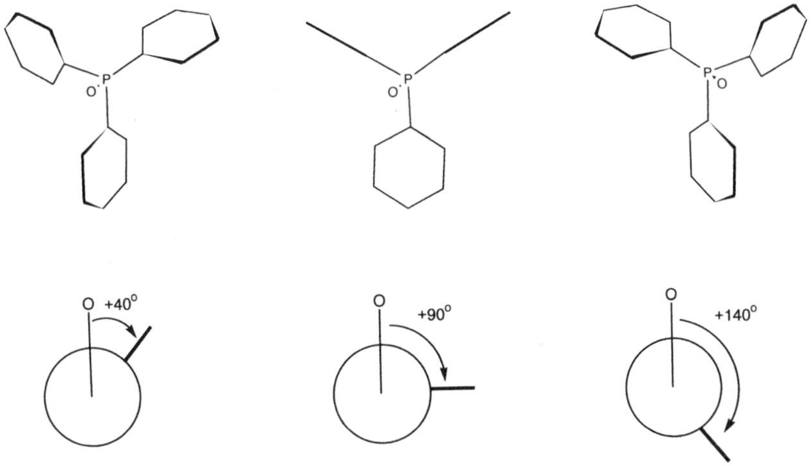

(b) 40°, 40°, 40°   (c) 90°, 10°, –10°   (d) 140°, –40°, –40°

FIGURE 13.11. Interconversions in triphenylphosphine oxide, $(C_6H_5)_3P=O$ (Refs. 46,47). (a) Diagram of the overall scheme of molecular rotation. O–P–C–C torsion angles 90°, 10°, –10°. (b), (c), and (d), The molecular shapes on interconversion. The molecule in (d) is the mirror image of that in (b). In the conversion of (c) to (b) or (d) the C–C–P=O torsion angles (in degrees), for rings A, B, and C, change respectively as:

Ring	A	B	C
(c) to (b)	–50	+30	+50
(c) to (d)	+50	–50	–30
(b) to (d)	+100	–80	+80

to cause them to be displaced in the same rather than in other directions. As a result the crystal lattice can be considered to vibrate in some way because the atoms are displaced in an interconnected way. Consider mica and graphite; these have layer structures, and vibration amplitudes would be expected to be larger perpendicular to the layers, that is, along the **c** direction (in their unit cells). As a result there will be diffuse intensity around the 0001 Bragg reflection. Thermal diffuse scattering causes a change in the energy (frequency) of the X rays, but the effect is very small. The effect is, however, much more noticeable for neutrons and is called "inelastic neutron scattering." Willis and Pryor[5] write: "It is because slow neutrons have *both* a wavelength comparable with interatomic spacings ⋯ that they constitute such a marvellous probe for examining the vibrational properties of crystals." For more information the reader is encouraged to study Reference 5.

13.3 Motion in proteins

Proteins, also, can vibrate in part or as a whole in the crystalline state. In contrast to crystals of small molecules, crystals of proteins almost invariably contain large numbers of molecules of solvent of crystallization, often corresponding to 50% or more of the unit cell volume. The extent to which these water molecules are ordered varies dramatically. Near the surface of the protein the water molecules may be well ordered. Beyond the first layer, water molecules typically show increasing levels of disorder. In addition, because of the high solvent content, there may be considerable motion and disorder in the protein molecule, particularly in the orientations of side chains. As a result, the number of independent Bragg reflections that can be measured is reduced, and this effectively reduces the resolution of the electron-density map.

In contrast to the case for small molecule crystal structure determinations, there are usually too few experimental data (Bragg reflections) to permit a refinement of anisotropic displacement parameters for protein structures. Often there are not even sufficient data to allow one to refine four parameters per atom (x, y, z, and B_{iso}); in fact, there may be fewer experimental observations than parameters to be determined. It might have been expected that lowering the temperature during data measurement would help by reducing atomic motion, leading to more Bragg reflections at high values of $\sin\theta/\lambda$ (see Figure 7.22, Chapter 7). The advantage of using low-temperature intensity data measurement for protein structures is that the crystal is stabilized. On the other hand, the effect of lowering the temperature on the intensities of Bragg reflections is small if the resolution is not very high, because the scattering factor curve has only been sampled at low values of $\sin\theta/\lambda$. Clearly, these limitations hinder the analysis of thermal motion and disorder for proteins.

Some protein structures can be determined to high resolution and the results of anisotropic refinements have, in a few instances, provided detailed information on the motions and flexibility of the molecules.[53-55] The observed motions in proteins have been grouped into three categories.[56,57] In the first category are proteins with flexibly linked globular domains, domains that are relatively independent of one another. The best examples are provided by antibody molecules. Crystallographic analyses of a large number of different antibodies have shown that the structures of the domains are well conserved, but that the orientation of the domains with respect to each other can vary considerably. This flexibility makes it possible for the antibodies to bind many different ligands with variable orientations. A second type of motion observed in proteins involves hinged globular domains. Again these are found in multidomain proteins. Essentially rigid globular domains move relative to one another but, in contrast to the motion in proteins with flexibly linked domains, hinged proteins appear in only two or so stable conformations. Hexokinase which is a monomer with two lobes separated by a distinct cleft, for example, appears to adopt two stable conformations.[58,59] When glucose is bound to this enzyme it induces a conformational change in the protein that alters the relative orientation of one lobe with respect to the other, a hinged-motion effect. The third type of motion involves order–disorder transitions. This type of motion is quite different from the previous two. Trypsinogen, the precursor of trypsin in the body, has an activation domain.[60] This domain consists of four segments of polypeptide chain, and it forms a patch on the surface of the protein. In the native structure this patch is disordered, and, as a result, the electron density is extremely diffuse and almost impossible to interpret. On the other hand, when specific ligands are bound to the zymogen, this activation domain becomes ordered and, as a result, can be observed in the electron density.

Techniques for the quantitative analysis of motion use principles derived from small molecule crystallography. In the course of refinement of a crystallographic model, one can adjust both the average coordinates and the disorder parameter for each atom so that it is possible to optimize the agreement between observed and calculated structure factors (see Chapter 10). In practice, as already discussed, the B value reflects vibrations in the molecules in the crystal and imperfections in the packing of molecules in the lattice. It is not practical to separate these two components of disorder in a protein refinement. This is because there are generally too few Bragg reflections measured to permit the determination of all the positional and disorder parameters with reasonable accuracy. The B value reported for a protein is generally a measure of the localized order of the molecule, rather than providing detailed information about atomic displacements. In addition, thermal parameters are affected seriously by both experimental errors in the data and by inadequacies in the current structural model. Small displacements, such as alternate side-chain conformations, can occasionally be identified in

proteins. As for small molecules, electron-density maps of proteins correspond to time averages of the structure over the lengths of the experiments and to space averages over all the molecules in the crystal. This
makes it difficult to distinguish between genuine motion and static disorder within the crystals. Figure 13.12, for example, shows the electron
density of tyrosine-62 in human lysozyme.[61] The electron density can be
interpreted in two ways: either the side chain adopts one conformation
in some of the molecules and a different conformation in others or spends
half of its time in each of the alternative conformations. An overall analysis of isotropic displacement parameters in proteins, however, can show
how lobes of an enzyme can be flexible[62] or that decreased motion of the
active site is achieved on substrate binding.[63] Comparisons of B values
for human and hen egg white lysozyme show a remarkable similarity.[61]
This suggests, at least in this instance, that the primary component of
thermal disorder arises from dynamic motions. In addition, the B values
of human lysozyme are larger in two "wings" of the active site. This
is shown in Figure 13.13 and implies relative ease of movement of these
portions of the molecule. For example,[61,64,65] for lysozyme, one of these

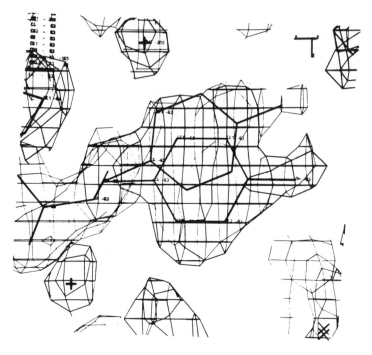

FIGURE 13.12. Electron density for tyrosine 62 in human leukemic lysozyme. Two
different conformations of the side chain fit the electron-density map, indicating disorder. (Courtesy D. C. Phillips)

FIGURE 13.13. Diagram of the human lysozyme molecule, with residues having displacements greater than 0.2 Å^2 outlined by parallel lines. The active site cleft runs almost vertically down the center of the drawing. (Courtesy D. C. Phillips)

modes involves the opening and closing of the active-site cleft. These are, in fact, two ways of saying the same thing either by giving refined displacement parameters or by describing the motion that would give these displacement parameters. A typical plot of B values versus amino-acid sequence number for a protein is given in Figure 13.14.

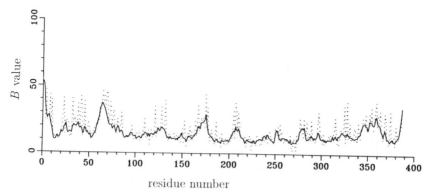

FIGURE 13.14. B values versus sequence number for amino acid residues in the protein D-xylose isomerase. Backbone C_α B values are joined by solid lines, those for the average of the side-chain B values (which are generally higher) are indicated by broken lines. (Courtesy H. L. Carrell).

13.4 Effects of displacements on molecular geometry

Thermal motion and disorder are two of the main factors limiting our ability to obtain accurate distances and angles by X-ray diffraction methods. As quoted earlier in this Chapter,[5] "··· if only the atoms kept still ··· " it would make the task much easier. An understanding of the effects of thermal motion and disorder is necessary so that they may be corrected for when determining bond distances. In order to obtain the most accurate set of bond distances, the X-ray diffraction experiment should be performed at as low a temperature as possible, thus reducing all atomic motion.

13.4.1 Libration and bond shortening

One of the effects of thermal motion is the apparent, but not real, shortening of bond lengths. This comes from oscillations of a bonded atom along an arc, rather than a straight line, an effect called **libration**. The distance between mean positions of atoms in a crystal must be distinguished from the average of the instantaneous separation of these atoms. The observed distance is usually shorter (never longer) than the actual average distance between instantaneous positions of atoms.

In the mid 1950s Durward W. J. Cruickshank[66-69] noted that atoms in, for example, a rotating molecule, are displaced towards the rotation axis. Rotational oscillations of molecules, such as found in crystalline benzene near its melting point, will cause an apparent displacement of atomic positions from their true positions because the best fit to the electron density should be curvilinear but, with the limitations of present-day techniques, is generally linear (Figure 13.15). If the root-mean-square amplitude of libration about an axis is ω (in radians), then the apparent (but not real) shortening of the bond, d, is:

$$\Delta d = d - d\cos\omega = d\omega^2/2. \tag{13.10}$$

For an r.m.s. libration amplitude of $5°$ (0.085 radians), the contraction in a distance of 1.5 Å is 0.005 Å, but if ω is $10°$ (0.17 radians), then the contraction[12] is approximately 0.025 Å.

13.4.2 Riding-motion corrections

Another effect of thermal motion is the riding motion of one atom on another, like that of a man (B) bouncing as he moves along riding a horse (A). In this model[45,70] there are two bonded atoms, A and B, one of which, B, is much lighter than the other, A. Atom B then "rides" on atom A in such a way that the translational motion of B contains all the motion of A plus an additional motion that is not correlated with that of A. It is assumed that the center of motion is located on atom A, and that the second bonded atom B is riding on that point.

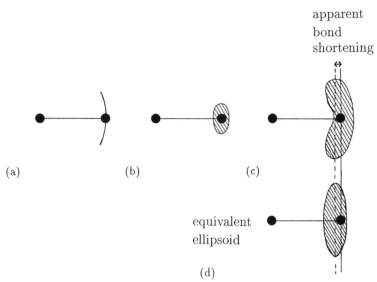

FIGURE 13.15. Libration causing apparent (not real) bond shortening. (a) Because the atom is vibrating but the bond length stays the same, the atom (b) vibrates in an arc (librates) (c). The electron density is interpreted as an ellipsoid (d), but its major axis is displaced as shown in that the bond appears shortened.

To determine a riding motion correction, the value of the mean-square displacement of atom A is subtracted from that of atom B (which is always greater). Bond lengths, corrected for riding motion, can then be computed. This model does not consider, however, any curvilinear component in the motion of A that is transferred to B. Hydrogen atoms (acting as atom B above) provide a good example of the importance of riding-motion correction. Riding-motion corrections are most appropriately applied to the anisotropic displacement parameters of hydrogen atoms if they have been obtained from neutron diffraction data, since X-ray diffraction studies do not give precise enough results for hydrogen atoms to warrant such an analysis.

13.4.3 Thermal motion analysis: TLS

The anisotropic rigid-body displacements of a molecule can be described[45,71] in terms of translation ($\mathbf{T}$, vibration along a straight-line path), libration ($\mathbf{L}$, vibration along an arc), and screw ($\mathbf{S}$, a combination of vibration and translation that may be regarded as vibration along a helical path). The mean-square amplitude of translational vibration is usually referred to as a system of Cartesian coordinates and unit vectors. $\mathbf{S}$ is the mean correlation between libration about an axis and translation parallel to this axis. Each of these three components can be expressed as tensors (20 in all plus one that is arbitrary). **T** is a symmetrical tensor with three perpendicular axes and expressed by six parameters. **L**, also a symmetrical tensor, is similarly expressed by six parameters, and is the mean-square amplitude of angular oscillation (libration) about three axes. **S** is an unsymmetrical tensor and is expressed by nine parameters, one of which is arbitrary. This **S** tensor is zero for centrosymmetric molecules.

The set of anisotropic displacement parameters, obtained from the least-squares refinement of the crystal structure (as described by Chapter 10) can be analyzed to obtain **T**, **L** and **S**. It has been assumed that there is no correlation between the motion of different atoms. Values of U_{ij} are analyzed (again by an additional least-squares analysis) in such a way that good agreement is obtained between the refined values and those predicted when constants have been obtained for the T, L, and S tensors. The total number of anisotropic displacement parameters (6 per atom) is the input, and a total of 12 parameters for a centrosymmetric structure, or 20 parameters for a noncentrosymmetric structure, is the output of this least-squares analysis. The results consist of the molecular translational (**T**), librational (**L**), and screw (**S**) tensors. This treatment leads to estimates of corrections that should be made to bond distances. On the other hand, this type of analysis cannot be used for intermolecular distances because the correlation between the motion of different molecules is not known.

13.5 Uses of anisotropic displacement parameters

Analyses of anisotropic displacement parameters give several pieces of information of interest to the chemist.

13.5.1 Studies of orientational disorder

When the determination of a crystal structure is difficult or refinement gives a strangely shaped molecule, it is possible that the molecule is disordered in a site in the crystal. This happens when the available space in the crystal is such as to accommodate two different orientations of the molecule. For example, in the crystal structure of 5-methylchrysene,[72] one of the two molecules in the asymmetric unit is disordered. It was difficult to solve the crystal structure until the nature of the disorder was realized. In a given site, this disordered molecule may be in one of two possible orientations. In some places in the disordered electron density map, atoms in the two orientations of the disordered molecule are near each other and their positions can be approximated (erroneously) by the use of highly anisotropic temperature parameters. The result is that the anisotropic temperature parameters on refinement do not make any sense until the nature of the disorder is understood.

A molecule that appears to be orientationally disordered in the crystalline state is cyclohepta[*de*]naphthalene (pleiadiene),[73] shown in Figure 13.16. The crystal structure is disordered at room temperatures, but ordered at 78 K. Four possible orientations of the molecule could be discerned in electron-density maps at the higher temperatures (100 K, 135 K, 200 K, and 295 K). The most populated arrangement, in which molecules pack in pairs about a center of symmetry [Figure 13.16(a)], varies in occurrence from 100% at 78 K to 41% at room temperature. As the temperature is raised, some reorientation by an in-plane rotation of 180° occurs [Figure 13.16(b)]. In addition, molecular orientations where rotations of 90° has occurred are observed. The 180° rotation causes pairs of double bonds to lie in the correct orientation for dimerization [marked by arrows in Figure 13.16(b)]. This explains why the dimerization can only occur at the higher temperatures, when such disorder is more prevalent. Azulene is another molecule that exhibits orientational disorder.[74] Interestingly, ^{1}H NMR indicates that the disorder in azulene is dynamic rather than static,[75] suggesting that the molecule can flip over in the crystal.

13.5.2 Thermal motion in aromatic hydrocarbons

The effects of correlated thermal motion on displacement parameters were recognized[66] soon after the introduction of the use of anisotropic displacement parameters. The reason for these differences puzzled Cruickshank until he realized that they could be explained by rotational oscillations of the molecules. They were analyzed in a study of crystalline benzene near its melting temperature[76,77] The bonding geometry from the crystal structure determination[76] did not agree with that obtained by Raman spectroscopists.[78] The spectroscopists had studied several deuterated benzenes and determined accurate bond lengths [1.397(1) Å for the C–C bonds] which differed markedly from the X-ray results [1.378(3) Å] at –3°C, very near the melting point (+5.5°C). The thermal motion of the benzene rings in the crystal at this temperature (–3°C) was appreciable, particularly in-plane libration, and was responsible for shortening the apparent bond lengths. A rigid-body thermal motion analysis[69] describing the rigid-body motion in terms of two tensors, $\mathbf{T}$ and $\mathbf{L}$ ($\mathbf{S} = 0$ because the molecule is centrosymmetric) showed r.m.s. amplitudes of translation of 0.24 Å at 270 K and an angular oscillation in the molecular plane about the sixfold axis of about 7.9°. These resulted in correction of the bond lengths to 1.392 Å (see Equation 13.10). Later, neutron diffraction studies[79,80] provided accurate hydrogen atom coordinates and resulted in carbon–carbon bond lengths of 1.398(8) Å, bringing the values determined by diffraction and spectroscopy into good agreement.

In naphthalene and other polycyclic aromatic hydrocarbons apparent bond lengths become shorter, especially when they are more distant from the molecular center. When corrected for effects of rigid-body libration, this trend disappears. Studies of naphthalene,[81] anthracene,[82] and 4-hydroxybiphenyl[83] illustrate this. In naphthalene

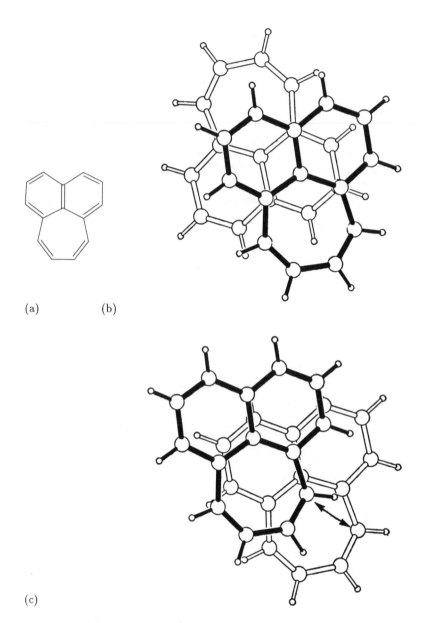

(a)   (b)

(c)

FIGURE 13.16. Orientational disorder in cyclohepta[*de*]naphthalene (Ref. 73). (a) Chemical formula. Shown are a pair of molecules approximately 3.5 Å apart, packing (b) head-to-tail, and (c) head-to-head. The arrangement in (c) can lead to dimerization in the solid state on exposure to X rays because of the proximity of the double bonds (indicated by a two-headed arrow).

and anthracene the principal axis of the **T** tensor lies, within experimental error, along the long axis of the molecule. Most of the atomic displacements are due to external modes of vibration (intermolecular modes), but the internal modes (intramolecular modes) were corrected for by calculations.[84,85] Values of the diagonal terms in the **T** and **L** matrices for naphthalene and anthracene are shown in Figure 13.17. The root-mean-square amplitudes and angles of oscillation are the square roots of these values. They are of the order of 0.1 to 0.2 Å and 3 to 5° at room temperature, and are much smaller at lower temperatures.

13.5.3 Estimation of barriers to rotation of groups

A TLS analysis can lead to data on force constants, frequencies and rotation barriers. For example, torsional amplitudes and quadratic force constants have been found for a series of groups.[86] The heights of the rotational energy barriers E_B in the crystal are much larger than those for free molecules. The increase is due to packing effects and can be reproduced by force-field calculations.

If the mean-square amplitude of libration $\langle \Phi^2 \rangle$ has been determined from a TLS analysis of the anisotropic displacement parameters, the force constant (for a harmonic oscillator) is

$$f = RT/\langle \Phi^2 \rangle, \tag{13.11}$$

where R is the gas constant. This equation can be used to find force constants and potential barriers for torsional motion.[86,87]

The potential energy $V(\Phi)$ is periodic during torsional rotation, giving n peaks and troughs during 360° rotation (n is an integer and is known from the molecular symmetry). This potential energy curve can be approximated by a cosine function of the form:

$$V(\Phi) = E_B(1 - \cos n\Phi)/2, \tag{13.12}$$

where E_B is the height of each potential energy barrier. If Φ is small, Equation 13.11 can be expanded as a Taylor series and truncated to give:

$$V(\Phi) = E_B n^2 \langle \Phi^2 \rangle /4. \tag{13.13}$$

Since $V(\Phi)$ also equals $RT/2$ (a classical Boltzmann distribution) and also equals $\kappa \langle \Phi^2 \rangle /2$ (a harmonic oscillator, Equation 13.5), it follows, using Equation 13.12, that:

$$E_B \approx 2RT/\left(n^2 \langle \Phi^2 \rangle\right) \approx 2\kappa/n^2, \tag{13.14}$$

although more precise formulæ have been developed and are now used.[22] Thus a value of the energy barrier E_B follows from the force constant, or the value of $\langle \Phi^2 \rangle$ from the TLS analysis.

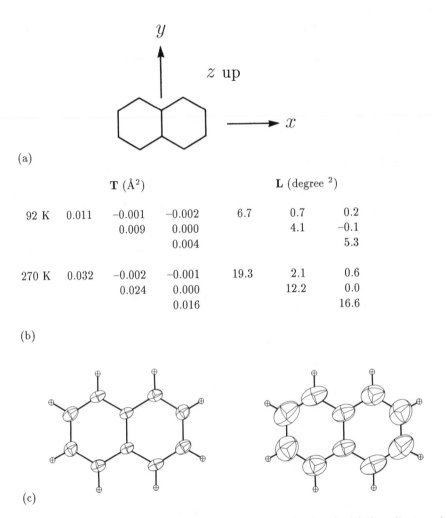

(a)

T (Å²)			L (degree ²)		
92 K	0.011	−0.001	−0.002	6.7	0.7	0.2

92 K   0.011   −0.001   −0.002   6.7   0.7   0.2
                0.009   0.000   4.1   −0.1
                         0.004   5.3

270 K 0.032   −0.002   −0.001   19.3   2.1   0.6
                0.024   0.000   12.2   0.0
                         0.016   16.6

(b)

(c)

FIGURE 13.17. Temperature effects in naphthalene (Ref. 81). (a) Coordinate axial system. (b) Values for **T** and **L** for naphthalene at 92 K and 270 K. Note the increase in numbers at the higher temperature. (c) Thermal ellipsoids at (left) 92 K and (right) 270 K. Note the increase in the sizes of the ellipsoids at the higher temperature where the atomic motion is higher.

Emily Maverick and Jack Dunitz used the anisotropic displacement parameters from crystal structure analyses of metallocenes at various temperatures to estimate the barriers for the rotation of an individual C_5H_5 ring in the crystal about its five fold axis.[22] The average mean-square amplitude of libration $\langle \Phi^2 \rangle$ is is found to be 28° at 101 K, corresponding to an energy barrier of approximately 2 kcal mol⁻¹ (9.3 kJ mol^{-1}) (the approximate Equation 13.13 would give about 3 kcal mol^{-1} = 12 kJ mol^{-1}). Once the potential energy barrier is known, the rate at which the molecule rotates from one side of the barrier to the other can be estimated. Values for rotation of cyclopentadienyl rings give 3 × 10^7 sec^{-1} at 100 K, agreeing well with results obtained by other physical chemical methods.[88]

The extent to which a phenyl group flips 180° in the crystalline state has been investigated by Keith Prout and co-workers.[89] Cross-polarized magic-angle spinning ^{13}C solid-state NMR spectra have been measured for some penicillin salts at a variety of temperatures and with a ^{13}C ^{1}H decoupling frequency of 70 kHz. For example, for the cesium salt, at 213 K there is no exchange, and sharp signals are obtained, but at 230–240 K, some exchange broadening occurs and a single peak is observed. The rate of flip of the phenyl group reaches the decoupling frequency (70 kHz) at 274 K, corresponding to 70,000 flips per second. As the speed of the flip exceeds the decoupling frequency when the temperature is raised further, two sharp lines are obtained again. The energy barrier is 40 kJ mol^{-1} (approximately 9.6 kcal mol^{-1}), but there does not seem to be sufficient space for a flip. Thus, in this crystal structure, a flip of the phenyl group is presumed to occur by an orchestrated motion of groups in that area of the crystal structure.

13.5.4 Distinction between the isotopes of hydrogen

The mean-square displacement amplitudes that result from an X-ray diffraction experiment have been used to distinguish isotopes of hydrogen. The zero-point vibration amplitude of an atom depends upon its mass, and therefore hydrogen and deuterium should have different displacement parameters, those for deuterium being smaller, particularly at low temperatures. This type of isotope effect has been used to probe the stereospecificity of an enzyme reaction[90] by distinguishing between hydrogen and deuterium. Crystal structure determinations for the reaction products of the enzymatic addition of D_2O and of H_2O to fumaric acid were studied; X-ray diffraction data were measured from crystals of phenylethylammonium hydrogen malate cooled to 95 K. The resulting isotropic U values of the hydrogen atoms in the deuterated malate versus those in the nondeuterated malate were compared (see Figure 13.18). A decrease was observed in the value of U for one of the hydrogen atoms in the enzymatically deuterated structure and this atom was assumed to be the deuterium atom added enzymatically. Although the difference in U values were small for this one atom, it indicated the correct absolute configuration, known from other methods. There is, however, a caveat to this type of experiment; it is at the limit of precision of the data and, unless done very carefully, can provide results that might be incorrect and therefore misleading. The method is introduced here because it illustrates an effect of isotopic variation in a structure.

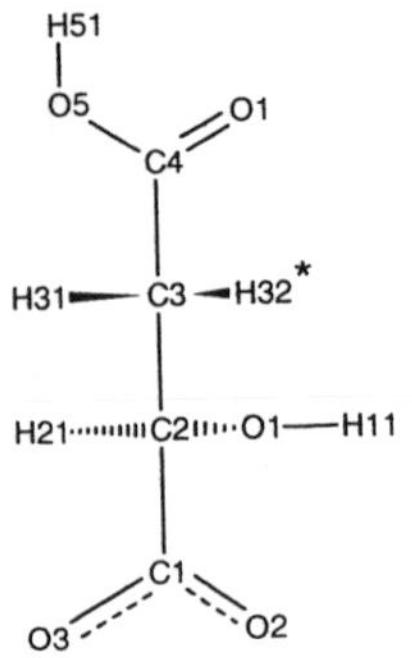

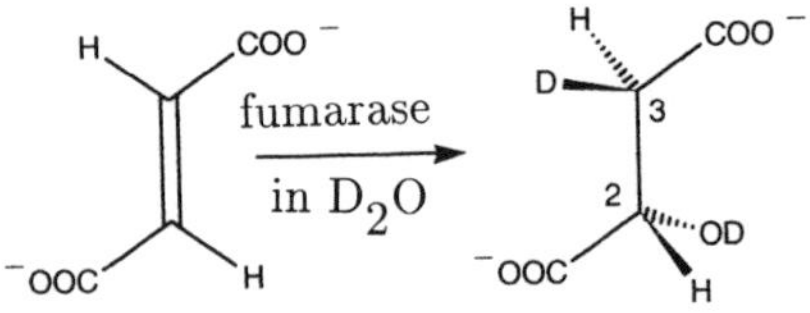

(a)

	monoammonium 2S-malate (all H)	2S-3-D-malate (one D)	difference in U values
H21	0.012(3)	0.011(3)	-0.001
H31	0.022(3)	0.022(4)	-0.000
H32	0.026(4)	0.021(4)	-0.005*

(b)

FIGURE 13.18. Absolute configuration of the product of an enzymatic reaction from displacement parameters (Ref. 90.). (a) The atomic numbering and the stereochemical course of the enzymatic reaction (from the U values listed) are shown. (b) U (in Å^2) values at 95 K. Note that H32 has the smallest U value in the product of enzymatic deuteration (malate), and the largest difference in U values. This implies that it is the hydrogen atom replaced enzymatically by D. Note, however, that the difference in U values is only 1.3σ.

13.5.5 Specific heat, characteristic temperature and entropy

Some measure of the variety of information available from displacement parameters can be illustrated by studies of cubic crystals of the elements. For example, the **specific heat** of an element is related to these displacement parameters. The **heat capacity** of an element or compound denoted C_v if measured at constant volume, is the quantity of heat required to raise the temperature of the atomic (or molecular) weight in grams of the material by 1°. The higher the value, the more heat that has to be put into the system. At low temperatures, the atomic heat capacities decrease, so at low enough temperatures they approach zero. The heat capacity, per gram atom, is observed to be approximately constant, 6 cal deg^{-1}, for all elements, a law put forward by Pierre Louis Dulong and Alexis Thérèse Petit.[91,92]

The value of M in Equation 13.1 may be expressed in terms of Θ, the Debye **characteristic temperature**[93] of the material. James[94] explains that the "characteristic temperature for a given crystal determines, roughly speaking, what temperatures may be considered as 'high' in dealing with the crystal." This quantity is obtained from measurements of the variation of specific heat with temperature. Debye writes[93] (this quotation italicized in the original translation): "The energy of a body is obtained if the Dulong Petit value is multiplied by a factor which is a universal function of the ratio T/Θ, that is, temperature T divided by characteristic temperature Θ." James[94] adds: "If Θ is properly chosen for each substance, the curve giving the specific heat as a function of T/Θ should be the same for all of them [cubic monatomic elements]. At very low temperatures C_v [the specific heat at vary as $(T/\Theta)^3$, and this is closely confirmed by experiment." For further details consult Chapter V of Reference 94.

Essentially the characteristic temperature is a measure of the temperature at which the atomic heat capacity is changing from zero to 6 cal deg^{-1}; for silver ($\Theta = 215$ K) this occurs around 100 K, but for diamond ($\Theta = 1860$ K) with a much more rigid structure, the atomic heat capacity does not reach 5 cal deg^{-1} until 900 K. Those elements that resist compression and that have high melting points have high characteristic temperatures.[94,95] Equations have been derived relating $\sqrt{\langle u^2 \rangle}$ to the characteristic temperature Θ. At room temperature diamond, with a characteristic temperature of 1860 K, has a root-mean-square amplitude of vibration, $\sqrt{\langle u^2 \rangle}$ of 0.02 Å, while copper and lead, with characteristic temperatures of 320 and 88 K, respectively, have values of 0.14 and 0.28 Å for $\sqrt{\langle u^2 \rangle}$.[96-98] Similar types of values are obtained for crystals with mixed atom (or ion) types.[99] For example, average values of $\sqrt{\langle u^2 \rangle}$ for Na$^+$ and Cl$^-$ in sodium chloride ($\Theta = 281$ K) are 0.14 Å at 86 K and 0.23 Å at 290 K.[100]

Cruickshank[101] used data on anthracene to calculate the entropy of naphthalene. By use of r.m.s. amplitudes of vibration of 3.9, 2.2, and 3.0° and moments of inertia (calculated from the refined atomic coordinates) of 1846, 391, and 2237 $\times 10^{-40}$ g. cm^{-2}, Cruickshank estimated the average rotational lattice frequencies as 81, 65, and 43 cm^{-1}. These may be compared with the values from Raman frequencies,[102] of 120(8), 68(5), and 48(4) cm^{-1} respectively. From such characteristic lattice frequencies, Cruickshank[101] estimated the entropy of naphthalene to be 40 and 12.5 e.u. at 298 and 90 K respectively.

13.5.6 Analysis of order–disorder phase transitions

Crystal structures themselves are temperature dependent. The simplest example of this is the expansion of the unit cell upon heating because the atoms move more and the average distance between them increases as the temperature is raised. For example, the unit cell dimensions of naphthalene have been measured at several temperatures; the values are plotted in Figure 13.19. In spite of these changes of the unit cell dimensions, while intermolecular distances may change, the fractional atomic coordinates (which may vary with temperature) indicate that the bond lengths do not vary appreciably with the temperature change.[81]

In addition, some crystals undergo a phase transition, that is, a change in crystal packing accompanied by a change in specific heat. This transition is commonly measured by differential scanning calorimetry, an illustration of which is given in Figure 13.20. The change in structure may or may not be reversible and may or may not dramatically affect the quality of the single crystal. There is a class of phase transitions, order–disorder phase transitions, that is particularly relevant to the topics discussed in this chapter because some attempts to analyze the nature of the disorder involve temperature-dependent studies and analyses of anisotropic displacement parameters.

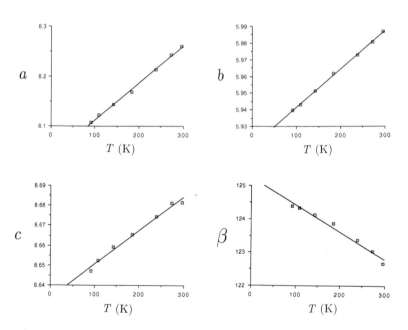

FIGURE 13.19. Variation of unit-cell dimensions of naphthalene with temperature (Ref. 81). Shown are variations in a, b, c, and β, for the monoclinic unit cell.

Order–disorder phase transitions are especially common in crystalline π-donor:acceptor complexes between planar polycyclic aromatic hydrocarbons and other organic compounds. The disordered phase can sometimes be characterized in terms of either a static- or a dynamic-disorder model, as shown in Figure 13.2. The dynamic-disorder model consists of disordered components in motion within the confines of a broad well in the potential energy curve, whereas the static disorder model requires that the disordered components be localized in two or more sites in the asymmetric unit, within one or another of the wells of a multiwell-potential energy curve. When the possible sites for static disorder are not resolved by the effective resolution of the data set, it is difficult to choose between these models. This turned out to be true for an anthracene-tetracyanobenzene complex studied at several temperatures above and below the transition temperature T_p of 206 K.[103,104] It was found[103–108] that the tetracyanobenzene molecules are "well ordered" at all temperatures, with the expected systematic decrease in the amplitudes of the anisotropic displacement parameters of the atoms as the temperature of the crystal is lowered. On the other hand, the anisotropic displacement parameters of the anthracene molecules change dramatically as

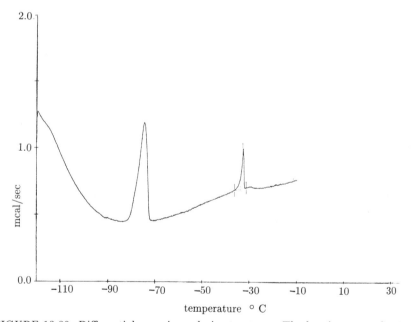

FIGURE 13.20. Differential scanning calorimetry trace. The heating curve for 1,1,2-trichloro-2,3,3-trifluorocyclobutane. The larger peak is due to the transition from the anisotropic crystalline phase to the plastic crystalline phase; the smaller peak is due to the transition from the plastic crystalline phase to the liquid phase. (Courtesy V. B. Pett and David L. Powell, The College of Wooster, Ohio.)

the temperature of the crystal is varied, and a systematic temperature-dependent reorientation of the anthracene molecules (as indicated by the angles between the long molecular axes of symmetry-related anthracene molecules) occurs in the low-temperature phase. This system is still under investigation.[104,108]

13.5.7 Intramolecular dynamic processes

Hans-Beat Bürgi[109,110] has demonstrated that results from a crystal structure determination of metal ion coordination compounds can be interpreted in terms of a variety of intramolecular dynamic processes. Using ΔU_{L-M}, that is, the differences in the anisotropic displacement parameters in the direction of the metal (M) to ligand (L) vector, Bürgi characterized bond stretching, spin crossover in a crystalline Fe(III) complex, and pseudodynamic Jahn–Teller deformations in octahedral crystalline Cu(II) and Mn(III) complexes.

Tri(dithiocarbamato)iron(III) complexes show a high-spin–low-spin equilibrium in the solid state.[111] This involves changes in the $Fe \cdots S$ bond length from about 2.30 Å in the low-spin state to one of 2.45 Å in the high-spin state. Seventeen Fe(III) complexes were studied using data extracted from the Cambridge Structural Database[112] (see Chapter 16). Anisotropic displacement parameters were listed for Fe and S in each structure. Values of $U_{Fe} - U_S = \Delta U_{Fe-S}$ varied from 0.0010 to 0.0096 Å along the direction of the $Fe \cdots S$ bond. While many of these ΔU values were not greater than $3\sigma(\Delta U)$, it was still possible to attach significance to the curve in Figure 13.21. The measurements in the crystalline state were shown to be a mixture of low-spin (LS) and high-spin (HS) distances (2.30 and 2.45 Å respectively, in that all measured $Fe \cdots S$ distances [d(Fe–S)] can be explained by Equation 13.15.

$$[d(Fe - S)] = p[d_{LS}] + (1 - p)[d_{HS}] \tag{13.15}$$

This may be converted to:

$$[d(Fe - S)] = 2.45 - 0.15p, \tag{13.16}$$

where p is the fractional population in the low-spin state.

$$\Delta U = p(\Delta U_{LS}) + p\cos\gamma(\Delta U_{LS})^2 + (1 - p)(\Delta U_{HS}) + (1 - p)\cos\gamma(\Delta U_{HS})^2 \tag{13.17}$$

Experimental data for Fe–S complexes could be fit by an equation of the form:

$$\Delta U_{obs} = 0.0071 + 0.0136\delta - 0.88\delta^2, \tag{13.18}$$

where $\delta = [d(Fe–S) - 2.375]$, the center of the observed distances. In crystal structures that are known to contain pure low-spin or pure high-

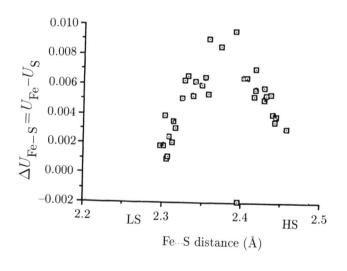

(a)

(b)

FIGURE 13.21. Variation in ΔU_{Fe-S} with Fe $\cdots$ S distance in some tri(dithiocarb-amato)iron(III) complexes. LS = low spin, HS = high spin complexes. (a) The reaction involved, and (b) a plot of ΔU versus Fe $\cdots$ S distances with experimental data from various crystal structure determinations.

spin complexes, experimental values for ΔU_{M-S} are reported to be approximately 0.0017 and 0.0035 Å², respectively. This (from Equation 13.17) leads to:

$$\Delta U_{calc} = 0.0017p + 0.0035(1 - p) + \cos \gamma (2.45 - 2.30)(p)(1 - p), \qquad (13.19)$$

which gives (since $p = 0.5 - 6.667\delta$ and $\delta = d(\text{Fe-S}) - 2.375$):

$$\Delta U_{calc} = 0.0078 - 0.0120\delta - 0.93\delta^2, \qquad (13.20)$$

where $\cos \gamma = 0.93$ accounts for the deviation of the direction of motion of the sulfur atom from the Fe–S bond. There is good agreement between the two equations, one (Equation 13.18) observed, and the other (Equation 13.20) calculated. This model therefore appears to be appropriate, implying that no intermediate spin state need be invoked.

Highly anisotropic displacement parameters can also indicate incipient reactions.[113] This has been illustrated by a reaction in which a boron–carbon bond is reversibly made and broken. The anisotropic displacement parameters and conformation of an intermediate give a good indication of the stereochemistry of the reaction. This will be discussed in Chapter 18.

13.6 Molecular dynamics of polypeptides and proteins

Motion in a protein may be modeled by computer using Isaac Newton's equation of motion, $f = ma$. This modeling requires the three-dimensional coordinates from an X-ray structure analysis as a starting point and some knowledge of interatomic potentials, so that only reasonable interatomic distances will be employed at all stages.[61,62,114–122] Such **molecular dynamics** calculations lead to a prediction of where atoms will move in a short period of time, and result in the calculation of a time-dependent trajectory of all atoms. Initially each atom is moved in the direction of the force on it from other atoms and then, as each atom moves, its trajectory may change to accommodate this. In addition, this method aids in protein structure refinement,[123–126] as was described in Chapter 10, although it is important to ensure that the model so refined still fits the electron density map.

Summary

1. The intensities of Bragg reflections at high scattering angles are decreased as a result of the vibration of atoms. This causes an exponential decrease in the scattering factor.
2. A similar effect is found if there is slight disorder from unit cell to unit cell in the crystal (static disorder).

3. In order to determine whether a decrease in scattering at high angles is due to vibration effects or to disorder, the data should be measured at a series of temperatures. Only the vibration effects should show a strong temperature dependence.

4. Displacements of atoms from their equilibrium positions can be anisotropic and are represented by anisotropic displacement parameters which, are refined by least-squares techniques together with the atomic coordinates (see Chapter 10). A further analysis of these anisotropic displacement parameters in terms of translation **T**, libration **L**, and screw **S** motions can give information on the nature of the molecular motion.

5. Molecular motion will affect the apparent molecular geometry and must be corrected for.

6. Detailed analyses of anisotropic displacement parameters can give information on barriers to the rotation of groups, on the characteristic temperature, and on entropy.

Glossary

Atomic displacement parameters: Atomic vibrations are displacements from equilibrium positions with periods that are typically smaller than 10^{-12} sec. Because *static* displacements of a given atom in a structure that vary in an essentially random fashion from one unit cell to another will simulate vibrations of that atom, the term displacement factors is preferred unless it is clear that the temperature-dependent vibrational displacements are not contaminated by static disorder. In the latter case the term temperature factors or vibration parameters may be correct.

Boltzmann distribution: The distribution of energies indicated by the formula

$$n = n_0 \exp(-E/kT)$$

where n_o is the number of particles with lowest energy, n is the number of particles with an energy E (above the minimum), k is the Boltzmann constant 1.380×10^{-16} erg/deg. and T is the absolute temperature ($0°C = 273.16°$ K).

Characteristic temperature : A quantity Θ obtained from a study of the variation of the specific heat of a material with absolute temperature T. At low temperatures the specific heat varies as $(T/\Theta)^3$. Materials with high melting points and that resist compression have high values of Θ.

Displacement parameters: (See atomic displacement parameters)

Equivalent isotropic displacement parameters: An expression representing the extent of thermal displacement of an atom, averaged over all directions:

$$U_{eq} = \frac{1}{3}[U_{11}(aa^*)^2 + U_{22}(bb^*)^2 + U_{33}(cc^*)^2$$
$$+2U_{12}aba^*b^*\cos\gamma + 2U_{13}aca^*c^*\cos\beta + 2U_{23}bcb^*c^*\cos\alpha$$

(see Refs. 5 and 26).

External (or lattice) modes of vibration: Modes of vibration involving the entire molecule moving with respect to others in the crystal structure.

Force constants: The constant κ in the equation (Hooke's Law) that relates the potential energy V of a displacement δ to that displacement. The equation is:

$$V = \kappa \delta^2 / 2.$$

Gas constant: The value of R in the equation of state of a perfect gas, $pv = RT$, where p, v and T are the pressure, volume and temperature ($^\circ$K). $R = 1.987$ calories per mole degree.

Harmonic vibration: A vibration that occurs under the influence of a force directly proportional to the displacement from equilibrium. The range of the vibration extends equal distances in either direction from an equilibrium position (the origin), and the acceleration is always toward the origin and directly proportional to the distance from it. For a Boltzmann distribution of particle energies, harmonic vibration leads to a probability distribution for the positions of the vibrating particles that is a Gaussian function of the displacements.

Heat capacity: The quantity of heat required to raise the temperature of a bod 1° (cf. Specific heat).

Internal (or molecular) modes of vibration: Ontramolecular modes of vibration such as stretching or bending of bonds.

Isotropic temperature parameter: An atomic displacement parameter (q.v.) that represents an equal amplitude of vibration of an atom in all directions through a crystal.

Libration: Oscillation along an arc rather than along a straight line. The lower end of a clock pendulum undergoes this type of motion.

Mean-square displacement amplitude: The average, in a given direction, of the square of the deviation of the instantaneous position of an atom from its average position. It is generally represented by $\overline{u^2}$ or $\langle u^2 \rangle$. The latter is used here.

Molecular dynamics: A quantitative study of the movement of systems over potential energy surfaces.

Nonpositive-definite matrix: A symmetric matrix M is positive definite if its determinant, each diagonal element, and each term $M_{ii}M_{kk} - M_{ik}M_{ki}$ is positive. If these conditions for a symmetric matrix do not hold, the matrix is said to be nonpositive definite.

Normal modes: In a normal vibration of a molecule, all atoms carry out simple harmonic motion at the same frequency and the same phase. Any vibration of the molecule can be represented as a superposition of these normal modes. Low-frequency modes are bending modes, while high-frequency modes are stretching modes.

Principal axes of thermal ellipsoids: Three mutually perpendicular directions, along two of which the amplitude of vibration of an atom, represented by an ellipsoid, is at a maximum and at a minimum. Each axis is characterized by an amplitude and a direction.

Probability density function: The probability that a random variable will take on a particular value in an infinitesimal time interval, divided by the length of the interval.

Rigid body: A group of particles that maintain a constant interparticle separation.

Rigid-body motion: Motion of a molecule (or part of a molecule) in which all atoms move in synchrony, as if completely rigid.

Specific heat: The quantity of heat necessary to raise the temperature of a unit of mass (such as one gram) by one degree of temperature (cf. Heat capacity).

Temperature factor: An exponential expression by which the scattering of an atom is reduced as a consequence of vibration (or a simulated vibration resulting from static disorder). For isotropic motion the exponential factor is $\exp(-B_{iso}\sin^2\theta/\lambda^2$), where B_{iso} is the isotropic temperature factor. It equals $8\pi^2\langle u^2\rangle$, where $\langle u^2\rangle$ is the mean-square displacement of the atom from its equilibrium position. For anisotropic motion the exponential expression usually contains six parameters, the anisotropic vibration or displacement parameters, which describe ellipsoidal rather than isotropic (spherically symmetrical) motion or average static displacements.

Temperature parameter: (See temperature factor or atomic displacement parameters.)

Vibration parameter: (See temperature factor or atomic displacement parameters.)

References

1. Dunitz, J. D. From crystal statics towards molecular dynamics. *Trans. Amer. Cryst. Assn.* **20**, 1–14 (1984).

2. Debye, P. Interferenz von Röntgenstrahlen und Wärmebewegung. [X-ray interference and thermal movement.] *Ann. Physik* **43**, 49–95 (1914). **English translation** in: *Collected Papers of Peter J. W. Debye.* pp. 3–39. Interscience: New York (1954).

3. Bragg, W. H. The intensity of reflection of X-rays by crystals. *Phil. Mag.* **27**, 881-899 (1914).

4. Wilson, A. J. C. Determination of absolute from relative X-ray intensity data. *Nature (London)* **150**, 152 (1942).

5. Willis, B. T. M., and Pryor, A. W. *Thermal Vibrations in Crystallography.* Cambridge University Press: Cambridge, UK (1975).

6. Dunitz, J. D., Schomaker, V., and Trueblood, K. N. Interpretation of atomic displacement parameters from diffraction studies of crystals *J. Phys. Chem.* **92**, 856–867 (1988).

7. Debije, P. [Debye, P.] Über den Einfluss der Wärmebewegung auf die Interferenzerscheinungen bei Röntgenstrahlen. [Influence of heat motion on the interference phenomena of Röntgen rays.] *Verhandlungen der Deutschen Physikalischen Gesellschaft* **15**, 678–689 (1913).

8. Debye, P. Über die intensitätsverteilung in den mit Röntgenstrahlen erzeugten Interferenz-bildern. [The intensity distribution in the interference produced by Röntgen rays.] *Verhandlungen der Deutschen Physikalischen Gesellschaft* **15**, 738–752 (1913).

9. Debije, P. [Debye, P.] Spektrale Zerlegung der Röntgenstrahlung mittels Reflexion und Wärmebewegung. [Spectral analysis of X rays by means of reflection and heat motion.] *Verhandlungen der Deutschen Physikalischen Gesellschaft* **17**, 857–875 (1913).

10. Waller, I. Zur Frage der Einwirkung der Wärmebewegung auf die Interferenz von Röntgenstrahlen. [The question of the influence of heat motion on the scattering of Röntgen rays.] *Z. Physik* **17**, 398–408 (1923).

11. Waller, I. Die Einwirkung der Wärmebewegung der Kristallatome auf Intensität, Lage und Schärfe der Röntgenspektrallinien. [The influence of thermal agitation of crystal atoms upon the intensity, position and sharpness of X-ray spectral lines.] *Annalen der Physik* **83**, 153–183 (1927).

12. Dunitz, J. D. *X-ray Analysis and the Structure of Organic Molecules.* Cornell University Press: London, Ithaca, NY (1979).

13. Ewald, P. P. The development of intensity interpretation in crystal x-ray diffraction. *Curr. Sci.*, pp. 11–13 (1937). Reprinted in: *Early Papers on the Diffraction of X-rays by Crystals.* Vol. 1. (**Eds.**, Bijvoet, J. M., Burgers, W. G., and Hägg, G.) Oosthoek: Utrecht (1969).

14. Ewald, P. P. The immediate sequels to Laue's discovery. In: *Fifty Years of X-ray Diffraction.* (**Ed.**, Ewald, P. P.) Chapter 5. pp. 57–80. Oosthoek: Utrecht (1962).

15. van der Helm, D., Glusker, J. P., Johnson, C. K., Minkin, J. A., Burow, N. E., and Patterson, A. L. X-ray crystal analysis of the substrates of aconitase. VIII. The structure and absolute configuration of potassium dihydrogen isocitrate isolated from *Bryophyllum calycinum. Acta Cryst.* **B24**, 578–592 (1968).

16. Karle, I. L., Karle, J., Mastropaolo, D., Camerman, A., and Camerman, N. [Leu^5]enkephalin: four cocrystallizing conformers with extended backbone that form an antiparallel β-sheet. *Acta Cryst.* **B39**, 625–637 (1983).

17. Taylor, M. R., Glusker, J. P., Gabe, E. J., and Minkin, J. A. The crystal structure of the antitumor agent 3-ethoxy-2-oxobutyraldehyde bis (thiosemicarbazonato) copper(II). *Bioinorg. Chem.* **3**, 189–205 (1974).

18. Punzi, J. S., Duax, W. L., Strong, P., Griffin, J. F., Flocco, M. M., Zacharias, D. E., Carrell, H. L., Tew, K. D., and Glusker, J. P. Molecular conformation of estramustine and two analogues. *Molecular Pharmacology* **41**, 569–576 (1992).

19. Lonsdale, K., and Milledge, H. J. Analysis of thermal vibrations in crystals: a warning. *Acta Cryst.* **14**, 59–61 (1961).

20. Seiler, P., and Dunitz, J. D. A new interpretation of the disordered crystal structure of ferrocene. *Acta Cryst.* **B35**, 1068–1074 (1979).

21. Seiler, P., and Dunitz, J. D. The structure of triclinic ferrocene at 101, 123 and 148 K. *Acta Cryst.* **B35**, 2020–2032 (1979).

22. Maverick, E., and Dunitz, J. D. Rotation barriers in crystals from atomic displacement parameters. *Molecular Physics* **62**, 451–459 (1987).

23. Savage, H. Water structure in vitamin B_{12} coenzyme crystals. II. Structural characteristics of the solvent networks. *Biophys. J.* **50**, 967–980 (1986).

24. Stezowski, J. J. Molecular motion in host-substrate complexes: the β-cyclodextrin:N-acetylphenylalanine methyl ester system. *Trans. Amer. Cryst. Assn.* **20**, 73–82 (1984).

25. Ringe, D., Kuriyan, J., Petsko, G. A., Karplus, M., Frauenfelder, H., Tilton, R. F., and Kuntz, I. D. Temperature dependence of protein structure and mobility. *Trans. Amer. Cryst. Assn.* **20**, 109–122 (1984).

26. Fischer, R. X., and Tillmanns, E. The equivalent isotropic displacement factor *Acta Cryst.* **C44**, 775–776 (1988).

27. Schomaker, V., and Marsh, R. E. E.s.d. of equivalent isotropic temperature factor *Acta Cryst.* **A39**, 819–820 (1983).

28. Johnson, C. K. ORTEP: a FORTRAN thermal-ellipsoid plot program for crystal structure illustrations. Report ORNL-3794. Oak Ridge National Laboratory, Oak Ridge, TN (1965). Report ORNL-5138. Oak Ridge National Laboratory, Oak Ridge, TN (1971).

29. Waser, J. The anisotropic temperature factor in triclinic coordinates. *Acta Cryst.* **8**, 731 (1955).

30. Busing, W. R., and Levy, H. A. Determination of the principal axes of the anisotropic temperature factor. *Acta Cryst.* **11**, 450–451 (1958).

31. Hamilton, W. C. On the isotropic temperature factor equivalent to a given anisotropic temperature factor. *Acta Cryst.* **12**, 609–610 (1959).

32. Hummel, W., Raselli, A., and Bürgi, H.-B. Analysis of atomic displacement parameters and molecular motion in crystals. *Acta Cryst.* **B46**, 683–692 (1990).

33. Hummel, W., Hauser, J., and Bürgi, H.-B. PEANUT: computer graphics program to represent atomic displacement parameters. *J. Molec. Graphics* **8**, 214–220 (1990).

34. Sands, D. E. *Vectors and Tensors in Crystallography.* Addison-Wesley: London, Amsterdam, Don Mills (Ontario), Sydney, Tokyo (1982).

35. Cyvin, S. J. (**Ed**.) *Molecular Vibrations and Mean Square Amplitudes.* Elsevier: Amsterdam (1968).

36. Zucker, U. H., and Schulz, H. Statistical approaches for the treatment of anharmonic motion in crystals. I. A comparison of the most frequently used formalisms of anharmonic thermal vibrations. *Acta Cryst.* **A38**, 563–568 (1982).

37. Zucker, U. H., and Schulz, H. Statistical approaches for the treatment of anharmonic motion in crystals. II. Anharmonic thermal vibrations and effective atomic potentials in the fast ionic conductor lithium nitride (Li_3N). *Acta Cryst.* **A38**, 568–576 (1982).

38. Johnson, C. K. Addition of higher cumulants to the crystallographic structure-factor equation: a generalized treatment for thermal-motion effects. *Acta Cryst.* **A25**, 187–194 (1969).

39. Johnson, C. K., and Levy, H. A. Thermal-motion analysis using Bragg diffraction data. In: *International Tables for X-ray Crystallography.* Vol. IV. *Revised and Supplementary Tables to Volumes II and III.* Section 5. (**Eds.**, Ibers, J. A., and Hamilton, W. C.) pp. 311–336 (1974).

40. Kuznetsov, P. I., Stratonovich, R. L., and Tikhonov, V. I. Quasi-moment functions in the theory of random processes. *Theory Prob. Appl.* **5**, 80–97 (1960). [English translation series.]

41. Dunitz, J. D., Maverick, E. F., and Trueblood, K. N. Atomic motion in molecular crystals from diffraction measurements. *Angew. Chem., Int. Ed. Engl.* **27**, 880–895 (1988).

42. Hirshfeld, F. L. Can X-ray data distinguish bonding effects from vibrational smearing. *Acta Cryst.* **A32**, 239–244 (1976).

43. Dunitz, J. D., and White, D. N. J. Non-rigid-body thermal-motion analysis. *Acta Cryst.* **A29**, 93–94 (1973).

44. Trueblood, K. N. Analysis of molecular motion with analysis for intramolecular torsion. *Acta Cryst.* **A34**, 950–954 (1978).

45. Johnson, C. K. Generalized treatments for thermal motion. In: *Thermal Neutron Diffraction.* (**Ed.**, Willis, B. T. M.) Ch. 9, pp. 132–160. Oxford University Press: London (1970).

46. Brock, C. P., Schweizer, W. B., and Dunitz, J. D. Internal molecular motion of triphenylphosphine oxide: analysis of atomic displacement parameters for orthorhombic and monoclinic crystal modifications at 100 and 150 K. *J. Amer. Chem. Soc.* **107**, 6964–6970 (1985).

47. Bye, E., Schweizer, W. B., and Dunitz, J. D. Chemical reaction paths. 8. Stereoisomerization path for triphenylphosphine oxide and related molecules: indirect observation of the structure of the transition state. *J. Am. Chem. Soc.* **104**, 5893–5898 (1982).

48. Luis, J., and Amorós, M. *Molecular Crystals. Their Transforms and Diffuse Scattering.* John Wiley: New York (1968).

49. Wooster, W. A. *Diffuse X-ray Reflections from Crystals.* Clarendon Press: Oxford (1962).

50. Laval, J. Diffusion of X-rays by crystals in directions other than those of selective reflection. *Comptes Rendues Acad. Sci. (Paris)* **208**, 1512–1514 (1939).

51. Zachariasen, W. H. A theoretical study of the diffuse scattering of X-rays by crystals. *Phys. Rev.* **57**, 597–602 (1940).

52. Epstein, J., and Welberry, T. R. Least-squares analyses of diffuse scattering from substitutionally disordered molecular crystals: application to 2,3-dichloro-6,7-dimethylanthracene. *Acta Cryst.* **A39**, 882–892 (1983).

53. Moss, D. S., Haneef, I., and Howlin, B. Anisotropic X-ray refinement of rigid group vibrations in protein structures. *Trans. Amer. Cryst. Assn.* **20**, 123–127 (1984).

54. Glover, I., Haneef, I., Pitts, J., Wood, S., Moss, D., Tickle, I., and Blundell, T. Conformational flexibility in a small globular hormone: X-ray analysis of avian pancreatic polypeptide at 0.98-Å resolution. *Biopolymers* **22**, 293–304 (1983).

55. Teeter, M. M., and Case, D. A. Harmonic and quasiharmonic descriptions of crambin. *J. Phys. Chem.* **94**, 8091–8097 (1990).

56. Huber, R. Conformational flexibility and its functional significance in some proteins. *Trends Biochem. Sci.* **4**, 271–283 (1979).

57. Huber, R., and Bennett, W. S. Functional significance of flexibility in proteins. *Biopolymers* **22**, 261–277 (1983).

58. Bennett, W. S., and Steitz, T. A. Glucose-induced conformational change in yeast hexokinase. *Proc. Natl. Acad. Sci. USA* **75**, 4848–4854 (1978).

59. Bennett, W. S., and Steitz, T. A. Structure of a complex between yeast hexokinase and glucose. *J. Molec. Biol.* **140**, 211–235 (1980).

60. Bode, W. The transition of bovine trypsinogen to a trypsin-like state upon strong ligand binding. *J. Molec. Biol.* **127**, 357–402 (1979).

61. Artymiuk, P. J., and Blake, C. C. F. Refinement of human lysozyme at 1.5 Å resolution. Analysis of nonbonded and hydrogen-bond interactions. *J. Molec. Biol.* **152**, 737–762 (1981).

62. Phillips, D. Protein structure and function. In: *Patterson and Pattersons.* (**Eds.**, Glusker, J. P., Patterson, B. K., and Rossi, M.) Chapter 7. pp. 117–144. Oxford University Press: Oxford (1987).

63. Poulos, T.L., Finzel, B. C., and Howard, A. J. Crystal structure of substrate-free *Pseudomonas putida* cytochrome P-450. *Biochemistry* **25**, 5314–5322 (1986).

64. Sternberg, M. J. E., Grace, D. E. P., and Phillips. D. C. Dynamic information from protein crystallography. An analysis of temperature factors from refinement of the hen egg-white lysozyme structure. *J. Molec. Biol.* **130**, 231–253 (1979).

65. McCammon, J. A., Gelin, B. R., Karplus, M., and Wolynes, P. G. The hinge-bending mode in lysozyme. *Nature (London)* **262**, 325–326 (1976).

66. Cruickshank, D. W. J. Errors in bond lengths due to rotational oscillations of molecules. *Acta Cryst.* **9**, 757–758 (1956).

67. Cruickshank, D. W. J. The determination of the anisotropic thermal motion of atoms in crystals. *Acta Cryst.* **9**, 747–753 (1956).

68. Cruickshank, D. W. J. The analysis of the anisotropic thermal motion of molecules in crystals. *Acta Cryst.* **9**, 754–756 (1956).

69. Cruickshank, D. W. J. The variation of vibration amplitudes with temperature in some molecular crystals. *Acta Cryst.* **9**, 1005–1009 (1956).

70. Busing, W. R., and Levy, H. A. The effect of thermal motion on the estimation of bond lengths from diffraction measurements. *Acta Cryst.* **17**, 142–146 (1964).

71. Schomaker, V., and Trueblood, K. N. On the rigid-body motion of molecules in crystals. *Acta Cryst.* **B24**, 63–76 (1968).

72. Kashino, S., Zacharias, D. E., Prout, C. K., Carrell, H. L., Glusker, J. P., Hecht, S. S., and Harvey, R. G. Structure of 5-methylchrysene, $C_{19}H_{14}$. *Acta Cryst.* **C40**, 536–540 (1984).

73. Hazell, A., Hazell, R. G., and Larsen, F. K. Orientational disorder in cyclo-hepta[*de*]naphthalene. Structure determination at 78, 100, 135, 200 and 295 K. *Acta Cryst.* **B42**, 621–626 (1986).

74. Robertson, J. M., Shearer, H. M. M., Sim, G. A., and Watson, D. G. A revision of the azulene structure. *Nature (London)* **182**, 177–178 (1958).

75. Fyfe, C. A., and Kupferschmidt, G. J. Molecular motion in azulene and the azulene-(*s*)-trinitrobenzene complex in the crystalline state. *Can. J. Chem.* **51**, 3774–3780 (1973).

76. Cox, E. G., and Smith, J. A. S. Crystal structure of benzene at −3°C. *Nature (London)* **173**, 75 (1954).

77. Cox, E. G., Cruickshank, D. W. J., and Smith, J. A. S. The crystal structure of benzene at −3°. *Proc. Roy. Soc. (London)* **A247**, 1–21 (1958).

78. Stoicheff, B. P. High resolution Raman spectroscopy of gases. II. Rotational spectra of C_6H_6 and C_6D_6 and internuclear distances in the benzene molecule. *Can. J. Phys.* **32**, 339–346 (1954).

79. Jeffrey, G. A., Ruble, J. R., McMullan, R. K., and Pople, J. A. The crystal structure of deuterated benzene. *Proc. Roy. Soc. (London)* **A414**, 47–57 (1987).

80. Bacon, G. E., Curry, N. A., and Wilson, S. A. A crystallographic study of solid benzene by neutron diffraction. *Proc. Roy. Soc. (London)* **A279**, 98–110 (1964).

81. Brock, C. P., and Dunitz, J. D. Temperature dependence of thermal motion in crystalline naphthalene. *Acta Cryst.* **B38**, 2218–2228 (1982).

82. Brock, C. P., and Dunitz, J. D. Temperature dependence of thermal motion in crystalline anthracene. *Acta Cryst.* **B46**, 795–806 (1990).

83. Brock, C. P., and Morelan, G. L. Relationships between molecular geometry, crystal packing, and thermal motion: temperature-dependent studies of three crystal forms of 4-hydroxybiphenyl. *J. Phys. Chem.* **90**, 5631–5640 (1986).

84. Gramaccioli, C. M., and Filippini, G. Lattice-dynamical evaluation of temperature factors in non-rigid molecular crystals. A first application to aromatic hydrocarbons. *Acta Cryst.* **A39**, 784–791 (1983).

85. Filippini, G. Thermal motion in [5]-, [6]- and [7]-circulene crystals: a harmonic lattice-dynamical calculation. *Acta Cryst.* **B46**, 643–645 (1990).

86. Trueblood, K. N., and Dunitz, J. D. Internal molecular motions in crystals. The estimation of force constants, frequencies and barriers from diffraction data. A feasibility study. *Acta Cryst.* **B39**, 120–133 (1983).

87. Maverick, E., Mirsky, K., Knobler, C. B., Trueblood, K. N., and Barclay, L. R. C. Rotation barriers in crystals from diffraction studies: 2,2′,4,4′,6,6′-hexa-*tert*-butylazobenzene. *Acta Cryst.* **B47**, 272–280 (1991).

88. Gardner, A. B., Howard, J., Waddington, T. C., Richardson, R. M. and Tomkinson, J. The dynamics of ring rotation in ferrocene, nickelocene and ruthenocene by incoherent quasi-elastic neutron scattering. *Chemical Physics* **57**, 453–460 (1981).

89. Fattah, J., Twyman, J. M., Heyes, S. J., Watkin, D. J., Edwards, A. J., Prout, K., and Dobson, C. M. Combination of CP/MAS NMR and X-ray crystallography: structure and dynamics in a low-symmetry molecular crystal, potassium penicillin V. *J. Amer. Chem. Soc.* **115**, 5636–5650 (1993).

90. Seiler, P., Martinoni, B., and Dunitz, J. D. Can X-ray diffraction distinguish between protium and deuterium atoms? *Nature (London)* **309**, 435–438 (1984).

91. Dulong, P. L., and Petit, A. T. Sur la mesure des températures et sur les lois de la communication de la chaleur. [On the measurement of temperatures and the laws of heat conduction.] *Ann. Chim. Phys.* **7**, 113–154 (1817).

92. Dulong, P. L., and Petit, A. T. Sur quelques points importans de la théorie de la chaleur. [On several important points concerned with the theory of heat.] *Ann. Chim. Phys.* **10**, 395–413 (1819).

93. Debye, P. Zur Theorie der spezifischen Wärmen. [On the theory of specific heats.] *Annalen der Physik* **39**, 789–839 (1912). **English translation** in: *Collected Papers of Peter J. W. Debye.* pp. 650–696. Interscience: New York (1954).

94. James, R. W. *Optical Principles of the Diffraction of X-rays.* Bell: London (1954).

95. Butt, N. M., Bashir, J., Willis, B. T. M., and Heger, G. Compilation of temperature factors of cubic elements. *Acta Cryst.* **A44**, 396–398 (1988).

96. Seitz, F. *Modern Theory of Solids.* McGraw-Hill: New York (1940).

97. Zener, C. Theory of the effect of temperature on the reflection of X-rays by crystals. II. Anisotropic crystals. *Phys. Rev.* **49**, 122–127 (1936).

98. Lonsdale, K. Vibration amplitudes of atoms in cubic crystals. *Acta Cryst.* **1**, 142–149 (1948).

99. Horning, R. D., and Staudenmann, J.-L. The Debye–Waller factor for polyatomic solids. Relationship between X-ray and specific-heat Debye temperatures. The Debye–Einstein model. *Acta Cryst.* **A44**, 136–142 (1988).

100. James, R. W., Waller, I., and Hartree, D. R. Existence of zero-point energy in the rock-salt lattice by an X-ray diffraction method. *Proc. Roy. Soc. (London)* **A118**, 334–350 (1928).

101. Cruickshank, D. W. J. The entropy of crystalline naphthalene. *Acta Cryst.* **9**, 1010–1011 (1956).

102. Fruhling, A. Recherches sur le spectre Raman de quelques monocristaux aromatiques. [Research on the Raman spectra of certain aromatic single crystals.] *Annu. Rev. Phys. (Paris)* **6**, 401–480 (1951).

103. Stezowski, J. Phase transition effects: a crystallographic characterization of the temperature dependency of the crystal structure of the 1:1 charge transfer complex between anthracene and tetracyanobenzene in the temperature range 297 to 119 K. *J. Chem. Phys.* **73**, 538–547 (1980).

104. Boyens, J. C. A., and Levendis, D. C. Static disorder in crystals of the anthracene-tetracyanobenzene charge transfer complex. *J. Chem. Phys.* **80**, 2681–2688 (1984).

105. Tschuiya, H., Marumo, F., and Saito, Y. The crystal structure of the 1:1 complex of anthracene and 1,2,4,5-tetracyanobenzene. *Acta Cryst.* **B28**, 1935–1941 (1972).

106. Stezowski, J. J. The crystal structure of a low temperature phase of the 1:1 charge transfer complex between anthracene and tetracyanobenzene. *J. Phys. Chem.* **83**, 550–551 (1979).

107. Möhwald, H., Erdle, E., and Thaer, A. Orientational phase transition in a charge-transfer crystal: triplet excitons as probes for lattice dynamics. *Chemical Physics* **27**, 79–87 (1978).

108. Stezowski, J. J. Temperature-dependent studies of disordered molecular crystals. *Trans. Amer. Cryst. Assn.* **17**, 59–69 (1981).

109. Bürgi, H.-B. Stereochemical lability in crystalline coordination compounds. *Trans. Amer. Cryst. Assn.* **20**, 61–71 (1984).

110. Chandrasekhar, K., and Bürgi, H.-B. Dynamic processes in crystals examined through difference vibration parameters ΔU: the low-spin–high-spin transition in tris(dithiocarbamato)iron(III) complexes. *Acta Cryst.* **B40**, 387–397 (1984).

111. Cambi, L., and Szegö, L. Über die magnetische Susceptibilität der komplexen Verbindungen. [The magnetic susceptibilities of the complex compounds.] *Ber. Deutsch. Chem. Gesel.* **B64**, 2591–2598 (1931).

112. Allen, F. H., Bellard, S., Brice, M. D., Cartwright, B. A., Doubleday, A., Higgs, H., Hummelink, T., Hummelink-Peters, B. G., Kennard, O., Motherwell, W. D. S., Rodgers, J. R., and Watson, D. G. The Cambridge Crystallographic Data Centre: computer-based search, retrieval, analysis and display of information. *Acta Cryst.* **B35**, 2331–2339 (1979).

113. Schmid, G., Meyer-Zaika, W., Boese, R., and Augart, N. Reversible coupling and opening of a boron-carbon bond. *Angew. Chem., Int. Ed. Engl.* **27**, 952–953 (1988).

114. Karplus, M., and McCammon, J. A. The internal dynamics of globular proteins. *CRC Crit. Rev. Biochem.* **9**, 293–349 (1981).

115. Levitt, M., Sandler, C., and Stern, P. S. Protein normal-mode dynamics; trypsin inhibitor, crambin, ribonuclease and lysozyme. *J. Molec. Biol.* **181**, 423–447 (1985).

116. Stuart, D. I., and Phillips, D. C. On the derivation of dynamic information from diffraction data. *Methods in Enzymology* **115**, 117–142 (1985).

117. Diamond, R. On the use of normal modes in thermal parameter refinement: theory and application to the bovine pancreatic trypsin inhibitor. *Acta Cryst.* **A46**, 425–435 (1990).

118. van Gunsteren, W. F., Berendsen, H. J. C., Hermans, J., Hol, W. G. J., and Postma, J. P. M. Computer simulation of the dynamics of hydrated protein crystals and its comparison with x-ray data. *Proc. Natl. Acad. Sci. USA* **80**, 4315–4319 (1983).

119. Levitt, M. Molecular dynamics of native protein. I. Computer simulation of trajectories. *J. Molec. Biol.* **168**, 595–657 (1983).

120. Singh, U. C., Brown, F. K., Bash, P. A., and Kollman, P. A. An approach to the application of free energy perturbation methods using molecular dynamics: applications to the transformations of $CH_3OH \rightarrow CH_3CH_3$, $H_3O^+ \rightarrow NH_4^+$, glycine $\rightarrow$ alanine, and alanine $\rightarrow$ phenylalanine in aqueous solution and to $H_3O^+(H_2O)_3 \rightarrow NH_4^+(H_2O)_3$ in the gas phase. *J. Amer. Chem. Soc.* **109**, 1607–1614 (1987).

121. van Eerden, J., Harkema, S., and Feil, D. Molecular-dynamics simulation of crystalline 18-crown-6: thermal shortening of covalent bonds. *Acta Cryst.* **B46**, 222–229 (1990).

122. Hagler, A. T., Osguthorpe, D. J., Dauber-Osguthorpe, P., and Hemoel, J. C. Dynamics and conformational energetics of a peptide hormone: vasopressin. *Science* **227**, 1309–1315 (1985).

123. Brünger, A. T., Kuriyan, J., and Karplus, M. Crystallographic R-factor refinement by molecular dynamics. *Science* **235**, 458–460 (1987).

124. Kuriyan, J., Brünger, A. T., Karplus, M., and Hendrickson, W. A. X-ray refinement of protein structures by simulated annealing: test of the method on myohemerythrin. *Acta Cryst.* **A45**, 396–409 (1989).

125. Postma, J. P. M., Parker, M. W., and Tsernoglou, D. Application of molecular dynamics in the crystallographic refinement of colicin A. *Acta Cryst.* **A45**, 471–477 (1989).

126. Fujinaga, M., Gros, P., and van Gunsteren, W. F. Testing the method of crystallographic refinement using molecular dynamics. *J. Appl. Cryst.* **22**, 1–8 (1989).

CHAPTER

14

Chirality and Absolute Structure

Chirality is concerned with objects and their mirror images. A **chiral** object is one that cannot be superimposed on its mirror image in such a way that the two appear identical. Right and left hands provide simple examples of chiral objects. Alan Bassindale wrote:[1] "A left hand and a right hand are, in appearance, nonsuperimposable. A right hand viewed in a mirror appears to be a left hand. The analogy with hands is often carried through to molecules. Chiral molecules are often said to have a 'handedness.' "

The term chirality is derived from the Greek (*cheir* = hand) and was introduced by William H. Thomson, Lord Kelvin in 1893,[2] and the analogy is illustrated in Figure 14.1. He wrote that the word should be used for an object "if its image in a plane mirror, ideally realized, cannot be brought into coincidence with itself." This definition stresses the importance of looking for reflection symmetry in an object.[3] If reflection symmetry is found, the object is not chiral and cannot be described as **achiral** (nonchiral). The apparently simple classification of molecules or structures as chiral or achiral is of utmost importance in nature since the vast majority of biologically active molecules are chiral and their chirality is essential to their function.[4] Alice in *Alice Through the Looking Glass*[5] said to her kitten "How would you like to live in Looking-glass House, Kitty? I wonder if they'd give you milk in there? Perhaps Looking-glass milk isn't good to drink ⋯" She was, of course, correct in realizing that Looking-glass milk would have different properties from ordinary milk.

The **absolute configuration or structure** of a molecule or crystal is its structure expressed in an absolute frame of reference, that is, with known directionalities of the axes.[6-9] This means that the structure parameters that define the structure contain not only the coordinates of each atom in the structure, but also a firm indication of the handedness of the structure.

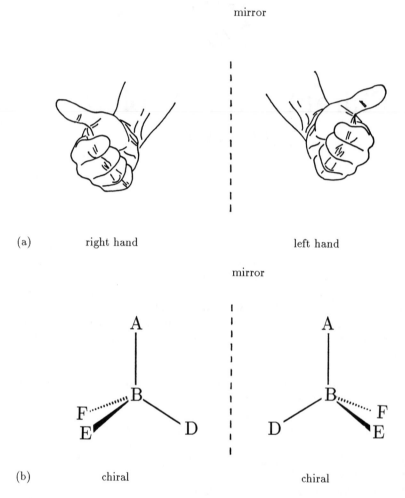

FIGURE 14.1. The importance of lack of mirror symmetry in defining a chiral object. (a) Right and left hands, and (b) two chiral molecules [B(ADEF)] are shown. One is the mirror image of the other, but they cannot be superimposed on each other. E and F, which are out of the plane of the page, cannot be superimposed on the mirror image of B(ADEF) by a rotation about A–B. F will lie on E, and E will lie on F after such a rotation.

Currently, the **absolute structure** (or **absolute configuration**) of a compound can be determined directly using a particular aspect of X-ray diffraction techniques, to be described in this Chapter. Other methods generally determine the absolute structure by correlating chemical or physical properties of an unknown molecule with those of one of known absolute structure.

14.1 Chirality of molecules

The relationship between a chiral object and its mirror-related object is called **enantiomerism**. A knowledge of the existence of enantiomers was one of the reasons that van't Hoff and Le Bel proposed, as described in Chapter 1, that the four valences of carbon are spatially directed to the corners of a regular tetrahedron.[10,11] The only difference between a pair of **enantiomers** is that, if one can be described as a left-handed form, the other will be a right-handed form; they have identical chemical formulæ. The major physical property that allows one to distinguish between enantiomers is the direction in which they, or their solutions, rotate the plane of polarized light, that is, when they are studied in the chiral environment provided by the polarized light. It is important to note that a molecule is not necessarily chiral just because it contains an asymmetric center, or that it is necessarily achiral because it lacks such an asymmetric center. These are not the criteria for molecular chirality. *The test for chirality in a molecule is the nonsuperimposability of the object on its mirror image.*

Since the use of a mirror plane or inversion center to test for chirality involves symmetry operations (see Chapter 4), the designation and description of chirality can be considered in the light of point-group or space-group theory. *A chiral object lacks a reflection plane or inversion center*; recall from the discussion in Chapter 4 that these two symmetry operators are referred to as improper symmetry operators or symmetry operators of the second kind (that convert a left-handed object into a right-handed object and vice versa). At most, a chiral object can only contain symmetry operations of the first kind, but it can *never* contain those of the second kind. Neither can a chiral molecule crystallize in space groups that contain the following types of symmetry elements: inversion centers, reflection planes, glide planes, or rotatory-inversion axes. Thus, proteins, which are chiral, cannot crystallize in space groups that contain these types of symmetry operations, and, as a result, the number of possible space groups for them is reduced from 230 (for both chiral and achiral objects) to only 65 (for chiral objects). These distinctions were listed in Table 4.6 (Chapter 4). It is found, as shown in this Table, that the most common space group for achiral small organic compounds is $P2_1/c$. By contrast, $P2_12_12_1$ is the most common space group for small chiral organic molecules.

Many compounds contain more than one chiral center. If a molecule contains n chiral centers, there will be a maximum of 2^n **stereoisomers**. If there is only one chiral center, there are two stereoisomers (enantiomers), while if there are two asymmetric carbon atoms, then there will be four isomers (two pairs of mirror-image-related molecules). While two chiral centers lead, theoretically, to four stereoisomers, these stereoisomers may not all be different. Some pairs may be identical, containing mirror-image symmetry because the two chiral centers are identical. The reason is that, while the two centers are chiral, if their substituents are the same, the rotation of the plane of plane-polarized light by one center is cancelled by the other (internally compensated). The molecules are described as **mesocompounds** if, in spite of the fact that they contain chiral centers, they are superimposable on their mirror images. A classic example of a molecule with two chiral centers but only three stereoisomers (instead of the possible maximum of four) is tartaric acid, diagrammed in Figure 14.2. Because one of the stereoisomers (in this case, mesotartaric acid) shows mirror symmetry, is superimposable on its mirror image, it is not chiral.

When a molecule contains more than one chiral center, there are pairs of molecules not related by mirror symmetry. These molecules are called **diastereomers**. They differ in the relative chiralities at pairs of chiral centers (see Figure 14.3) and their chemical and physical properties are not identical. In this they differ from enantiomers, which have identical chemical and physical properties except those connected with their interaction with chiral detectors — plane-polarized light or enzymes, for example. Members of a pair of diastereoisomers that differ only in the configuration at one carbon atom are called **epimers**.

In this Chapter we will describe chiral molecules and their representations, then ways of determining absolute structure, and finally extensions of the experimental methods used.

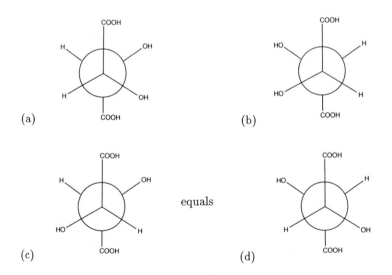

FIGURE 14.2. The stereoisomers of tartaric acid molecules (a) and (b) are enantiomers, related by miror symmetry. Molecules (c) and (d) are equivalent, termed mesocompounds, because they have two identical chiral centers. Thus there are only three stereoisomers [(a), (b), and (c) = (d)].

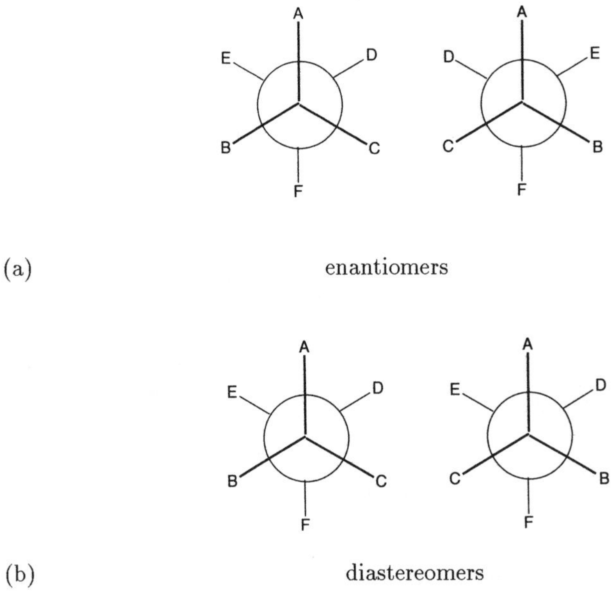

(a) enantiomers

(b) diastereomers

FIGURE 14.3. Comparison of (a) enantiomers that are mirror images of each other, and (b) diastereomers, which are not.

14.1.1 Examples of chiral molecules

We discuss here chirality in a variety of molecular structures. This may help show the importance of nonsuperimposability of mirror images rather than the presence of an asymmetric carbon atom. Thus, if a molecule is chiral it lacks some, but not necessarily all, symmetry elements.

A carbon atom with four different substituents around it (e.g., bromochlorofluoromethane, CHFClBr) has a **chiral center** (also called a **stereocenter** or an **asymmetric carbon atom**). This molecule, Figure 14.4(a), does not have reflection symmetry and therefore is chiral. 2-Iodobutane, Figure 14.4(b), also is chiral, and the carbon atom to which the iodine is attached is the chiral center (also called an asymmetric carbon atom). Similarly, a sulfur atom with three different groups (and one lone pair) also is chiral, Figure 14.4(c).[12]

Biphenyl has a **chiral axis**, as shown in Figure 14.5(a). Its nonplanar conformation is a result of the steric repulsion of the *ortho* hydrogen atoms; this steric overcrowding is enough to offset the resonance energy lost by such nonplanarity. If the *ortho* hydrogen atoms are replaced by bulkier groups, the energy barrier to rotation about the single bond is increased and the resulting molecules can, if the steric overcrowding is

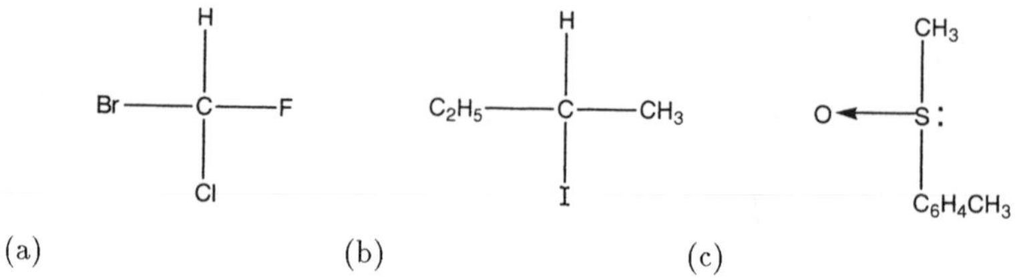

FIGURE 14.4. Some chiral molecules. (a) Bromochlorofluoromethane, (b) 2-iodobutane, and (c) methyl *p*-tolyl sulfoxide (Ref. 12).

extensive enough, exist as enantiomeric pairs. For example, the barrier to rotation of biphenyl-2,2'-disulfonic acid is greater than 100 kJ/mol (25 kcal mol^{-1}, and it exists in two enantiomeric forms, as diagrammed in Figure 14.5(b). Another set of chiral molecules are certain substituted allenes. Allene $CH_2=C=CH_2$ is not planar, but has two terminal H–C–H groups in planes that are oriented 90° with respect to each other. A substituted allene of the type shown in Figure 14.6 (but not allene itself) is chiral. The relationship between Fischer projection and Newman diagrams of asymmetric molecule (potassium isocitrate)[13] was shown in Figure 12.3 (Chapter 12). It can be seen in this Figure that groups that are adjacent in the Fischer representation are actually *trans* to each

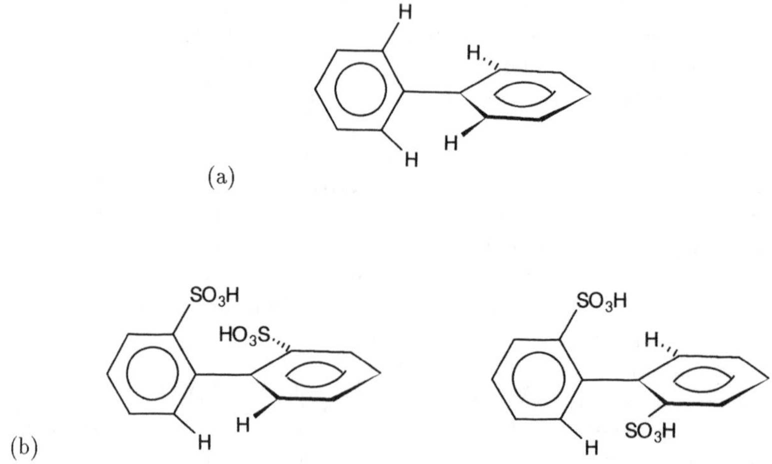

(a)

(b)

FIGURE 14.5. Biphenyl, a chiral molecule. (a) Biphenyl, showing steric overcrowding with respect to hydrogen atoms, and (b) the two enantiomers of its derivative 2,2'-disulfonic acid.

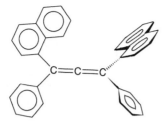

FIGURE 14.6. A 1,3-disubstituted allene, a chiral molecule. Note that, while the substituents on the left-hand side are in the plane of the paper, those on the right-hand side lie in a plane perpendicular to that of the paper.

other. An enzyme, HIV-1 protease, has been prepared containing only D, rather than L amino acids.[14] The peptide substrate specificity is the mirror image of that of the normal enzyme, implying that the folding of proteins composed of D- or L-amino acids gives mirror-image proteins.

14.1.2 The R/S System

A method for describing the structure of a molecule in an absolute frame of reference, the **R/S system**, was put forward by Robert S. Cahn, Christopher K. Ingold and Vladimir Prelog[15,16] (see Figure 14.7). It is used to specify the configuration of a chiral center in an unambiguous way. Each atom around the chiral center is assigned a priority, depending on its atomic number, the atoms with the higher atomic numbers receiving higher priorities. For example, bromochlorofluoromethane (CHFClBr)

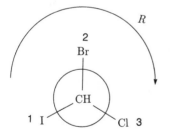

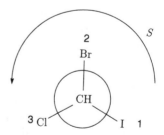

clockwise: Latin, *rectus* = right   counterclockwise: Latin, *sinister* = left

FIGURE 14.7. The *R/S* system for a carbon atom. The substituents are numbered with the highest priority for the highest atomic weight elements. Then one proceeds from high to low priority (atomic weight) with the lowest priority atom obscured by the carbon atom. If the progression is clockwise, the configuration is denoted *R*, if it is counterclockwise it is denoted *S*.

would have Br = 1 (highest), Cl = 2, F = 3, H = 4 (lowest) assigned as priorities for the atom around the asymmetric carbon atom. Often it is necessary to check on the priorities of adjacent atoms if two atoms directly attached to the chiral center are the same (methyl and ethyl, for example, where ethyl with carbon and two hydrogen atoms, has a higher priority than methyl with three hydrogen atoms). If there is a multiple bond, replica atoms of the same type are added. Once the assignments of priorities has been made, the molecule is viewed from the side opposite the atom of lowest priority so that this lowest-priority atom is behind the chiral atom (carbon) looking down the threefold axis (Figure 14.7). The direction in which the assigned priorities *decrease* is then noted. If this direction is clockwise, the configuration is designated R (Latin *rectus* = right) and if it is counterclockwise, it is designated S (Latin *sinister* = left). As heraldic experts know well, the more correct pairing[17] would have been *dexter* and *sinister*, since the meaning of *rectus* as "right" is in the sense of "correct" or "proper." R, however, is presumably preferred over D because the latter is also used for the older term, dextrorotatory. Another convention, the D/L system, still remains in the literature for the naturally occurring L-amino acids and the D-sugars. It is now more appropriate to use the R/S system, because it is a three-dimensional way of describing a structure. Therefore the R/S system gives an excellent way of describing the absolute configurations of molecules found in X-ray diffraction analyses. Its use in a description of the stereoisomers of tartaric acid is shown in Figure 14.8.

The doubly bonded carbon atom, with four groups around the pair of atoms, can be treated in a similar way using the **E/Z system** of nomenclature.[15] The priority sequence of the two substituents on each carbon atom are derived as in the R/S system. Then the terms Z (on the same side, from the German *zusammen*) and E (on opposite sides, from the German, *entgegen*) are used, depending on whether the higher priority group on each carbon atom is on the same or opposite side of the double bond. Two examples are shown in Figure 14.9.

Care must be taken to use the R/S system correctly. For example, if hydrogen in a molecule is replaced by fluorine, an atom with a priority far lower than that of carbon has been replaced by one with a priority higher than oxygen. Therefore the R/S designation of the molecule may change,[18,19] as diagrammed in Figure 14.10. If there are two or more chiral centers in a molecule, and their relative configurations have been determined, it is usual to designate the formula either by the term *rel* (for *relative*), or to affix asterisks to the R or S designation.

14.1.3 Prochirality

Methane, methyl chloride (CH_3Cl) and bromochloromethane (CH_2ClBr) are each achiral. Bromochloromethane, however, has the property that further substitution of one of the two hydrogen atoms (by other than

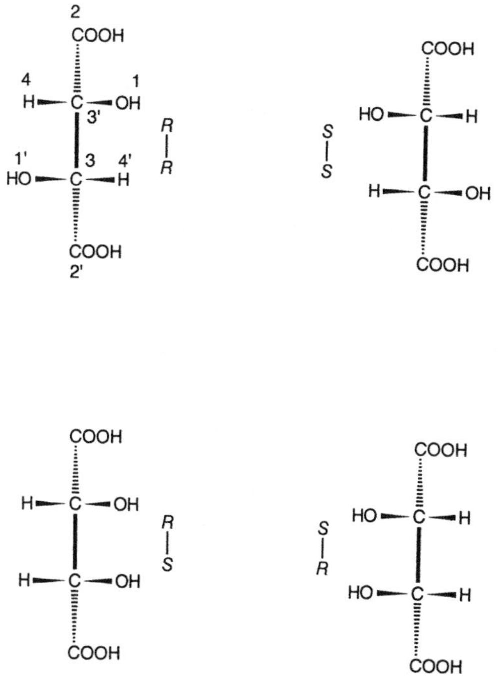

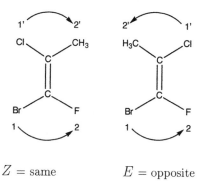

FIGURE 14.8. *R/S* designations for tartaric acid (compare with Figure 14.2). The priorities 1, 2, 3, and 1′, 2′, and 3′ are indicated for the upper and lower asymmetric carbon atoms, respectively, in one enantiomer (compare with Figure 14.7).

Z = same   *E* = opposite

FIGURE 14.9. The *E/Z* system. Priorities of substituents on the C=C group are assigned, as in the *R/S* system (1 = highest atomic number = highest priority, 4 = lowest atomic number = lowest priority). If the atoms with the higher priority on each carbon atom lie on the same side of the double bond, the compound is designated *Z*, if they lie on opposite sides the compound is designated *E*.

bromine or chlorine) creates a chiral product. Therefore, the carbon atom in bromochloromethane is described as "*prochiral.*" **Prochiral** is the term used to describe the situation in which an achiral atom, upon a single substitution, produces a chiral product.[20,21] The two hydrogen atoms of bromochloroethane are described as "enantiotopic." Substitution of one gives a product with an *R* configuration, while substitution of the other gives a product with an *S* configuration. If, however, these hydrogen atoms yield two indistinguishable achiral products upon such substitution, they are described as "homotopic." If diasteroisomers are produced upon substitution, the atoms are called "diastereotopic."

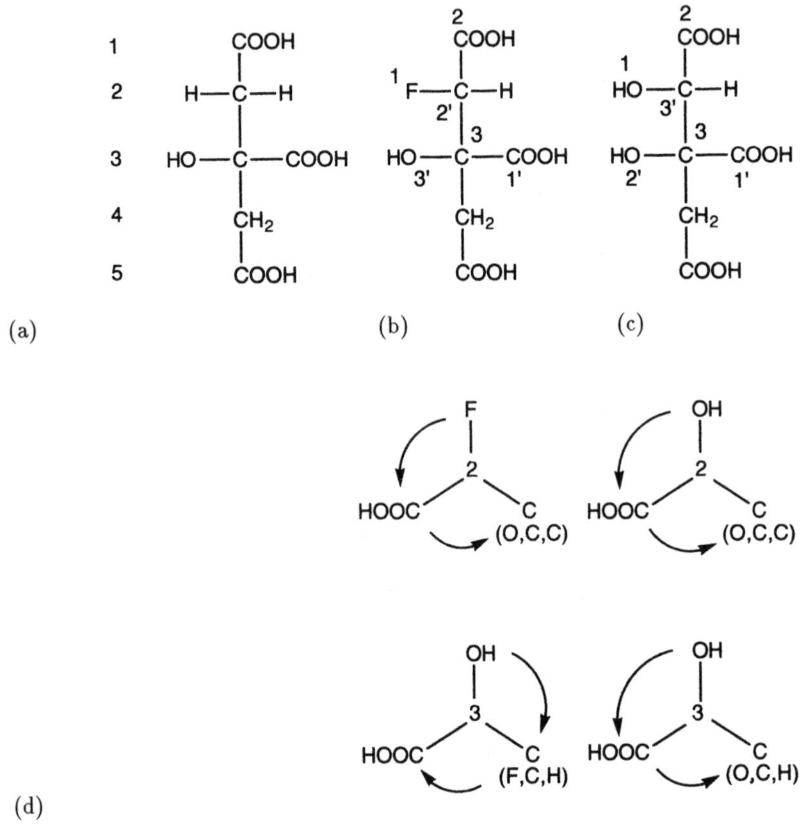

(a)   (b)   (c)

(d)

FIGURE 14.10. Effect of changing a substituent on the designation of a molecule in the *R/S* system. (a) Citric acid. The asterisked atom is the *pro-2S,3R* hydrogen atom. (b) (2*S*,3*R*)-2-fluorocitric acid and (c) (2*S*,3*S*)-2-hydroxycitric acid. (d) When fluorine is introduced, it has a higher priority than oxygen, and the overall designation of the molecule is changed. In the hydroxycitric acid it is (2*S*,3*S*) while when OH is replaced by F it is (2*S*,3*R*).

The two atoms or groups containing the same atoms can be distinguished, if the atom of interest is given higher priority than its mate, it can, by application of the rules for R/S assignment, be denoted *pro R*, or *pro S*. The four atoms attached to the C–C group are assigned priorities as in the R/S system. The arrows point from high to low atomic numbers. If they point in the same direction the molecule is denoted Z; if they point in the opposite direction the molecule is denoted E. A biochemical example is provided by the citrate ion which contains two – CH_2COOH groups and no asymmetric carbon atom. As shown in Figure 14.10, these two groups can be distinguished as *pro R* and *pro S*. If one of the methylene group hydrogen atoms is replaced by F or OH,[22] four stereoisomers are possible.

A similar terminology, the **re/si system**[20] exists for reactions on opposite sides of a planar system, such as a π-system. If addition of chemical groups to the different faces of the plane produces enantiomers, then the faces are described as "enantiotopic faces." The planar group of interest is viewed onto the face. The term *re* (*rectus*) is used if the priority sequence (high to low) of the ligands is clockwise. When the ligands are arranged in a counterclockwise manner, the face is designated *si* (for *sinister*). This is shown in Figure 14.11.

While a *pro*-chiral molecule is achiral, its two equivalent groups can be distinguished if it interacts with a chiral receptor. This was noted by Alexander Ogston,[23] who explained why the two identical $-CH_2-COO^-$ groups in *pro*-chiral citrate are distinguished by the enzyme aconitase. One $-CH_2-COO^-$ group of citrate is derived from oxaloacetate, while the

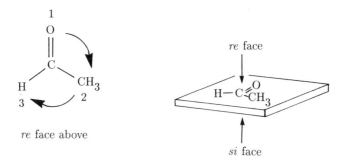

FIGURE 14.11. The *re/si* system. A planar arrangement of groups around an sp^2-hybridized carbon atom. Priorities are assigned to groups as in the R/S system. The *re* face of the plane through the carbon atom and its three substituents is the one in which the priorities follow a clockwise sense. This sequence is counterclockwise for the *si* face.

other is derived from acetate. It has been shown by isotopic labelling experiments[24,25] that these two groups remain distinct in the Krebs cycle (which involves a series of enzymatic reactions). No scrambling of the citrate molecule occurs, contrary to expectations. The reason for this must involve some unique citrate–enzyme interaction.

Ogston showed that **three-point attachment** of the *pro*-chiral substrate to the chiral enzyme must cause the two equivalent groups on citrate to have different environments;[26] either both are bound at different chiral sites on the enzyme, or only one is bound and the other is free. In each case the two groups are differentiable after binding, as shown in Figure 14.12. Thus the inherent lack of symmetry in the citric acid molecule is recognized and exploited by the enzyme.[27] The situation in which *pro*-chiral methylaminomalonate binds to a cobalt complex[28] is shown in Figure 14.13. Only two sites on the cobalt(III) coordination site were available for binding, but the smaller molecule bound at three sites, one via a hydrogen bond to the ligand. It is evident in this figure that the two carboxyl groups have different environments. In fact, asymmetric decarboxylation occurs with this complex, as shown, because of the different environments of the two carboxyl groups and their accessibilities to chemical agents.[28]

14.1.4 Racemates

When equimolar quantities of enantiomeric pairs of molecules are mixed together, the mixture does not show any optical activity, even though it is composed of two chiral compounds. Such a mixture may crystallize in more than one way. If the crystals are a **racemic mixture** they have crystallized giving two types of crystals each containing one of the two enantiomeric compounds. In effect the mixture has been "resolved" into its two enantiomers. As we shall see later in this Chapter, such a mixture of crystals was used by Louis Pasteur to separate the two enantiomers of tartaric acid.[29] If the crystals are **racemates** they have crystallized giving only one type of crystal, one that has equal numbers of both enantiomers within its crystal structure. In this case the crystal structure itself will generally contain a mirror plane or a center of symmetry relating the two enantiomers to each other. For every left-handed structure (molecule, conformer) there is also a right-handed structure in the crystal, and no optical activity will be observed.

14.2 Optical activity and chiral molecules

A compound is considered optically active if it rotates the plane of plane-polarized light, as described in Chapter 5. An explanation for optical activity[30] is that, upon entering a crystal, plane-polarized light is split into two circularly polarized waves, one right-handed and the other left-handed. Since the crystal has a right-handed or left-handed character, the two circularly polarized waves travel through it with different velo-

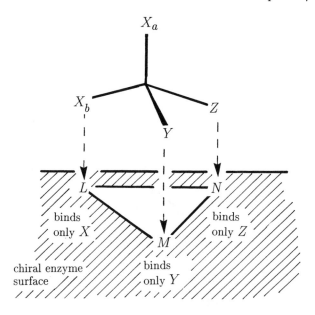

(a) Binding

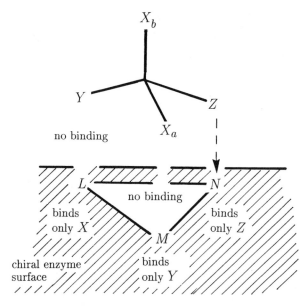

(b) No binding

FIGURE 14.12. Three-point attachment of a *prochiral* molecule CX_2YZ to a chiral receptor (hatched area). Since the receptor binds the groups X, Y, and Z at specific sites only, it must bind X_b but cannot bind X_a. Note that, after binding, X_a (free in space) would be presumed to have different chemical properties from X_b (bound to the receptor). (a) Binding, and (b) no binding.

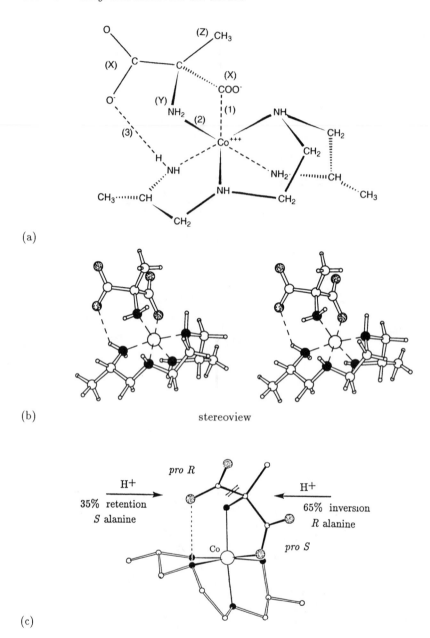

(a)

(b)   stereoview

(c)

FIGURE 14.13. Mode of three-point attachment of *pro*chiral methylaminomalonate to a chiral surface (a cobalt(III) complex). (a) Chemical formula of complex, (b) stereoview of complex, and (c) stereospecific decarboxylation. The carboxyl group that is hydrogen bonded to the complex is lost rather than the metal-bound carboxylate.

cities, and will be out of phase when they exit. They recombine in such a way that the rotation of the plane of polarization is half the difference in relative phase of the two components.

Such optical activity or rotatory dispersion,[30-33] was first described for quartz crystals by François Arago in 1811.[31] Similar observations of optical activity for solutions, for example those of tartaric acid (a byproduct of wine production), and for some organic liquids, were made by Jean Baptiste Biot.[34-38] This led Biot to suggest that if optical activity is observed in solution it indicates that individual molecules, rather than their aggregates (as in quartz crystals), are chiral. In 1821, John Frederick William Herschel[39] at Cambridge, England, noticed small, additional faces (secondary facets) on crystals of quartz (see Figure 5.16, Chapter 5). He further noted that the number of these small faces was only half that expected by the full symmetry of the crystal. Therefore they were called hemihedral, as opposed to holohedral when all the expected faces are present. The important point, however, was that the hemihedral property imparted some asymmetry to the crystals. As a result, crystals of quartz could be divided into two enantiomorphous groups, depending on the handedness of the hemihedral faces.

Herschel also realized that there was a relationship between the sense of these hemihedral faces and the direction of rotation of the plane of polarization of light. A right-handed quartz crystal is one that gives a right-handed optical rotation along the optic axis. One enantiomorphous group of quartz crystals rotated the plane of polarized light counterclockwise, while those of the other group rotated it clockwise. Thus, he linked crystalline form and optical rotation and wrote[39] in 1822: "It may lead us to pay minuter attention to those seemingly capricious truncations on the edges and angles of crystals ⋯ It is not improbable that an accurate examination of them may afford us evidence of the operation of forces of which we have at present no suspicion."

Eilhard Mitscherlich in 1844 reported that solutions of commercial tartaric acid rotated the plane of polarized light in a clockwise manner, while solutions of "racemic acid" had no effect on plane-polarized light.[40] This finding was surprising because the chemical formulæ of these two acids are the same, $HOOC–CH(OH)–CH(OH)–COOH$, and their crystals appeared identical. Mitscherlich posed this problem as "Mitscherlich's riddle."

The answer to Mitscherlich's riddle was found by Louis Pasteur[41-46] in 1848. Because of his great contributions to medical science, it is not always appreciated that he started his scientific career as a crystallographer, and that he solved the enigma posed by Mitscherlich. Pasteur's work on tartaric acid and racemic acid is a testimony to his fine powers of observation, a faculty that stood him in great stead in later years. He wrote at this stage (translated), "I am very happy. I shall soon publish a paper on crystallography."[45]

Pasteur[44,45] continued the studies of Jöns Jakob Berzelius[47] on the crystallization of tartrates and of Biot[34-38] on their optical rotatory power. Berzelius had coined the word **isomeric** to describe, among other cases, the relationship between tartaric and racemic acids.[47] Pasteur noticed something missed by Mitscherlich and Biot. He observed that sodium ammonium tartrate crystals had small asymmetrically placed little faces on some of their edges, that is, they were hemihedral; this hemihedry was always in the same sense (Figure 14.14). Fortunately Pasteur worked in a cold French laboratory in which the temperature was below 26°C, and he obtained a useful crystalline form.[29] Above that temperature a racemic form is obtained, and no resolution is possible. Pasteur had expected that crystals of the racemate salts would be more symmetrical than the hemihedral tartrate crystals, but, to his surprise, they were not. He wrote (translated),[45] "I saw that the tartrate of soda and ammonia carried the little faces that betrayed its asymmetry; but when I passed to the examination of the crystal form of the paratartrate [racemic tartrate] my heart lost a beat; all the crystals bore the facets of asymmetry." Thus, the racemate crystals also had the hemihedry, but not all the crystals were hemihedral in the same sense. Pasteur now understood what was happening. "The crystals are often hemihedral to the left, often to the right, that is the difference between the two salts." The answer to Mitscherlich's riddle was that the tartaric acid that rotated the plane of polarized light in a clockwise manner contained one enantiomer, while "racemic acid" contained both the left- and right-handed forms of the tartrate ions. These enantiomers of sodium ammonium tartrate had been separated on crystallization. At this point Pasteur, in his excitement, rushed out and embraced the surprised curator of the building. Later, he used the the phrase "idée preconnue," that is, "prepared mind"

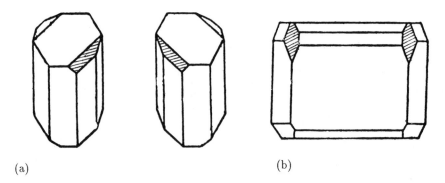

(a)   (b)

FIGURE 14.14. (a) Hemihedral faces (shaded) of sodium ammonium tartrate compared (b) with the holohedral faces (shaded) of the racemate. The hemihedral faces in (a) were used by Pasteur to separate left-handed and right-handed crystals.

and quickly pointed out that he would not have noticed the hemihedry if Gabriel Delafosse[48] had not stressed it so much in his lectures.

Pasteur was able to use tweezers to separate racemic sodium ammonium tartrate crystals into two groups depending on the handedness of the hemihedry. One group when dissolved in water, gave a levorotatory solution, while the other gave a dextrorotatory solution. A present day inspection of such crystals of sodium ammonium tartrate[49,50] demonstrates the acuteness of Pasteur's powers both of observation and deduction. These little hemihedral faces that define left- or right-handedness are not easy to see. For the two mirror image forms, the specific rotations were exactly equal but opposite in sign. Since this effect was observed in solution, Pasteur concluded that it was due to molecules (i.e., mirror image forms) rather than to their arrangement in crystals, as was the case for quartz, as studied by Herschel.

While selection of crystals on the basis of their hemihedry may sometimes be employed as a means of resolving optically active compounds, an alternate method — complexation with a previously resolved optically active substance, a chiral **resolving agent** — is more generally used. This method[45] was introduced by Pasteur in 1853, and it involves the preparation of a complex of two optically active compounds (to give what is called a diastereoisomeric complex). One component of this complex is the racemate or racemic mixture of interest, and the other is a resolving agent, itself chiral. Upon complexation of these two components, a pair of diastereomers is formed which may then be be separated by physical means. The resolving agent is then removed. In this way the two enantiomers are separated. This method was used to establish the absolute configuration of the biochemically active isomer of fluorocitrate, produced from fluoroacetate and oxaloactate by the enzyme citrate synthase.[51,52] Diethyl fluorocitrate was synthesized chemically and resolved by crystallization of a complex with a substituted amine of known absolute configuration [(–)-Sα-methylbenzylamine],[53,54] (also known as phenylethylamine). Therefore, when the crystal structure was determined, it was possible to assign the directions (positive or negative) of the unit-cell axes to correspond to the known absolute configuration of the amine. In this way the absolute configuration of the citrate derivative was deduced. This method did not involve the use of anomalous dispersion, but merely the standard methods of structure determination. It gave, as shown in Figure 14.15, the absolute configuration of the active isomer of fluorocitrate as (2R,3R), a result that agreed with subsequent biochemical studies.[55]

Similarly, the absolute **stereochemistry** of the conversion of fumarate to malate by the enzyme fumarase was confirmed by a neutron-diffraction structure determination of the structure of the (+)-R-α-methylbenzylammonium salt of malate (see Figure 14.16).[56] The malate was prepared enzymatically with the use of heavy water (D_2O) so that

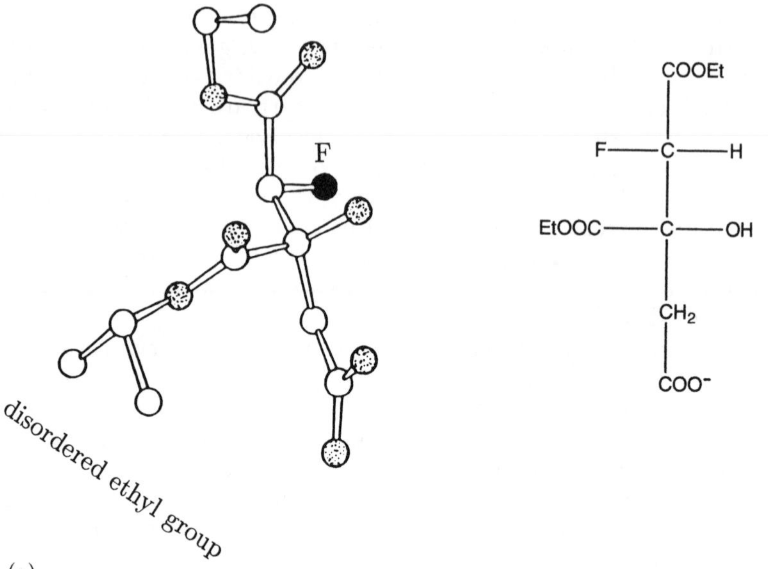

F
disordered ethyl group
COOEt
F—C—H
EtOOC—C—OH
CH2
COO-

(a)

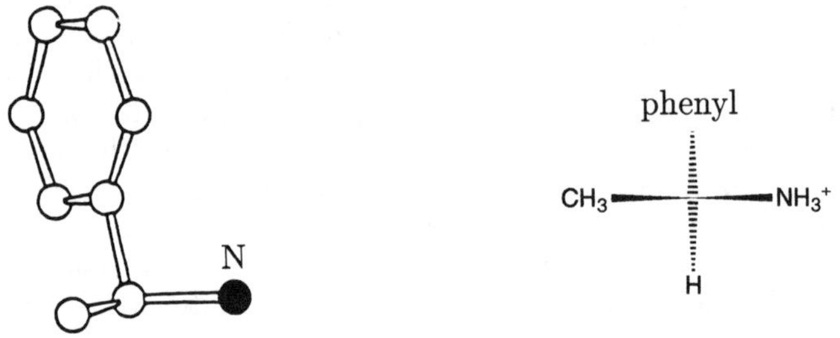

N
phenyl
CH3
NH3+
H

(b)

FIGURE 14.15. Relative and absolute structure of the methylbenzylamine salt of fluorocitrate. Hydrogen atoms omitted for clarity. (a) X-ray diffraction results showing that the absolute configuration of the fluorocitrate studied was 2*S*,3*S*. This is the mirror image of the inhibitory isomer, 2*R*,3*R*. (b) Absolute configuration of methylbenzylamine and its appearance in a crystal structure of the complex with fluorocitrate, drawn with its correct absolute configuration.

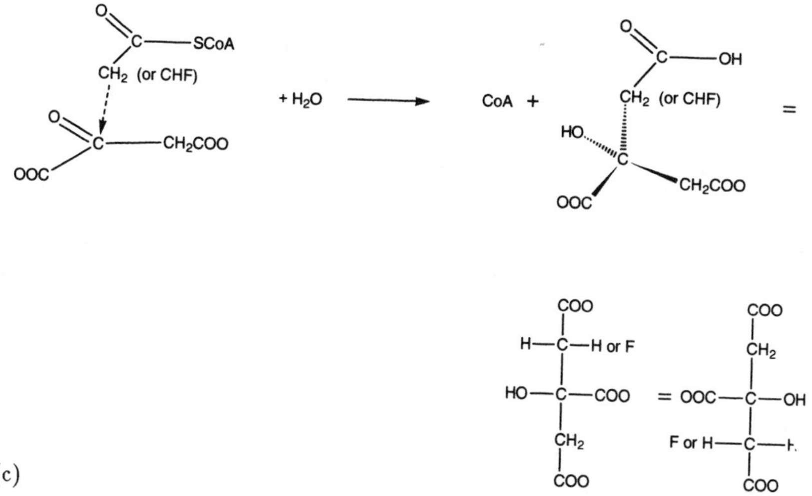

(c)

FIGURE 14.15 (cont'd). (c) Deduced stereochemical course of the reaction catalyzed by the enzyme citrate synthase.

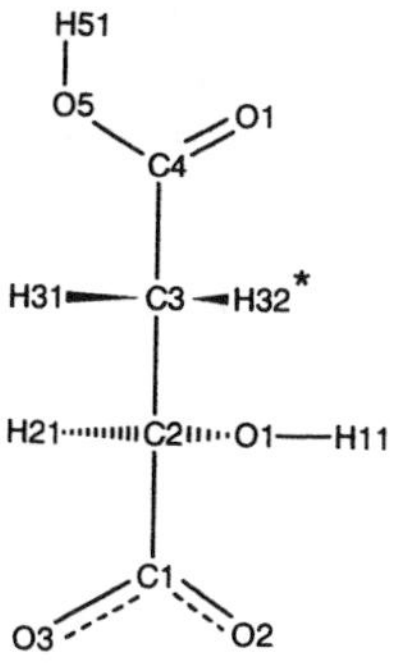

implying

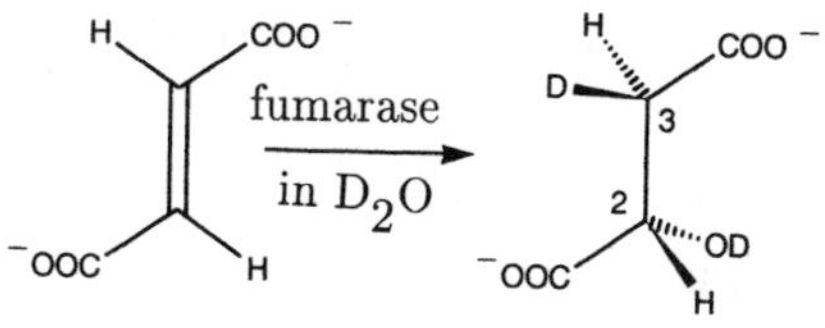

FIGURE 14.16. Stereochemistry of the enzymatic conversion of fumarate to malate (Ref. 56). Shown are (a) the formula with H32 which is converted to D, and (b) the deduced steric course of the enzymatic reaction.

the malate contained the CHD group, the D from the water and the H from the fumarate ion. The analysis demonstrated how neutron diffraction can be used to distinguish hydrogen from deuterium; this method can also distinguish between carbon and nitrogen more clearly than can X-ray diffraction.

14.3 Anomalous dispersion measurements

The experimental method used to determine the chirality or absolute structure of a molecule or crystal structure involves the use of the anomalous dispersion of X-rays by one or more atoms in the structure. We will now describe this effect and how Bijvoet used it to determine the absolute configuration of (+)-tartaric acid from the differences in the intensities of the hkl and $\overline{hkl}$ Bragg reflections.

Dispersion is the difference in refractive index of different wavelengths of light. It is the reason why light is split into colors by a prism or a crystal, as diagrammed for diamond in Figure 5.5 (Chapter 5). In this Figure 5.5 it is shown that the violet light is more deviated from the direction of the incident beam than is the red light. Occasionally an anomalous effect, called **anomalous dispersion**, will occur for the refractive index as the wavelength of the radiation passes through an absorption band. In this case the variation of the refractive index with respect to wavelength reverses sign and the dispersion is termed "anomalous" because the red light is now more deviated than the violet. This effect was noted with the dye fuchsin and with iodine vapor; both have absorption bands in the visible region. Glass and quartz, however, also show such anomalous dispersion in the infrared and ultraviolet regions. Therefore, it appears that this dispersion is not really "anomalous," but its name has remained.

In this experimental method, X-ray scattering by an atom near its absorption edge is investigated. An example of an absorption edge was diagrammed in Figure 6.23 (Chapter 6). Wavelengths that are at or below the absorption edge, the energy of the X rays (inversely proportional to the wavelength) is sufficient either to excite an electron in the strongly absorbing atom to a higher quantum state or to eject the electron completely from this atom. This causes a phase change on scattering, diagrammed in Figure 14.17, and leads to anomalous scattering. The theory of the atomic scattering factor that results from an anomalous scatterer was developed by Helmut Hönl.[57-59] His studies laid the groundwork for the subsequent studies on absolute configuration and absolute structure.

14.3.1 Anomalous scattering of X rays

The scattering factor for the anomalously scattering atom, j, is complex and is expressed as:

$$f_{\text{anom}} = f_j + \Delta f'_j + i\Delta f''_j, \tag{14.1}$$

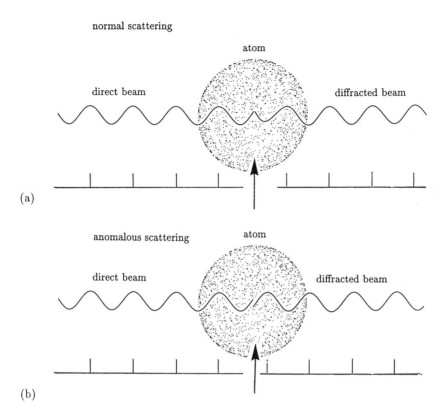

FIGURE 14.17. Phase change on anomalous scattering. Compare this Figure with Figure 3.8 (Chapter 3). (a) Normal, and (b) anomalous scattering.

where $i = \sqrt{-1}$, and $\Delta f'_j$ and $\Delta f''_j$ vary with the wavelength of the incident radiation. The value of $\Delta f''_j$ is largest when the wavelength of the radiation is near the absorption edge of the atom, j. The unperturbed scattering factor, f_j, which is the Fourier transform of the electron density of an atom, is a function of $\sin\theta/\lambda$ but otherwise is independent of wavelength. On the other hand, $\Delta f'$ and $\Delta f''$ are highly dependent on the value of the wavelength, λ, but not very sensitive to 2θ (Figure 14.18). Carol H. Dauben, David H. Templeton, Donald T. Cromer and David Liberman have determined and listed values of $\Delta f'$ and $\Delta f''$ for X rays.[59-63] More detailed studies of the variation of these quantities with wavelength have been made by David H. and Lieselotte K. Templeton.[64-68]

If an atom absorbs X radiation in this way (even moderately), the result will be a phase change in the X rays scattered by that atom, relative to the X rays scattered by the other atoms in the structure (which do not scatter anomalously). This phase change is equivalent to a path length

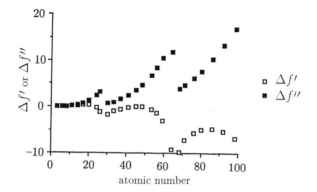

FIGURE 14.18. $\Delta f'$ and $\Delta f''$ versus atomic number for Cu $K\alpha$ radiation. Note how large $\Delta f''$ may be.

change in the scattered radiation (as if there had been a "gulp" before scattering, thereby delaying or advancing the wave). Then, by constructions like those shown in Figure 14.19, which involves a change in path lengths, it can be shown that the apparent path differences of two reflections (hkl) and $(\overline{hkl})$, related by Friedel's law (Equation 4.1, Chapter 4), are not the same, so that their intensities are different. Georges Friedel[69] had assumed that all that was needed to compute phase differences between scattered waves was a knowledge of their path differences on scattering. But, if the waves are affected by an anomalous phase change, the phase difference will depend on the relationship of the scattering vector to the absolute atomic arrangement, as shown in Figure 14.19. Thus, if a structure is noncentrosymmetric, the intensities of $I(hkl)$ and $I(\overline{hkl})$ will differ. If the structure is centrosymmetric there will not be an intensity change but there may be a phase change, so that phase angles will no longer be precisely 0° or 180°.

14.3.2 EXAFS

Another physical technique that involves studies near absorption edges is **EXAFS** (extended X-ray absorption fine structure).[70-72] This consists of a study of the variations in absorption at frequencies above an absorption edge that arise from interference effects of photoelectrons scattered by neighboring atoms. The variation of absorption with energy, shown in Figure 6.23 (Chapter 6), is examined experimentally from the absorption edge to about 1 keV of this edge. There is a ripple in this area, called the Kronig fine structure, and Fourier transform techniques can be used to give an approximate radial distribution function around the atom of interest. From this it is possible to determine the number of atoms in the first coordination sphere of a metal ion and the distances to these atoms. Thus, have proved useful in studies of metalloproteins.[73]

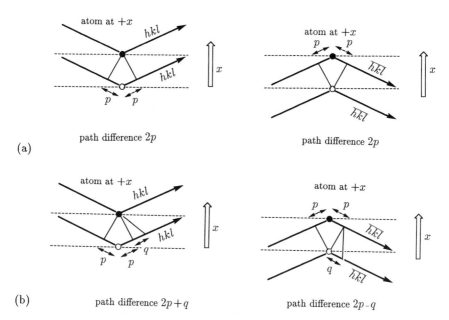

(a)   path difference $2p$ path difference $2p$

(b)   path difference $2p+q$ path difference $2p-q$

FIGURE 14.19. The phase difference depends on the absolute structure. (a) In normal diffraction, with no anomalous dispersion, for the hkl Bragg reflection, the path difference (PD) = $2p$. (b) For the $\overline{hkl}$ Bragg reflection, the path difference also equals $2p$ in normal diffraction. Therefore $I(hkl) = I(\overline{hkl})$. (c) If there is an anomalous scattering atom in the crystal structure, the PD = $2p + q$ for thre hkl Bragg reflection, and (d) the PD = $2p - q$ for the $\overline{hkl}$ Bragg reflection. Therefore $I(hkl) \neq I(\overline{hkl})$.

14.3.3 The "polar" character of zinc blende

Polarity or **polar character** is one-dimensional chirality; for example, a spear or arrow has direction and it is always clear which is the "head" of the arrow. One of the first uses of the breakdown of Friedel's law as a result of anomalous dispersion was in the determination of the polarity of zinc blende. In this crystal structure layers of zinc atoms and layers of sulfur atoms are arranged in pairs through the crystal. The polarity of zincblende is expressed with respect to some observable physical property (for example, the appearance of crystal faces at different ends of the crystal). The question is whether the zinc or the sulphur layers are on the shiny-face side of these pairs. Anomalous scattering of Au Lα X rays was used to determine the polarity of the arrangement of these layers.

Zinc blende (sphalerite, ZnS) has a diamond-type structure. The space group is $F\overline{4}3m$ for a cubic unit cell with $a = 5.42$ Å. The structure is illustrated in Figure 14.20. Parallel to the (100) face of zinc blende

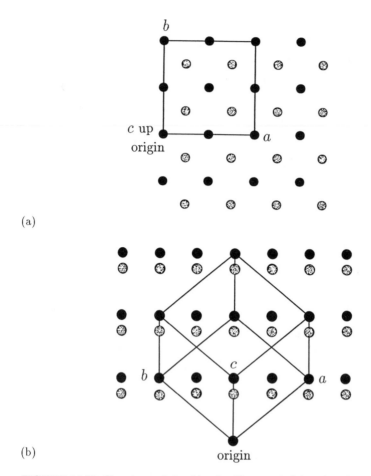

FIGURE 14.20. Structure of zinc blende. Shown are (a) a view down a unit-cell axis, and (b) a view onto the 111 plane, showing the layers of atoms of zinc (filled circles) and sulfur (stippled circles).

there are alternate layers of zinc and sulfur atoms, and parallel to the (110) face each layer contains both zinc and sulfur atoms. Parallel to the (111) faces, however, there are pairs of faces of zinc and sulfur atoms separated by only one quarter of the spacing in that direction; that is, alternate layers of atoms are separated either by $a/4$ or by $3a/4$, as shown in Figure 14.21. Therefore the direction *perpendicular* to the (111) face has "polar character" since either zinc layers or sulfur layers are in front, depending on the direction along this perpendicular being viewed. The aim was to find a relationship between the directionality of the layers of atoms separated by $a/4$ and some externally observed features of the crystals.

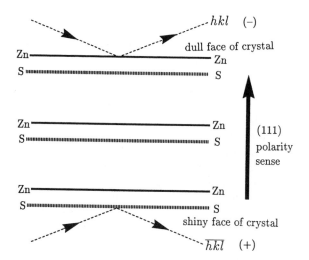

FIGURE 14.21. Polar character of zinc blende. The results of the X-ray diffraction experiment show that the shiny faces have sulfur atoms on their surfaces, and that the dull faces have zinc atoms on their surfaces.

Opposite faces [(111) and ($\overline{1}\overline{1}\overline{1}$)] of zinc blende have different appearances; those on one side are shiny and those on the other side are dull or matte. These two types of faces have zinc atoms on one face and sulfur atoms on the other. One face is (111) and the other is ($\overline{1}\overline{1}\overline{1}$), but which is which? This question was investigated by Shoji Nishikawa[74] in Japan in 1928 and by Dirk Coster, Kornelis S. Knol, and Jan Albert Prins[75,76] in 1930. It was assumed that cleavage does not occur between the closely arranged pairs of zinc and sulfur planes but between planes that are more separated ($3a/4$ rather than $a/4$). Au Lα radiation was used in the anomalous dispersion experiment; this has an $\alpha_1\alpha_2$ doublet with wavelengths Au Lα_1 = 1.276 Å and Au Lα_2 = 1.288 Å, on either side of the absorption edge of zinc (1.283 Å). Only the Au Lα_1 radiation is scattered anomalously by the zinc atoms. The observed intensity differences between the 111 Bragg reflection and the $\overline{1}\overline{1}\overline{1}$ Bragg reflections give the information that the atoms on the surfaces of the larger shiny faces are sulfur and those on the surfaces of the smaller rough (matte) faces are zinc. When pressure is applied in a direction perpendicular to the (111) face in piezoelectric experiments (see Chapter 5), the shiny faces with sulfur on the exterior become relatively positively charged and the matte faces with zinc on the surface become relatively negatively charged.

14.3.4 The absolute structure of (+)-tartaric acid

The experiment with zinc blende just described involved polarity, which is chirality in one dimension. The principles used in the analysis of zinc blende were appreciated and extended to three dimensions by Johannes M. Bijvoet, Antonius F. Peerdeman and Adrianus J. van Bommel[77–79] in 1951. Bijvoet had been working on methods of phase determination by multiple isomorphous replacement methods.[80] Peerdeman wrote[81] that, on examining the "half-forgotton zinc sulphide experiment of Coster, Knol, and Prins," Bijvoet "realized that absolute configuration, phase determination and polarity sense in zinc sulphide were but three varieties of one and the same problem and that the solution of the Zn–S polarity also applied to the other." Bijvoet, Peerdeman, and van Bommel demonstrated that the anomalous scattering of X rays from noncentrosymmetric crystals of the sodium rubidium salt of dextrorotatory (+)-tartaric acid (Figure 14.22) by Zr Kα X radiation could be used to find the absolute structure of the tartrate. The wavelength of Zr radiation lies near the absorption edge of rubidium, and the absolute configuration of sodium rubidium tartrate was established by the measurement of the effect of anomalous scattering of zirconium radiation by a rubidium atom in this noncentrosymmetric structure (Table 14.1).

Bijvoet's method, used by the X-ray crystallographer to measure the absolute configuration of a chiral structure, involves measurement of Bragg reflection data using X rays of a wavelength that is near the absorption edge for one or more of the atoms in the crystal structure. The anomalous dispersion effect will cause a breakdown of Friedel's Law,[69] so that the intensities of the two Bragg reflections, $I(hkl)$ and $I(\overline{hkl})$, normally equal, will now be different. The absolute magnitude of these differences, $I(hkl) - I(\overline{hkl})$, can be used to determine the absolute configuration of the molecule in the crystal structure. Thus, when the atomic positions in a chiral crystal have been determined, the absolute frame of reference is established because, for one enantiomer, the *hkl* Bragg re-

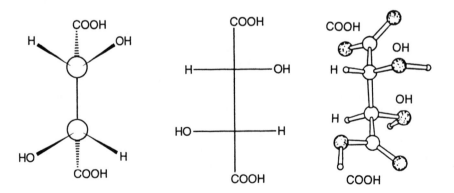

FIGURE 14.22. Absolute structure of (+)-tartaric acid.

TABLE 14.1. Some intensities of rubidium sodium tartrate (Ref. 78).

h	k	l	$I(hkl)$	$I(\overline{hkl})$		h	k	l	$I(hkl)$	$I(\overline{hkl})$	
1	4	1	361	377	?	2	6	1	828	817	+
1	5	1	337	313	?	2	7	1	18	8	+
1	6	1	313	241	+	2	8	1	763	716	+
1	7	1	65	78	−	2	9	1	170	166	?
1	8	1	185	148	+	2	10	1	200	239	−
1	9	1	65	46	+	2	11	1	159	149	?
1	10	1	248	208	+	2	12	1	324	353	−
1	11	1	27	41	−						

Note that ? implies that the difference in intensities is within the expected error and is probably not significant.

flection will be more intense than the $\overline{hkl}$ Bragg reflection, while for the other enantiomer the opposite is true.

Bijvoet, Peerdeman, and van Bommel wrote "The result is that Emil Fischer's *convention* ⋯ *appears to answer to reality.*" Emil Fischer,[82,83] in the nineteenth century had arbitrarily assigned an absolute configuration to the asymmetric carbon atom in (+)-glyceraldehyde. He had just a 50% chance of being right (he could equally well have chosen the mirror image of the correct absolute configuration). The anomalous dispersion experiment of Bijvoet and his co-workers, however, combined with chemical information relating the absolute configurations of (+)-tartaric acid and (+)-glyceraldehyde, showed that Fischer had chosen correctly. Chemistry texts did not have to be revised in light of the 1951 experiment! By use of this method many X-ray determinations of chirality (absolute configuration) have been made.[54]

The method of determining absolute configurations, suggested by Bijvoet, was confirmed by an independent test following a challenge that proved to be incorrect. In 1972 Jiro Tanaka[84−86] reported a disagreement between the assignment of absolute configuration from circular dichroism techniques and crystal structure analysis by anomalous dispersion methods. It further appeared that his findings were corroborated by more detailed investigations into the quantum theory of anomalous scattering. He suggested that the sign of $\Delta f''$ used by Bijvoet and co-workers was incorrect and should be reversed. This would have the effect of reversing the sense of chirality of all crystal structures determined until that time by X-ray anomalous dispersion studies. An independent experimental proof of the validity of the original Bijvoet assignment of absolute configuration arrived from an unexpected method, noble gas ion reflection mass spectrometry on sphalerite (ZnS) and CdS crystals.[87] Beams of positive ions of known energies (He+ or Ne+, for example) were reflected from the opposite (111) and ($\overline{1}\overline{1}\overline{1}$) faces of these crystals. The energy loss of the ions on reflection from a crystal face is a function of the atomic masses of the atoms that they have collided with. This experiment clearly verified the earlier results of Coster, Knol, and Prins[75] that the sulfur layers are exposed on the surface of the shiny (111) faces and zinc layers on the matte ($\overline{1}\overline{1}\overline{1}$) faces. Thus, the sense of polarity so determined was in agreement with that derived by the anomalous X-ray scattering method. The agreement between circular dichroism measurements and X-ray analysis was subsequently restored upon more correct interpretations[87] of the chiroptical technique in the case studied by Tanaka.[84-86]

Since the first experiment by Bijvoet and co-workers[79] the absolute structures of many compounds have been determined. First was an amino acid, so that the relationships between the absolute configurations of groups of molecules, already worked out by organic chemists could be used with the now known absolute structures of (+)-tartaric acid[79] and D-(−)-isoleucine.[88] Then followed many other determinations of absolute structure, some of which provided information on a previously unknown absolute configuration and others that confirmed results from other methods, thereby, in some cases, establishing the steric courses of chemical and biochemical reactions.[89-94] If a crystal is chiral, it is a simple matter to do the necessary measurements to determine the absolute configuration. It is part of the overall scheme of intensity measurements.[95] As stressed here, in all cases the result of an absolute configuration determination must be related to some other effect that the crystal has such as the sense of its optical activity, its crystal hemihedry, or its biochemical reactivity. Otherwise there is no general relevance to the observation.

14.3.5 Physical tests for chiral structures

The 230 space groups may be divided into three categories. Those that are made up of only one enantiomer of a chiral molecule are necessarily noncentrosymmetric (see Table 4.6, Chapter 4). Proteins and nucleic acids fall in this category. A second category consists of pairs of enantiomers or of achiral molecules, crystallizing in a centrosymmetric space group, so that for every atom at x, y, z in the unit cell there is another at $-x$, $-y$, $-z$. There is, however, a third category, somewhat confusing to many. It is possible for achiral objects (such as racemates) to crystallize in noncentrosymmetric space groups (Table 4.6, Chapter 4). Thus, the fact that a space group is noncentrosymmetric does not mean that the component molecules are chiral. In this category the space group contains no center of symmetry, only mirror planes and glide panes. Therefore, although the crystal structure is noncentrosymmetric, the component molecules exist as equal numbers of left-handed and right-handed (chiral) molecules or they are (achiral) molecules with no inherent asymmetry.

Some space groups are enantiomers of others. There are 11 such pairs, listed in Table 4.6 (Chapter 4) and Table 14.2.[96,97] If the (+) isomer of a chiral molecule crystallizes in one of these space groups, the (−) isomer will crystallize in the enantiomorphous space group. The systematic absences in the Bragg reflections are the same for both members of these pairs of space groups, but anomalous dispersion can aid in distinguishing between them. Several proteins have crystallized in enantiomorphous space groups.

There are several tests for chirality in a crystal.

1. Crystals of pure chiral objects, such as proteins or nucleic acids, cannot crystallize in a space group that contains mirror planes, glide planes, or inversion centers. Therefore, if a crystal is grown and its space groups contains one or more of these symmetry operations, the crystal is not chiral.

2. Anomalous X-ray dispersion effects will cause intensity differences if the structure is noncentrosymmetric and contains an atom that scatters anomalously. Friedel's Law, which states that the X-ray diffraction pattern of a crystal is centrosymmetric even if the crystal structure is not, will not be obeyed. This means that, in the presence of anomalous dispersion effects, $I(hkl) \neq I(\overline{hkl})$. If the structure is centrosymmetric, then $I(hkl) = I(\overline{hkl})$, but if there is an anomalously scattering atom in the structure then the phases angles will no longer be precisely 0° or 180°. Thus, if anomalous dispersion is observed and $I(hkl) \neq I(\overline{hkl})$, then the crystal structure must be chiral.

3. Certain physical phenomena, such as piezoelectric or pyroelectric effects, or second harmonic generation, are found only for noncentrosymmetric structures [see Table 4.6 (Chapter 4) and Chapter 5].

4. The statistics of the intensities will usually tell if a crystal structure is centrosymmetric or noncentrosymmetric[98] (see Chapter 8).

It has already been pointed out that the test for noncentrosymmetric space groups will not tell if a molecule is chiral or not, only that if the space group is centrosymmetric the molecule cannot be chiral. If, however, the molecule is chiral and only one enantiomer is present in the crystal, then anomalous dispersion can be used to determine which enantiomer is present.

An example of tests for chirality is provided by studies of mandelic acid crystals (Figure 14.23).[99] These crystals are polar and noncentrosymmetric, space group $P2_1$, but no hemihedral faces develop and therefore there are no external indications that allow one to distinguish the two ends of its polar axis **b**. Other techniques have to be used. In order to differentiate between the two ends of the hexagonally-shaped crystals (which were shown, by X-ray diffraction studies to have the **c** axis along the unique axis of the crystal).[100] The directions of the **a** and **b** axes with respect to crystal habit were also established by X-ray diffraction studies.

TABLE 14.2. Enantiomorphic pairs of space groups (Ref. 96).

(a) Threefold screw axes 3_1 or 3_2

   $P3_1$ and $P3_2$
   $P3_112$ and $P3_212$
   $P3_121$ and $P3_221$

Fourfold screw axes 4_1 or 4_2

   $P4_1$ and $P4_2$
   $P4_122$ and $P4_322$
   $P4_12_12$ and $P4_32_12$
   $P4_132$ and $P4_322$

(c) Sixfold screw axes 6_1 and 6_5, or 6_2 and 6_4

   $P6_1$ and $P6_5$
   $P6_2$ and $P6_4$
   $P6_122$ and $P6_522$
   $P6_222$ and $P6_422$

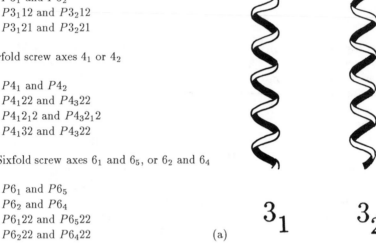

(a) 3_1 3_2

Note how these pairs involve the three-, four-, and sixfold screw axes.

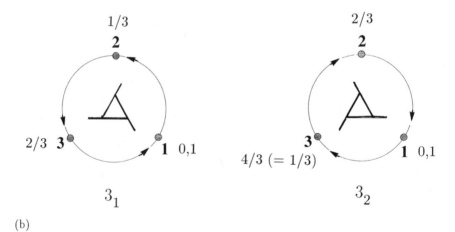

(b)

Shown are two threefold screw axes 3_1 and 3_2, viewed (a) sideways, and (b) down the screw axis. Heights of points 1, 2, and 3 are indicated to show the difference between 3_1 and 3_2

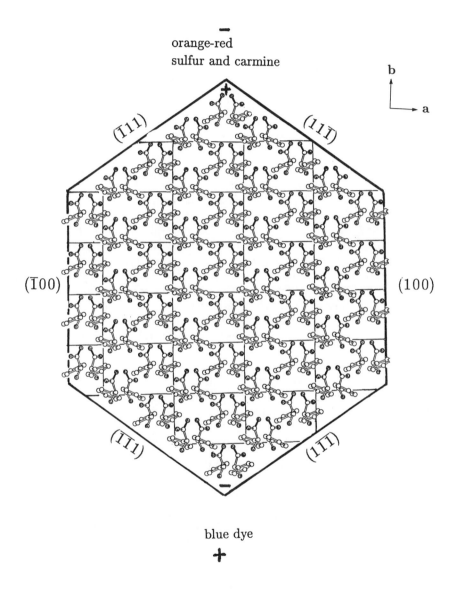

FIGURE 14.23. Determination of the sense of chirality of mandelic acid crystals (Ref. 99). (a) Molecular packing in a crystal.

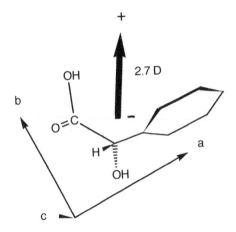

FIGURE 14.23 (cont'd). Determination of the sense of chirality of mandelic acid crystals (Ref. 99). (b) Calculated molecular dipole moment.

Two powders were sprayed onto a single crystal. One of these powders, lycopodium powder (dyed with methylene violet), develops a positive charge, while a mixture of sulfur and carmine powders develops a negative charge (compare with Figure 5.18, Chapter 5). These two powders were then attracted to the negative and positive ends of the polar axis of the heated crystal and showed which end was which. In both (R)-$(-)$- and (S)-$(+)$-mandelic acid crystals, the carboxylic hydroxyl group vectors are directed towards the positive end of the crystal. The dipole moments for the enantiomers (using crystal structure coordinates) were calculated using *ab initio* methods. From a knowledge of the crystal morphology and the absolute configuration of the crystals, it was possible to determine the direction of the electric dipole. This example (Figure 14.23) shows what can be done when the enantiomeric crystals are not easily distinguishable.

14.3.6 Experimental measurements of sense of chirality

The effect of anomalous dispersion on the Bragg intensities is very small, but it is important and useful. A. Lindo Patterson[101] derived the equations involved in anomalous dispersion, initially in order to determine how to eliminate this effect on the intensities so that the electron-density map will not contain electron density that is unreal (imaginary, from an equation involving $i=\sqrt{-1}$). He considered the structure factors $F(hkl)$ and $F(\overline{hkl})$ to be made up of two components: one represents the anomalously dispersing atom and is here subscripted d (A_d, B_d, f_d, Δf_d, and $\Delta f_d''$), and the other representing the remaining parts of the structure factor here subscripted n (A_n, B_n, f_n). He then defined two quantities, $2S$, the sum of the squares of the structure factors of the hkl and $\overline{hkl}$ Bragg reflections, and $2D$, their difference:

$$2S = \text{sum} = \left[\mid F(hkl) \mid^2 + \mid F(\overline{hkl}) \mid^2 \right]$$

$$= A_n^2 + B_n^2 + (2\Delta f_d'/f_d)(A_n A_d + B_n B_d)$$

$$+ \left[(\Delta f_d')^2 + (\Delta f_d'')^2/f_d^2 \right](A_n A_d^2 + B_n B_d^2) \qquad (14.2)$$

$$2D = \text{difference} = \left[\mid F(hkl) \mid^2 - \mid F(\overline{hkl}) \mid^2 \right]$$

$$= -2 \left[(\frac{\Delta f_d''}{f_d}(A_n B_d - B_n A_d) \right] \qquad (14.3)$$

where A_n and B_n are the structure factor components for the atoms that do not scatter anomalously, and A_d and B_d are those for the anomalously dispersing atom. These equations, especially Equation 14.3, are useful for deriving the magnitude of the effect. It depends on the values of A_n, B_n, A_d, B_d, f, $\Delta f_d'$ and $\Delta f_d''$ is the difference in two large but nearly equal numbers. Dorothy Hodgkin wrote:[102] "It is very curious how blind we often are. Today we look at X-ray photographs of crystals containing large molecules and anomalously scattering atoms and we recognise, with immediate delighted conviction, small differences in Friedel related pairs of reflections. With the help of these differences we confidently place the heavy atoms and determine phase angles and crystal structures. Yet before 1949, we must often have neglected to see similar differences on the photographs we took of many heavy-atom-containing crystals." Like many others she "thought the effects would be too small to be useful. A year or two later ⋯ the differences became suddenly glaringly obvious to us. We could never again not see them." The values of the anomalous portions of the scattering factors ($\Delta f_d''$) thus contribute to the structure factors, as shown in Figure 14.24. This information can be used, as shown in Figure 8.26 (Chapter 8), in the determination of relative phases by anomalous dispersion.

In practice, values of $[I(hkl) - I(\overline{hkl})]$ are calculated for a model of the structure using Equations 14.3. A comparison is then made between observed and calculated values made for a selection of Bragg reflections. If the signs of the differences so calculated agree with the measured values, then the model has the correct absolute configuration. If the signs are wrong, then the absolute configuration of the model must be reversed by converting the signs of all x,y,z to $-x,-y,-z$. Examples are given in Table 14.3 and Figure 14.25.[103] It is important in all such experiments to keep h, k, and l and x, y, and z in axial systems with the same handedness (conventionally, right-handed). Such consistency in the indexing of these quantities is essential to the validity of the results.[104]

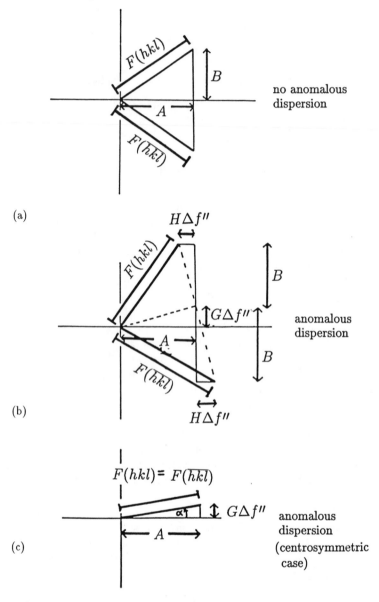

FIGURE 14.24. Effects of $\Delta f'$ on the relative phase of a structure factor. (a) The structure factor with no anomalous dispersion has components $A = A(hkl)$ and $B = B(hkl)$. (b) When anomalous dispersion occurs, as shown in Equations 14.2 and 14.3, there are perturbations. $H\Delta f'' = (B_d/f_d)\Delta f''_d$ and $G\Delta f'' = (A_d/f_d)\Delta f''_d$ (c) When the crystal structure is noncentrosymmetric, $B = 0°$, but $G\Delta f''$ does not equal zero. Therefore the relative phase angle is not $0°$ or $180°$.

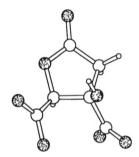
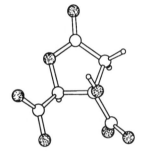

(a)

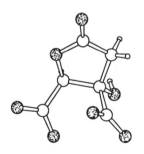

(b)   stereoview Molecule A

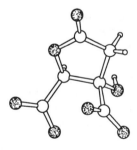

(c)   stereoview Molecule B

FIGURE 14.25. Absolute configurations of two hydroxycitrate lactones (Table 14.3 and Ref. 103). Shown are (a) the chemical formulæ, and (b) and (c) stereoviews.

TABLE 14.3. Some intensity data leading to absolute structure determination (Ref. 103). Ratios of fractional intensity differences in Friedel-related pairs of Bragg reflections. Values are given for two chiral compounds, A and B. The listed ratio, $RI = ([I(hkl)-I(\overline{hkl})]/[I(hkl)+I(\overline{hkl})] = $ (difference)/(sum) [D/S in Equations 14.2 and 14.3] for hkl and $\overline{hkl}$ Bragg reflections. Note that all the ratios have the calculated symmetry, showing that the model used to calculate D and S has the correct absolute configuration. The two structures are shown in Figure 14.25.

h	k	l	Molecule A $(D/S)_{obs}$	Molecule A $(D/S)_{calc}$	h	k	l	Molecule B $(D/S)_{obs}$	Molecule B $(D/S)_{calc}$
4	5	2	-0.12	-0.15	1	2	1	-0.22	-0.22
5	6	2	-0.11	-0.13	1	3	4	$+0.32$	$+0.36$
6	3	2	$+0.18$	$+0.22$	1	5	4	-0.32	-0.37
6	8	3	-0.13	-0.14	2	3	2	$+0.13$	$+0.20$
9	1	4	-0.12	-0.07	2	4	1	$+0.05$	$+0.06$
9	3	3	-0.23	-0.11	2	4	2	$+0.16$	$+0.24$
9	10	3	-0.16	-0.22	2	5	1	-0.31	-0.34

Since the result of the assignment of absolute configuration is important to chemists and biochemists, it is essential that it be correct. The assignment of absolute configuration depends on the analysis of small differences between measurements of two fairly large numbers ($I(hkl)$ and $I(\overline{hkl})$). Therefore, both $I(hkl)$ and $I(\overline{hkl})$ should be measured precisely and under as nearly the same conditions as possible. Only those differences in intensity that are considered experimentally significant [e.g., differences greater than $2.7\sigma(I)$] should be used. It is important to measure the intensities of the pairs of Bragg reflections using the same crystal, to measure them close together in time so that no crystal deterioration occurs between the two measurements, to make sure that these pairs of Bragg reflections trace essentially the same path length through the crystal, and to ensure that the chosen Bragg reflections are reasonably intense. Weak intensity data should not be used because they are prone to considerable errors in measurement.

There are three general experimental methods used to find the absolute configuration of molecules in a crystal. One involves direct measurements of $I(hkl)$ and $I(\overline{hkl})$ when this difference is considered highly significant with respect to the e.s.d. values of the measurements. This should be done for several, say 20, pairs of Bragg reflections and was the method used by Bijvoet in his study of the absolute configuration of the tartrate ion.[79] All that is needed to determine an absolute configuration is a list of signs of the differences [which one is larger?, $I(hkl)$ or

$I(\overline{hkl})$?], for both observed and calculated data. The reasons of preci- sion *the difference in intensities should be much larger than three times their e.s.d. values* if the difference is to be taken into account. If most or, better still, all of the observed and calculated differences agree with respect to their signs, it is concluded that the model has the correct ab- solute configuration. If most of the differences in observed signs are the opposite of those calculated, then the wrong absolute configuration has been chosen for the model used to calculate the intensities. As a result, the signs of all x, y and z of the model will have to be reversed (i.e., every x, y, z must be replaced by $-x, -y, -z$). If there is not a very clear result, positive or negative, the data are not sufficiently precise for an absolute configuration determination, or the model needs adjustment, or the crystal is not chiral.

A second method for the determination of absolute configurations, due to Walter C. Hamilton,[105] involves a computation of R values for both the trial structure and its mirror image. The ratio of these two R values will show which structure is correct and will allow an assessment to be made of the reliability of the choice. Small and possibly insignificant differences in $I(hkl)$ and $I(\overline{hkl})$ are included in this, however, and a few reflections with large differences should also be checked as described above in the first method.

A third method involves use of $\Delta f''$, or its sign, as a variable in the least-squares refinement. This method was first used by Carroll K. Johnson,[13,106] rediscovered by Donald Rogers[107] and improved on by Gérald Bernardinelli and Howard D. Flack.[108-110] Johnson included $\Delta f''$ as a parameter, considered independent of $\sin\theta/\lambda$, for the least-squares refinement of potassium dihydrogen isocitrate.[13] He used Cr $K\alpha$ radia- tion ($\lambda = 2.2896$ Å) to measure the intensities of some Bragg reflections and determined that $\Delta f''$ has values 2.38(8) for potassium, 0.19(5) for oxygen and 0.07(5) for carbon. Later an independent parameter, whose reliability can be assessed, together with its e.s.d., that defines the sense of the absolute structure (approximately +1 or –1 which can be applied to $\Delta f''$) was introduced into the least-squares refinement of data. It then defines the chirality (polarity) of crystals, and is a way of assessing the confidence of an absolute structure determination.

Any reasonably heavy atom can serve as an anomalous scatterer if the appropriate wavelength of X rays is used. For neutrons, however, there are fewer atoms that scatter anomalously[56] (for example, ^{6}Li, ^{10}B, ^{113}Cd, ^{144}Sm, ^{151}Eu, ^{157}Gd) and their high absorption may limit their use. If X-ray diffraction intensity data are sufficiently carefully measured, it is possible to determine absolute configurations even if there are no heavy atoms in the structure, although values of $\Delta f''$ are very small for the light elements. Results of determinations of absolute structure must come from laboratories with an excellent reputation for careful intensity measurement. For example, Håkon Hope has been able to assign absolute configurations on the basis of the anomalous scattering of oxygen.[111] He reported $\Delta f''$ of 0.041(4) for oxygen for Cu $K\alpha$ X rays, noting, however, that it is necessary to have very precise values for $\Delta f''$ for this method to work. It is better, if at all possible, to have an atom present in the crystal structure that scatters anomalously to as large an extent as possible, and then to measure pairs of Bragg reflections $I(hkl)$ and $I(\overline{hkl})$, and study the differences in intensity within the pairs.

When the absolute structure has been determined, the result must be correlated with some physical property of the crystal, otherwise the result has no use to the chemist. The obvious correlation is with the direction of rotation of the plane of plane-polarized light, that is, whether the compound or crystal is dextrorotatory or levorotatory. Another correlation can be made with crystal appearance; this was shown for zinc blende with its matte and shiny faces, and for silica and sodium ammonium tartrate crystals for the disposition of their hemihedral faces. If such data are not available, it may be necessary to list physical properties of diastereomers made with chiral complexing agents. Then, whenever this same compound is encountered by a chemist, its absolute structure is well known.

14.4 Uses of anomalous dispersion

The use of anomalous dispersion has led to many results of interest in chemistry and biochemistry such as the steric course of certain chemical reactions.[112,113] The uses for detailed studies of chirality and for protein relative phase determination will now be discussed.

14.4.1 Studies of chirality in biochemical reactions

Both neutron and X-ray diffraction were used to determine the absolute configuration of the isomer of deuterated glycolate that is produced biochemically.[114] In this way the stereochemistry of this particular biochemical pathway was established, as shown in Figure 14.26. Deuterated glycolic acid, HO–CHD–COOH, was prepared enzymatically, and the structure of lithium glycolate was determined by X-ray diffraction methods.[115] Since hydrogen and deuterium have the same atomic number, they were not distinguished by this X-ray diffraction study. These two isotopes of hydrogen, however, have different signs for their neutron scattering factors (see Figure 3.13, Chapter 3). Crystals of lithium glycolate, prepared using lithium hydroxide enriched with ^6Li, were then studied by neutron diffraction. The differing scattering amplitudes of hydrogen and deuterium $[-0.378$ and $+0.65$ $(\times 10^{-12}$ cm), respectively] made them distinguishable. In addition, the fact that neutrons are anomalously scattered by ^6Li $(0.18 + 0.025i) \times 10^{-12}$ cm led to the determination of the absolute configuration of the crystal structure from neutron diffraction. As a result, the steric course of the biochemical process was elucidated, as shown in Figure 14.26.

hkl	calc[a]	obs[b]	hkl	calc[a]	obs[b]
021	−5.3	− 4.8 (0.7)	$\overline{2}23$	+ 2.6	+ 3.9 (0.8)
041	−9.4	−12.2 (1.6)	$\overline{3}31$	−15.8	−20.2 (3.8)
131	+7.0	+ 5.8 (1.3)	422	+ 4.8	+ 5.4 (1.9)
220	−2.3	− 3.3 (0.6)	430	− 4.4	− 6.6 (1.6)
$\overline{2}21$	+8.9	+10.2 (1.5)	$\overline{5}21$	− 2.2	− 3.2 (1.1)

[a] The coherent scattering amplitude used for 96% enriched ^{6}Li was $b = (0.16 + 0.024i) \times 10^{-12}$ cm. The glycolate ion was assumed to have 100% deuterium at the position indicated. [b] Quantities in parentheses are the e.s.d. values based on counting statistics alone.

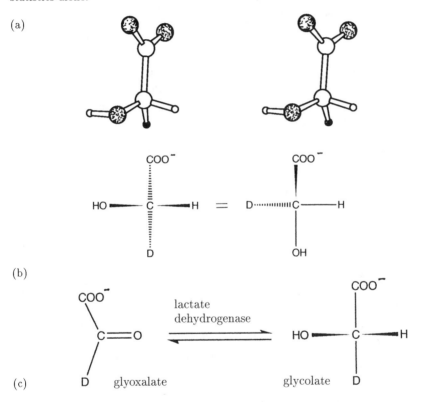

(a)

(b)

(c)

FIGURE 14.26. Absolute structure of deuterated lithium glycolate and its biochemical implications (Ref. 114). (a) Some neutron dispersion data are listed for lithium-6 *S*-glycolate-2-*d*. Values are given for $100(I_+ - I_-)/(I_+ + I_-) = 100$ difference/sum = 100 D/S (see Equations 14.2 and 14.3). Calc. = calculated value from the absolute structure found, and Obs. = measured value from the Bragg reflections. (b) The structure of the glycolate ion, together with its chemical formula, and (c) the steric course of the enzymatic reaction deduced from this X-ray diffraction study.

14.4.2 Facilitation of protein structure determination

Anomalous dispersion can also be used as an aid in the determination of phase angles. It was realized early on that anomalous scattering from the heavy atom in a derivative could be used to resolve the phase ambiguities if a single heavy-atom derivative is all that is available.[116-120] In proteins containing heavy-atom substituents, it is possible to use the magnitude of the anomalous dispersion effect, together with data on isomorphous replacement, to solve the phase problem (see Figure 14.24). This is done by a variation of the intersecting-circle method used to determine phases by isomorphous replacement. The construction is shown in Figure 8.26(c) (Chapter 8). This method is particularly powerful with neutron or synchrotron radiation sources, for which any required wavelength may be selected. Intensity data can then be determined in the presence and absence of anomalous dispersion,[121,122] although the signal-to-noise ratio may be low. Anthony C. T. North then showed that differences due to anomalous scattering should be weighted more heavily than the isomorphous differences.[123] Michael G. Rossmann proposed the Bijvoet-difference Patterson synthesis.[124] Joseph Kraut proposed using Bijvoet differences as coefficients and experimental phases to locate the heavy atoms in proteins.[125] Methods of deriving phases directly, by use of anomalous scattering, are being worked on by Jerome Karle, Herbert Hauptmann, and others.[126,127]

Anomalous scattering can also be used directly if the protein is small and a suitable anomalous scatterer can be used. The three-dimensional structure of the small protein, crambin, was determined by Wayne A. Hendrickson and Martha Teeter by the use of anomalous dispersion measurements.[128] This protein contains 45 amino acid residues and diffracts to 0.88 Å resolution. It crystallizes with 72 water and four ethanol molecules per protein molecule. Since there is a sulfur atom in the protein molecule, the use of its anomalous scattering was made. The nearest absorption edge of sulfur lies at 5.02 Å, but for Cu $K\alpha$ radiation, wavelength 1.5418 Å, values of $\Delta f'$ and $\Delta f''$ for sulfur are 0.3 and 0.557, respectively. Friedel-related pairs of reflections were measured to 1.5 Å resolution, and sulfur atom positions were computed from difference Patterson maps. The structure is now fully refined and a portion of an α helix was shown in Figure 12.27 (Chapter 12).

14.4.3 Sine-Patterson maps

The intensity differences between the Bragg reflections hkl and $\overline{hkl}$ suggested to Yoshi H. Okaya and Raymond Pepinsky[129] their use in a sine function (rather than the cosine function of the normal Patterson map) to give an analogue of the Patterson map. This sine-Patterson map has some interesting properties (see Figure 14.27). Peaks in the map represent vectors from the anomalously scattering atom, A, to all other atoms,

X, but now the vectors between atoms have a distinguishable direction. When peaks in the sine-Patterson map are positive they represent $A \to X$ vectors, while the holes (troughs) in the map represent $X \to A$ vectors.

normal Patterson map

$$P(uvw) = \frac{1}{V} \sum_{h,k,l} \sum \sum |F(hkl)|^2 \cos(hu + kv + lw) \qquad (8.3)$$

sine Patterson map

$$P_S(uvw) = \frac{1}{V} \sum_{h,k,l} \sum \sum \left[F^2(hkl) - F^2(\overline{hkl}) \right] \sin(hu + kv + lw). \qquad (14.4)$$

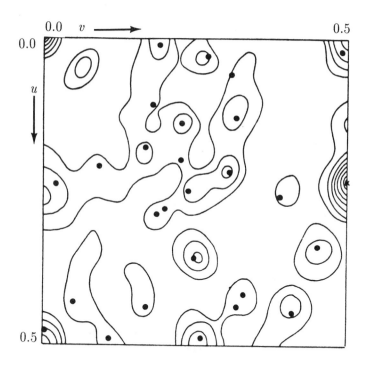

FIGURE 14.27. (a) The cosine (normal) and (b) the sine Patterson maps of potassium isocitrate (Ref. 13). Peaks in these maps are marked with filled circles if they are positive and open circles if they are negative. Note that the high $K \cdots K$ peaks at $u = 0.25$, $v = 0.50$ and $u = 0.50$, $v = 0.0$ (present in the cosine Patterson map), are missing in the sine Patterson map. Note that each vector in the sine Patterson map has a sign (positive or negative).

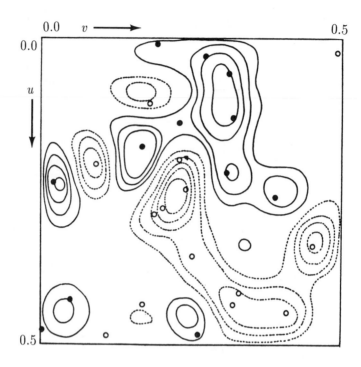

(b)

FIGURE 14.27 (cont'd).

Thus, by comparing peaks in a normal Patterson map with those in the sine-Patterson function, a sign for each peak can be found that directly gives the absolute structure of the crystal. This is shown in the example in Figure 14.27.

Summary

1. A molecule or crystal is *chiral* if it cannot be superimposed on its mirror image.
2. Chiral molecules of one handedness can only crystallize in space groups lacking any symmetry operations that would convert a left-handed molecule into a right-handed molecule.
3. The R/S system is used for a description of chiral molecules. Designations also exist for *pro*-chiral molecules, (where replacement of one group will produce a chiral center), *pro-R, pro-S*, as well as faces of planes containing atoms, *re* and *si*.
4. Crystals with an asymmetric habit may be separated into groups of left-handed and right-handed crystals. These two groups rotate the plane of polarized light in opposite directions.

5. If a solution of dissolved crystals also rotates the plane of plane-polarized light, the molecules in the crystal must be chiral, so that crystals contain only molecules of one chirality. If the rotation of plane-polarized light occurs only in the crystalline state and not in solution, then this property depends on the manner in which the individual molecules pack in the crystal structure, and not on the chirality of individual molecules.

6. Differences in the intensities of pairs of Bragg reflections (hkl and $\overline{hkl}$) give information on the absolute structure of the crystal. Normally these pairs of Bragg reflections should be equal (Friedel's Law). The predicted and experimental differences are compared and, if they have the same sign, the model used to calculate the differences has the correct absolute structure. If they have opposite signs the x, y, and z of all of the atoms in the model have to be reversed.

7. Anomalous dispersion measurements have proved to be very useful as an aid in phase determination in protein structure determination.

Glossary

Absolute structure, absolute configuration: The structure of a crystal or molecule expressed in an absolute frame of reference. A chiral object or structure is nonsuperimposable on its mirror image. This absolute structure or configuration can often be correlated with other optical phenomena that depend on chirality, such as the direction of rotation of the plane of polarization of light by a solution of the crystal.

Achiral: Lacking chirality.

Anomalous dispersion: A discontinuity in the plot of refractive index against wavelength (i.e., the dispersion). This effect occurs at wavelengths in the vicinity of absorption edges of the absorbing substance.

Asymmetric carbon atom: A carbon atom that has four different groups attached to it. The presence or absence of a chiral center in a molecule is not a criterion of chirality of the molecule as a whole.

Chiral axis: An asymmetric axis such that the two ends can be differentiated.

Chiral center: An atom that has no symmetry in the arrangement of groups around it. The term is often used for a chiral carbon atom that has four different groups attached to it, also called an asymmetric carbon atom (q.v.). The presence or absence of a chiral center in a molecule is not a criterion of chirality of the molecule as a whole.

Diastereomers: Stereoisomers that are not mirror images of each other.

Enantiomer: One of a pair of chiral objects related to the other by mirror symmetry.

Enantiomerism: The property of having two enantiomers.

Epimers: Two molecules with several chiral centers are said to be epimers if they differ only at one of these chiral centers.

EXAFS (extended X-ray absorption fine structure: A method of determining the geometry of atoms surrounding a dispersing atom by an analysis of the fine structure in the absorption edge of the dispersing atom.

E/Z (from *entgegen/zusammen* = on opposite sides/on the same side): Affixes to the name of a compound designating steric relations around a double bond and assigned by use of the sequence-rule procedure. For example, for a C=C double bond, the priorities of the two subtituents on the carbon atoms of the double bond are assigned (see *R/S* system). If the higher priority on each carbon is on the same side of the double bond the specification is *Z*, if it is on the opposite side it is *E*.

Isomeric: Having the same atomic composition but a different arrangement of these atoms.

Mesocompound: A compound for which the molecules are superimposable on their mirror images in spite of the presence of chiral centers, because these are "internally compensated."

Polar character: A direction is called polar if its two directional senses are geometrically or physically different.

Plane-polarized light: Electromagnetic radiation in which the electric vectors of the waves are confined to a single plane.

Pro-chiral: An achiral object is *pro*-chiral if replacement of one ligand by a ligand different from others present produces chirality. For example, the molecule $CXYH_2$ is prochiral because it contains two hydrogen atoms. If one of these hydrogen atoms is replaced by iodine, the resulting molecule is chiral. If one hydrogen atom is replaced, and the resulting chiral molecule can be described as (*R*), this replaced hydrogen atom is called "*pro-R*." For similar reasons, the other hydrogen atom is called "*pro-S*."

Racemate: Any homogeneous phase containing equimolar amounts of enantiomeric molecules.

Racemic mixture: A mixture composed of equal amounts of dextrorotatory and levorotatory forms (enantiomorphs) of the same compound present as separate solid phases. It displays no optical rotatory power.

re/si system: A method for designating the faces of the plane, particularly one containing a C = C bond. The substituents of the carbon atom are assigned priorities (see *R/S* system). The planar group is viewed onto its face. The term *re* (*rectus*) is used if the priority sequence (high to low) of the ligands is clockwise. When the ligands are arranged in an counterclockwise manner, the face is designated *si* (*sinister*).

R/S system: A method for designating the absolute configuration about a chiral atom in a molecule. For example, four atoms around an asymmetric carbon atom are assigned sequence priorities, highest for the atom with the highest atomic number. If necessary the identities of the next bonded atoms must be taken into account. The asymmetric atom is viewed with the substituent of lowest priority directly behind it. If the sequence of priorities of the other three substituents, from the highest to the lowest is clockwise the configuration of the asymmetric carbon is *R* (*rectus*), and if the sequence is anticlockwise, the configuration is *S* (*sinister*).

Resolving agent: A compound that causes a racemic mixture to be resolved into its optically pure components.

Stereocenter: Also called a **chiral center** (q.v.) or an **asymmetric carbon atom** (q.v.).

Stereochemistry: Relating to isomerism due to differences in the spatial arrangement of bonds without any differences in connectivity or bond multiplicity between the isomers.

Stereoisomer: Isomers that differ only in the arrangement of their atoms in space and are not superimposable by rotations about single bonds.

Three-point attachment: Attachment of a *pro*-chiral compound by three points to a chiral compound will serve to differentiate the two equivalent groups of the *pro*-chiral compound.

References

1. Bassindale, A. *The Third Dimension in Organic Chemistry.* John Wiley: Chichester, New York, Brisbane, Toronto, Singapore (1984).

2. Lord Kelvin [Thomson, W.H.]. *Baltimore Lectures on Molecular Dynamics and the Wave Theory of Light.* pp. 436 and 619. C. J. Clay and Son: London (1904).

3. Buda, A. B., Auf der Heyde, T., and Mislow, K. On quantifying chirality. *Angew. Chem. Intl. Ed. Engl.* **31**, 989–1007 (1992).

4. Sayers, D., and Eustace, R. *The Documents in the Case.* Avon: New York (1930).

5. Carroll, L. *Through the Looking-Glass and What Alice Found There.* see Woollcott, A. *The Complete Illustrated Lewis Carroll.* Galley Books: New York (1991).

6. Glazer, A. M. *The Structure of Crystals.* Adam Hilger: Bristol, UK (1987); Taylor and Francis: Philadelphia, PA (1987).

7. Glazer, A. M., and Stadnicka, K. The use of the term 'absolute' in crystallography. *Acta Cryst.* **A45**, 234–238 (1989).

8. Heilbronner, E., and Dunirz, J. D. *Reflections on Symmetry in Chemistry · · · and Elsewhere.* VHCA: Basel, and VCH: Weinheim (1993).

9. Mislow, K. *Introduction to Stereochemistry.* W. A. Benjamin: New York, Amsterdam (1965).

10. van't Hoff, J. H. *Voorstel tot uitbreiding der tegen woordig in de scheikunde gebruike structuur-formules in de ruimte, benevens een daarme samehangende vermogen en chemische constitutie van organische verbindingen.* [Proposal for the extension of the structural formulæ now in use in chemistry into space, together with a related note on the relation between the optical active power and the chemical constitution of organic compounds.] J. Greven: Utrecht (1874). **French translation:** Sur les formules de structure dans l'espace. *Archives Néerlandaises des Sciences Exactes et Naturelles,* **9**, 445–454. **English translation:** *Arrangement of Atoms in Space.* Longmans, Green: London (1898).

11. Le Bel, J. A. Sur les relations qui existent entre les formules atomiques des corps organiques, et le pouvoir rotatoire de leurs dissolutions. [On the relationships that exist between the atomic formulæ of organic materials and the rotatory power of their solutions.] *Bulletin de la Société chimique de France Bulletin,* **22**, 337–347 (1874). **English translation:** Benfy, O. T. In: *Classics in the Theory of Chemical Combinations.* pp. 151–171. Dover: New York (1963).

12. Hope, H., del la Camp, U., Homer, G. D., Messing, A. W., and Sommer, L. H. Stereochemical course of Grignard reactions at asymmetric sulfur. *Angew. Chem., Intl. Edn. Engl.* **8**, 612–613 (1969).

13. van der Helm, D., Glusker, J. P., Johnson, C. K., Minkin, J. A., Burow, N. E., and Patterson, A. L. X-ray crystal analysis of the substrates of aconitase. VIII. The structure and absolute configuration of potassium dihydrogen isocitrate isolated from *Bryophyllum calycinum. Acta Cryst.* **B24**, 578–592 (1968).

14. Milton, R. C. deL., Milton, S. C. F., and Kent, S. B. H. Total synthesis of a D-enzyme: the enantiomers of HIV-1 protease show demonstration of reciprocal chiral substrate specificity. *Science* **256**, 1445–1448 (1992).

15. Cahn, R. S., Ingold, C. K., and Prelog, V. The specification of asymmetric configuration in organic chemistry. *Experientia* **12**, 81–124 (1956).

16. Prelog, V. Chirality in science. *Science* **193**, 17–24 (1976).

17. Koga, G. Configurational symbols. *Chem. & Eng. News.* p. 3. April 3 (1988).

18. Peters, R. A. Biochemical light upon an ancient poison: a lethal synthesis. *Endeavour* **13**, 147–154 (1954).

19. Glusker, J. P., and Srere, P. A. Citrate enzyme substrates and inhibitors: depiction of their absolute configurations. *Bioorg. Chem.* **2**, 301–310 (1973).

20. Hanson, K. R. Applications of the sequence rule. I. Naming the paired ligands g,g at a tetrahedral atom Xggij. II. Naming the two faces of a trigonal atom Yghi. *J. Amer. Chem. Soc.* **88**, 2731–2742 (1966).

21. Hirschmann, H., and Hanson, K. R. Elements of stereoisomerism and prostereoisomerism. *J. Org. Chem.* **36**, 3293–3306 (1971).

22. Sullivan, A. C., Singh, M., Srere, P. A., and Glusker, J. P. Reactivity and inhibitor potential of hydroxycitrate isomers with citrate synthase, citrate lyase and ATP citrate lyase. *J. Biol. Chem.* **252**, 7583–7590 (1977).

23. Ogston, A. G. Interpretation of experiments on metabolic processes, using isotopic tracer techniques. *Nature (London)* **162**, 963 (1948).

24. Wood, H. G., Werkman, C. H., Hemingway, A., and Nier, A. O. Mechanism of fixation of carbon dioxide in the Krebs cycle. *J. Biol. Chem.* **139**, 483–484 (1941).

25. Evans, E. A. Jr., and Slotin, L. Carbon dioxide utilization by pigeon liver. *J. Biol. Chem.* **141**, 439–450 (1941).

26. Job, R., Kelleher, P. J., Stallings, W. C., Jr., Monti, C. T., and Glusker, J. P. Three-point attachment of citrate to a cobalt(III) tetramine complex. *Inorg. Chem.* **21**, 3760–3764 (1982).

27. Alworth, W. L. *Stereochemistry and its Application in Biochemistry. The Relation between Substrate Symmetry and Biological Stereospecificity.* Wiley-Interscience: New York, London, Sydney, Toronto (1972).

28. Glusker, J. P., Carrell, H. L., Job, R., and Bruice, T. C. Mechanism for chiral recognition of a prochiral center, and for amino acid complexation to a Co(III) tetramine. The crystal structure, absolute configuration and circular dichroism of $\Lambda(-)_{436}$-β-[(2S;9S)-2,9-diamino-4,7-diazadecanecobalt(III)aminomethylmalonate] perchlorate monohydrate. *J. Amer. Chem. Soc.* **96**, 5741–5751 (1974).

29. Kauffman, G. B., and Myers, R. D. The resolution of racemic acid. *J. Chem. Ed.* **52**, 777–781 (1975).

30. Nye, J. F. *Physical Properties of Crystals. Their Representation by Tensors and Matrices.* Paperback edn. of 1957 book, revised. Clarendon Press: Oxford (1985).

31. Arago, D. F. J. Mémoire sur une modification remarquable qu'éprouvent les rayons lumineux dans leur passage à travers certain corps diaphanes, et sur quelques autre nouveaux phénomènes d'optique. [On an interesting effect shown by light rays on their passage through certain transparent materials, and some other new optical phenomena.] *Mémoires de la Classe des Sciences Mathématiques et Physiques de l'Institut Impérial de France. Part I. (Paris)* **XII**, 93–134 (1811).

32. Glazer, A. M., and Stadnicka, K. On the origin of optical activity in crystal structures. *J. Appl. Cryst.* **19**, 108–122 (1986).

33. Devarajan, V., and Glazer, A. M. Theory and computation of optical rotatory power in inorganic crystals. *Acta Cryst.* **A42**, 560–569 (1986).

34. Biot, J. B. Mémoire sur un nouveau genre d'oscillation que les molécules de la lumière éprouvent en traversant certains cristaux. [On a new type of oscillation exhibited by light traveling through certain crystals.] *Mémoires de la Classe des Sciences Mathématiques et Physiques de l'Institut Impérial de France. Part I. (Paris)* **II**, 1–372 (1812).

35. Biot, J. B. Mémoire sur les rotations que certains substances impriment aux axes de polarisation des rayons lumineux. [The rotation of the axes of polarization of light by certain substances.] *Mémoires de la Classe des Sciences Mathématiques et Physiques de l'Institut Impérial de France. Part I. (Paris)* **II**, 41–136 (1812).

36. Biot, J. B. Sur la decouverte d'une propriété nouvelle dont jouissent les forces polarisantes des certains cristaux. [On the discovery of a new property shown by polarizing forces in certain crystals.] *Mémoires de la Classe des Sciences Mathématiques et Physiques de l'Institut Impérial de France. (Paris)* **II**, 19–30 (1812).

37. Biot, J. B. Sur l'emploi de la lumiere-polarisée pour manifester les differences des combinaisons isomeriques. [On the use to polarized light to demonstrate the difference between isomeric compounds.] *Annales de Chimie* **69**, 27 (1838).

38. Biot, J. B. Chemie optique: sur le degré de précision des charactères optiques dans leur application á l'analyse des matières sucrées, et dans leur emploi comme charactere distinctif des corps. [Optical chemistry: on the degree of precision of optical characteristics in their use in the analysis of sugars and for identifying materials.] *Comptes Rendus, Acad. Sci. (Paris)* **15**, 693–712 (1842).

39. Herschel, J. F. W. On the rotation impressed by plates of rock crystal on the planes of polarisation of the rays of light, as connected with certain peculiarities in its crystallisation. *Trans. Camb. Phil. Soc.* **1**, 43–52 (1821).

40. Biot, J. B. Communication d'une note de M. Mitscherlich. [Communication of a letter from Mr. Mitscherlich.] *Comptes Rendus, Acad. Sci. (Paris)* **19**, 720–722 (1844).

41. Pasteur, L. Mémoire sur la relation qui peut exister entre la forme cristalline et la composition chimique, et sur la cause de la polarisation rotatoire. [Note on the relationship of crystalline form to chemical composition, and on the cause of rotatory polarization.] *Comptes Rendus, Acad. Sci. (Paris)* [seance du 15 mai, 1848] **26**, 535–538 (1848).

42. Pasteur, L. Recherches sur les relations qui peuvent exister entre la forme cristalline, la composition chimique, et le sens de la polarisation rotatoire. [Research on the relationships between crystalline form, chemical composition, and the sense of rotatory polarization.] *Annales de Chimie* **24**, 442–459 (1848).

43. (Pasteur) *Leçons chimie professées en 1860 par M. M. Pasteur, Cahours, Wurtz, Berthelot, Saint–Claire Deville, Barral et Dumas.* [Chemistry courses taught in 1860.] Paris (1861).

44. Pasteur, L. Recherche sur les propriétés spécifiques des deux acides qui composent l'acide racémique. [Research on the specific properties of the two acids that comprise racemic acid.] *Ann. Chim. Physique* **28**, 56–99 (1850).

45. Vallery-Radot, R. *Œuvres de Pasteur.* [*The Works of Pasteur.*] **English translation:** Devonshire, R. L. Doubleday, Page & Co: New York (1923).

46. Dubois, R. J. *Louis Pasteur, Free Lance of Science.* Little, Brown: Boston, MA (1950).

47. Berzelius, J. J. Composition de l'acide tartarique et de l'acide racémique (Traubensaure). [The composition of tartaric and racemic acids.] *Annales de Chimie* **46**, 113–147 (1831).

48. Delafosse, G. Recherches relatives á la cristallisation considérée sous les rapports physiques et mathématiques. [Research on the physical and mathematical aspects of crystallization.] *Comptes Rendus, Acad. Sci. (Paris)* **11**, 394–400 (1840).

49. Patterson, T. S., and Buchanan, C. Historical and other considerations regarding the crystal form of sodium ammonium *d*- and *l*-tartrate, potassium *d*- and *l*-tartrate, potassium ammonium *d*- and *l*-, and potassium racemate. I. *Annals of Science* **5**, 288–295 (1945).

50. Patterson, T. S., and Buchanan, C. Historical and other considerations regarding the crystal form of sodium ammonium *d*- and *l*-tartrate, and potassium ammonium *d*- and *l*- tartrate, and potassium racemate. *Annals of Science* **5**, 317–324 (1945).

51. Carrell, H. L., Glusker, J. P., Villafranca, J. J., Mildvan, A. S., Dummel, R. J., and Kun, E. Fluorocitrate inhibition of aconitase: relative configuration of inhibitory isomer by X-rays. *Science* **170**, 1412–1414 (1970).

52. Stallings, W. C., Monti, C. T., Belvedere, J. F., Preston, R. K., and Glusker, J. P. Absolute configuration of the isomer of fluorocitrate that inhibits aconitase. *Arch. Biochem. Biophys.* **203**, 65–72 (1980).

53. Bush, M. A., Dullforce, T. A., and Sim, G. A. Orientation of $Cr(CO)_3$ and $Mn(CO)_3$ groups in tricarbonylchromiumbenzenecarboxylate and tricarbonyl-manganesecyclopentadienylcarboxylate anions. *J. Chem. Soc., Chem. Commun.* 1491–1493 (1969).

54. Klyne, W., and Buckingham, J. *Atlas of Stereochemistry.* Oxford University Press: New York (1974).

55. Marletta, M. A., Srere, P. A., and Walsh, C. Stereochemical outcome of processing fluorinated substrates by ATP citrate lyase and malate synthase. *Biochemistry* **20**, 3719–3723 (1981).

56. Bau, R., Brewer, I., Chiang, M. Y., Fujita, S., Hoffman, J., Watkins, M. I., and Koetzle, T. F. Absolute configuration of a chiral CHD group via neutron diffraction: confirmation of the absolute stereochemistry of the enzymatic formation of malic acid. *Biochem. Biophys. Res. Commun.* **115**, 1048–1052 (1983).

57. Hönl, H. Zur Dispersionstheorie der Röntgenstrahlen. [On the theory of dispersion of X rays.] *Z. Physik* **84**, 11–16 (1933).

58. Hönl, H. Atomfaktor für Röntgenstrahlen als Problem der Dispersiontheorie (*K*-Schale). [Atom form factors for X rays by dispersion theory *K*-shell).] *Annalen der Physik* **18**, 625–655 (1933).

59. James, R. W. *Optical Principles of the Diffraction of X-rays; The Crystalline State, Vol. 2.* Bell: London (1958).

60. Cromer, D. T., and Liberman, D. Relativistic calculation of anomalous scattering factors for X rays. *J. Chem. Phys.* **53**, 1891–1898 (1970).

61. Cromer, D. T. Calculation of anomalous scattering factors at arbitrary wavelengths. *J. Appl. Cryst.* **16**, 437 (1983).

62. Ibers, J. A. and Hamilton, W. C. (**Eds.**) *International Tables for X-ray Crystallography. Volume IV. Revised and Supplementary Tables to Volumes II and III.* 2.3. Dispersion effects. pp. 148–151 (1974).

63. Dauben, C. H., and Templeton, D. H. A table of dispersion corrections for X-ray scattering of atoms. *Acta Cryst.* **8**, 841–842 (1955).

64. Templeton, L. K., Templeton, D. H., Phizackerley, R. P., and Hodgson, K. O. L_3-edge anomalous scattering by gadolinium and samarium measured at high resolution with synchrotron radiation. *Acta Cryst.* **A38**, 74–78 (1982).

65. Templeton, D. H., and Templeton, L. K. X-ray dichroism and polarized anomalous scattering of uranyl ion. *Acta Cryst.* **A38**, 62–67 (1982).

66. Templeton, D. H., and Templeton, L. K. Polarized X-ray absorption and double refraction in vanadyl bisacetylacetonate. *Acta Cryst.* **A36**, 237–241 (1980).

67. Templeton, D. H., and Templeton, L. K. Tensor X-ray optical properties of the bromate ion. *Acta Cryst.* **A41**, 133–142 (1985).

68. Templeton, D. H., and Templeton, L. K. X-ray dichroism and anomalous scattering of potassium tetrachloroplatinate(II). *Acta Cryst.* **A41**, 365–371 (1985).

69. Friedel, G. Sur les symétries cristallines que peut reveler la diffraction des rayons Röntgen. [Concerning crystal symmetry revealed by X-ray diffraction.] *Comptes Rendus, Acad. Sci. (Paris)* **157**, 1533–1536 (1913).

70. Lytle, F. W., Sayers, D. E., and Stern, E. A. Extended X-ray-absorption fine-structure technique. II. Experimental practice and selected results. *Phys. Rev.,* **B11**, 4825–4835 (1975).

71. Stern, E. A., Sayers, D. E., and Lytle, F. W. Extended X-ray-absorption fine-structure technique. III. Determination of physical parameters. *Phys. Rev.* **B11**, 4836–4846 (1975).

72. Stern, E. A. The analysis of materials by X-ray absorption. *Sci. Amer.* **234(4)**, 96–103 (1976).

73. Diakun, G. P., Fairall, L., and Klug, A. EXAFS study of the zinc-binding sites of the protein transcription factor IIIA. *Nature (London)* **324**, 698–699 (1986).

74. Nishikawa, S., and Matukawa, K. Hemihedry of zinc blende and X-ray reflection. *Proc. Imperial Academy (Japan)* **4**, 96–97 (1928).

75. Coster, D., Knol, K. S., and Prins, J. A. Unterschiede in der Intensität der Röntgenstrahlenreflexion an den beiden 111 Flächen der Zinkblende. [Differences in intensity between the X-ray reflections from the two 111 faces of zinc blende.] *Z. Physik* **63**, 345–369 (1930). **English translation:** Stezowski, J. J. In: *Structural Crystallography in Chemistry and Biology.* (**Ed.**, Glusker, J. P.) pp. 158–160. Hutchinson Ross: Stroudsburg, PA, Woods Hole, MA (1981).

76. Prins, J. A. Epilogue to "differences in the intensities of the X-ray reflections from the two (111) faces of zinc blende." In: *Structural Crystallography in Chemistry and Biology.* (**Ed.**, Glusker, J. P.) pp. 158–160. Hutchinson Ross: Stroudsburg, PA, Woods Hole, MA (1981).

77. Bijvoet, J. M. Phase determination in direct Fourier-synthesis of crystal structures. *Proc. Koninklijke Nederlandse Akademie van Wetenschappen [Proc. Roy. Acad. (Amsterdam)]* **52**, 313–314 (1949).

78. Peerdeman, A. F., van Bommel, A. J., and Bijvoet, J. M. Determination of absolute configuration of optically active compounds by means of X-rays. *Proc. Koninklijke Nederlandse Akademie van Wetenschappen [Proc. Roy. Acad. (Amsterdam)]* **B54**, 16–19 (1951).

79. Bijvoet, J. M., Peerdeman, A. F., and van Bommel, A. J. Determination of the absolute configuration of optically active compounds by means of X-rays. *Nature (London)* **168**, 271–272 (1951).

80. Bokhoven, C., Schoone, J. C., and Bijvoet, J. M. On the crystal structure of strychnine sulfate and selenate. III. [001] projection. *Proc. Koninklijke Nederlandse Akademie van Wetenschappen [Proc. Roy. Acad. (Amsterdam)]* **52**, 120–121 (1949).

81. Peerdeman, A. F. Use of anomalous scattering in structure analysis. In: *Anomalous Scattering.* (**Eds.**, Ramaseshan, S. and Abrahams, S. C.) pp. 3–11. Munksgaard: Copenhagen (1975).

82. Fischer, E. Synthesen in der Zuckergruppe I. [Syntheses of sugars. I.] *Ber. Deutsch. Chem. Gesell.* **23**, 2114–2141 (1890).

83. Fischer, E. Synthesen in der Zuckergruppe II. [Syntheses of sugars. II.] *Ber. Deutsch. Chem. Gesell.* **27**, 3189–3232 (1894).

84. Tanaka, J. The quantum theory of X-ray scattering. *Acta Cryst.* **A28**, S229 (1972).

85. Tanaka, J., Ogura, F., Kuritani, M., and Nakagawa, M. Circular dichroism and the absolute configuration of 2, 7-disubstituted triptycenes. *Chimia* **26**, 471–472 (1972).

86. Tanaka, J., Ozeki-Minakata, K., Ogura, F., and Nakagawa, M. Absolute configuration of a coupled chromophore system by exciton analysis of circular dichroism. *Nature, Phys. Sci. (London)* **241**, 22–23 (1973).

87. Brongersma, H. H., and Mul, P. M. Absolute configuration assignment of molecules and crystals in discussion. *Chem. Phys. Lett.* **19**, 217–220 (1973).

88. Trommel, J., and Bijvoet, J. M. Crystal structure and absolute configuration of the hydrochloride and hydrobromide of D(–)-isoleucine. *Acta Cryst.* **7**, 703–709 (1954).

89. Peerdeman, A. F. The absolute configuration of natural strychnine. *Acta Cryst.* **9**, 824 (1956).

90. Saito, Y., Nakatsu, K., Shiro, M., and Kuroya, H. Determination of the absolute configuration of optically active complex ion $[Co\ en_3]^{3+}$ by means of X-rays. *Acta Cryst.* **8**, 729–730 (1955).

91. Beecham, A. F., Hurley, A. C., Mathieson, A. McL., and Lamberton, J. A. Absolute configuration by X-ray and circular dichroism methods of calycanthine. *Nature, Phys. Sci. (London)* **244**, 30–32 (1973).

92. de Vries, A. Determination of the absolute configuration of α-quartz. *Nature (London)* **181**, 1193 (1958).

93. Cornforth, J. W., Reichard, S. A., Talalay, P., Carrell, H. L., and Glusker, J. P. Determination of the absolute configuration at the sulfonium center of S-adenosylmethionine: correlation with the absolute configuration of the diastereomeric S-carboxymethyl-(S)-methionine salts. *J. Amer. Chem. Soc.* **99**, 7292–7300 (1977).

94. Speckhard, D. C., Pecoraro, V. L., Knight, W. B., and Cleland, W. W. Determination of the absolute configurations of the isomers of triamminecobalt(III) adenosine triphosphate. *J. Amer. Chem. Soc.* **108**, 4167–4171 (1986).

95. Jones, P. G. Absolute structure and how not to determine it. In: *Crystallographic Computing 3: Data Collection, Structure Determination, Proteins, and Databases.* (**Eds.**, Sheldrick, G. M., Krüger, C., and Goddard, R.) pp. 260–263. Clarendon Press: Oxford.

96. Dunitz, J. D. *X-ray Analysis and the Structure of Organic Molecules.* Cornell University Press: London, Ithaca, NY, (1979).

97. Hahn, T. (**Ed.**) *International Tables for Crystallography. Volume A. Space-group Symmetry.* D. Reidel: Dordrecht, Boston (1983).

98. Wilson, A. J. C. The probability distribution of X-ray intensities. *Acta Cryst.* **2**, 318–321 (1949).

99. Patil, A. O., Pennington, W. T., Paul, I. C., Curtin, D. Y., and Dykstra, C. E. Reactions of crystalline (R)-(–)- and (S)-(+)-mandelic acid with amines. Crystal structure and dipole moment of (S)-mandelic acid. A method of determining absolute configuration of chiral crystals. *J. Amer. Chem. Soc.* **109**, 1529–1535 (1987).

100. Lang, S. B. *Sourcebook of Pyroelectricity.* Gordon and Breach: New York (1974).

101. Patterson, A. L. Treatment of anomalous dispersion in X-ray diffraction data. *Acta Cryst.* **16**, 1255–1256 (1963).

102. Hodgkin, D. Foreword. In: *Anomalous Scattering*. (**Eds.**, Ramaseshan, S. and Abrahams, S. C.) Munksgaard: Copenhagen (1975).

103. Glusker, J. P., Minkin, J. A., Casciato, C. A., and Soule, F. B. Absolute configurations of the naturally occurring hydroxycitric acids. *Arch. Biochem. Biophys.* **132**, 573–575 (1969).

104. Peerdeman, A. F., and Bijvoet, J. M. The indexing of reflexions in investigations involving the use of the anomalous scattering effect. *Acta Cryst.* **9**, 1012–1015 (1956).

105. Hamilton, W. C. Significance tests on the crystallographic *R*-factor. *Acta Cryst.* **18**, 502–510 (1965).

106. Johnson, C. K. Discussion. In: *Crystallographic Computing. Proceedings of the 1969 International Summer School on Crystallographic Computing.* p. 197. (**Ed.**, Ahmed, F. R.). Munksgaard: Copenhagen (1970).

107. Rogers, D. On the application of Hamilton's ratio test to the assignment of absolute configuration and an alternative test. *Acta Cryst.* **A37**, 734–741 (1981).

108. Flack, H. On enantiomorph-polarity estimation. *Acta Cryst.* **A39**, 876–881. 1983).

109. Bernardinelli, G., and Flack, H. D. Least-squares absolute-structure refinement. Practical experience and ancillary calculations. *Acta Cryst.* **A41**, 500–511 (1985).

110. Watkin, D. J. Crystallographic determination. *Chem. Britain* **25**, 32 (1989).

111. Hope, H., and de la Camp, U. Anomalous scattering by oxygen: measurements on (+)-tartaric acid. *Acta Cryst.* **A28**, 201–207 (1972).

112. Kryger, L., Rasmussen, S. E., and Danielsen, J. Walden inversion. I. The crystal structure and absolute configuration of (−)-chlorosuccinic acid. *Acta Chem. Scand.* **26**, 2339–2348 (1972).

113. Kryger, L., and Rasmussen, S. E. Walden inversion. II. The crystal structure and absolute configuration of cobalt(II) (−)-malate trihydrate. *Acta Chem. Scand.* **26**, 2349–2359 (1972).

114. Johnson, C. K., Gabe, E. J., Taylor, M. R., and Rose, I. A. Determination by neutron and X-ray diffraction of the absolute configuration of an enzymatically formed α-monodeuterioglycolate. *J. Amer. Chem. Soc.* **87**, 1802–1804 (1965).

115. Gabe, E. J., and Taylor, M. R. The crystal structure of anhydrous lithium glycolate, ^{6}Li-(S)-glycolate-2-d. *Acta Cryst.* **21**, 418–421 (1966).

116. Blow, D. M., and Rossmann, M. G. The single isomorphous replacement method. *Acta Cryst.* **14**, 1195–1202 (1961).

117. Hendrickson, W. A. Phase information from anomalous-scattering measurements. *Acta Cryst.* **A35**, 245–247 (1979).

118. Hendrickson, W. A., Smith, J. L. and Sheriff, S. Direct phase determination based on anomalous scattering. *Methods in Enzymology* **115**, 41–55 (1985).

119. Langs, D. A. Unique solutions for the SIR and SAS phase problems and the use of partial structural information in phase refinement. *Acta Cryst.* **A42**, 362–368 (1986).

120. Wallace, B. A., Hendrickson, W. A., and Ravikumar, K. The use of single-wavelength anomalous scattering to solve the crystal structure of a gramicidin A/cæsium chloride complex. *Acta Cryst.* **B46**, 440–446 (1990).

121. Chapuis, G., Templeton, D. H., and Templeton, L. K. Solving crystal structures using several wavelengths from conventional sources. Anomalous scattering by holmium. *Acta Cryst.* **A41**, 274–278 (1985).

122. Guss, J. M., Merritt, E. A., Phizackerley, R. P., Hedman, B., Murata, M., Hodgson, K. O., and Freeman, H. C. Phase determination by multiple-wavelength X-ray diffraction: crystal structure of a basic "blue" copper protein from cucumbers. *Science* **241**, 806–811 (1988).

123. North, A. C. T. The combination of isomorphous replacement and anomalous scattering data in phase determination of non-centrosymmetric reflexions. *Acta Cryst.* **18**, 212–216 (1965).

124. Rossmann, M. G. The position of anomalous scatterers in protein crystals. *Acta Cryst.* **14**, 383–388 (1961).

125. Kraut, J. Bijvoet-difference Fourier function. *J. Molec. Biol.* **35**, 511–512 (1968).

126. Karle, J. Triplet phase invariants from single isomorphous replacement or one-wavelength anomalous dispersion data, given heavy-atom information. *Acta Cryst.* **A42**, 246–253 (1986).

127. Weeks, C. M., DeTitta, G. T., Miller, R., and Hauptman, H. A. Applications of the minimal principle to peptide structures. *Acta Cryst.* **D49**, 179–181 (1993).

128. Hendrickson, W. A., and Teeter, M. Structure of the hydrophobic protein crambin determined directly from the anomalous scattering of sulphur. *Nature (London)* **290**, 107–113 (1981).

129. Okaya, Y., and Pepinsky, R. New formulation and solution of the phase problem in X-ray analysis of noncentric crystals containing anomalous scatterers. *Phys. Rev.* **103**, 1645–1647 (1956).

15

Packing in Crystals

A crystal consists of an arrangement of molecules or ions packed in a regular manner such that the total free energy of the system is a minimum. This packing is determined by the forces between atoms, expressed by the sizes, shapes, charges, dipoles, and hydrophobicities of the individual molecules or ions. Ions are packed so that their charges balance locally as well as throughout the crystal as a whole. For example, in the sodium chloride structure[1] (Figure 1.6, Chapter 1), positively charged Na^+ ions are surrounded by six negatively charged Cl^- ions and vice versa, so that each individual charge is essentially neutralized by surrounding ions of the opposite charge.

Nonionic compounds pack in a great variety of ways. If the functional groups are capable of accepting or donating hydrogen bonds, such hydrogen bonds will be formed wherever possible.[2,3] The numbers of potential hydrogen bond donors and acceptors in a crystal are, ideally, equal. If this situation is difficult to achieve, water (or some other solvent), which can donate two and accept either one or two hydrogen bonds, can help settle the balance of donor and acceptor groups.[4] **Hydrophobic** groups in a molecule tend to pack near hydrophobic groups of other molecules in the unit cell. If a molecule contains both hydrophobic and **hydrophilic** areas, these will pack separately, as far apart as possible. X-ray diffraction analyses have provided information on each of these types of interaction.

15.1 Forces holding crystals together

The free energy in a crystal is a minimum with respect to the arrangement of molecules and ions within it. A useful but approximate indication of the forces between components in a crystal is given by the energy needed to evaporate crystals into their separate molecules or ions. Typical values are the following: for the ionic crystals lithium fluoride and sodium chloride, the energies required to break up the crystal into component anions and cations are 247 and 186 kcal mol^{-1} (1033 and 778 kJ mol^{-1}) respectively; for covalent crystals, such as diamond, 170 kcal mol^{-1} (715 kJ mol^{-1}) are required; while for metallic crystals, such as those of iron, 99 kcal mol^{-1} (415 kJ mol^{-1}) are required to make gaseous iron. Finally, molecular crystals, such as those of argon and carbon dioxide, evaporate with absorption of 1.7 or 6 kcal mol^{-1} (7 or 24 kJ mol^{-1}) of energy respectively.[5] The strengths of ionic interactions are thus of the same order of magnitude as those of covalent bonds. James E. Huheey[6] writes: "The common notion that ionic bonds are considerably stronger than covalent bonds results from mistaken interpretations of melting-point and boiling-point phenomena ···" Thus covalent and ionic crystals require considerably more energy if they are to break apart, while molecular crystals require much less energy. This implies that there are weaker forces between molecules in molecular crystals than between atoms in covalent crystals or ions in ionic crystals (Table 15.1).

The packing of ions in crystalline salts takes into account the sizes as well as the charges of the ions. If both cations and anions are similar in size, the crystal will have a sodium chloride-like arrangement of ions shown in Figure 1.6 (Chapter 1), but if the sizes are very different, other arrangements will be found in the crystal structure. Covalent crystals are characterized by a covalent-bond network throughout the crystal, a feature reminiscent of ionic crystals, since in both cases no single "molecule" can be found in the crystal structure. The physical properties of covalent crystals resemble those of ionic crystals; both are extremely hard and have high melting points.

Molecular crystals contain molecules that have aggregated by virtue of **van der Waals** interactions,[7-9] hydrogen bonding, and other forces. If two molecules are **polar**, that is, they have a **dipole moment**, they

---

TABLE 15.1. Types of intermolecular forces in crystal structures.

---

1. *Electrostatic forces (Coulombic).*
   (*a*) *Ion-ion interactions between cations and anions* (*e.g.*, $Na^+ + Cl^-$).

2. *Forces between neutral molecules (van der Waals).*

   (*a*) *Dipole/dipole interactions, This category includes hydrogen bonding* (*e.g.*, $HCl + HCl$, *or* $H_2O + H_2O$).

   (*b*) *Dipole/induced-dipole interactions. A polar molecule interacts with a polarizable molecule and induces a dipole in the latter* (*e.g.*, $HCl + C_6H_6$).

   (*c*) *Induced dipole/induced dipole interactions (London dispersion forces). A non-polar molecule may induce a small fluctuating instantaneous dipole in another nearby non-polar molecule* (*e.g.*, $C_6H_6 + C_6H_6$).

---

interact in a head-to-tail arrangement, the one with the lowest energy. If one molecule is polar and the other is nonpolar, but polarizable, the polar molecule may induce a dipole in the nonpolar molecule. These two dipoles, one permanent and the other induced, will interact in a manner similar to that for two dipoles. The strength of this interaction depends on the magnitude of the permanent dipole moment of the polar molecule, and on the polarizability of the second molecule. Even if the two molecules are nonpolar, there can be attractive, low-energy molecular interactions between them, albeit weaker than those between polar molecules. These weak interactions result from induced dipoles aligning with other induced dipoles. These different types of interactions are listed in Table 15.1. Because dipole–dipole interactions are weak, crystals in which these are the strongest forces are generally soft and have low melting points. For example, the magnitudes at 25°C for the average interaction energy for pairs of molecules with a dipole moment $\mu = 1$ D (such as chloroform) is about 0.33 kcal mol^{-1} (1.4 kJ mol^{-1}) when the separation between molecules is 3 Å; the energy is negative, indicating lower energy on interaction and a need to absorb energy to separate molecules or ions, as mentioned earlier. When a polarizable molecule with no permanent dipole moment, such as benzene, interacts with a polar one, such as chloroform, 3 Å away, the average energy of interaction is about 0.2 kcal mol^{-1} (0.8 kJ mol^{-1}). Two methane molecules (which have no permanent dipole moment because of their symmetry) lying 3 Å apart have an interaction energy of 0.12 kcal mol^{-1} (0.5 kJ mol^{-1}). In a crystal a multitude of such individual interactions occur.

15.1.1 Physical properties of minerals

The atomic arrangement and interatomic interactions within a crystal affect its chemical and physical properties in a wide variety of ways. Physical properties of common minerals, for example, display this. Silicates comprise more than 90% of the minerals in the earth's crust, and their chemistry can be explained structurally in terms of SiO_4 tetrahedra and the manner in which these tetrahedra are arranged and interconnected.[10] Examples are diagrammed in Figure 15.1.

In the simple orthosilicates, such as olivine, the SiO_4 tetrahedra are discrete [Figure 15.1(a)]. They are held together in the crystal structure by divalent cations (Mg^{2+} and Fe^{2+}). Pyrosilicates, in which the tetrahedra have one corner in common, giving $Si_2O_7^{6-}$ units [Figure 15.1(b)], are comparatively rare. Three tetrahedra can give cyclic structures $(Si_3O_9)^{6-}$, as shown in Figure 15.1(c), and these are found in minerals such as benitoite. There are many examples of pyroxene-type silicates in which single, infinite, zigzag chains of tetrahedra occur, each tetrahedron having two corners in common with its neighbors [Figure 15.1(d)]. In such chains, all the tetrahedra are pointed along a threefold axis and the chain formula is $(SiO_3)_n^{2-}$. Continuation of this process leads to the

(a) $[SiO_4]^{2-}$

Simple orthosilicate

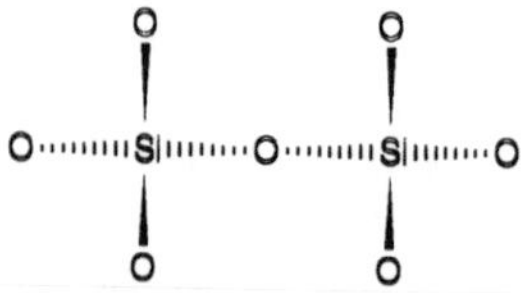

(b) $[Si_2O_7]^{6-}$

One oxygen atom shared

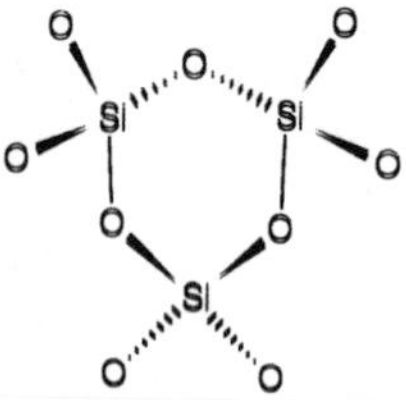

(c) $[Si_3O_9]^{6-}$

Three oxygen atoms, each shared by two silicon atoms

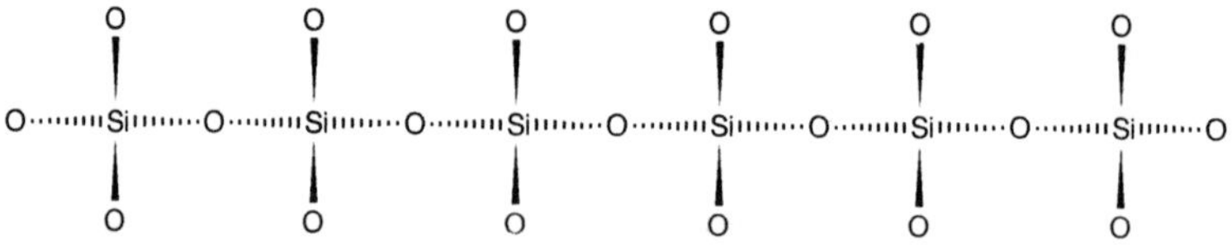

(d) $[SiO_3)_n]^{2n-}$

Long chain with one oxygen atom shared by two silicon atoms   Pyroxenes

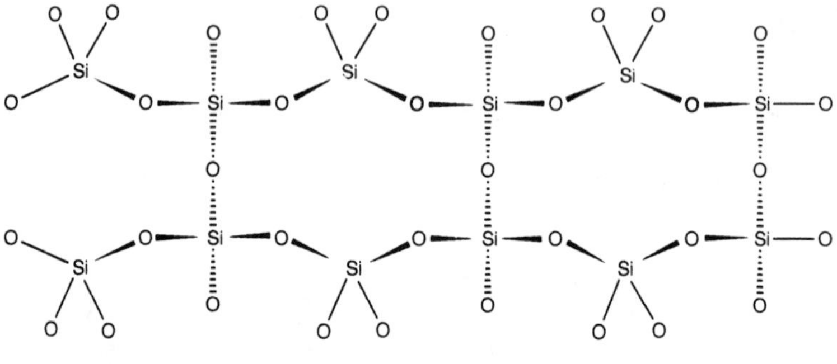

(e) $[(Si_4O_{11})_n]^{6n-}$

Long double chains   Amphiboles

FIGURE 15.1. Arrangements of SiO_4 tetrahedra. (a) Simple orthosilicates $[SiO_4]^{2-}$, (b) one oxygen atom shared $[Si_2O_7]^{6-}$, (c) three oxygen atoms each shared by two silicon atoms, (d) long chain with one oxygen atom shared by two silicon atoms, as in pyroxenes $[(SiO_3)_n]^{2n-}$, and (e) long double chains, as in amphiboles $[(Si_4O_{11})_n]^{6n-}$.

infinite, two-dimensional sheet structures of the micas, the sheet composition being $(Si_2O_5)_n^{2-}$. These anionic chains or sheets are connected by countercations, commonly alkali metal or alkaline earth ions. Further interconnections are possible [see Figure 15.1(e)]. Results of crystal structure determinations are important for an understanding of the composition and properties of minerals It is usual to see the silicate groups represented as tetrahedra in drawings of mineral crystal structures.

Micas, which are composed of infinite two-dimensional sheets of silicate tetrahedra, cleave, like graphite, into thin sheets because interatomic interactions are weaker between than within the sheets. On the other hand, minerals that crystallize with a three-dimensional network do not cleave so readily. Feldspars, such as orthoclase, $KAlSi_3O_8$, provide an example of this. The silicate tetrahedra are linked at all four corners and so give a strong three-dimensional network, like that of diamond.

15.1.2 Hydrogen bonds

One very important determinant of the nature of the packing of molecules or ions is whether there is a possibility for the formation of hydrogen bonds. The hydrogen bond, usually characterized as $A–H \cdots B$, is a weak electrostatic interaction that forms between a hydrogen atom (H) [covalently bonded to an electronegative atom (A)] and another atom (B) that is basic, that is one that has lone pairs of electrons and is fairly electronegative. The strength of a hydrogen bond is intermediate between that of a weak covalent bond and a van der Waals interaction.[11] The more readily the hydrogen is removed in the form of H^+ from AH, the stronger is the hydrogen bond $H \cdots B$. The terminology for a hydrogen bond is confusing with respect to the terms "donor" and "acceptor." Atoms A and B are either considered in terms of donors and acceptors of lone pairs (B the donor and A the acceptor) or as proton donors (A the donor and B the acceptor). When reading articles on hydrogen bonding it is important to be sure which definition of donor and acceptor is used. In this book we use the convention for a hydrogen bond $A–H \cdots B$ that A is the proton donor, and B is the the proton acceptor.

The hydrogen bond was first invoked to explain both the weakness of trimethylammonium hydroxide as a base[12] and the low reactivity to bases of aromatic carbonyl compounds with an *ortho*-hydroxyl rather than to bases of aromatic carbonyl compounds with a *meta*-hydroxyl group.[13] The low reactivity was taken to indicate some interaction (Figure 15.2) between the *ortho*-hydroxyl group and the carbonyl group. Wendell M. Latimer and Worth H. Rodebush[14] then showed how many physical properties of compounds, such as the tendency of acetic acid to form dimers, could also be explained by hydrogen bonding. Now that hydrogen atoms can be well located by neutron diffraction or good X-ray diffraction studies, there is a large amount of information available on hydrogen bonding (see examples in Table 15.2).[3,15−10]

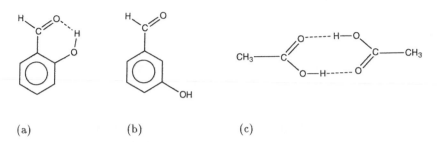

(a)   (b)   (c)

FIGURE 15.2. Hydrogen bonding, represented by broken lines, in some simple organic molecules. (a) *o*-hydroxybenzaldehyde, (b) *m*-hydroxybenzaldehyde, and (c) acetic acid dimers. Note the hydrogen bonds in (a) and (c), but not in (b).

The hydrogen bond may be described in terms of the angle between the two interactions that it makes, that is, H–O and H ··· O in Figure 15.2(c). The O–H ··· O system may be nearly linear [Figure 15.3(a)], but is often considerably bent from 180°, to as low a value as 120°. It has been usual to take the sum of the van der Waals radii as a criterion for hydrogen bonding; if the distance is less than the sum of the van der Waals radii, the interaction is presumed to be a hydrogen bond, provided a hydrogen atom is available to form such an interaction. George A. Jeffrey[3] points out, however, that the hydrogen bond is primarily electrostatic and far-ranging in effect. As a result, longer distances should be taken into consideration. For example, the hydrogen atom may be attracted by two electronegative atoms and a three-center hydrogen bond may be formed, even though distances from the hydrogen atom may be longer than the sum of the van der Waals radii of the atoms involved.[20] In such a case, all four atoms involved are in approximately the same plane, and the two distances between the hydrogen atom and the acceptor atoms often differ appreciably. This arrangement of atoms, illustrated in Figure 15.3(b), occurs for 20–30% of all O–H ··· O and N–H ··· O hydrogen bonds.[3] On the other hand, it can be argued that the directionality of hydrogen bonds excludes its description as a purely electrostatic interaction.

Extended networks of hydrogen bonds are found in many crystal structures and, as they run continuously through the crystal, they act **cooperatively**[3,19,21] with lower energy than expected from the sum of the energies of the individual hydrogen bonds. Hydrogen-bonding schemes in crystal structures can be analyzed readily, provided that all the hydrogen atoms have been located and included in the refined crystal structure. Neutron diffraction data give more reliable information on hydrogen atom positions than do X-ray data. The interatomic distances and angles between hydrogen-bond donors and acceptors are determined

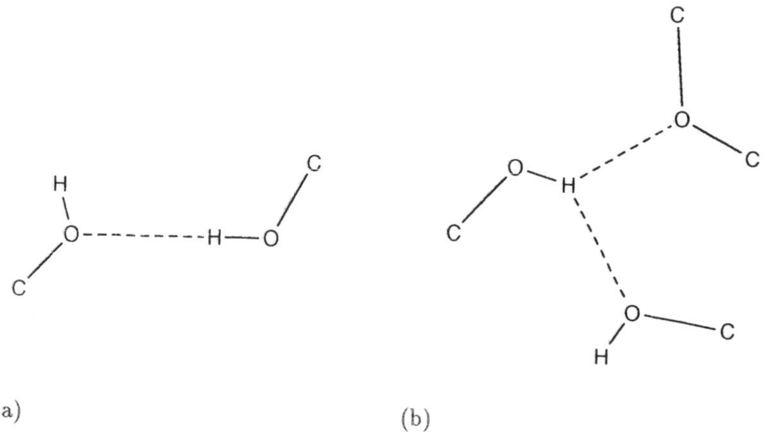

(a)   (b)

FIGURE 15.3. The hydrogen bond (a) linear and (b) threecentered. The O–H · · · O angle describes the extent of linearity of the hydrogen bond.

keeping in mind the periodicity and space-group symmetry of the crystal structure. Peggy Etter[2] was able, from extensive studies of crystal structure determination results, to derive some rules for hydrogen bonding. Among these was her finding that the best proton donors and the best proton acceptors will bind to one another. If possible, all proton donors will form hydrogen bonds. In general all good proton acceptors will also form hydrogen bonds, although this rule is not rigorously observed.

Hydrogen bonds between symmetry-related molecules in a crystal frequently replace those between the molecule and solvent in solution. If the crystal retains water molecules from the solution as water of crystallization, it is not uncommon for such water molecules to have less than a full complement (two donor and two acceptor interactions) of hydrogen

TABLE 15.2. Hydrogen bonds in crystal structures (Ref. 3).

A–H · · · B	H · · · B distance (Å)
O–H · · · COO$^-$	1.35 – 1.85
N–H$^+$ · · · O	1.70 – 2.22
O–H · · · OH$^-$	1.70 – 2.15
O$_W$–H · · · O*	1.75 – 2.06
O–H · · · Cl$^-$	1.95 – 2.15
N–H$^+$ · · · Cl$^-$	2.05 – 2.22

* O$_W$ = oxygen of water.

bonds. When this occurs the water molecules are often not as well ordered in the crystal as the molecules to which they bind.

Graph-set analyses have been used[22] for describing hydrogen bonding. A hydrogen-bonding pattern is described as $G_d^a(r)$, where G describes the type (S for self or intramolecular, D for noncyclic finite pattern such as a dimer, R for rings, and C for chains). The sub- and superscripts d and a, refer respectively to the number of proton donors and acceptors, and r is the size of the pattern.[2,22] This scheme highlights extended systems of hydrogen bonds and similarities and differences of patterns of hydrogen bonding in crystals of related compounds. For example, the cyclic system in Figure 15.2(a) would be described as $S_1^1(6)$, and that is in Figure 15.2(c) as $R_2^2(8)$.

Weaker interactions that are analogous to hydrogen bonds, such as C–H$\cdots$O or C–H$\cdots$N interactions were first described by D. June Sutor.[23] They have been the subject of much discussion.[24-26] Are they hydrogen bonds or not? If not, why do they appear to have directionality? These interactions appear to play a significant role in the packing of, for example, polynucleotides. The distances involved, if short or the order of the sum of the van der Waals radii, are generally attractive rather than repulsive, and have come to be regarded as very weak hydrogen bonds. If many such interactions are formed per unit cell, their contribution to the overall energy may be substantial.

15.1.3 Metal ion coordination

Another important determinant of packing in crystals is the tendency for metal ions to gather negatively charged atoms, often from different ions or molecules, around them. Metal ions are generally positively charged and therefore act as electrophiles, seeking to share electron pairs of other atoms or ions so that a bond or a charge–charge interaction can be formed and electrical neutrality thereby attained. By analogy, as shown in Equations 15.1 and 15.2, the binding of a metal ion M^{n+} to its surrounding ligands :L can be considered as a Lewis acid–Lewis base interaction, in which an electron-poor **Lewis acid** A (the metal ion M) accepts the partial donation of an electron pair from electron-rich **Lewis base** B (the ligand L).

$$A + \, :B \longrightarrow A:B \tag{15.1}$$

$$M + \, :L \longrightarrow M:L \tag{15.2}$$

The work of Alfred Werner[27] led to a considerable understanding of the stereochemistries of arrangements of ligands. This understanding was reinforced by X-ray diffraction studies of metal complexes, for example, in the determination of the structures of ammonium hexachloroplatinate and chloropalladite,[28,29] shown in Figure 15.4. In these there are six groups around the metal ion, at an equal distance, rather like a small portion of the sodium chloride structure.

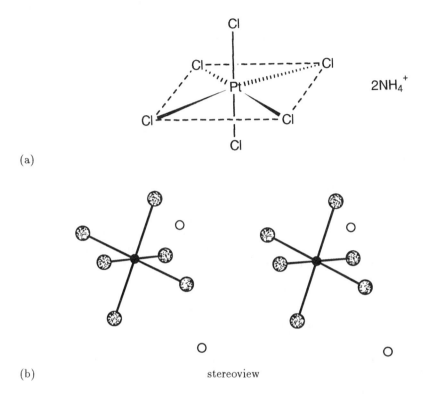

(a)

(b)   stereoview

FIGURE 15.4. Metal coordination in ammonium hexachloroplatinate (Ref. 28). (a) Chemical formula, and (b) stereoview.

The crystal structures of a vast number of metal complexes have now been determined so that plenty of data on metal–ligand interatomic distances are available. The shortest distance between contiguous ions of opposite sign is presumed to be the sum of the radii of the cation and the anion, assumed to be hard spheres:

$$r_{\text{exptl}} = (r^+) + (r^-). \qquad (15.3)$$

Since, however, only values of the sum, r_{exptl}, are obtained experimentally, it is necessary to assume a value for the ionic radius of either r^+ or r^- in order to derive the ionic radius of the other. How is this "size" determined, since only distances between centers of the atoms or ions can be measured? The method used to determine **ionic or atomic radii** is to record the closest distance that two identical nonbonded atoms or ions approach each other in crystal structures. The atomic or ionic radii are considered to be half this distance. For example, the minimum anion–anion $O^{2-} \cdots O^{2-}$ and $Cl^- \cdots Cl^-$ distances in crystal structures are approximately 2.8 and 3.9 Å, respectively.[30] This leads to the assumption[31,32] that ionic radii are 1.40 Å for O^{2-} and 1.95 Å for Cl^-. By this method the values for ionic radii[31-40] listed in Table 15.3 were obtained. Then cation radii can be found from metal ion-oxide or metal ion-chloride distances by subtraction.

Ions tend to pack in such a way that the number of cations about an anion and anions around a cation is a maximum, subject to the condition that the central ion is in contact with each of its neighbors. It is energetically unfavorable to have a central ion "rattling around" in a cavity made by surrounding ions of the opposite sign,[41] and therefore this does not commonly happen. Allowance must also be made for repulsion between ions of like charge, for example, as they touch each other in the surrounding arrangement of anions about the metal ion.

The number of ions or atoms that pack around a metal ion in a crystal is referred to as the **coordination number** of the metal ion. Its value is determined to a large extent by the ratio of the radius of the cation to that of the anion that packs around it.[42,43,50,51]

$$\text{radius ratio} = (r^+)/(r^-). \tag{15.4}$$

A coordination number of 6 is common for a metal ion. Small, highly charged cations, such as Be^{2+} and Al^{3+}, have low coordination numbers, although these coordination numbers are usually higher in crystals than in gases or liquids. For example, beryllium chloride, $BeCl_2$, in the gas phase, exists as an isolated linear molecule,[44] with coordination number 2 (not truly ionic); in crystals, however, it exists as a polymeric, bridged structure[45] in which the beryllium has the preferable coordination number of 4 (Figure 15.5). Ionic radii vary somewhat with coordination number, and are shorter if the coordination number is smaller.[39] For example, a cation with a tetrahedral coordination of 4 anions has 93–95% the radius

TABLE 15.3. Ionic radii (in Å) (Ref. 42).

Ion	Radius (Å)	Ion	Radius (Å)	Ion	Radius (Å)		
Be^{2+}	0.31	Sn^{4+}	0.71	K^+	1.33	Cl^-	1.81
Al^{3+}	0.50	Sc^{3+}	0.81	Ba^{2+}	1.33	S^{2-}	1.84
Li^+	0.60	Pb^{4+}	0.84	F^-	1.36	Br^-	1.95
Mg^{2+}	0.65	Y^{3+}	0.93	O^{2-}	1.40	Se^{2-}	1.98
		Na^+	0.95	Rb^+	1.48	I^-	2.16
		Ca^{2+}	0.99	Cs^+	1.69	Te^{2-}	2.21
		Sr^{2+}	1.13				

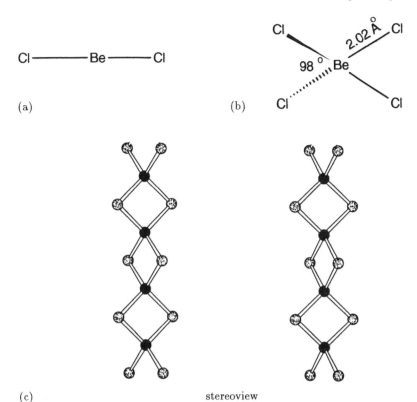

(a)

(b)

(c)   stereoview

FIGURE 15.5. Beryllium chloride (a) in the gaseous state, and (b) in a crystal. In the gaseous state the coordination number of the beryllium is 2, in the crystal it is 4, giving a continuous series of $BeCl_4^{2-}$ tetrahedra that share edges. (c) A stereoview of the coordination in the crystal structure [diagrammed in (b)] (Refs. 44 and 45).

of the same cation with a coordination number of 6. On the other hand, if the coordination number of the cation is 8, the radius for coordination number 6 must be multiplied by 1.03. Anions also appear to have different ionic radii when their coordination numbers are different. Values for the oxide anion varies from 1.35 Å (coordination number 2), to 1.40 Å (coordination number 6), to 1.42 Å (coordination number 8).

The spatial arrangements of anions or atoms around a cation, found in X-ray diffraction studies, are referred to as **coordination polyhedra**.[46] They are the solid figures obtained by joining all ligand atoms L that are directly bonded to the metal cation. Examples of polyhedra commonly found are illustrated in Figure 15.6. These different polyhedra are generally formed as a result of different cation-to-anion radius ratios. Essentially they depend on the holes formed when different numbers of anions pack together. For example, while a linear arrange-

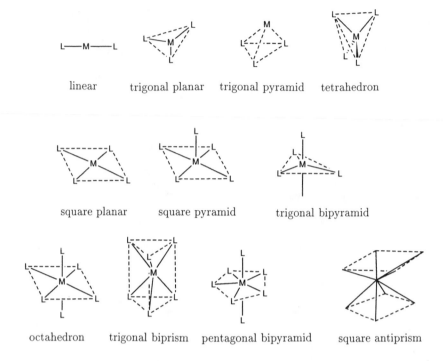

FIGURE 15.6. Some coordination polyhedra. M = metal ion, L= ligand. Shown are polyhedra for metal ions coordination number 2 (linear), 3 (trigonal planar, trigonal pyramidal), 4 (tetrahedral, square planar), 5 (square pyramidal, trigonal bipyramidal), 6 (octahedral, trigonal biprismatic), 7 (pentagonal bipyramidal) and 8 (square antiprismatic).

ment $L-M-L$ shows no limit on cation-to-anion radius ratio, in a trigonal planar arrangement, ML_3 the ratio of radii of M to L generally lies between 0.225 and 0.155. If this ratio is greater than 0.225, a tetrahedral arrangement, ML_4, is more likely. As the cation-to-anion radius ratio increases, there is room for more anions around the cation, and more possibilities for different types of coordination polyhedra. For example, the coordination polyhedron ML_5 may be either a square pyramid or a trigonal bipyramid; there are several $L-M-L$ angles of 90° in both of these polyhedra. The radius ratios are the same for both and, more importantly, the same as for an octahedron. It is found that a coordination number of five is not as common as one of six.

If a cation, charge 1+, gathers n equivalent ligands around it, M^+L_n, then each of these anions contributes a partial negative charge of $-1/n$ to the area around the cation, so that local electroneutrality results. For example, if a magnesium ion Mg^{2+} is surrounded by six oxygen atoms, each oxygen atom will experience $\frac{1}{6}$ of the charge of the Mg^{2+} ion $(= \frac{1}{3}^+)$ and, in turn, each Mg ion will experience $\frac{1}{3}$ of an electron ($= \frac{1}{3}^-$) from each oxygen atom. This value of $\frac{1}{3}$ is called a **bond valence**,[47-49] denoted S_{ij}. This provides a value for the strength of each interaction between a cation i and various anions j. For example, in magnesium citrate decahydrate[50] there is one magnesium cation surrounded by six water molecules with approximately equal $Mg^{2+} \cdots O$ distances, 2.06 to 2.08 Å. A second magnesium cation binds the citrate ion with $Mg^{2+} \cdots O$ distances that vary from 2.02 to 2.12 Å (Figure 15.7). The shorter $Mg^{2+} \cdots O$ interactions would be assumed to be stronger, with a greater partial charge donated by the oxygen atom to the metal ion in order to attain electroneutrality. Such analyses require that the sum of the partial negative charges contributed by each ligand around a cation must be the total cation charge, the valence sum rule.

$$\text{charge on cation} = \sum \text{bond valences of surrounding anions}$$

$$= \sum S_{ij}. \tag{15.5}$$

With a sufficient number of data, one can work backwards and, by measuring $M \cdots L$ distances, obtain an estimate of the bond valence. Equation 15.5 can be applied and will give values of partial charges on anions in view of the known cation charge (see Table 15.4).[42,43]

TABLE 15.4. Values of S. calculated for citrate and water O atoms in magnesium citrate (Ref. 50). r is the interatomic distance. S is the bond strength and is computed (for coordination distance r) as $S = (0.333)(r/2.076)^{-5.2}$ for $Mg^{2+} \cdots O$ distances and $S = (r/1.184)^{-2.2}$ for $H^+ \cdots O$ distances (in hydrogen bonds). The total gives an indication of the charge on each O atom.

		Mg(1)	Mg(2)	H···O	Bond valences
O(1)	carboxylate		2.072	2.016	0.633
O(2)	carboxylate		2.019	2.299	0.501
O(3)	carboxylate			1.913,1.891	0.348
O(4)	carboxylate		2.077	2.119	0.471
O(5)	carboxylate		2.082	2.071	0.474
O(6)	carboxylate			1.980,2.109,1.974	0.465
O(7)	hydroxyl		2.118		0.300
O(8)	water		2.030		0.374
O(9)	water	2.081			0.329
O(10)	water	2.080			0.330
O(11)	water	2.061			0.346

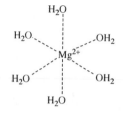

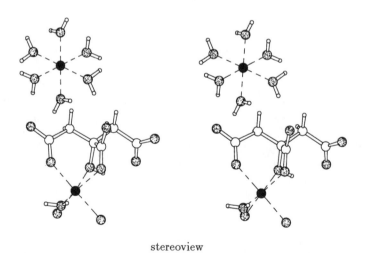

(a)

(b)   stereoview

FIGURE 15.7. Magnesium citrate decahydrate (Ref. 50). (a) Chemical formula, and (b) stereoview of the surroundings of both Mg^{2+} ions (filled circles).

15.1.4 Van der Waals forces in crystals

The crystal structures of hydrocarbons, such as polycyclic aromatic hydrocarbons, are mainly determined by van der Waals interactions.[7,8] Individually these interactions are small, but so many are formed in a crystal that the total effect per unit cell may be appreciable. These interactions have less directionality than the other interactions described earlier. The crystal structures of polycyclic aromatic hydrocarbons can be described by packing principles introduced in a geometric fashion by Alexandr Isaakovitch Kitaigorodskii.[52,53] He considered the packing in terms of fairly constant van der Waals radii for the individual atoms. Carbon atoms, for example, rarely approach each other closer than 3.4 Å, and hydrogen atoms rarely approach each other closer than 2.0 Å. In hydrocarbons, van der Waals interactions predominate because there are no groups capable of forming hydrogen bonds or ionic interactions. As a result, the crystal structures of polycyclic aromatic hydrocarbons can be considered as the nesting of solid figures with slight positive charges on the hydrogen atoms and corresponding slight negative charges around the carbon atoms. Each molecule is surrounded in a plane by six molecules, a close-packing situation. These layers of molecules can stack giving a total of 10 to 14 molecules around each molecule.

Often there are different ways in which this close packing may be achieved, and, as a result, a compound may crystallize in two or more different space groups, although each crystal has approximately the same density. There are only a few space groups that allow for efficient close packing of organic molecules that are somewhat irregularly shaped.[52-55] These space groups, $P1$, $P2_1$, $P2_1/c$ (or $P2_1/a$), $Pca2$, $Pna2_1$, and $P2_12_12_1$ are found for compounds that have no inherent molecular symmetry. The only molecular symmetry element present that can be retained with efficient packing is an inversion center. In such cases, centrosymmetric molecules achieve close packing in the space groups $P\bar{1}$, $P2_1/c$, $C2/c$, and $Pbca$. Examples are provided by benzene ($Pbca$) and naphthalene ($P2_1/a$). Crystals of asymmetric or chiral molecules (which cannot contain centers of symmetry or symmetry planes) in which only one optical isomer is present are generally found to be in one or another of the close-packed space groups, $P2_1$ or $P2_12_12_1$ showing (as described in Chapter 4) that a translation (involved in screw axes in these two examples) aids in the nesting of asymmetric molecules. Racemates, on the other hand, can crystallize in space groups that have glide planes or inversion centers. Substituents in aromatic molecules, such as chlorine atoms, play an important role in aligning molecules in crystals,[56,57] as shown in Figure 15.8.

There may be more than one molecule in the asymmetric unit. This appears to happen more frequently when the molecules have awkward shapes that do not accommodate well to simple packing principles. An example is provided by the crystal structure of 5-methylchrysene[58] in

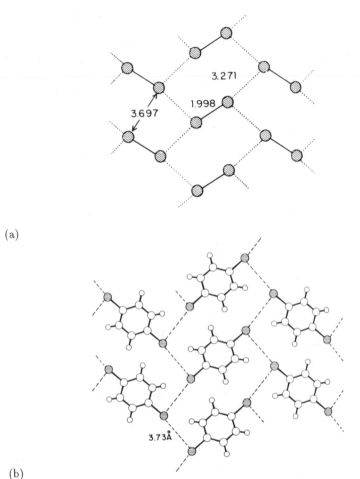

(a)

(b)

FIGURE 15.8. The role of chlorine in aligning molecules (Ref. 18). The similarities in the crystal structures of (a) chlorine and (b) monoclinic 1,4-dichlorobenzene are shown. The two structures are held together in similar ways by Cl · · · Cl interactions. (Reproduced with the permission of G. R. Desiraju and Elsevier Science Publishers BV, Academic Publishing Division, Amsterdam, The Netherlands.)

which there are two molecules in the asymmetric unit, one ordered and the other disordered. In the case of 2-keto-3-ethoxybutyraldehyde bis(thiosemicarbazone)[59,60] an interaction is found between the two different molecules in the asymmetric unit so that this dimer may be considered the motif that crystallized. Sometimes there is almost a symmetry axis of some type between the two molecules of the asymmetric unit, this is particularly noticeable in protein crystal structures.

15.1.5 Packing in polypeptides and proteins

Proteins are made from many amino acids with hydrophilic and hydrophobic side chains. The main principle is that amino acids with hydrophobic side chains tend to be in the core of a protein, while the hydrophilic side chains lie on the exterior. The hydrophilic tendencies of carbonyl and amino groups in the backbone of the protein are neutralized by hydrogen bonding, such as in α helices (see Chapter 12).

The crystal structure of an aralkylated tripeptide[61] is presented in Figure 15.9. This molecule contains regions that are hydrophobic, while others are hydrophilic. These areas segregate in the crystal structure as shown, and water adds in order to stabilize the hydrophilic areas. Often, however, the distinction between hydrophobic and hydrophilic interactions is not so clear.

The effect of crystal packing on the conformations of molecules is of great interest because it is relevant to the question of whether the conformation in the crystal is the same as a low-energy form (not necessarily the *lowest* energy form) in solution or in the gas phase. Packing effects generally have only modest influence on the conformation of a molecule, but packing energy may favor one conformer over another, especially if a substituent is highly flexible. Nevertheless, it is likely that, under the conditions of crystallization, the conformation that is incorporated into the crystal is also present in solution, even though it is not necessarily the most populous conformer. Many examples have been studied. For example, a cyclic pentapeptide that is an analogue of an antineoplastic agent (dolastatin 3) crystallizes with four molecules per asymmetric unit together with seven molecules of toluene and two molecules of methanol.[62]

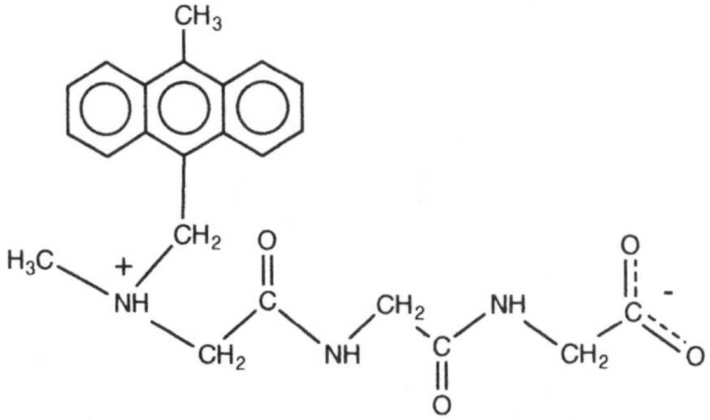

(a)

FIGURE 15.9. (a) Chemical formula of an aralkylated peptide (sarcosylglycylglycine) for which the crystal packing is shown in (b) and (c).

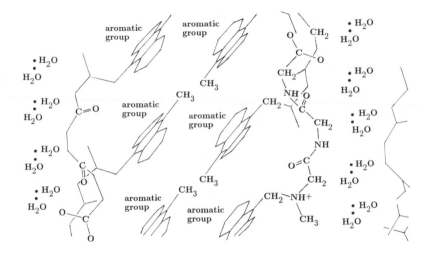

(b)

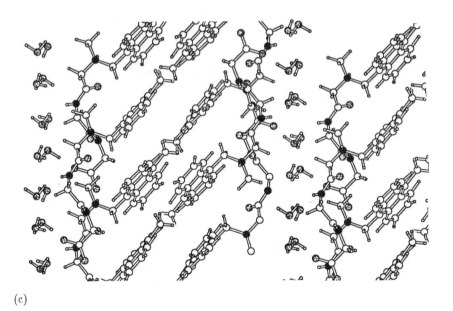

(c)

FIGURE 15.9 (cont'd). Crystal structure of an aralkylated peptide showing segregation of hydrophobic and hydrophilic portions of molecules into different regions as they pack. (b) Diagram of the molecular packing, and (c) view of the crystal structure. Crystals contain two molecules of water per alkylated peptide molecule (Ref. 61).

The conformations of the four peptide molecules are found to be nearly identical except for some very minor variations in the orientations of the valine and leucine side chains, as shown in Figure 15.10. This indicates the variability of conformation in this crystal. Toluene molecules are distributed in channels throughout the crystal and the methanol molecules are "encapsulated" in small holes in the crystal structure. Upon standing in air, the crystals disintegrate, probably as a result of loss of solvent of crystallization. The association of peptide molecules with toluene and methanol appears necessary to achieve a stable packing arrangement. Studies of this dolastatin 3 analogue in solution also supports this interpretation.[62]

The complex surface of folded proteins and the internal packing of atoms in their crystal structures have been analyzed by the procedure of Byungkook Lee and Frederic M. Richards.[63] Each atom of the protein is depicted as a sphere that has the appropriate van der Waals radius. When the atoms are covalently bonded, these spheres will overlap and therefore are truncated at the point of overlap. The complex surface that results is referred as the "van der Waals surface." It has a strictly defined

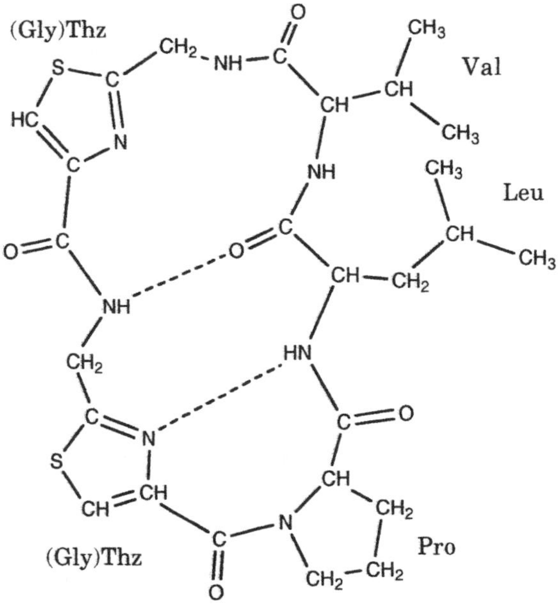

(a)

FIGURE 15.10. A cyclic peptide with four different molecules in the asymmetric unit of the crystal. (a) The chemical formula of one asymmetric unit. The unusual amino acid is designated (Gly)Thz.

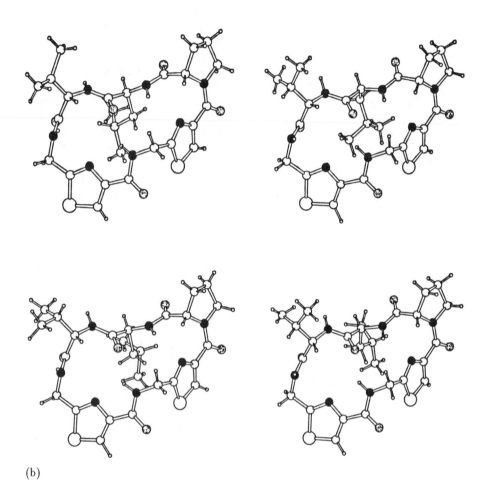

(b)

FIGURE 15.10 (cont'd). Conformations of the four different molecules of a cyclic pentapeptide [formula in (a)], per asymmetric unit (Ref. 62). (b) Views of the four symmetry-independent molecules in the asymmetric unit. Note the different orientations of the Leu and Val side chains.

surface area and encloses a finite volume. A spherical probe, representative of an atom or selected group of atoms, is used to investigate possible points of contact of the protein. That part of the van der Waals surface that makes contact with the surface of a probe is designated the "contact surface." The re-entrant contact surface is a series of disconnected patches showing areas accessible to the probe. When the probe is simultaneously in contact with more than one atom, it defines, in an analogous manner, the re-entrance surface. The contact surface and the re-entrant surface together define a continuous area called the "molecular surface" which can then be analyzed. Protein crystallographers indicate in a visual manner the distinction between hydrophilic and hydrophobic regions of such a surface by use of color-coded dot surfaces (called "Connolly surfaces").[64]

15.1.6 Stacking of flat aromatic organic molecules

The types of packing found in the crystal structures of polycyclic aromatic hydrocarbons (PAHs) were categorized by John Monteath Robertson.[65] His scheme is shown in Figure 15.11. Small PAHs, such as benzene, naphthalene, anthracene, phenanthrene, and chrysene crystallize with one unit cell dimension of about 5.5 Å. The packing appears to result from interactions of the slightly positively charged hydrogen atoms of one aromatic molecule with the electron-rich π-electron system of another molecule, yielding the so-called herring-bone structure. Larger molecules such as pyrene, perylene, and benzo[g,h,i]perylene crystallize in pairs with the π-electron systems interacting. Even larger molecules, such as

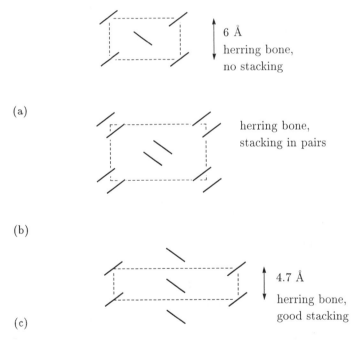

FIGURE 15.11. Packing arrangements for PAHs. The PAH molecule is viewed end on along the long axis of the molecule. (a) Crystal packing of small PAHs such as benzene, naphthalene, anthracene, phenanthrene and chrysene. (b) Crystal packing of larger PAHs such as pyrene, perylene, and benzo[g,h,i]perylene. (c) Crystal packing of large PAHs, such as coronene and ovalene.

coronene and ovalene, crystallize with a vertical stacking of molecules. Each molecule lies inclined at about 45° to a short axis of 4.7 Å so that the interplanar spacing is 3.4 Å ($\approx 4.7 \cdot \cos 45°$).

The dominant intermolecular interactions in crystals of the smaller PAHs are between hydrogen atoms and carbon atoms, giving, as just described, a herring-bone structure. As the PAH becomes larger, the carbon-to-hydrogen ratio increases and intermolecular carbon–carbon interactions become relatively more important. The result is that the PAH molecules stack one above the other. These findings may be expressed by the potential energy expression:[66,67]

$$V_{jk} = \sum_{j,k} [-A_{jk}r_{jk}^{-6} + B_{jk}\exp(-C_{jk}r_{jk}) + q_j q_k r_{jk}^{-1}] \tag{15.6}$$

where r_{jk} is a nonbonded interatomic distance between atoms j and k, q is the point electrostatic charge on an atom, and A_{jk}, B_{jk}, and C_{jk} are adjustable parameters that can be deduced from experimental measurements of unit cell dimensions, interatomic distances and packing arrangements. A_{jk} represents the coefficient of the **London dispersion attraction** term between atoms j and k, while B_{jk} and C_{jk} are short-range repulsive energy terms. The summation is over all interactions (all j and all k). Equations such as 15.6 make is possible to predict packing arrangements for PAHs, particularly unit cell dimensions, which can then be compared with experimental values. In this way the constants in Equations 15.6 may be refined. Values are listed in Table 15.5. Parameters for systems involving other types of atoms, such as oxygen, are currently being worked on, but the equations are then much more complex and the results less satisfactory to date.

15.2 Two or more species in the same crystal

Several types of crystals contain two or more types of molecules per unit cell, the hydrates providing a very simple example, in which the solvent molecules take part in the crystal structure and help satisfy requirements for hydrogen bonding and metal coordination. Sometimes solid solutions of two compounds with similar structures may form. This situation can cause problems in refining a crystal structure, especially if it is not real-

---

TABLE 15.5. Potential parameters for polycyclic aromatic hydrocarbons, expressed in kJ mol^{-1} and Å. These may be used in Equation 15.6 (see Ref. 67).

A_{HH}	126	A_{HC}	552	A_{CC}	2420	q	0.159
B_{HH}	9816	B_{HC}	63410	B_{CC}	409600		
C_{HH}	3.74	C_{HC}	3.67	C_{CC}	3.60		

ized that the molecule under study may be replaced by a slightly different molecule, sometimes a precursor in the chemical preparation. This was true in crystals of 6-mercaptopurine.[68,69] Two well-refined structures gave slightly different results, which could be attributed to contaminant molecules in at least one of the crystal preparations.

The types of structures containing more than one species that will be described here are π- and **charge-transfer complexes**, clathrates and host–guest complexes.

15.2.1 π-complexes and charge-transfer complexes

Flat aromatic molecules such as polycyclic aromatic hydrocarbons (PAHs) are rich in delocalized electrons. They form molecular complexes with other flat molecules containing electron-poor ring systems.[70] The resulting complexes, which involve interactions between the π-electron systems of the two types of molecules, are generally called π-complexes. They are characterized by short intermolecular distances perpendicular to the stacking direction, like graphite. Therefore it is common to find crystals growing as needles, elongated along the stacking direction. Examples of these types of complexes are provided by the trinitrobenzene complexes of PAHs,[71,72] such as that shown in Figure 15.12.

The two molecules forming such a π-complex consist of a donor molecule with a low ionization potential so that an electron can be readily lost (a **delocalized π-electron** of the PAH) and an acceptor molecule with a high affinity for electrons (the aromatic ring of trinitrobenzene which is electron-poor because the three nitro groups have pulled electrons out of it). As a result, stacks of alternating donor and acceptor molecules are found in the crystal. The relative orientations within the parallel planes of these donor and acceptor molecules are determined, to a considerable extent, by the orientations of charge distributions of the highest occupied molecular orbital (HOMO) of the polycyclic aromatic hydrocarbon from which an electron will come, and the lowest unoccupied molecular orbital (LUMO) of the trinitrobenzene, to which the electron will go.[73-75] From such considerations it is possible to predict how the molecules will be aligned in alternate stacks. The structures of many other such complexes of PAHs have been reported.[71-78]

In some complexes of this type, charge transfer can occur between the two molecular components. In such charge-transfer complexes the interplanar spacing of flat molecules is smaller than usual. Thus, while the normal spacing is 3.4 Å, it can decrease to 3.1 Å in a charge-transfer complex. The theory of charge-transfer complexes was developed by Robert Mulliken, who considered that such donor–acceptor complexes are formed as a result of mixing (resonance) between a "no-bond" structure and one in which charge transfer (i.e., transference of an electron from donor to acceptor, as in salt formation) has occurred.[79,80] An example is the charge-transfer complex between iodine and benzene. The no-bond structure predominates in the ground state, but there is transfer of charge in the first excited state, with benzene as the electron donor, and iodine as the electron acceptor. A "charge-transfer band" is formed in the visible or ultraviolet spectrum (at 300 nm). This band is observed in addition to the normal absorption bands of the two component molecules. The wavelength of such a charge-transfer band is inversely related to the first ionization potential of the electron donor (benzene). Thus, the usual test for the formation of a charge-transfer complex is to examine the spectrum for any evidence of a charge-transfer band. If a charge-transfer band lies in the visible region, the solution will become colored or change color as the charge-transfer complex forms. The stack-

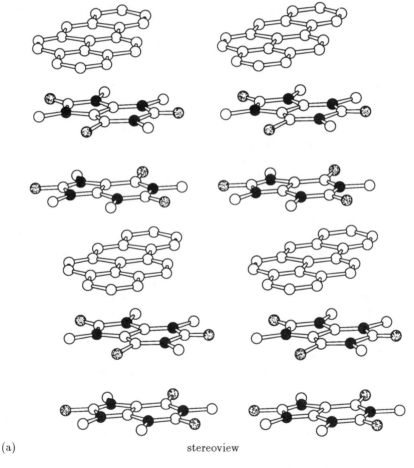

(a)   stereoview

FIGURE 15.12.   Some PAH–trinitrobenzene crystalline complexes.   (a) 1,3,7,9-tetramethyluric acid and benzo[a]pyrene (Ref. 72).

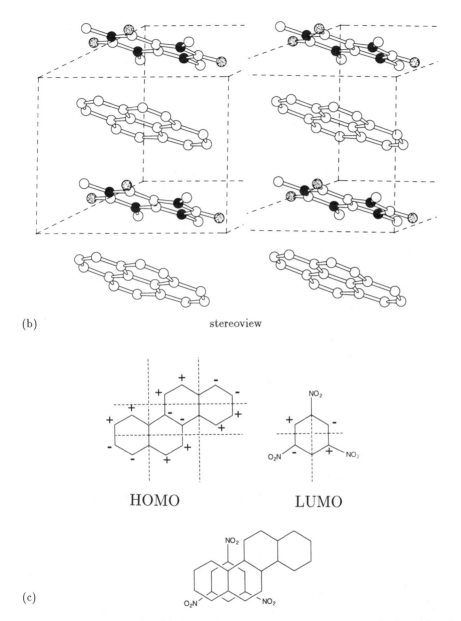

(b) stereoview

HOMO LUMO

(c)

FIGURE 15.12 (cont'd). (b) Crystal structure of 1,3,7,9-tetramethyluric acid and pyrene. (c) HOMO/LUMO overlaps for PAH–TNB complexes. Shown are signs of the wave functions (not electronic charge) for the HOMO and LUMO, and the overlap of molecules observed in the crystalline state (as predicted from the HOMO/LUMO overlap of positive with positive wave functions).

ing of planar molecules in parallel planes does not, in itself, demonstrate that a charge-transfer complex has been formed, but a color change may indicate this. For example, when (colorless) solutions of naphthalene and pyromellitic dianhydride are mixed, a deep red solution results, from which crystals may separate if the conditions are appropriate to give the complex of these two components.

15.2.2 Host–guest complexes

Molecules may pack so that there are cavities in the crystal structure. These cavities may be of an appropriate size, shape and hydrophobicity to accommodate a different type of molecule that has the size and character to fill the cavity. A particularly interesting example is provided by molecular complexes called **choleic acids**.[81] These were discovered when it was noted that fatty acids are tenaciously retained during the purification of bile acids from the body. These complexes, isolated from bile, consist of approximately 92% deoxycholic acid and 8% stearic or palmitic acid. Several choleic acids were studied by X-ray techniques in the early 1930s and were found to have similar powder diffraction patterns, with unit cell dimensions almost independent of the nature of the guest molecule (stearic or palmitic acid, for example). These unit cell dimensions[82] are $a = 13.4$–13.6, $b = 7.2$–7.3, and $c = 25.8$–26.8 Å. It was found that a framework of deoxycholic acid molecules is formed upon crystallization and that this contains channels running parallel to the b axis. The guest molecules, if small enough, can then lie in these channels and may, if necessary, extend through more than one unit cell.[82,83] This was shown to be true for the phenanthrene complex,[84] in which disordered phenanthrene molecules lie in such channels. Phenanthrene molecules interact closely with methyl groups on two deoxycholic acid molecules in such a way that these methyl groups point to the centers of the outermost rings of the phenanthrene molecules.

Another such host molecule is β-cyclodextrin. This is the main product obtained when a starch solution, pretreated with α-amylase, is acted on by the enzyme cyclodextrin transglycosylase. The cyclodextrins so produced have 6, 7, or 8 pyranose units and are named α, β, and γ cyclodextrins, respectively. For example, β cyclodextrin is a torus-shaped molecule composed of seven glucopyranose units, each connected through C1 and C4. All of the glucosyl oxygen-atom bridges point to the center of the torus, which is hydrophobic, while the exterior of the torus is hydrophilic. Therefore an organic molecule that prefers a hydrophobic environment will enter the cavity in the center of the torus.[85–87] The crystal structure of β cyclodextrin·11H$_2$O has been studied by X-ray diffraction at room temperature[88] and by neutron diffraction at 120 K.[89] At room temperature only three water molecules are ordered. The remaining molecules are distributed over 16 sites. At 120 K all water sites but one are fully occupied. Cyclodextrin complexes are studied not only for their binding properties, but can also be engineered to assist in catalytic reactions.[90]

Another example of crystals containing more than one component is provided by the **clathrates**. In clathrates of β–quinol, three quinol molecules are hydrogen bonded together to form an approximately spherical cavity of radius 4 Å (Figure 15.13).[91] Any molecule of appropriate size such as oxygen, nitrogen, krypton, xenon, methane, sulfur dioxide, or methyl alcohol can be trapped, and if it is not disordered within the clathrate, its location and orientation can be determined in the crystalline state by X-ray diffraction methods. In most cases, when a clathrate is

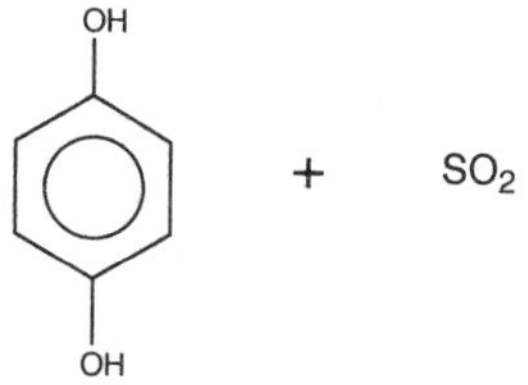

(a)

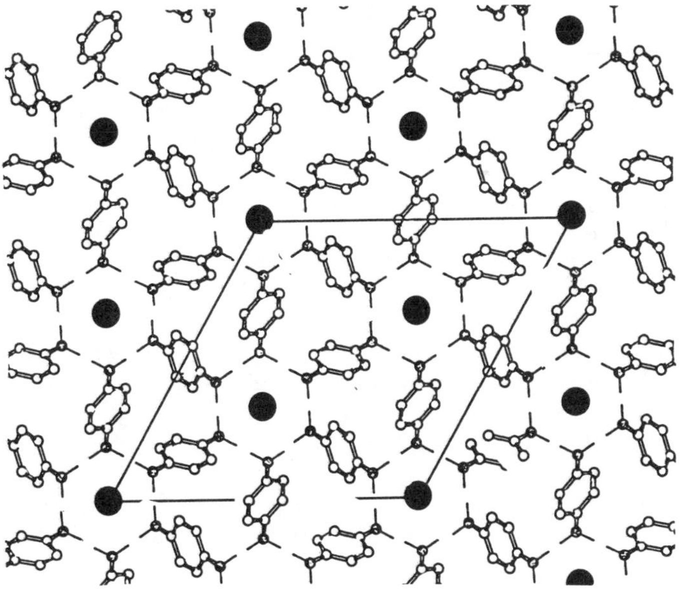

(b)

FIGURE 15.13. A quinol-sulfur dioxide clathrate (Ref. 91). (a) Molecular components, and (b) the crystal structure of the complex. The sulfur dioxide molecule is disordered and is indicated by a solid circle.

dissolved or melted, the guest molecules escape. The host structure is generally less stable without the trapped molecules than with them, since the guest molecules contribute van der Waals energy and so help to stabilize the whole structure.

The name "clathrate," from the Greek *klathron* ($\kappa\lambda\alpha\theta\rho o\nu$, "bolt" or "lock," since the volatile guest compound is locked into the crystal),[92] was coined by Herbert M. Powell,[91,93] who studied many of them. Examples of clathrates are provided by the gas hydrates, first identified by Sir Humphrey Davy, who prepared chlorine hydrate by bubbling chlorine[94] into cool water. This hydrate was shown[95] to have the chemical formula $8Cl_2 \cdot 46H_2O$. The anesthetic properties of chloroform have been attributed to the formation of such gas hydrates in brain tissue.[96]

A full description of the three-dimensional structures of such gas hydrates is provided by X-ray diffraction studies.[97,98] The basic motif of water in the gas hydrates is a regular pentagon formed by hydrogen bonding. The H–O$\cdots$H angle is approximately 108° in such pentagons. These pentagons can assemble to form polyhedra.[96] Regular pentagons assemble in three dimensions to give a dodecahedron (12-hedron, a 12-faced figure) with 20 vertices, each the position of a water molecule. If additional hexagonal faces are added, it is possible to obtain larger polyhedra. The 12-hedra contain only pentagons, while the 14-, 15-, and 16-hedra contain, respectively, 2, 3, and 4 additional hexagonal faces. These polyhedra contain voids of specific sizes so that small guest molecules, such as krypton, xenon, methyl chloride, or sulfur dioxide, can fit into them.[97] The forces between the host and guest molecules are generally only van der Waals interactions.

In the type I hydrates there are two 12-hedra (composed of 20 water molecules) and six 14-hedra (composed of 24 molecules) in a cubic unit cell. These polyhedra share pentagonal faces. The 14-hedra share 8 faces with other 14-hedra and 4 faces with other 12-hedra. As a result, there are 46 water molecules per unit cell. The clathrate formula, $6X \cdot 2Y \cdot 46H_2O$, is explained by this number of polyhedra; six guest molecules, X, fit in the six 14-hedra, and two guest molecules, Y, fit in the two 12-hedra (Figure 15.14). Chlorine can fit in both types of polyhedra (so that $X = Y$) giving the compound $8Cl_2 \cdot 46H_2O$ prepared by Davy. Larger molecules can only fit into the 14-hedra so that such clathrates have the formula $6X \cdot 46H_2O$.

In the type II hydrates, the host framework is composed of 12- and 16-hedra, the latter composed of 27 water molecules, to give a unit cell $8X \cdot 16Y \cdot 136H_2O$, where X fits in the eight 12-hedra and Y fits in the sixteen 16-hedra. Larger molecules, such as *n*-butane, do not readily form clathrate hydrates but can do so in the presence of methane (a "hilf-gas," helper-gas[97]) giving a type II hydrate in which X = methane and Y = *n*-butane. The methane has served to help stabilize this type of clathrate.

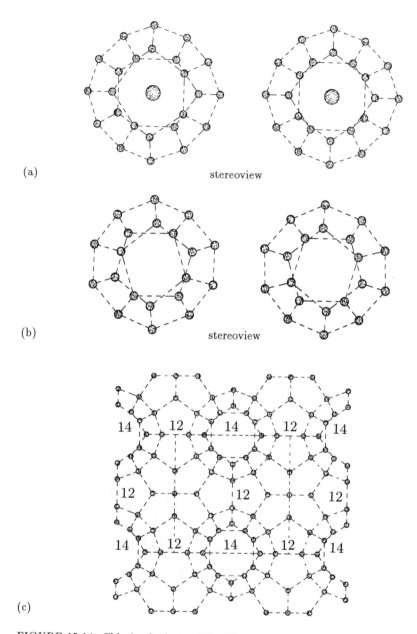

(a)

stereoview

(b)

stereoview

(c)

FIGURE 15.14. Chlorine hydrate, $8Cl_2 \cdot 46H_2O$. Hydrogen atoms omitted for clarity. (a) Stereoviews of a 14-hedron consisting of 24 water molecules. (b) Stereoviews of a 12-hedron consisting of 20 water molecules. (c) Molecular packing of the whole complex.

Clathrates provide cavities of a specific size and shape and therefore they can be used very effectively for separating gases with different sizes of molecules. For example, urea clathrates have been used to separate linear from branched hydrocarbons. The hydroquinone clathrate can be used to store and deliver radioactive krypton. In addition, if the host is chiral, there can be chiral discrimination so that one enantiomer of a guest is enclosed in the clathrate structure in preference to the other enantiomer. The trapped guest molecules can be liberated, for example, by solution in an apolar solvent, at the convenience of the user.

15.3 Polymorphism

Polymorphism is the existence of more than one crystalline form of the same chemical substance. These forms differ in their unit cell dimensions and atomic arrangements. The term **allotropy** is used for the polymorphism of a chemical element. Polymorphs can be described in terms of their relative thermodynamic stabilities; the phase with the lowest free energy is the most stable at a given temperature and pressure, although the differences in free energies of different polymorphs may be very small (1–2 kcal mol^{-1}, 4–8 kJ mol^{-1}). Under conditions of temperature and pressure where this phase is no longer the lowest in free energy, a transition to a more stable form may occur. For example, under high pressure sodium chloride can be converted from a less dense NaCl-type crystal to a denser CsCl-type structure (Figure 15.15).[99,100]

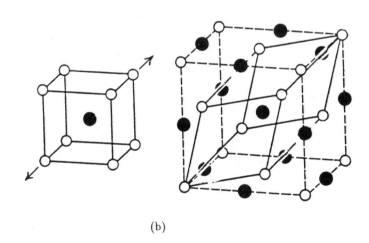

(a)   (b)

FIGURE 15.15. The transformation of (a) a CsCl-type structure to (b) an NaCl-type structure. This occurs at 465°C when the arrangement of ions found in CsCl crystals distorts to give the arrangement of ions found in NaCl crystals. $\Delta H = 2.90$ kJ mol^{-1} and $\Delta V = 17\%$ (Ref. 100). The filled circles represent Cs ions.

The phenomenon of polymorphism was noted by Martin Heinrich Klaproth[101] in 1798, when he proposed that the minerals calcite and aragonite must have the same chemical composition, $CaCO_3$. Calcite forms a rhombohedral uniaxial crystal and is the stable form under normal conditions, with a density of 2.71 g/ml. Its metastable polymorph, aragonite, is an orthorhombic biaxial crystal with a density of 2.94 g/ml. This work was continued by Louis Jacques Thernard, Jean Baptiste Biot, and Eilhard Mitscherlich.[102-104] Mitscherlich, for example, reported on it in his studies of phosphates and arsenates.[104] The transition from calcite to aragonite has been studied at different pressures.[105,106]

There is considerable interest in the factors that cause one (rather than another) polymorph to be formed. It is common first to obtain a metastable form during crystallization and then, later, to find that it may become transformed into a more stable form. Moritz Ludwig Frankenheim,[107] in 1839, reported a polymorphic transition in potassium nitrate crystals. At room temperature potassium nitrate forms calcite-like rhombohedra, but, after a while, dendritic aragonite-like prisms form and destroy the rhombohedra, which are then converted to crystallites. Above 110° only the calcitelike form is obtained. The aragonitelike form is then the metastable form. Similarly, mercuric iodide, when it is first precipitated, is yellow but then immediately changes to red.[108] The red, tetragonal form is the stable one, while the yellow, orthorhombic form is metastable. A third form can be obtained by cooling the vapor by sudden expansion.[109] The high-temperature modification of a substance often forms more symmetrical crystals than does the low-temperature form. The red form can be reversibly converted to the yellow form[110] by heating above 126°C. Another interesting set of polymorphs[111] consists of quartz, tridymite, and cristobalite which can each exist as β (low-temperature) and α (high-temperature) modifications.

Each polymorph contains the same chemical contents of the respective unit cells. If the chemical contents differ, for example by the presence of different amounts of solvent, they are called **pseudopolymorphs**. Polymorphs may differ with respect to physical properties such as melting points, or solubilities, as also may pseudopolymorphs. Their existence often presents a serious problem in the pharmaceutical industries since physical properties of crystals are often used as criteria for quality control and thereby the effectivity of a given preparation.[112] Polymorphs and pseudopolymorphs are usually obtained when crystals are grown under different conditions. For example, **metastable** crystals of the π-donor:acceptor complex between biphenylene and pyromellitic dianhydride were obtained when crystals were grown by sublimation at high temperatures, whereas a different polymorph, stable at room temperature, was grown by the same method at a lower temperature.[113]

An analysis of the crystal structures of pseudopolymorphs can also lead to a better understanding of interrelationships between structure and chemical and biological properties.[114] Single crystals of the free base of oxytetracycline, a medicinally important tetracycline antibiotic, were grown from aqueous methanol and also from anhydrous toluene. Both crystal structures were determined at 120 K. The crystals differed not only in solvent content (dihydrate for the crystals from aqueous methanol and anhydrous from toluene), but also in the tautomeric form and conformation of the oxytetracycline molecules. The tetracycline derivative was found to be zwitterionic in the dihydrate crystals and nonionized in the anhydrous crystals.

There is good reason to encourage the chemist and biologist setting out to grow crystals for crystal structure determinations to try a wide spectrum of crystallization techniques. The probability is high that if sufficiently different conditions are tried, then more than one crystal modification may be obtained. Dunitz[54] wrote: "Polymorphism appears to be a widespread phenomenon, and one might conjecture that almost all compounds are polymorphic if the right conditions for crystallizing the different forms could be found."

15.3.1 Ice polymorphs

In ice each oxygen atom is surrounded by two near and two far hydrogen atoms, but the total arrangement is disordered, even near absolute zero. The concept of orientational disorder of hydrogen atoms in hydrogen bonds in ice was introduced by Linus Pauling.[115] He showed that there is a statistical arrangement with short-range but no long-range order so that the configurational entropy could be calculated, and the result agreed with experiment.[116] In 1957 the predicted structure of ice was established by a single-crystal neutron diffraction study of ice.[117] The hydrogen atoms show 50% occupancy of each possible site, as shown in Figure 15.16. Hugh Savage[118,119] showed that a major factor in determining the orientation of water molecules in crystals is the distance between $H(2)$ and $O(2)$ in the hydrogen-bond sequence $H(2)-O(1)-H(1) \cdots O(2)$ (see Figure 15.17). This value is found to be in the range of 2.9–3.1 Å.

Eleven different forms of crystalline ice have been found, and they are numbered "Ice I" through "Ice XI." In each there are four nearest oxygen neighbors at 2.75–2.9 Å from a given oxygen atom. Denser forms of ice have been studied at low temperatures.[120] Ice II was shown to have a structure with ordered hydrogen atoms.[121] Some of the ice structures are listed in Table 15.6.

15.3.2 Tetraacetylribofuranose

A very interesting case of **dimorphism** is provided by 1,2,3,5-tetra-*O*-acetyl-*β*-D-ribofuranose.[122–125] This compound crystallizes in two forms. Form *A* melts at 329–331 K, while form *B* melts at a higher temperature, 358 K. Form *A* has a density that is 1.5% higher than that of form *B*.

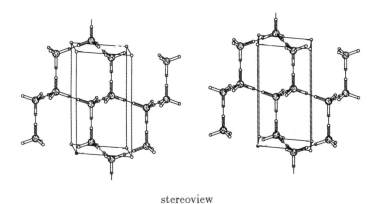

stereoview

FIGURE 15.16. The crystal structure of ice (hexagonal unit cell, $a = b = 4.48$ Å, $c = 7.31$ Å. Water molecules are drawn with four hydrogen atoms representing disordered positions.

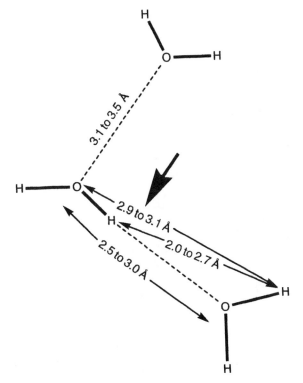

FIGURE 15.17. Interactions around water (Ref. 119).

TABLE 15.6. Polymorphs of Ice

Form	System	Cell Vol. ($\mathring{A}^3$)	Density (g cm^{-3})	Space Group	Cell Dimensions* (a,b,c,angle)
I	Hexagonal	128	0.931	$P6_3/mmc$	4.50,4.50,7.32
Ic	Cubic	256	0.93	$Fd3m$	6.35
II	Rhombohedral	304	1.18	$R\bar{3}$	7.79, $\alpha = 113.1°$
III	Tetragonal	(309)	(1.16)	$P4_12_12$	(6.73,6.73,6.83)
V	Monoclinic	680	1.23	$A2/a$	9.22,7.54,10.35, $\beta = 109.2°$
VI'	Orthorhombic	228	1.31	$Pmmn$	6.27,6.27,5.79
VI	Tetragonal	(228)	(1.31)	($P4_2/nmc$)	(6.27,6.27,5.79)
VII	Cubic	(40)	(1.49)	($Pn3m$)	3.43
VIII	Tetragonal	161	1.49	$I4_1/amd$	4.80,4.80,6.99
IX	Tetragonal	309	1.16	$P4_12_12$	6.73,6.73,6.83

* Less certain values in parentheses (for Ice III and Ice VI, both tetragonal).

Generally, but by no means always, the density and melting points are higher for the more stable polymorph. In this set of polymorphs, however, while form A has a lower melting point, it is more dense. The molecular conformations are similar except for torsion about two of the four acetyl groups. The conformation in form A is approximately 3.9 kcal mol^{-1} more stable than that of form B. On the other hand, the packing is such that crystals of form A are less stable than those of form B, and the less stable form A is converted into form B if the laboratory becomes contaminated with form B. For a while laboratories in which crystals of form A had been prepared communicated by telegram to avoid contamination, but A. Lindo Patterson described how, when form B was finally introduced into his laboratory in Philadelphia, the form A did not change until form B was powdered and sprinkled over it.[122] From that time on no form A could be grown in that laboratory, an example of a "disappearing crystal form."[126,127] The crystal structure of the metastable form A was determined in Hungary.[123] Recently it has been shown that in general cases, with care, less stable polymorphic forms can be recovered.[127–129]

15.4 Phase changes and transitions

Polymorphs are converted into one another by phase transformations. Martin Buerger[130] wrote "The polymorphs of a substance are, in a sense, nodes, and the **phase transformations** comprise the network which connects them." There is an energy barrier to any polymorphic transition since an activation energy is necessary. External physical effects such as increasing the temperature may cause a change to a different polymorph. The potential energy barrier and the proportion of atoms with that energy will affect the rate of the transformation. For example, phase transitions will occur in response to pressure changes,[131-134] giving the more dense polymorph at the higher pressure. Phase transitions can be readily detected by calorimetric methods, and a scanning differential calorimeter trace, which provides an example, is shown in Figure 13.20 (Chapter 13).

The rate of a polymorphic transition depends on the mobility of atoms in the solid state and varies with temperature. Slow rates may be very convenient; for example, when steel is hardened or tempered by quenching, it is suddenly cooled far below its melting point so that the rate of transition to the stable phase at room temperature becomes negligible. Another interesting example, already discussed in Chapter 11, is white tin, which is converted to gray tin below 18°C, although the transition rate is very slow. The gray tin so formed has lower density (5.7 versus 7.3 g cm^{-1}) and therefore expands giving a powder.

A polymorphic transition has also been noted for biphenyl. The conformation of biphenyl results from a competition between conjugation of double bonds in the two rings and the steric repulsion of the *ortho*-hydrogen atoms. In the gas phase the torsion angle is 44.4°. In crystals there is an equilibrium between these two tendencies, one involving a twisted molecule and the other a planar molecule (which favors the packing of phenyl groups of different molecules). In the crystalline state, above a transition temperature, the crystal structure contains an averaged planar structure (representing two or more nonplanar units). Below 38 K the molecule is twisted by 10°.[131] In this case thermal motion is not great enough for the molecule to surmount the potential barrier to the conversion of one twisted form to the other. The nonplanar form is more stable than the planar form by about 1.5 kcal mol^{-1} (near 6 kJ mol^{-1}).

15.5 Variations from true crystallinity

Some of the experimental aspects of crystal structure analyses can now be applied to compounds that are not truly crystalline, or that are variants of the perfectly ordered crystals we have been describing so far.

15.5.1 Liquid crystals

The term **liquid crystal** was first applied to compounds that, unlike most compounds that melt in a single step at a definite temperature, show one or more well-defined phases between the solid and the true liquid. Otto Lehmann in 1888 was contacted by Friedrich Reinitzer who had observed that crystalline cholesteryl benzoate, on heating, seemed to have two melting points.[135-137] At 145.5° C a cloudy liquid forms that, on further heating, changes sharply to a clear liquid at 178.5° C. On cooling the reverse order of phases was found. The turbid liquid is doubly refracting, like the anisotropic crystals described in Chapter 5, hence the name liquid crystals (German: *fliessende Kristalle*) that Lehmann gave to them. They are also, sometimes, described as mesomorphic phases, "in-between phases". In each liquid–crystal phase (a mesophase) there is some molecular ordering, less than that in a crystal but more than that found in true liquids. Thus, liquid crystals exhibit aspects of both the liquid and solid states, but also have properties not found in either of these states. Many compounds, such as *p*-azoxyanisole, are found to form liquid crystals.[138] This is a molecule that is much longer than its width or thickness, contains central aromatic units linked by planar trigonal atoms and has tetrahedral terminal or wing groups. Peter J. Collins writes:[137] "A good model of a typical liquid crystal molecule is therefore a short pencil with a short piece of cooked spaghetti attached to each end."

Liquid crystals may be classified by the method used to modify order in the solid state. Thermotropic liquid crystals are obtained by varying the temperature, while lyotropic liquid crystals are formed by adding a solvent, such as water or oil. There are two main structural groups of liquid crystals. In nematic liquid crystals (from the Greek, *threadlike*), the long axes of the molecules are somewhat parallel, but the centers of gravity of the molecules are randomly arranged and can vary as the molecules rotate about their axes or move up and down or side to side. In smectic liquid crystals (from the Greek, *soaplike*), molecules are arranged approximately in layers that can slide over each other. The molecules within each layer, however, while aligned in a parallel fashion, can only move sideways, not up or down from one layer to another.[138,139] This smectic phase is more ordered than the nematic phase, and therefore it is formed at the lower temperature for compounds that form both types of phases. A third classification, cholesteric, resembles the nematic phase, but the molecules composing it are chiral. In studies of crystals of compounds related to liquid crystals, it has been possible to gain some understanding of their nature.[140] Robert F. Bryan has described in detail the relationship between crystal structures and liquid crystallinity.[141] Domains of liquid crystals can be aligned by electric or magnetic fields, and are of technological importance.

In a similar way, long-chain *n*-alkyl pyranosides were noted by Emil Fischer[142] to have, apparently, a double melting point. In other words, these amphiphilic molecules, with moieties that have different physical properties, form liquid crystals.[143] Thermotropic liquid crystals melt in two stages because of differences in the cohesive forces in different parts of the molecules. In the case of the long-chain alkyl pyranosides, the hydrocarbon chains disengage first in the crystal structure and "melt." The hydrogen-bonded carbohydrate portions have stronger cohesive forces and melt at a higher temperature when an isotropic homogeneous liquid forms. In a similar way, a difference in solubilities of two major portions of a molecule may cause the formation of lyotropic liquid crystals.[139] Bryan M. Craven studied the crystal structures of several esters of cholesterol, such as cholesteryl myristate,[144] which forms smectic and cholesteric phases. The solid crystalline phase (Figure 15.18) consists of stacked bilayers, 50.7 Å thick. The C_{17} chain ends, which lie at the bilayer surface, are described as almost liquid in character. In the crystal structure, cholesterol packs with cholesterol, and myristate packs with myristate within a bilayer. The different inner X-ray diffraction rings of the cholesteric and smectic A phases can be explained by features in the crystal structure, implying that crystallinity is not totally lost in these mesophases.

Dichroic dye molecules absorb light in an anisotropic way and show different colors in different directions. If such dyes are dissolved in a liquid crystal, they form a guest–host-type interaction, and are oriented by the host liquid crystal molecule. Application of an electric field will reorient the liquid crystal and dye; this effect is used for "liquid-crystal" displays.[145,146]

15.5.2 Polytypism

Polytypism[147–151] is a special kind of one-dimensional polymorphism that is found for certain close-packed and layer structures. It occurs when an element or compound can crystallize in several different modifications, each of which may be regarded as being built up by different stacking arrangements of layers of the same structure and composition. The modifications differ only in their stacking sequence. For any successive layer, there are two or more possible positions relative to its predecessor. Periodic (ordered) as well as nonperiodic (disordered) stackings are found to occur.

15.5.3 Defect structures

While the arrangement of atoms from unit cell to unit cell is usually consistently repeated throughout the crystal, perturbations to this exact regularity are encountered in practice. These **defects** can be of great technological importance because of the physical properties they impart to the crystal.[152] The simplest of these defects are those in which atoms or ions become displaced throughout the crystal structure. Such defects have been observed by electron microscopy techniques.

Schottky defects[153] occur when sites that are normally occupied by atoms or ions are left vacant. In order that the crystal structure maintain its electrical neutrality, for every cation-site vacancy there must be an anion-site vacancy. At room temperatures, one in 10^{15} sites is typically vacant, but this adds up to 10^4 Schottky defects in a 1 mg crystal.[154] A less commonly observed defect is a **Frenkel defect**,[155] in which an atom or ion is displaced from its site to an interstitial site that is normally unoccupied. In so doing the number of nearest neighbors of one component of the crystal is changed. This type of defect is seen in

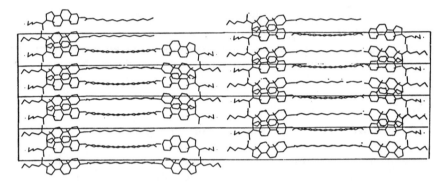

(a)

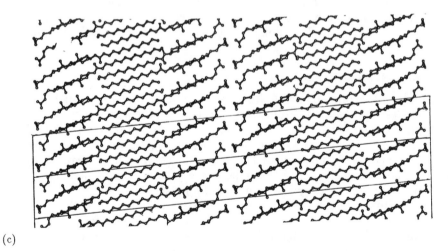

(b)

(c)

FIGURE 15.18. The structure of cholesterol myristate (Ref. 144). Some unit-cell edges are indicated. (a) Chemical formula, and (b) and (c) two views of the molecular packing in the crystal. (b) View onto the steroid ring systems, and (c) view along them. The equilibria, with temperatures, are:

crystal $\Leftarrow^{73.7°}\Rightarrow$ smectic A $\Leftarrow^{79.7°}\Rightarrow$ cholesteric $\Leftarrow^{85.5°}\Rightarrow$ isotropic liquid

silver chloride crystals where the cations and anions have very different sizes. The silver cation may move from a site with four neighbors to an interstitial site where it has eight neighbors. A third type of defect, involving electrons trapped in vacant holes, is referred to as an **F center** (from the German, *Farbenzentrum,* color center). Sodium chloride, which forms this type of defect, will absorb sodium atoms when heated in sodium vapor. In so doing it takes on a greenish-yellow color. The sodium fills cation sites as sodium ions. The additional electrons derived in formation of the cation are then trapped in vacant anion sites and give the sodium chloride crystal some interesting electrical and optical properties.[156] Other types of defects, readily observed by electron microscopy, are dislocations, of which a specific type, the screw dislocation, was described in Chapter 2.

An important feature of defects in crystals is that molecules trapped in defect sites have different environments and therefore slightly different energies than those in the bulk of the crystal. In addition, there may be more space at the defect site. As a result, defects may be the sites of chemical reactions (such as certain photochemical reactions), and the nature of the product will depend on the orientation of molecules at the defect sites.

15.5.4 Modulated structures

Deviation from a regular packing of molecules is found for nonstoichiometric compounds and for modulated structures. The term **modulated structure** is used for "any periodic, or partially periodic perturbation of a crystal structure with a repetition distance appreciably greater than basic unit cell dimensions."[157] Changes to a pseudostructure are modulated over many unit cells. These reflect a change in the true lattice that is usually larger than the original pseudocell. If the deviations from perfect periodicity are not too large, they may be expressed as a modulation of the undistorted structure by an additional function of different periodicity.

$$\rho_o(x) = \sum_h A_h (\cos 2\pi h x/a) \tag{15.7}$$

then becomes:

$$\rho_o(x) = \rho_o(x)[1 + p\cos(2\pi x/(Qa))] \tag{15.8}$$

where Qa is the period of modulation, detected experimentally by the presence of weak diffraction spots on either side of the main diffraction spot.[157] If there are many such modulations and they have large periodicities, then a diffuse halo will result around each Bragg reflection. A wider definition covers incommensurate periodic perturbations, that is, those in which the ratio of the imposed periodicity to that of the unit cell is irrational. For more information consult Reference 157.

15.5.5 Quasicrystalline materials

An alloy of manganese and aluminum was found in 1984 to show sharp diffraction spots that could not be indexed on any known crystal lattice. The diffraction pattern showed icosahedral $53m$ point group symmetry, together with long-range orientational order.[158] Pauling suggested multiple **twinning** of a cubic phase had occurred.[159] On the other hand, a Penrose tiling pattern[160-162] can provide a two-dimensional explanation of the existence of this icosahedral phase. The subject is still under active investigation.[163,164]

15.6 Macromolecular assembly

There are interesting similarities between the energetics that govern crystal packing and those that govern protein folding. Secondary structure, important in protein folding, provides conformational motifs that must be accommodated in the tertiary structure of the protein.[165-167] In crystal packing, some aspects of molecular conformation are also so favored energetically that they, too, must be accommodated by the packing mechanism. As in protein secondary structure, these conformational motifs may involve intramolecular hydrogen bonding. For example, continuing the analogy between the packing of large and small molecules, it is found that enolic diketones frequently form intramolecular hydrogen bonds that favor a planar conformation. Hydrophobic effects play important roles both in determining protein folding and in crystal packing.[168] Where possible, hydrophobic components of small molecules will often come into close contact to give a "compact" conformation. If the structure does not permit such interactions, hydrophobic moieties of different molecules will generally "associate" as molecules pack in the crystal. The parallel with protein folding is the association of subunits to give oligomers as the stable (and often active) form of the protein.

This discussion of the similarities of crystal packing and protein folding can be carried further to gain some feeling for the magnitude of packing energy in crystals of organic molecules. Protein folding involves the interactions (primarily hydrogen bonding and hydrophobic effects, with most of the charged groups distributed on the surface) of hundreds of amino acids.[169] Because a single hydrogen bond may release approximately 5 to 10 kcal mol^{-1} (21–42 kJ mol^{-1}) in bonding energy, one might expect the folding energy of proteins to be on the order of hundreds of kcal mol^{-1}. This is not so; rather the folding energy is typically less than 25 kcal mol^{-1} (105 kJ mol^{-1}. Many of the intramolecular hydrogen bonds found in the folded protein were previously present in the unfolded state as intermolecular hydrogen bonds with water molecules. Most of the folding free energy can be accounted for by the increased entropy resulting from release of water molecules as the protein folds. Intermolecular hydrogen bonds formed in the crystal are probably also present in the mother liquor. As in protein folding, they may involve different species, but the enthalpic contribution is essentially unchanged. As the molecules associate to form the crystal, solvation molecules are released, which increases entropy.

The packing density of a protein is the ratio of its van der Waals volume divided by the actual space occupied, which can be calculated by use of **Voronoï polyhedra**.[170] A low packing density would imply that the recognition between amino acids is poor and that the association is similar to that which occurs in micelles. Small molecules pack very efficiently within the unit cell; for example, graphite has a packing density of 0.89, while crystals of organic molecules have packing densities of 0.65–0.80. To establish a frame of reference, the close packing of spheres has a packing density of approximately 0.74, while for close-packed cylinders it is 0.91. By comparison, the atoms in proteins such as ribonuclease and lysozyme have packing densities between 0.72 and 0.76, demonstrating that the interiors of proteins are well organized and similar to the packing of small molecules.

15.6.1 Packing of helices and sheets

When proteins fold to form their tertiary structures, secondary structural units, such as α helices and β sheets, come into contact, frequently forming highly organized **supersecondary structures**. How do these secondary structural units pack within the protein?[171] Prior to the first X-ray structure determination of a protein, a model of α-helical packing within a protein was described by Francis Crick using a simple geometrical model of an idealized helical net that consists of knobs and holes. The knobs represent the positions of the side chains, and the holes are the spaces between the side chains. Now we can imagine taking a model of an α helix, slicing it down the back and opening it so that it can lay flat, as shown in Figure 15.19(a). The side-chain positions of the α helix form a kind of lattice arrangement on this opened-up model. Crick recognized that the packing of two α helices could be considered in terms of the packing of two lattices.[172] The object here is to minimize the space between the two helical nets. Packing the first lattice onto the second lattice is equivalent to bringing the exteriors of the two α helices into contact. Crick showed that the amino-acid residue or knob at some arbitrary position i on the first α helix must fit between residues j, $j + 3$, $j + 4$, and $j + 7$ of the second α helix. A knob from one of the helices sits in a hole created by four residues on the other helix. Two distinct solutions satisfy the packing criterion of filling space. One occurs when α helices are antiparallel, and the other occurs when they are perpendicular to one another. From a number of crystal structures it was observed that pairs of α helices pack in three distinct fashions rather than just two. To account for this third packing arrangement, Cyrus Chothia, using the same graphical representation as Crick, described the packing problem in terms of ridges and grooves, rather than knobs or holes.[173,174] All three

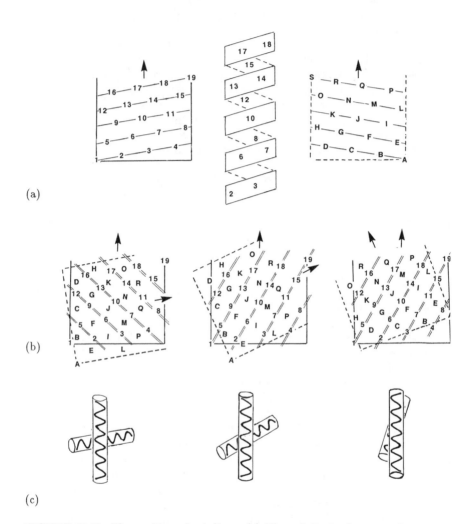

(a)

(b)

(c)

FIGURE 15.19. The packing of α helices. (a) The α helix in the center is cut open and laid out to give the diagram on the left. Amino-acid residues are numbered. On the right is a similar plot for an α helix that will interact with the first. In this diagram the amino-acid residues are represented in order by alphabetical letters. The two α helices interact by a narrow line down the middle of each flat diagram. (b) Three ways that the two α helices can interact without amino-acid side chains causing steric problems. Not that the situation on the left pertains if the side chains arew small (Gly). (c) The stacking of α helices that results. Left: Side chain Gly, angle near 80°; center: side chain Ala, Val, Ile, or Ser, angle near 60°; and right: side chain Leu or Thr, angle near 20°.

observed classes are accounted for by positioning the surface ridges in the first α helix into the grooves on the second α helix, and vice versa. These three classes are shown in Figure 15.19(b). Chothia showed that an amino-acid residue j in an α helix has neighbors $j + 3$, $j + 4$ from above and $j - 2$ and $j - 4$ from below. One row is formed by residues j, $j + 3$, $j + 6$ and another by j, $j + 4$, $j + 8$. The model for helix–helix packing requires that the surface ridges in the first α helix pack into the grooves between the ridges on the second α helix and vice versa.

Timothy J. Richmond and Frederic M. Richards[175,176] extended these ideas, examining the residue preference and the nature of helix interactions. They observed a strong residue preference that depends on the three classes, defined by the angle between the two α helices. For example, the Class 1 type interactions, where the angle is approximately 80°, has a central residue that can only be a glycine residue. At the point of contact of the two helices, main-chain atoms of both helices are also in contact. Class 2 types of interactions, characterized by an interhelical angle of approximately 60°, have side-chain residues such as alanine, valine, serine, or threonine. The third observed class, which has a small interhelical angle of 20°, can accommodate bulkier side chains. Note how for interhelical angles near 90°, the contact residues are small, but increase in size as the α helices become more parallel. The specifications of the contact residues are based upon size and polarity. Large residues will not allow close penetration of the α helices, and polar residues will resist being buried in a nonpolar environment.

Similar types of analyses were performed to examine the packing of β sheets. An analysis of a large number of protein structures revealed that β sheets are not flat, but have variable extents of twist.[177–179] When the twists of the β sheets are negligible, then the two sheets are complementary to one another and can associate to form a compact folding unit. But how do β sheets with different degrees of twist and residue compositions associate so that the central portion of each β sheet is close-packed? Chothia, using the same types of arguments described for α helices, showed that two β sheets having different degrees of twist, can associate when one β sheet is rotated with respect to the other so that the protruding amino-acid side chains do not overcrowd each other. A rotation of the axis of one sheet with respect to another optimizes the amount the two sheets are overlapped. When the two β sheets are rectangular in shape, but have different degrees of twist, then portions of the one sheet will not overlay the other, as shown in Figure 15.20(a). In real structures, the first and last strands are usually shortened relative to the middle strands, which makes double sheets appear elliptical.

A third type of supersecondary structure results from the packing of α helices on to β sheets.[180] Two parallel β strands form a sheet with a particular handedness. The linker between these β structures can either be an α helix, a coil, or another strand. By examining several protein

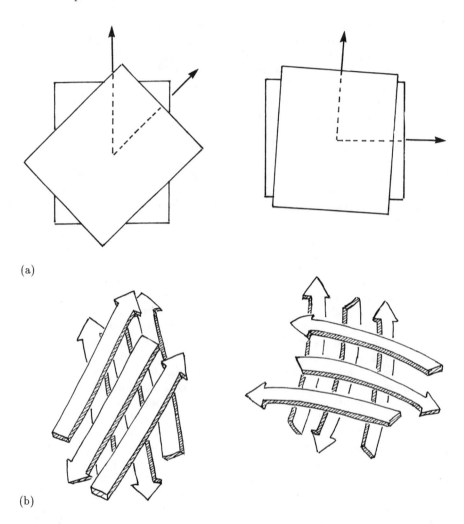

(a)

(b)

FIGURE 15.20. Packing of two independent β sheets. As for α helices, the optimal packing depends on the bulkiness of the side chains. In addition, since β sheets have a right-handed twist, this also affect the geometry of the packing. Two β sheets with the same twist pack well with parallel strands. If the twists are different, the β strands will not be parallel, but rotated with respect to each other. (a) Representation of β sheets by rectangles. (b) Representation of the packing of β sheets. In both (a) and (b) the left-hand diagram shows nearly aligned sheets, while the right-hand diagram shows nearly orthogonal sheets.

structures, it was observed that the connection between β sheets is always right handed. The most common type of connection between β strands is an α helix. This forms the third class of supersecondary structures referred to as the $\beta\alpha\beta$ unit. Analysis of these structures has shown that the amino-acid residues of the helix that interact with β sheets at positions that can be described as i, $i + 1$, $i + 4$, $i + 5$, and $i + 8$ have an angle of approximately 20° between the α helix and the β sheet.

15.6.2 Assembly from subunits

When proteins form a higher-ordered structure they associate in very specific ways. Identical subunits called monomers form higher-ordered structures called oligomers. By contrast, monomers that are not identical associate into protomers. All protomers are monomers but not all monomers are protomers. The dimeric enzyme alcohol dehydrogenase (ADH) is an oligomer consisting of two monomers or protomers.[181,182] Hemoglobin, however, is an oligomer consisting of two protomers $\alpha_1\beta_1$ and $\alpha_2\beta_2$, where each chain of the protomer α_1 and α_2 are monomers.[183–185] The structure of a dimer of the enzyme D-xylose isomerase is illustrated in Figure 15.22.

The interaction of subunits with other subunits produces the folded complete protein. Each subunit folds into apparently independent globular structures which then interact with other folded subunits. This quaternary structure of a protein is often vital to its function. Identical subunits are usually related by either cyclic, dihedral, or icosahedral point–group symmetry. Molecules related by cyclic symmetry can have n subunits that are arranged in a plane with n-fold symmetry. There are many examples of proteins that associate with cyclic symmetry, forming dimers and trimers. When the interaction surfaces are identical, a closed structure with a two-fold axis of symmetry results. On the other hand, the interaction surfaces may be different, in which case an infinite helical structure is produced. For example, a single site mutation in the hemoglobin molecule near the contact site results in a long helical structure which is known as sickle-cell hemoglobin.[186–188] Dihedral symmetry frequently is used to relate four molecules by using two perpendicular axes.

The most impressive assemblies are the spherical viruses with icosahedral symmetry. Sixty copies of a fundamental unit are arranged on the surface of a shell. These structures have 532 symmetry meaning that there are twofold, threefold, and fivefold axes of symmetry. Icosahedral symmetry is the most efficient arrangement for a closed shell and as such is analogous to the geodesic dome.

15.6.3 Virus assembly

Viral capsid structures are built with high efficiency and economy from a single fundamental unit and as such are analogous to geodesic domes.[189] There are five regular polyhedra, a tetrahedron, a cube, an octahedron, a dodecahedron and an icosahedron. These polyhedra are often referred to as **Platonic solids** because of their significance in Plato's natural

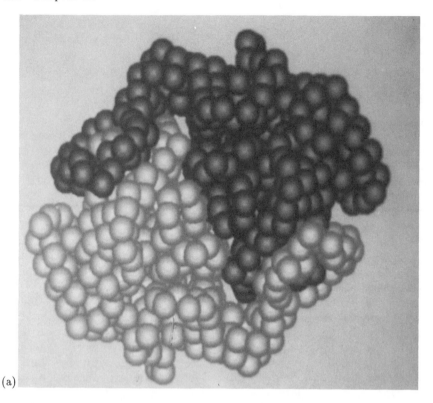

(a)

FIGURE 15.22. Assembly of subunits in proteins. Shown are two subunits of the enzyme D-xylose isomerase, forming a dimer by a "one-armed embrace." (a) Line drawing connecting C_α carbon atoms, and (b) computer-generated view of the molecule with atoms with van der Waals radii on each C_α carbon atom. No side chains are shown in either diagram. In (a) one monomer is illustrated with black spheres, and the other with white spheres, in order to differentiate between them. (Courtesy H. L. Carrell)

philosophy. Platonic figures are built from regular polygons and as such the polyhedra are also regular.[190] Polyhedra have faces that are triangles for tetrahedra and icosahedra, squares for cubes and regular pentagons for dodecahedra. The faces of the polyhedral structures are related to one another by a pure rotational symmetry. Crick and Watson[191] put forward the idea that the virus capsid must have rotational symmetry of either a tetrahedron (23), an octahedron (432), or an **icosahedron** (532) (see Figure 15.23) requiring 12, 24, or 60 subunits, respectively, all in identical environments.

Donald L. D. Caspar[192] showed that a particular virus, tomato bushy stunt virus (TBSV), has a capsid structure with icosahedral symmetry.

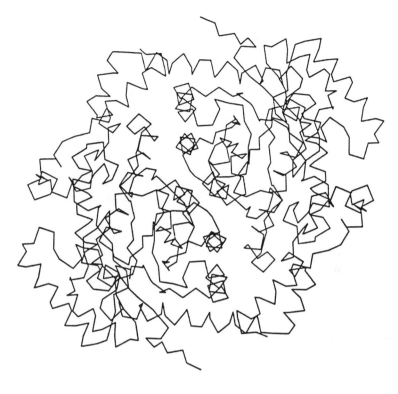

(b)

FIGURE 15.22 (cont'd). (b) Line drawing of a dimer of the enzyme D-xylose iso-merase.

The icosahedron has the lowest surface-to-volume ratio, most nearly approximating a sphere. A regular three-dimensional icosahedron can be generated from an extended planar hexagonal net of equilateral triangles. It is possible to put 60 equivalent objects on the surface of an icosahedron in such a way that each is identically situated and related to the others by a rotational operation, yet many virus structures have a capsid with more than 60 units. To account for the additional subunits, it is necessary to relax the requirement that each subunit have the same environment, a situation described[193] as "quasi-equivalence." Writing about the geodesic dome, Donald L. D. Caspar and Aaron Klug[193] point out that "In the omni-triangulated geodesic dome ⋯ the complete sphere would consist of 720 triangular units, but although they are actually of 12 different types, they are all very similar. The asymmetric unit of this radome consists of these 12 symmetrically distinct triangles, but the physical subunit may be considered as one "average triangle."" They then explain that "molecular structures are not built to conform to exact mathematical concepts but, rather, to satisfy the condition that the system be in a minimum

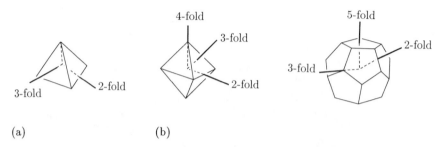

(a)   (b)

FIGURE 15.23. Platonic figures found in virus structures, and their symmetries (see Ref. 191). (a) Tetrahedron, (b) octahedron, and (c) icosahedron. Rotation axes are indicated.

energy configuration." The reader is urged to read Reference 193 for more details.

Summary

1. The molecules and ions in crystals are held together by ionic forces, van der Waals interactions, and dipole–(induced-) dipole interactions. Hydrogen bonds and metal–ligand interactions are prominent in the packing, which is such as to minimize the total energy.
2. π complexes and charge-transfer complexes generally form good crystals, and the relative orientations of the electron-donor and -acceptor molecules is generally that expected from molecular orbital theory.
3. Some molecules aggregate in crystals, generally by hydrogen bonding, to give crystal structures that contain channels and cavities that can accommodate foreign molecules. Examples are provided by the choleic acids, the gas hydrates, and the clathrates.
4. Polymorphism is the existence of more than one crystalline form of the same chemical substance. If there are only two forms, the phenomenon is dimorphism; if the materials are elements, it is allotropy; if the forms differ by solvent of crystallization, they are called pseudopolymorphs. Different polymorphs have different relative stabilities, but these may be varied by changing temperature, pressure, and other conditions.
5. Phase changes may be detected calorimetrically, and the differences of the phases are revealed by structure determinations.
6. Assembly in proteins involves packing of α helices and β sheets. Together they play important roles in protein folding. Assembly of subunits gives the active protein. In spherical viruses such assembly involves identical subunits, and the symmetry found in the resulting coat is of interest.

Glossary

Allotropy: Polymorphism of chemical elements.

Bond valence: The strength of an interaction between two atoms expressed in terms of electrostatics. It is used particularly for a description of metal coordination. In coordination, both shared electrons in a bond are provided by one of the two atoms.

Charge-transfer complex: A complex formed as a result of electron transfer from an electron donor to an electron acceptor. The charge transfer band lies at near-UV wavelength (290-300 nm) and is typically very intense. Such a band indicates that the first excited state has almost complete electron transfer.

Clathrate: An inclusion complex formed when molecules of one kind are trapped in the crystalline network formed by molecules of a different kind.

Choleic acid: A specific complex between deoxycholic acid, which crystallizes with channels throughout the structure, and various organic molecules, such as hydrocarbons or fatty acids, that can fit in these channels.

Cooperative hydrogen bonding: An interaction between neighboring hydrogen bonds such that the energy to form such a bond is less than the average of the energies of the different hydrogen bonds individually.

Coordination number of a cation: The number of ligands around a metal ion in a complex.

Coordination polyhedron: The polyhedron obtained by joining adjacent ligands coordinated to a metal cation.

Defect: A feature of a structure that spoils its regularity.

Delocalized electrons: Electrons that are not specific to a particular chemical bond but have a probability of existing anywhere throughout an area specified as having delocalized electrons.

Dipole moment: The magnitude of either charge in a dipole (consisting of separated charges) multiplied by the distance between the two charges.

F center: A crystal defect in which electrons are trapped in vacant holes, resulting in a colored crystal.

Frenkel defect: A crystal defect in which an interstitial atom is found in a new site, leaving vacant the site from which it has moved.

Hydrophilicity: The tendency of a group of atoms to become solvated by water. An affinity for water (Greek *hydros* = water, *philos* = love)

Hydrophobicity: The tendency of a group of atoms to resist becoming solvated by water. (Greek hydros=water, phobos=fear)

Icosahedron: A symmetrical polyhedron with 12 vertices and 20 faces, each an equilateral triangle.

Lewis acid: A substance that can accept an electron pair to form a covalent bond (an electron-pair acceptor).

Lewis base: A substance that can supply an electron pair to form a covalent bond (an electron-pair donor).

Liquid crystal: A phase that has the mobility of a liquid and some of the internal order of a crystal. It is also called a mesophase.

London dispersion forces: Dispersion forces that are attractive and due to induced-dipole–induced-dipole interactions.

Metastable: A condition of limited stability that generally eventually changes to one that is more stable.

Modulated structure: A regular structure that is modified by a periodic or partially periodic perturbation. This is revealed by additional haloes or spots around Bragg reflections in the photograph of the diffraction pattern.

Phase transformation: The change from one (crystalline) phase to another.

Platonic solid: Any one of five regular three-dimensional solids – the tetrahedron, the cube, the octahedron, the dodecahedron, and the icosahedron.

Polar molecule: One with a permanent dipole moment, indicating a separation of charge across the molecule.

Polytypism: The existence of different forms of a crystal that differ in the manner that layers of atoms are stacked.

Pseudopolymorphs: Polymorphs that differ by solvent of crystallization.

Radii, ionic or atomic: The radius of a sphere about an atomic nucleus within which the sphere around any other atom or ion may not penetrate.

Schottky defect: A crystal defect in which an atom has migrated to the surface of the crystal leaving a vacant site.

Supersecondary structure: A pattern of protein structure that is not an entire domain but is at a higher level than secondary structure. Examples are β barrels and Greek-key structures.

Twinning: A nonparallel intergrowth of separate crystals related by a symmetry not possessed by the substance.

van der Waals interactions: Forces, ranging from weak to strong, between atoms and molecules as a result of various types of dipole–dipole interactions. The components are a dispersion effect (induced dipole–induced dipole), an induction effect (dipole–induced dipole), and an orientation effect (dipole–dipole interactions).

Voronoï polyhedra: The Voronoï polyhedron is a geometrical construction that allows an objective assignment of volume to each atom in a structure. The volume associated with an atom can be calculated by constructing a set of planes normal to the vectors between the atom of interest and its nearest neighbors. The volume inside the polyhedron is a good approximation to the volume associated with the particular atom.

References

1. Bragg, W. L. The structure of some crystals as indicated by their diffraction of X rays. *Proc. Roy. Soc. (London)* **A89**, 248–277 (1913).
2. Etter, M. C. Encoding and decoding hydrogen-bond patterns of organic crystals. *Acc. Chem. Res.* **23**, 120–126 (1990).
3. Jeffrey, G. A. The nanometer world of hydrogen bonds. In: *Patterson and Pattersons. Fifty Years of the Patterson Function.* (**Eds.**, Glusker, J. P., Patterson, B. K., and Rossi, M.) Ch. 10, pp. 193–221. International Union of Crystallography/Oxford University Press: New York (1987).
4. Desiraju, G. R. Hydration in organic crystals: prediction from molecular structure. *J. Chem. Soc., Chem. Commun.* 426–428 (1991).
5. Rigby, M., Smith, E. B., Wakeham, W. A., and Maitland, G. C. *The Forces between Molecules.* Clarendon Press: Oxford (1986).
6. Huheey, J. E. *Inorganic Chemistry.* 3rd edn. Harper and Row: New York (1983).

7. van der Waals, J. D. *Die Kontinuität des gasförmigen und flüssigen Zustandes.* [The continuity of the gaseous and liquid states.] J. A. Barth: Leipzig (1881).

8. London, F. Über einige Eigenschaften und Anwendungen der Molekularkräfte. [Concerning certain properties and effects of molecular forces.] *Z. Phys. Chem.* **Abt. B**, 222–251 (1930).

9. London, F. Zur Theorie und Systematik der Molekularkräfte. [On the theory and systematics of molecular forces.] *Z. Physik* **63**, 245–279 (1930).

10. Bragg, W. L., and Claringbull, G. F. *Crystal Structures of Minerals. The Crystalline State - Vol. IV.* G. Bell: London (1965).

11. Allen, L. C. A simple model of hydrogen bonding. *J. Amer. Chem. Soc.* **97**, 6921–6940 (1975).

12. Moore, T. S., and Winmill, T. F. State of amines in aqueous solution. *J. Chem. Soc.* **101**, 1635–1676 (1912).

13. Pfeiffer, P., Fischer, P., Kuntner, J., Monti, P., and Pros, Z. Zur Theorie der Farblacke, II. [Theory of dye lakes.] *Justus Liebigs Annalen der Chemie* **398**, 137–196 (1913).

14. Latimer, W. M., and Rodebush, W. H. Polarity and ionization from the standpoint of the Lewis theory of valence. *J. Amer. Chem. Soc.* **42**, 1419–1433 (1920).

15. Pimentel, G. C., and McClellan, A. L. *The Hydrogen Bond.* Freeman: San Francisco (1960).

16. Taylor, R., Kennard, O., and Versichel, W. The geometry of the N–H $\cdots$ O=C bond. 3. Hydrogen bond distances and angles. *Acta Cryst.* **B40**, 280–288 (1984).

17. Görbitz, C. H. Hydrogen-bond distances and angles in the structures of amino acids and peptides. *Acta Cryst.* **B45**, 390–395 (1989).

18. Desiraju, G. R. *Crystal Engineering. The Design of Organic Solids.* Elsevier: Amsterdam (1989).

19. Jeffrey, G. A., and Saenger, W. *Hydrogen Bonding in Biological Structures.* Springer-Verlag: Berlin, New York (1991).

20. Parthasarathy, R. Crystal structure of glycylglycine hydrochloride. *Acta Cryst.* **B25**, 509–518 (1969).

21. Steiner, T., Mason, S. A., and Saenger, W. Cooperative O–H $\cdots$ O hydrogen bonds on β-cyclodextrin-ethanol-octahydrate at 15 K: a neutron diffraction study. *J. Amer. Chem. Soc.* **112**, 6184–6190 (1990).

22. Etter, M. C., MacDonald, J. C., and Bernstein, J. Graph-set analysis of hydrogen-bond patterns in organic crystals. *Acta Cryst.* **B46**, 256–262 (1990).

23. Sutor, D. J. Evidence for the existence of C–H $\cdots$ O hydrogen bonds in crystals. *J. Chem. Soc.*, 1105–1110 (1963).

24. Donohue, J. Selected topics in hydrogen bonding. In: *Structural Chemistry and Molecular Biology.* (**Eds.**, Rich, A., and Davidson, N.) pp. 443–465. W. H. Freeman: San Francisco, London (1968).

25. Taylor, R., and Kennard, O. Crystallographic evidence for the existence of C–H $\cdots$ O, C–H $\cdots$ N and C–H $\cdots$ Cl hydrogen bonds. *J. Amer. Chem. Soc.* **104**, 5063–5070 (1982).

26. Desiraju, G. R. The C–H ⋯ O hydrogen bond in crystals? What is it? *Acc. Chem. Res.* **24**, 290–296 (1991).

27. Werner, A. Beitrag zur Konstitution anorganischer Verbindungen. [Contribution on the constitution of inorganic compounds.] *Z. Anorg. Allgemeine Chemie* **3**, 267–342 (1893). **English translation** in: *Classics in Coordination Chemistry, Part 1: The Selected Papers of Alfred Werner.* (**Ed.**, Kauffman, G. B.) pp. 5–88. Dover: New York (1968).

28. Wyckoff, R. W. G., and Posnjak, E. The crystal structure of ammonium chloroplatinate. *J. Amer. Chem. Soc.* **43**, 2292–2309 (1921).

29. Dickinson, R. G. The crystal structures of potassium chloroplatinite and of potassium and ammonium chloropalladites. *J. Amer. Chem. Soc.* **44**, 2404–2411 (1922).

30. Pauling, L., and Huggins, M. L. Covalent radii of atoms and interatomic distances in crystals containing electron-pair bonds. *Z. Krist.* **87**, 205–238 (1934).

31. Pauling, L. *The Nature of the Chemical Bond.* Cornell University Press: Ithaca, New York. 1st edn., 1939; 2nd edn., 1940; 3rd edn., 1960.

32. Pauling, L. The sizes of ions and the structure of ionic crystals. *J. Amer. Chem. Soc.* **49**, 765–790 (1927).

33. Barlow, W., and Pope, W. J. The relation between the crystalline form and chemical constitution of simple inorganic substances. *Trans. Chem. Soc. (London)* **91**, 1150–1214 (1907).

34. Wasastjerna, J. A. On the radii of ions. *Societas Scientiarum Fennica Commentationes Physico-Mathematicae I.* **38**, 1–25 (1923).

35. Goldschmidt, V. M. (1926). Geochemische Verteilungsgesetze, VII: Die Gesetze der Krystallochemie (nach Untersuchungen gemeinsam mit T. Barth, G. Lunde, and W. H. Zachariasen). [Geochemical distribution principles. The rules of crystal chemistry (studies with others).] *Shrifter Norske Videnskaps-Akademi i Oslo Arbok (Mat. Natl. Kl.)* **2**, 90–117 (1926).

36. Goldschmidt, V. M. Crystal structure and chemical constitution. *Trans. Faraday Soc.* **25**, 253–283 (1929).

37. Pauling, L. The principles determining the structure of complex ionic crystals. *J. Amer. Chem. Soc.* **51**, 1010–1026 (1929).

38. Zachariasen, W. H. A set of empirical crystal radii for ions with inert gas configuration. *Z. Krist.* **80**, 137–153 (1931).

39. Baur, W. H. Bond length variation and distorted coordination polyhedra in inorganic crystals. *Trans. Amer. Cryst. Assn.* **6**, 129–155 (1970).

40. Brown, I. D., and Shannon, R. D. Empirical bond strength–bond-length curves for oxides. *Acta Cryst.* **A29**, 266–282 (1973).

41. Orgel, L. G. *An Introduction to Transition-metal Chemistry. Ligand-field Theory.* 2nd edn. Methuen: London/ Wiley: New York (1966).

42. Brown, I. D. What factors determine cation coordination numbers? *Acta Cryst.* **B44**, 545–553 (1988).

43. Pauling, L. The coordination theory of the structure of ionic crystals. In: *Probleme der Modernen Physik, Arnold Sommerfeld Festschrift.* pp. 11–17. Verlag Hirzel: Leipzig (1928).

44. Rahlfs, O., and Fischer, W. Thermal properties of halides. VI. Vapor pressures and vapor densities of beryllium and zirconium halides. *Z. Anorg. Allgem. Chem.* **211**, 349–367 (1933).

45. Rundle, R. E., and Lewis, P. H. Electron deficient compounds. VI. The structure of beryllium chloride. *J. Chem. Phys.* **20**, 132–134 (1952).

46. Wells, A. F. *Structural Inorganic Chemistryu.* Clarendon Press: Oxford (1962).

47. Brown, I. D. Bond valences – a simple structural model for inorganic chemistry. *Chem. Soc. Rev.* **7**, 359–376 (1978).

48. Brown, I. D. A structural model for Lewis acids and bases. An analysis of the structural chemistry of acetate and trifluoroacetate ions. *J. Chem. Soc., Dalton,* 1118–1123 (1980).

49. Brese, N. E., and O'Keefe, M. Bond-valence parameters for solids. *Acta Cryst.* **B47**, 192–197 (1991)

50. Johnson, C. K. (1965). X-ray crystal analysis of the substrates of aconitase. V. Magnesium citrate decahydrate $[Mg(H_2O)_6][MgC_6H_5O_7(H_2O)]_2.2H_2O$. *Acta Cryst.* **18**, 1004–1018 (1965).

51. Rossi, M., Rickles, L. F., and Glusker, J. P. Trilithium citrate pentahydrate, $C_6H_5O_7Li_3·5H_2O$. *Acta Cryst.* **C39**, 987–990 (1983).

52. Kitaigorodskii, A. I. *Organic Chemical Crystallography.* (**English translation:** Consultants Bureau: New York) (Russian edition, 1955. Academy of Sciences Press: Moscow) (1961).

53. Kitaigorodsky, A. I. *Molecular Crystals and Molecules.* Academic Press: New York, London (1973).

54. Dunitz, J. D. *X-ray Analysis and the Structure of Organic Molecules.* Cornell University Press: London, Ithaca, NY (1979).

55. Haisa, M. The origin of the crystallographic pedigree. *Acta Cryst.* **A38**, 443–453 (1982).

56. Sarma, J. A. R. P., and Desiraju, G. R. The chloro-substituent as a steering group: a comparative study of non-bonded interactions and hydrogen bonding in crystalline chloroaromatics. *Chem. Phys. Lett.* **117**, 160–164 (1985).

57. Sarma, J. A. R. P., and Desiraju, G. R. The role of Cl $\cdots$ Cl and C–H $\cdots$ O interactions in the crystal engineering of 4 Å-short axis structures. *Acc. Chem. Res.* **19**, 222–228 (1986).

58. Kashino, S., Zacharias, D. E., Prout, C. K., Carrell, H. L., Glusker, J. P., Hecht, S. S., and Harvey, R. G. Structure of 5-methylchrysene, $C_{19}H_{14}$. *Acta Cryst.* **C40**, 536–540 (1984).

59. Taylor, M. R., Gabe, E. J., Glusker, J. P., Minkin, J. A., and Patterson, A. L. The crystal structures of compounds with antitumor activity. 2-Keto-3-ethoxybutyraldehyde bis(thiosemicarbazone) and its cupric complex. *J. Amer. Chem. Soc.* **88**, 1845–1846 (1966).

60. Gabe, E. J., Taylor, M. R., Glusker, J. P., Minkin, J. A., and Patterson, A. L. The crystal structure of 2-keto-3-ethoxybutyraldehyde-bis(thiosemicarbazone). *Acta Cryst.* **B25**, 1620–1631 (1969).

61. Glusker, J. P., Carrell, H. L., Berman, H. M., Gallen, B., and Peck, R. M. Alkylation of a tripeptide by a carcinogen: the crystal structure of sarcosylglycylglycine, 9-methyl-10-chloromethylanthracene, and their reaction product. *J. Amer. Chem. Soc.* **99**, 595–601 (1977).

62. Stezowski, J. J., Pöhlmann, H. W., Haslinger, E., Kalchauser, H., Schmidt, U., and Pozolli, B. The conformation of cyclo[L-Pro–L-Leu–L-Val–(gly)Thz–(gly)Thz], a dolastatin 3 analog, in the crystalline and solution states. *Tetrahedron* **43**, 3923–3930 (1987).

63. Lee, B. K., and Richards, F. M. The interpretation of protein structures: estimation of static accessibility. *J. Molec. Biol.* **55**, 379–400 (1971).

64. Connolly, M. L. Analytical molecular surface calculation. *J. Appl. Cryst.* **16**, 548–558 (1983).

65. Robertson, J. M. The measurement of bond lengths in conjugated molecules of carbon centres. *Proc. Roy. Soc. (London)* **A207**, 101–110 (1951).

66. Kitaigorodskii, A. I. Interaction curve of non-bonded carbon and hydrogen atoms and its applications. *Tetrahedron* **14**, 230–236 (1961).

67. Williams, D. E. Computer calculations of the structure and physical properties of crystalline hydrocarbons. *Trans. Amer. Cryst. Assn.* **6**, 21–33 (1970).

68. Sletten, E., Sletten, J., and Jensen, L. H. The crystal and molecular structure of 6-mercaptopurine monohydrate. *Acta Cryst.* **B25**, 1330–1338 (1969).

69. Brown, G. M. The crystal and molecular structure of 6-mercaptopurine monohydrate. A second, independent X-ray diffraction determination. *Acta Cryst.* **B25**, 1338–1353 (1969).

70. Weil-Malherbe, H. The solubilization of polycyclic aromatic hydrocarbons by purines. *Biochem. J.* **40**, 351–363 (1946).

71. Damiani, A., De Sanctis, P., Giglio, E., Liquori, A. M., Puliti, R., and Ripamonti, A. The crystal structure of the 1:1 molecular complex between 1,3,7,9-tetramethyluric acid and pyrene. *Acta Cryst.* **19**, 340–348 (1965).

72. Damiani, A., Giglio, E., and Liquori, A. M. The crystal structure of the 2:1 molecular complex between 1,3,7,9-tetramethyluric acid and 3,4-benzpyrene. *Acta Cryst.* **23**, 675–681 (1967).

73. Herbstein, F. H. Crystalline π-molecular compounds: chemistry, spectroscopy, and crystallography. In: *Perspectives in Structural Chemistry.* (**Eds.**, Dunitz, J.D., and Ibers, J.A.) pp. 166–395. Wiley: New York, London, Sydney, Toronto (1971).

74. Mayoh, B., and Prout, C. K. Molecular complexes. Part 13.– Influence of charge transfer interactions on the structure of π–π^* electron donor-acceptor molecular complexes. *J. Chem. Soc., Faraday Trans.* 1072–1082 (1972).

75. Zacharias, D. E., Prout, K., Myers, C. B., and Glusker, J. P. (1990) Structure and molecular orbital studies of potentially mutagenic methylchrysenes and their π–π^* electron donor-acceptor complexes. *Acta Cryst.* **B47**, 97–107 (1991).

76. Munnoch, P. J., and Wright, J. D. Crystal structure of the 1:1 molecular complex of chrysene and 7,7,8,8-tetracyanoquinodimethane. *J. Chem. Soc., Perkin Trans. II*, 1397–1400 (1974).

77. Munnoch, P. J., and Wright, J. D. Crystal structure of the 1:1 molecular complex of chrysene and tetrafluoro-*p*-benzoquinone (fluoranil). *J. Chem. Soc., Perkin Trans. II*, 1071–1074 (1975).

78. Foster, R., Iball, J., Scrimgeour, S. N., and Williams, B. C. Crystal structure and nuclear magnetic resonance spectra of a 1:1 complex of benz[*a*]anthracene and pyromellitic dianhydride (benzene-1,2,4,5-tetracarboxylic dianhydride). *J. Chem. Soc., Perkin Trans. II*, 682–685 (1976).

79. Mulliken, R. Structures of complexes formed by halogen molecules with aromatic and with oxygenated solvents. *J. Amer. Chem. Soc.* **72**, 600–608 (1950).

80. Mulliken, R. Molecular compounds and their spectra. III. The interaction of electron donors and acceptors. *J. Phys. Chem.* **56**, 801–822 (1952).

81. Wieland, H., and Sorge, H. Untersuchungen über die Gallensäuren. II. Mitteilung. Zur Kenntnis der Choleinsäure. [Research on bile acids. On the identification of choleic acid.] *Hoppe-Seyler's Z. Physiol. Chem.* **96**, 1–27 (1916).

82. Giacomello, G. von, and Kratky, O. Röntgenographische Studien an Cholansäurenäthylestern. [X-ray studies of choleic acid ethyl esters.] *Z. Krist.* **A95**, 459–464 (1936).

83. Herndon, W. C. The structure of choleic acids. *J. Chem. Educ.* **44**, 724–728 (1967).

84. De Sanctis, S. C., Giglio, E., Pavel, V., and Quagliata, C. A study of the crystal packing of the 2:1 and 3:1 canal complexes between deoxycholic acid and *p*-diiodobenzene and phenanthrene. *Acta Cryst.* **B28**, 3656–3661 (1972).

85. Jogun, K. H., and Stezowski, J. J. Metastable crystals of β-cyclodextrin complexes and the membrane diffusion model. *Nature (London)* **278**, 667–668 (1979).

86. Saenger, W. Circular hydrogen bonds. *Nature (London)* **279**, 343–344 (1979).

87. Lindner, K., and Saenger, W. OH clusters with homodromic circular arrangement of hydrogen bonds. *Angew. Chem., Int. Ed. Engl.* **19**, 398–399 (1980).

88. Betzel, C., Saenger, W., Hingerty, B. E., and Brown, G. D. Circular and flip-flop hydrogen bonding in β-cyclodextrin undecahydrate: a neutron diffraction study. *J. Amer. Chem. Soc.* **106**, 7545–7557 (1984).

89. Zabel, V., Saenger, W., and Mason, S. A. Neutron diffraction study of the hydrogen bonding in β-cyclodextrin undecahydrate at 120 K: from dynamic flip-flops to static homodromic chains. *J. Amer. Chem. Soc.* **108**, 3664–3673 (1986).

90. Bender, M. L., asnd Komiyama, M. *Cyclodextrin Chemistry.* Springer-Verlag (1978).

91. Powell, H. M., and Riesz, P. β-quinol: an example of the firm union of molecules without the formation of chemical bonds between them. *Nature (London)* **161**, 52–53 (1948).

92. Mylius, F. Zur Kenntniss des Hydrichinons und der Ameisensäure. [On the identification of hydroquinone and formic acid.] *Ber. Deutsch. Chem. Gesell.* **19**, 999–1009 (1886).

93. Palin, D. E., and Powell, H. M. Structure of molecular compounds. III. Crystal structure of addition complexes of hydroquinone with certain volatile compounds. *J. Chem. Soc.* 208–221 (1947).

94. Davy, H. On a combination of oxymuriatic gas and oxygene gas. *Phil. Trans. Roy. Soc.* (*London*) **101**, 155–162 (1811).

95. Faraday, M. On hydrate of chlorine. *Quant. J. Sci. Let. Arts* **15**, 71–74 (1823).

96. Pauling, L., and Marsh, R. E. The structure of chlorine hydrate. *Proc. Natl. Acad. Sci. USA* **38**, 112–118 (1952).

97. Tsoucaris, G. Clathrates. In: *Organic Solid State Chemistry.* (**Ed.**, Desiraju, G. R.) Ch. 7, pp. 207–270. Elsevier: Amsterdam, Oxford, New York, Tokyo (1987).

98. Jeffrey, G. A. Water structure in organic hydrates. *Acc. Chem. Res.* **2**, 344–352 (1969).

99. Evdokimova, V. V., and Vereschagin, L. F. Polymorphic transitions in NaCl under pressure. *J. Exptl. Theoret. Phys.* (*USSR*) **43**, 1208–1212 (1962).

100. Eysel, W., Geyer, A., and Wies, S. Polymorphic transision of CsCl. (Abstract PS–07.01.09) *Acta Cryst.* **A46** (Supplement) p. C-242 (1990).

101. Mallard, E. Explication des phénomènes optiques anomaux qui présentent un grand nombre des substances cristalisées. [Explanation of the anomalous optical phenomena shown by many crystalline materials.] *Ann. Mines* **10**, 60–196 (1876).

102. Thernard, L. J., and Biot, J. B. Mémoire sur l'analyse comparée de l'arragonite et du carbonate de chaux rhomboidal. [On the comparative analyses of aragonite and rhombohedral calcium carbonate.] *Mem. Phys. II. Soc. d'Arcueil* **2**, 176–206 (1809).

103. Mitscherlich, E. Sur la relation qui existe entre la forme cristalline et les proportions chimiques. I. Mémoire sur les arseniates et les phosphates. [On the relationship between crystalline form and chemical composition. I. Note on arsenates and phosphates.] *Ann. Chim. Phys.* **19**, 350–419 (1822).

104. Mitscherlich, E. Sur la rapport qui existe entre les proportions chimiques et la forme cristalline. III. Mémoire sur les corps qui affectent deux formes cristallines différentes. [On the relationship between chemical composition and crystalline form. III. On the substances that affect two different crystalline forms.] *Ann. Chim. Phys.* **24**, 264–271 (1823).

105. MacDonald, G. J. F. Experimental determination of calcite-aragonite equilibrium relations at elevated temperatures and pressures. *Amer. Mineralogist* **41**, 744–756 (1956).

106. Jamieson, J. C. Introductory studies of high-pressure polymorphism to 24,000 bars by X-ray diffraction with some comments on calcite II. *J. Geol.* **65**, 334–343 (1957).

107. Frankenheim, M. L. Über die Isomerie. [On isomerism.] *Erdm. J. Prakt. Chem.* **16**, 1–14 (1839).

108. Gernez, D. Sur la température de transformation des deux variétés quadratique et orthorhombique de l'iodure mercurique. [On the temperature of transformation of two forms, tetragonal and orthorhombic mercuric iodide.] *Comptes Rendus, Acad. Sci.* (*Paris*) **129**, 1234–1236 (1899).

109. Tammann, G. Über eine farblose Form des Quecksilberjodides. [A colorless modification of mercuric iodide.] *Nachrichten von der Königlichen Gesellschaft der Wissenschaften zu Göttingen. Mathematisch-physikalische Klasse* **2**, 292–293 (1916)

110. Jeffrey, G. A., and Vlasse, M. On the crystal structures of the red, yellow and orange forms of mercuric iodide. *Inorg. Chem.* **6**, 396–399 (1967).

111. Bergerhoff, G., and Brown, I. D. Inorganic crystal structure database. In: *Crystallographic Databases. Information Content, Software Applications, Scientific Applications.* Section 2.2. pp. 77–95. International Union of Crystallography: Bonn, Cambridge, Chester (1987).

112. Byrn, S. R. *Solid-state Chemistry of Drugs.* Academic Press: New York (1982).

113. Stezowski, J. J., Stigler, R-D., and Karl, N. Crystal structure and charge transfer energies of complexes of the donor biphenylene with the acceptors TCNB and PMDA. *J. Chem. Phys.* **84**, 5162–5170 (1986).

114. Hughes, L. J., Stezowski, J. J., and Hughes, R. E. Chemical-structural properties of tetracycline derivatives. 7. Evidence for the coexistence of the zwitterionic and nonionized forms of the free base in solution. *J. Amer. Chem. Soc.* **101**, 7655–7657 (1979).

115. Pauling, L. The structure and entropy of ice and of other crystals with some randomness of atomic arrangement. *J. Amer. Chem. Soc.* **57**, 2680–2684 (1935).

116. Nagle, J. F. Lattice statistics of hydrogen bonded crystals. I. The residual entropy of ice. *J. Math. Phys.* **7**, 1484–1491 (1966).

117. Peterson, S. W., and Levy, H. A. A single-crystal neutron diffraction study of heavy ice. *Acta Cryst.* **10**, 70–76 (1957).

118. Savage, H. F. J. Water structure in crystalline solids: ices to proteins. In: *Water Science Reviews.* (**Ed.**, Franks, F.) Ch. 2, pp. 67–148. Cambridge University Press: Cambridge (1986).

119. Savage, H. F. J., Finney, J. L. Repulsive regulations of water structure in ices and crystalline hydrates. *Nature (London)* **322**, 717–720 (1986.)

120. Kamb, B. Crystallography of water structures. In: *Crystallography in North America.* (**Eds.**, McLachlan, D., Jr., and Glusker, J. P.) Section F. Applications to various sciences. Chapter 3. pp. 336–342. American Crystallographic Association: New York (1983).

121. Bernal, J. D., and Fowler, R. H. A theory of water and ionic solution, with particular reference to hydrogen and hydroxyl ions. *J. Chem. Phys.* **1**, 515–548 (1933).

122. Patterson, A. L., and Groshens, B. P. A solid state transformation in tetraacetyl-D-ribofuranose. *Nature (London)* **173**, 398–399 (1954).

123. Czugler, M., Kalman, A., Kovacs, J., and Pinter, I. Structure of the unstable monoclinic 1,2,3,5-tetra-*O*-acetyl-β-D-ribofuranose. *Acta Cryst.* **B37**, 172–177 (1981).

124. James, V. J., and Stevens, J. D. 1,2,3,5-Tetra-*O*-acetyl-β-D-ribofuranose, $C_{13}H_{18}O_9$. *Cryst. Struct. Commun.* **2**, 609–612 (1973).

125. Poppleton, B. J. 1,2,3,5-Tetra-*O*-acetyl-β-D-ribofuranose. *Acta Cryst.* **B32**, 2702–2705 (1976).

126. Woodward, G. D., and McCrone, W. C. Unusual crystallization behaviour. *J. Appl. Cryst.* **8**, 342 (1975).

127. Bernstein, J. Conformational Polymerism. In: *Organic Solid State Chemistry.* (**Ed.**, Desiraju, G. R.) Ch. 13. pp. 471–518. Elsevier: Amsterdam, Oxford, New York, Tokyo (1987).

128. Jacewicz, V. W. and Nayler, J. C. Can metastable forms "disappear"? *J. Appl. Cryst.* **12**, 396–397 (1979).

129. Bar, I., and Bernstein, J. Molecular conformation and electronic structure. VI. The structure of *p*-methyl-*N*-(*p*-methylbenzilidene)aniline (Form I). *Acta Cryst.* **B38**, 121–125 (1982).

130. Buerger, M. J. Crystal-structure aspects of phase transformations. *Trans. Amer. Cryst. Assn.* **7**, 1–23 (1971).

131. Busing, W. R. Modeling the phase change in crystalline biphenyl by using a temperature-dependent potential. *Acta Cryst.* **A39**, 340–347 (1983).

132. Rao, C. N. R. Phase transitions and the chemistry of solids. *Acc. Chem. Res.* **17**, 83–89 (1984).

133. Brock, C. P., and Minton, R. P. Systematic effects of crystal-packing forces: biphenyl fragments with H atoms in all four ortho positions. *J. Amer. Chem. Soc.* **111**, 4586-4593 (1989).

134. Almenningen, A., Bastiansen, O., Fernholdt, L., Cyvin, B. N., Cyvin, S. J., and Samdal, S. Structure and barrier of internal rotation of biphenyl derivatives in the gaseous state. Part I. The molecular structure and normal coordinate analysis of normal biphenyl and perdeuterated biphenyl. *J. Molec. Struct.* **128**, 59–76 (1985).

135. Reinitzer, F. Beiträge zur Kenntniss des Cholesterins. [Contribution on the identification of cholesterol.] *Sitzgsber. d. Akad. d. Wissensch.* **94**, 420–441 (1888)

136. Lehmann, O. Über fliessende Krystalle. [On liquid crystals.] *Z. Phys. Chem.* **4**, 462–473 (1889).

137. Collings, P. J. *Liquid Crystals. Nature's Delicate Phase of Matter.* Princeton University Press: Princeton, NJ (1990).

138. Vorländer, D. Einfluss der molekularen Gestalt auf den krystallinisch-flüssigen Zustand. [Influence of molecular form on the liquid crystalline state.] *Ber. Deutsch. Chem. Gesell.* **40**, 1970–1972 (1908).

139. Friedel, G. Les états mésmorphes de la matiére. [The mesomorphic states of matter.] *Annal. Phys.* **18**, 273–474 (1922).

140. de Vries, A. Liquid crystals. In: *Crystallography in North America.* (**Eds.**, McLachlan, D., Jr., and Glusker, J. P.) Section F. Applications to various sciences. Ch. 2. pp. 325–335. American Crystallographic Association: New York (1983).

141. Bryan, R. F. Crystal structure and liquid-crystallinity. In: *Proceedings. Pre-Congress Symposium on Organic Crystal Chemistry. Poznań – Dymaczewo, 30 July - 2 August 1978.* (**Eds.**, Kałuski, Z., and Kosturkiewicz, Z.) pp. 105–154. Adam Mickiewicz University: Poznań (1978).

142. Fischer, E., and Helferich, B. Über neue synthetische Glucoside. [New synthetic glycosides.] *Justus Liebig's Annalen der Chemie.* **383**, 68–91 (1911).

143. Jeffrey, G. A. Carbohydrate liquid crystals. *Acc. Chem. Res.* **19**, 168–173 (1986).

144. Craven, B. M., and DeTitta, G. T. Cholesteryl myristate: structure of the crystalline solid and mesophases. *J. Chem. Soc., Perkin Trans. II*, 814–822 (1976).

145. Templer, R., and Attard, G. The world of liquid crystals. The fourth state of matter. *New Scientist* **1767** 25–29 (1991).

146. Gibbons, W. M., Shannon, P. J., Sun, S-T., and Swetlin, B. J. Surface-mediated alignment of nematic crystals with polarized light. *Nature (London)* **353**, 49–50 (1991).

147. Backhaus, K. O., Grell, K. O., Grell, H., and Fichtner, K. Database of OD (order-disorder) structures. In: *Crystallographic Databases. Information Content. Software Applications. Scientific Applications.* Section 3.4, pp. 178–181. Published by the Data Commission of the International Union of Crystallography: Bonn, Cambridge, Chester (1987).

148. Dornberger-Schiff, K. *Grundzüge einer Theorie der OD-Strukturen aus Schichten.* [Fundamentals of a theory of order-disorder layer structures.] Abh. d. Deutsch Akademie der Wiss. zu Berlin: Akademie-Verlag (1964).

149. Dornberger-Schiff, K. *Lehrgang über OD-Strukturen.* [Course of instruction on order-disorder structures.] Akademie-Verlag: Berlin (1966).

150. Dornberger-Schiff, K. On order-disorder structures (OD-structures). *Acta Cryst.* **9**, 593–601 (1956).

151. Verma, A. R., and Krishna, P. *Polymorphism and Polytypism in Crystals.* J. Wiley: New York (1966).

152. Wagner, C., and Schottky, W. Theorie der geordneten Mischphasen. [Theory of ordered mixed phases.] *Z. Phys. Chem.* **B11**, 163–210 (1930).

153. Schottky, W. Über den Mechanismus der Ionenbewegung in festen Elektrolyten. [The mechanism of ionic motion in solid electrolytes.] *Z. Phys. Chem.* **B29**, 335–355 (1935).

154. West, A. R. *Solid State Chemistry and its Applications.* Wiley: Chichester (1987).

155. Frenkel, Y. I. Über die Wärmebewegung in festen and flüssigen Körpern. [Thermal agitation in solids and liquids.] *Z. Physik* **35**, 652–669 (1926).

156. Rees, A. L. G. *Chemistry of the Defect Solid State.* Methuen: London (1954).

157. Cowley, J. M., Cohen, J. B., Salamon, M. B., and Wuensch, B. J. (**Eds.**) *Modulated Structures – 1979. AIP Conference Proceedings.* **53**. American Institute of Physics: New York (1979).

158. Shechtman, D., Blech, I., Gratias, D., and Cahn, J. W. Metallic phase with long-range orientational order and no translational symmetry. *Phys. Rev. Lett.* **53**, 1951–1953 (1984).

159. Pauling, L. Apparent icosahedral symmetry is due to directed multiple twinning of cubic crystals. *Nature (London)* **317**, 512–514 (1985).

160. Cahn, J. W., and Gratias, D. A structural determination of the Al–Mn icosahedral phase. *Journal de Physique (Paris)* **47**, C3-415–C3-424 (1986).

161. Stephens, P. W., and Goldman, A. I. The structure of quasicrystals. *Sci. Amer.* **264(4)** 44–53 (1991).

162. Nelson, D. R. Quasicrystals. *Scientific American* **255 (2)** 43–51 (1986).

163. Dubost, B., Lang, J-M., Tanaka, M., Saintfort, P., and Audier, M. Large Al-CuLi single quasicrystals with triacontahedral solidification morphology. *Nature (London)* **324**, 48–50 (1986).

164. Pierrot, M. (**Ed.**) *Structure and Properties of Molecular Crystals.* Elsevier: Amsterdam, Oxford, New York, Tokyo (1990).

165. Voet, D., and Voet, J. G. *Biochemistry.* John Wiley: New York (1990).

166. Levitt, M., and Chothia, C. Structural patterns in globular proteins. *Nature (London)* **261**, 552–558 (1976).

167. Jaenicke, R. Protein folding: local structures, domains, subunits, and assemblies. *Biochemistry* **30**, 3147–3161 (1991).

168. Pakula, A. A., and Sauer, R. T. Reverse hydrophobic effects relieved by amino-acid substitutions at a protein surface. *Nature (London)* **344**, 363–364 (1990).

169. Baker, E. N., and Hubbard, R. E. Hydrogen bonding in globular proteins. *Prog. Biophys. Molec. Biol.* **44**, 97–179 (1984).

170. Coxeter, H. S. M. *Introduction to Geometry.* Wiley: New York (1961).

171. Nagano, K. Logical analysis of the mechanism of protein folding IV. Super secondary structures. *J. Molec. Biol.* **109**, 235–250 (1977).

172. Crick, F. H. C. The packing of α-helices: simple coiled-coils. *Acta Cryst.* **6**, 689–697 (1953).

173. Chothia, C., Levitt, M., and Richardson, D. Helix to helix packing in proteins. *J. Molec. Biol.* **145**, 215–250 (1981).

174. Chothia, C. Principles that determine the structure of proteins. *Annu. Rev. Biochem.* **53**, 537–572 (1984).

175. Efimov, A. V. Packing of α-helices in globular proteins. Layer-structure of globin hydrophobic cores. *J. Molec. Biol.* **134**, 23–40 (1979).

176. Richmond, T. J., and Richards, F. M. Packing of α-helices: geometrical constraints and contact areas. *J. Molec. Biol.* **119**, 537–555 (1978).

177. Sternberg, M. J. E., and Thornton, J. M. On the conformation of proteins: the handedness of the connection between parallel β-strands. *J. Molec. Biol.* **110**, 269–283 (1977).

178. Sternberg, M. J. E., and Thornton, J. M. On the conformation of proteins: hydrophobic ordering of strands in β-pleated sheets. *J. Molec. Biol.* **115**, 1–17 (1977).

179. Chothia, C. Conformation of twisted β-pleated sheets in proteins. *J. Molec. Biol.* **75**, 295–302 (1973).

180. Chothia, C., Levitt, M., and Richardson, D. Structure of proteins: packing of α-helices and pleated sheets. *Proc. Natl. Acad. Sci. USA* **74**, 4130–4134 (1977).

181. Brändén, C.-I., Eklund, H., Nordström, B., Boiwe, T., Söderlund, G., Zeppeza-uer, E., Ohlsson, I. and Åkeson, Å. Structure of liver alcohol dehydrogenase at 2.9-Å resolution. *Proc. Natl. Acad. Sci. USA* **70**, 2439–2442 (1973).

182. Eklund, H., Nordström, B., Zeppezauer, E., Söderlund, G., Ohlsson, I., Boiwe, T., Soderberg, B. O., Tapia, O., and Brändén, C.-I. Three-dimensional structure of horse liver alcohol dehydrogenase at 2.4 Å resolution. *J. Molec. Biol.* **102**, 27–59 (1976).

183. Kendrew, J. C., Dickerson, R. E., Strandberg, B. E., Hart, R. G., Davies, D. R., Phillips, D. C., and Shore, V. C. Structure of myoglobin, a three-dimensional Fourier synthesis at 2 Å resolution. *Nature (London)* **185**, 422–427 (1960).

184. Perutz, M. F., Muirhead, H., Cox, J. M., and Goaman, L. C. G. Three-dimensional Fourier synthesis of horse oxyhæmoglobin at 2.8 Å resolution: the atomic model. *Nature (London)* **219**, 131–139 (1968).

185. Frier, J. A., and Perutz, M. F. Structure of human deoxyhæmoglobin. *J. Molec. Biol.* **112**, 97–112 (1977).

186. Magdoff-Fairchild, B., Swerdlow, P. H., and Bertles, J. F. Intermolecular organization of deoxygenated sickle haemoglobin determined by X-ray diffraction. *Nature (London)* **239**, 217–219 (1972).

187. Dykes, G., Crepeau, R. H., and Edelstein, S. J. Three-dimensional reconstruction of the fibres of sickle cell hemoglobin. *Nature (London)* **272**, 506–510 (1978).

188. Padlan, E. A., and Love, W. E. Refined crystal structure of deoxyhemoglobin S. II. Molecular interactions in the crystal. *J. Biol. Chem.* **260**, 8280–8291 (1985).

189. Marks, R. W. *The Dymaxion World of Buckminster Fuller.* Reinhold: New York (1960).

190. Coxeter, H. S. M. *Regular Polytopes.* 3rd edn. Dover: New York (1973).

191. Crick, F. H. C., and Watson, J. D. The structure of small viruses. *Nature (London)* **177**, 473–475 (1956).

192. Caspar, D. L. D. Structure of tomato bushy stunt virus. *Nature (London)* **177**, 476–477 (1956).

193. Caspar, D. L. D., and Klug, A. Physical principles in the construction of regular viruses. *Cold Spring Harbor Symp. Quant. Biol.* **27**, 1–24 (1962).

CHAPTER
16

Comparisons of Structures

Information on those features of the three-dimensional structure of a molecule (small or large) that are important with respect to their chemical or biochemical reactivities can often be obtained by comparisons of analogous molecular geometries in various different crystal structures. Effective techniques for comparing three-dimensional structures are therefore crucial to these types of analyses. If the structure of, for example, the citrate ion is compared as it occurs in several different crystal structures,[1] some measure will be obtained of the possible conformational variability of this ion, as shown in Figure 16.1. Each such conformation determined from a crystal structure analysis represents a low-energy form of the ion.

If the variability in conformation from crystal structure to crystal structure is extremely small, it is probably due to molecular vibrations and rotations, that is, the dynamics of molecular motion within the crystal. If the variability is large, however, this is not likely to be the case. There may be appreciable potential energy barriers between the various possible conformers, so that each of these conformers can exist as a separate entity in the crystal, as described in Chapter 13. Some measure of the probability of the relative proportions of the various conformers in solution may be derived from the percentage of times each conformer is seen in the total of crystal structures of that compound or ion. This argument hinges, however, on the assumption that the sampling in the crystal structures studied is random, an assumption that merits investigation in each analysis.

One might also wish to compare the structures and conformations of substrates and inhibitors of an enzyme. Citrate and isocitrate are two ions with slightly different formulæ and different conformations. As they are both substrates of the same enzyme (the Krebs cycle enzyme aconitase), a comparison between them would be expected to lead to information on how each is bound in the active site of the enzyme. This comparison, shown in Figure 16.2, provided some insight into the reasons

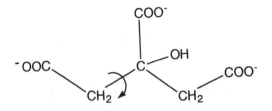

(a)

(b)

(c)

FIGURE 16.1. Conformational variability of the citrate ion. Shown are (a) the chemical formula, (b) a diagram indicating atom types in (c) which is a line drawing of several superposed citrate structures. The main variability is at the terminal carboxyl groups and rotation about one CH_2–C group. Alternate conformers are indicated in (b) by solid bonds for the main conformer and broken lines for another possible conformer.

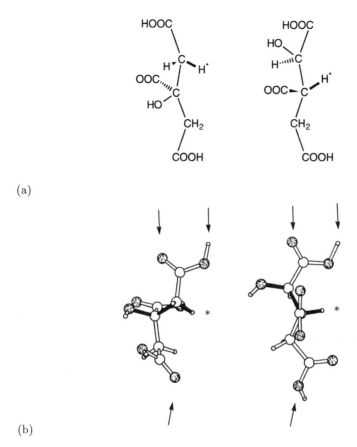

(a)

(b)

FIGURE 16.2. Comparison of citrate and isocitrate (Ref. 2) with respect to the reaction catalyzed by the enzyme aconitase. The asterisk shows the hydrogen atom abstracted by the enzyme. The arrows indicate possible sites of binding to the enzyme (via carboxyl groups of the anions).

for the stereospecificity of the reaction catalyzed by the enzyme aconitase which is now being investigated by structural studies of the enzyme with bound citrate or isocitrate or their analogues.[3]

But we have a dilemma. When comparing three-dimensional structures, what should be compared to what? In the case of direct comparisons of citrate ions, there was a central tetrahedral carbon atom in the molecule and all citrate ions could be laid on top of each other, as shown in Figure 16.1, with the central three carbon atoms superimposed. When, however, citrate and isocitrate are compared, it is less clear how the comparison should be made. In the example of aconitase substrates, it was made by reasoning that the enzyme probably "holds on" to the terminal carboxyl groups and somehow the central carboxyl group is held by other groups on the enzyme,[2] as shown in Figure 16.2. But this is just conjecture. Analyses of appropriate aconitase-bound substrates will be needed to determine the true binding mode.[3]

There are two major requirements for such relevant comparisons of three-dimensional structures to be made:

1. ready access to data from all the appropriate crystal structure determinations that contain the features chosen for comparison and
2. good, easy-to-use analysis techniques.

Fortunately, the need to amass coordinates of atoms in reported crystal structures has already been appreciated, and efficient computer-based databases are now available. As a result, it is not necessary to type large sets of numbers into a computer, because they can be derived in computer-ready form from a database. In addition to the ready access to atomic coordinate tables, there are many excellent computer-graphics, geometrical, and statistical programs available for comparisons of structures. In practice, however, it is additionally advantageous to build molecular models either of the ball-and-stick variety or the space-filling variety, if chemical or biochemical insight is required from a comparison of molecular structures observed by crystal structure analyses.

Two quotations[4] may help the reader appreciate the significance to science of the preparation and maintenance of **databases**. J. D. Bernal, who envisioned the need for and who planned the first computerized crystallographic database, wrote[5] about the information explosion: "However large an array of facts, however rapidly they accumulate, it is possible to keep them in order and to extract from time to time digests containing the most generally significant information while indicating how to find those items of specialized interest." This has been done and the crystallographic databases, which are the subject of this chapter, have enabled the crystallographer to keep up with the huge amount of data generated yearly. Not only can the data be stored, sorted and accessed, but it can also provide food for thought. Bragg is reported to have stated[6] that "the important thing in science is not so much to obtain new facts, as to discover new ways of thinking about them."

16.1 Crystallographic databases and their uses

A very early database on crystallographic measurements was a huge compendium[7] containing 3342 drawings and diagrams of crystals, together with information on measurements on nearly 10,000 crystalline substances, published by Paul Heinrich Ritter von Groth, Professor of Mineralogy in Munich, Germany in the period 1906 – 1919. Groth classified crystals by chemical composition and site of occurrence, categories of interest to chemists and mineralogists, respectively. The first of these classifications was innovative for that time, and Ewald wrote of Groth[8] that "in particular he fought for the wide introduction of crystallographic methods in organic chemistry;" Groth's text *Physikalische Kristallographie* was widely used, and he founded the journal *Zeitschrift für Kristallographie und Mineralogie* in 1877, a journal currently still in existence. Groth's works provided the major sources of information for the crystallographer of the time. From these publications arose additional compendia of crystallographic data: *Strukturbericht,*[9] *Structure Reports,*[10] *The Barker Index of Crystals*[11] and *Crystal Data Determinative Tables.*[12]

Now that computers are available for data handling, the information from crystal structures can be stored and readily accessed in the computer databases. There are several crystallographic databases, described below, each covering different aspects of crystallography.[13] These are maintained so that they are completely up to date with respect to the published literature. From these databases it is possible to find out which structures have been determined, their precision and the resulting unit cell dimensions, space groups, chemical formulæ, atomic coordinates, and atomic connectivities.

16.1.1 The Cambridge Structural Database

The Cambridge Structural Database (CSD)[14-27] contains information on approximately 100,000 three-dimensional crystal structure determinations that have been studied by X-ray or neutron diffraction (as of the middle of 1993). All crystallographic structure determinations of carbon-containing compounds, excluding some inorganic compounds such as carbonates, are included. The Database was initiated by Olga Kennard and was built up by her, together with Frank H. Allen, David G. Watson, W. D. Samuel Motherwell, Robin Taylor, and many other experts on database management, computing, chemistry, and crystallography. This Database is maintained in Cambridge, England.

Information that is put into this Database is derived from published reports of crystal structure determinations. The data extracted from the scientific literature in this way include the atomic coordinates, information on the space group, chemical connectivity, and the literature reference to each structure determination. Each compound listed in the Database is identified by a six-letter code (the REFCODE), unique to each crystal structure determination. Duplicate structures and remeasurements of the same crystal structure are identified by an additional two digits after the REFCODE. Scientific journals are scanned regularly by the Database staff for reports of crystal structure determinations, and the data are then entered into this Database. Structural data are also deposited by journals, for example, *Chemical Communications*, that publish articles, but do not have space for atomic coordinates. All crystallographic data reported in the literature are tested by the Database staff for internal consistency, precision, and chemical reasonableness. In this way errors in reported structures can be corrected, usually with the assistance of the original authors. This Database is particularly valuable for providing structural information suitable for comparison (Figure 16.3). The **software** that comes with this Database allows one to view the chemical formula of each compound accessed.

The Cambridge Structural Database may be searched in several ways. It is possible to find reported three-dimensional structural data on all compounds in a given class (such as steroids or peptides). It is also possible to search for compounds by name or portion of name. For example,

---

CAMBRIDGE STRUCTURAL DATABASE

Sample search (morphine)

```
save 3
tl *refc morphm
ques tl
--------+---------+---------+---------+---------+---------+---------+---------+
*REFC=MORPHM // (-)-Morphine monohydrate // *FORM=C17 H19 N1 O3,H2 O1 // *AUTH=E
.Bye // *CODE=306(Acta Chem.Scand.Ser.B) // *VOLU= 30 // *PAGE= 549 // *YEAR=197
6 // *FSKY=C 17H 19N 1O 3 H 2O 1 .... // *PREF=MORPHM // *ADAT=761225 // *MDAT=8
51129 // *MSDB=958014 // *CASN=0 // *NBSI=517887 // *CDRE=5 // *BATC=0 // *BCLA=
58 // *TOLE=.40 // *COOR=43 // *SPGN=19 // *SPAC=P212121 // *RFAC=.0450 // *TEMP
=295 // *MAXA=8 // *ZVAL=4.00 // *DENM=1.310 // *DENX=1.320 // *DENC=1.321 // *R
CP1=7.43799 // *RCP2=13.75099 // *RCP3=14.90099 // *RCP4=90.0000 // *RCP5=90.000
0 // *RCP6=90.0000 // *SIGF=2 // *MATF=3 // *INTF=3 // *CATF=3 // *METR=32 // *B
RV2=10 // *BRV1=10 // *RCVO=1524 // *SCOR=0 // *NW01=136 // *NW02=83 // *NW03=32
5 //
```

(a)

```
#MORPHM  33761225        22  9  0   0  0  4  4 43   0  0 5313220000002000000000076
   7438 13751 14901      90    90     90333000 1 3 3 0 0 0131132 19P212121    440
R=0.0450
211 0121 0112 0211 6101 6110 0011 0121 6110 6011 6101 0112 6
H   23C  68N  68O  68
C1     30680  28210   8110 C10   13240  44420   8400 C11   17960  34390  11960
C12    11390  31790  20230 C13    -720  37730  26230 C14    4560  48430  24680
C15   -20700  36400  23950 C16  -24940  39980  14490 C17  -23320  53450   4050
C2     36630  20040  12660 C3    30720  17770  21340 C4    17950  24010  24920
C5      4340  33440  35540 C6    17470  39790  40780 C7    29740  45690  34840
C8     23680  49990  27590 C9     1310  50680  14730 H1    35800  29900   2100
H10    22300  32000  50800 H11   48100  26400  42800 H12   47400  15700  10000
H13     8000  44400  44000 H14   31900  54800  24300 H15    8400  44200   2700
H16    -2700  52900  27900 H17  -24000  28900  24700 H18  -20400  34700   9900
H19   -36800  54800   4400 H2    -7000  32100  38600 H20   29200   6300  30100
H21    52900  19500  36600 H3    42600  46500  36800 H4     4200  58300  13700
H5     25000  47800   7400 H6   -27600  40500  28400 H7   -38600  40000  13600
H8    -19000  49300  -1000 H9   -19200  60100   3400 N1   -18200  49950  13050
O1     38380  10400  26070 O2    27920  34510  47030 O3    12310  23850  33740
O4     55440  22600  40470
 3  3  4  5 616  5 728 110  4 5131415  2 14143101416  2  6  7  8 01340431517  2  7  8  9  9 811
1412 0 617 939111213421739
```

(b)

FIGURE 16.3. Input and output from the Cambridge Structural Database. Morphine has been chosen as an example. (a) Search for the crystal structure of morphine. This can be done by name, atomic connectivity, or REFCODE (as in this example). The input was simply the first three lines shown above. The output follows (at REFC*) and includes the journal reference, space group, unit-cell dimensions, R factor, etc. (b) Output of atomic coordinates from the input at the top of (a). Courtesy the Cambridge Structural Database.

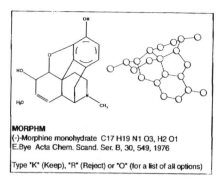

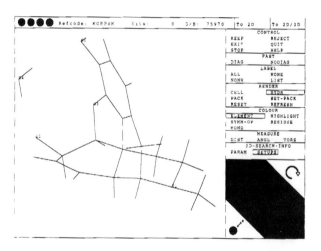

(c)

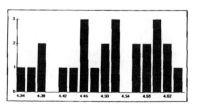

(d)

FIGURE 16.3 (cont'd). Output from the CSD. (c) Molecular formula and three-dimensional diagrams generated from search. (d) Geometric searches leading to, for example, histograms of comparisons of numerical output. Courtesy the Cambridge Structural Database.

one might want to search for all acetates by "ACETATE" (or the truncated version "ACETAT," as for *Chemical Abstracts*); this latter search should give a longer list of compounds than the search on ACETATE. The names of authors or any item in the literature reference, such as year or journal name, may also be used as search codes, and it is possible to search in several of these categories at the same time. These are referred to as **bibliographic searches**. Finally, of particular use are the **connectivity searches**. One can search for a small group of atoms, either a full molecule or a part (fragment) of a chemical structure, with the bonding precisely defined by the code used in the search. Such searches can be made to be highly specific (with an exact description in the input code of the number of multiple bonds, and of the hydrogen atoms bonded to a given atom). Alternatively, the searches can be kept general by specifying only the minimum number of bond types (C–C, C–N, C–O, etc.) to a given atom. The example of the crystal structure of morphine monohydrate is given in Figure 16.3. When groups of atoms that meet the required specifications have been extracted from the Database, tables of required geometrical data may be generated, or the structures may be displayed on a computer graphics screen. In this way comparisons of the three-dimensional geometries of the molecular fragment chosen for study can be made for all other entries in the Database containing that fragment.

16.1.2 The Protein Data Bank

The Protein Data Bank (PDB),[28,29] started in 1971, is a computerized archive for data on the three-dimensional structures of biological macromolecules. It is organized and maintained by Thomas F. Koetzle, Frances C. Bernstein and Enrique E. Abola, and many others, at Brookhaven National Laboratory. At present there are structural entries (atomic coordinates, journal references, and, in many cases, experimentally observed structure factors) for about 1200 biological macromolecules, over 550 with atomic coordinates (see Figure 16.4). Also available is a list of proteins that have been crystallized and the experimental conditions used.[30] In the Protein Data Bank each protein structure reported has an identifying code (IDCODE), a header record containing useful information on the protein such as the name and source of the protein, a series of references to published articles on the protein, and the resolution of the data. Data are included on the refinement, such as the programs used, the R value, the number of Bragg reflections, the r.m.s. deviations of the bond lengths and angles from ideality, and the number of water molecules that have been located. Then follows a description of the protein, its amino acid sequence, including an analysis of which parts are helix, sheet, or turn. The main entry is a list of atomic coordinates (ATOM) and information on three-dimensional coordinates of metals, substrates, and inhibitors bound to the protein (HETATM). In addition,

```
HEADER    ISOMERASE(INTRAMOLECULAR OXIDOREDUCTSE) 11-OCT-90   7XIA    7XIA    2
COMPND    D-*XYLOSE ISOMERASE (E.C.5.3.1.5)                           7XIA    3
SOURCE    (STREPTOMYCES $RUBIGINOSUS)                                 7XIA    4
AUTHOR    H.L.CARRELL,J.P.GLUSKER                                     7XIA    5
REVDAT   1   15-OCT-91 7XIA       0                                   7XIA    6
SPRSDE       15-OCT-91 7XIA       2XIA                                7XIA    7
JRNL        AUTH   H.L.CARRELL,J.P.GLUSKER,V.BURGER,F.MANFRE,D.TRITSCH 7XIA   8
JRNL        AUTH J.-*F.BIELLMANN                                      7XIA    9
JRNL        TITL   X-RAY ANALYSIS OF D-XYLOSE ISOMERASE AT 1.9        7XIA   10
JRNL        TITL 2 ANGSTROMS: NATIVE ENZYME IN COMPLEX WITH SUBSTRATE 7XIA   11
JRNL        TITL 3 AND WITH A MECHANISM-DESIGNED INACTIVATOR          7XIA   12
JRNL        REF    PROC.NAT.ACAD.SCI.USA       V.  86  4440 1989      7XIA   13
JRNL        REFN   ASTM PNASA6  US ISSN 0027-8424                040  7XIA   14
REMARK   1                                                           7XIA   15
REMARK   1 REFERENCE 1                                               7XIA   16
REMARK   1  AUTH   K.HENRICK,D.M.BLOW,H.L.CARRELL,J.P.GLUSKER         7XIA   17
REMARK   1  TITL   COMPARISON OF BACKBONE STRUCTURES OF GLUCOSE       7XIA   18
REMARK   1  TITL 2 ISOMERASE FROM STREPTOMYCES AND ARTHROBACTER       7XIA   19
REMARK   1  REF    PROTEIN ENG.                V.   1   467 1987      7XIA   20
REMARK   1  REFN   ASTM PRENE9  UK ISSN 0269-2139                859  7XIA   21
REMARK   1 REFERENCE 2                                               7XIA   22
REMARK   1  AUTH   H.L.CARRELL,B.H.RUBIN,T.J.HURLEY,J.P.GLUSKER       7XIA   23
REMARK   1  TITL   X-RAY CRYSTAL STRUCTURE OF D-*XYLOSE ISOMERASE AT  7XIA   24
REMARK   1  TITL 2 4-*ANGSTROMS RESOLUTION                           7XIA   25
REMARK   1  REF    J.BIOL.CHEM.                V. 259  3230 1984      7XIA   26
REMARK   1  REFN   ASTM JBCHA3  US ISSN 0021-9258                071  7XIA   27
REMARK   2                                                           7XIA   28
REMARK   2 RESOLUTION. 1.9 ANGSTROMS.                                7XIA   29
REMARK   3                                                           7XIA   30
REMARK   3 REFINEMENT. BY THE RESTRAINED LEAST SQUARES PROCEDURE OF J. 7XIA  31
REMARK   3 KONNERT AND W. HENDRICKSON (PROGRAM *PROLSQ*) WITH FFT     7XIA   32
REMARK   3 ACCELERATION IMPLEMENTED BY M. KNOSSOW, M. LEWIS, D. REES, 7XIA   33
REMARK   3 I. WILSON, J. SKEHEL AND D. WILEY.  THE R VALUE IS 0.131   7XIA   34
REMARK   3 FOR 28313 REFLECTIONS IN THE RESOLUTION RANGE 8.0 TO 1.9   7XIA   35
REMARK   3 ANGSTROMS WITH INTENSITY GREATER THAN 1.5*SIGMA(I).  THE   7XIA   36
REMARK   3 RMS DEVIATION FROM IDEALITY OF THE BOND LENGTHS IS 0.020   7XIA   37
REMARK   3 ANGSTROMS.   THE RMS DEVIATION FROM IDEALITY OF THE BOND   7XIA   38
REMARK   3 ANGLE DISTANCES IS 0.041 ANGSTROMS.  ATOMS WITH THERMAL    7XIA   39
REMARK   3 FACTORS WHICH CALCULATE LESS THAN 5.00 ARE ASSIGNED THIS   7XIA   40
REMARK   3 VALUE.  THIS IS THE LOWEST VALUE ALLOWED BY THE REFINEMENT 7XIA   41
REMARK   3 PROGRAM.                                                  7XIA   42
REMARK   4                                                           7XIA   43
REMARK   4 THE AMINO ACID SEQUENCE WAS TAKEN FROM INTERNATIONAL PATENT 7XIA  44
REMARK   4 APPLICATION NUMBER:PCT/US88/02765, INTERNATIONAL          7XIA   45
REMARK   4 PUBLICATION NUMBER:WO89/01520.  IN THIS PATENT RESIDUE 41  7XIA   46
REMARK   4 IS SPECIFIED AS ARG.  HOWEVER, RESIDUE 41 APPEARS TO BE GLN 7XIA  47
REMARK   4 BASED ON THE CRYSTALLOGRAPHIC STUDY.  IN THIS ENTRY RESIDUE 7XIA  48
REMARK   4 41 IS PRESENTED AS GLN.                                   7XIA   49
REMARK   5                                                           7XIA   50
REMARK   5 THE CARBOXYL TERMINAL RESIDUE GLY 388 WAS NOT DEFINED BY   7XIA   51
REMARK   5 THE ELECTRON DENSITY AND, THEREFORE, NO COORDINATES ARE    7XIA   52
REMARK   5 GIVEN FOR THIS RESIDUE IN THIS ENTRY.                     7XIA   53
REMARK   6                                                           7XIA   54
REMARK   6 THE STRUCTURE OF THE MONOMER IS AN EIGHT-FOLD ALPHA-BETA   7XIA   55
REMARK   6 BARREL WITH AN EXTENDED C-TERMINAL LOOP WHICH FACILITATES  7XIA   56
REMARK   6 AGGREGATION OF MONOMERS TO TETRAMERS.  TETRAMERS ARE       7XIA   57
REMARK   6 POSITIONED ON THE 222 SYMMETRY SITE AT THE ORIGIN OF THE   7XIA   58
REMARK   6 CELL.                                                     7XIA   59
REMARK   7                                                           7XIA   60
REMARK   7 NOTE THAT SOME WATERS ARE NOT IN THE SAME ASYMMETRIC UNIT  7XIA   61
REMARK   7 AS THE PROTEIN.                                           7XIA   62
SEQRES   1   388  MET ASN TYR GLN PRO THR PRO GLU ASP ARG PHE THR PHE 7XIA  63
SEQRES   2   388  GLY LEU TRP THR VAL GLY TRP GLN GLY ARG ASP PRO PHE 7XIA  64
SEQRES   3   388  GLY ASP ALA THR ARG ARG ALA LEU ASP PRO VAL GLU SER 7XIA  65
SEQRES   4   388  VAL GLN ARG LEU ALA GLU LEU GLY ALA HIS GLY VAL THR 7XIA  66
SEQRES   5   388  PHE HIS ASP ASP ASP LEU ILE PRO PHE GLY SER SER ASP 7XIA  67
SEQRES   6   388  SER GLU ARG GLU GLU HIS VAL LYS ARG PHE ARG GLN ALA 7XIA  68
SEQRES   7   388  LEU ASP ASP THR GLY MET LYS VAL PRO MET ALA THR THR 7XIA  69
SEQRES   8   388  ASN LEU PHE THR HIS PRO VAL PHE LYS ASP GLY GLY PHE 7XIA  70
SEQRES   9   388  THR ALA ASN ASP ARG ASP VAL ARG ARG TYR ALA LEU ARG 7XIA  71
SEQRES  10   388  LYS THR ILE ARG ASN ILE ASP LEU ALA VAL GLU LEU GLY 7XIA  72
SEQRES  11   388  ALA GLU THR TYR VAL ALA TRP GLY GLY ARG GLU GLY ALA 7XIA  73
SEQRES  12   388  GLU SER GLY GLY ALA LYS ASP VAL ARG ASP ALA LEU ASP 7XIA  74
SEQRES  13   388  ARG MET LYS GLU ALA PHE ASP LEU LEU GLY GLU TYR VAL 7XIA  75
SEQRES  14   388  THR SER GLN GLY TYR ASP ILE ARG PHE ALA ILE GLU PRO 7XIA  76
SEQRES  15   388  LYS PRO ASN GLU PRO ARG GLY ASP ILE LEU LEU PRO THR 7XIA  77
```

(a)

FIGURE 16.4. Portions of an output file from the Protein Data Bank. (a) Bibliographic information, comments on the structure determination, and the amino-acid sequence. Shown are journal references, information on the resolution, and the amino-acid sequence. Courtesy the Protein Data Bank.

```
SEQRES   16   388   VAL GLY HIS ALA LEU ALA PHE ILE GLU ARG LEU GLU ARG   7XIA   78
SEQRES   17   388   PRO GLU LEU TYR GLY VAL ASN PRO GLU VAL GLY HIS GLU   7XIA   79
SEQRES   18   388   GLN MET ALA GLY LEU ASN PHE PRO HIS GLY ILE ALA GLN   7XIA   80
SEQRES   19   388   ALA LEU TRP ALA LYS LEU PHE HIS ILE ASP LEU ASN   7XIA   81
SEQRES   20   388   GLY GLN ASN GLY ILE LYS TYR ASP GLN ASP LEU ARG PHE   7XIA   82
SEQRES   21   388   GLY ALA GLY ASP LEU ARG ALA ALA PHE TRP LEU VAL ASP   7XIA   83
SEQRES   22   388   LEU LEU GLU SER ALA GLY TYR SER GLY PRO ARG HIS PHE   7XIA   84
SEQRES   23   388   ASP PHE LYS PRO PRO ARG THR GLU ASP PHE ASP GLY VAL   7XIA   85
SEQRES   24   388   TRP ALA SER ALA ALA GLY CYS MET ARG ASN TYR LEU ILE   7XIA   86
SEQRES   25   388   LEU LYS GLU ARG ALA ALA ALA PHE ARG ALA ASP PRO GLU   7XIA   87
SEQRES   26   388   VAL GLN GLU ALA LEU ARG ALA SER ARG LEU ASP GLU LEU   7XIA   88
SEQRES   27   388   ALA ARG PRO THR ALA ASP GLY LEU GLN ALA LEU LEU   7XIA   89
SEQRES   28   388   ASP ASP ARG SER ALA PHE GLU GLU PHE ASP VAL ASP ALA   7XIA   90
SEQRES   29   388   ALA ALA ALA ARG GLY MET ALA PHE GLU ARG LEU ASP GLN   7XIA   91
SEQRES   30   388   LEU ALA MET ASP HIS LEU LEU GLY ALA ARG GLY   7XIA   92
                                                                      7XIA   93
FTNOTE    1                                                           7XIA   94
FTNOTE    1   RESIDUE PRO 187 IS A CIS PROLINE.                       7XIA   95
HET      MN   390        1   MANGANESE                                7XIA   96
HET      MN   391        1   MANGANESE                                7XIA   97
FORMUL    2   MN      2(MN1)                                          7XIA   98
FORMUL    3   HOH    *306(H2 O1)                                      7XIA   99
CRYST1  93.900   99.700  102.900   90.00   90.00   90.00 I 2 2 2    8  7XIA  100
ORIGX1   1.000000  0.000000  0.000000        0.00000                 7XIA  101
ORIGX2   0.000000  1.000000  0.000000        0.00000                 7XIA  102
ORIGX3   0.000000  0.000000  1.000000        0.00000                 7XIA  103
SCALE1   0.010650  0.000000  0.000000        0.00000                 7XIA  104
SCALE2   0.000000  0.010030  0.000000        0.00000                 7XIA  105
SCALE3   0.000000  0.000000  0.009718        0.00000                 7XIA  106
ATOM      1   N    MET   1      43.489  41.271  90.198  1.00 62.94    7XIA  107
ATOM      2   CA   MET   1      42.838  41.978  89.056  1.00 62.19    7XIA  108
ATOM      3   C    MET   1      42.935  40.973  87.898  1.00 60.26    7XIA  109
ATOM      4   O    MET   1      43.876  40.975  87.073  1.00 60.26    7XIA  110
ATOM      5   CB   MET   1      43.418  43.328  88.737  1.00 66.16    7XIA  111
ATOM      6   CG   MET   1      44.904  43.427  89.011  1.00 69.71    7XIA  112
ATOM      7   SD   MET   1      45.198  45.150  89.653  1.00 73.34    7XIA  112


ATOM     29   N    GLN   4      38.217  36.651  84.178  1.00 30.22    7XIA  134
ATOM     30   CA   GLN   4      37.772  35.288  83.883  1.00 27.55    7XIA  135
ATOM     31   C    GLN   4      36.448  35.276  83.143  1.00 23.84    7XIA  136
ATOM     32   O    GLN   4      35.495  35.822  83.635  1.00 22.32    7XIA  137
ATOM     33   CB   GLN   4      37.572  34.525  85.216  1.00 34.50    7XIA  138
ATOM     34   CG   GLN   4      38.371  33.292  85.399  1.00 43.53    7XIA  139
ATOM     35   CD   GLN   4      37.709  31.996  84.990  1.00 49.39    7XIA  140
ATOM     36   OE1  GLN   4      37.807  30.952  85.705  1.00 53.47    7XIA  141
ATOM     37   NE2  GLN   4      37.019  31.944  83.849  1.00 50.63    7XIA  142
ATOM     38   N    PRO   5      36.451  34.637  81.998  1.00 21.53    7XIA  143
ATOM     39   CA   PRO   5      35.207  34.551  81.233  1.00 21.15    7XIA  144
ATOM     40   C    PRO   5      34.279  33.499  81.913  1.00 22.43    7XIA  145
ATOM     41   O    PRO   5      34.766  32.533  82.510  1.00 23.18    7XIA  146
ATOM     42   CB   PRO   5      35.628  33.839  79.954  1.00 20.58    7XIA  147
ATOM     43   CG   PRO   5      37.118  33.576  80.016  1.00 20.93    7XIA  148
ATOM     44   CD   PRO   5      37.610  33.989  81.365  1.00 21.70    7XIA  149
ATOM     45   N    THR   6      33.004  33.692  81.754  1.00 21.78    7XIA  150
ATOM     46   CA   THR   6      32.020  32.703  82.161  1.00 23.93    7XIA  151
ATOM     47   C    THR   6      31.067  32.608  80.939  1.00 24.39    7XIA  152
ATOM     48   O    THR   6      30.919  33.622  80.187  1.00 24.37    7XIA  153
ATOM     49   CB   THR   6      31.193  33.208  83.415  1.00 22.53    7XIA  154
ATOM     50   OG1  THR   6      30.547  34.422  83.093  1.00 23.37    7XIA  155
ATOM     51   CG2  THR   6      32.197  33.456  84.555  1.00 28.21    7XIA  156


ATOM    3043  CD   ARG  387     41.875  59.456  83.323  1.00 34.74    7XIA 3148
ATOM    3044  NE   ARG  387     41.715  60.754  84.035  1.00 37.25    7XIA 3149
ATOM    3045  CZ   ARG  387     42.678  61.696  83.865  1.00 39.98    7XIA 3150
ATOM    3046  NH1  ARG  387     43.798  61.486  83.173  1.00 39.51    7XIA 3151
ATOM    3047  NH2  ARG  387     42.404  62.915  84.315  1.00 40.80    7XIA 3152
HETATM  3054  MN   MN   390     39.205  36.647  58.172  0.50 15.96    7XIA 3153
HETATM  3055  MN   MN   391     34.689  35.426  59.602  0.50 23.84    7XIA 3154
HETATM  3056  O    HOH    1     27.818  75.112  70.549  1.00 21.93    7XIA 3155
HETATM  3057  O    HOH    2     42.765  44.700  64.337  1.00  6.00    7XIA 3156
HETATM  3058  O    HOH    3     23.938  22.494  76.046  1.00 24.35    7XIA 3157
```

(b)

FIGURE 16.4 (cont'd). Portions of an output file from the Protein Data Bank. (b) The end of the amino-acid sequence, the presence of metal ions, and orthogonal coordinates of each atom together with its displacement factor. Courtesy the Protein Data Bank.

coordinates of oligonucleotides and their complexes with dyes and drugs are contained in the PDB. Coordinates of a protein or nucleic acid can readily be extracted in a suitable form for use with a computer-graphics system. Portions of an output file are shown in Figure 16.4. From information in this Database it has been possible to map side-chain–side-chain interactions in proteins.[31]

16.1.3 The Inorganic Crystal Structure Database

The Inorganic Crystal Structure Database[32-38] is maintained through a collaboration of the Fachinformationszentrum Energie Physik Mathematik, Karlsruhe, Germany, and the Institute for Inorganic Chemistry of the University of Bonn, Germany, and is managed by Günter Bergerhoff and I. David Brown. This database contains information on all compounds containing at least one nonmetallic element but no C–C or C–H bonds (because these are covered by the Cambridge Structural Database). Each reported crystal structure has a separate entry. Information provided in the database includes the chemical name, phase designation, unit cell dimensions, density, space group and the oxidation state of the elements. Also listed are R value, temperature, pressure, method of measurement and the full journal reference. Atomic information includes coordinates and displacement parameters. Errors, such as a charge sum that is not zero or a calculated formula that does not agree with the atomic coordinates, are corrected, as for the Cambridge Structural Database, by consultation with the authors. If the error cannot be corrected, a warning message is included in the database entry. Again the extracted coordinates can be analyzed in any appropriate way by the chemist or biochemist.

16.1.4 NRCC Metals Crystallographic Data File

NRCC Metals Crystallographic Data File (CRYSTMET)[39] is a database of crystallographic and bibliographic data on intermetallic phases. Also included are some hydrides and binary oxides. The database contains about 11,000 entries. It was started by Don T. Cromer at Los Alamos Scientific Laboratories, William Burton Pearson at Waterloo University, and Lauriston D. Calvert at The National Research Council of Canada (NRCC) and is currently maintained by the Canadian Institute for Scientific and Technical Information (CISTI).

16.1.5 The Powder Diffraction File

The Powder Diffraction File[40-43] contains data on single-phase X-ray powder diffraction patterns. These are stored as lists of interplanar spacings d_{hkl} and relative intensities characteristic of each compound, as shown in Figure 16.5. Unit cell dimensions, space groups and densities are also listed.

41-1476 ★

KCl							
			d Å	**Int**	**hkl**	**d Å**	**Int**	**hkl**

KCl		d Å	Int	hkl	d Å	Int	hkl
Potassium Chloride	Sylvite, syn	3.633	1	111			
		3.146	100	200			
		2.2251	37	220			
Rad. CuKα₁ λ 1.54056 **Filter** Mono. **d-sp** Diff.		1.8972	<1	311			
Cut off 15.0 **Int.** Diffractometer **I/I**cor.		**1.8169**	10	222			
Ref. Welton, J., McCarthy, G., North Dakota State University, Fargo, North Dakota, USA, *JCPDS Grant-in-Aid Report*, (1989)		1.5730	5	400			
		1.4071	9	420			
		1.2839	5	422			
Sys. Cubic **S.G.** Fm3m (225)		1.1121	1	440			
a 6.2917(3) **b** **c** **A** **C**		1.0485	2	600			
α β γ **Z** 4 **mp** 790 C		0.9948	2	620			
Ref. Copper, M., Rouse, K., *Acta Crystallogr., Sec. A*, **29** 514 (1973)		0.9485	1	622			
Dₓ 1.99 **D**m 1.99 **SS/FOM** F₁₅ = 88(.009,20)		0.9081	<1	444			
		0.8725	1	640			
εα **n**ωβ 1.4904 εγ **Sign** **2V**		0.8408	1	642			
Ref. Winchell, A., Winchell, H., *Microscopic Character of Artificial Inorg. Solid Sub.*, 15 (1964)							

Color White
Peak height intensities. Sample from Mallinckrodt. Lot analysis showed sample as 99.9 + % pure. Sample recrystallized from 50/50 ethanol water solvent system and heated at 600 C for 72 hours. Merck Index, 8th Ed., p. 853. σ(I_obs) = ± 0.07. Halite group, halite subgroup. Silicon used as internal standard. PSC: cF8. To replace 4-587, and validated by calculated patterns 26-920 and 26-921.

2-326

KClO₄		d Å	Int	hkl
		4.34	40	111
Potassium Chlorate		**3.75**	100	200
		2.66	90	220
		2.26	20	311
Rad. CuKα λ 1.5418 **Filter** **d-sp**		**2.16**	60	222
Cut off **Int.** **I/I**cor.		1.876	20	400
Ref. Herrmann, L., Ilge, Z. *Kristallogr., Kristallgeom., Kristall-phys., Kristallchem.*, **75** 47 (1930)		1.676	60	420
		1.533	60	422
Sys. Cubic **S.G.** F̄43m (216)		1.327	40	440
a 7.505(2) **b** **c** **A** **C**		1.251	40	600
α β γ **Z** 4 **mp**		1.185	20	620
Ref. Braekken, Harang. Z. *Kristallogr., Kristallgeom., Kristallphys., Kristallchem.*, **75** 540 (1930)				
Dₓ 2.177 **D**m **SS/FOM** F₁₁ = 14(.056,14)				

Pattern at 340 C. CAS#: 3811-04-9. Specimen above 300 C. a = 7.51 at 310 C. Cell parameter generated by least squares refinement. Reference reports: a = 7.49. PSC: cF24.

FIGURE 16.5. Data for two salts from the Powder Diffraction File. The crystal system, unit-cell dimensions, space group and bibliographic references are given together with experimental details. Listed on the right are d_{hkl}, $I(hkl)$ and hkl for the most intense Bragg reflections. Courtesy, JCPDS — International Centre for Diffraction Data (ICDD), publishers of the Powder Diffraction File.

This Database is particularly useful for the identification of materials, either pure chemical substances or mixtures, such as the deposit scraped from the inside of the smokestack of an industrial factory. A certain number of these diffraction peaks in the diffraction pattern of the compound or deposit are compared, both with respect to scattering angle and relative intensity with the substances already in the Powder Diffraction File. Often good analyses of mixtures can be made in this way.

16.2 Some methods for comparing structures

Correlations of three-dimensional structure with the chemical, physical, and biological properties of a compound require comparisons of bonding geometry, of conformation, and/or of interactions of the compounds with their respective environments in the crystal structure. Lists of results from calculations of bonding geometry can be used, but graphical methods are frequently more effective, and easier for the reader.

The structure of a molecule in a crystal is not always the same as that of the free molecule *in vacuo*. The environment of the molecule in the crystal may favor a conformational change because of intermolecular interactions. The geometrical parameters of a molecule that may be changed on crystallization depend on the strengths (stiffness) of the various molecular parameters. Bond lengths will be only slightly affected while torsion angles may be more affected. Comparisons of structures give us a large amount of information on the types of deformations that can occur in molecules, and hence the way that they might react to an attacking reagent. We should, however, beware of the erroneous conclusion that a study of such conformations give information on energy. Bürgi and Dunitz[44] warn us, "Although low-energy regions of molecular potential energy surfaces can certainly be recognized and mapped from distributions of observed structures, the derivation of quantitative energy relationships from statistical analysis alone is not possible without introducing arbitrary and unwarranted assumptions." They emphasize (in italics) *"An ensemble of structural parameters obtained from chemically different compounds in different crystal structures does not even remotely resemble a closed system at thermal equilibrium and does not therefore conform to the conditions necessary for the application of the Boltzmann distribution."* This means that it is not possible to derive relative energies from purely geometrical data. On the other hand,[44] any observed distribution of structural parameters in different crystal structures give some qualitative information on the shapes of low-energy regions of potential energy surfaces (see Figure 16.6).

16.2.1 Bar graphs/histograms

When data files are used to extract data, it is often useful to illustrate the results by means of **bar graphs** or **histograms** containing information

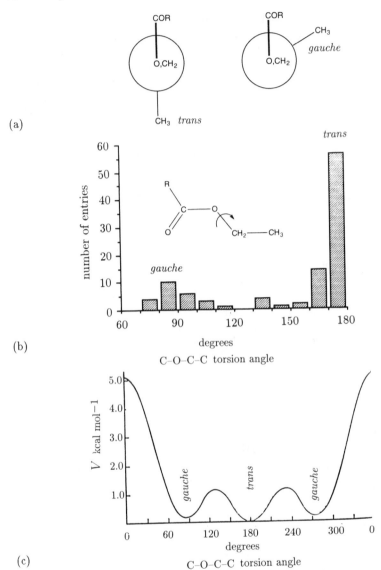

FIGURE 16.6. Conformations of ethyl esters from the CSD. (a) View down the C–O bond showing *trans/ gauche* designations. (b) Frequencies with which different values are found for the C–O–C–C torsion angle for ethyl esters in the CSD. Shown are the number of entries in the Database with torsion angles in 10° ranges. (c) Variation of conformation (rotation about the C–O bond) with potential energy for gaseous ethyl formate (Refs. 45 and 46). Note that the peaks in (b), corresponding to many entries in the CSD, lie at the same torsion angles as valleys in (c), corresponding to low-energy conformers.

on the number of structures studied that have a bond length, interbond angle or torsion angle of a particular value. What, in fact, happens is that, for a given molecular geometrical parameter (such as a torsion angle), most values are near that of the lowest-energy conformation. Sometimes the molecule is overcrowded in some way, and a less favorable conformation must necessarily occur. As a consequence, a given torsion angle may adopt a conformation that deviates from the "preferred" conformation (where there were no steric restrictions). For example, the frequency of this occurrence seems, in the structures studied to date, to parallel the probability implicit in the potential energy curve for rotation of that torsion angle.

The conformations of ethyl esters about their C–O bonds [Figure 16.6(a)] has been studied by Peter Murray-Rust.[4] A plot of the torsion angle about this bond versus the frequency of occurrence of this torsion angle among all the crystal structures studied to date is shown in Figure 16.6(b). The results may be compared with those from microwave studies by E. Bright Wilson,[45,46] which show a similar angular variation of energy versus conformation angle pattern [Figure 16.6(c)]. Of course, those ethyl esters that have torsion angles near 90° probably do so for some steric reason since the most common value is near 180°, but the general similarities of Figures 16.6(b) and (c), where the conformations of lowest energy are the most common, suggest that this comparison is justified, remembering the caveat (given above) that absolute values for energies cannot be directly obtained in this way.

16.2.2 Superpositions

A simple method for comparing conformations of molecules in different crystals is to superimpose two or more molecular structures upon each other to give a **superposition diagram**, such as that in Figure 16.1. The molecules are treated as rigid bodies defined by their crystallographic coordinates. A computer program then rotates and translates one (or more) of these rigid molecules to obtain a best fit between selected portions of the molecules under study, often by least-squares techniques. Such a fit can be made to a particular reference structure or within a group of structures. If a single structure stands out, for example, by its high reactivity, the former method, using this structure as a reference, may be preferred. A computer-generated plot of the coordinate set, transformed to one with the best fit to the standard structure, can be visually inspected. Such plots can be drawn as stereoscopic projections to aid visual perception of similarities and differences in the conformations of the molecules studied.

Superposition diagrams can also be constructed using depictions of space-filling models. A van der Waals surface depiction for each molecule is constructed, the diagrams so obtained are superimposed on each other. Then one van der Waals diagram is subtracted from the other.[47] The resultant diagram (called an **excluded volume** plot) clearly defines areas in space where one molecule has electron density and the other does not, and vice versa (Figure 16.7). Such plots can be particularly useful when comparing molecules with different kinds of pharmacological activities. For example, **antagonists** and **agonists** may be found to have electron densities in different regions in space. A newly designed drug may have a shape resembling one or another of these. It may not bind appropriately if it is too bulky in certain areas; these will be positive areas in an excluded volume plot. Observations of this type provide insight into interactions at the molecular level with an unknown receptor. The three-dimensional shape of the active site of the interaction in an enzyme or on a receptor is mapped in this way.

16.2.3 Distance matrices

A method for comparing the backbone folding in protein structures is to calculate a matrix (table) of distances from each α-carbon atom to every other α-carbon atom in a protein. This **distance matrix** is indexed in the consecutive order of the amino acid residues in the protein. The matrix of such $C\alpha \cdots C\alpha$ distances is then contoured so that it is evident where these distances are short and where they are long. The plot, an example of which is shown in Figure 16.8 for a small molecule (to illustrate the principles), has symmetry across the diagonal. This method, suggested by David C. Phillips,[48] led to many interesting analyses of protein shape. The advantage is that no decision has to be made about what part of the molecule has to be compared with what. For example, α helices and β pleated sheets have different contours in the distance **matrix,** and the comparisons of distance matrices of proteins can show

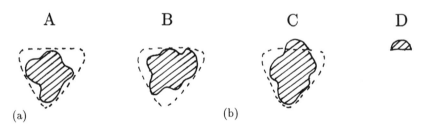

FIGURE 16.7. Diagram illustrating excluded volume. (a) Since active drugs A and B work, they are presumed to fit in the required space in the enzyme or receptor. Their shapes give an indication of the combined volume available for active binding (the excluded volume shown by broken lines). (b) Drug C does not work, and when its shape is compared with those of active drugs, it is evident that one portion (D) fills space that should not be filled if the enzyme or receptor is to be active. This is called the "essential volume," shown as D. If this space is filled by ligands, they will not be active (see Ref. 47).

	O1	O2	O3	O4	O5	O6	O7	C1	C2	C3	C4	C5	C6
O1		2.23 / 2.30	6.95 / 6.47	6.71 / 5.43	4.58 / 3.12	4.75 / 3.02	4.26 / 3.43	1.30 / 1.32	2.35 / 2.36	3.73 / 2.94	4.83 / 4.37	6.17 / 5.36	4.23 / 2.73
O2			6.18 / 6.68	5.56 / 6.35	4.00 / 4.79	3.13 / 4.52	2.87 / 3.44	1.24 / 1.23	2.40 / 2.42	2.87 / 3.46	4.36 / 4.72	5.30 / 5.90	3.10 / 4.14
O3				2.21 / 2.23	4.76 / 4.76	4.97 / 4.77	3.47 / 3.60	5.92 / 5.96	4.64 / 4.75	3.50 / 3.55	2.33 / 2.36	1.31 / 1.32	4.34 / 4.26
O4					3.39 / 3.19	3.35 / 3.09	3.17 / 3.32	5.54 / 5.38	4.48 / 4.40	3.00 / 2.94	2.42 / 2.39	1.21 / 1.20	2.98 / 2.77
O5						2.23 / 2.25	3.57 / 3.58	3.71 / 3.57	2.97 / 2.94	2.37 / 2.37	2.83 / 2.92	3.50 / 3.46	1.32 / 1.33
O6							2.70 / 2.65	3.57 / 3.54	3.40 / 3.45	2.41 / 2.41	3.52 / 3.48	3.75 / 3.57	1.20 / 1.20
O7								3.02 / 2.88	2.45 / 2.41	1.41 / 1.43	2.35 / 2.38	2.76 / 2.89	2.41 / 2.40
C1									1.50 / 1.50	2.58 / 2.51	3.87 / 3.84	5.08 / 5.03	3.11 / 3.02
C2										1.54 / 1.54	2.49 / 2.50	3.85 / 3.86	2.52 / 2.54
C3											1.53 / 1.53	2.52 / 2.52	1.54 / 1.53
C4												1.51 / 1.51	2.53 / 2.54
C5													3.09 / 2.98
C6													

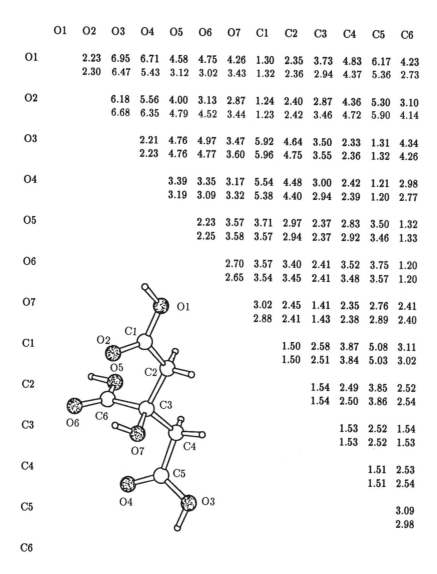

FIGURE 16.8. An example of a distance matrix for anhydrous citric acid (Refs. 49,50), and its monohydrate (Ref. 51). Both are numbered in the same way. Shown is half the matrix (which is symmetrical with all diagonal terms equal to zero), i.e., $A_{ij} = A_{ji}$ and $A_{ii} = 0$. Distances (in Å) are given for anhydrous citric acid (upper entry), and citric acid monohydrate (lower entry). Note the difference in the orientation of the carboxyl group on C1 in each.

similarities and differences in the folding of proteins. It canalso readily show when one protein has a deletion or additional piece of a polypeptide chain within the backbone.[52-54]

Another use of distance matrices has been to look at the changes in protein conformation that accompany binding of a small molecule, for example, a substrate or an inhibitor.[52,53] If, instead of distances, the differences between distances in the enzyme and enzyme with bound substrate are calculated, the distance matrix will more clearly show movements of portions of the protein upon binding of ligands. On binding substrate, for example, a particular α helix may move relative to another.

On the other hand, when this method is applied to small molecules, even with groups rather than single atoms used for the distance computation, the amount of data generated is large and may be complicated to analyze. As an example, a series of corrins was analyzed in this way as part of a project aimed at determining how vitamin B_{12} and its coenzymes work.[55] The major variation in such corrin structures is a "butterfly-like" folding (described later in this Chapter). Since the corrin ring is fairly flat, changes in distance across the corrin ring were not very sensitive to the folding angle. On the other hand, distances from axial substituents did reveal the nature of any variation, and indicated not only the region of maximum flexibility in the molecule, but also the nature of the conformational change.

16.2.4 Analyses of data when there are many variables

Most experiments have inherent in them a large number of variables. The experimenter usually ignores most of these variables, assuming on a subjective basis that they are not important. The problem has been stated by Thomas Auf der Heyde and Hans-Beat Bürgi[56] as a twofold one: "First, how can we avoid subjectively carving up the data into two-dimensional subsets, i.e., how can all n variables be analyzed *simultaneously* in order to reveal correlations between them? Second, how can the dimensionality of the problem be *objectively* reduced in order to interpret and visualize these correlations?" Two main methods are presently employed to try to resolve this problem. These methods are: **cluster analysis,** which sorts the data into groups (clusters) so that the data can be classified, and **factor analysis,** which finds those variables (factors) that account most successfully for the sample variance, that is, the differences or similarities between the various sets of geometric data.

In cluster analysis[57-66] each variable is considered as a point in a separate dimension, and the points so obtained are analyzed in multidimensional space. Similar points identified within this multidimensional space are grouped together. For example, in a study of five-coordinate d^8 metal complexes,[56,67,68] Bürgi and Auf der Heyde showed that the coordination of the metal in crystal structures can be described by three data clusters. These are representations of a **trigonal bipyramid** and

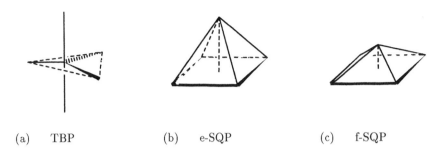

(a) TBP	(b) e-SQP	(c) f-SQP

FIGURE 16.9. Results of a cluster analysis of five-coordinate derivatives of d^8 metals. The structures can be separated into three clusters. (a) Trigonal bipyramid (TBP), (b) elevated square pyramid (e-SQP), and (c) flattened square pyramid (f-SQP).

two types of **square-planar pyramids** (one elevated and the other flattened, see Figure 16.9).

Factor analysis[58,59,69–71] is a method by which correlations among all variables are searched for simultaneously. The overall aim (in a crystallographic context) is to analyze molecular geometry from a set of similar structures, some of which are of limited precision, and find out if there are any significant effects that stand out from the noise due to experimental errors. In the example of five-coordinate complexes of d^8 metals by Bürgi and Auf der Heyde (described above), a factor analysis indicated that the difference between the elevated and flattened pyramids is significant.

Factor analysis involves the transformation of n orthogonal axes that represent the n variables into n new axes that lie along the directions of maximum variance (that is, differences between the various pieces of data). This transformation is usually illustrated in a two-dimensional example, by the scatter of points about a straight line, as shown in Figure 16.10. The two axes are rotated in this Figure so that one lies along the direction of most points. Two factors, $\mathbf{f}_1$ and $\mathbf{f}_2$, result, where the direction of greatest variance is $\mathbf{f}_1$, while $\mathbf{f}_2$ is perpendicular to this direction. Customarily, the factors finally chosen are linear combinations of the original variables.

$$\mathbf{f}_1 = a_1 x_1 + a_2 x_2 + a_3 x_3 + \cdots + a_n x_n, \text{ etc.,} \tag{16.1}$$

where x_i are the original variables and a_i, the "loadings," represent the relative importance of each original variable to this particular factor, $\mathbf{f}_1$. Essentially the factor analysis involves an **eigenanalysis** of the covariance or correlation matrix. The factor with the largest eigenvalue is the most important. It is named the principal **factor** and will lie along the axis of maximum variance of the data. Computer programs have been written to do this type of analysis.[59]

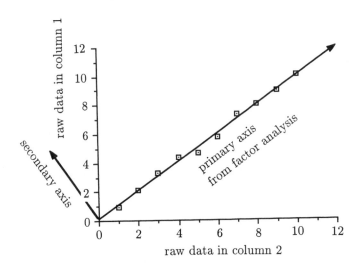

FIGURE 16.10. Two-dimensional plot showing factors. Data in "column 1" are plotted against data in "column 2." The result is approximately a straight line, which, since this line is only one-dimensional, can be considered as just one factor.

The philosophy of a factor analysis is illustrated, in a book by Edmund R. Malinowski and Darryl G. Howery,[71] by an analysis of grades that college students obtain for a laboratory report they have written. "··· the same laboratory report will receive different grades from different professors because of variations in marking criteria ···" In other words, each Professor will view a selected report in different ways, depending on his or her particular style and interests (see Figure 16.11). The grade received can therefore be expressed as:

$$\text{grade}_{ik} = \sum_{j=1}^{n} (\text{true score}_{ij} \times \text{loading}_{jk}), \qquad (16.2)$$

where the "true score" is the grade that students i should obtain in factor j, while the "loading" is the importance that Professor k gives to factor j (neatness, percent yield, completeness of report, etc.). If all the grades that all students received for all reports are available, then the data matrix:

grades : rows = students / columns = professors

can be expressed as the product of two matrices, one the "true score matrix:"

rows = students / columns = factors

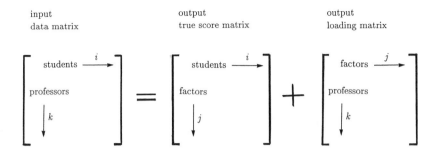

FIGURE 16.11. Matrices in the example used here.

and the other the "loading matrix:"

$$rows = factors / columns = professors.$$

"The purpose of factor analysis $\cdots$ is to extract the student score index [the true score matrix] from the data matrix in order to determine the students' true abilities in each subject, in effect removing the professors' prejudices [the loadings] from the grades." The starting point is the data matrix, and the end point is a series of factors with (hopefully) some physical meaning (Figure 16.11). The method for doing this can be complex and mathematical. For more information see Reference 71.

A cluster analysis can be used to identify discrete conformational subgroups of a chemical fragment, while a factor analysis gives an indication of significant components in each cluster. These types of analysis are used to interpret a large amount of data of great complexity, and provide a simplification in the description of the system. It answers the questions: how many factors were involved in making up the observed data, and how can these factors be described in terms of a physical model. For example, in the analysis of crystal structure results, a significant use of factor analysis is the determination of factors that account for a slight conformational variability between structures being compared.

16.3 Examples of cluster and factor analyses

Comparisons of the types just described can provide information of great chemical or biochemical interest. A few examples, some already alluded to, follow.

16.3.1 Thioesters

The ester bond –CO–O–R contains both a C=O and a C–O–R group. It is found that the electrons are delocalized within this group, resulting in resonance stabilization of the type for benzene:

$$R_3C-\overset{\overset{\displaystyle O}{\|}}{C}-O-CR'_3 \rightleftharpoons R_3C-\overset{\overset{\displaystyle O^-}{|}}{C}=\overset{+}{O}-CR'_3$$

On the other hand, infra red spectroscopic studies[73] indicate that this resonance stabilization does not occur with thioesters.

$$R_3C-\overset{\overset{\displaystyle O}{\|}}{C}-S-CR'_3 \;\cancel{\rightleftharpoons}\; R_3C-\overset{\overset{\displaystyle O^-}{|}}{C}=\overset{+}{S}-CR'_3$$

This was further analyzed (Figure 16.12), by a comparison of the X-ray crystal structures of thioesters with the normal *O*-esters.[73] The extent

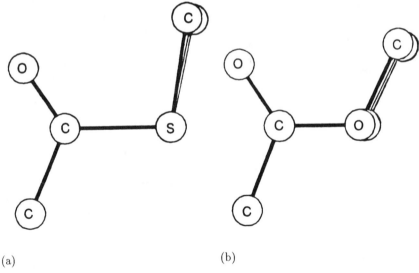

(a)   (b)

FIGURE 16.12. Comparison of *O*-esters and thioesters (*S*-esters) (Ref. 73). (a) Comparison of the averaged alkyl vinyl/aryl sulfide with an averaged thioester. These two are superimposed at the carbonyl carbon atom and aligned along the C(carbonyl)–S bond. Note that the sulfur atom is in the same place in the superposition. (b) An analogous comparison of an averaged alkyl vinyl/aryl ether and an averaged ethyl ester. Note that the oxygen atom of the C–O–C group is not in the same position for esters and ethers, indicating a shortening (due to resonance) in the ester.

of delocalization of the electrons within the two types of ester linkage, a thioester and an *O*-ester, was investigated (see Figure 16.12). The lengths of the C–O and C–S bonds could serve as indicators of which species (charged or uncharged) predominates.

An ester carbonyl group, however, involves an sp^2 carbon atom while the other carbon atom attached to the doubly bonded oxygen atom is sp^3. The sp^2 bonding results in a shortening of the C–(O,S) bond, and this has to be taken into account in the analysis.

The Cambridge Structural Database was used for the analysis and *O*-ester geometries were compared with those of alkyl vinyl/aryl ethers.[71] Similarly geometries of thioesters were compared with those of alkyl vinyl/aryl sulfides. The scheme is shown in Figure 16.12. Note that the C–O bond is shorter in *O*-esters than in alkyl vinyl or alkyl aryl ethers, but that the C–S bond length is the same in thioesters and in alkyl vinyl or alkyl aryl ethers. The results of this analysis imply that there is more resonance in *O*-esters than in thioesters. These results agree with the spectroscopic finding that the carbonyl group in acetyl-coenzyme A is ketone-like.[74] Since thioesters have little or no double-bond character in the C–S bond, they have a higher free energy of hydrolysis than do *O*-esters, which have appreciable double-bond character in the C–O bond. The lack of resonance stabilization of thioesters compared with *O*-esters provides a driving force in biological systems for *O*-ester or amide formation via acyl transfer.

16.3.2 Corrins

Vitamin B_{12} is the "anti-pernicious anemia factor (see Figure 1.13, Chapter 1), and it acts in the body via its adenosyl or methyl derivatives, called B_{12} coenzymes, which are cofactors for certain enzymes. The chemical formula is given in Figure 16.13. Both types of coenzyme contain a Co – C bond and therefore may be described as naturally occurring organometallic compounds. The coenzymes appear to act by cleavage of the Co–C bond; this cleavage is believed to be homolytic, that is, one electron goes to the cobalt and one to the carbon atom when the Co–C bond is broken.[75–78] Most of the enzymes that utilize a B_{12} coenzyme in their activity catalyze the vicinal exchange of hydrogen and another functional group; the remaining B_{12}-utilizing enzymes are involved in nucleic acid synthesis via a ribonucleotide reductase.

The crystal structure of vitamin B_{12} was determined by Dorothy Hodgkin and co-workers in 1954, and this established the chemical formula of the vitamin (Figure 16.13, compare with Figure 1.13, Chapter 1). It was shown to contain a ring structure that had resemblances to a porphyrin ring system, but with one bridge atom missing and with all β positions on the pyrrole rings saturated. This ring system was later named a "corrin," where the first three letters imply *core*, not *cobalt*, so that complexes with other metals can also be called corrins. Vitamin B_{12}

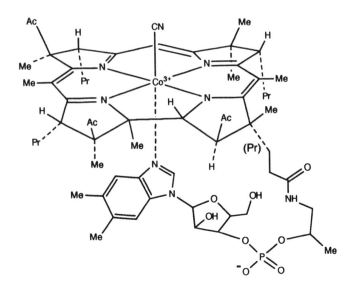

(a)

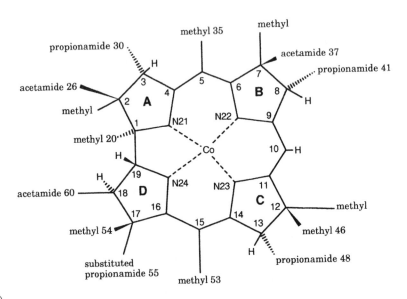

(b)

FIGURE 16.13. Formula for vitamin B$_{12}$. (a) Simplified formula where Me = CH$_3$, Ac = CH$_2$CONH$_2$, Pr = CH$_2$CH$_2$CONH$_2$. (b) Details of the numbering in the corrin ring system.

and its coenzymes have axially bound 5,6-dimethylbenzimidazole, connected to an amide side chain through a linkage of ribose and phosphate. The coenzymes have either adenosyl or methyl groups bound by a Co–C bond on the opposite side of the corrin ring system from the axial Co–N bond involving 5,6-dimethylbenzimidazole. The vitamin has cyanide in place of the adenosyl or methyl groups. The vitamin is a heptaamide with one acetamido or propionamido side chain on each β position, except for one that merely has a methyl group on it. These side chains were shown by the structure analyses to project axially from the β positions, because of their bulk. They seriously limit the flexibility of the molecule. They serve to protect the chemically reactive Co–C bond in the coenzymes from external attack, thereby controlling the fate of the free radical when the Co–C bond is broken homolytically.

The crystal structures of many corrins have been determined. Therefore it was questioned whether flexing of the corrin ring system can, in any way, assist in the cleavage of the Co–C bond. In a factor analysis of twelve corrin crystal structures[55] the principal factor involved variation perpendicular to the corrin ring, as shown in Figure 16.14, which represents the limits of variation. While atoms near C5 and C15 move up, those near C10 move down and vice versa. This can also be described as a butterfly-like folding of the corrin ring about a line through the cobalt atom and C10. A distance matrix analysis (described above) was also used. Since this type of analysis is more sensitive to variations with respect to out-of-ring atoms than in-plane atoms, the methyl group, C20, was used as a reporter group. There are 110 atoms in an average corrin molecule (excluding hydrogen atoms), and this number makes for immense distance matrices. The analysis was therefore simplified by partitioning the matrix. The molecule was divided into segments, principally the five-membered rings and the bridges between them. Distances from all atoms in one such segment to all atoms in a second segment were summed and used as a single entry in the matrix. Differences in such values for various B_{12} derivatives were calculated. The analysis showed, as did the factor analysis, a flexing that was mainly about the Co···C10 direction, but it also indicated that most of the flexing occurs on the C5 side of the molecule, as expected, because this is the side sterically pushed on by the axial 5,6-dimethylbenzimidazole.

When the Co–C bond is broken in the adenosyl coenzyme, the adenosyl group must move and transfer the free radical to the substrate site that is probably not directly on the cobalt atom. This analysis of corrin ring flexibility also included information on the accessibility of various portions of the molecule. Such **steric accessibility** indicates the availability of that part of the molecule to attack by other molecules. Random points (50 points per Å^3, for example) were generated around each atom at its van der Waals radius. If any of these points around an atom was within the van der Waals radius of another atom in the structure, the

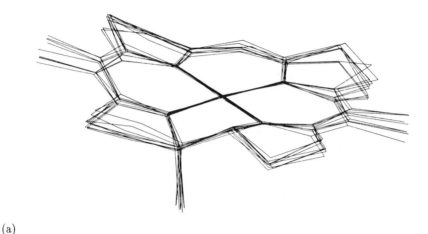

(a)

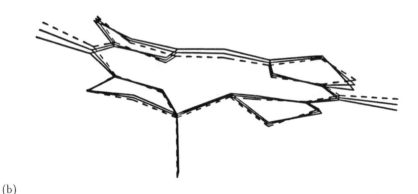

(b)

FIGURE 16.14. Results of a factor analysis for corrins. (a) Experimental variation in conformations of a series of corrins, fit to the five central atoms, and (b) limits of displacement in a factor analysis with broken lines at one limit and solid lines at the other. Note that the ring system folds in a butterfly-like manner.

atom was considered buried, otherwise it was considered exposed. The ratio of exposed to total points was computed for each atom.[55] Such an analysis showed that the area around C10 is the most accessible part of the corrin ring, while the area around C1 and C19 is the least accessible.

Further analyses, by direct sectioning with van der Waals radii above the corrin ring plane, showed that the cobalt atom is accessible, after loss of the adenosyl group, from the adenosyl side in B_{12} coenzyme (adenosylcobalamin) by three routes: one route is directly onto the cobalt in a path perpendicular to the corrin ring plane; of the other two routes, one is over C10 and the other is over C5. It was proposed from these comparative studies of crystal structure determinations that the adenine group may remain somewhat fixed in the enzyme and that rotation about the glycosidic bond of the adenosyl group, previously on the coenzyme, may occur (Figure 16.15). In this way the free radical can be moved to the active site of the B_{12}-utiizing enzyme.

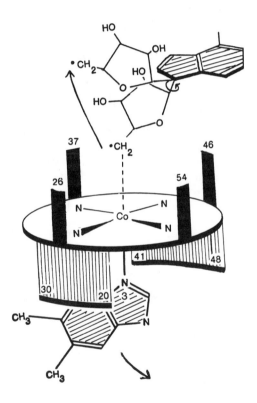

FIGURE 16.15. Proposed movement of a free radical when the Co–C bond is broken in a B_{12} coenzyme (Ref. 78). The Co^{3+}–C bond is broken and the adenosyl radical moves to the active site of the enzyme bound to the B_{12} cofactor. The lower axial ligand, the 5,6-dimethylbenzimidazole, can also swing away, but is constrained in its pathway by the sidechains 20, 30, 41, and 48. In a similar way the path of the adenosyl group is controlled by the sidechains 26, 37, 46, and 54.

16.3.3 Five-coordinate d^8 metal complexes

The analyses of five-coordinate d^8 metal complexes, where the metal is Ni(II), Pd(II), Pt(II), Rh(I) or Ir(I) described earlier, have thrown some interesting light on the stereochemistry of reactions of these complexes.[67] The chemical reaction involved is

$$XML_3 + Y \rightarrow [XML_3Y] \rightarrow X + ML_3Y \qquad (16.3)$$

where M is the atom of interest, and X and L represent any ligand atom and Y is either X, L, or solvent. The investigation has given information on the likelihood that the complexes could undergo a **Berry pseudorotation mechanism** (Figure 16.16). The CSD was searched for five-coordinate complexes of the d^8 metals. In addition, four-coordinate complexes were reviewed to determine if they could alternately be classified as five-coordinate or nearly five-coordinate. Five-coordinate complexes generally have a trigonal bipyramidal or square-based pyramidal structure or slightly distorted forms of either. R. Stephen Berry, however, showed that trigonal bipyramidal (TBP) and square-based pyramidal (SQP) structures are interconvertible.[79]

These two types of five-coordinate structures, TBP and SQP, can be described by two sets of nonredundant symmetry coordinates leading to two 12-dimensional spaces spanned by symmetry coordinates. These spaces were termed T space for TBP and S space for SQP. Since several different metals were involved in this study, the data on metal–ligand

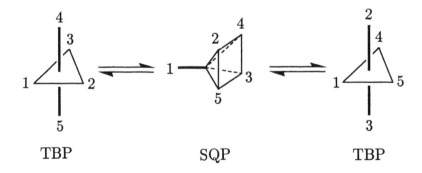

TBP   SQP   TBP

FIGURE 16.16. Five-coordinate d^8 complexes and the Berry pseudorotation mechanism. In a trigonal bipyramid (TBP) there are three equatorial ligands (**1, 2, 3**) at 120° to each other, and two axial ligands (**4, 5**) at 90° to the equatorial ligands. By slight movements of the ligands, such that the interligand angles change between 90° and 120°, the identities of the axial and equatorial ligands can be changed. The intermediate is a square-based pyramid (SQP), as shown.

distances were put on a common basis by comparing the distances to standard values. The resulting distributions of points in the two 12-dimensional spaces were then probed by cluster analysis. In both cases, four clusters of points were obtained. The analysis in T space gave a trigonal bipyramid and three slightly distorted square pyramids with *trans*–basal angles of 161° and 169°, respectively. The analysis in S space gave a flattened square pyramid (denoted f-SQP) with *trans*–basal angles of 171° and an apical bond 0.7 Å in excess of the standard M–L length and an elevated square pyramid (denoted e-SQP) with *trans*–basal angles of 163° and an apical bond only 0.03 Å longer than the standard M–L length (denoted e-SQP). However, the square pyramids in the T space were shown to be divided between these two types (f-SQP and e-SQP). Thus the two cluster analyses give the same result, that there are three main conformations of d^8 complexes: a trigonal bipyramid (TBP), a flattened square pyramid (f-SQP), and an elevated square pyramid (e-SQP). A factor analysis was then used to map the types of distortions from these three conformations that are actually found in crystals. The results of this factor analysis showed that the structures of the d^8 complexes can be divided into three groups: platinum and palladium complexes (mainly f-SQP), rhodium and iridium complexes (mainly TBP and e-SQP), and nickel complexes (slight preference for e-SQP over TBP and f-SQP). This is shown in Figure 16.17. Interestingly, f-SQP conformations do not show a Berry-type distortion, while e-SQP conformations do.

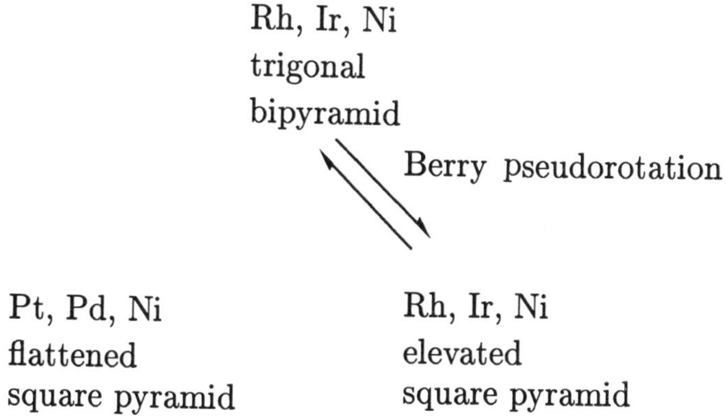

Rh, Ir, Ni
trigonal
bipyramid

Berry pseudorotation

Pt, Pd, Ni
flattened
square pyramid

Rh, Ir, Ni
elevated
square pyramid

FIGURE 16.17. The three types of five-coordinate d^8 complexes and their relationship to the Berry pseudorotation mechanism (see Figure 16.9). It appears that five-coordinate complexes of Rh, Ir, and Ni can undergo pseudorotation between a TBP and e-SQP structure. Other five-coordinate complexes of Ni, plus those of Pt and Pd, form f-SQP structures and do not undergo pseudorotation.

Thus the Berry coordinate represents a viable option for intramolecular exchange in rhodium and iridium complexes, in contrast to platinum and palladium complexes. Nickel complexes, on the other hand, can adopt either tetrahedral or square-planar conformations in the four-coordinate structures, and therefore the fact that these complexes can take on any of the three conformations is not surprising. This analysis is described in detail in Reference 67.

16.3.4 β/α barrels in proteins

A common structural motif in proteins is the (β/α-**barrel structure** [($\beta\alpha)_8$- or TIM barrel],mentioned in Chapter 12.[80−86] It is constructed of eight β sheets surrounded by eight α helices (Figure 16.18), and was first found in the crystal structure of the glycolytic enzyme triose phosphate isomerase(abbreviated to TIM) from chicken muscle.[81] Since that time it has been found in many other proteins[81] such as pyruvate kinase,[83] KDPG aldolase,[84] taka-amylase,[87], glycolate oxidase,[88] and D-xylose isomerase.[89] Detailed comparisons of the differing shapes of these various β/α barrels have been made.[80,90] The variety of biochemical reactions involved also lead to analyses of the mechanisms and the locations on the active sites where these reactions occur.[91]

The motif is a hydrophobic β strand with conserved residues at its C terminus, followed by a loop of conserved residues that lead to an amphipathic α helix. A final loop also appears to tolerate amino-acid deletions or mutations.[92,93] This motif is repeated eight times and is packed to give a cylindrical, barrel-like structure with the eight β sheets in the interior and the eight α helices outside them. The β strands are tilted approximately $-36°$ to the axis of the barrel, while the axes of the α helices are approximately parallel to the strands. Since $\beta\alpha\beta$ units generally have a right-handed connection,[94] the direction of the strands around the barrel is the same in all such structures. The β sheets are staggered as shown in Figure 16.19 for TIM. The average twist between successive strands is $26°$ as found in open β sheets.

Variations in the shape of the barrel, the number of residues involved, and the disposition of the α helices have been studied by superpositions of all such structures.[92−99] A comparison of the shapes of β/α barrels in nine proteins[95] showed that the mean radius of the barrel is fairly constant but the axial ratio can vary from 1.0 to 1.5. There appear to be at least two classes of β/α barrels,[96] depending on the number of residues that the N-terminal strand and the other odd-numbered strands contribute. The interior of the barrel is filled mainly by amino acid side chains. By slicing through a space-filling model of the protein at 1 Å intervals, it was possible to show that the total volume of amino acid residues in the interior of the barrel is that normally expected for the interior of proteins. The active site of the enzyme is generally found to lie at the C-terminal end of the β strands.[92]

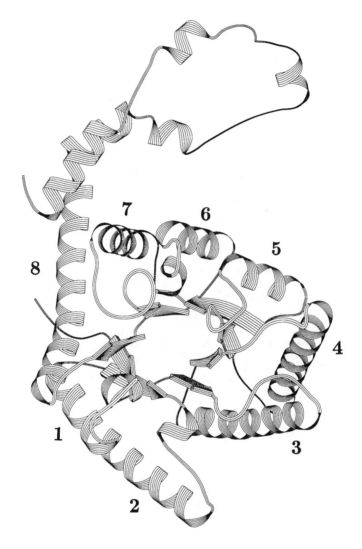

(a)

FIGURE 16.18. (a) Overall diagram of a (β/α) barrel (D-xylose isomerase) (Ref. 89). The β strands form a lining for the barrel, and the α helices, numbered sequentially from the N terminus, pack around the β lining and provide a stable structure.

16.3.5 Further studies

The number of reported comparisons of sets of molecules for crystal sheets is large, and there is not space here to include each. A selection of references of additional analyses of small molecules is included for the reader who is interested in this subject. These studies include comparisons of

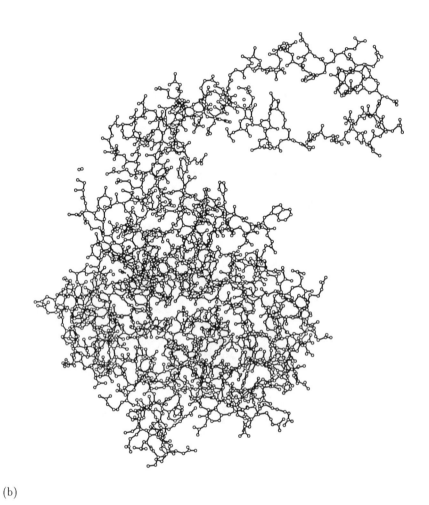

(b)

FIGURE 16.18 (cont'd). (b) Atomic connectivity of the protein molecule represented by "ribbons" in (a).

carboxyl and related groups,[100-104] nucleosides and nucleotides,[105-108] small rings,[109-114] and tertiary-butyl groups.[115,116]

Finally, a warning should be added here reminding the reader that in comparing structures the estimated significances of compared structural data must be taken into account. An introduction to this was given in Chapter 11.

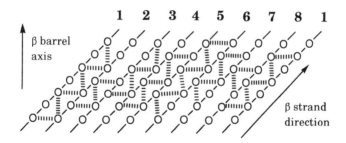

FIGURE 16.19. Details of β-sheet folding in the lining of a $(\beta\alpha)_8$ barrel. Amino-acid residues are shown as open circles, joined by solid lines; these indicate the backbone. Hydrogen bonds are indicated by thick broken lines. These bind the various strands (**1** to **8**) into a β sheet. The direction of the strands in the sheet is inclined at an angle (as shown) to the $(\beta\alpha)_8$ barrel axis.

Summary

1. The results of crystal structure determinations are stored in accessible form in crystallographic databases. Extraction of this information is made simple by those who designed the databases. The following databases contain information on crystal structures:

   (a) the Cambridge Structural Database contains crystallographic data on all small organic compounds for which published results of crystal structure analyses are available,

   (b) the Protein Data Bank contains crystallographic data on all biological macro-molecules, mainly proteins and nucleic acids, for which published results of crystal structure analyses are available,

   (c) the Inorganic Crystal Structure Database contains crystallographic data on all inorganic compounds for which published results of crystal structure analyses are available,

   (d) the NRCC Metals Crystallographic Data File contains crystallographic data on all metals, together with such data on some hydrides and binary oxides, for which published results of crystal structure analyses are available, and

   (e) the Powder Diffraction File contains information on unit cell dimensions, space groups and intensities of diffraction patterns of any powdered solid materials. Its major use is for comparing the powder diffraction pattern of a material with those in the Powder Diffraction File. It is accessed in order to compare diffraction patterns, usually in order to establish the identity of a material.

2. Comparisons of molecules may be done by a variety of techniques:

   (a) by preparing lists of analogous bond lengths, interbond angles, and torsion angles for all the compounds being compared,

(b)   by use of bar graphs for illustrating the frequency with which a given value of a geometric parameter is encountered in a comparison among similar compounds,

(c)   by superpositions of groups of atoms, usually by least-squares methods, minimizing deviations from a selected group of atoms,

(d)   by preparation and comparison of distance matrices, which represent all interatomic distances in an ordered fashion, such as along the backbone of a protein,

(e)   by cluster analysis which is used to sort the geometric data (of molecules being compared) into groups, and

(f)   by factor analysis which is used to extract factors that account for the variability between geometric parameters in the various structures being compared.

Glossary

Agonist: Agonists bind to a biological receptor and elicit an appropriate biological response.

Antagonist: A drug that binds to a biological receptor, but is unable to elicit the normally expected response. It prevents an agonist from binding to the same sits as that normally occupied by the agonist (competition), or at another site (allosterism).

Bar graph: A chart that represents a frequency distribution (a histogram). Vertical rectangles are of equal width, and their heights are proportional to the frequencies of their values.

Berry pseudorotation mechanism: In five-coordinate complexes, trigonal bipyramidal and square-based pyramidal conformations have similar energies and can be interconverted by small and simple angle deformations in which *axial* and *equatorial* vertices of the trigonal bipyramid are interchanged. A square-based pyramid is an intermediate in the interconversion of one trigonal bipyramid to another. The resultant trigonal bipyramidal conformation is identical to a rotated version of the original conformer, but, because this rotation is apparent but not real, it is called pseudorotation.

Bibliographic search: A search (by computer) for the journal reference(s) to a selected crystal structure or a set of crystal structures.

Cluster analysis: A statistical analysis of data with respect to certain characteristics in order to find out whether those data fall into recognizable groups (clusters).

Connectivity search: A search (by computer) for crystal structures containing atoms that are connected together in a manner defined by the input to the search program.

Database: A collection of data on a particular subject, such as unit cell dimensions, space group, and atomic coordinates of a class of compounds. These data can be retrieved by computer data techniques.

Distance matrix: A matrix of distances, originally defined as from each α-carbon atom in a protein to every other α-carbon atom. The matrix is indexed in the consecutive order of the amino acid residues in the protein. The term is also used now more generally.

Eigenanalysis: An analysis to determine the characteristic vectors (eigenvectors) of a matrix. These are a measure of the principal axes of the matrix.

Excluded volume: The difference between the shapes of a compound that binds to, for example, an enzyme and one that does not. Some of the volume areas in a model of the inactive compound, with van der Waals radii around atoms, will not fit in the active site of the enzyme. These areas should be avoided when designing new inhibitors.

Factor: A parameter that contributes to the production of a result.

Factor analysis: A method of simultaneously searching for correlations among all variables in a data set. A factor is a component not initially completely defined. The aim is to characterize it by analyses of all the available data.

Histogram: A graphical representation of a frequency distribution. Rectangles are used. Their heights represent the frequencies, and their widths represent the class interval.

Software: The specifications provided to a computer (the hardware) that enables it to perform desired calculations. The software includes assemblers, translators, and programs.

Square-based pyramid: A polygon with a square base and the other four faces composed of triangles that share a vertix in a pyramidal manner.

Steric accessibility: A measure of the probability that a (very small) molecule can reach a site on the surface of a different (large) molecule.

Superposition diagrams: The results of laying all the structures on top of each other, using a selected group of atoms for as exact an overlap as possible.

Trigonal bipyramid: Two pyramids that share a common triangular base and extend on either side of it.

References

1. Glusker, J. P. Citrate conformation and chelation: enzymatic implications. *Acc. Chem. Res.* **13**, 345–352 (1980).
2. Glusker, J. P. Mechanism of aconitase action deduced from crystallographic studies of its substrates. *J. Molec. Biol.* **38**, 149–162 (1968).
3. Lauble, H., Kennedy, M. C., Beinert, H., and Stout, C. D. Crystal structures of aconitase with isocitrate and nitroisocitrate bound. *Biochemistry* **31**, 2735–2748 (1992).
4. Murray-Rust, P. Chemical information from crystal structures. In: *Patterson and Pattersons. Fifty Years of the Patterson Function.* (Eds., Glusker, J. P., Patterson, B. K., and Rossi, M.) Ch. 6, pp. 86–116. International Union of Crystallography/Oxford University Press: New York (1987).
5. Bernal, J. D. *Science in History.* p. 943. 3rd edn. Hawthorn: New York (1965).
6. Hydén, H. Afterthoughts. In: *Beyond Reductionism: New Perspectives in the Life Sciences. (The Alpbach Symposium 1968.)* (Eds., Koestler, A. and Smithies, J. R.) pp. 114–115. Hutchinson: London (1968).
7. Groth, P. *Chemische Kristallographie.* [Chemical crystallography.] Engelmann: Leipzig (1906–1919).

8. Ewald, P. P. Laue's discovery of X-ray diffraction by crystals. In: *Fifty Years of X-ray Diffraction.* (**Ed.**, Ewald, P. P.) Ch. 4, pp. 31–56. Oosthoek: Utrecht (1962).

9. *Strukturbericht.* Vols. 1–7. Akademische Verlagsgesellschaft. Becker and Erler: Leipzig (1931–1943). Reprinted by Edward Brothers, Inc: Ann Arbor, MI (1943).

10. *Structure Reports. A. Metals and Inorganic. B. Organic (including organometallic compounds).* 60 year indices available (1913–1973 with supplements for 1974–1975) Section A complete through 1985. Section B complete through 1984. Oosthoek (IUCr): Utrecht (1913 to date).

11. Porter, M. W., and Spiller, R. C. *The Barker Index of Crystals.* Vol. I (1951), Vol. II (1956). Cambridge, England (1951, 1956).

12. *Crystal Data Determinative Tables.* All crystallographic phases except proteins for which unit cell dimensions have been determined. Vols. 1,3,5,6 Organic and organometallic compounds through 1981. Vols. 2,4 inorganic compounds through 1969. 3rd edn. (**Eds.**, Mighell, A. D., and Stalick, J. K.) U.S. Department of Commerce, NBS/ JCPDS–International Centre for Diffraction Data, Swarthmore, PA (1972–1983).

13. *Crystallographic Databases. Information Content. Software Applications. Scientific Applications.* Published by the Data Commission of the International Union of Crystallography: Bonn, Cambridge, Chester (1987).

14. Allen, F. H., Bellard, S., Brice, M. D., Cartwright, B. A., Doubleday, A., Higgs, H., Hummelink, T., Hummelink-Peters, B. G., Kennard, O., Motherwell, W. D. S., Rodgers, J. R., and Watson, D. G. The Cambridge Crystallographic Data Centre: computer-based search, retrieval, analysis and display of information. *Acta Cryst.* **B35**, 2331–2339 (1979).

15. Kennard, O., Watson, D. G., Allen, F. H., Motherwell, W., Town, W., and Rodgers, J. Crystal clear data. *Chem. Britain* **11**, 213–216 (1975).

16. Wilson, S. R., and Huffman, J. C. Cambridge Data File in Organic Chemistry. Applications to transition-state structure, conformational analysis, and structure/activity studies. *J. Org. Chem.* **45**, 560–566 (1980).

17. Allen, F. H., Kennard, O., and Taylor, R. Systematic analysis of structural data as a research technique in organic chemistry. *Acc. Chem. Res.* **16**, 146–153 (1983).

18. Bürgi, H. B., and Dunitz, J. D. From crystal statics to chemical dynamics. *Acc. Chem. Res.* **16**, 153–161 (1983).

19. Schweizer, W. B. Practical applications of data bases. In: *Crystallographic Computing 3.* (**Eds.**) Sheldrick, G. M., Kruger, C., and Goddard, R.) pp. 119–127. Clarendon Press: Oxford (1985).

20. Murray-Rust, P., and Motherwell, W. D. S. Computer retrieval and analysis of molecular geometry. I. General principles and methods. *Acta Cryst.* **B34**, 2518–2526 (1978).

21. Murray-Rust, P., and Bland, R. Computer retrieval and analysis of molecular geometry. II. Variance and its interpretation. *Acta Cryst.* **B34**, 2527–2533 (1978).

22. Taylor, R., and Kennard, O. The estimation of average molecular dimensions from crystallographic data. *Acta Cryst.* **B39**, 517–525 (1983).

23. Murray-Rust, P., and Raftery, J. Computer analysis of molecular geometry. Part VI. Classification of differences in conformation. *J. Molec. Graphics* **3**, 50–59 (1985).

24. Murray-Rust, P., and Raftery, J. Computer analysis of molecular geometry. Part VII. The identification of chemical fragments in the Cambridge Structural Data File. *J. Molec. Graphics* **3**, 60–68 (1985).

25. Taylor, R. The Cambridge Structural Database in molecular graphics: techniques for the rapid identification of conformational minima. *J. Molec. Graphics* **4**, 123–131 (1986).

26. Murray-Rust, P. Automatic analysis of crystal and molecular geometry on data files. In: *Methods and Applications in Crystallographic Computing.* (**Eds.**, Hall, S. R., and Ashida, T.) pp. 387–396. Clarendon Press: Oxford (1984).

27. Murray-Rust, P. How useful are X-ray studies of conformation. In: *Molecular Structure and Biological Activity.* (**Eds.**, Griffin, J. F., and Duax, W. L.) pp. 117–130. Elsevier Biomedical: New York (1982).

28. Bernstein, F. C., Koetzle, T. F., Williams, G. J. B., Meyer, E. F., Jr., Brice, M. D., Rodgers, J. R., Kennard, O., Shimanouchi, T., and Tasumi, M. The Protein Data Bank: a computer-based archival file for macromolecular structures. *J. Molec. Biol.* **112**, 535–542 (1977).

29. Abola, E. E., Bernstein, F. C., Bryant, S. H., Koetzle, T. F., and Weng, J. Protein Data Bank. Section 2.4. pp. 107–132. In: *Crystallographic Databases. Information Content, Software Applications, Scientific Applications.* International Union of Crystallography: Bonn, Cambridge, Chester (1987).

30. Gilliland, G. L. A biological macromolecule crystallization database: a basis for a crystallization strategy. *Crystal Growth* **90**, 51–59 (1988).

31. Singh, J., and Thornton, J. M. *Atlas of Protein-Side Chain Interactions.* 2 vols. IRL Press: Oxford, New York, Tokyo (1992).

32. Bergerhoff, G., Hundt, R., Sievers, R., and Brown, I. D. The Inorganic Crystal Structure Database. *J. Chem. Inf. Comput. Sci.* **23**, 66–69 (1983).

33. Bergerhoff, G., and Brown, I. D. Inorganic Crystal Structure database. In: *Crystallographic Databases. Information Content, Software Applications, Scientific Applications.* Section 2.2. pp. 77–95. International Union of Crystallography: Bonn, Cambridge, Chester (1987).

34. Brown, I. D. The standard crystallographic file structure. *Acta Cryst.* **A39**, 216–224 (1985).

35. Brown, I. D. Standard crystallographic file structure — 84. *Acta Cryst.* **A41**, 399 (1983).

36. Brown, I. D., Bradley, S. M., and Altermatt, D. Using the Inorganic Crystal Structure Database for the systematic examination of inorganic crystal structures. In: *Molecular Structure: Chemical Reactivity and Biological Activity.* (**Eds.**, Stezowski, J. J., Huang, J-L., and Shao, M-C.) Ch. 52, pp. 513–519. Oxford University Press: Oxford (1988).

37. Altermatt, D., and Brown, I. D. The automatic searching from chemical bonds in inorganic crystal structures. *Acta Cryst.* **B41**, 240–244 (1985).

38. Brown, I. D., and Altermatt, D. Bond-valence parameters obtained from a systematic analysis of the Inorganic Crystal Structure Database. *Acta Cryst.* **B41**, 244–247 (1985).

39. Rodgers, J. R., and Wood, G. H. NRCC Metals Crystallographic Data File (CRYSTMET). In: *Crystallographic Databases. Information Content, Software Applications, Scientific Applications.* Section 2.3. pp. 96–106. International Union of Crystallography: Bonn, Cambridge, Chester (1987).

40. Jenkins, R., and Smith, D. K. Powder diffraction file. In: *Crystallographic Databases. Information Content, Software Applications, Scientific Applications.* Section 3.3. pp. 158–175. International Union of Crystallography: Bonn, Cambridge, Chester (1987).

41. Mighell, A. D., and Himes, V. L. Compound identification and characterization using lattice-formula matching techniques. *Acta Cryst.* **A42**, 101–105 (1986).

42. Parrish, W. History of the X-ray powder method in the U.S.A. In: *Crystallography in North America.* (**Eds.**, McLachlan, D., Jr., and Glusker, J. P.) Section D. Apparatus and methods. Ch. 1, pp. 201–214. American Crystallographic Association: New York (1983).

43. Hanawalt, J. D. History of the Powder Diffraction File PDF. In: *Crystallography in North America.* (**Eds.**, McLachlan, D., Jr., and Glusker, J. P.) Section D. Apparatus and methods. Ch. 2, pp. 215–219. American Crystallographic Association: New York (1983).

44. Bürgi, H.-B., and Dunitz, J. D. Can statistical analysis of structural parameters from different crystal environments lead to quantitative energy relationships? *Acta Cryst.* **B44**, 445–448 (1988).

45. Riveros, J. M., and Wilson, E. B. Microwave spectrum and rotational isomerism of ethyl formate. *J. Chem. Phys.* **46**, 4605–4612 (1967).

46. Wilson, E. B. Conformational studies on small molecules. *Chem. Soc. Rev.* **1**, 293–318 (1972).

47. Marshall, G. R., Barry, C. D., Bosshard, H. E., Dammkoehler, R. A., and Dunn, D. A. The conformational parameter in drug design: the active analog approach. In: *Computer-assisted Drug Design.* (**Eds.**, Olson, E. C., and Christofferson, R. E.) Ch. 9, pp. 205–226. ACS Symposium Series 112. American Chemical Society: Washington (1979).

48. Phillips, D. C. The development of crystallographic enzymology. In: *British Biochemistry Past and Present.* (**Ed.**, Goodwin, T. W.) pp. 11–28. Academic Press: London (1970).

49. Nordman, C. E., Weldon, A. S., and Patterson, A. L. X-ray crystal analysis of the substrates of aconitase. II. Anhydrous citric acid. *Acta Cryst.* **13**, 418–426 (1960).

50. Glusker, J. P., Minkin, J. A., and Patterson, A. L. X-ray crystal analysis of the substrates of aconitase. IX. A refinement of the structure of anhydrous citric acid. *Acta Cryst.* **B25**, 1066–1072 (1969).

51. Roelofsen, G., and Kanters, J. A. Citric acid monohydrate, $C_6H_8O_7 \cdot H_2O$. *Cryst. Struct. Commun.* **1**, 23–26 (1972).

52. Liebman, M. N. Correlations of structure and function in biologically active small molecules and macromolecules by distance matrix partitioning. In: *Molecular Structure and Biological Activity.* (**Eds.**, Griffin, J. F., and Duax, W. L.) pp. 193–212. Elsevier Biomedical: New York, Amsterdam, Oxford (1982).

53. Liebman, M. Quantitative analysis of structural domains in proteins. *Biophys. J.* **32**, 213–215 (1980).

54. Nishikawa, K., Ooi, T., Isogai, Y., and Saito, N. Tertiary structure of proteins. I. Representation and computation of the conformations. *J. Phys. Soc. Japan* **32**, 1331–1337 (1972).

55. Pett, V. B., Liebman, M. N., Murray-Rust, P., Prasad, K., and Glusker, J. P. Conformational variability of corrins: some methods of analysis. *J. Amer. Chem. Soc.* **109**, 3207–3215 (1987).

56. Auf der Heyde, T. P. E., and Bürgi, H.-B. Molecular geometry of d^8 five-coordination. 2. Cluster analysis, archetypal geometries, and cluster statistics. *Inorg. Chem.* **28**, 3970–3981 (1989).

57. Massart, D. L., and Kaufman, L. *The Interpretation of Analytical Chemical Data by the Use of Cluster Analysis.* John Wiley: New York, Chichester, Brisbane, Toronto (1983).

58. Auf der Heyde, T. P. E. Analysing chemical data in more than two dimensions. *J. Chem. Educ.* **67**, 461–469 (1990).

59. Dixon, W. J. (**Ed.**) *BMPD Statistical Software, 1983.* University of California Press: Berkeley, Los Angeles, London (1985).

60. Harvey, D. T., and Bowman, A. Factor analysis of multicomponent samples. *J. Chem. Educ.* **67**, 470–472 (1990).

61. Chatfield, C., and Collins, A. J. *Multivariate Analysis.* Chapman and Hall: London, New York (1980).

62. Allen, F. H., Doyle, M. J., and Taylor, R. / Automated conformational analysis from crystallographic data. 1. A symmetry-modified single-linkage clustering algorithm for three-dimensional pattern recognition. *Acta Cryst.* **B47**, 29–40 (1991).

63. Allen, F. H., Doyle, M. J., and Taylor, R. Automated conformational analysis from crystallographic data. 2. Symmetry-modified Jarvis-Patrick and complete-linkage clustering algorithms for three-dimensional pattern recognition. *Acta Cryst.* **B47**, 41–49 (1991).

64. Allen, F. H., Doyle, M. J., and Taylor, R. Automated conformational analysis from crystallographic data. 3. Three-dimensional pattern recognition within the Cambridge Structural Database System: implementation and practical examples. *Acta Cryst.* **B47**, 50–61 (1991).

65. Allen, F. H., and Johnson, O. Automated conformational analysis from crystallographic data. 4. Statistical descriptors for a distribution of torsion angles. *Acta Cryst.* **B47**, 62–67 (1991).

66. Everitt, B. *Cluster Analysis.* 2nd edn. Halstead Heinemann: London (1980).

67. Auf der Heyde, T. P. E., and Bürgi, H.-B. Molecular geometry of d^8 five-coordination. 3. Factor analysis, static deformations, and reaction coordinates. *Inorg. Chem.* **28**, 3982–3989 (1989).

68. Norskov-Lauritsen, L., and Bürgi H.-B. Cluster analysis of periodic distributions; application to conformational analysis. *J. Comput. Chem.* **6**, 216–228 (1985).

69. Domenicano, A., Murray-Rust, P., and Vaciago, A. Molecular geometry of substituted benzene derivatives. IV. Analysis of variance in monosubstituted benzene rings. *Acta Cryst.* **B39**, 457–468 (1983).

70. Murray-Rust, P. Computer analysis of molecular geometry. V. Symmetry aspects of factor analysis. *Acta Cryst.* **B38**, 2765–2771 (1982).

71. Malinowski, E. R., and Howery, D. G. *Factor Analysis in Chemistry.* John Wiley: New York, Chichester, Brisbane, Toronto (1980).

72. Harman, H. H. *Modern Factor Analysis.* 1st edn., revised. University of Chicago Press: Chicago (1976).

73. Zacharias, D. E., Murray-Rust, P., Preston, R. M., and Glusker, J. P. The geometry of the thioester group and its implications for the chemistry of acyl coenzyme A. *Arch. Biochem. Biophys.* **222**, 22–34 (1983).

74. El-Aasar, A. M. M., Nash, C. P., and Ingraham, L. L. Infrared and Raman spectra of S-methyl thioacetate: toward an understanding of the biochemical reactivity of esters of coenzyme A. *Biochemistry* **21**, 1972–1976 (1982).

75. Hodgkin, D. C., Pickworth, J., Robertson, J. H., Trueblood, K. N., Prosen, R. J., and White, J. G. The crystal structure of the hexacarboxylic acid derived from B$_{12}$ and the molecular structure of the vitamin. *Nature (London)* **176**, 325–328 (1955).

76. Lenhert, P. G. (1968) The structure of vitamin B$_{12}$. VII. The X-ray analysis of the vitamin B$_{12}$ coenzyme. *Proc. Roy. Soc. (London)* **A303**, 45–84 (1968).

77. Glusker, J. P. X-ray crystallography of B$_{12}$ and cobaloximes. In: *B$_{12}$. Vol. 1. Chemistry.* (**Ed.**, Dolphin, D.) pp. 23–106. Wiley Interscience: New York (1982).

78. Rossi, M., and Glusker, J. P. Vitamin B$_{12}$ and its coenzymes: structures and reactivities. In: *Environmental Influences and Recognition in Enzyme Activity.* (**Eds.**, Liebman, J. F., and Greenberg, A.) Chap. 1, pp. 1–58. VCH: New York, Weinheim (1988).

79. Berry, R. S. Correlation of rates of intramolecular tunneling processes, with application to some group V compounds. *J. Chem. Phys.* **32**, 933–938 (1960).

80. Muirhead, H. Triose phosphate isomerase, pyruvate kinase and other α/β-barrels. *Trends Biochem. Sci.* **8**, 326–330 (1983).

81. Alber, T., Banner, D. W., Bloomer, A. C., Petsko, G. A., Phillips, D. C., Rivers, P. C., and Wilson, I. A. On the three-dimensional structure and catalytic mechanism of triose phosphate isomerase. *Phil. Trans. Roy. Soc. (London)* **B293**, 159–171 (1981).

82. Chothia, C. Protein structure – the 14th barrel rolls out. *Nature (London)* **333**, 598–599 (1988).

83. Stuart, D. I., Levine, M., Muirhead, H., and Stammers, D. K. Crystal structure of cat muscle pyruvate kinase at a resolution of 2.6 Å resolution. *J. Molec. Biol.* **134**, 109–142 (1979).

84. Mavridis, I. M., Hatada, M. H., Tulinsky, A., and Lebioda, L. Structure of 2-keto-3-deoxy-6-phosphogluconate aldolase at 2.8 Å resolution. *J. Molec. Biol.* **162**, 419–444 (1982).

85. Chothia, C., and Finkelstein, A. V. The classification and origin of protein folding patterns. *Annu. Rev. Biochem.* **59**, 1007–1039 (1990).

86. Jaenicke, R. Protein folding: local structures, domains, subunits and assemblies. *Biochemistry* **30**, 3147–3161 (1991).

87. Matsuura, Y., Kusunoki, M., Harada, W., Tanaka, N., Iga, Y., Yasuoka, N., Toda, H., Narita, K., and Kakudo, M. Molecular structure of Taka-amylase A. I. Backbone chain folding at 3 Å resolution. *J. Biochem. (Tokyo)* **87**, 1555–1558 (1980).

88. Lundqvist, Y., and Brändén, C-I. Structure of glycolate oxidase from spinach at a resolution of 5.5 Å. *J. Molec. Biol.* **143**, 201–211 (1990).

89. Carrell, H. L., Rubin, B. H., Hurley, T. J., and Glusker, J. P. X-ray crystal structure of D-xylose isomerase at 4-Å resolution *J. Biol. Chem.* **259**, 3230–3236 (1984).

90. Lebioda, L., Hatada, M. H., Tulinsky, A., and Mavridis, I. Comparison of the folding of 2-keto-3-deoxy-6-phosphogluconate aldolase, triose phosphate isomerase and pyruvate kinase. Implications in molecular evolution. *J. Molec. Biol.* **162**, 445–458 (1982).

91. Rose, I. A. Chemistry of proton abstraction by glycolytic enzymes (aldolase, isomerases and pyruvate kinase). *Phil. Trans. Roy. Soc. (London)* **B293**, 131–143 (1981).

92. Niermann, T., and Kirschmann, K. Improving the prediction of secondary structure of 'TIM-barrel' structures. *Protein Engineering* **4**, 359–370 (1991).

93. Wilmanns, M., Hyde, C. C., Davies, D. R., Kirschner, K., and Jansonius, J. N. Structural conservation in parallel β/α-barrel enyzmes that catalyze three sequential reactions in the pathway of tryptophan biosynthesis. *Biochemistry* **30**, 9161–9169 (1991).

94. Richardson, J. S. Handedness of crossover connections in β sheets. *Proc. Natl. Acad. Sci. USA* **73**, 2619–2623 (1976).

95. Lasters, I., Wodak, S. J., Alard, P., and van Cutsem, E. Structural principles of parallel β-barrels in proteins. *Proc. Natl. Acad. Sci. USA* **85**, 3338–3342 (1988).

96. Lesk, A. M., Brändén, C-I., and Chothia, C. Structural principles of α/β barrel proteins: the packing of the interior of the sheet. *Proteins: Structure, Function, and Genetics* **5**, 139–148 (1989).

97. Urfer, R., and Kirschner, K. The importance of surface loops for stabilizing an eightfold $\beta\alpha$ barrel protein. *Protein Science* **1**, 31–45 (1992).

98. Chou, K.-C., and Carlacci, L. Energetic approach to the folding of $(\alpha\beta)$ barrels. *Protein Struct. Funct. Genet.* **9**, 280–285 (1990).

99. Luger, K., Szadkowski, H., and Kirschner, K. An 8-fold $\beta\alpha$-barrel protein with redundant folding possibilities. *Protein Engineering* **3**, 249–258 (1990).

100. Borthwick, P. W. Some relationships between bond lengths and angles in $-COO^-$, $-COOH$ and $-COOCH_3$ groups. *Acta Cryst.* **B36**, 628–632 (1980).

101. Schweizer, W. B., and Dunitz, J. D. Structural characteristics of the carboxylic ester group. *Helv. Chim. Acta* **65**, 1547–1554 (1982).

102. Chakrabarti, P., and Dunitz, J. D. Structural characteristics of the carboxylic amide group. *Helv. Chim. Acta* **65**, 1555–1562 (1982).

103. Jeffrey, G. A., Houk, K. N., Paddon-Row, M. N., Rondan, N. G., and Mitra, J. Pyramidalization of carbonyl carbons in asymmetric environments: carboxylates, amides, and amino acids. *J. Amer. Chem. Soc.* **107**, 321–326 (1985).

104. Norskov-Lauritsen, L., Bürgi, H.-B., Hoffmann, P., and Schmidt, H. R. 9. Bond angles in lactams. *Helv. Chim. Acta.* **68**, 76–83 (1985).

105. Murray-Rust, P., and Motherwell, W. D. S. Computer retrieval and analysis of molecular geometry. III. Geometry of the β-1'-aminofuranoside fragment. *Acta Cryst.* **B34**, 2534–2546 (1978).

106. Taylor, R., and Kennard, O. The molecular structures of nucleosides and nucleotides. Part 1. The influences of protonation on the geometries of nucleic acid constituents. *J. Molec. Struct.* **78**, 1–28 (1982).

107. Taylor, R., and Kennard, O. Molecular structures of nucleosides and nucleotides. 2. Orthogonal coordinates for standard nucleic acid base residues. *J. Amer. Chem. Soc.* **104**, 3209–3212 (1982).

108. Sheldrick, B., and Akrigg, D. Rigid-body coordinates of pyranose rings. *Acta Cryst.* **B36**, 1615–1621 (1980).

109. Allen, F. H. The geometry of small rings. I. Substituent-induced bond-length asymmetry in cyclopropane. *Acta Cryst.* **B36**, 81–96 (1980).

110. Allen, F. H. The geometry of small rings. II. A comparative geometrical study of hybridization and conjugation in cyclopropane and the vinyl group. *Acta Cryst.* **B37**, 890–900 (1981).

111. Allen, F. H. The geometry of small rings. III. The effect of small-ring fusion on the geometry of benzene. *Acta Cryst.* **B37**, 900–906 (1981).

112. Allen, F. H. The geometry of small rings — IV Molecular geometry of cyclopropene and its derivatives. *Tetrahedron* **38**, 645–655 (1982).

113. Norrestam, R., and Schepper, L. Prediction of molecular geometries of aromatic six-membered rings. *Acta Chem. Scand.* **A35**, 91–103 (1981).

114. Allen, F. H. The geometry of small rings. VI. Geometry and bonding in cyclobutane and cyclobutene. *Acta Cryst.* **B40**, 64–72 (1984).

115. Nachbar, R., Johnson, C. A., and Mislow, K. Structure and internal dynamics of systems containing two or more tert-butyl groups in close proximity. *J. Org. Chem.* **47**, 4829–4833 (1982).

116. van Koningsveld, H., and Jansen, J. C. Review of the preferred rotational orientation of the carboxyl and tert-butyl groups. Structure of the trans-4-tert-butyl-1-cyclohexanecarboxylic acid, $C_{11}H_{20}O_2$. *Acta Cryst.* **B40**, 420–424 (1984).

Recognition and Receptors

In order to form a crystal, molecules must aggregate in an orderly manner. This implies that intermolecular interactions have occurred in specific ways. It therefore follows that the crystal structure *per se* contains information on preferred modes of binding between the molecules in the crystalline state. In this Chapter we show how information on the most likely stereochemistries of interactions between functional groups in different molecules can be extracted from the three-dimensional coordinates of atoms listed in reports of crystal structure determinations. Three-dimensional structural data on binding stereochemistry may also be obtained from X-ray diffraction studies of the binding of small molecules to crystalline proteins and other macromolecules. These two types of information can be used, for example, to predict how drugs will interact with their biological receptors.

17.1 Surroundings of functional groups

The orientations of the interactions between different molecules in the solid state can be found readily from crystal structure analyses; such data are more difficult to obtain by other methods. They provide information on modes of recognition between different groups on separate molecules. Hydrogen bonding is an important intermolecular recognition interaction, but other types of forces are also significant. When direct evidence on how a specific small molecule interacts with a biological macromolecule is not readily obtainable, model building, based on the most likely distances and directions between functional groups in the complex, must be used. Such model building necessarily requires good information on how functional groups would be expected to interact with each other. The methods used to obtain this information are described in this Chapter.

The observed interactions between molecules in a crystal are a consequence of the minimization of the free energy for the entire atomic arrangement, as described in Chapter 15. Therefore intermolecular interactions in one crystal structure may be biased by the requirements for packing of the rest of the crystal structure. To overcome this problem, structural data from a large number of different crystal structures, extracted from one of the databases listed in Chapter 16, are analyzed statistically. Such analyses can lead to some general principles for the modeling of ligand–receptor interactions, for the design of new biologically active agents, and for an understanding, on an atomic scale, of the stereochemistry of some biologically significant processes, such as macromolecular assembly and catalytic reactions.

Such analyses are of use to pharmacologists who are interested in how drugs interact with receptors. The **pharmacophore**[1] has been defined as the three-dimensional arrangement of functional groups on a small molecule that is essential for both recognition (by binding) and activation (to give a reaction) of a macromolecular receptor. Thus the pharmacophore constitutes the significant portions of a drug molecule that are involved in pharmacological activity. Studies of the preferred orientations of intermolecular reactions can lead to a prediction of the stereochemistry of the biological receptor of a pharmacophore.

17.1.1 Scatterplots

Nonbonded surroundings of a functional group can be analyzed to determine if there are preferred directions in which the group binds its neighbors. A functional group of interest, such as a carboxyl group[2] or an epoxide group[3] is selected. All crystal structures containing this chosen group may be accessed by searching a database, such as the Cambridge Crystallographic Database (CSD).[4] The positions of atoms (such as metal cations, or the hydrogen atoms of hydroxyl groups) that pack in each crystal structure around the selected group are calculated from the three-dimensional coordinates of the crystal structure stored in the database. Then, with the functional group in a fixed position and orientation for each crystal structure, the positions of atoms packed around it are superimposed to give a **scatterplot**, such as that shown in Figure 17.1(a). This scatterplot is then examined for any specific directionalities of packing, evidenced by an area of high density of points in the scatterplot.

The various points in the scatterplot, are sometimes hard to analyze objectively. For example, outlying points may unintentionally be given undue significance by the observer. To circumvent this problem it has been found advantageous to contour the scatterplot[5] by putting a **Gaussian-like peak** on each point of the scatterplot, and contouring the result, in the same way that an electron density map is contoured [see Figures 9.3, Chapter 9, and 17.1(b) and (c)]. This has the effect of smearing out the probability plot and highlighting any directional preferences of binding, which should appear as the highest peaks in the plot.

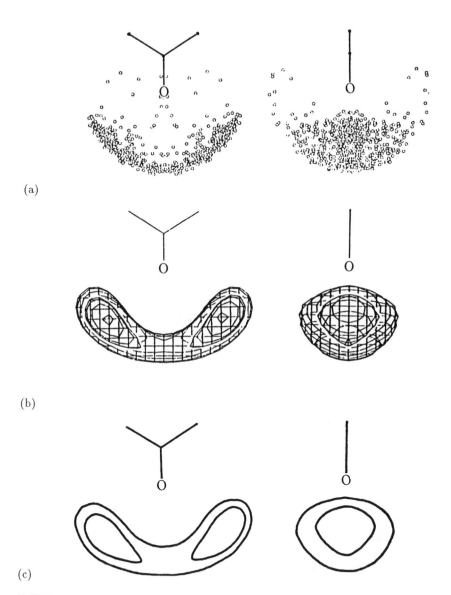

(a)

(b)

(c)

FIGURE 17.1. Three-dimensional scatterplots (as points) for the OH, and NH groups surrounding the functional oxygen atom of a carbonyl (C=O) group. Left hand side: view onto the C–CO–C plane. Right hand side: view along the plane. (a) A scatterplot, (b) the result of contouring the scatterplot, and (c) a simplified version of (b) (Ref. 3).

17.1.2 Surrounding of C–S groups

Early work on the directional preferences of binding was done in the laboratory of Rengachary Parthasarathy. He studied the surroundings of divalent sulfur bonded to carbon (C–S–C),[6] as shown in Figure 17.2. Electrophiles such as metal cations or hydrogen bond donors approach in a direction that is from 50° to 90° from the sulfide (C–S–C) plane. This suggests, by frontier orbital theory,[7] that electrophiles are interacting with the highest occupied molecular orbital (HOMO) which is a lone-pair orbital nearly perpendicular to the C–S–C plane. On the other hand, nucleophiles such as negatively charged groups or ions, approach the sulfur within 30° of the sulfide C–S–C plane. These interactions tend to lie along the extension of one of the C–S bonds, the direction predicted for the lowest unoccupied molecular orbital (LUMO). Therefore electrophilic attack of divalent sulfur would be expected to occur in a direction perpendicular to the sulfide plane, while nucleophilic attack would be expected to occur in the direction of an S–C bond.

Such relationships have also been examined in proteins.[8] Of the two sulfur-containing amino acids (methionine and cysteine), cysteine, is the only one that could be a hydrogen bond donor.[9] Hydrogen bonds are formed from its S–H group to the carbonyl C=O groups of the $(i - 4\text{th})$ residue. These interactions are observed in "C-terminal capping," in which a cysteine side chain donates a hydrogen bond to an unsatisfied C=O near the end of the helix. Such capping may be important in

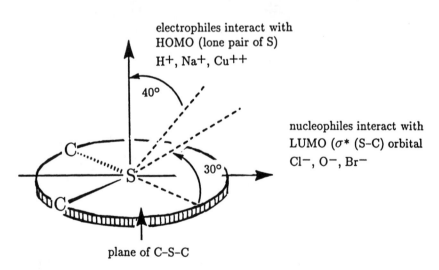

electrophiles interact with
HOMO (lone pair of S)
H+, Na+, Cu++

40°

nucleophiles interact with
LUMO ($\sigma*$ (S–C) orbital
Cl−, O−, Br−

30°

plane of C–S–C

FIGURE 17.2. Directional preferences of surroundings of divalent sulfur (Ref. 6). From an analysis of the environments of C–S–C groups in crystal structures it was concluded that electrophiles approach at an angle of 50–90° from the C–S–C plane, while nucleophiles interact within 0–30° of the C–S–C plane.

The probability of O···H–O hydrogen bond formation is significantly higher than that for the formation of a hydrogen bond involving a C–F group. One of the crystal structures in which the C–F group takes part in hydrogen bonding involves a three-centered (bifurcated) hydrogen bond, with the hydrogen atom shared between an oxygen atom and a fluorine atom (Figure 17.4). The fluorine atom in a C–F bond, however, can only act as a proton acceptor in a hydrogen bond, while the oxygen atom in the C–OH bond can act both as a proton donor and acceptor. The result for oxygen is a cooperative effect — oxygen (as O–H) can give and accept hydrogen bonds, while fluorine can only accept them.

Fluorine is the most electronegative element.[15] Why, then, does the C–F bond form hydrogen bonds so rarely, especially since former when H···F⁻ forms the shortest hydrogen bonds measured (H···F⁻ $= 1.15$ Å)?[16] The strength of a hydrogen bond $(D–H \cdots A)$[17,18] appears to be a complicated balance of various factors[19] which include:

1. the interactions between fractional charges that have developed on D, A, and H,
2. the deformability (polarizability) of the electron cloud around the acceptor atom A so that it can make its lone pairs available to the proton (the softness of A),
3. the transfer of charge from A to H (σ-bond transfer).
4. the electronic repulsion between D and A,
5. how readily a hydrogen-bond donor atom D will lose its covalently bound hydrogen atom as H⁺ (related to the electronegativity of D and the strength of the $D–H$ bond), and
6. how readily the hydrogen bond acceptor A can accept the H⁺ (the electronegativity of A),

Of these, the important attractive forces are electrostatic (1.), polarization effects (2.), and charge-transfer interactions (3.). The electrostatic component (1.) falls off less rapidly as a function of distance than do the others, and therefore at longer distances it is the most important.[18] These energy components are balanced by repulsive forces (4.) which become important at short distances. Thus the the hydrogen atom in a C–H bond is held firmly by the carbon atom and retains some of its electron cloud, which would be significantly repelled by the hydrogen-bond acceptor atom A if the distance were short. Therefore the C–H group does not readily form a hydrogen bond, and when it does, the H $\cdots A$ distance is relatively long. While some charge transfer also occurs, it is weak. The fluorine atom in a C–F group does not, apparently, develop much negative charge, and, in addition, it is not readily polarized. Thus it appears that the C–F group only forms very weak hydrogen bonds (as the hydrogen-bond acceptor). While fluorine is more electronegative than oxygen, oxygen is somewhat more polarizable. On the other hand,

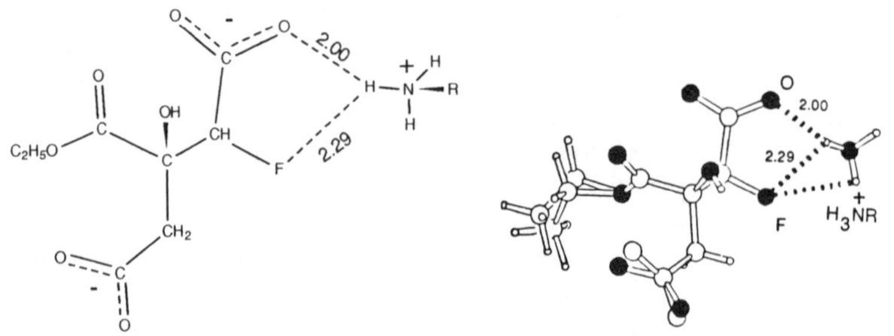

(a)   (b)

FIGURE 17.4. Examples of C–F $\cdots$ H–O and C–F $\cdots$ H–N interactions. (a) Chemical formula of a fluorocitrate ester, and (b) a ball-and-stick model of the results of the crystal structure analysis of this ester. Note the different H $\cdots$ O and H $\cdots$ F distances (2.0 Å and 2.3 Å, respectively) (Ref. 13).

the reason that the fluoride ion can form short H $\cdots$ F$^-$ hydrogen bonds in hydrofluoric acid, is presumably the result of a proton resting between two negatively charged ions, a powerful electrostatic effect.

An interaction between a C–F bond and a metal ion (C–F $\cdots$ M^{n+}) has been observed so far by X-ray diffraction studies only when the metal ion is an alkali metal ion. Examples are shown in Figures 17.4 and 17.5. The coordination of carbon-bound oxygen and fluorine atoms around the metal ion can be analyzed in terms of the requirement that the sum of the bond valences should be near 1.0 for a monovalent cation. This sum is only achieved if the $M^{n+} \cdots$ F–C interaction is taken into account as well as those involving oxygen atoms. The fluorine atom contributes to the local neutralization of the charge of the cation, and is, indeed, a significant member of the first coordination sphere of the metal ion.

17.1.5 Surroundings of carboxylate groups

The carboxylate ion has one delocalized negative charge, and each oxygen atom has two lone pairs disposed at 120° to the C–O bond. The directionalities of the oxygen lone pairs are important for metal–ion binding. It is found in various crystal structures that a given carboxylate group may bind up to six cations, and that it may, in certain circumstances, share the metal cation between both oxygen atoms, as has been found for calcium ions.[20]

The geometry of the interactions of a carboxyl group has been investigated by Julius Rebek, Jr.[21–24] He designed, prepared, and studied many compounds in which the carboxylic acid groups are forced, by steric effects, to approach each other in a controlled way. The directions of the lone-pair electrons in a carboxyl group are designated *syn* and *anti* (il-

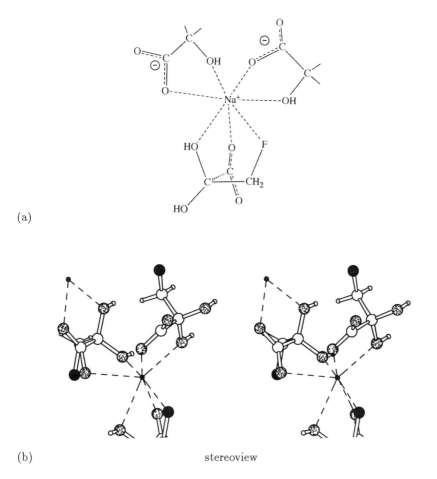

(a)

(b)   stereoview

FIGURE 17.5. Surroundings of a sodium ion in hydrated sodium fluoropyruvate (Ref. 25). (a) Chemical formula, and (b) stereoview. Note that the fluorine of a C–F group is part of the coordination sphere of the metal ion. $Na^+ \cdots O$ distances 2.306, 2.384, 2.407, 2.426, 2.611, 2.681 Å, $Na^+ \cdots F$ distance 2.470 Å (with e.s.d. values of 0.001 Å for each).

lustrated in Figure 17.6). The proton on another molecule can approach in either of these directions. In the *syn* conformation (Z form) the proton is on the same side of the C–O bond as the other C–O bond; this is the conformation found when carboxyl groups dimerize by forming two hydrogen bonds. On the other hand, in the *anti* conformation (E form) the proton is on the opposite side of the C–O bond from the other C–O bond. *Ab initio* quantum chemical studies of formic acid indicate that the *syn* (Z) conformation is more stable than the *anti* (E) conformation by 4.5 kcal mol^{-1} (19 kJ mol^{-1}), implying that the *syn* lone pairs are more basic (and therefore bind metal ions more readily) than do the *anti* lone pairs.[26] Richard Gandour[27] noted that the carboxylates in active sites of enzymes generally employ the more basic *syn* lone pairs for metal chelation rather than the less basic *anti* lone pairs. He estimated that *syn* protonation is 10^4-fold more favorable than *anti* protonation (since 1.4 kcal mole^{-1} corresponds approximately to a tenfold increase in rate). The carboxylate ion is therefore a weaker base when constrained to accept a proton in the *anti* (E) direction.

An analysis of the directions in which metal ions approach a carboxyl group in crystal structures[2] showed that the most likely arrangements of metal cations are *syn*, *anti*, and direct (see Figure 17.6). Alkali metal and some alkaline earth metals, which ionize readily and form strong bases, and have less specific directions of binding, and show extensive out-of-plane interactions (see Figure 17.7). Most other metal ions bind to the carboxyl group in its plane. In Figure 17.8 a plot of the type of bonding (*syn*, *anti*, or direct) as a function of $M^{n+} \cdots O$ range is shown. Direct bonding, in which the metal ion is equidistant from the two oxygen atoms of the carboxylate ion, is preferred in the range $M^{n+} \cdots O = 2.3–$

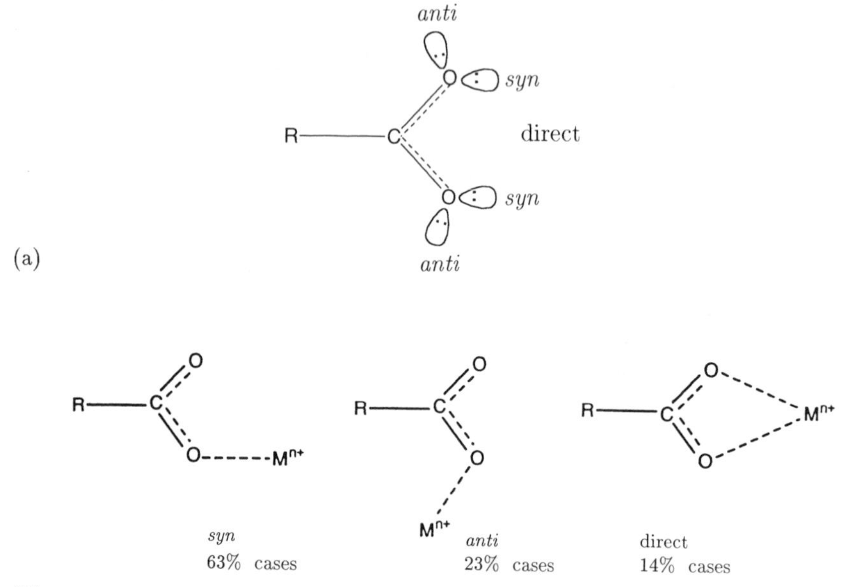

FIGURE 17.6. (a) Dispositions of the *syn* and *anti* lone pairs of a carboxyl group. (b) Percentages of metal ion binding to the lone pairs, and shared ("direct") (Ref. 2).

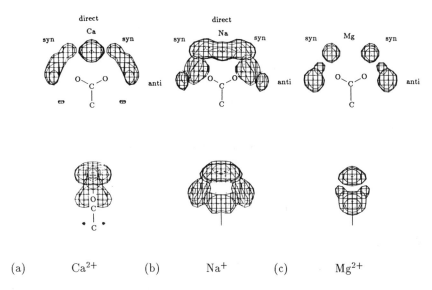

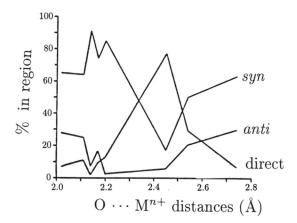

FIGURE 17.7. Examples of scatterplots of the surroundings of carboxyl groups. (a) Ca^{2+}, (b) Na^+, and (c) Mg^{2+}. The upper portion of each diagram is the view onto the plane of the carboxyl group, while the lower portion is a view along the carboxyl group, which points up vertically. Note that the Ca^{2+} ion is involved in "direct binding" and that the Na^+ ion binds out of the plane as well as in the plane.

FIGURE 17.8. Plot of type of bonding versus $M^{n+} \cdots O$ distances (Ref. 2). Note that the peak for direct bonding lies near 2.4–2.5 Å.

2.6 Å; otherwise *syn* is generally preferred, except when the $M^{n+} \cdots O$ distances are short (less than 1.95 Å). Elements with a reasonably high percentage of "direct" bonding are calcium(II), cadmium(II), mercury(II), samarium(III), thallium(III), cerium(IV), and uranium(VI). At other $M \cdots O$ distances the "direct bonding" is rare. These results are a function of the carboxylate "bite" size (2.2 Å) and of a need to keep $O \cdots M^{n+} \cdots O$ angles reasonably large (larger than approximately 60°). The arrangement with the metal in the *syn* position is much more common than that for which it is *anti*, possibly due to steric constraints. Exceptions are found at very short and very long $M^{n+} \cdots O$ distances, that is, for very small and very large cations. When the carboxyl group is not ionized, it is found, from detailed neutron diffraction studies, that the hydroxyl portion of the carboxyl group, while a powerful hydrogen-bond donor, does not accept hydrogen bonds.[28,29]

17.1.6 Surroundings of α-hydroxycarboxylate groups

Chelating groups that bind metals are found in iron-binding hydroxamates,[30,31] some of which are powerful enough to extract iron from stainless steel. Other examples of metal-chelating groups are provided by α-hydroxycarboxylate groups, $HO-CR_2-COO^-$, found in many important biochemical compounds such as citrates and malates.[32] The geometry of their metal binding has been investigated by X-ray crystallographic techniques (Figure 17.9). If the cation is of a suitable size, the entire chelating group is approximately planar, even though the $O \cdots O$ distance is shorter than the sum of the van der Waals radii for oxygen

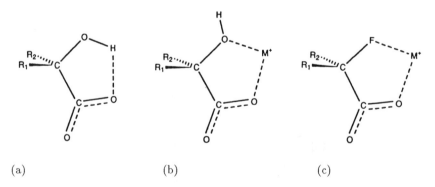

(a)   (b)   (c)

FIGURE 17.9. Metal binding to an α-substituted carboxyl group. (a) Hydrogen bonding in a carboxylic acid. (b) Metal binding, where the metal replaces the hydrogen ion in (a). (c) The fluoro analogue has no hydrogen atom analogous to (a) and is observed to bind a metal ion. In each case there is a proton or metal ion chelated by the α-substituted carboxyl group.

atoms [Figure 17.9(b)]. A study of one form of potassium citrate revealed,[33] surprisingly, that the potassium ion did not chelate the α-hydroxycarboxylate group. The hydroxyl hydrogen atom, however, formed an internal hydrogen bond, as shown in Figure 17.9(a). This was verified by a neutron diffraction study.[33] When the central hydroxyl group of citric acid is replaced by fluorine, the α-fluorocarboxylate group is still a good chelating group, but no hydrogen atom is available to form an internal hydrogen bond in salts such as the potassium salt. As a result, in dipotassium 3-fluorodeoxycitrate, a potassium cation is chelated by the α-fluorocarboxylate group [Figure 17.9(c)]. It appears that an α-hydroxycarboxylate group will either chelate a cation or form an internal hydrogen bond via the hydroxyl hydrogen atom, with the hydrogen atom acting as a substitute for the metal ion. If no such hydrogen atom is available to form a hydrogen bond, the group will chelate a cation, as for dipotassium 3-fluorodeoxycitrate.

17.1.7 Surroundings of imidazole and histidine groups

Imidazole groups in histidyl side chains bind in proteins to metal ions, but the metal cation nearly always lies in or very near the plane of the imidazole group. One imidazole can bind one or two metal ions, depending on the ionization state and suitability of the metal ion (Zn^{2+} and Cu^{2+} being common binders). A scatterplot of metal–imidazole interactions is shown in Figure 17.10. In proteins,[34,35] the metal ion shows a preference

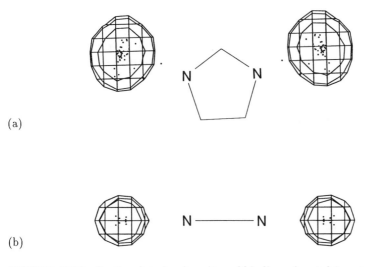

(a)

(b)

FIGURE 17.10. Scatterplot showing sites of binding of metal ions to an imidazole group in small molecule-crystal structures. (a) View onto the plane of the ring, and (b) along the plane.

for binding to the nitrogen atom furthest from the C_α of the histidine. If the metal binds to the other nitrogen atom, the histidine is rotated from the value normally found in proteins. A hydrogen bond to the nitrogen atom that is not coordinated by metal serves to stabilize the orientation of the ring.

Metal ions are found to lie in the plane of the imidazole group even if the oxidation state or coordination number of the metal ion changes. Rotation of the imidazole ring about the $M^{n+} \cdots N$ (imidazole) bond is controlled by the steric requirements of the rest of the histidyl residue, and by hydrogen bonding to the nitrogen atom of the other ring. The flexibility of imidazole–metal bonds may be a reason for the effectiveness of histidine side chains as metal-binding sites in proteins. Several other systems of chemical and biochemical interest have been investigated and directional preferences of binding of certain functional groups have been established.[36–42] Space does not allow their inclusion here, but the reader is encouraged to read References 36–42.

17.1.8 Metal-binding sites in proteins

The experimental observation that there is a high probability that the metal ion lies in the plane of a neighboring carboxylate ion provides a mechanism for searching for metal-binding positions in proteins. This test is complementary to that suggested by David Eisenberg and co-workers,[43] which probes for the existence of a hydrophilic area in the protein (negatively-charged oxygen atoms) with a hydrophobic area immediately behind it (carbon atoms of the carboxylate group). The probes used are diagrammed in Figure 17.11. The method was used successfully for the enzyme xylose isomerase from *Streptomyces rubiginosus*.[44,45] Two metal sites were located by use of these probes. One metal-binding site involves three carboxylates (aspartate and glutamate), one histidine, and

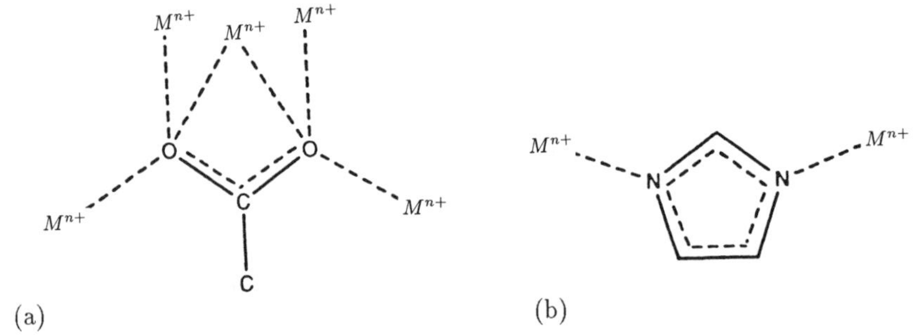

FIGURE 17.11. Probes of metal-binding sites in proteins, deduced from studies of crystal structures of small molecules. Shown are likely in-plane sites for a metal ions with respect to a carboxyl group or an imidazole.

water and the other involves four carboxylate groups and water (Figure 17.12). In proteins[46] the *syn* lone pair of an oxygen atom is preferred, possibly as a site for metal–ion binding. A metal ion-containing five-membered ring (six if the hydrogen is included) has been noted with water in the other *syn* position, as shown in Figure 17.12.

17.1.9 C–H ⋯ O interactions in molecular recognition

The interaction between a C–H group and a neighboring oxygen atom can be considered as a very weak hydrogen bond,[47,48] about one third the energy of an O–H ⋯ O hydrogen bond. If, however, there are many such interactions, they can have a significant effect on the total energy of an interacting system. For example, the importance of C–H ⋯ O interactions in aligning molecules has been shown by the design, by Gautam R. Desiraju and co-workers, of dibenzylidene ketones that interact by way of C–H ⋯ O interactions with 1,3,5-trinitrobenzene.[49] Use was made of the great acidity of the hydrogen atoms in 1,3,5-trinitrobenzene, and the propensity of nitro groups to interact with hydrogen atoms bound to carbon atoms. It was shown by X-ray crystallographic studies that the C–H ⋯ O interactions have a significant aligning capability, and that

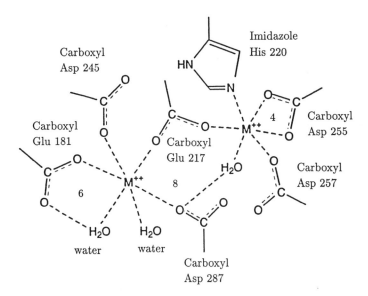

FIGURE 17.12. Binding of carboxylate and histidyl groups and water molecules to the two metal ions in the active site of the enzyme D-xylose isomerase (Ref. 45). Note the carboxylate–water–metal ion chelate at the left, and the direct metal–carboxylate bonds to the right. The number of atoms in each closed cycle is shown.

there is a similarity of binding of dibenzylidene acetone, cyclopentanone, and cyclohexanone (Figure 17.13). These complexes are analogous to the designed complexes containing O–H ⋯ O and N–H ⋯ O hydrogen bonds, but in the complex described here only C–H ⋯ O interactions could be formed.

17.2 Electrostatic potentials

Deformation density maps of crystal structures were described in Chapter 9 (see Figures 9.16 and 9.17). Methods for refining data to obtain charge information was also described. From the experimental charge parameters so derived it is possible to map the **electrostatic potential** in the crystal,[50–53] or for a molecule or group of atoms removed from the crystal.[54] These electrostatic potential maps show which areas around a molecule are electronegative and which are electropositive.

The method for calculating the electrostatic potential map is outlined in Reference 54, and an example of the electrostatic potential around a nucleic-acid base pair is shown in Figure 17.14. The electron density at the distance r from an atomic nucleus may be represented, as in Chapter 9, Equation 9.4, by

$$\rho_{atom} = \rho_c + P_v\rho_v(\kappa r) + \sum_{lmn} P_{lmn} R_n(r) Y_{lm}(\theta, \phi) \tag{17.1}$$

where κ is a constant, and ρ_c and ρ_v are core and valence electron densities, repectively. P_v is a population parameter and the right-hand term

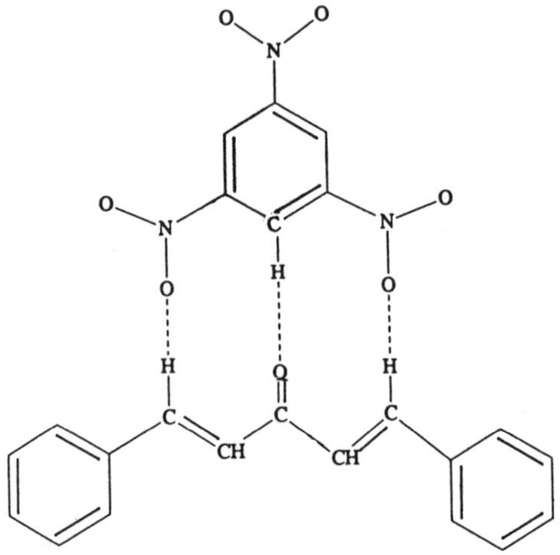

FIGURE 17.13. C–H ⋯ O interactions that align two molecules in a crystal structure via the nitro groups and a carbonyl group (Ref. 49)

represents the asphericity of the electron density $P_{lmn}R_n(r)Y_{lm}(\theta,\phi)$ where P_{lmn} is a population parameter obtained from the fit to the data, l, m, and n are integers (n is 0 for hydrogen, 2 for first row atoms, $m < l \leq n$), $R_n(r)$ is a radial density function (of the Slater type), $Y_{mn}(\theta,\phi)$ is a angular function (a spherical harmonic), and (r,θ,ϕ) are the spherical coordinates with respect to the atomic nucleus. The constants are derived by least-squares refinement. The pseudoatom potential can be derived from the electron density $\rho(r)$ as

$$\Phi(r) = \int \rho(r') \mid R - r' \mid^{-1} d^3 r' \qquad (17.2)$$

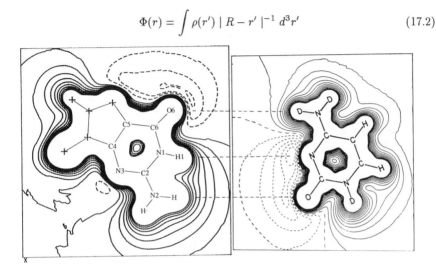

FIGURE 17.14. The electrostatic potential complementarity between guanine and cytosine. This montage is made up from two different crystal structure determinations. The experimental data were fit using a multipole model with atomic deformation terms extending to the octapole level. As a result the valence shell charge density distribution for each carbon, nitrogen or oxygen atom can be described as the sum of 16 terms (see Equation 17.1). With these terms, together with others of similar type for describing the invariant K-shell density, the experimental charge density can be mapped for the molecule lifted out of its crystal structure, and with thermal vibrations removed. The electrostatic potential is then obtained using Equation 17.2 (see Ref. 54). Note that although the molecules are removed from their crystal environment, they retain the polarization (e.g., hydrogen bonding) effects that occur in the crystal, and they differ from truly isolated molecules such as would be studied by *ab initio* calculations. In the maps shown here for a base pair of guanine and cytosine, the contours are at intervals of 0.05 Å with solid contours for electropositive regions and dashed contours for electronegative regions. The maps for the two molecules are additive. At proper hydrogen-bonding distances, electropositive and electronegative potentials would superpose and almost cancel in the central region between molecules. The cytosine map is from Ref 55. Figures courtesy J. Luo and B. M. Craven, and the International Union of Crystallography. Text courtesy B. M. Craven.

The contribution of each atom to the total potential at a point r_P in space is the sum of: (1.) the contribution of the nucleus, (2.) the electron density in the shell extending from the nucleus and having a radius r_P, and (3.) the electron density in the shell from radius r_P to infinity.

Robert F. Stewart and Bryan M. Craven[54,56] show that the molecular electrostatic potential can be represented approximately as the sum of atomic contributions

$$V(r) = (q/r) + \exp(-\beta r^2) \tag{17.3}$$

where r (in Å) is a distance greater than the van der Waals radius, and q is a net atomic charge given by $(P_v - Z_v)$, where P_v is from Equation 17.1, and Z_v is the number of valence electrons in the neutral atom. They found $\beta = 1.47$ Å^{-1} for carbon atoms, $\beta = 1.66$ Å^{-1} for nitrogen atoms, and $\beta = 1.83$ Å^{-1} for oxygen atoms for r greater than 1.2 Å. The contributions of pseudoatom higher terms can be neglected at distances greater than 2.5 Å from atomic nuclei. For hydrogen atoms the approximate equation

$$V(r, \phi) = (0.165 \cos \phi)/(r^2 + 0.615) \tag{17.4}$$

may be used, where ϕ is the angle between the direction of r and the X–H bond.

17.3 Surface recognition and chirality

The effects of certain impurities on the habit of a growing crystal was described in Chapter 2. This is another aspect of molecular recognition. If an impurity that is similar to the molecules in the bulk of the crystal, it is adsorbed on a specific crystal face. The impurity molecule then, however, presents to the exterior a side onto which further molecules of host material cannot adsorb. As a result, crystal growth in that direction is terminated. This is diagrammed in Figure 17.15 for the effect of cinnamide as an impurity in the crystallization of cinnamic acid. As a result of the adsorption of a layer of impurity molecules, the growth of that face will be inhibited, and the face will then be prominent in the habit of the resulting crystal. This effect was described and illustrated earlier in Chapter 2 (Figure 2.8).

The substitution of chiral molecules onto the crystal surfaces has been investigated by Mein Lahav, Lia Addadi, Leslie Leiserowitz and co-workers.[57–68] The chiral molecule is "recognized" by the molecules already oriented in the crystal lattice. This can be true for either an enantiomorphous crystal composed of molecules with only one chirality sense, or for a centrosymmetric racemic crystal in which a face that is composed only of one enantiomer interacts with a chiral agent.

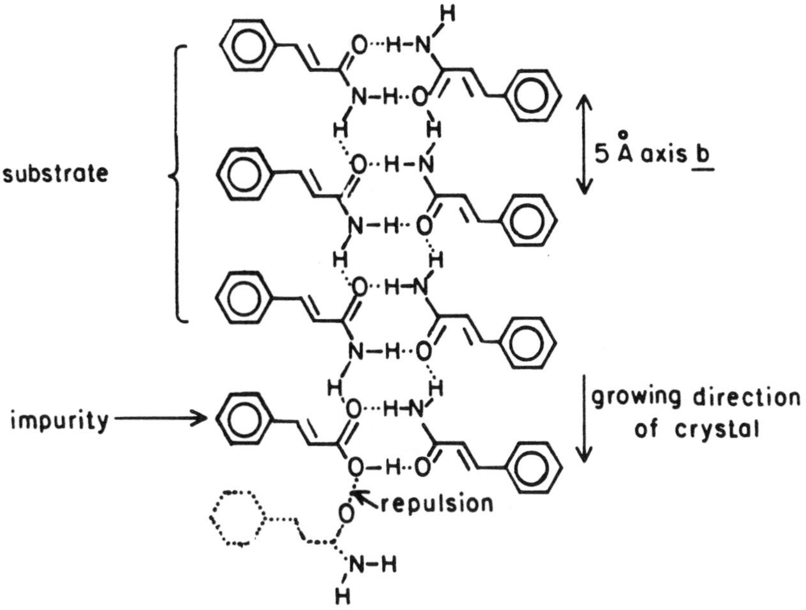

FIGURE 17.15. The effect of absorption of an impurity on the surface of a crystal. In this case a carboxylic acid (cinnamic acid) is bound to the crystal face as an impurity on crystals of an amide (cinnamide). As a result, crystal growth in the **b** direction is stopped (compare with Figure 2.8, Chapter 2, for the consequences of this). (Reproduced from Reference 69 with the permission of G. R. Desiraju and Elsevier Science Publishers BV, Academic Publishing Division, Amsterdam, The Netherlands.)

We showed in Chapter 14 how anomalous dispersion could be used to determine the absolute structure of chiral molecules in noncentrosymmetric crystals. Centrosymmetric crystals, however, contain equal numbers of right- and left-handed molecules, and the determination of absolute structure is not possible. The disposition of functional groups on the various faces of the crystal may, however, be chiral. The requirement for recognition of asymmetry in a racemic crystal (containing both R and S enantiomers) is that a specific functional group in each molecule in the crystal point towards, say, hkl faces in an R molecule, and not towards $\overline{hkl}$ faces, while the reverse is true for S molecules. An example is provided by the crystal structure of racemic (R,S) serine, shown in Figure 17.16. Both C–H bonds in the methylene group of serine have components along the unique **b** axis. If a specific one of these hydrogen atoms (in serine) is replaced by a methyl group (to give threonine), indicated by large filled circles in Figure 17.16, then, when threonine binds to a growing crystal of serine, the methyl group of (R)-threonine (side-chain chirality S) will replace the *pro-S* hydrogen atom of (*R*)-serine, and will inhibit further growth of the crystal in the +**b** direction. Similarly the methyl group of (*S*)-threonine will replace the *pro-R* hydrogen atom of (*S*)-serine and inhibit growth along −**b**. This has been verified experimentally.[65,70−72]

The molecules in the original crystal need not be chiral; they can be *pro*-chiral and still give an analogous effect. Glycine, which lacks an asymmetric carbon atom, crystallizes in the α form as bipyramids with well-developed [110] and [011] sets of faces.[73] These crystals are mono-

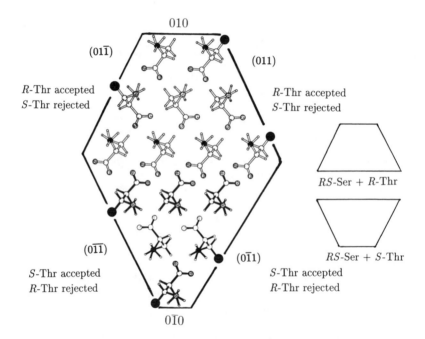

FIGURE 17.16. The crystal packing of *RS* serine is shown, viewed down the **a** axis. *R*- and *S*-threonine and -serine are D- and L-threonine and -serine, respectively. Threonine impurity molecules (with their methyl groups represented by large filled circles) are added stereospecifically. They then prevent further addition of serine molecules on the labelled crystal faces. Remember, from Chapter 2, that the faces that are observed on a crystal are those that have grown most slowly. Therefore, when threonine adds to a growing crystal, the face to which it adds becomes prominent.

clinic and centrosymmetric, space group $P2_1/n$. The crystals grow with a change in habit on addition of the resolved amino acid (see Figure 17.17), giving pyramidal crystals with a large [010]-type face.

D-amino acids induce large (010) faces, while L-amino acids induce large (0$\bar{1}$0) faces. Racemates (D- and L-amino acids) give plate-like crystals. Just as groups could be distinguished with respect to direction in DL-serine, so can the *pro-R* and *pro-S*-hydrogen atoms of glycine be distinguished in the crystal structure of glycine.[67,68] The additional R side chain of the impurity, replacing hydrogen in glycine, emerges out of the face in the direction of the C–H pointing in the +**b** or –**b** direction. Further growth is inhibited and the face becomes prominent. Only D-amino acids [(R)-amino acids] can readily replace glycine at growing (010) faces while only L-amino acids [(S)-amino acids] can grow on (0$\bar{1}$0) faces.

Crystals of α-glycine float on the surface of the solution, with one of the two flat (010) or (0$\bar{1}$0) faces exposed to solution. If one enantiomer of a chiral hydrophobic amino acid (leucine or phenylalanine, for example) is now added to the solution, the crystal will become oriented so that the face that occludes amino acids of opposite chirality is exposed to solution. D-amino acids are only incorporated in crystals with the 010 face exposed to solution, while L-amino acids only interact with the exposed (0$\bar{1}$0) face. Therefore, if a floating crystal with the (010) face exposed to solution is presented with a solution of DL amino acids, the crystal will bind D-amino acids and will enrich the solution with respect to L-amino acids. If, for example, the underside of a film of a resolved amino acid with a long hydrophobic side chain [such as (S)-α-aminooctanoic acid] is exposed to a solution of a racemic mixture of amino acids, some optical resolution will occur by the principles just described (see Reference 64).

17.4 Ionophores and other metal–binding molecules

Ionophores or "ion bearers" are compounds designed by nature to carry small ions across lipid membrane barriers.[74] The first such compound to be characterized was valinomycin [see Figure 17.18(a)][75] which is a cyclic dodecadepsipeptide with a large ring of 36 atoms. Its crystal structure,[76] shown in Figure 17.18(b), reveals a cagelike structure with the carbonyl oxygen atoms 2.7–2.8 Å from the potassium cation that it is specifically engineered to bind. The hydrogen bonding in the valinomycin cage prevents it from contracting to bind the smaller sodium ion, and therefore this ionophore has a high K^+:Na^+ selectivity.[74] The structure of the potassium salt is shown in Figure 17.18(c). Another interesting ionophore is enniatin,[77] which is a hexadepsipeptide with a smaller ring of only 18 atoms. In the potassium complex all carbonyl oxygen atoms are oriented to the interior in two triads, one 1.5 Å above the mean ring plane and the other the same distance below it.

In **guest–host chemistry** the shapes of the participating molecules are fundamental to the process of co-aggregation. Charles J. Pedersen[78]

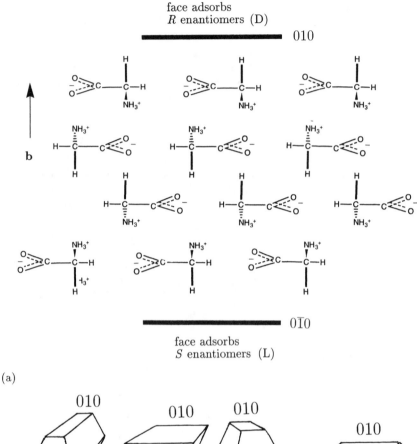

(a)

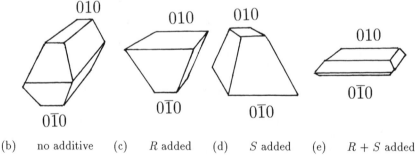

(b)   no additive   (c)   *R* added   (d)   *S* added   (e)   *R* + *S* added

FIGURE 17.17. Crystals of α-glycine grown in the presence of other amino acids. The important feature is the C–H bond aligned along the +b or −b direction. (a) Diagram of the molecular packing of crystalline glycine. The crystal structure is centrosymmetric and glycine does not contain an asymmetric carbon atom. Note the C–H vectors, perpendicular to the marked crystal faces [(010) at the top, (0$\bar{1}$0) below). (b) Pure glycine crystals. (c) *R* amino acid added, (d) *S* amino acid added, and (e) both enantiomers added.

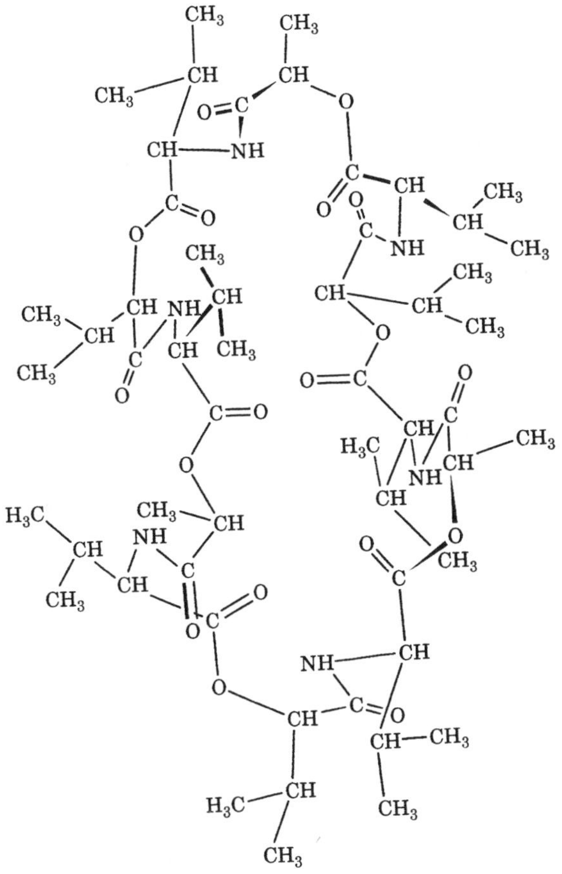

(a)

FIGURE 17.18. The ionophore valinomycin. (a) Chemical formula of valinomycin.

was trying to find means of controlling the action of a vanadium catalyst in a scheme to make synthetic rubber. He found, unexpectedly, during this study that some cyclic molecules, **crown ethers**, could accommodate a metal ion within them. The structures of several of these have been determined. For example, dibenzo-18-crown-6 holds the metal ion approximately in the plane of the oxygen atoms,[79,80] as shown in Figure 17.19(c). Donald J. Cram and Jean-Marie Lehn[81–83] extended this work and made much more complex molecules, designing them to perform certain functions with the specific aims of mimicking selective transport across biological membranes and certain aspects of enzyme chemistry.

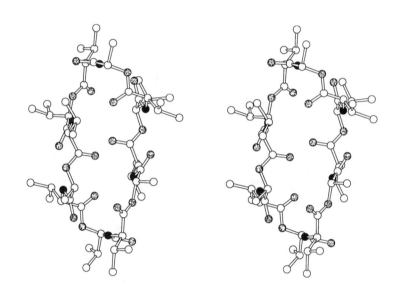

(b)   stereoview

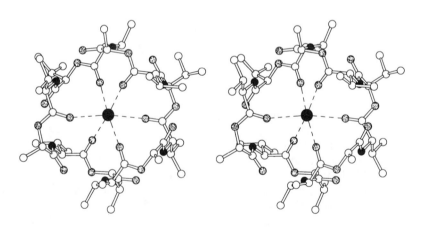

(c)   stereoview

FIGURE 17.18 (cont'd). (b) Stereoview of valinomycin (Ref. 76), and (c) stereoview of its potassium derivative (Refs. 84 and 85).

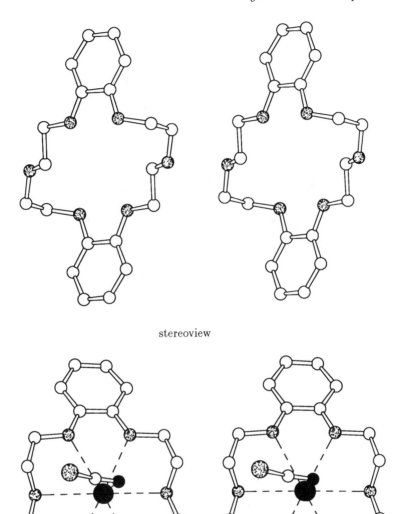

(a)   stereoview

(b)   stereoview

FIGURE 17.19. Stereoviews of (a) dibenzo-18-crown-6 (Refs. 80 and 84), and (b) its rubidium complex (Ref. 80). Hydrogen atoms are omitted for clarity.

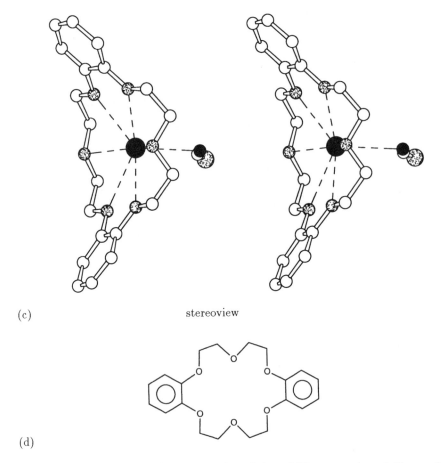

(c)   stereoview

(d)

FIGURE 17.19 (cont'd). (c) Another view of the rubidium complex of dibenzo-18-crown-6 (Ref. 80). (d) The chemical formula of this complexing agent.

One metal-binding motif found in proteins is the helix–loop–helix motif, also known as an *EF* hand.[86–89] A comparison is made with a right hand with an index finger (the *E* helix), a curled second finger (the loop), and a thumb (the *F* helix). This metal-binding motif has a strong affinity for calcium ions. This is generally an octahedral arrangementwith one site shared by two of the oxygen atoms of a glutamic acid side chain. The vertices of the octahedron are designated X, Y, Z, $-X$, $-Y$, $-Z$ in a right-handed system, as shown in Figure 17.20. The position $-Y$ is filled by a conserved backbone carbonyl group and $-Z$ by the shared glutamic acid group. Generally in proteins with this calcium-binding site X is aspartic acid, $-X$ a water molecule. The ordered sequence of amino acids in the protein requires that positions $-X$ and $-Y$ be inverted, as

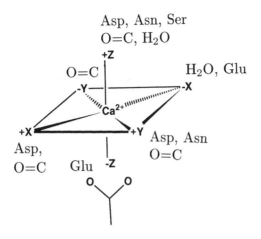

(a)

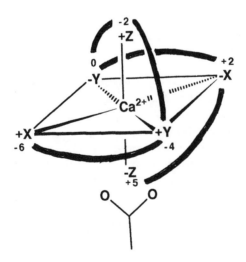

(b)

FIGURE 17.20. An *EF* hand, a calcium-binding motif in proteins. (a) Ligands binding to each coordination position around Ca^{2+}, and (b) the sequence numbering (starting at −6 and proceeding to +5) of the protein as it binds to Ca^{2+}

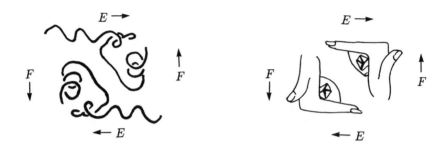

(c)

FIGURE 17.20 (cont'd). (c) An *EF* hand, designating the directions of the helices.

shown in Figure 17.20. Nature has utilized this calcium-specific binding motif in several proteins, such as troponin C, calmodulin, parvalbumin, and lactalbumin.

An interesting metal-binding motif in proteins is the **zinc finger**, which contains a β ribbon–turn–α helix motif that binds a zinc ion via four sulfur atoms of cysteine residues, or via two such sulfur atoms and two nitrogen atoms (from histidine).[90,91] This motif, first identified by NMR studies,[92] has been found in the crystal structures of several protein-nucleic acid complexes.[93]

17.5 Drug–protein and drug–nucleic acid interactions

The binding of drugs to biological macromolecules involves a fit between the drug and its receptor that depends not only on their respective shapes, but also on the distribution of electronic charge on the surfaces of both molecules so that a complementary fit may be obtained. Molecules with similar, but not identical formulae may bind to the same receptor in the same or different ways, depending on the nature of the functional groups in each.

The challenge of how to represent the "receptor" that is deduced from these results is greatly helped by the use of computer graphics systems. For example, it was used to show the complementarity of molecular surfaces in the crystal structure of the complex between an antibody and lysozyme.[94] The contact area is about 20×20 Å^2. Side chains from one, such as Gln 121 of lysozyme, fit in a depression on the surface of the other protein. Interactions with small antigens, such as phosphorylcholine,[95] have also demonstrated the specificity of the interactions between antibodies and antigens.

17.5.1 Binding of drugs to hemoglobin

The protein hemoglobin is composed of two α chains, each 141 amino acids long, and two β chains, 146 amino acids long. Each has one bound heme group, four in all.[96,97] This tetramer has pseudotetrahedral symmetry with true twofold axes between pairs of α subunits and between pairs of β subunits. Hemoglobin is allosteric and is in equilibrium with two alternative structures. These are deoxyhemoglobin (T structure) with a high affinity for oxygen, and oxyhemoglobin (R structure) with a low affinity for oxygen. Sickle-cell hemoglobin results from a mutation in which a glutamic acid on the exterior of the protein is replaced by valine. As a result this hemoglobin forms filaments.[98]

An analysis of the ways that various drugs bind to hemoglobin has been made by soaking compounds into crystalline human deoxyhemoglobin.[99] The compounds used include ethacrynic acid (a diuretic), bezafibrate (an antihyperlipoproteinemia drug), succinyl-L-tryptophan-L-tryptophan, and p-bromobenzyloxyacetic acid. The structures of such drug–hemoglobin complexes (Figure 17.21) show that the drugs seek out cavities in the protein and bind in accord with the shape of the cavity and with the maximum possible number of electrostatic interactions.

Van der Waals forces also play a significant role in binding. When the drugs bind to hemoglobin, small distortions in the protein occur, and these affect the solubility of deoxyhemoglobin S (sickle-cell hemoglobin). They may change the oxygen affinity of the hemoglobin.

17.5.2 Binding of drugs to cytochrome P-450

When drugs and other foreign chemicals, such as polycyclic aromatic hydrocarbons enter the body, they are oxidized to hydroxylated products,

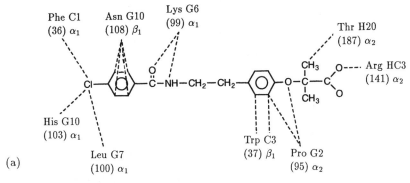

(a)

FIGURE 17.21. Binding of a drug (bezafibrate, an antihyperlipoproteinemia drug) to hemoglobin. (a) Functional groups in crystalline hemoglobin that are involved in binding the drug.

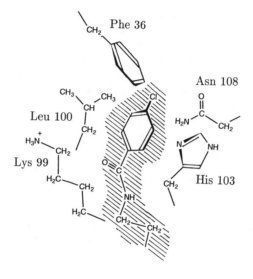

(b)

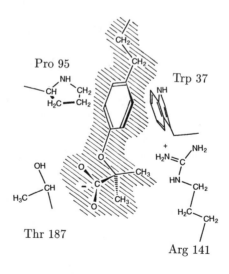

(c)

FIGURE 17.21 (cont'd).   (b) and (c).   Diagrams demonstrating the three-dimensionality of the binding of bezafibrate to crystalline hemoglobin that was shown in (a). (b) Binding of the chlorobenzene and amide portions of bezafibrate, and (c) binding of the phenoxy and isobutyrate portions of bezafibrate.

which are more water soluble and therefore more readily excreted. The enzyme system involved, known as aryl hydrocarbon hydroxylase (AHH), consists of at least three components:[100]

1. A monooxygenase enzyme (cytochrome P-450 or its equivalent) that uses molecular dioxygen to oxidize the substrate;
2. NADPH-cytochrome P-450 reductase, a flavoprotein that transfers electrons from NADH to cytochrome P-450, reducing the iron of the P-450 hemoprotein from the Fe(III) to the Fe(II) state so that molecular oxygen will bind to it in the first stages of its action; and
3. a lipid fraction.

The process for a PAH (here designated RH) is:

$$RH + NADPH + H^+ + {}^{18}O_2 \rightarrow R^{18}OH + NADP^+ + H_2^{18}O. \tag{17.5}$$

Cytochrome P-450 can activate molecular oxygen (O_2) so that one of the two oxygen atoms is incorporated into the substrate and the other oxygen atom is reduced by two electrons to water. Unfortunately, hydroxylation by P-450 enzymes, a reaction that was meant to cause detoxification, may, under certain conditions, cause a conversion of a hydrocarbon to a carcinogenic agent.

The cytochrome P-450 enzyme system that has been best characterized structurally is the cytochrome P-450$_{cam}$ which oxidizes the monoterpene camphor (2-bornanone) (Figure 17.22). This enzyme takes part in the first step in a catabolic process that allows the organism to use camphor [Figure 17.23(a)] as a sole carbon source. Cytochrome P-450$_{cam}$, more precisely called cytochrome P-450 camphor 5-*exo*-hydroxylase, has a molecular weight of 45,000 and contains a single protoporphyrin IX group [Figure 17.23(b)]. The crystal structure of this P-450$_{cam}$, for which the amino-acid sequence is known,[101] has been determined to 2.2 Å res-

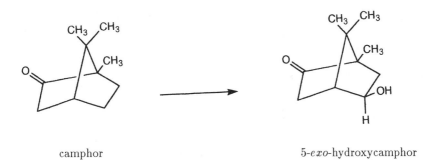

camphor   5-*exo*-hydroxycamphor

FIGURE 17.22. The reaction catalyzed by cytochrome P-450$_{cam}$.

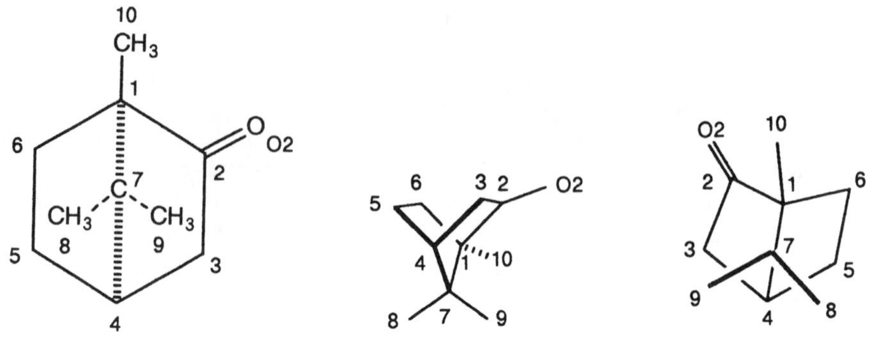

(a)

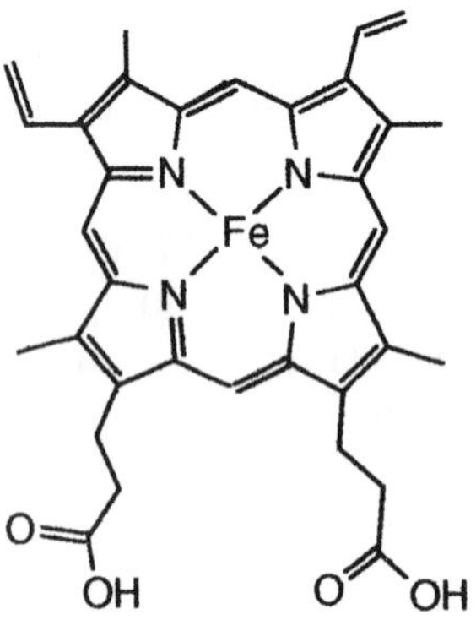

(b)

FIGURE 17.23. (a) Chemical formula of camphor, drawn in various orientations. (b) Chemical formula of protoporphyrin IX. The upper axial ligand is dioxygen (O_2) and the lower axial ligand is a thiolate (S^-) side chain on the enzyme (Cys 357).

olution in the absence of bound substrate and to 1.6 Å resolution in the presence of camphor.[102–106] The camphor–P-450$_{cam}$ complex[103] contains a pentacoordinated iron atom with the sulfur atom of Cys-357 (Figure 17.24) as one axial ligand, as suggested by spectroscopic studies.[107] The camphor is located in the active site, at a minimum distance of 4 Å above the protoporphyrin IX.[108.109]

The heme is sandwiched between two parallel α helices. Cys-357, the axial thiolate ligand to the iron of the protoporphyrin is protected in a pocket of Phe-350, Leu-358, and Gln-360. It extends from the N-terminal end of a helix on the molecular surface.

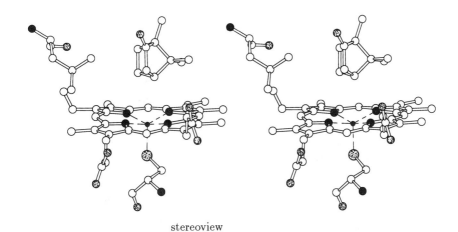

stereoview

FIGURE 17.24. Active site of cytochrome P-450$_{cam}$ showing Cys-357 and camphor in the active site. The sulfur atom of Cys-357 is indicated by a large stippled circle.

The active site of the enzyme is lined with the hydrophobic residues. The substrate, camphor, is held in place by a hydrogen bond to the hydroxyl group of Tyr-96.[110] Contacts of Phe-87, Val-247, and Val-295 with the methyl groups of the camphor are hydrophobic. Norcamphor, which lacks the three methyl groups of camphor, is much less rigidly held by the enzyme and binds 0.9 Å further from the oxygen-binding site.[111] On binding of substrate, large decreases in thermal motion are observed in the areas of Tyr-96, Thr-185, and Asp-251, implying increased rigidity of structure ion binding. These amino-acid side chains are involved in substrate access and binding in the active site. The orientation of the camphor, shown in Figure 17.24, implies that a heme-bound "activated" oxygen atom can add only from the direction that gives 5-*exo*-hydroxycamphor, the product that is actually found.[99,112] Thus the stereospecific nature of the oxidation is demonstrated in the crystal structure. Abstraction of hydrogen can occur from either the *exo* or *endo* positions C5 of camphor, but oxygen adds stereospecifically only to the *re* face to give only the *exo* compound (Figure 17.22).

Crystal structures have been reported for the enzyme with several bound drugs. Metyrapone- and 1-, 2- and 4-phenylimidazole-inhibited complexes of cytochrome P-450$_{cam}$ have each been refined to 2.1 Å resolution.[113] Except in the 2-phenylimidazole complex, each complex forms an N $\cdots$ Fe interaction to the heme iron atom. In the 2-phenylimidazole complex, water or hydroxide coordinates the heme iron atom and the inhibitor binds in the camphor pocket. Details of the inhibitor binding are shown in Figure 17.25. Eukaryotic cytochrome P-450 is membrane bound and has a different structure from the soluble

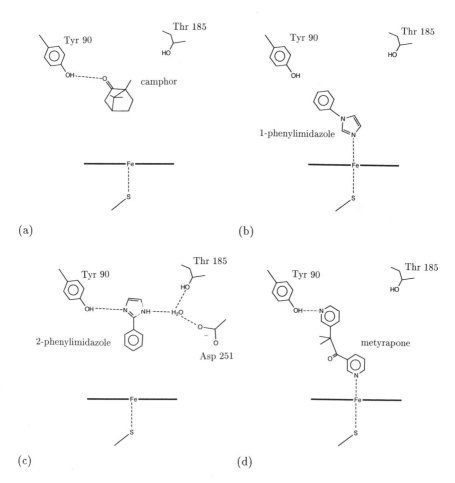

FIGURE 17.25. Interactions of cytochrome P-450$_{cam}$ with (a) camphor, (b) 1-phenylimidazole. (c) 2-phenylimidazole, and (d) metyrapone. Note that the enzyme-bound Fe does not interact with a ring nitrogen atom in (c) for steric reasons.

bacterial cytochrome P-450$_{cam}$ since it must span a membrane.[114,115] Studies of other cytochrome P-450 enzymes from the known amino acid sequences and the known crystal structure of cytochrome P-450$_{cam}$ are now in progress.[112]

17.5.3 Binding of drugs to dihydrofolate reductase

An extensively studied enzyme–inhibitor system involves the protein dihydrofolate reductase,[116–122] and demonstrates an important feature of drug–enzyme interactions – the fact that an inhibitor drug may not bind in the same way as substrate even though both have similar chemical formulæ. Crystal structure data are available for the enzyme and its complexes with both substrates and inhibitors. This enzyme catalyzes, as shown in Figure 17.26, the reduction of 7,8-dihydrofolate to 5,6,7,8-tetrahydrofolate, an essential coenzyme used in the synthesis of thymidylate, inosinate, and methionine. Methotrexate is an antimetabolite that is a folic acid analogue. It inhibits dihydrofolate reductase by binding tightly and thereby causing a cellular deficiency of thymidylate

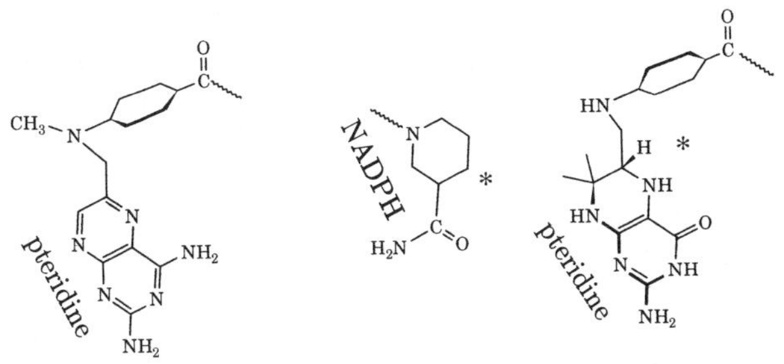

(a) methotrexate in *E. coli* enzyme

(b) methotrexate and NADPH in *L. casei* enzyme

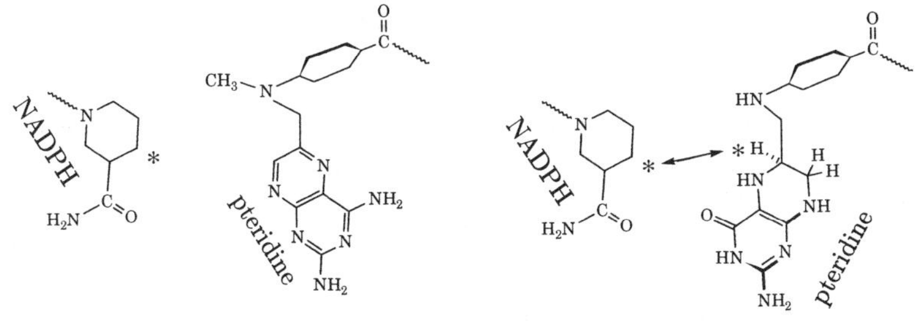

(c) model of tetrahydrofolate in the same orientation as methotrexate

(d) tetrahydrofolate rotated so that H can add from the NADPH side

FIGURE 17.26. Binding of (a) methotrexate to dihydrofolate reductase and (b) a comparison with a model of tetrahydrofolate bound in the same orientation. (c) The binding mode observed for NADPH and methotrexate in the crystalline state. It is seen in (c) that when tetrahydrofolate is bound in the same way as methotrexate, the hydrogen atom cannot be transferred. (d) Therefore the tetrahydrofolate must be bound in a different manner from methotrexate for a reaction to occur. The two orientations of the two ligands are related by a rotation. This has been verified by an X-ray structure determination (Ref. 122). Thus tetrahydrofolate and methotrexate do not bind in the same manner to the enzyme.

("thymine-less death"). This drug is used as a cancer chemotherapeutic agent. Crystallographic studies of folic acid, trimethoprim, and several other antifolates have been determined. Methotrexate is found in a cavity 15 Å deep in the *E. coli* enzyme complex. The pteridine ring of methotrexate lies nearly perpendicular to the aromatic ring of the *p*-amino benzoyl glutamate group. The entire inhibitor molecule is bound by at least 13 amino acid residues to the enzyme by a system of hydrogen bond and hydrophobic interactions. A similar binding is found for the *L. casei* enzyme complex.

The absolute configuration of tetrahydrofolate has been established[123,124] to be *S* at C6, but if dihydrofolate is bound in the same orientation as methotrexate, the configuration should be *R*. When dihydrofolate (or tetrahydrofolate) is bound in the enzyme in the same way as methotrexate, the hydrogen atom that is transferred to NADPH (asterisk in Figure 17.26) is pointing in the wrong direction. Thus dihydrofolate and methotrexate must bind to the enzyme with different orientations as established by X-ray crystallographic studies. The binding of folate and methotrexate are shown in Figure 17.26. The recognition of the site in the enzyme is governed mainly by hydrogen bonding, and not simply by molecular shape.

17.5.4 Binding of drugs to nucleic acids

Antitumor agents that interact with DNA do so in one of three ways.

1. *Intercalation* between the nucleic acid bases of DNA. This mode of binding is found for anthracyclines, actinomycin D and acridines. The DNA backbone flexes so that the bases, normally separated by 3.4 Å, are moved to approximately 7 Å apart. As a result, a flat aromatic molecule, can slip between them.
2. *Non-intercalative groove binding.* The drug lies in one of the **grooves in DNA**, following the helical contour of DNA as it binds. Netropsin and distamycin bind in this way.
3. *Covalent bond formation.* This occurs with mitomycin C.

Many **intercalation** complexes have been studied by X-ray diffraction techniques. The first model for the interaction of actinomycin D with a nucleic acid was based on the crystal structure of an actinomycin–guanine complex.[125] Many other such intercalative complexes have been studied; these involve self-complementary dinucleoside phosphates or short oligonucleotides that can dimerize by hydrogen bond formation to give a portion with Watson–Crick base pairing. These crystal structures demonstrate the types of conformational changes that the phosphodiester backbone has undergone on binding of the drug, and the hydrogen bonding that the inserted molecule can make to the phosphate or other nucleic acid groups.

The anthracyclines are aminoglycosidic derivatives of tetracyclic compounds that contain an anthraquinone chromophore. Originally, daunomycin and adriamycin, its hydroxy analogue, were isolated from cultures of *Streptomyces peucetius*.[126] The main target of these drugs appears to be DNA, most likely by formation of an intercalation complex. The drugs also affect topoisomerase II, an enzyme which interacts with the DNA that has been modified by intercalation.[126]

Daunomycin and adriamycin have a planar chromophore with an amino sugar extending out of the plane. In the crystal structure of the complex of adriamycin with a self-complementary hexadeoxynucleotide.[127,128] The oligonucleotide forms a right-handed helix with two molecules of adriamycin intercalated in the d(CpG) sequences at each end. Adriamycin is intercalated with its long axis perpendicular to those of the base pairs, as shown in Figure 17.27(b). The hydroxyl group forms two hydrogen bonds to an adjacent guanine. The amino sugar lies in the minor groove of the distorted helix, but does not form bonds to the nucleic acid. Thus ring A anchors adriamycin in DNA, and the nucleic-acid backbone is distorted by the interactions formed.

Minor groove binders are generally A·T specific and fit snugly in the narrow minor groove. Such drugs are replacements for well-ordered water molecules found in the absence of drugs. Netropsin and distamycin A, which are are pyrrole-amidine oligopeptide antibiotics, bind very tightly to A·T regions of duplex DNA. These compounds are relatively cytotoxic. In the crystal structure of netropsin bound to a dodecamer with six A·T base pairs in the middle (Figure 17.28), the six base pairs have high propeller twists.[129–131] As a result, the amino group of adenine is located half-way between the carbonyl groups of two adjacent thymine residues and some three-center hydrogen bonds are formed. The additional hydrogen bond in A·T base pairs may help stabilize the DNA. An-

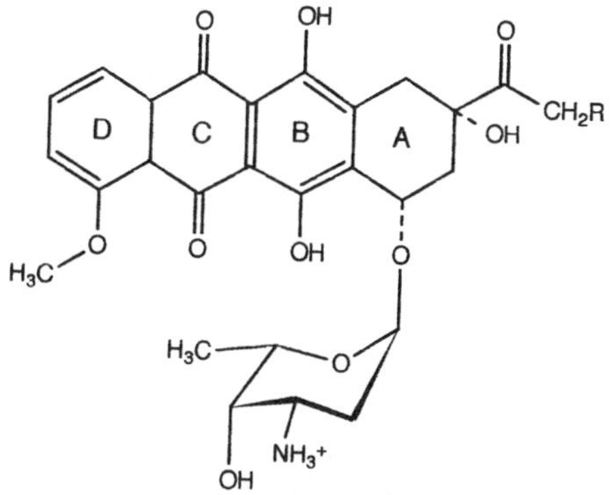

(a)

FIGURE 17.27. Binding of adriamycin to an oligonucleotide (Ref. 128). (a) Chemical formulæ. R = H for daunomycin, and R = OH for adriamycin.

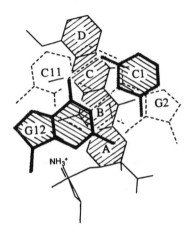

(b)

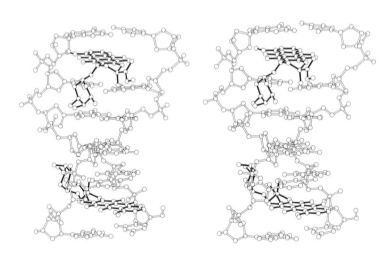

(c)   stereoview

FIGURE 17.27 (cont'd). (b) Diagram of the intercalation of the chromophore of adriamycin between base pairs. View down the helix axis. (c) Stereoview of the binding of adriamycin (black bonds) to an oligonucleotide (open bonds). The chromophore is intercalated between base pairs and the amino sugar extends in the minor groove.

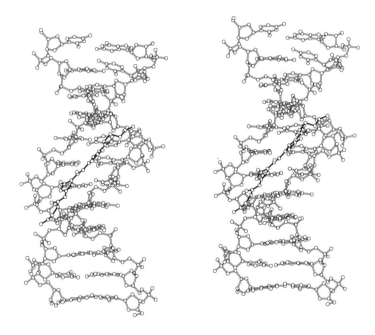

(a)

(b)   stereoview

FIGURE 17.28. Binding of netropsin to a polynucleotide. (a) Chemical formula of netropsin, and (b) stereoview of the binding of netropsin in the minor groove of an oligonucleotide (Ref. 129).

other minor-groove binder is Hoechst dye 33258 which is a DNA stain. The structures of this dye 33258 bound both to d(CGCGATATCGCG) and to d(CGCGAATTCGCG) have been determined.[132-134] In each, the DNA undergoes alterations in the helical twist at ApT and TpA steps and rotation of some DNA bases if they form hydrogen bonds to the drug. Thus it appears that DNA can be quite flexible.

17.5.5 Protein-nucleic acid interactions

Proteins can interact with nucleic acids by means of a helix-turn-helix motif[135] in the protein structure, by a zinc finger[93] or other type of domain formed by zinc binding, or by a leucine zipper[136] in which leucine side chains that are seven amino-acid residues apart on along each of two α helices interact and force a protein conformation that is suitable for binding to a nucleic acid. The recognition of the bases in DNA by hydrogen bonding involves patterns diagrammed in Figure 17.29. This shows that each combination of base pairs in a nucleic acid — GC, AT, CG, and TA — can be differentiated in the major groove by the hydrogen-bond pattern. The recognition is less specific in the minor groove.[137] The side chains of proteins available for such recognition are arginine with two hydrogen-bond donor groups in the required orientation, glutamic or aspartic acid with two hydrogen-bond acceptors, and asparagine and glutamine which contain one donor and one acceptor. Many crystal structures that involve protein-nucleic acid interactions have been published in the scientific literature recently. Some show this recognition scheme of Figure 17.29, as in a zinc finger protein-nucleic acid complex,[92] but others do not. Their interactions are more varied and apparently less specific. This is an exciting area of research and more results are being published each month. There is still a need to establish unequivocally what the recognition rules are for protein-nucleic acid interactions, and this requires further structural studies.

17.5.6 Binding of drugs to protein receptors

The general modes of binding of drugs to receptors have been reported extensively in the literature.[138-143] These studies often involve conformational studies and a deduction of the conformation of the active species. Molecular modelling is also used.[144] In this way pharmaceutical drugs can be rationally designed with due attention to the stereochemistry as well as the electronic features of the binding.[145] Information is now becoming available on the actual binding modes of drugs from crystal structures of the drug-receptor complex, of the type shown for dihydrofolate reductase. The design of specific molecules to recognize specific targets is now underway and can be very effective. C_{60}, buckminsterfullerene, has been modelled into HIV-1 protease and predicted to act as an inhibitor. This was verified experimentally.[146]

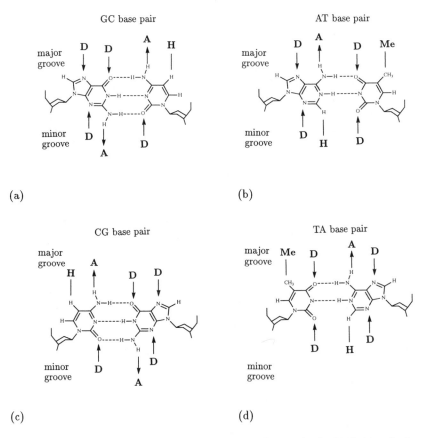

FIGURE 17.29. Hydrogen bonding pattern of the base pairs in DNA. **D** = hydrogen-bond donor, **A** = hydrogen-bond acceptro, and Me = methyl groups. Note that the pattern in the major groove allows one to discriminate between the four possible base-pair arrangements (Ref. 137).

The inhibition of the enzyme purine nucleoside phosphorylase has been studied by crystallographic techniques. Inhibition of this enzyme has been targeted as possible therapy for arthritis, and some other diseases. An inhibitor was designed to fit snugly into the active site of the enzyme. The new inhibitor contains a 9-deazaguanine group, a chlorinated benzene ring to fill the sugar-binding site, and an acetate group to fill the phosphate-binding site. It is much more effective than previously designed inhibitors.[145] This is a rapidly evolving area of research.

Summary

In the study of crystal structures molecular recognition is evident in may ways:

1. The act of crystallization is the recognition of one molecule by a like molecule. At least one interaction, weak or strong, leads to the formation of a crystal.
2. Surroundings of functional groups in the molecules in crystals can be analyzed from crystal structure determinations. Scatterplots, preferably contoured, highlight any preferred directions of binding. This information can be used in model building of drugs binding to their receptors.
3. The recognition of crystal faces by impurity molecules can be exploited in various ways. It is possible to resolve enantiomorphic compounds even when the original crystal structure is centrosymmetric.
4. Certain molecules, such as naturally occurring ionophores and synthetic crown ethers, can recognize and bind metal ions, often fairly specifically. Enzymes are proteins that fold in a way that accomplishes this stereospecific binding of required molecules.
5. Information on molecular recognition is also found in the crystal structures of proteins or oligonucleotides complexed with small molecules.

Glossary

Crown ether: A synthetic cyclic polyether with three or more binding sites that can bind a guest, often a metal cation, in a central position.

Electrostatic potential: Electrostatics is concerned with the phenomena associated with an electric charge at rest. The electrostatic potential at a point r is the energy required to bring a unit positive charge from infinity to that point. Values may ne positive or negative.

Gaussian-like peak: A bell-shaped curve described by the equation $Y = A \exp -(x^2/B)$. See Chapter 10 for more details.

Groove in DNA: The structures of DNA have grooves between the lines of phosphodiester backbone. In the Watson–Crick model (B-DNA) the major groove is deep and wide and the minor groove is shallow and narrow. Different functional groups on the bases are accessible in the two different grooves of the same DNA molecule.

Host–guest complex: A guest molecule trapped in a host crystal structure.

Intercalation: A process whereby a flat aromatic molecule, such as an acridine, becomes inserted between two adjacent stacked bases in double-stranded DNA. Similarly, cations can become intercalated between the sheets of atoms in graphite.

Ionophore: A lipid-soluble compound that increases the permeability of a membrane to metal ions, which are then transported across the membrane. Thus, ions are transferred from an aqueous medium into a hydrophobic phase (a membrane, for example) (see Ref. 84).

Pharmacophore: A three-dimensional arrangement of functional groups in a drug that is essential for its recognition and activations.

Scatterplot: A plot of data as points in an orthogonal coordinate system.

Zinc finger: A DNA-recognizing motif in a protein. The zinc ion is coordinated by cysteine and/or histidine amino-acid side chains. The zinc finger contains antiparallel β ribbon and an α helix.

References

1. Marshall, G. R., Barry, C. D., Bosshard, H. E., Dammkoehler, R. A. and Dunn, D. A. The conformational parameter in drug design: the active analog approach. In: *Computer-assisted Drug Design.* (**Eds.**, Olson, E. C., and Christofferson, R. E.) Ch. 9, pp. 205–226. ACS Symposium Series 112. American Chemical Society: Washington (1979).

2. Carrell, C. J., Carrell, H. L., Erlebacher, J., and Glusker, J. P. Structural aspects of metal ion–carboxylate interactions. *J. Amer. Chem. Soc.* **110**, 8651–8656 (1988).

3. Murray-Rust, P., and Glusker, J. P. Directional hydrogen bonding to sp^2- and sp^3-hybridized atoms and its relevance to ligand–macromolecule interactions. *J. Amer. Chem. Soc.* **106**, 1018–1025 (1984).

4. Allen, F. H., Bellard, S., Brice, M. D., Cartwright, B. A., Doubleday, A., Higgs, H., Hummelink, T., Hummelink-Peters, B. G., Kennard, O., Motherwell, W. D. S., Rodgers, J. R., and Watson, D. G. The Cambridge Crystallographic Data Centre: computer-based search, retrieval, analysis and display of information. *Acta Cryst.* **B35**, 2331–2339 (1979).

5. Rosenfield, R. E., Jr., Swanson, S. M., Meyer, E. F., Jr., Carrell, H. L., and Murray-Rust, P. Mapping the atomic environment of functional groups: turning 3D scatter plots into pseudo-density contours. *J. Molec. Graphics* **2**, 43–46 (1984).

6. Rosenfield, R. E., Jr., Parthasarathy, R., and Dunitz, J. D. Directional preferences of nonbonded atomic contacts with divalent sulfur. 1. Electrophiles and nucleophiles. *J. Amer. Chem. Soc.* **99**, 4860–4862 (1977).

7. Fukui, K., Yonezawa, T., and Shingu, H. A molecular orbital theory of reactivity in aromatic hydrocarbons. *J. Chem. Phys.* **20**, 722–725 (1952).

8. Chakrabarti, P. Geometry of interaction of metal ions with sulfur-containing ligands in proteins. *Biochemistry* **28**, 6081–6085 (1989).

9. Gregoret, L. M., Rader, S. D., Fletterick, R. J., and Cohen, F. E. Hydrogen bonds involving sulfur atoms in proteins. *Proteins: Structure, Function, and Genetics.* **9**, 99–107 (1991).

10. Presta, L. G., and Rose, G. D. Helix signals in proteins. *Science* **240**, 1632–1641 (1988).

11. Vallee, B. L., and Auld, D. S. Zinc coordination, function, and structure of zinc enzymes and other proteins. *Biochemistry* **29**, 5647–5659 (1990).

12. Vedani, A., and Huhta, D. W. A new force field for modelling metalloproteins. *J. Amer. Chem. Soc.* **112**, 4759–4767 (1990).

13. Murray-Rust, P., Stallings, W. C., Monti, C. T., Preston, R. M., and Glusker, J. P. Intermolecular interactions of the C–F bond: the crystallographic environment of fluorinated carboxylic acids and related structures. *J. Amer. Chem. Soc.* **105**, 3206–3214 (1983).

14. Ramasubbu, N., Parthasarathy, R., and Murray-Rust, P. Angular preferences of intermolecular forces around halogen centers: Preferred directions of approach of electrophiles and nucleophiles around the carbon–halogen bond. *J. Amer. Chem. Soc.* **108**, 4308–4314 (1986).

15. Pauling, L. *The Nature of the Chemical Bond.* Cornell University Press: Ithaca, New York. 1st edn., 1939; 2nd edn., 1940; 3rd edn., 1960.

16. Carrell, H. L., and Donohue, J. Interatomic distances in the bifluoride ion. *Isr. J. Chem.* **10**, 195–200 (1972).

17. Allen, L. C. A simple model of hydrogen bonding. *J. Amer. Chem. Soc.* **97**, 6921–6940 (1975).

18. Jeffrey, G. A., and Saenger, W. *Hydrogen Bonding in Biological Structures.* Springer-Verlag: Berlin, New York (1991).

19. Umeyama, H., and Morokuma, K. The origin of hydrogen bonding. An energy decomposition study. *J. Amer. Chem. Soc,* **99**, 1316–1332 (1977).

20. Einspahr, H., and Bugg, C. E. The geometry of calcium–carboxylate interactions in crystalline complexes. *Acta Cryst.* **B37**, 1044–1052 (1981).

21. Rebek, J., Jr., Marshall, L., Wolak, R., Parris, K., Killoran, M., Askew, B., Nemeth, D., and Islam, N. Convergent functional groups: synthetic and structural studies. *J. Amer. Chem. Soc.* **107**, 7476–7481 (1985).

22. Rebek, J., Jr., Duff, R. J., Gordon, W. E., and Parris, K. Convergent functional groups provide a measure of stereoelectronic effects at carbonyl oxygen. *J. Amer. Chem. Soc.* **108**, 6068–6069 (1986).

23. Rebek, J., Jr. Molecular recognition with model systems. *Angew. Chem. Int. Ed. Engl.* **29**, 245–255 (1990).

24. Rebek, J., Jr. Binding forces, equilibria, and rates: new models for enzymic catalysis. *Acc. Chem. Res.* **17**, 258–264 (1984).

25. Hurley, T. J., Carrell, H. L., Gupta, R. K., Schwartz, J., and Glusker, J. P. The structure of sodium β-fluoropyruvate: a *gem*-diol. *Arch. Biochem. Biophys.* **193**, 478–486 (1979).

26. Petersen, M. R., and Csizmadia, I. G. Determination and analysis of the formic acid conformational hypersurface. *J. Amer. Chem. Soc.* **101**, 1076–1079 (1979).

27. Gandour, R. On the importance of orientation in general base catalysis by carboxylate. *Bioorg. Chem.* **10**, 169–176 (1981).

28. Chidambaram, R., and Ramanadham, M. Hydrogen bonding in biological molecules – an update. *Physica B* **174**, 300–305 (1991).

29. Ramanadham, M., Jakkal, V. S., and Chidambaram, R. Carboxyl group hydrogen bonding in X-ray protein structures analyzed using neutron studies on amino acids. *FEBS Letters* **323**, 203–206 (1993).

30. van der Helm, D., Jalal, M. A. F., and Hossain, M. B. The crystal structures, conformations and configurations of siderophores. In: *Iron Transport in Microbes, Plants and Animals.* (**Eds.**, Winkelmann, G., van der Helm, D., and Neilands, J. B.) pp. 135–165. VCH: Weinheim (1987).

31. Hossain, M. B., Jalal, M. A. F., Benson, B. A., Barnes, C. L., and van der Helm, D. Structure and conformation of two coprogen-type siderophores: neocoprogen I and neocoprogen II. *J. Amer. Chem. Soc.* **109**, 4948–4954 (1987).

32. Glusker, J. P. Citrate conformation and chelation: enzymatic implications. *Acc. Chem. Res.* **13**, 345–352 (1980).

33. Carrell, H. L., Glusker, J. P., Piercy, E. A., Stallings, W. C., Zacharias, D. E., Davis, R. L., Astbury, C., and Kennard, C. H. L. Metal chelation versus internal hydrogen bonding of the α-hydroxy carboxylate group. *J. Amer. Chem. Soc.* **109**, 8067–8071 (1987).

34. Chakrabarti, P. Geometry of interaction of metal ions with histidine residues in protein structures. *Protein Engineering* **4**, 57–63 (1990).

35. Carrell, A. B., Shimoni, L., Carrell, C. J., Bock, C. W., Murray-Rust, P., and Glusker, J. P. The stereochemistry of the recognition of nitrogen-containing heterocycles by hydrogen bonding and by metal ions. *Receptor* **3**, 57–76 (1993).

36. Taylor, R., Kennard, O., and Versichel W. Geometry of the N–H···O=C hydrogen bond. 1. Lone-pair directionality. *J. Amer. Chem. Soc.* **105** 5761–5766 (1983).

37. Vedani, A., and Dunitz, J. D. Lone-pair directionality in hydrogen bond potential functions for molecular mechanics calculations: the inhibition of human carbonic anhydrase II by sulfonamides. *J. Amer. Chem. Soc.* **107**, 7653–7658 (1985).

38. Britton, D., and Dunitz, J. D. Directional preferences of approach of nucleophiles to sulfonium ions. *Helv. Chim. Acta* **63**, 1068–1073 (1980).

39. Rosenfield, R. E., and Murray-Rust, P. Analysis of the atomic environment of quaternary ammonium groups in crystal structures, using computerized data retrieval and interactive graphics: modeling acetylcholine–receptor interactions. *J. Amer. Chem. Soc.* **104**, 5427–5430 (1982).

40. Chakrabarti, P., and Dunitz, J. D. Directional preferences of ether O-atoms towards alkali and alkaline earth cations. *Helv. Chim. Acta* **65**, 1482–1488 (1982).

41. Cody, V., and Murray-Rust, P. Iodine···X(O, N, S) intermolecular contacts: models of thyroid hormone–protein binding interactions using information from the Cambridge Crystallographic Data Files. *J. Molec. Struct.* **112**, 189–199 (1984).

42. Gould, R. O., Gray, A. M., Taylor, P., and Walkinshaw, M. D. Crystal environments and geometries of leucines, isoleucine, valine, and phenylalanine provide estimates of minimum nonbonded contact and preferred van der Waals interaction. *J. Amer. Chem. Soc.* **107**, 5921–5927 (1985).

43. Yamashita, M., Wesson, L., Eisenman, G., and Eisenberg, D. Where metal ions bind in proteins. *Proc. Natl. Acad. Sci. USA* **87**, 5648–5652 (1990).

44. Carrell, H. L., Rubin, B. H., Hurley, T. J., and Glusker, J. P. X-ray crystal structure of D-xylose isomerase at 4-Å resolution *J. Biol. Chem.* **259**, 3230–3236 (1984).

45. Carrell, H. L., Glusker, J. P., Burger, V., Manfre, F., Tritsch, D., and Biellmann, J-F. X-ray analysis of D-xylose isomerase at 1.9 Å: native enzyme in complex with substrate and with a mechanism-designed inactivator. *Proc. Natl. Acad. Sci. USA* **86**, 4440–4444 (1989).

46. Chakrabarti, P. Interaction of metal ions with carboxylic and carboxamide groups in protein structures. *Protein Engineering* **4**, 49–56 (1990).

47. Sutor, D. J. Evidence for the existence of C–H $\cdots$ O hydrogen bonds in crystals. *J. Chem. Soc.* 1105–1110 (1963).

48. Burley, S. K., and Petsko, G. A. Weakly polar interactions in proteins. *Advances in Protein Chemistry* **39**, 125–189 (1988).

49. Biradha, K., Sharma, C. V. K., Panneerselvam, K., Shimoni, L., Carrell, H. L., Zacharias, D. E., and Desiraju, G. R. Design of a C–H $\cdots$ O hydrogen bonded receptor for the recognition of 1,3,5-trinitrobenzene. *Chem. Commun.* 1473–1475 (1993).

50. Stewart, R. F. On the mapping of electrostatic properties from Bragg diffraction data. *Chem. Phys. Lett.* **65**, 335–342 (1979).

51. Coppens, P., Guru Row, T. N., Leung, P., Stevens, E. D., Becker, P. J., and Yang, Y. W. Net atomic charges and molecular dipole moments from spherical-atom X-ray refinements and the relation between atomic charge and shape. *Acta Cryst.* **A35**, 63–72 (1979).

52. Swaminatham, S., and Craven, B. M. Electrostatic properties of phospho-rylethanolamine at 123K from crystal diffraction data. *Acta Cryst.* **B40**, 511–518 (1984).

53. Su, Z., and Coppens, P. On the mapping of electrostatic properties from the multipole description of the charge density. *Acta Cryst.* **A48**, 188–197 (1992).

54. Stewart, R. F., and Craven, B. M. Molecular electrostatic potentials from crystal diffraction: the neurotransmitter γ-aminobutyric acid. *Biophys. J.* **65**, 998–1005 (1993).

55. Weber, H.-P., and Craven, B. M. Electrostatic properties of cytosine monohy-drate from diffraction data. *Acta Cryst.* **B46**, 532–538 (1990).

56. Stewart, R. F. Electron population analysis with rigid pseudoatoms. *Acta Cryst.* **A32**, 565–574 (1976).

57. Berkovitch-Yellin, Z., Addadi, L., Idelson, M., Leiserowitz, L., and Lahav, M. Absolute configuration of chiral polar crystals. *Nature (London)* **296**, 27–34 (1982).

58. Weissbuch, I., Addadi, L., Leiserowitz, L., and Lahav, M. Total asymmetric transformations at interfaces with centrosymmetric crystals: role of hydrophobic and kinetic effects in the crystallization of the system glycine/ α-amino acids. *J. Amer. Chem. Soc.* **110**, 561–567 (1988).

59. Lahav, M., Addadi, L., and Leiserowitz, L. Chemistry at the surfaces of organic crystals. *Proc. Natl. Acad. Sci. USA* **84**, 4737–4738 (1987).

60. Wolf, S. G., Leiserowitz, L., Lahav, M., Deutsch, M., Kjaer, K., and Als-Nielsen, J. Elucidation of the two-dimensional structure of an α-amino acid surfactant monolayer on water using synchrotron X-ray diffraction. *Nature (London)* **328**, 63–66 (1987).

61. Shimon, L. J. W., Lahav, M., and Leiserowitz, L. Stereoselective etchants for molecular crystals. Resolution of enantiomorphs and assignment of absolute structure of chiral molecules and polar crystals. *Nouv. J. Chim.* **10**, 723–737 (1986).

62. Addadi, L., Berkovitch-Yellin, Z., Weissbuch, I., Lahav, M., and Leiserowitz, L. A link between macroscopic phenomena and molecular chirality: crystals as probes for the direct assignment of absolute configuration of chiral molecules. *Topics in Stereochemistry* **16**, 1–85 (1986).

63. Weissbuch, I., Zbaida, D., Addadi, L., Leiserowitz, L., and Lahav, M. Design of polymeric inhibitors for the control of crystal polymorphism. Induced enantiomeric resolution at racemic histidine by crystallization at 25°C. *J. Amer. Chem. Soc.* **109**, 1869–1871 (1987).

64. Landau, E. M., Levanon, M., Leiserowitz, L., Lahav, M., and Sagiv, J. Transfer of structural information from Langmuir monolayers to three-dimensional growing crystals. *Nature (London)* **318**, 353–356 (1985).

65. Weissbuch, I., Shimon, L. J. W., Addadi, L., Berkovitch-Yellin, Z., Weinstein, S., Lahav, M., and Leiserowitz, L. Stereochemical discrimination at organic crystal surfaces. 1. The systems serine/threonine and serine/allothreonine. *Isr. J. Chem.* **25**, 353–361 (1985).

66. Berkovitch-Yellin, Z., Van Mil, J., Addadi, L., Idelson, M., Lahav, M., and Leiserowitz, L. Crystal morphology engineering by "tailor-made" inhibitors: a new probe to fine intermolecular interactions. *J. Amer. Chem. Soc.* **107**, 3111–3122 (1985).

67. Weissbuch, I., Addadi, L., Berkovitch-Yellin, Z., Gati, E., Lahav, M., and Leiserowitz, L. Spontaneous generation and amplification of optical activity in α-amino acids by enantioselective occlusion into centrosymmetric crystals of glycine. *Nature (London)* **310**, 161–164 (1984).

68. Weissbuch, I., Addadi, L., Berkovitch-Yellin, Z., Gati, E., Weinstein, S., Lahav, M., and Leiserowitz, L. Centrosymmetric crystals for the direct assignment of the absolute configuration of chiral molecules. Application to the α-amino acids by their effect on glycine crystals. *J. Amer. Chem. Soc.* **105**, 6615–6621 (1983).

69. Desiraju, G. R. *Crystal Engineering. The Design of Organic Solids.* Elsevier: Amsterdam (1989).

70. Addadi, L., Berkovitch-Yellin, Z., Weissbuch, I., Lahav, M., Leiserowitz, L., and Weinstein, S. The use of "enantiopolar" directions in centrosymmetric crystals for direct assignment of absolute configuration of chiral molecules: application to the system serine/threonine. *J. Amer. Chem. Soc.* **104**, 2075–2077 (1982).

71. Frey, M. N., Lehman, M. S., Koetzle, T. F., and Hamilton, W. C. Precision neutron diffraction structure determination of protein and nucleic acid components.

XI. Molecular configuration and hydrogen bonding of serine in the crystalline amino acids L-serine monohydrate and DL-serine. *Acta Cryst.* **B29**, 876–884 (1973).

72. Miles, F. D. The apparent hemihedrism of crystals of lead chloride and some other salts. *Proc. Roy. Soc. (London)* **A132**, 266–281 (1931).

73. Albrecht, G. and Corey, R. B. The crystal structure of glycine. *J. Amer. Chem. Soc.* **61**, 1087–1103 (1939).

74. Pressman, B. C. Alkali metal chelators – the ionophores. In: *Inorganic Biochemistry.* (**Ed.**, Eichhorn, G. L.) Ch. 6, pp. 203–226. Elsevier: Amsterdam, Oxford, New York (1975).

75. Brockmann, H. and Schmidt-Kastner, G. Valinomycin. I, XXVII. Mitteil. Über Antibiotica aus Actinomyceten. *Chem. Berichte* **88**, 57–61 (1955).

76. Duax, W. L., Hauptman, H., Weeks, C. M., and Norton, D. A. Valinomycin crystal structure determination by direct methods. *Science* **176**, 911–914 (1972).

77. Kratky, C. and Dobler, M. The crystal structure of enniatin B. *Helv. Chim. Acta* **68**, 1798–1803 (1985).

78. Pedersen, C. J. Crystalline salt complexes of macrocyclic ethers. *J. Amer. Chem. Soc.* **92**, 386–391 (1970).

79. Bright, D. and Truter, M. R. Crystal structure of a cyclic polyether complex of alkali metal thiocyanate. *Nature (London)* **225**, 176–177 (1970).

80. Bright, D. and Truter, M. R. Crystal structure of complexes between alkali-metal salts and cyclic polyethers. Part 1. Complex formed between rubidium sodium isothiocyanate and 2,3,11,12-dibenzo-1,4,7,10,13,16-hexaoxocyclo-octadeca-2,11-diene ('dibenzo-18-crown-6'). *J. Chem. Soc. (B)* 1544–1550 (1970).

81. Cram, D. J. Molecular container compounds. *Nature (London)* **356**, 29–36 (1992).

82. Cram, D. J., Tanner, M. E., Keipert, S. J., and Knobler, C. B. Two chiral [1.1.1]orthocyclophane units bridged by three biacetylene units as a host which binds medium-sized organic guests. *J. Amer. Chem. Soc.* **113**, 8909–8916 (1991).

83. Lehn, J-M. Supramolecular chemistry — scope and perspectives. Molecules, supermolecules, and molecular devices. (Nobel Prize Lecture). *Angew. Chem., Int. Ed. Engl.* **27**, 90–112 (1988).

84. Dobler, M. *Ionophores and their Structures.* Wiley-Interscience: New York, Chichester, Brisbane, Toronto (1981).

85. Neupert-Laves, K. and Dobler, M. The crystal structure of a K^+ complex of valinomycin. *Helv. Chim. Acta* **58**, 432–442 (1973).

86. Kretsinger, R. H., and Nockolds, C. E. Carp muscle calcium-binding protein. II. Structure determinations and general description. *J. Biol. Chem.* **248**, 3313–3326 (1973).

87. Szebenyi, D. M. E., and Moffat, K. The refined structure of vitamin D-dependent calcium-binding protein from bovine intestine. Molecular details, ion binding, and implications for the structure of other calcium-binding proteins. *J. Biol. Chem.* **261**, 8761–8777 (1986).

88. Strynadka, N. C. J., and James, M. N. G. Crystal structures of the helix–loop–helix calcium-binding proteins. *Annu. Rev. Biochem.* **58**, 951–998 (1989).
89. Snyder, E. E., Buoscio, B. W., and Falke, J. J. Calcium(I) site specificity: effect of size and charge on metal ion binding to an EF-hand-like site. *Biochemistry* **29**, 3937–3943 (1990).
90. Miller, J., McLachlan, A. D., and Klug, A. Repetitive zinc-binding domains in the protein transcription factor IIIA from Xenopus oocytes. *EMBO J.* **4**, 1609–1614 (1985).
91. Klug, A., and Rhodes, D. 'Zinc fingers:' a novel protein motif for nucleic acid recognition. *Trends Biochem. Sci.* **12**, 464–469 (1987).
92. Párraga, G., Horvath, S. J., Eisen, A., Taylor, W. E., Hood, L., Young, E. T., and Klevit, R. E. Zinc-dependent structure of a single-finger domain of yeast ADR1. *Science* **241**, 1489–1492 (1988).
93. Pavletich, N. P., and Pabo, C. O. Zinc finger–DNA recognition: crystal structure of a Zif268–DNA complex at 2.1 Å resolution. *Science* **252**, 809 – 817 (1991).
94. Amit, A. G., Mariuzza, R. A., Phillips, S. E. V., and Poljak, R. J. Three-dimensional structure of an antigen-antibody complex at 2.8 Å resolution. *Science* **233**, 747–753 (1986).
95. Padlan, E. A., Davies, D. R., Rudikoff, S., and Potter, M. Structural basis for the specificity of phosphorylcholine-binding immunoglobulins. *Immunochemistry* **13**, 945–949 (1976).
96. Perutz, M. F. Stereochemistry of cooperative effects in hæmoglobin. *Nature (London)* **228**, 726–734 (1970).
97. Perutz, M. F., Fermi, G., Luisi, B., Shaanan, B., and Liddington, R. C. Stereochemistry of cooperative mechanisms in hemoglobin. *Acc. Chem. Res.* **20**, 309–321 (1987).
98. Magdoff-Fairchild, B., Swerdlow, P. H., and Bertles, J. F. Intermolecular organization of deoxygenated sickle haemoglobin determined by X-ray diffraction. *Nature (London)* **239**, 217–219 (1972).
99. Perutz, M. F., Fermi, G., Abraham, D. J., Poyart, C., and Bursaux, E. Hemoglobin as a receptor of drugs and peptides: X-ray studies of the stereochemistry of binding. *J. Amer. Chem. Soc.* **108**, 1064–1078 (1986).
100. Alfred, L. J., and Gelboin, H. V. Benzpyrene hydroxylase induction by polycyclic hydrocarbons in hamster embryonic cells grown in vitro. *Science* **157**, 75–76 (1967).
101. Haniu, M., Armes, L. G., Yasunobu, K. T., Shastry, B. A., and Gunsalus, I. C. Amino acid sequence of the *Pseudomonas putida* cytochrome P-450. II. Cyanogen bromide peptides, acid cleavage peptides, and the complete sequence. *J. Biol. Chem.* **257**, 12664–12671 (1982).
102. Poulos, T. L., Perez, M., and Wagner, G. C. Preliminary crystallographic studies on cytochrome P-450$_{cam}$. *J. Biol. Chem.* **257**, 10427–10429 (1982).
103. Poulos, T. L., Finzel, B. C., Gunsalus, I. C., Wagner, G. C., and Kraut, J. The 2.6-Å crystal structure of *Pseudomonas putida* cytochrome P-450. *J. Biol. Chem.* **260**, 16122–16130 (1985).

104. Poulos, T. L., Finzel, B. C., and Howard, A. J. High-resolution crystal structure of cytochrome P450cam. *J. Molec. Biol.* **195**, 687–700 (1987).

105. Poulos, T. L. The crystal structure of cytochrome P-450$_{cam}$. In *Cytochrome P450; Structure, Mechanism, and Biochemistry.* (**Ed.**, Ortiz de Montellano, P. R.) Ch. 30, pp. 505–523. Plenum: New York, London (1986).

106. Poulos, T. L., Finzel, B. C., and Howard, A. J. Crystal structure of substrate-free *Pseudomonas putida* cytochrome P-450. *Biochemistry* **25**, 5314–5322 (1986).

107. Dawson, J. H. Probing structure-function relations in heme-containing oxygenases and peroxidases. *Science* **240**, 433–439 (1988).

108. Ortiz de Montellano, P. R., Kunze, K. L., and Beilan, H. S. Chiral orientation of prosthetic heme in the cytochrome P450 active site. *J. Biol. Chem.* **258**, 45–47 (1983).

109. Martinis, S. A., Atkins, W. M., Stayton, P. S., and Sligar, S. G. A conserved residue of cytochrome P-450 is involved in heme-oxygen stability and activation. *J. Amer. Chem. Soc.* **111**, 9252–9253 (1989).

110. Atkins, W. M., and Sligar, S. G. Tyrosine-96 as a natural spectroscopic probe of the cytochrome P-450$_{cam}$ active site. *Biochemistry* **29**, 1271–1275 (1990).

111. Raag, R., and Poulos, T. L. The structural basis for substrate-induced changes in redox potential and spin equilibrium in cytochrome P-450$_{CAM}$. *Biochemistry* **28**, 917–922 (1989).

112. Gelb, M. H., Heimbrook, D. C., Mälkönen, P., and Sligar, S. G. Stereochemistry and deuterium isotope effects in camphor hydroxylation by the cytochrome *P*450$_{cam}$ monoxygenase system. *Biochemistry* **21**, 370–377 (1982).

113. Poulos, T. L., and Howard, A. J. Crystal structures of metyrapone- and phenylimidazole-inhibited complexes of cytochrome P-450$_{cam}$. *Biochemistry* **26**, 8165–8174 (1987).

114. Hudeček, J., and Anzenbacher, P. Secondary structure prediction of liver microsomal cytochrome *P*-450; proposed model of spatial arrangement in a membrane. *Biochim. Biophys. Acta* **955**, 361–370 (1988).

115. Furuya, H., Shimizu, T., Hirano, K., Hatano, M., Fujii-Kuriyama, Y., Raag, R., and Poulos, T. L. Site-directed mutageneses of rat liver cytochrome P-450$_{d}$: catalytic activities toward benzphetamine and 7-ethoxycoumarin. *Biochemistry* **28**, 6848–6857 (1989).

116. Zvelebil, M. J. J. M., Wolf, C. R., and Sternberg, M. J. E. A predicted three-dimensional structure of human cytochrome P450: implications for substrate specificity. *Protein Engineering* **4**, 271–282 (1991).

117. Matthews, D. A., Alden, R. A., Bolin, J. T., Freer, S. T., Hamlin, R., Xuong, N., Kraut, J., Poe, M., Williams, M., and Hoogsteen, K. Dihydrofolate reductase: X-ray structure of the binary complex with methotrexate. *Science* **197**, 452–455 (1977) .

118. Matthews, D. A., Alden, R. A., Bolin, J. T., Filman, D. J., Freer, S. T., Hamlin, R., Hol, W. G. J., Kisliuk, R. L., Pastore, E. J., Plante, L. T., Xuong, N.-h.,

and Kraut, J. Dihydrofolate reductase from *Lactobacillus casei*. X-ray structure of the enzyme-methotrexate-NADPH complex. *J. Biol. Chem.* **253**, 6946–6954 (1978).

119. Bolin, J. T., Filman, D. J., Matthews, D. A., Hamlin, R. C., and Kraut, J. Crystal structures of *Escherichia coli* and *Lactobacillus casei* dihydrofolate reductase refined at 1.7 Å resolution. I. General features and binding of methotrexate. *J. Biol. Chem.* **257**, 13650–13662 (1982).

120. Filman, D. J., Bolin, J. T., Matthews, D. A., and Kraut, J. Crystal structures of *Escherichia coli* and *Lactobacillus casei* dihydrofolate reductase refined at 1.7 Å resolution. II. Environment of bound NADPH and implications for catalysis. *J. Biol. Chem.* **257**, 13663–13672 (1982).

121. Matthews, D. A., Bolin, J. T., Burridge, J. M., Filman, D. J., Volz, K. W., Kaufman, B. T., Beddell, C. R., Champness, J. N., Stammers, D. K., and Kraut, J. Refined crystal structures of *Escherichia coli* and chicken liver dihydrofolate reductase containing bound trimethoprim. *J. Biol. Chem.* **260**, 381–391 (1985).

122. Oefner, C., D'Arcy, A., and Winkler, F.K. Crystal structure of human dihydrofolate reductase complexed with folate. *Eur. J. Biochem.* **174**, 377–385 (1988).

123. Fontecilla-Camps, J. C., Bugg, C. E., Temple, C. Jr., Rose, J. D., Montgomery, J. A. and Kisliuk, R. L. Absolute configuration of biological tetrahydrofolates. A crystallographic determination. *J. Amer. Chem. Soc.* **101**, 6114–6115 (1979).

124. Armarego, W. L. F., Waring, P. and Williams, J. W. Absolute configuration of 6-methyl-5,6,7,8-tetrahydropterin produced by enzymatic reduction (dihydrofolate reductase and NADPH) of 6-methyl-7,8-dihydropterin. *J. Chem. Soc., Chem. Commun.*, 334–336 (1980).

125. Sobell, H. M. How actinomycin binds to DNA. *Sci. Amer.* **231** (2), 82–91 (1974).

126. Grein, A., Spalla, C., Di Marco, A., and Canevazzi, G. Descrizione e classificazione di un attinomicete (*Streptomyces peucetius* sp. *nova*) produttoce di una sostaviza ad attivata antitumorale – la daunomicina. [Description and classification of an actinomyces production of an active antitumor agent — daunomycin.] *Giorn. Microbiol.* [*J. Microbiol.*] **11**, 109–118 (1963).

127. Quigley, G.J., Wang, A. H.-J., Ughetto, G., Van der Marel, G., Van Boom, J. H., and Rich, A. Molecular structure of an anticancer drug-DNA complex: daunomycin plus d(CpGpTpApCpG). *Proc. Natl. Acad. Sci. USA* **77**, 7204–7208 (1980).

128. Frederick, C. A., Williams, L. D., Ughetto, G., van der Marel, G. A., van Boom, J. H., Rich, A., and Wang, A. H.-J. Structural comparison of anticancer drug-DNA complexes: adriamycin and daunomycin. *Biochemistry* **29**, 2538–2549 (1990).

129. Kopka, M. L., Yoon, C., Goodsell, D., Pjura, P., and Dickerson, R. E. Binding of an antitumor drug to DNA. Netropsin and C-G-C-G-A-A-T-T-^{Br}C-G-C-G. *J. Molec. Biol.* **183**, 553–563 (1985).

130. Kopka, M. L., Yoon, C., Goodsell, D., Pjura, P., and Dickerson, R. E. The molecular origin of DNA-drug specificity in netropsin and distamycin. *Proc. Natl. Acad. Sci. USA* **82**, 1376–1380 (1985).

131. Coll, M., Aymani, J., van der Marel, G., van Boom, J. H., Rich, A., and Wang, A. H.-J. Molecular structure of the netropsin-d(CGCGATATCGCG) complex: DNA conformation in an alternating AT segment. *Biochemistry* **28**, 310–320 (1989).

132. Pjura, P. E., Grzeskowiak, K., and Dickerson, R. E. Binding of Hoechst 33258 to the minor groove of B-DNA. *J. Molec. Biol.* **197**, 257–271 (1987).

133. Teng, M.-K., Usman, N., Frederick, C. A., and Wang, A. H.-J. The molecular structure of the complex of Hoechst 33258 and the DNA dodecamer d(CGCGAATTCGCG). *Nucleic Acids Res.* **16**, 2671–2690 (1988).

134. Carrondo, M. A. A. F. de C. T., Coll, M., Aymani, J., Wang, A. H.-J., van der Marel, G. A., van Boom, J. H., and Rich, A. Binding of a Hoechst dye to d(CGCGATATCGCG) and its influence on the conformation of the DNA fragment. *Biochemistry* **28**, 7849–7859 (1989).

135. Anderson, W. F., Ohlendorf, D. H., Takeda, Y., and Matthews, B. W. Structure of the cro repressor from bacteriophage I and its interaction with DNA. *Nature* (*London*) **290**, 754–758 (1981).

136. Ellenberger, T. E., Brandl, C.-I., Struhl, K., and Harrison, S. C. The GCN4 basic region leucine zipper binds DNA as a dimer of uninterrupted α helices: crystal structure of the protein–DNA complex. *Cell* **71**, 1223–1237 (1992).

137. Seeman, N. C., Rosenberg, J. M., and Rich, A. Sequence-specific recognition of double helical nucleic acids by proteins. *Proc. Natl. Acad. Sci. USA* **73**, 804–808 (1976)..

138. Kuntz, I. D., Blaney, J. M., Oatley, S. J., Langridge, R., and Ferrin, T. E. A geometric approach to macromolecule-ligand interactions. *J. Molec. Biol.* **161**, 269–288 (1982).

139. Goodford, P. J. Drug design by the method of receptor fit. *J. Med. Chem.* **27**, 557–564 (1984).

140. Luger, P., Griss, G., Hurnaus, R., and Trummlitz, G. The α_2-adrenoceptor agonists B-HT 920, B-HT 922 and B-HT 958, a comparative X ray and molecular mechanics study. *Acta Cryst.* **B42**, 478–490 (1986).

141. Laver, W. G., and Air, G. M. (**Eds.**) *Use of X-ray Crystallography in the Design of Antiviral Agents.* Academic Press: San Diego (1989).

142. Ealick, S. E., and Armstrong, S. R. Pharmacologically relevant proteins. *Current Opinion in Structural Biology* **3**, 861–867 (1993).

143. Navia, M. A., and Murcko, M. A. Use of structural information in drug design. *Current Opinion in Structural Biology* **2**, 202–210 (1992).

144. Marshall, G. R. Computer-aided drug design. *Annu. Rev. Pharmacol. Toxicol.* **27**, 193–213 (1987).

145. Bugg, C. E., Carson, W. M., and Montgomery, J. A. Drugs by design. *Scientific American* **269 (6)**, 92–98 (1993).

146. Friedman, S. H., DeCamp, D. L., Sijbesma, R. P., Srdanov, G., Wudl, F., and Kenyon, G. L. Inhibition of the HIV-1 protease by fullerene derivatives; model building studies and experimental verification. *J. Amer. Chem. Soc.* **115**, 6506–6509 (1993).

18

Structure-Activity Results

In this Chapter we discuss the ways in which X-ray crystallographers have been able to study some stereochemical aspects of chemical and biochemical reactions in the solid state. Several types of experiments will be described. The first is the observation of reactions that occur within a crystal on exposure to radiation, and the second is the study of reactions that occur on the surfaces of crystals. The third involves the analyses of structures that can be considered representative of points on reaction pathways, and the fourth is the study of enzyme reactions from the point of view of substrate and inhibitor binding and deductions from site-directed mutagenesis studies. Finally the observation of a change in the diffraction pattern of a mixture as a chemical reaction occurs and the ways in which this effect can be monitored are described.

Ideally the chemist interested in structure would like to obtain the multidimensional geometry of a compound at each step along a reaction pathway during the course of a chemical reaction. This difficult task has been tackled by many a chemist and biochemist. The transition state of a reaction is represented by the saddle point or high pass in the multi-dimensional plot of energy versus coordinates. If there are troughs near the transition state in this plot, then the structures of compounds with geometries that are very similar to that of the transition state may be investigated. Carefully chosen crystal structures may be considered to be instant snapshots of the shapes of molecules at points on a low-energy pathway on the potential energy surface that links reactants to products.

18.1 Solid-state reactions

Many chemical reactions take place in crystals. Detailed analyses of the stereo- and electronic effects are useful for an understanding of chemical reactions in general.[1] Such solid-state reactions can be studied by X-ray diffraction techniques when the participating molecules are close together in the crystal and there is room for the reaction to take place, so that the accompanying readjustments of molecular geometry do not destroy the crystal.

18.1.1 Solid-state photoreactions between double bonds

Crystalline *trans*-cinnamic acid forms cyclobutane derivatives upon irradiation of the crystal when C=C double bonds in neighboring molecules interact with each other to give cyclobutane derivatives, as illustrated in Figure 18.1.[2–4] In solution or in the molten state, the only reaction observed is isomerization from *trans*- to *cis*-cinnamic acid, but in the crystalline state the stereospecific reaction to form cyclobutane derivatives may occur.

The development of the principles of solid-state reactions of substituted cinnamic acids was pioneered by Gerhard M. J. Schmidt. The *trans*-acid was found to be polymorphic, and three different crystal forms (designated α, β, and γ) were identified by him.[5,6] The finding that the nature of the cyclobutane derivatives formed by the solid-state photochemical reaction on crystals of *trans*-cinnamic acid depend on which polymorph is irradiated was of great interest. The products of the photo-

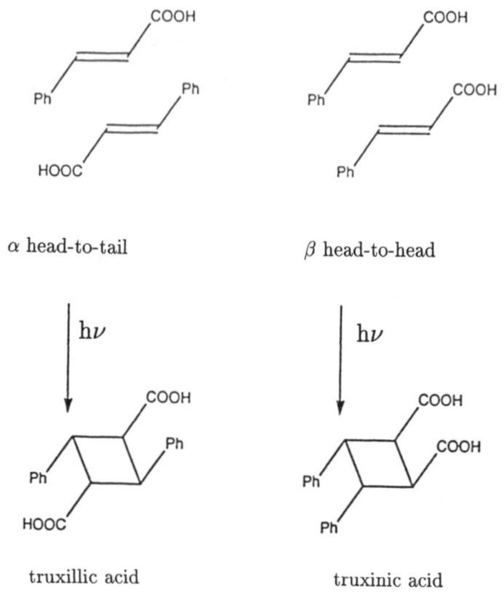

FIGURE 18.1. Solid-state reactions that occur on irradiation of cinnamic acids. The reaction products are truxillic or truxinic acids, depending on the relative orientations of the molecules in the crystal. The distance between C=C bonds in the solid state must lie in the range 3.6–4.1 Å.

chemical reaction are classified as truxillic acids when the carboxyl groups are attached to opposite carbon atoms, and truxinic acids when the carboxyl groups and phenyl groups lie on adjacent carbon atoms of the cyclobutane ring (Figure 18.1). Crystals of the α form of cinnamic acid and its derivatives contain neighboring molecules packed in a head-to-tail arrangement with antiparallel C=C double bonds 3.6–4.1 Å apart, and related by a center of symmetry. Upon irradiation, the cinnamic acid molecules are transformed to truxillic acids with *trans* stereochemistry. In the β polymorph, cinnamic acid molecules are packed in the crystal in a head-to-head and tail-to-tail arrangement. As a result, the C=C double bonds, which are also 3.6–4.1 Å apart, pack with parallel double bonds. Upon irradiation, the photochemical reaction gives molecules with *cis* stereochemistry, that is, truxinic acids. In the third form, the γ form, the distances between the nearest pairs of C=C bonds are greater than 4.1 Å, too far apart for a reaction to occur. This crystalline form is stable to irradiation.

Photocyclodimerization of crystalline olefinic compounds (giving cyclobutane derivatives) thus requires that the C=C bonds be aligned parallel (or antiparallel) to each other and about 4 Å apart. It also requires that the formation of new bonds should cause a minimal movement of atoms. This type of solid-state reaction is called a **topochemical reaction**. It is essentially diffusionless[7] since all that is required is a minor reorganization of atoms rather than a diffusion so that two reactant molecules can approach each other in a suitable manner for reaction. For example, *cis*-4a,5,8,8a-dimethyltetrahydronaphthoquinone,[8–11] forms dimers stereospecifically in the crystalline state upon irradiation (Figure 18.2). By contrast, intramolecular cyclization occurs in solution.

Certain crystal structures, however, that appear to fit the criteria just listed for α or β cinnamic acids do not, in fact, produce photodimers. For example, 3,4-dihydroxy-*trans*-cinnamic acid is photostable.[12] It is suggested that this is because symmetry-related molecules are held together by strong hydrogen bonding which does not permit the molecular flexibility that requires the 4.0 Å interaction to be reduced to approximately 1.5 Å. It appears that there needs to be sufficient space in the crystal "reaction" cavity for the reaction to take place, and sufficient flexibility in the overall crystal packing for the required atomic spatial reorganization to occur.

If the C=C bonds are more than 4.6 Å apart, the reaction does not take place. As molecules need to have double bonds aligned and about 4 Å apart for such photoreactions, it is sometimes possible to force the required molecule to crystallize with a unit cell dimension of this length. For example, mercuric chloride crystallizes with one unit cell dimension of 4.33 Å, and its complex with coumarin maintains this unit-cell length.[13] The required 4 Å packing distance can also be obtained by use of chlo-

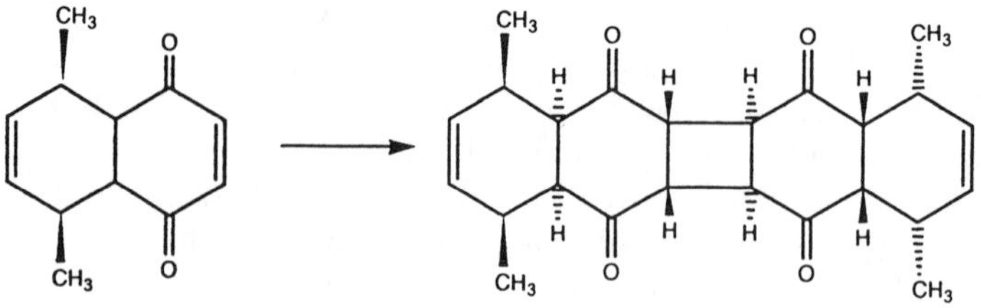

(a)

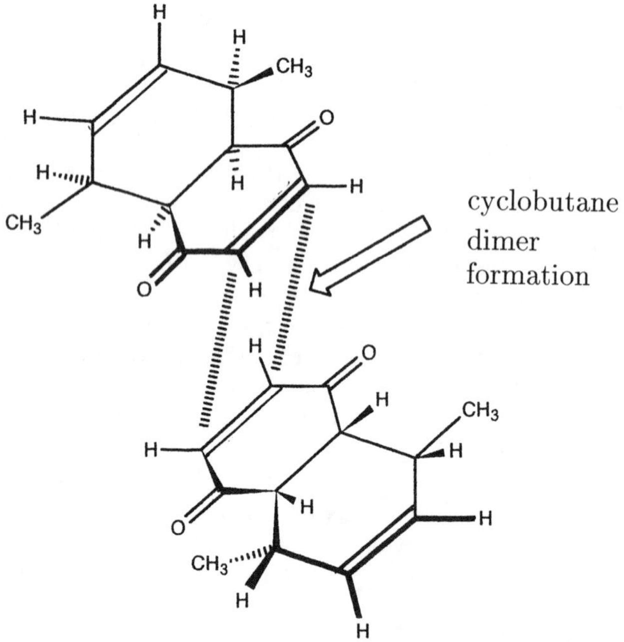

cyclobutane dimer formation

(b)

FIGURE 18.2. The solid-state polymerization (at wavelengths greater than 340 nm). (a) The reaction occurring in the solid state, and (b) a diagram of the solid-state reaction. The ene dione double bonds of adjacent molecules in the crystal have a C ··· C distance (indicated by hatched lines) of 3.62 Å. Cyclobutane dimerization occurs across a center of symmetry. (Refs. 10 and 14).

rine atoms in the structure, at least for a host molecule. The presence of two chlorosubstituents in a molecule with an aromatic group and double bonds tends to result[15] in a unit cell dimension of 4.0(2) Å. It is even possible to "engineer" required products by use of mixed crystals of related compounds, since the crystal packing essentially controls the stereochemistry of the product.[16]

If the molecules each contain more than one double bond, oligomerization throughout the crystal is a possibility. For example, the crystal structure determination of a partially photooligomerized (i.e., polymerized) crystal of 2,5-distyryl pyrazine (DSP)[17-21] demonstrates another photochemical reaction in the crystalline state. The reaction involved is shown in Figure 18.3. When the crystal structure was determined, a difference electron-density map (Figure 18.4) contained four peaks in the positions expected for a cyclobutane-like adduct (right-hand side of Equation 18.1). There were also peaks in this difference electron-density map that could be interpreted as peaks resulting from the reorientation of the "central ring" on oligomerization.

18.1.2 Cobaloximes

Cobaloximes are bis(dimethylglyoximato)cobalt(III) complexes that consist of a planar (equatorial) chelating agent for Co(III) consisting of two monoanions of dimethylglyoxime hydrogen bonded to each other, and two axial positions on the cobalt ion that are available for binding. The coordination of the Co(III) is octahedral. Frequently the axial ligands of cobaloximes are an alkyl derivative and a nitrogenous base such as pyridine. In this way cobaloximes have been used successfully to model vitamin B_{12} compounds.[22]

The axial Co–C bonds in cobaloximes are weak. Several complexes studied by Yuji Ohashi and co-workers,[23-27] contain an (R)-1-cyanoethyl group and a heterocyclic nitrogenous base as the other axial ligand (Figure 18.5). These complexes show, upon irradiation with X rays during diffraction data measurement, a slow racemization of the optically active cyanoethyl group. This solid-state racemization occurs without any decomposition of the crystals, although the space group is sometimes changed by it. This implies that there is sufficient space in the crystal to accommodate a cyanoethyl group in an alternate configuration, after cleavage of the Co–C bond, as in B_{12} coenzyme. The concept of a "reaction cavity" in the crystal structure is established by this experiment, since it appears that the reaction rate is directly correlated with the amount and dimensions of free space surrounding the cyanoethyl group in the crystal. If this cavity is large and of the appropriate shape, the reacting group can move freely, otherwise the reaction cannot take place. The inversion reaction is slow, so that several X-ray diffraction data sets can be measured to monitor the course of the reaction. An examination of electron-density maps obtained from diffraction data measurements

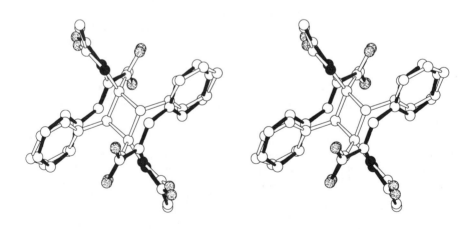

(a)

(b)   stereoview

FIGURE 18.3. The formation of cyclobutane dimers. (a) Chemical formula of dimerization. (b) Stereoview of the monomers (black bonds) and the cyclobutane adduct (open bonds). This diagram shows the amount of movement of C=C carbon atoms needed to form a cyclobutane ring. Ph = phenyl (Ref. 28).

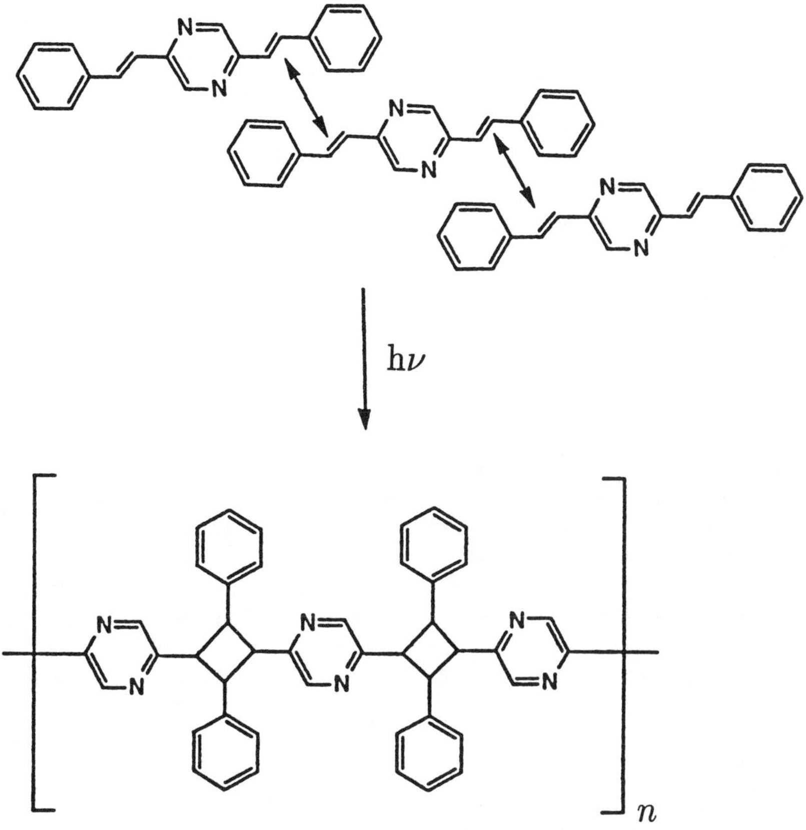

FIGURE 18.4. Photooligomerization of a pyrazine derivative. The presence of two double bonds in each molecule makes it possible, with the crystallographic alignment, to form an oligomer (Ref. 21).

at different stages of the reaction shows that, in the first step, the weak Co-C bond is broken, allowing the cyanoethyl group to rotate and give product.

Several types of photochemical reactions have been observed in chiral crystalline cyanoethyl cobaloximes in which there is sufficient space for an inversion of configuration. In one type, crystals contain one molecule per asymmetric unit, and 50% of the cyanoethyl groups are rotated during the photochemical reaction. Loss of crystallinity is not observed because the space occupied by the rotated cyanoethyl group is similar to that occupied by it in the original structure. Even the unit cell dimensions of

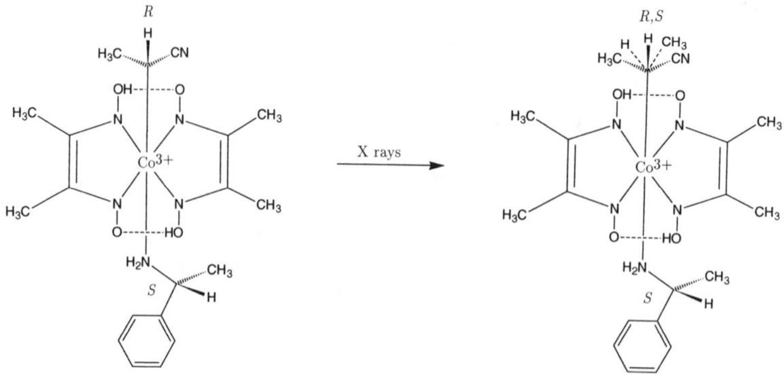

FIGURE 18.5. Solid-state X-ray induced reaction of a cobaloxime. The Co–C bond is broken, and racemization of the upper axial substituent occurs, as shown.

the enantiomer and its racemate are similar, although the racemic crystal has a disordered structure. In a second type, there are two independent molecules in the asymmetric unit. These are related by a pseudocenter of symmetry and all of the cyanoethyl groups of the second molecule show a change in configuration from R to S, so that the center of symmetry becomes real and the resultant crystal is ordered. In a third type, there are also two independent molecules in the asymmetric unit of the crystal, but the cyanoethyl groups in both molecules are inverted in a random manner.

For example, a racemic mixture of (1-cyanoethyl)(piperidine)-cobaloxime, which contains the chiral 1-cyanoethyl group, crystallizes in the chiral space group $P2_12_12_1$ with R and S enantiomers in each asymmetric unit. Irradiation with X rays at 343 K converts the S enantiomers to R while the R enantiomer is not affected. This is due to different constraints in the two crystallographic environments. The respective volumes for the two enantiomers are 7.49 and 11.57 Å^3 respectively. The critical volume[22] for such racemization is 11.5 Å^3. In agreement with this estimate it is found that the molecule with the small cavity does not show evidence of racemization, while the one with the large reaction cavity does, and a disordered cyanoethyl group is observed as a peak in two positions in the electron-density map. This reaction is illustrated in Figure 18.6 which shows the crystal structure at 293 K (where no reaction occurs) and at 333 K (where reaction is only observed in one of the two cavities).

18.2 Reactions of crystal surfaces with gases

Another way to study reactions is to investigate heterogeneous catalysis at crystal surfaces. Low-energy electron diffraction has been used[29,30] to

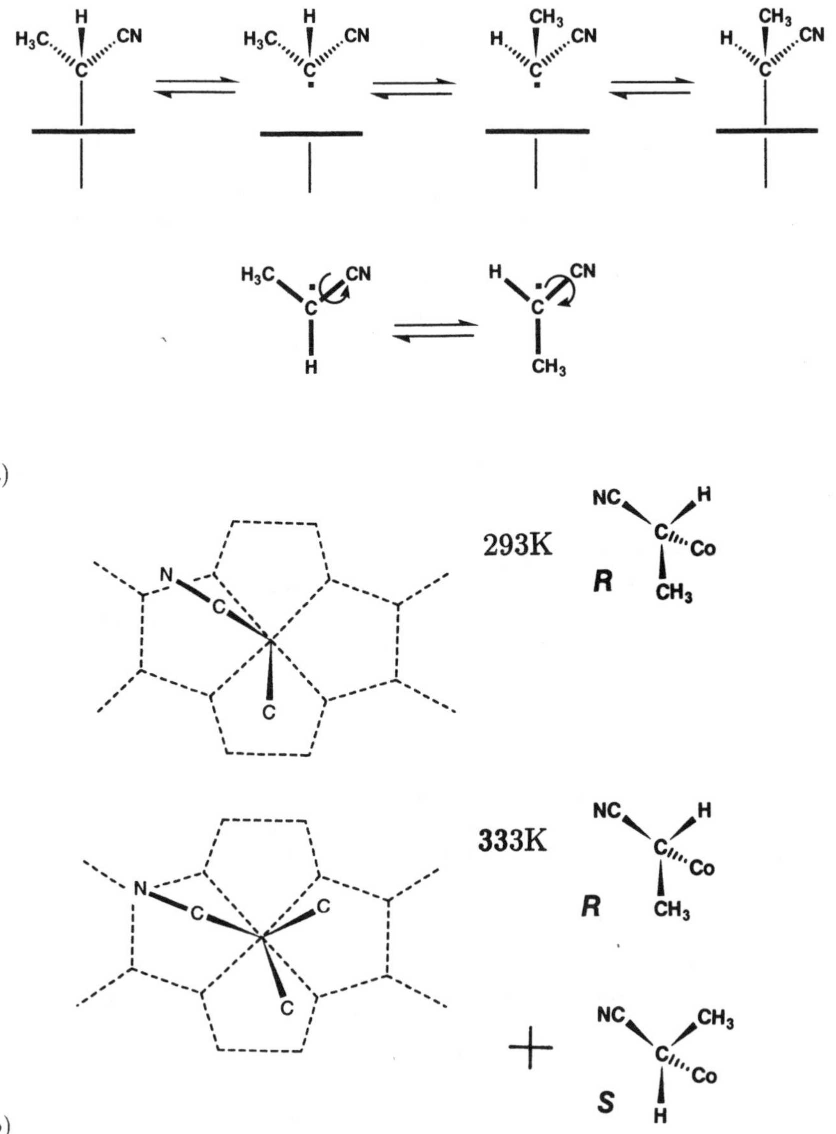

(a)

(b)

FIGURE 18.6. Cobaloxime reaction. (a) Details of the reaction in Figure 18.5. (b) Difference electron-density maps show that, above 293K, the racemization occurs, giving two possible positions for the methyl group.

determine the nature of the packing on the surface of platinum crystals cut along various planes. It was shown that selective bond-breaking ability (C–C, C–H, or H–H bond breaking) could be correlated with the coordination number of the metal at surface sites.

Iain C. Paul, David Y. Curtin, and Rodger S. Miller[31–33] have studied reactions between gases and crystalline compounds as a function of which crystalline faces are exposed to the gas. This work showed the importance of the relative orientation of molecules on different crystal faces, as illustrated for sodium and chloride ions in Figure 2.15 (Chapter 2). Single crystals of 4-chlorobenzoic acid can be converted quantitatively to the ammonium salt by reaction with gaseous ammonia. The crystal becomes opaque as ammonium 4-chlorobenzoate is formed in the reaction. Observed changes in the crystals demonstrated that the reaction is anisotropic. The (100) face remained clear throughout the reaction while opacity, indicating crystal disorder caused by the formation of ammonium 4-chlorobenzoate, moved uniformly through the other faces.

As salt forms, a disorientation of a layer of structure occurs sufficient to allow ammonia gas to diffuse to the next layer of acid molecules where the process is repeated. The presence of carboxyl groups at the surface of the crystal is a necessary but not sufficient condition for this solid-gas reaction to occur. The carboxyl groups must be arranged so that reaction of one carboxyl group exposes the next to ammonia. Thus carboxyl groups, layer by layer throughout the crystal, must be sequentially exposed to the attacking agent at the time that those in the upper layer are converted to the ammonium salt. Disorder resulting from the reaction permits penetration of the gaseous reactant (ammonia) into the interior of the crystal. Examination of the crystal structure of 4-chlorobenzoic acid, illustrated in Figure 18.7, shows that each of the faces has carboxyl groups exposed to attack except for the (100) faces. On these (100) faces the carboxyl groups are shielded from the attacking ammonia molecules by the 4-chlorophenyl groups. In this way the experimental observations can be explained by a knowledge of the crystal structure. This is an experimental area for further study.

18.3 Inferences on chemical reactions

We would like to be able to take a series of instantaneous snapshots of the shapes of molecules during a chemical reaction. In this way we can measure the distortions from ground-state geometry that occur at different stages along the reaction path. Unfortunately this is generally not possible by diffraction methods. Therefore it is necessary to study the structures of molecules that, for various reasons, are distorted in some way corresponding to points along the anticipated chemical reaction path.

Hans-Beat Bürgi and Jack D. Dunitz[34] and their co-workers have postulated that if the mode of restructuring of compounds as a result of a chemical reaction is known, then a study of the detailed geometries of

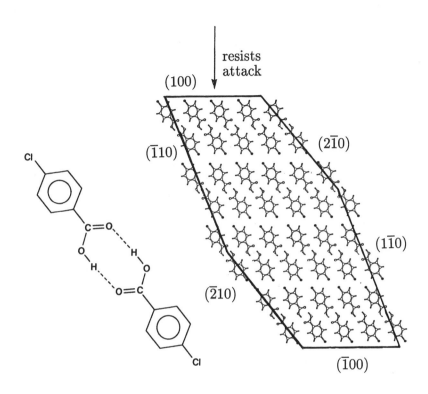

FIGURE 18.7. Reaction of ammonia with 4-chlorobenzoic acid to give ammonium 4-chlorobenzoate.

such compounds may provide significant information on the three-dimensional geometry of the reaction pathway. This method is analogous to the study of nonhindered and somewhat-hindered esters mentioned earlier providing a plot of the conformational variability in ethyl esters (see Figure 16.6, Chapter 16). Some examples follow.

18.3.1 S_N1 reactions

Unimolecular **nucleophilic** substitution reactions, S_N1, of tetrahedral molecules follow a pathway (see Figure 18.8) that involves the dissociation of a leaving group X and the production of a planar intermediate which

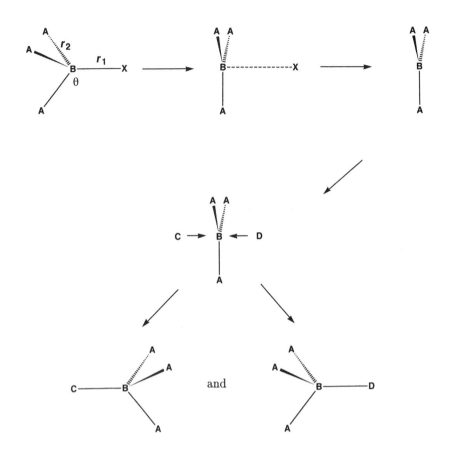

FIGURE 18.8. Example of an S_N1 reaction. The molecule A_3BX loses the group X as r_1 increases from a bond length to infinity. As r_1 (the $A-X$ bond length) increases, θ (the $A-B-X$ angle) changes from 109.5° (tetrahedral) to 120.0° (trigonal-planar), and the intermediate A_3B is planar. Atoms C or D may add to either side of the planar A_3B group with inversion (C) or retention (D) of configuration with respect to A_3BX (Ref. 35).

does not have the leaving group X attached to it. The intermediate can be attacked from either side by a nucleophilic group, to give a mixture of stereoisomers. The system studied can be represented by AB_3X, where A is the central atom of the molecule with C_{3v} (threefold) symmetry. In an S_N1 reaction the $A-X$ bond (where X is the leaving group) is elongated and then breaks. The resulting planar intermediate AB_3 can be attacked by C from one side (inversion of configuration), or D from the other side (retention of configuration). The $A-C$ or $A-D$ bond that is formed becomes shortened from a van der Waals to a bond distance. Peter Murray-Rust, with Bürgi and Dunitz,[35] studied a series of compounds, particularly aluminum, sulfur, phosphorus, tin, silicon, and germanium derivatives, each with a threefold symmetry axis, i.e., AB_3X). Bond lengths $A–X$ or $A–B$ in compounds of formulae AX_4 or AB_4 respectively were used as standards. Three parameters were then defined: Δr_1 is the $A–B$ bond length in AB_3X minus the $A–B$ value in AB_4; Δr_2 is the $A–X$ bond length in AB_3X minus the $A–X$ value in AX_4. The angle θ is the average angle between axial $(A–X)$ and basal $(A–B)$ bonds.

To relate bond lengths to their bond number n (single, double, or triple) Pauling[26] suggested a relationship, where d is the observed interatomic distance and d_o the assigned single bond distance. The equation is

$$d - d_o = \Delta d = -c \log n. \tag{18.1}$$

where n is the Pauling bond number for the $A–B$ bond,[36] and c is a constant (found to be near 0.5). From the valence sum rule (Chapter 15)

$$n_2 + 3n_1 = 4 \text{ (the valence of atom } A) \tag{18.2}$$

where the subscripts 1 and 2 refer to B and X, respectively. Values of Δr_1 and Δr_2 were tabulated for 200 crystal structure determinations found in the CSD.[37] A plot of Δr_1 and and Δr_2 versus θ, was shown, as expected, to be described by:

$$\Delta r_1 = -0.5 \log\left(\frac{4}{3} - 3\cos^2\theta\right) \tag{18.3}$$

$$\Delta r_2 = -0.5 \log\left(9\cos^2\theta\right). \tag{18.4}$$

since $n_1 = (4/3) - 3\cos^2\theta$ and $n_2 = 9\cos^2\theta$ The two curves intersect at $\theta = 109.5°$ where Δr_1 and $\Delta r_2 = 0$. In the transition state $\theta = 90°$, $\Delta r_1 = -0.08$ Å, and Δr_2 is very large (i.e., the leaving group has left). All molecules deform along the same path, maintaining the threefold symmetry.

18.3.2 $S_N 2$ reactions

In a bimolecular nucleophilic substitution reaction ($S_N 2$), the $B–Y$ bond from the central carbon atom to the leaving group (Y) is broken, and a new bond, $B–X$ (see Figure 18.9), is formed with the nucleophile (X). The intermediate in this, the transition state, may be envisioned as a trigonal bipyramid with the leaving group, Y, and the nucleophile, X, in axial positions. The result of this reaction is that the leaving group leaves from one side and the nucleophile adds to the other side a trigonal group so that an inversion of configuration, the Walden inversion, occurs (Figure 18.9).

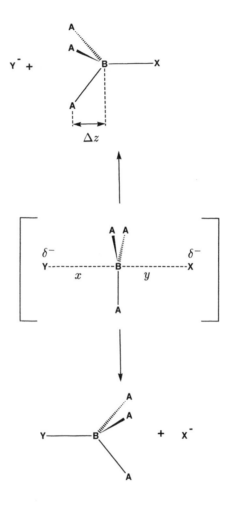

FIGURE 18.9. Example of an S_N2 reaction. Δz = displacement of B from the plane through the three A atoms (Ref. 34).

No compound with five-coordinate carbon as the central atom has yet been crystallized. Bürgi,[34] however, has studied a series of trigonal bipyramidal complexes of divalent cadmium with three equatorial sulfur ligands and with axial ligands iodide, hydroxyl, or sulfur groups (*Y* and *X*).

First the deviation Δz (in Å) of the cadmium atom from the plane of the three sulfur atoms was studied. Since the nature of X and Y varied among the compounds selected, it was necessary to convert all axial distances to values that could be directly compared. This was done by subtracting the sum of the covalent radii (from Pauling's *The Nature of the Chemical Bond*)[36] for each bond length, from each Cd–X or Cd–Y bond length to give ΔX and ΔY, the deviations of these values for the nucleophile and leaving groups, respectively. As the leaving group Lv departs from the cadmium, ΔY is increased while ΔX is decreased. The nucleophile Nu becomes more tightly bound, as shown by the decrease in ΔX. Values of Δz, plotted against ΔX or ΔY (with the sign convention that Δz is positive when displaced toward Y, and negative when displaced toward B–X). An ideal CdS$_4$ tetrahedron is formed when Δz is a maximum 0.84 Å. At the transition state, where $\Delta z = 0$, the geometry is trigonal bipyramidal. It was found that at the point of the transition state both ΔX and ΔY had absolute values of 0.32 Å. Since the ionic radii of the atoms involved are approximately 0.3 Å greater than their covalent radii, the absolute values of ΔX and ΔY of 0.32 Å imply that the transition state has ionic character.

An inspection of Figure 18.9 shows that, for the cadmium complexes:

$$\Delta Y = f(\Delta z) = -1.05 \log[(\Delta z + 0.84)/(2 \times 0.84)] \text{ Å} \qquad (18.5)$$

$$\Delta X = f(-\Delta z) = -1.05 \log[(-\Delta z + 0.84)/(2 \times 0.84)] \text{ Å}, \qquad (18.6)$$

where the bond orders are:

$$n_Y = (\Delta z + 0.84)/(2 \times 0.84) = 0.5 + c_1 \Delta z \qquad (18.7)$$

$$n_X = (-\Delta z + 0.84)/(2 \times 0.84) = 0.5 - c_1 \Delta z. \qquad (18.8)$$

where $c_1 = 0.6$. These equations show how the bond orders change with the displacement Δz.

The interbond angles will also change during the reaction. In the tetrahedron, CdS$_4$, all angles are 109.5°. In the trigonal bipyramidal state, the angles Y–Cd–S and X–Cd–S are 90°. As X approaches a tetrahedral CdYS_4 molecule the X–Cd–S angle becomes 70.5°, compared with the Y–Cd–S angle of 109.5°. Thus, to approach the transition state the first angle in the structure must increase while the second must decrease.

18.3.3 Nucleophilic addition to a carbonyl group

Another type of chemical reaction that has been investigated is the addition of a nucleophile to a carbonyl center, illustrated for the addition of ammonia to a carbonyl group in Figure 18.10. Nucleophilic addition is an important feature of many biochemical reactions and appears to involve a tetrahedral intermediate. Bürgi, Dunitz, and Eli Shefter[38,39] studied molecules with both a carbonyl group and a tertiary amino group that were separated by varying numbers of carbon atoms. They measured,

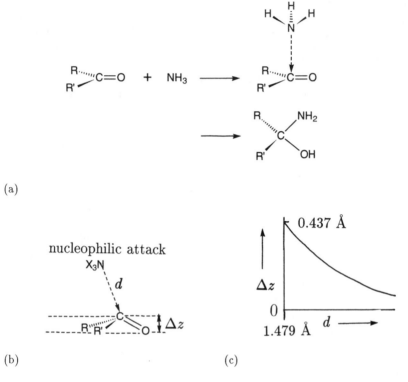

(a)

(b)   (c)

FIGURE 18.10. Nucleophilic addition to a carbonyl group. Δz = deviation of C from the plane of O, R, and R. (a) An example of addition of ammonia to a carbonyl group is illustrated. (b) The geometry of the intermediate, leading to Equation 18.9. (c) A plot of experimental values for d_1 versus Δz (Ref. 38).

by an analysis of crystal structures in the CSD, the distances between the nitrogen atom of the amino group and the carbon atom of the carbonyl group (an N$\cdots$C=O interaction), and found that the C$\cdots$N distance varied from 2.91 Å in methadone[38] to 1.49 Å in N-brosylmitomycin A.[40] Thus the distance varies from a van der Waals distance to one near a covalent bond. Therefore the geometry of N $\cdots$ C interactions was used as an indication of the pathway of approach of the nitrogen atom to the carbonyl carbon as a bond is formed. The carbon atom, as the **electrophilic** center, becomes more pyramidal in shape as the nitrogen atom approaches it (Figure 18.10). The deviation Δz of the carbonyl carbon atom from the plane of its three neighbors R_1, R_2, and O increases as the

N···C distance (d_1) decreases. At the same time the carbonyl C–O bond length (d_2) increases in value. The experimental results for the N ··· C distance (in Å) may be expressed as

$$d_1 = -1.70 \log \Delta + 1.479 \text{ Å} \qquad (18.9)$$

where $\Delta = \Delta z/\Delta_{max}$ with Δ_{max} equal to 0.437 Å and 1.479 Å as the standard C–N bond length. The C–O bond length is then expressed as:

$$d_2 = -0.71 \log(2 - \Delta) + 1.426 \text{ Å}, \qquad (18.10)$$

where 1.426 Å is the single-bond length of a C–O bond and 0.71 is a value chosen to fit the CSD data in the best possible manner.

The nucleophile approaches the carbonyl center at an angle that is approximately 107°. This is a strong indication for a preferred orientation of nucleophilic attack. The closer the nitrogen atom comes to the carbonyl carbon atom, the less planar the RR′C=O system becomes.[39] The C–O bond lengthens and R_1 and R_2 bend away. The lone pair of electrons on the nitrogen atom is consistently oriented toward the carbonyl carbon atom in a direction roughly parallel to the N ··· C vector and at an angle of about 109° to the C=O bond. A tetrahedral bond is then readily formed to give product. Quantum-mechanical calculations of the addition of a hydride ion to formaldehyde gave similar results.[41] Several other analyses of chemical pathways have been studied.[42–49] For example, a combination of crystallographic and kinetic data gives a linear relationship between C–O extension and the free energy of activation for heterolytic fission of the bond.[50]

18.4 Protein-catalyzed reactions

Enzymes catalyze an enormous variety of biochemical reactions.[51–54] They serve to regulate the rate of these specific reactions, for which they have been uniquely designed. Like any other catalyst, they alter only the rate of a reaction; their chemical structure is not altered by the reaction. They do not alter the equilibrium between the reactants and products but merely increase the rate at which that equilibrium is attained. Enzymes may, however, participate in the reaction, transiently changing the chemical structure, but are quickly regenerated to their original form. In order to understand how an enzyme works, it is necessary to know its three-dimensional structure and, more importantly, the structure of the enzyme complex involving substrates, intermediates and products of the reaction.

Enzymes bind their substrates with a high degree of specificity. The reactions that a particular enzyme catalyzes are also specific and take place at a particular location in the enzyme, its active site. This active site is made up of amino acid side chains arranged in such a way that they are complementary to the substrate with respect to both shape and charge. Molecules that do not have this complementarity cannot bind to the enzyme. Most binding sites in enzymes are preformed and immediately available for substrate binding. The best ligand for an enzyme is, as was pointed out by Linus Pauling,[55] the arrangement of atoms that represents the transition-state of the reaction being catalyzed. Therefore the use of ligands that approximate this transition-state geometry has yielded useful information on the mechanism of action of several enzymes.[56]

The first crystal structure of an enzyme, that of lysozyme, was determined by David C. Phillips and coworkers[57] in 1965. The most striking feature in the three-dimensional structure of lysozyme is a prominent cleft that traverses one face of the molecule. The X-ray structure of lysozyme complexed with a three-residue oligosaccharide showed that this cleft was, indeed, the substrate-binding site. The crystal structure of this complex provided the first three-dimensional model for how enyzmes work.

The cleft really binds a penta- or hexasaccharide (*ABCDEF*) but data were only available at that time for a bound trisaccharide (*ABC*). By model building Phillips proposed, as shown in Figure 18.11, that glutamic acid 35 transfers its proton to the bridging oxygen atom between sugar rings *D* and *E*. The C–O bond then breaks, leaving a positive charge at C1 that is stabilized by the ionized carboxylate aspartate 52. The *EF* portion of the polysaccharide then leaves, and the carbonium ion interacts with a hydroxyl group, regenerating the active enzyme. Phillips proposed that the *D* ring is distorted to a half-chair conformation, more like the transition state. Others have suggested that loss of water on substrate binding increases the ability of aspartate 52 to stabilize the transition state. Details of this mechanism are still under investigation.[58-64]

18.4.1 Mechanism of catalysis by chymotrypsin

Proteolytic enzymes, such as the serine proteases, are among the best characterized of all enzymes.[65-71] They are important in digestive processes because they break down proteins. They each catalyze the same type of reaction, that is. the breaking of peptide bonds by hydrolysis. The crystal structures of several serine proteases have been determined, and the mechanism of hydrolysis is similar for each. The specificity of each enzyme is, however, different and is dictated by the nature of the side chains flanking the scissile peptide bond (the bond that is broken in catalytic mechanism. Chymotrypsin is one of the best characterized of these serine proteases. The preferred substrates of chymotrypsin have bulky aromatic side chains. The crystal structure determination of the active site of chymotrypsin, illustrated in Figure 18.12, has provided much of the information used to elucidate a plausible mechanism of action of the enzyme. In the first step of any catalyzed reaction, the enzyme and substrate form a complex, ES, the Michælis complex. The hydrolysis of the peptide bond by chymotrypsin involves three amino acid residues,

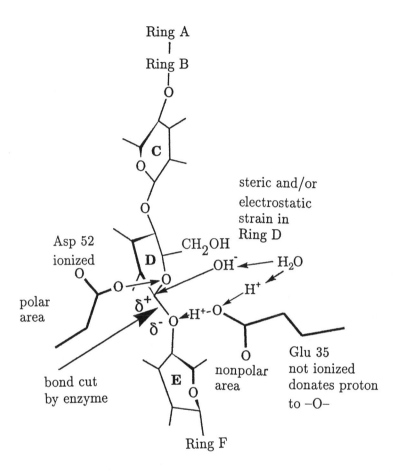

Ring A
|
Ring B
|
O

C

steric and/or
electrostatic
strain in
Ring D

Asp 52
ionized
O

polar
area

CH₂OH

D

OH⁻ ⟵ H₂O

H⁺

δ⁺

δ⁻ O ⟵H⁺-O

H⁺-O

O

E

nonpolar
area

bond cut
by enzyme

Glu 35
not ionized
donates proton
to –O–

Ring F

FIGURE 18.11. Proposed mechanism of action of lysozyme (Ref. 57).

histidine 57, serine 195, and aspartic acid 102, which form a hydrogen-bonded grouping called a "catalytic triad" that is very important in the mechanism. Aspartic acid 102 is ionized but unsolvated, and is buried in a deep pocket, inaccessible to solvent. It is hydrogen bonded to histidine 57.

A serine protease has four major requirements. The catalytic triad Asp, His, and Ser, are essential for the chemical mechanism. The oxyanion hole contains side chains that form hydrogen bonds to the oxygen atom that has developed a negative charge. The specificity pocket pro-

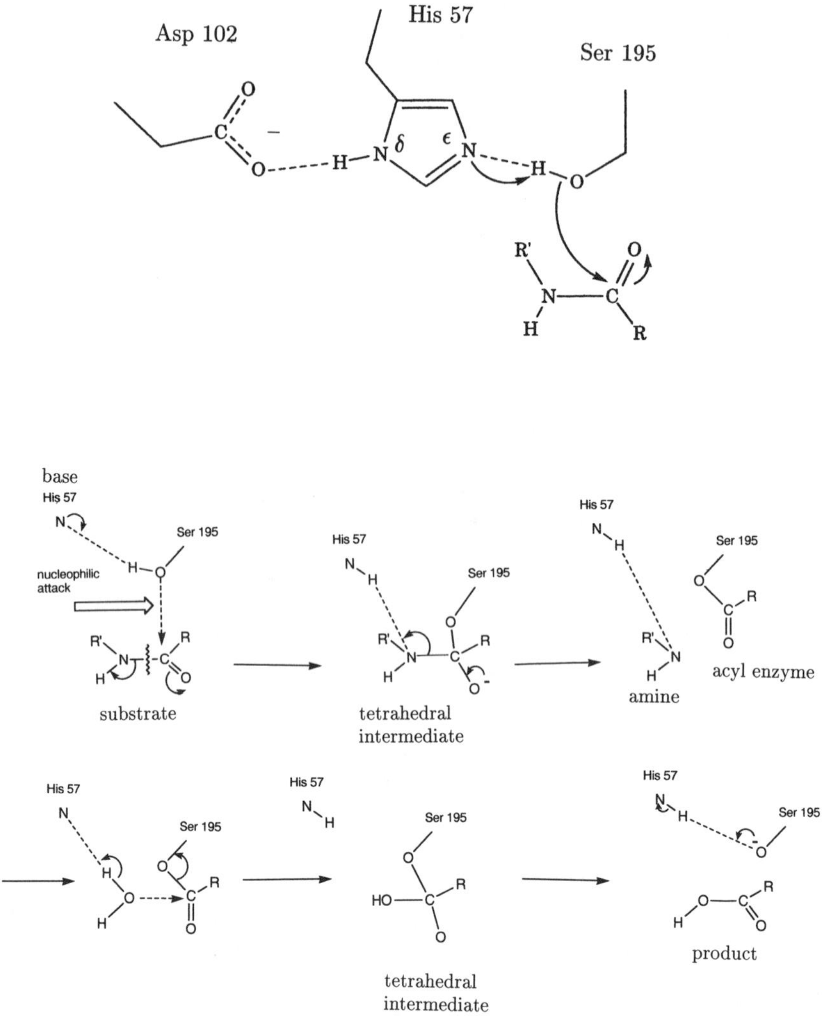

(a)

(b)

FIGURE 18.12. Proposed mode of action of chymotrypsin. (a) The polypeptide substrate binds, and (b) the carbon atom of the C=O group undergoes nucleophilic attack by the side-chain oxygen atom of Ser 195 to give a tetrahedral intermediate. This splits off NH_2R', replacing it with water which attacks the Ser 195-bound carbonyl group. The bond from Ser 195 to the substrate is then broken, giving a new C-terminus for the polypeptide chain. The enzyme is now ready for another catalytic cycle.

vides appropriate binding for the R side chain. Main-chain substrate binding is aided by a short region of antiparallel β sheet on the enzyme, These form hydrogen bonds to the C terminus of the substrate.

In the catalyzed reaction the stable amide linkage of the substrate is converted to an unstable acylated enzyme. This is then hydrolyzed, regenerating active enzyme. The shape of the active site of chymotrypsin is complementary to the aromatic side chain of its substrates that fit snugly into the hydrophobic pocket near the catalytic triad. Serine 195 is then in an ideal position and orientation to carry out the required nucleophilic attack on the **scissile** peptide **bond**. The imidazole ring of histidine 57 accepts the proton from aspartic acid 102 forming a general base. The scissile bond is then cleaved and a tetrahedral acyl-enzyme intermediate is formed (bound through serine 195), as shown in Figure 18.12. The new N terminus of the substrate (now product) is released from the enyzme and replaced by a water molecule. The acyl-enzyme is then converted back into its native state by the second half of the mechanism which is a deacylation. A proton is abstracted from a water molecule and the resulting hydroxide ion attacks the carbonyl group of the acyl group that is covalently attached to serine 195. Histidine 57 donates a proton to the oxygen, the new C terminus of the product diffuses away, and the enzyme is restored to its original active state. The importance of enzyme-bound water to this mechanism has been pointed out.[68]

18.4.2 Site-directed mutagenesis

Data derived from structural and enzymatic studies of proteins containing a mutation in one amino acid residue provide complementary information that aids in the elucidation of the mechanism of enzymes. As previously described, serine proteases have three residues that are essential for catalysis. Structural studies of several serine proteases have provided detailed information on the spatial relationship of these amino acids. The catalytic role of serine 195 and histidine 57 is well established, but the function of aspartic acid 102 is less well known. This aspartic acid may either stabilize the conformation of the histidine, maintain the correct tautomer, or stabilize the positive charge that forms during the reaction. A mutant serine protease has been studied in the form of a genetically engineered trypsin with asparagine in place of aspartic acid at position 102. The activity of the mutant enzyme toward a variety of substrates was found to be reduced several orders of magnitude relative to that of the native enzyme. This decrease in activity presumably resulted from the lower nucleophilicity of serine 195. A crystallographic analysis of the asparagine mutant clearly demonstrated the role of aspartic acid in the native structure. This implies that aspartic acid 102 plays the critical role by providing hydrogen bond stabilization of the functional tautomer and serves to keep the catalytic site (the histidine and serine) correctly oriented for action.[72,73] The effect of replacing histidine by asparagine in the enzyme D-xylose isomerase[74] is shown in Figure 18.13. The overall crystal structure is the same for the mutant and wild-type enzyme, but an additional water molecule is incorporated in the mutant enzyme, as shown. This fills the space left by the larger histidine side chain.

Mutational studies can also complement structural analysis in elucidating substrate specificity. By genetically altering residues in the substrate binding pocket of α-lytic protease, another serine protease, the specificity can be altered. Enzymes generally have a limited range of substrates, making each enzyme highly specific. Replacing an active site methionine with an alanine enlarged the binding pocket, which in turn enabled the mutant enzyme to accommodate large side chain substrates.[75] One must not, however, assume that a mutant enzyme necessarily acts by exactly the same mechanism as that of the **wild-type enzyme**.

18.5 Monitoring chemical reactions by diffraction

Chemical and biochemical reactions can be viewed by diffraction methods provided the reactions are slow and the techniques for measuring them are rapid. For example, Lonsdale and coworkers studied the conversion of a photo-oxide of anthracene on exposure to Cu $K\alpha$ or Mo $K\alpha$ X rays at room temperature. A single mixed crystal of anthraquinone and anthrone is formed which still shows crystallinity.[76] The crystal does not change in appearance, and the space group remains $P2_1/a$, but the unit-cell dimensions change:

initial: $a = 15.94$, $b = 5.86$, $c = 11.43$ Å, $\beta = 108.2°$, $Z = 4$,

product: $a = 15.8$, $b = 4.0$, $c = 7.9$ Å, $\beta = 102°$, $Z = 2$.

The product is actually a twinned crystal and the molecules are disordered and simulate anthraquinone.

The stages of disorder, decomposition, and recrystallization were followed by X-ray studies as a function of time. It appears that the photo-oxide first forms chains of molecules parallel to (010) and these chains then break down.

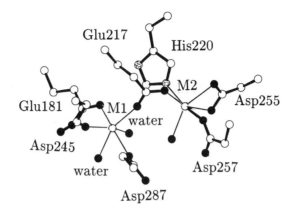

(a)

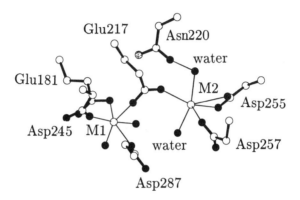

(b)

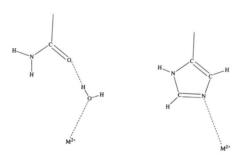

(c)

FIGURE 18.13. Crystallographic analysis of a mutant enzyme (Ref. 74). Crystal structures are shown in approximately the same orientation for (a) and (b). (a) "Wild-type" enzyme with His 220, and (b) the mutant enzyme with His 220 mutated to Asn 220. (c) Formulæ of the changes found. Note that histidine has been replaced by asparagine and a water molecule.

18.5.1 Chemical reactions viewed by powder diffraction

Powder diffraction techniques can be used to study solid-state reactions as a function of time.[77,78] This is done by scanning 2θ as a function of diffraction intensity at regular intervals of time, and plotting the result in a manner such as that shown in Figures 18.14 and 18.15. Powder neutron diffraction methods have been used to study orientational order–disorder transitions as a function of time.[77]

The reaction of a 1:1 host-guest complex of 1,1,2,2-tetraphenylethane-1,2-diol with various lutidines (dimethylpyridines) has been investigated by Luigi Nassimbeni and coworkers.[79,80] The complexes, which are inclusion compounds,[81] were prepared by dissolving the host diol in a minimum amount of diethyl ether, adding an excess of the lutidine, and evaporating the solution over a week or so. The crystal structure of the host is shown in Figure 18.16(a), and that of the inclusion complex of lutidine in Figure 18.16(b). The hydroxyl groups of host molecules are disordered, presumably a result of two or more possible orientations of the molecules in the crystal. The phenyl groups appear ordered. Low-

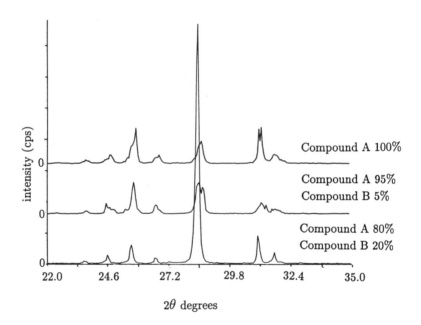

FIGURE 18.14. Powder diffraction patterns of a compound (A), and small amounts of an added compound (B). Note the high intensity of a Bragg reflection near 28° 2θ. The intensity scale is shifted vertically for the upper traces. This type of analysis can be used to study mixtures or the time course of chemical reactions (if A were converted to B). Courtesy F. Caruso.

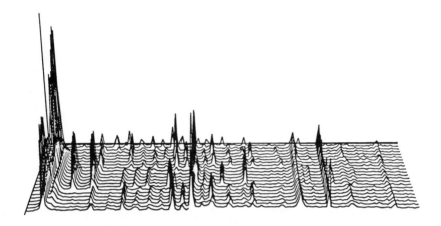

(a)

FIGURE 18.15. (a) Changes in the powder diffraction pattern with respect to time for a zeolite that loses water on heating. An intensity/ temperature / Bragg angle diagram showing the course of the reaction as a function of time (the axis into the paper). Each line drawing represents a separate powder diffraction pattern at a given time. Courtesy of Glover A. Jones, DuPont Company, Central Research and Development, C S and E.

temperature data were measured, but the disorder persisted. The host-guest complex, shown in Figure 18.16(b), shows host-host and host-guest hydrogen bonding. Thermogravimetric analyses and differential calorimetric analyses were done. The thermal analysis showed experimental weight loss in line with that expected. The differential calorimetric curve showed a skewed endotherm corresponding to the weight loss of the guest shown in the thermogravimetric curve. The melting endotherm of the host appears at 198°C. X-ray powder diffraction photographs served to characterize the host and host-guest phases. They show that the inclusion complex (β) can revert to the nonporous α phase.

18.5.2 Reactions in the crystal: glycogen phosphorylase b

The elucidation of catalytic mechanisms of enzymes by diffraction methods is very difficult, even when the three-dimensional structure has been determined to a reasonably high resolution. With conventional X-ray sources, diffraction data collection for macromolecules usually takes days, while chemical reactions that are catalyzed by enzymes may occur in fractions of a second. Nonetheless, it has been possible to study catalysis by

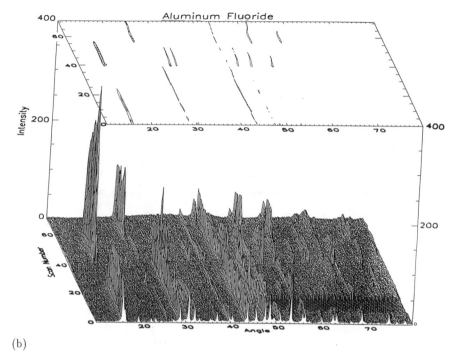

(b)

FIGURE 18.15 (cont'd). (b) Three-dimensional plot of a transformation of aluminum fluoride in the temperature interval 25–750°C. The scan number corresponds to the temperature (in equal increments). This plot shows how fast the reaction occurs, and which Bragg reflections to use when studying the course of the reaction. Courtesy Glover A. Jones and M. R. Short, E. I. du Pont de Nemours and Company, Central Research and Development, C S and E.

host

guest

(a)

FIGURE 18.16. Inclusion of lutidine by a diol. (a) Chemical formulæ of host and guest molecules.

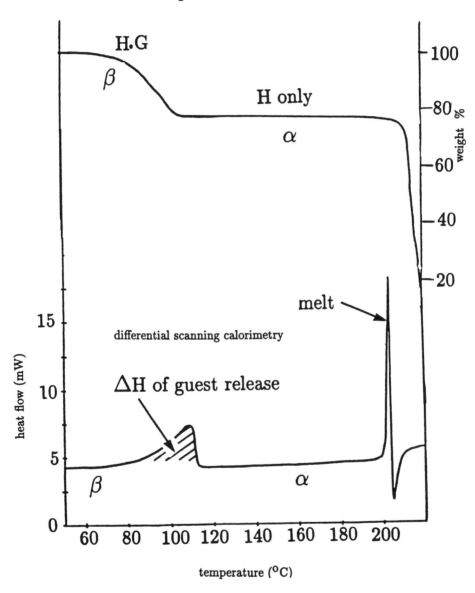

(b)

FIGURE 18.16 (cont'd). Inclusion of lutidine by a diol. (b) Approximate diagrams of results of TG and DSC experiments.

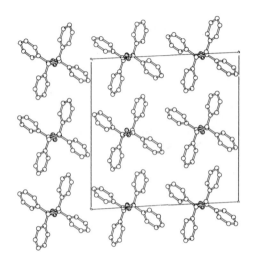

(c)

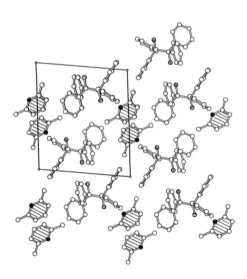

(d)

FIGURE 18.16 (cont'd). Inclusion of lutidine by a diol. (c) Crystal packing in the host compound with disorder in the center of each molecule (Ref. 79). Molecules are hydrogen-bonded by hydroxyl groups in ribbons perpendicular to the plane of the paper. (d) Crystal packing in the host-guest complex (Ref. 80). One hydroxyl group of the host forms a hydrogen bond to the ring nitrogen atom of the guest and the other hydroxyl group forms a hydrogen bond to the hydroxyl group of another host molecule. Guest molecules are shaded.

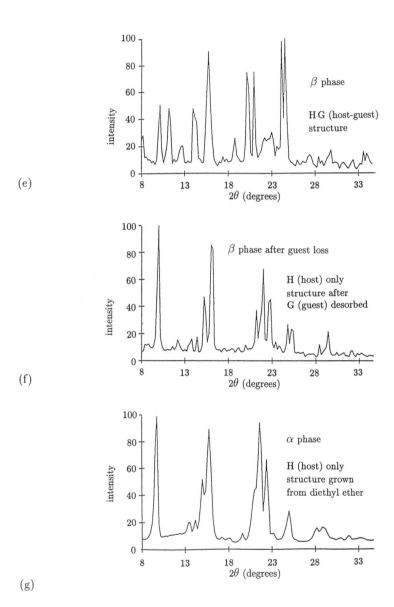

(e)

(f)

(g)

FIGURE 18.16 (cont'd). Inclusion of lutidine by a diol. Powder diffraction photographs of (e) the host-guest complex, (f) the host structure after guest molecules have been desorbed (dried at 120° for 24 hours), and (g) the host structure grown from diethyl ether before guest has been introduced. Courtesy L. Nassimbeni.

altering the experimental conditions (e.g., pH, ionic strength, or temperature) such that substrate is much more slowly converted to product. Alternatively, the chemical nature of the substrate and/or enzyme may be altered. As a result, the enzyme–substrate complex can be viewed crystallographically.

The recent development of very high-intensity X-ray sources (synchrotron radiation) has resulted in a dramatic reduction in the time needed for diffraction data collection. For particular problems, transient events that occur on the multisecond time scale can now be investigated. With fast data collection methods, it is possible to obtain a set of crystal structures that represent various stages along the enzyme-catalyzed reaction coordinate. Methods are being developed that will enable the investigator to obtain a direct visualization of the conversion of native enzyme–substrate complex to enzyme–product complex as a function of time. This is done by use of Laue methods,[82] in which large numbers of Bragg reflection intensity data are recorded simultaneously (see Figure 7.13, Chapter 7).

Glycogen phosphorylase b is an enzyme that catalyzes the phospholytic breakdown of glycogen to glucose-1-phosphate (Equation 18.11 and Figure 18.17). Crystals of the phosphorylase are catalytically active, but the reaction is too fast to study directly by diffraction methods. A glycosylic substrate analogue, heptenitol, is slowly converted to heptulose-2-phosphate, presumably by the same mechanism.

$$\text{heptenitol} + P_i \rightarrow \text{heptulose} \cdot 2 \cdot \text{phosphate}, \tag{18.11}$$

This substrate analogue provides a useful model system for following the enzyme-catalyzed reaction which can be monitored by crystallographic methods. The rate-limiting step in this reaction is the conversion of

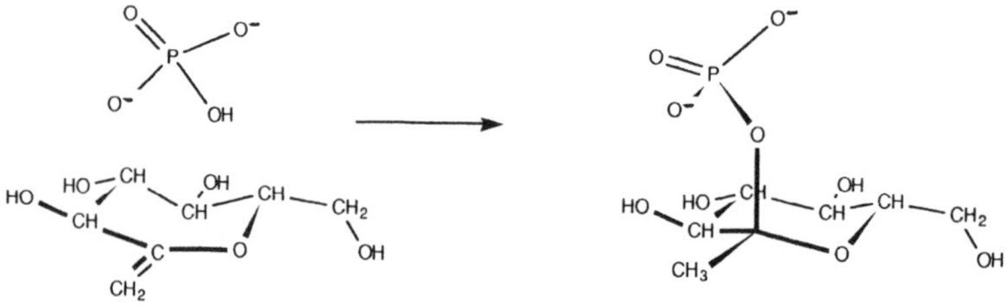

FIGURE 18.17. The use of X-ray diffraction to follow the course of the reaction of glycogen phosphorylase. Shown is the catalytic reaction.

the Michælis complex to product, so that the enzyme–substrate complex accumulates transiently in the crystals. Several X-ray diffraction data sets were recorded as a function of time. Each data set then provided a snapshot of the reaction depicting the formation and transformation of the enzyme–substrate complex into product.[83]

The crystal was mounted in a flow cell,[84] substrate solution flowed over the crystal for about 10 minutes, and Laue photographs were taken with a synchrotron source of white radiation. Since the source of X rays is so intense, it was possible to measure over 100,000 reflections per second.[83] Data sets of one second duration were taken before, during, and after the initiation of the reaction. The site of binding had already been established by structural work with monochromatic radiation, so difference Fourier techniques were used to follow the small changes as a function of the time (Figure 18.18). Unfortunately, if the lifetime of the intermediate is very short, less than 3 seconds, other methods must be used. These are currently being investigated.

18.6 Concluding Remarks

The techniques of determining molecular structure by X-ray diffraction analyses have been initiated, developed, and expanded during the twentieth century, and in this book we have tried to show how the method works and what type of results can be expected.

As we proceed to the twentyfirst century, it is evident that structural studies in chemical and biochemical systems will continue. Higher resolution structures will be determined, more will be learned about the electron distribution in molecules, and, with shorter times of measurement, reactions can be followed. In the twentyfirst century, scientists will be able to concentrate their studies on the mechanism of action of reactions. This will be possible by use of a variety of methods, of which X-ray crystallography will be an important but not sole member. NMR studies will give information on conformations in solution is assignments can be made correctly, as they are for most proteins. Microscopic studies will give information in the surface structure. Other spectroscopic methods will give information in the gaseous state. Many other physical methods will converge together with theoretical methods and will give information on chemical and biochemical reactions, the bases of these sciences.

Summary

1. Photochemical reactions between neighboring molecules are found to occur in some crystals in which C=C double bonds are within 4 Å of each other and in which there is sufficient space for the molecules to undergo the reaction. Similarly, groups of atoms of some molecules in which a metal–carbon bond is broken photochemically may rearrange if the cavity is of an appropriate shape and size.

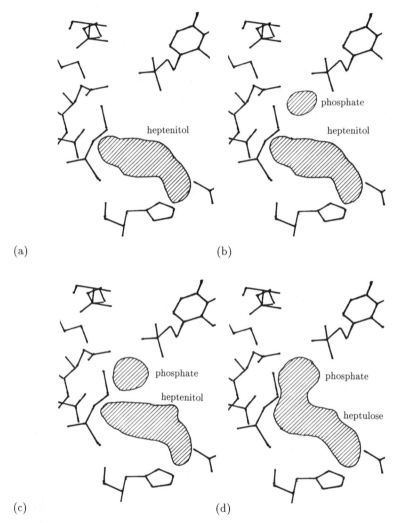

(a)

(b)

(c)

(d)

FIGURE 18.18. Snapshots of the reaction of heptenitol plus phosphate to give heptulose 2-phosphate in crystals of glycogen phosphorylase *b*. This reaction was followed by Laue diffraction studies. (a) Crystalline enzyme plus heptenitol alone, showing binding site (no reaction). (b) Results of diffraction of a crystal after 10 minutes soak in phosphate and heptenitol (60 minutes data collection), (c) 15 minute soak with heptenitol and phosphate plus maltoheptose (45 minute data collection), and (d) 50 hour soak with heptenitol and phosphate and maltoheptose (2.5 hour data collection). *p*H 6.7, temperature 20°C. The difference electron density is contoured at one third the maximum height. The amino-acid residues are shown as in the unliganded enzyme (Ref. 83).

2. Reactions may occur on the surfaces (faces) of crystals, and the different presentation of the molecules on different faces leads to differing reactivities of these faces.

3. Inferences on the changes in geometry accompanying a chemical reaction may be made from the geometry of various related molecules and the interatomic distances within them. Such analyses may lead to information on the nature of the transition state of the reaction being studied.

4. Enzyme-catalyzed reactions may be analyzed from structural studies of the active site of the enzyme and its perturbation and mode of binding of substrates and inhibitors. These analyses may also be aided by structure determinations of enzymes in which appropriate functional side chains have been mutated.

5. Chemical reactions may be followed by observing changes in the diffraction pattern as a function of time. This may be seen in solids via powder diffraction, or in biological macromolecules by use of Laue photographs (in which large amounts of data are measured in a small amount of time). The availability of synchrotron radiation has greatly facilitated this.

Glossary

Electrophile: An ion or molecule that accepts a pair of electrons in order to form a covalent bond.

Nucleophile: An ion or molecule that donates a pair of electrons to an electron-deficient atom so that a new covalent bond is formed.

Scissile bond: A bond that is broken in a chemical or biochemical reaction.

Topochemical reaction: A reaction that occurs in the solid state with a minimum atomic or molecular movement. The term was originally introduced by Gerhardt Schmidt and applied to some photochemical reactions.

Transition-state analogue: A molecule that resembles the geometry and charge distribution of the arrangement of atoms in the transition state of a chemical or biochemical reaction. Such an analogue should bind particularly well to the active site of an enzyme (which has generally been designed by Nature to facilitate the formation of a transition state to the reaction it catalyzes).

Wild-type enzyme: An enzyme in the form found naturally, before mutation.

References

1. Desiraju, G. R. (**Ed.**) *Organic Solid State Chemistry.* Elsevier: Amsterdam (1987).

2. Liebermann, C. Ueber die γ- und δ-Isatropasäure. [On the γ- and δ-isatropic acids.] *Ber. Deutsch. Chem. Gesell.* **22**, 124–130 (1889).

3. Liebermann, C., and Bergami, O. Ueber die Einwirkung der Schwefelsäure auf γ- und δ-Isatropasäure. [On the effect of sulphuric acid on γ- and δ-isatropic acids.] *Ber. Deutsch. Chem. Gesell.* [*Chem. Berichte*] **22**, 782–786 (1889).

4. Kaupp, G. Photodimerization of cinnamic acid in the solid state: new insights on application of atomic force microscopy. *Angew. Chem., Intl. Ed. Engl.* **31**, 592–595 (1992).

5. Schmidt, G. M. J. Photodimerization in the solid state. *Pure Appl. Chem.* **27**, 647–678 (1971).

6. Schmidt, G. M. J. Topochemistry. Part III. The crystal chemistry of some trans-cinnamic acids. *J. Chem. Soc.*, 2014–2021 (1964).

7. Thomas, J. M. Diffusionless reactions and crystal engineering. *Nature (London)* **289**, 633–634 (1991).

8. Dzakpasu, A. A., Phillips, S. E. V., Scheffer, J. R., and Trotter, J. Intramolecular photochemical hydrogen abstraction reactions in the solid state. Correlation with X-ray crystal structure data. *J. Amer. Chem. Soc.* **98**, 6049–6051 (1976).

9. Trotter, J. Structural aspects of the solid-state photochemistry of tetrahydron-aphthoquinones. *Acta Cryst.* **B39**, 373–381 (1983).

10. Scheffer, J. R., Bhandari, K. S., Gayler, R. E., and Wostradowski, R. A. Solution photochemistry. XIII. Hydrogen abstraction reactions proceeding through five-membered transition states. Mechanistic studies indicating conformational control. *J. Amer. Chem. Soc.* **97**, 2178–2189 (1975).

11. Theocharis, C. R., and Jones, W. Topotactic and topochemical photodimerization of benzylidenecyclopentanones. In *Organic Solid State Chemistry*. (**Ed.**, Desiraju, G. R.) Ch. 2., pp. 47–68. Elsevier: Amsterdam, Oxford, New York, Tokyo (1987).

12. Garcia-Granda, S., Beurskens, G., Beurskens, P. T., Krishna, T. S. R., and Desiraju, G. R. Structure of 3,4-dihydroxy-*trans*-cinnamic acid (caffeic acid) and its lack of solid-state topochemical reactivity. *Acta Cryst.* **C43**, 683–685 (1987).

13. Bregman, J., Osaki, K., Schmidt, G. M. J., and Sonntag, F. I. Topochemistry. Part IV. The crystal chemistry of some *cis*-cinnamic acids. *J. Chem. Soc.*, 2021–2030 (1964).

14. Phillips, S. E. V., and Trotter, J. The crystal and molecular structure of *cis*-4a,5,8,8a-tetrahydro-1,4-naphthoquinone and comparison with some of its derivatives. *Acta Cryst.* **B33**, 996–1003 (1977).

15. Sarma, J. A. R. P., and Desiraju, G. R. The role of Cl ··· Cl and C–H ··· O interactions in the crystal engineering of 4-Å short-axis structures. *Acc. Chem. Res.* **19**, 222–228 (1986).

16. Green, B. S., Lahav, M., and Rabinovich, D. Asymmetric syntheses via reactions in chiral crystals. *Acc. Chem. Res.* **12**, 191–197 (1979).

17. Iguchi, M., Nakanishi, H., and Hasegawa, M. Crystals of polymers derived from divinyl compounds by photoradiation in the solid state. *J. Polymer Sci.* **A6**, 1055–1057 (1968).

18. Hasegawa, M., Suzuki, Y., Suzuki, F., and Nakanishi, H. Four-center type photopolymerization in the solid state. I. Polymerization of 2,5-distyrylpyrazine and related compounds. *J. Polymer Sci.* **A7**, 743–752 (1968).

19. Braun, H.-G., and Wegner, G. New experiments concerning the mechanism of solid-state photopolymerization of 2,5-distyrylpyrazine. *Makromol. Chem.* **184**, 1103–1119 (1983).

20. Peachey, N. M., and Eckhardt, C. J. Electronic structure and photochemistry of the 2,5-distyrylpyrazine molecular crystal. *Chem. Phys. Lett.* **188**, 462–466 (1992).

21. Stezowski, J. J., Peachey, N. M., Goebel, P., and Eckhardt, C. J. Structural and lattice dynamical investigation of models for reactions in organic crystals. *J. Amer. Chem. Soc.* **115**, 6499–6505 (1993).

22. Ohashi, Y. Dynamical structure analysis of crystalline-state racemization. *Acc. Chem. Res.* **21**, 268–274 (1988).

23. Ohashi, Y., and Sasada, Y. X-ray analysis of Co–C bond cleavage in the crystalline state. *Nature (London)* **267**, 142–144 (1977).

24. Danno, M., Uchida, A., Ohashi, Y., Sasada, Y., Ohgo, Y., and Baba, S. Crystalline-state reactions of cobaloxime complexes by X-ray exposure. XIV. Uneven racemization at the independent reaction sites. *Acta Cryst.* **B43**, 266–271 (1987).

25. Danno, M., Uchida, A., Ohashi, Y., Sasada, Y., Ohgo, Y., and Baba, S. Crystalline-state reaction of cobaloxime complexes by X-ray exposure. XIV. Uneven racemization at the independent reaction sites. *Acta Cryst.* **B43**, 266–271 (1987).

26. Osano, Y. T., Uchida, A., and Ohashi, Y. Optical enrichment of a racemic chiral crystal by X-ray irradiation. *Nature (London)* **352**, 510–512 (1991).

27. Osano, Y. T., Danno, M., Uchida, A., Ohashi, Y., Ohgo, Y., and Baba, S. Crystalline-state reactions of cobaloxime complexes by X-ray exposure. 15. Different reactivity between two crystallographically independent molecules. *Acta Cryst.* **B47**, 702–707 (1991).

28. Iwamoto, T., and Kashino, S. Topochemical studies. XVI. Direct observation of the solid-state photoreaction of α-(acetylamino)cinnamic acid dihydrate by single crystal X-ray diffraction. *Bull. Chem. Soc. Japan* **66**, 2190–219 (1993).

29. Somorjai, G. A., and Blakely, D. W. Mechanism of catalysis of hydrocarbon reactions by platinum surfaces. *Nature (London)* **258**, 580–583 (1975).

30. Lebrilla, C. B., and Maier, W. F. C–H activation on platinum, a mechanistic study. *J. Amer. Chem. Soc.* **108**, 1606–1616 (1986).

31. Miller, R. S., Curtin, D. Y., and Paul, I. C. Reactions of molecular crystals with gases. I. Reactions of solid aromatic carboxylic acids and related compounds with ammonia and amines. *J. Amer. Chem. Soc.* **96**, 6329–6334 (1974).

32. Miller, R. S., Paul, I. C., and Curtin, D. Y. Reactions of molecular crystals with gases. II. The X-ray structure of crystalline 4-chlorobenzoic acid and the anisotropy of its reaction with ammonia gas. *J. Amer. Chem. Soc.* **96**, 6334–6339 (1974).

33. Miller, R. S., Curtin, D. Y., and Paul, I. C. Reactions of molecular crystals with gases. III. The relationship of anisotropy to crystal structure in reactions of carboxylic acids and anhydrides with ammonia gas. *J. Amer. Chem. Soc.* **96**, 6340–6349 (1974).

34. Bürgi, H.-B., and Dunitz, J. D. From crystal statics to chemical dynamics. *Acc. Chem. Res.* **16**, 153–161 (1983).

35. Murray-Rust, P., Bürgi, H.-B., and Dunitz, J. D. Chemical reaction paths. V. The S_N1 reaction of tetrahedral molecules. *J. Amer. Chem. Soc.* **97**, 921–922 (1975).

36. Pauling, L. *The Nature of the Chemical Bond.* Cornell University Press: Ithaca, New York. 1st edn., 1939; 2nd edn., 1940; 3rd edn. (1960).

37. Allen, F. H., Bellard, S., Brice, M. D., Cartwright, B. A., Doubleday, A., Higgs, H., Hummelink, T., Hummelink-Peters, B. G., Kennard, O., Motherwell, W. D. S., Rodgers, J. R., and Watson, D. G. The Cambridge Crystallographic Data Centre: computer-based search, retrieval, analysis and display of information. *Acta Cryst.* **B35**, 2331–2339 (1979).

38. Bürgi, H.-B., Dunitz, J. D., and Shefter, E. Geometrical reaction coordinates. II. Nucleophilic addition to a carbonyl group. *J. Amer. Chem. Soc.* **95**, 5065–5067 (1973).

39. Bürgi, H.-B., Dunitz, J. D., and Shefter, E. Chemical reaction paths. IV. Aspects of O···C=O interactions in crystals. *Acta Cryst.* **B30**, 1517–1527 (1974).

40. Tulinsky, A., and van den Hende, J. H. The crystal and molecular structure of N-brosylmitomycin A. *J. Amer. Chem. Soc.* **89**, 2905–2911 (1967).

41. Bürgi, H.-B., Lehn, J. M., and Wipff, G. An *ab initio* study of nucleophilic addition to a carbonyl group. *J. Amer. Chem. Soc.* **96**, 1956–1957 (1974).

42. Auf der Heyde, T. P. E., and Nassimbeni, L. R. Reaction pathways from structural data: dynamic stereochemistry of nickel compounds. *Inorg. Chem.* **23**, 4525–4532 (1984).

43. Auf der Heyde, T. P. E., and Nassimbeni, L. R. Reaction pathways from structural data: dynamic stereochemistry of zinc(II) compounds. *Acta Cryst.* **B40**, 582–590 (1984).

44. Britton, D., and Dunitz, J. D. Chemical reaction paths. 7. Pathways for S_N2 and S_N3 substitution at Sn(IV). *J. Amer. Chem. Soc.* **103**, 2971–2979 (1981).

45. Bye, E., Schweizer, W. B., and Dunitz, J. D. Chemical reactions paths. Stereoisomerization path for triphenylphosphine oxide and related molecules: indirect observation of the structure of the transition state. *J. Amer. Chem. Soc.* **104**, 5893–5898 (1982).

46. Gilli, G., Bertolasi, V., Bellucci, F., and Ferretti, V. Stereochemistry of the $R_1(X=)C(sp^3)R_2R_3$ fragment. Mapping of the cis-trans isomerization path by rotation around the C–N bond from crystallographic structural data. *J. Amer. Chem. Soc.* **108**, 2420–2424 (1986).

47. Chandrasekhar, K., and Bürgi, H.-B. Chemical reactions paths. 9. Conformational interconversions of Wilkinson's catalyst and of related square-planar $XM(PR_3)_3$ compounds as determined from systematic analysis of solid-state structural data. *J. Amer. Chem. Soc.* **105**, 7081–7093 (1983).

48. Allen, F. H. and Kirby, A. J. Bond length and reactivity. Variable length of the C–O single bond. *J. Amer. Chem. Soc.* **106**, 6197–6200 (1984).

49. Jones, P. G. and Kirby, A. J. Simple correlation between bond length and reactivity. Combined use of crystallographic and kinetic data to explore a reaction coordinate. *J. Amer. Chem. Soc.* **106**, 6207–6212 (1984).

50. Bürgi, H.-B., and Dunitz, J. D. Fractional bonds: relations among their lengths, strengths, and stretching force constants. *J. Amer. Chem. Soc.* **109**, 2924–2926 (1987).

51. Voet, D., and Voet, J. G. *Biochemistry*. John Wiley: New York (1990).

52. Lipscomb, W. N. Acceleration of reactions by enzymes. *Acc. Chem. Res.* **15**, 232–238 (1982).

53. Lienhard, G. E. Enzymatic catalysis and transition state theory. *Science* **180**, 149–154 (1973).

54. Wolfenden, R. Analogue approaches to the structure of the transition state in enzyme reactions. *Acc. Chem. Res.* **5**, 10–18 (1972).

55. Pauling, L. Molecular architecture and biological reactions. *Chem. Eng. News* **24**, 1375–1377 (1946).

56. Wilson, D. K., Rudolph, F. B., and Quiocho, F. A. Atomic structure of adenosine deaminase complexed with a transition-state analog: understanding catalysis and immunodeficiency mutations. *Science* **252**, 1278–1284 (1991).

57. Phillips, D. C. The three-dimensional structure of an enzyme molecule. *Sci. Amer.* **215** (5), 75–80 (1966).

58. Ford, L. O., Johnson, L. N., North, A. C. T., Phillips, D. C., and Tijan, R. Crystal structure of a lysozyme-tetrasaccharide lactone complex. *J. Molec. Biol.* **88**, 349–371 (1974).

59. Secemski, I. I., Lehrer, S. S., and Lienhard, G. E. A transition state analogue for lysozyme. *J. Biol. Chem.* **247**, 4740–4748 (1972).

60. Schindler, M., Ascaf, Y., Sharon, N., and Chipman, D. M. Mechanism of lysozyme catalysis: role of ground-state strain in subsite D in hen egg and human lysozymes. *Biochemistry* **16**, 423–431 (1977).

61. Warshel, A., and Levitt, M. Theoretical studies of enzymatic reactions; dielectric, electrostatic and steric stabilization of the carbonium ion in the reaction of lysozyme. *J. Molec. Biol.* **103**, 227–249 (1976).

62. Kirby, A. J. Mechanism and stereoelectronic effects in the lysozyme reaction. *CRC Crit. Rev. Biochem.* **22**, 283–315 (1987).

63. Post, C. B., and Karplus, M. Does lysozyme follow the lysozyme pathway? An alternative based on dynamic, structural, and stereoelectronic considerations. *J. Amer. Chem. Soc.* **108**, 1317–1319 (1986).

64. Kuroki, R., Weaver, L. H., and Matthews, B. W. A covalent enzyme-substrate intermediate with saccharide distortion in a mutant T4 lysozyme. *Science* **262**, 2030–2033 (1993).

65. Blow, D. M. Structure and mechanism of chymotrypsin. *Acc. Chem. Res.* **9**, 145–152 (1976).

66. Stroud, R. M. A family of protein-cutting proteins. *Sci. Amer.* **231** (2), 74–88 (1974).

67. Sprang, S., Standing, T., Fletterick, R. J., Stroud, R. M., Finer-Moore, J., Xuong, N.-H., Hamlin, R., Rutter, W. J., and Craik, C. S. The three-dimensional structure of Asn[102] mutant of trypsin: role of Asp[102] in serine protease catalysis. *Science* **237**, 905–909 (1987).

68. Meyer, E., Cole, G., Radhakrishnan, R. and Epp, O. Structure of native porcine pancreatic elastase at 1.65 Å resolution. *Acta Cryst.* **B44**, 26-28 (1988).

69. Corey, D. R., and Craik, C. S. An investigation into the minimum requirements for peptide hydrolysis by mutation of the catalytic triad of trypsin. *J. Amer. Chem. Soc.* **114**, 1784–1790 (1992).

70. Perona, J. J., Craik, C. S., and Fletterick, R. J. Locating the catalytic water molecule in serine proteases. *Science* **261**, 620–621 (1993).

71. Singer, P. T., Smalås, Carty, R. P., Mangel, W. F., and Sweet, R. M. Locating the catalytic water molecule in serine proteases. *Science* **261**, 621–622 (1993).

72. Craik, C. S., Roczniak, S., Largman, C., and Rutter, W. J. The catalytic role of the active site aspartic acid in serine proteases. *Science* **237**, 909–913 (1987).

73. Bone, R., Silen, J. L., and Agard, D. A. Structural plasticity broadens the specificity of an engineered protease. *Nature (London)* **339**, 191–195 (1989).

74. Cha, J., Cho, Y., Whitaker, R. D., Carrell, H. L., Glusker, J. P., Karplus, A., and Batt, C. A. Perturbing the metal site in D-xylose isomerase: the effect of mutations of His-220 on enzyme stability. *J. Biol. Chem.* in press.

75. Warshel, A., Naray-Szabo, G., Sussman, F., and Hwang, J.-K. How do serine proteases really work? *Biochemistry* **28**, 3639–3637 (1989).

76. Lonsdale, K., Nave, E., and Stephens, J. F. X-ray studies of a single crystal chemical reaction: photooxide of anthracene to (anthraquinone, anthrone). *Phil. Trans. Roy. Soc. (London)* **A26**, 1–31 (1966).

77. Corbin, D. R., Abrams, L., Jones, G. A., Eddy, M. M., Harrison, W. T. A., Stucky, G. D., and Cox, D. E. Flexibility of the zeolite RHO framework. *In situ* X-ray and neutron powder structural characterization of divalent cation-exchanged zeolite RHO. *J. Amer. Chem. Soc.* **112**, 4821–4830 (1990).

78. Cockroft, J. K., Fitch, A. N., and Simon, A. Powder neutron diffraction studies of orientational order–disorder transitions in molecular and molecular-ionic solids: use of symmetry-adapted spherical harmonic functions in the analysis of scattering density distributions arising from orientational disorder. In: *Collected Papers. Summerschool on Crystallography and its Teaching. Tianjin, China. Sept 15-24, 1988.* (**Ed.**, Miao, F.-M.) p. 427. Tianjin: Tianjin Normal University (1988).

79. Bond, D. R., Bourne, S. A., Nassimbeni, L. R., and Toda, F. Complexation with diol host compounds. Part 2. Structures of 1,1,2,2,-tetraphenylethane-1,2-diol and its 1:2 molecular inclusion complex with dimethylsulphoxide. *J. Crystallographic and Spectroscopic Research* **19**, 809–822 (1989).

80. Bourne, S. A., Nassimbeni, L., and Toda, F. Complexation with diol host compounds. Part 7. Structures and thermal analysis of 1,1,2,2,-tetraphenylethane-1,2-diol with lutidine guests. *J. Chem. Soc., Perkin Trans. 2* 1335–1341 (1991).

81. Atwood, J. L., Davies, J. E., and MacNicol, D. D. (**Eds.**) *Inclusion Compounds.* Vols. I – III. Academic Press: London (1984).

82. Moffat, K., Szebenyi, D., and Bilderback, D. X-ray Laue diffraction from protein crystals. *Science* **223**, 1423–1425 (1984).

83. Hajdu, J., Machin, P. A., Campbell, J. W., Greenhough, T. J., Clifton, I. J., Zurek, S., Gover, S., Johnson, L. N. and Elder, M. Millisecond X-ray diffraction and the first electron density maps from Laue photographs of a protein crystal. *Nature (London)* **329**, 178–181 (1987).

84. Wyckoff, H. W., Doscher, M., Tsernoglou, D., Inagami, T., Johnson, L. N., Hardman, K. D., Allewell, N. N., Kelly, D. M., and Richards, F. M. Design of a diffractometer and flow cell system for X-ray analysis of crystalline proteins with applications to the crystal chemistry of ribonuclease-S. *J. Molec. Biol.* **27**, 563–578 (1967).

APPENDIX

Strategies for determining crystal structures

Small molecules

1. Grow a suitable crystal, 0.1-0.3 mm in each dimension.
2. Glue the crystal to a fine glass fibre and place the fiber in a goniometer head. If the crystal is unstable it will need to be mounted in a glass capillary.
3. Take some preliminary X-ray diffraction photos in order to:
   a) establish if the crystal diffracts
   b) estimate the quality of the crystal
   c) obtain preliminary values for the unit-cell dimensions.
4. Measure the density of flotation in a mixture of liquids. Check the weight of the unit-cell contents.
5. If the crystal is single and diffracts well, align it in an X-ray diffractometer and measure unit-cell dimensions.
6. Measure three-dimensional diffraction data with appropriate diffractometer settings for scanning each reflection and recording its direction and intensity. Select 3–5 standard reflections and measure them at regular intervals. If possible, the data are measured at low temperatures.
7. Examine the diffraction data to establish the space group of the crystal structure.
8. Correct the experimental data for Lorentz, polarization, absorption, and other factors, and the results to relative values of the structure amplitudes $F(hkl)$.
9. Use direct methods to solve the crystal structure.
10. If problems arise, calculate the Patterson map to see if information on the atomic arrangement can be obtained.
11. Check the geometry of the derived crystal structure, particularly searching for anomalies in intermolecular packing.
12. Refine the structure by least squares methods.
13. Calculate a difference map to check for any possible hydrogen atoms, or for errors in the structure determination.

14. Refine the entire structure by least-squares methods.
15. Calculate the intramolecular geometry.
16. Calculate any intermolecular interactions.

Macromolecular structures

1. Set out matrices with respect to pH, ionic strength, temperature, etc. to find the best experimental conditions for growing crystals. Then grow suitable crystals.
2. Mount crystal in capillary.
3. Measure some preliminary data on a crystal so obtained to see
    a) if it diffracts,
    b) the quality of the crystal, and the ultimate resolution of the data, and
    c) what the unit-cell dimensions are.
    d) If the first two are poor, start again.
4. Measure the crystal density and check the unit-cell dimensions.
5. Measure X-ray diffraction data (the "native" data set).
6. Reduce the data by applying appropriate correction factors to the measured data.
7. Soak complexes of heavy atoms of various kinds into a crystal.
8. Screen diffraction data from heavy-atom derivatives of the protein for differences in intensities from those in the native data set.
9. Calculate difference Patterson maps and locate the positions of the heavy atoms in the unit cell.
10. Determine relative phases by isomorphous replacement methods and calculate an electron-density map. Alternatively use molecular replacement methods to determine the structure and calculate an electron-density map.
11. Start fitting polyalanine to the map ("tracing the chain").
12. Use the known amino-acid sequence to fit the amino-acid side chains to the electron-density map.
13. Refine the structure.
14. Measure data to higher resolution.
15. Extend the relative phases to include the higher resolution data.
15. Soak in substrate or inhibitor molecules into the protein crystals.
16. Measure data on the crystal with bound small molecules.
17. Calculate difference maps to identify binding sites.
18. Crystallize mutant enzymes.
19. Check if the crystals are isomorphous with the original protein crystals (same space group and unit-cell dimensions).
20. Measure data on the mutant enzyme. Bind molecules if so desired.
21. Calculate difference maps to identify changes in the mutant enzyme.
22. If the macromolecule is an enzyme, determine how it carries out the catalytic mechanism.

GLOSSARY ITEMS (BOLDFACE IN TEXT)

α helix, 510
Ångström unit, 24
Absolute scale, 267
Absolute structure, absolute
   configuration, 615
Absorption correction, 267
Absorption edges, 220
Accuracy, 408
Achiral, 615
Acicular crystals, 64
Aggregation, 64
Agonist, 722
Allotrope, 447
Allotropy, 674
Amphipathic helix, 510
Amplitude, 24
Anisotropy, 176
Anomalous dispersion, 615
Anomalous scattering, 220
Antagonist, 722
Anti, 510
Anticlinal conformation (*ac*),
   512
Antiperiplanar conformation
   (*ap*), 512
Area detector, 267
Asymmetric carbon atom, 615
Asymmetric unit, 136
Atomic and ionic refractivities,
   176
Atomic coordinates, 448
Atomic displacement parame-
   ters, 563
Atomic parameters, 448
Atomic scattering factor, 100
Automated diffractometer, 267
Axial bond (*a*), 512
Axis of symmetry, 136
Azimuth angle, 267
Azimuthal scan, 268

β barrel, 512
β sheet, 512

$\beta\alpha\beta$ motif, 512
Bar graph, 722
Berry pseudorotation
   mechanism, 722
Best plane, 512
Biaxial crystals, 176
Bibliographic search, 722
Birefringence, 176
Boat conformation, 512
Boltzmann distribution, 563
Bond length, 448
Bond order, 448
Bond valence, 675
Bragg reflection, 101
Bragg's Law, the Bragg
   equation, 100
Bragg, 24
Bravais lattice, 136
Brilliance, 176

Calculated phase angle, 333
Cartesian coordinate, 448
Cauchy–Schwarz inequality,
   333
Center of symmetry (or center
   of inversion), 136
Centrosymmetric crystal
   structure, 333
Chair conformation, 512
Characteristic temperature,
   563
Characteristic X rays, 268
Charge-transfer complex, 675
Chiral axis, 615
Chiral center, 615
Chiral objects, 136
Choleic acid, 675
Cis, 512
Clathrate, 675
Cleavage, 176
Cluster analysis
Coherent scattering, 101
Configuration, 512

825

INDEX